Newnes Mechanical Engineer's Pocket Book

Newnes Mechanical Engineer's Pocket Book

Third edition

Roger L. Timings

AMSTERDAM • BOSTON • HEIDELBERG • LONDON • NEW YORK
OXFORD • PARIS • SAN DIEGO • SAN FRANCISCO • SINGAPORE
• SYDNEY • TOKYO

ELSEVIER

Newnes is an imprint of Elsevier

Newnes is an imprint of Elsevier
The Boulevard, Langford Lane, Kidlington, Oxford, OX5 1GB, UK
30 Corporate Drive, Suite 400, Burlington, MA 01803, USA

First edition 1990
Reprinted 1992, 1993, 1995 (twice)
Second edition 1997
Reprinted 2001, 2002
Third edition 2006
Reprinted 2008, 2009, 2010

British Library Cataloguing in Publication Data
A catalogue record for this book is available from the British Library

Library of Congress Cataloging-in-Publication Data
A catalog record for this book is available from the Library of Congress

ISBN: 978-0-7506-6508-7

For information on all Newnes publications
visit our website at www.elsevierdirect.com/newnes

Printed and bound by CPI Group (UK) Ltd, Croydon, CR0 4YY

Transferred to digital print 2013

Contents

Foreword

It is now 14 years since the first edition of the Mechanical Engineer's Pocket Book was published in its present format. Although a second edition was published some 7 years ago to accommodate many updates in the British Standards incorporated in the text, no changes were made to the structure of the book.

During the 14 years since the first edition of the Mechanical Engineer's Pocket Book was published, the British Engineering Industry has undergone many changes, with the emphasis moving from manufacture to design and development. At the same time manufacturing has been largely out-sourced to East Europe, Asia and the Far East in order to reduce operating costs in an increasingly competitive global market.

Therefore, before embarking on this third edition, the Publishers and the Author have undertaken an extensive market research exercise. The main outcomes from this research have indicated that:

- The demand has changed from an engineering *manufacturing* driven pocket book to an engineering *design* driven pocket book.
- More information was requested on such topic areas as roller chain drives, pneumatic systems and hydraulic systems in the section on power transmission.
- More information was requested on selected topic areas concerning *engineering statics*, *dynamics* and *mathematics*.
- Topic areas that have fallen out of favour are those related to cutting tools. Cutting tool data is considered less important now that the emphasis has move from manufacture to design. Further, cutting tool data is now widely available on the web sites of cutting tool manufacturers. A selection of useful web sites are included in Appendix 4.

Roger L. Timings (2005)

Preface

As stated in the foreword, this new edition of the Mechanical Engineer's Pocket Book reflects the changing nature of the Engineering Industry in the UK and changes in the related technology. Many of the British Standards (BS) quoted in the previous editions have been amended or withdrawn and replaced by BS EN and BS EN ISO standards. The British Standards catalogue is no longer available in hard copy but is published on a CD.

Definitions of entries

Catalogue entries are coded as in the example shown below. The various elements are labelled A–J. The key explains each element. Not all the elements will appear in every Standard.

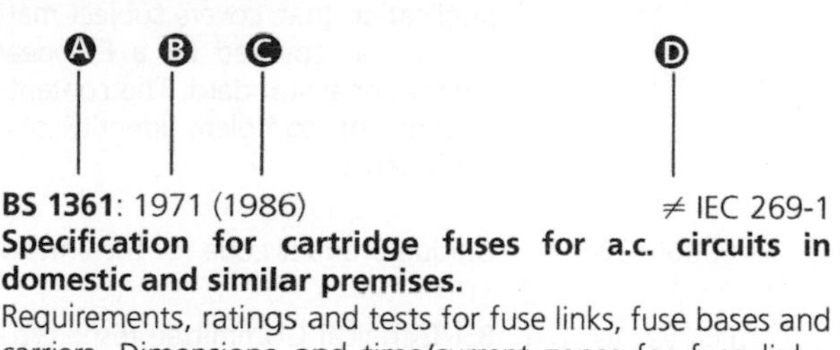

BS 1361: 1971 (1986) ≠ IEC 269-1

Specification for cartridge fuses for a.c. circuits in domestic and similar premises.

Requirements, ratings and tests for fuse links, fuse bases and carriers. Dimensions and time/current zones for fuse links. Type 1-rate 240 V and 5–45 A for replacement by domestic consumers; Type II-rated 415 V and 60, 80 or 100 A for use by the supply authority in the incoming unit of domestic and similar premises.

AMD 4171, January 1983 (Gr 0) R 00000821
AMD 4795, January 1985 (Gr 2) R 00000833
AMD 6692, January 1991 (FOC) 00000845

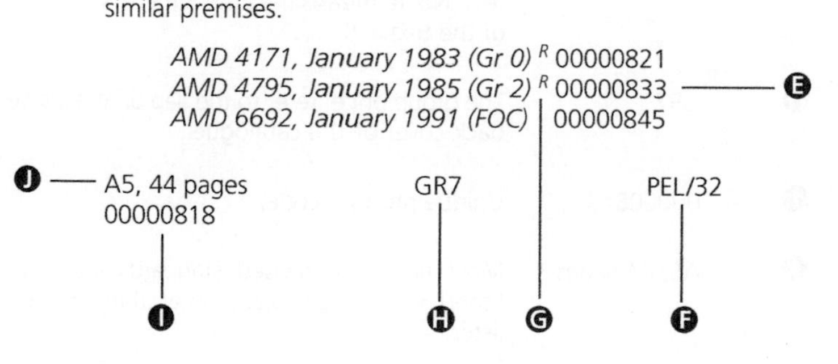

A	**BS 1361**	Product identifier.
B	1971	Original publication date.
C	(1986)	Confirmed in 1986, indicating the continuing currency of the standard without full revision.
D	≡	An identical standard: a BSI publication identical in every detail with a corresponding European and/or international standard.
or	=	A (technically) equivalent standard: a BSI publication in all technical respects the same as a corresponding European and/or international standard, though the wording and presenting may differ quite extensively.
or	≠	A related but not equivalent standard: a BSI publication that covers subject matters similar to that covered by a European and/or international standard. The content however is short of complete identity or technical equivalence.
E	00000833	Unique product code for the amendment.
F	PEL/32	BSI Technical Committee responsible for this publication.
G	*R*	Amendment incorporated in the reprinted text. No 'R' means the amendment is not part of the text.
H	GR7	The group price: refer to the flap on the inside back cover of the catalogue.
I	00000818	Unique product code.
J	A5, 44 pages	Most new and revised standards are published in A4 size. Sizes other than A4 are listed.

Amendments

All separate amendments to the date of despatch are included with any main publication ordered. Prices are available on application. The amendment is then incorporated within the next reprint of the publication and the text carries a statement drawing attention to this, and includes an indication in the margin at the appropriate places on the amended pages.

Review

The policy of the BSI is for every standard to be reviewed by the technical committee responsible not more than 5 years after publication, in order to establish whether it is still current and, if it is not, to identify and set in hand appropriate action. Circumstances may lead to an earlier review.

When reviewing a standard a committee has four options available:

- *Withdrawal*: indicating that the standard is no longer current.
- *Declaration of Obscelence*: indicating by amendment that the standard is not recommended for use in new equipment, but needs to be retained for the servicing of existing equipment that is expected to have a long service life.
- *Revision*: involving the procedure for new projects.
- *Confirmation*: indicating the continuing currency of the standard without full revision. Following confirmation of a publication, stock of copies are overstamped with the month and year of confirmation.

The latest issue of standards should always be used in new product designs and equipments. However many products are still being manufactured to obsolescent and obsolete standards to satisfy a still buoyant demand. This is not only for maintenance purposes but also for current manufacture where market forces have not yet demanded an update in design. This is particularly true of screwed fasteners. For this reason the existing screw thread tables from the previous editions have been retained and stand alongside the new BE EN ISO requirements.

The new standards are generally more detailed and prescriptive than there predecessors, therefore there is only room to include the essential information tables in this Pocket Book as a guide. Where further information is required the full standard should always be consulted and a list of libraries holding up-to-date sets of standards is included in an appendix at the end of this book. The standards quoted in the Pocket Book are up-to-date at the time of publication but, in view of the BSI's policy of regular reviews, some may become subject to revision within the life of this book.

For this reason the validity of the standards quoted should always be verified. The BSI helpline should be consulted.

This Pocket Book has been prepared as an aid to mechanical engineers engaged in the design, development and manufacture of engineering products and equipment. It is also a useful source book for others who require a quick, day-to-day reference of engineering information. For easy reference this book is divided into 8 main sections, namely:

1. Mathematics
2. Engineering statics
3. Engineering dynamics
4. Fasteners
5. Power transmission
6. Materials
7. Linear and geometrical tolerancing
8. Computer aided engineering.

Within these main sections the material has been assembled in a logical sequence for easy reference and numbered accordingly. This enables the reader to be lead directly to the item required by means of a comprehensive list contents and the inclusion of a comprehensive alphabetical index.

This Pocket Book is not a text book but a compilation of useful information. Therefore in the sections concerned with mathematics, statics and dynamics worked examples are only included where anomalies might otherwise occur. The author is indebted to the British Standards Institution and to all the industrial and commercial companies in the UK and abroad who have co-operated in providing up-to-date data in so many technical areas. Unfortunately, limitations of space have allowed only abstracts to be included from amongst the wealth of material provided. Therefore the reader is strongly recommended to consult the complete standards, industrial manuals, design manuals or catalogues after an initial perusal of the tables of data found in this book. To this end, an appendix is provided with the names and addresses of the contributors to this book. They all have useful web sites where additional information may be found.

The section on computer-aided manufacture has been deleted from this edition as it was too brief to be of much use and could not be expanded with in the page count available. Specialist texts are available on application from the publishers of this Pocket Book on such topic areas as:

- Computer numerical control
- Computer-aided drawing and design
- Industrial robots
- Flexible manufacturing systems

- Programmable logic controllers
- Manufacturing management
- Project management.

Within the constraints of commercial viability, it is the continuing intention of the author and publisher to update this book from time to time. Therefore, the author would appreciate (via the publishers) suggestions from the users of this book for additions and/or deletions to be taken into account when producing new editions.

Roger L. Timings (2005)

Acknowledgements

We would like to thank all the companies who have kindly given permission for their copyright material to be used in this edition:

British Fluid Power Association (BFPA) (Sections 5.6.9–5.6.18, 5.6.20).
ContiTech United Kingdom Ltd. (Sections 5.2.4–5.2.18).
David Brown Engineering Ltd. (Section 5.1.1).
Henkel Loctite Adhesives Ltd. (Sections 4.5.1–4.5.9).
IMI Norgren Ltd. (Section 5.6.19).
National Broach & Machine Co. (Sections 5.1.4–5.1.17).
Renold plc. (Sections 5.3–5.4, 5.3.10–5.3.13, 5.3.16, 5.3.18–5.3.19, 5.3.21–5.3.22, 5.3.27–5.3.30).
Emhart Teknologies (Tucker Fasteners Ltd.) (Sections 4.2.8–4.2.18).
Butterworth Heinemann for allowing us to reproduce material from Higgins, R. A., *Properties of Materials* in Sections 6.2.2, and 6.2.3.
Butterworth Heinemann for allowing us to reproduce material from Parr, Andrew, *Hydraulics and Pneumatics* in Sections 5.6.1–5.6.4.
Newnes for allowing us to reproduce material from Stacey, Chris, *Practical Pneumatics* in Section 5.6.3.
Pearson Education Ltd., for allowing us to reproduce material from Timings, R. L., *Materials Technology volumes 1* and *2* in Sections 6.1.1–6.1.21 and 7.1–7.5.
Extracts from British Standards are reproduced with permission of BSI. Complete copies of these documents can be obtained by post from BSI Sales, Linford Wood, Milton Keynes, Bucks., MK14 6LE.

Further information concerning the above companies can be found in the Appendix 3 at the end of this Pocket Book.

1 Engineering Mathematics

1.1 The Greek alphabet

Name	Symbol		Examples of use
	Capital	Lower case	
alpha	A	α	Angles, angular acceleration, various coefficients
beta	B	β	Angles, coefficients
gamma	Γ	γ	Shear strain, surface tension, kinematic viscosity
delta	Δ	δ	Differences, damping coefficient
epsilon	E	ε	Linear strain
zeta	Z	ζ	
eta	H	η	Dynamic viscosity, efficiency
theta	Θ	θ	Angles, temperature, volume strain
iota	I	ι	
kappa	K	κ	Compressibility
lambda	Λ	λ	Wavelength, thermal conductivity
mu	M	μ	Poisson's ratio, coefficient of friction
nu	N	ν	Dynamic viscosity
xi	Ξ	ξ	
omicron	O	o	
pi	Π	π	Mathematical constant
rho	P	ρ	Density
sigma	Σ	σ	Normal stress, standard deviation, sum of
tau	T	τ	Shear stress
upsilon	Y	υ	
phi	Φ	ϕ	Angles, heat flow rate, potential energy
chi	X	χ	
psi	Ψ	ψ	Helix angle (gears)
omega	Ω	ω	Angular velocity, solid angle (ω) electrical resistance (Ω)

1.2 Mathematical symbols

Is equal to	$=$	Is not equal to	$\neq$
Is identically equal to	$\equiv$	Is approximately equal to	$\approx$
Approaches	$\rightarrow$	Is proportional to	$\propto$
Is smaller than	$<$	Is larger than	$>$
Is smaller than or equal to	$\leqslant$	Is larger than or equal to	$\geqslant$
Magnitude of a	$\lvert a \rvert$	a raised to power n	a^n
Square root of a	$\sqrt{a}$	nth root of a	$\sqrt[n]{a}$
Mean value of a	$\bar{a}$	Factorial a	$a!$
Sum	Σ	Product	Π
Complex operator	i, j	Real part of z	Re z
Imaginary part of z	Im z	Modulus of z	$\lvert z \rvert$
Argument of z	arg z	Complex conjugate of z	z^*

a multiplied by *b*	$ab, a \cdot b, a \times b$
a divided by *b*	$a/b, \frac{a}{b}, ab^{-1}$
Function of *x*	$f(x)$
Variation of *x*	δx
Finite increment of *x*	Δx
Limit to which *f*(*x*) tends as x approaches *a*	$\lim_{x \to a} f(x)$
Differential coefficient of *f*(*x*) with respect to *x*	$\frac{df}{dx}$, df/dx, $f'(x)$
Indefinite integral of *f*(*x*) with respect to *x*	$\int f(x)dx$
Increase in value of *f*(*x*) as *x* increases from a to *b*	$[f(x)]_a^b$
Definite integral of *f*(*x*) from $x = a$ to $x = b$	$\int_a^b f(x)dx$
Logarithm to the base 10 of *x*	$\lg x$, $\log_{10}x$
Logarithm to the base *a* of *x*	$\log_a x$
Exponential of *x*	$\exp x$, e^x
Natural logarithm	$\ln x$, $\log_e x$
Inverse sine of *x*	arcsin *x*
Inverse cosine of *x*	arccos *x*
Inverse tangent of *x*	arctan *x*
Inverse secant of *x*	arcsec *x*
Inverse cosecant of *x*	arccosec *x*
Inverse cotangent of *x*	arccot *x*
Inverse hyperbolic sine of *x*	arsinh *x*
Inverse hyperbolic cosine of *x*	arcosh *x*
Inverse hyperbolic tangent of *x*	artanh *x*
Inverse hyperbolic cosecant of *x*	arcosech *x*
Inverse hyperbolic secant of *x*	arsech *x*
Inverse hyperbolic cotangent of *x*	arcoth *x*
Vector	**A**
Magnitude of vector **A**	$\|\mathbf{A}\|$, *A*
Scalar products of vectors **A** and **B**	$\mathbf{A} \cdot \mathbf{B}$
Vector products of vectors **A** and **B**	$\mathbf{A} \times \mathbf{B}$, $\mathbf{A} \wedge \mathbf{B}$

1.3 Units: SI

1.3.1 Basic and supplementary units

The International System of Units (SI) is based on nine physical quantities.

Physical quantity	Unit name	Unit symbol
Length	metre	m
Mass	kilogram	kg
Time	second	s
Plane angle	radian	rad
Amount of substance	mole	mol
Electric current	ampere	A
Luminous intensity	candela	cd
Solid angle	steradian	sr
Thermodynamic temperature	kelvin	K

1.3.2 Derived units

By dimensionally appropriate multiplication and/or division of the units shown above, derived units are obtained. Some of these are given special names.

Physical quantity	Unit name	Unit symbol	Derivation
Electric capacitance	farad	F	$(A^2 s^4)/(kg\,m^2)$
Electric charge	coulomb	C	A s
Electric conductance	siemens	S	$(A^2 s^3)/(kg\,m^2)$
Electric potential difference	volt	V	$(kg\,m^2)/(A\,s^3)$
Electrical resistance	ohm	Ω	$(kg\,m^2)/(A^2 s^3)$
Energy	joule	J	$(kg\,m^2)/s^2$
Force	newton	N	$(kg\,m)/s^2$
Frequency	hertz	Hz	1/s
Illuminance	lux	lx	$(cd\,sr)/m^2$
Inductance	henry	H	$(kg\,m^2)/(A^2 s^2)$
Luminous flux	lumen	lm	cd sr
Magnetic flux	weber	Wb	$(kg\,m^2)/(A\,s^2)$
Magnetic flux density	tesla	T	$kg/(A\,s^2)$
Power	watt	W	$(kg\,m^2)/s^3$
Pressure	pascal	Pa	$kg/(m\,s^2)$

Some other derived units not having special names.

Physical quantity	Unit	Unit symbol
Acceleration	metre per second squared	m/s^2
Angular velocity	radian per second	rad/s
Area	square metre	m^2
Current density	ampere per square metre	A/m^2
Density	kilogram per cubic metre	kg/m^3
Dynamic viscosity	pascal second	Pa s
Electric charge density	coulomb per cubic metre	C/m^3
Electric field strength	volt per metre	V/m
Energy density	joule per cubic metre	J/m^3
Heat capacity	joule per kelvin	J/K
Heat flux density	watt per square metre	W/m^2
Kinematic viscosity	square metre per second	m^2/s
Luminance	candela per square metre	cd/m^2
Magnetic field strength	ampere per metre	A/m
Moment of force	newton metre	N m
Permeability	henry per metre	H/m
Permittivity	farad per metre	F/m
Specific volume	cubic metre per kilogram	m^3/kg
Surface tension	newton per metre	N/m
Thermal conductivity	watt per metre kelvin	W/(m K)
Velocity	metre per second	m/s
Volume	cubic metre	m^3

1.3.3 Units: not SI

Some of the units which are not part of the SI system, but which are recognized for continued use with the SI system, are as shown.

Physical quantity	Unit name	Unit symbol	Definition
Angle	degree	°	$(\pi/180)$ rad
Angle	minute	′	$(\pi/10\,800)$ rad
Angle	second	″	$(\pi/648\,000)$ rad
Celsius temperature	degree Celsius	°C	K − 273.2 (For K see 1.3.1)
Dynamic viscosity	poise	P	10^{-1} Pa s
Energy	calorie	cal	≈4.18 J
Fahrenheit temperature	degree Fahrenheit	°F	$\left(\frac{9}{5}\right)$°C + 32

continued

Section 1.3.3 (*continued*)

Physical quantity	Unit name	Unit symbol	Definition
Force	kilogram force	kgf	≈9.807 N
Kinematic viscosity	stokes	St	10^{-4} m²/s
Length	inch	in.	2.54×10^{-2} m
Length	micron	μm	10^{-6} m
Mass	pound	lb	≈0.454 kg
Mass	tonne	t	10^3 kg
Pressure	atmosphere	atm	101 325 Pa
Pressure	bar	bar	10^5 Pa
Pressure	millimetre of mercury	mm Hg	≈133.322 Pa
Pressure	torr	torr	≈133.322 Pa
Thermodynamic temperature	degree Rankine	°R	°F + 459.7
Time	minute	min	60 s
Time	hour	h	3600 s
Time	day	d	86 400 s

1.3.4 Notes on writing symbols

(a) Symbols should be in roman type lettering: thus cm, not *cm*.
(b) Symbols should remain unaltered in the plural: thus cm, not cms.
(c) There should be a space between the product of two symbols: thus N m, not Nm.
(d) Index notation may be used: thus m/s may be written as m s^{-1} and W/(m K) as W m^{-1} K^{-1}.

1.3.5 Decimal multiples of units

For quantities which are much larger or much smaller than the units so far given, decimal multiples of units are used.

Internationally agreed multiples are as shown.

For small quantities

Multiple	Prefix	Symbol
10^{-1}	deci	d
10^{-2}	centi	c
10^{-3}	milli	m
10^{-6}	micro	μ
10^{-9}	nano	n
10^{-12}	pico	p
10^{-15}	femto	f
10^{-18}	atto	a

For large quantities

Multiple	Prefix	Symbol
10	deca	da
10^2	hecto	h
10^3	kilo	k
10^6	mega	M
10^9	giga	G
10^{12}	tera	T
10^{15}	peta	P
10^{18}	exa	E

Notes

(a) A prefix is used with the gram, not the kilogram: thus Mg, not kkg.
(b) A prefix may be used for one or more of the unit symbols: thus kN m, N mm and kN mm are all acceptable.
(c) Compound prefixes should not be used: thus ns, not mμs.

1.4 Conversion factors for units

The conversion factors shown below are accurate to five significant figures where FPS is the foot-pound-second system.

1.4.1 FPS to SI units

Acceleration	
1 ft/s^2	= 0.304 80 m/s^2
Angular velocity	
1 rev/min	= 0.104 72 rad/s
Area	
1 in.^{2}	= 6.4516 cm^2
1 ft^2	= 0.092 903 m^2
1 yd^2	= 0.836 13 m^2
1 acre	= 0.404 69 ha
Density	
1 lb/ft^3	= 16.018 kg/m^3
Energy	
1 ft pdl	= 0.042 140 J
1 ft lbf	= 1.355 82 J
1 k W h	= 3.6000 MJ
1 therm	= 0.105 51 GJ
Force	
1 pdl	= 0.138 26 N
1 lbf	= 4.4482 N
Length	
1 in.	= 2.5400 cm
1 ft	= 0.304 80 m
1 yd	= 0.914 40 m
1 mi	= 1.6093 km
Mass	
1 oz	= 28.350 g
1 lb	= 0.453 59 kg
1 cwt	= 50.802 kg
1 ton	= 1.0161 tonne
Moment of force	
1 lbf ft	= 1.3558 N m
Plane angle	
1°	= 0.017 45 rad
Power	
1 ft lbf/s	= 1.3558 W
1 hp	= 0.745 70 kW
Pressure and stress	
1 in. Hg	= 33.864 mbar
1 lbf/in.^{2}	= 6.8948 kPa
1 tonf/in.^{2}	= 15.444 N/mm^2
Specific heat capacity	
1 Btu/(lb°F)	= 4.1868 kJ/(kg°C)
Velocity	
1 ft/s	= 0.304 80 m/s
1 mi/h	= 1.6093 km/h
Volume	
1 in.^{3}	= 16.387 cm^3
1 ft^3	= 0.028 317 m^3
1 yd^3	= 0.764 56 m^3
1 pt	= 0.568 26 l
1 gal	= 4.5461 l

1.4.2 SI to FPS units

1 m/s^2	= 3.2808 ft/s^2
1 rad/s	= 9.5493 rev/min
1 cm^2	= 0.155 00 in.^{2}
1 m^2	= 10.764 ft^2
1 m^2	= 1.1960 yd^2
1 ha	= 2.4711 acre
1 kg/m^3	= 0.062 428 lb/ft^3
1 J	= 23.730 ft pdl
1 J	= 0.737 56 ft lbf
1 MJ	= 0.277 78 k W h
1 GJ	= 9.4781 therm
1 N	= 7.2330 pdl
1 N	= 0.224 81 lbf
1 cm	= 0.393 70 in.
1 m	= 3.2808 ft
1 m	= 1.0936 yd
1 km	= 0.621 37 mi
1 g	= 0.035 274 oz
1 kg	= 2.2046 lb
1 kg	= 0.019 684 cwt
1 tonne	= 0.984 21 ton
1 N m	= 0.737 56 lbf ft
1 rad	= 57.296°
1 W	= 0.737 56 ft lbf/s
1 kW	= 1.3410 hp
1 mbar	= 0.029 53 in. Hg
1 kPa	= 0.145 04 lbf/in.^{2}
1 N/mm^2	= 0.064 749 tonf/in.^{2}
1 kJ/(kg °C)	= 0.238 85 Btu/(lb °F)
1 m/s	= 3.2808 ft/s
1 km/h	= 0.621 37 mi/h
1 cm^3	= 0.061 024 in.^{3}
1 m^3	= 35.315 ft^3
1 m^3	= 1.3080 yd^3
1 l	= 1.7598 pt
1 l	= 0.219 97 gal

1.5 Preferred numbers

When one is buying, say, an electric lamp for use in the home, the normal range of lamps available is 15, 25, 40, 60, 100 W and so on. These watt values approximately follow a geometric progression, roughly giving a uniform percentage change in light emission between consecutive sizes. In general, the relationship between the sizes of a commodity is not random but based on a system of *preferred numbers*.

Preferred numbers are based on *R numbers* devised by Colonel Charles Renard. The principal series used are R5, R10, R20, R40 and R80, and subsets of these series. The values within a series are approximate geometric progressions based on common ratios of $\sqrt[5]{10}$, $\sqrt[10]{10}$, $\sqrt[20]{10}$, $\sqrt[40]{10}$ and $\sqrt[80]{10}$, representing changes between various sizes within a series of 58% for the R5 series, 26% for the R10, 12% for the R20, 6% for the R40 and 3% for the R80 series.

Further details on the values and use of preferred numbers may be found in BS 2045:1965. The rounded values for the R5 series are given as 1.00, 1.60, 2.50, 4.00, 6.30 and 10.00; these values indicate that the electric lamp sizes given above are based on the R5 series. Many of the standards in use are based on series of preferred numbers and these include such standards as sheet and wire gauges, nut and bolt sizes, standard currents (A) and rotating speeds of machine tool spindles.

1.6 Mensuration

1.6.1 Plane figures

Square

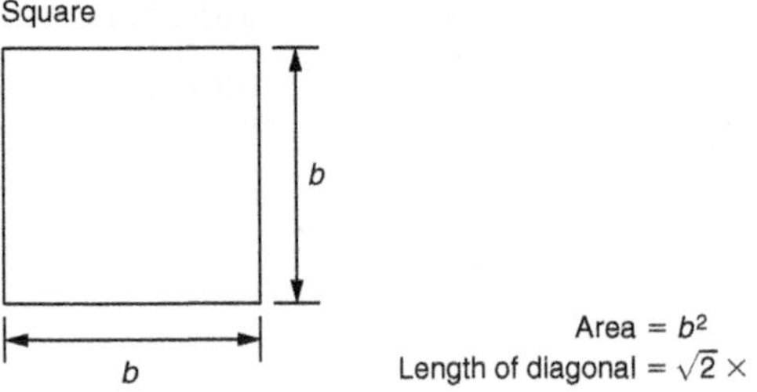

Area $= b^2$

Length of diagonal $= \sqrt{2} \times b$

Rectangle

Area $= b \times h$

Length of diagonal $= \sqrt{b^2 + h^2}$

Parallelogram

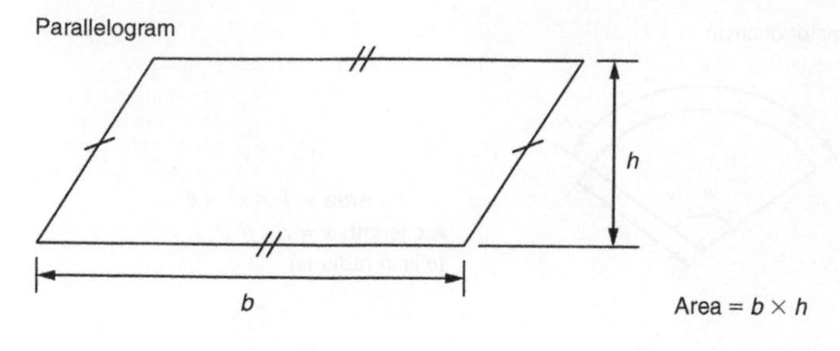

Area = $b \times h$

Trapezium

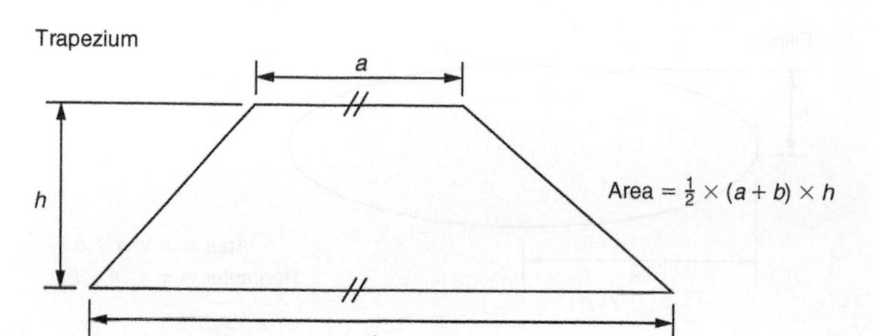

Area = $\frac{1}{2} \times (a + b) \times h$

Triangle

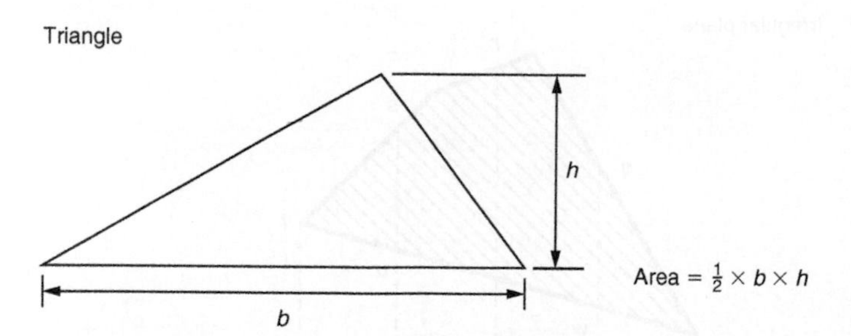

Area = $\frac{1}{2} \times b \times h$

Circle

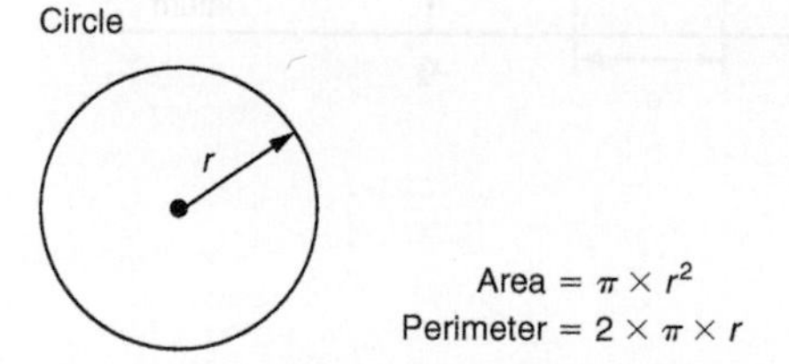

Area = $\pi \times r^2$

Perimeter = $2 \times \pi \times r$

Sector of circle

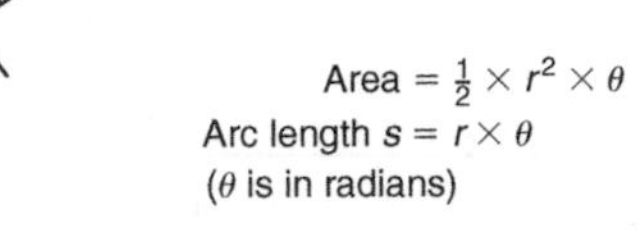

Area $= \frac{1}{2} \times r^2 \times \theta$
Arc length $s = r \times \theta$
(θ is in radians)

Ellipse

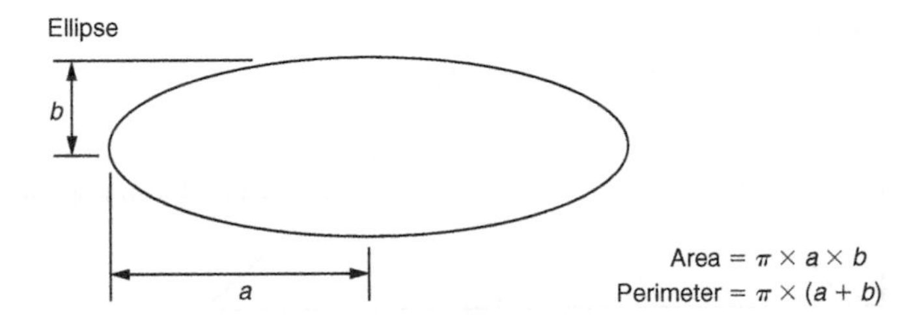

Area $= \pi \times a \times b$
Perimeter $= \pi \times (a + b)$

Irregular plane

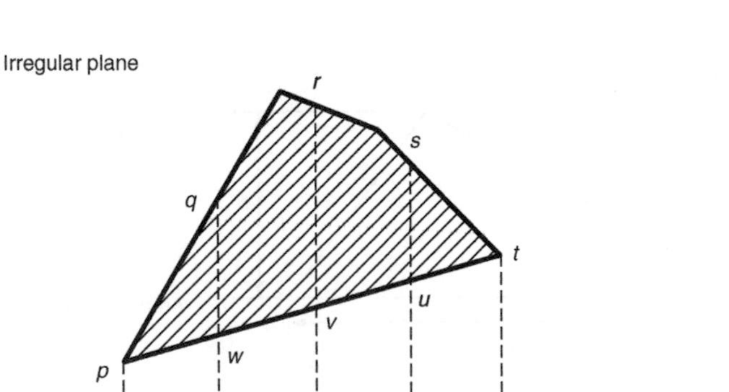
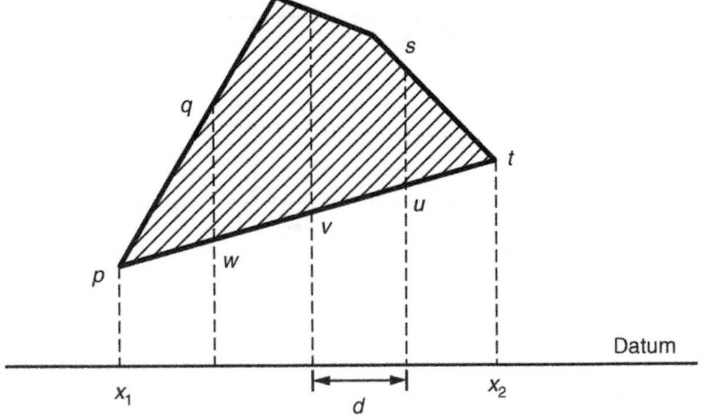

Several methods are used to find the shaded area, such as the mid-ordinate rule, the trapezoidal rule and Simpson's rule. As an example of these, Simpson's rule is as shown. Divide x_1x_2 into an even number of equal parts of width d. Let $p, q, r, \ldots$ be the lengths of vertical lines measured from some datum, and let A be the approximate area of the irregular plane, shown shaded. Then:

$$A = \frac{d}{3}[(p + t) + 4(q + s) + 2r] - \frac{d}{3}[(p + t) + 4(u + w) + 2v$$

In general, the statement of Simpson's rule is:

Approximate area = (d/3) × [(first + last) + 4(sum of evens) + 2(sum of odds)]

where first, last, evens, odds refer to ordinate lengths and d is the width of the equal parts of the datum line.

1.6.2 Solid objects

Rectangular prism

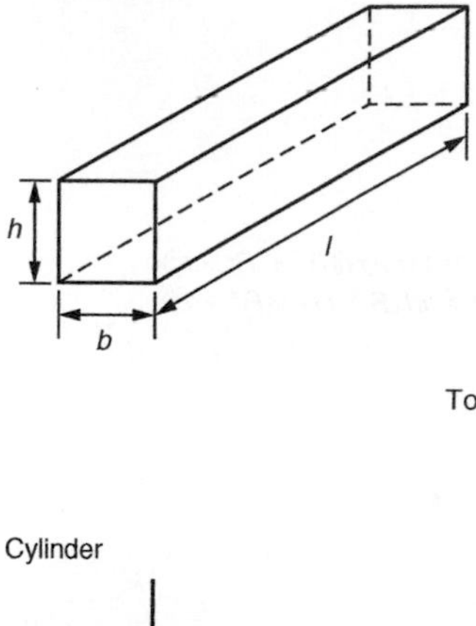

Volume = bhl

Total surface area = $2(bh + hl + lb)$

Cylinder

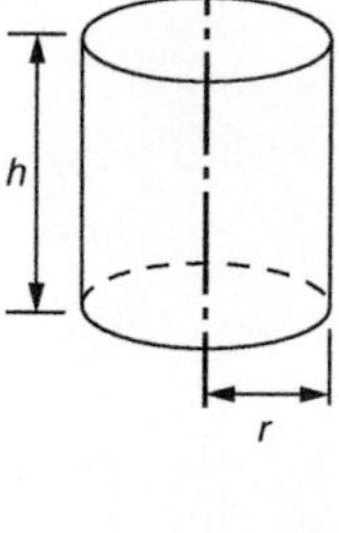

Volume = $\pi r^2 h$

Total surface area = $2\pi r(r + h)$

Cone

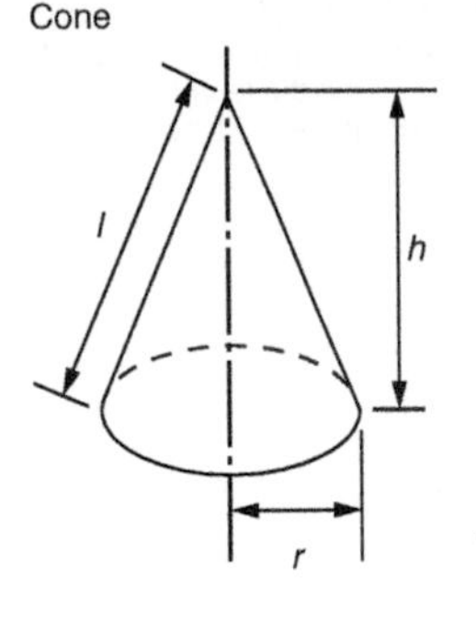

Volume = $(1/3)\pi r^2 h$

Total surface area = $\pi r(l + r)$

Frustrum of cone

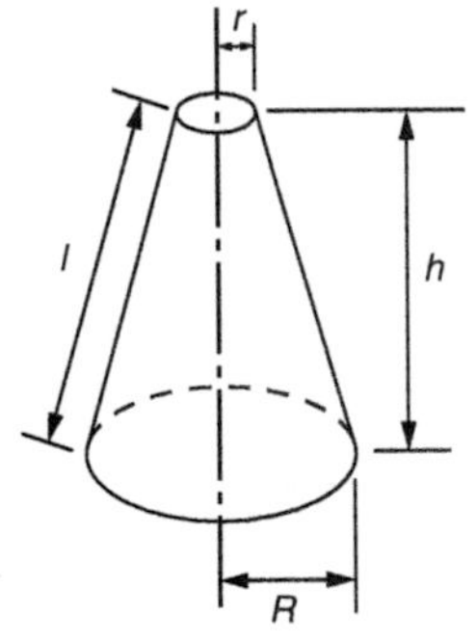

Volume = $(1/3)\pi h(R^2 + Rr + r^2)$

Total surface area = $\pi l(R + r) + \pi(R^2 + r^2)$

Sphere

r

Volume = $(4/3)\ \pi r^3$

Total surface area = $4\pi r^2$

Zone of sphere

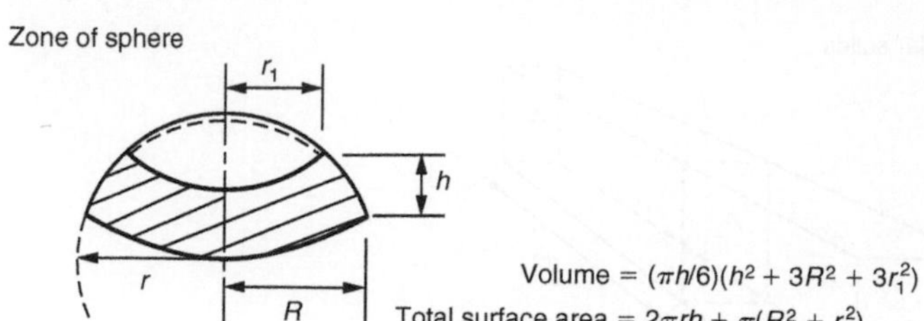

Volume $= (\pi h/6)(h^2 + 3R^2 + 3r_1^2)$
Total surface area $= 2\pi rh + \pi(R^2 + r_1^2)$
Where r is the radius of the sphere

Pyramid

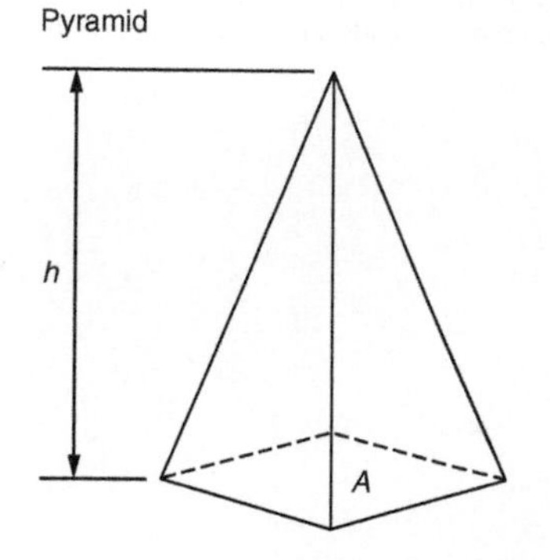

Volume $= (1/3)Ah$
Where A is the area of the base
and h is the perpendicular height

Regular solids

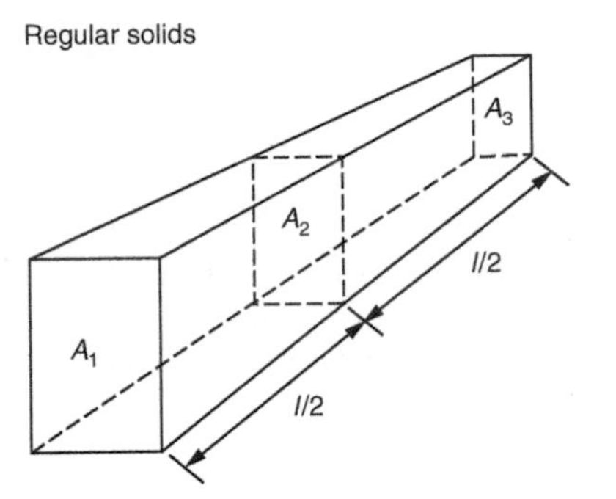

The volume of any regular solid can be found by using the prismoidal rule. Three parallel planes of areas A_1, A_3 and A_2 are considered to be at the ends and at the centre of the solid, respectively. Then:

$$\text{Volume} = (l/6)(A_1 + 4A_2 + A_3)$$

Where:
l is the length of the solid.

Irregular solids

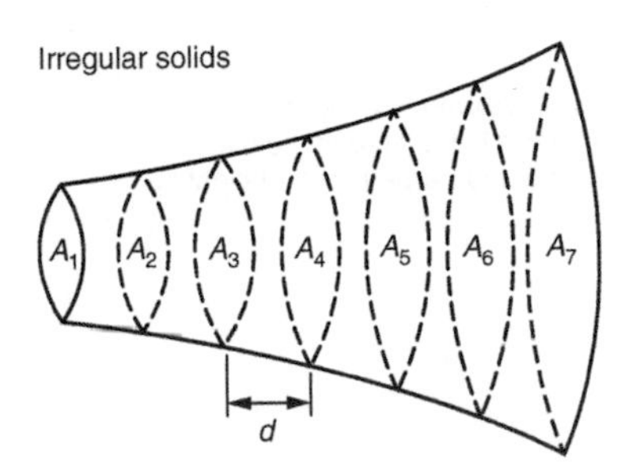

Various methods can be used to determine volumes of irregular solids; one of these is by applying the principles of Simpson's rule (see earlier this section). The solid is considered to be divided into an even number of sections by equally spaced, parallel planes, distance d apart and having areas of A_1, A_2, A_3, ... Assuming, say, seven such planes, then approximate volume = $(d/3)\,[(A_1 + A_7) + 4(A_2 + A_4 + A_6) + 2(A_3 + A_5)]$.

1.7 Powers, roots and reciprocals

n	n^2	$\sqrt{n}$	$\sqrt{10n}$	n^3	$(n)^{1/3}$	$(10n)^{1/3}$	$(100n)^{1/3}$	$1/n$
1	1	1.000	3.162	1	1.000	2.154	4.642	1.000 00
2	4	1.414	4.472	8	1.260	2.714	5.848	0.500 00
3	9	1.732	5.477	27	1.442	3.107	6.694	0.333 33
4	16	2.000	6.325	64	1.587	3.420	7.368	0.250 00
5	25	2.236	7.071	125	1.710	3.684	7.937	0.200 00
6	36	2.449	7.746	216	1.817	3.915	8.434	0.166 67
7	49	2.646	8.367	343	1.913	4.121	8.879	0.142 86
8	64	2.828	8.944	512	2.000	4.309	9.283	0.125 00
9	81	3.000	9.487	729	2.080	4.481	9.655	0.111 11
10	100	3.162	10.000	1 000	2.154	4.642	10.000	0.100 00
11	121	3.317	10.488	1 331	2.224	4.791	10.323	0.090 91
12	144	3.464	10.954	1 738	2.289	4.932	10.627	0.083 33
13	169	3.606	11.402	2 197	2.351	5.066	10.914	0.076 92
14	196	3.742	11.832	2 744	2.410	5.192	11.187	0.071 43
15	225	3.873	12.247	3 375	2.466	5.313	11.447	0.066 67
16	256	4.000	12.649	4 096	2.520	5.429	11.696	0.062 50
17	289	4.123	13.038	4 913	2.571	5.540	11.935	0.058 82
18	324	4.243	13.416	5 832	2.621	5.646	12.164	0.055 56
19	361	4.359	13.784	6 859	2.668	5.749	12.386	0.052 63
20	400	4.472	14.142	8 000	2.714	5.848	12.599	0.050 00
21	441	4.583	14.491	9 261	2.759	5.944	12.806	0.047 62
22	484	4.690	14.832	10 648	2.802	6.037	13.006	0.045 45
23	529	4.796	15.166	12 167	2.844	6.127	13.200	0.043 48
24	576	4.899	15.492	13 824	2.884	6.214	13.389	0.041 67
25	625	5.000	15.811	15 625	2.924	6.300	13.572	0.040 00
26	676	5.099	16.125	17 576	2.962	6.383	13.751	0.038 46
27	729	5.196	16.432	19 683	3.000	6.463	13.925	0.037 04
28	784	5.292	16.733	21 952	3.037	6.542	14.095	0.035 71
29	841	5.385	17.029	24 389	3.072	6.619	14.260	0.034 48
30	900	5.477	17.321	27 000	3.107	6.694	14.422	0.033 33
31	961	5.568	17.607	29 791	3.141	6.768	14.581	0.032 26

continued

Section 1.7 (*continued*)

n	n^2	$\sqrt{n}$	$\sqrt{10n}$	n^3	$(n)^{1/3}$	$(10n)^{1/3}$	$(100n)^{1/3}$	$1/n$
32	1024	5.657	17.889	32 768	3.175	6.840	14.736	0.031 25
33	1089	5.745	18.166	35 937	3.208	6.910	14.888	0.030 30
34	1156	5.831	18.439	39 304	3.240	6.980	15.037	0.029 41
35	1225	5.916	18.708	42 875	3.271	7.047	15.183	0.028 57
36	1296	6.000	18.974	46 656	3.302	7.114	15.326	0.027 78
37	1369	6.083	19.235	50 653	3.332	7.179	15.467	0.027 03
38	1444	6.164	19.494	54 872	3.362	7.243	15.605	0.026 32
39	1521	6.245	19.748	59 319	3.391	7.306	15.741	0.025 64
40	1600	6.325	20.000	64 000	3.420	7.368	15.874	0.025 00
41	1681	6.430	20.248	68 921	3.448	7.429	16.005	0.024 39
42	1764	6.481	20.494	74 088	3.476	7.489	16.134	0.023 81
43	1849	6.557	20.736	79 507	3.503	7.548	16.261	0.023 26
44	1936	6.633	20.976	85 184	3.530	7.606	16.386	0.022 73
45	2025	6.708	21.213	91 125	3.557	7.663	16.510	0.022 22
46	2116	6.782	21.448	97 336	3.583	7.719	16.631	0.021 74
47	2209	6.856	21.679	103 823	3.609	7.775	16.751	0.021 28
48	2304	6.928	21.909	110 592	3.634	7.830	16.869	0.020 83
49	2401	7.000	22.136	117 649	3.659	7.884	16.985	0.020 41
50	2500	7.071	22.361	125 000	3.684	7.937	17.100	0.020 00
51	2601	7.141	22.583	132 651	3.708	7.990	17.213	0.019 61
52	2704	7.211	22.804	140 608	3.733	8.041	17.325	0.019 23
53	2809	7.280	23.022	148 877	3.756	8.093	17.435	0.018 87
54	2916	7.348	23.238	157 464	3.780	8.143	17.544	0.018 52
55	3025	7.416	23.452	166 375	3.803	8.193	17.652	0.018 18
56	3136	7.483	23.664	175 616	3.826	8.243	17.758	0.017 86
57	3249	7.550	23.875	185 193	3.849	8.291	17.863	0.017 54
58	3364	7.616	24.083	195 112	3.871	8.340	17.967	0.017 24
59	3481	7.681	24.290	205 379	3.893	8.387	18.070	0.016 95
60	3600	7.746	24.495	216 000	3.915	8.434	18.171	0.016 67
61	3721	7.810	24.698	226 981	3.936	8.481	18.272	0.016 39
62	3844	7.874	24.900	238 328	3.958	8.527	18.371	0.016 13
63	3969	7.937	25.100	250 047	3.979	8.573	18.469	0.015 87
64	4096	8.000	25.298	262 144	4.000	8.618	18.566	0.015 62

65	4225	8.062	25.495	274 625	4.021	8.662	18.663	0.015 38
66	4356	8.124	25.690	287 496	4.041	8.707	18.758	0.015 15
67	4489	8.185	25.884	300 763	4.062	8.750	18.852	0.014 93
68	4624	8.246	26.077	314 432	4.082	8.794	18.945	0.014 71
69	4761	8.307	26.268	328 509	4.102	8.837	19.038	0.014 49
70	4900	8.367	26.458	343 000	4.121	8.879	19.129	0.014 29
71	5041	8.426	26.646	357 911	4.141	8.921	19.220	0.014 08
72	5184	8.485	26.833	373 248	4.160	8.963	19.310	0.013 89
73	5329	8.544	27.019	389 017	4.179	9.004	19.399	0.013 70
74	5476	8.602	27.203	405 224	4.198	9.045	19.487	0.013 51
75	5625	8.660	27.386	421 875	4.217	9.086	19.574	0.013 33
76	5776	8.718	27.568	438 976	4.236	9.126	19.661	0.013 16
77	5929	8.775	27.749	456 533	4.254	9.166	19.747	0.012 99
78	6084	8.832	27.928	474 552	4.273	9.205	19.832	0.012 82
79	6241	8.888	28.107	493 039	4.291	9.244	19.916	0.012 66
80	6400	8.944	28.284	512 000	4.309	9.283	20.000	0.012 50
81	6561	9.000	28.460	531 441	4.327	9.322	20.083	0.012 35
82	6724	9.055	28.636	551 368	4.344	9.360	20.165	0.012 20
83	6889	9.110	28.810	571 787	4.362	9.398	20.247	0.012 05
84	7056	9.165	28.983	592 704	4.380	9.435	20.328	0.011 90
85	7225	9.220	29.155	614 125	4.397	9.473	20.408	0.011 76
86	7396	9.274	29.326	636 056	4.414	9.510	20.488	0.011 63
87	7569	9.327	29.496	658 503	4.431	9.546	20.567	0.011 49
88	7744	9.381	29.665	681 472	4.448	9.583	20.646	0.011 36
89	7921	9.434	29.833	704 969	4.465	9.619	20.724	0.011 24
90	8100	9.487	30.000	729 000	4.481	9.655	20.801	0.011 11
91	8281	9.539	30.166	753 571	4.498	9.691	20.878	0.010 99
92	8464	9.592	30.332	778 688	4.514	9.726	20.954	0.010 87
93	8649	9.644	30.496	804 357	4.531	9.761	21.029	0.010 75
94	8836	9.695	30.659	830 584	4.547	9.796	21.105	0.010 64
95	9025	9.747	30.822	857 375	4.563	9.830	21.179	0.010 53
96	9216	9.798	30.984	884 736	4.579	9.865	21.253	0.010 42
97	9409	9.849	31.145	912 673	4.595	9.899	21.327	0.010 31
98	9604	9.899	31.305	941 192	4.610	9.933	21.400	0.010 20
99	9801	9.950	31.464	970 299	4.626	9.967	21.472	0.010 10

1.8 Progressions

A set of numbers in which one number is connected to the next number by some law is called a *series* or a *progression*.

1.8.1 Arithmetic progressions

The relationship between consecutive numbers in an arithmetic progression is that they are connected by a *common difference*. For the set of numbers 3, 6, 9, 12, 15, …, the series is obtained by adding 3 to the preceding number; that is, the common difference is 3. In general, when a is the first term and d is the common difference, the arithmetic progression is of the form:

Term	1st	2nd	3rd	4th	last
Value	a	$a + d$	$a + 2d$	$a + 3d, \dots$	$a + (n - 1)d$

where:
n is the number of terms in the progression.

The sum S_n of all the terms is given by the average value of the terms times the number of terms; that is:

$$S_n = [(\text{first} + \text{last})/2] \times (\text{number of terms})$$
$$= [(a + a + (n - 1)d)/2] \times n$$
$$= (n/2)[2a + (n - 1)d]$$

1.8.2 Geometric progressions

The relationship between consecutive numbers in a geometric progression is that they are connected by a *common ratio*. For the set of numbers 3, 6, 12, 24, 48, …, the series is obtained by multiplying the preceding number by 2. In general, when a is the first term and r is the common ratio, the geometric progression is of the form:

Term	1st	2nd	3rd	4th	last
Value	a	ar	ar^2	$ar^3, \dots$	ar^{n-1}

where:
n is the number of terms in the progression.

The sum S_n of all the terms may be found as follows:

$$S_n = a + ar + ar^2 + ar^3 + \dots + ar^{n-1} \quad (1)$$

Multiplying each term of equation (1) by r gives

$$rS_n = ar + ar^2 + ar^3 + \dots + ar^{n-1} + ar^n \quad (2)$$

Subtracting equations (2) from (1) gives:

$$S_n(1 - r) = a - ar^n \quad (3)$$
$$S_n = [a(1 - r^n)]/(1 - r)$$

Alternatively, multiplying both numerator and denominator by −1 gives:

$$S_n = [a(r^n - 1)]/(r - 1) \quad (4)$$

It is usual to use equation (3) when $r < 1$ and (4) when $r > 1$.

When $-1 > r > 1$, each term of a geometric progression is smaller than the preceding term and the terms are said to *converge*. It is possible to find the sum of all the terms of a converging series. In this case, such a sum is called the sum to infinity. The term $[a(1 - r^n)]/(1 - r)$ can be rewritten as $[a/(1 - r)] - [ar^n/(1 - r)]$. Since r is less than 1, r^n becomes smaller and smaller as n grows larger and larger. When n is very large, r^n effectively becomes zero, and thus $[ar^n/(1 - r)]$ becomes zero. It follows that the *sum to infinity of a geometric progression* is $a/(1 - r)$, which is valid when $-1 > r > 1$.

1.8.3 Harmonic progressions

The relationship between numbers in a harmonic progression is that the *reciprocals* of consecutive terms form an arithmetic progression. Thus for the arithmetic progression 1, 2, 3, 4, 5, …, the corresponding harmonic progression is 1, 1/2, 1/3, 1/4, 1/5, ….

1.9 Trigonometric formulae

1.9.1 Basic definitions

In the right-angled triangle shown below, a is the side *opposite* to angle A, b is the *hypotenuse* of the triangle and c is the side *adjacent* to angle A. By definition:

$$\sin A = \text{opp/hyp} = a/b$$
$$\cos A = \text{adj/hyp} = c/b$$
$$\tan A = \text{opp/adj} = a/c$$
$$\operatorname{cosec} A = \text{hyp/opp} = b/a = 1/\sin A$$
$$\sec A = \text{hyp/adj} = b/c = 1/\cos A$$
$$\cot A = \text{adj/opp} = c/a = 1/\tan A$$

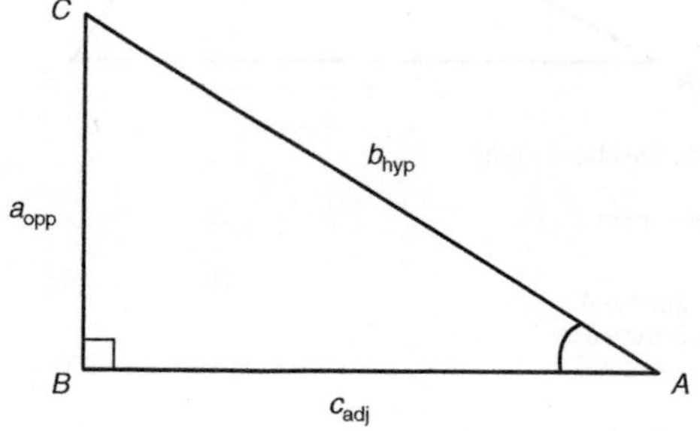

1.9.2 Identities

$\sin^2 A + \cos^2 A = 1$
$1 + \tan^2 A = \sec^2 A$
$1 + \cot^2 A = \mathrm{cosec}^2 A$
$\sin(-A) = -\sin A$
$\cos(-A) = \cos A$
$\tan(-A) = -\tan A$

1.9.3 Compound and double angle formulae

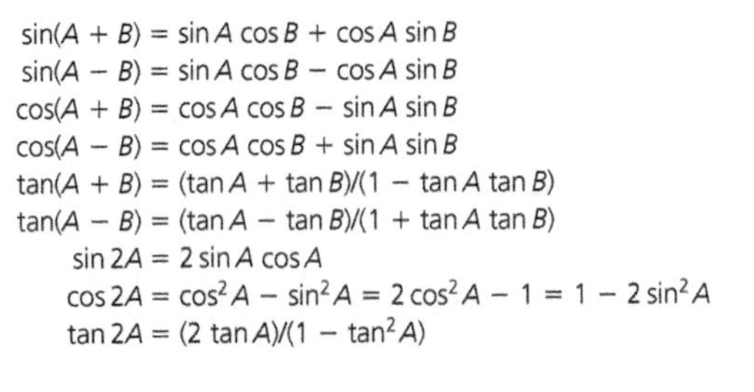

$\sin(A + B) = \sin A \cos B + \cos A \sin B$
$\sin(A - B) = \sin A \cos B - \cos A \sin B$
$\cos(A + B) = \cos A \cos B - \sin A \sin B$
$\cos(A - B) = \cos A \cos B + \sin A \sin B$
$\tan(A + B) = (\tan A + \tan B)/(1 - \tan A \tan B)$
$\tan(A - B) = (\tan A - \tan B)/(1 + \tan A \tan B)$
$\sin 2A = 2 \sin A \cos A$
$\cos 2A = \cos^2 A - \sin^2 A = 2\cos^2 A - 1 = 1 - 2\sin^2 A$
$\tan 2A = (2 \tan A)/(1 - \tan^2 A)$

1.9.4 'Product to sum' formulae

$\sin A \cos B = \frac{1}{2}[\sin(A + B) + \sin(A - B)]$
$\cos A \sin B = \frac{1}{2}[\sin(A + B) - \sin(A - B)]$
$\cos A \cos B = \frac{1}{2}[\cos(A + B) + \cos(A - B)]$
$\sin A \sin B = -\frac{1}{2}[\cos(A + B) - \cos(A - B)]$

1.9.5 Triangle formulae

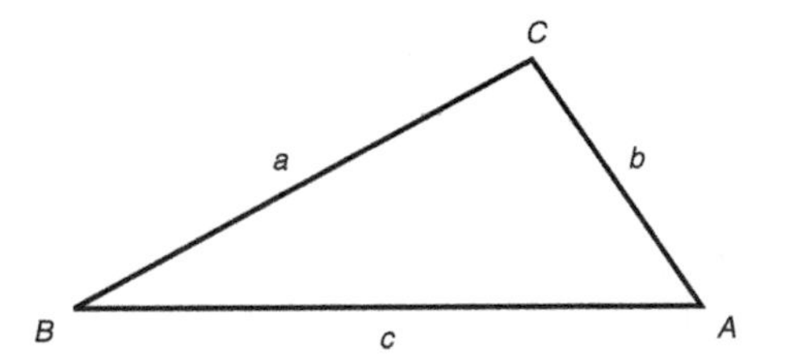

With reference to the above figure:
Sine rule:
$a/\sin A = b/\sin B = c/\sin C$

Cosine rule:
$a^2 = b^2 + c^2 - 2bc \cos A$
$b^2 = c^2 + a^2 - 2ca \cos B$
$c^2 = a^2 + b^2 - 2ab \cos C$

Area:
Area $= \frac{1}{2} ab \sin C = \frac{1}{2} bc \sin A = \frac{1}{2} ca \sin B$

Also:

Area $= \sqrt{s(s-a)(s-a)(s-c)}$

where:

s is the semi-perimeter, that is, $(a + b + c)/2$.

1.10 Circles: some definitions and properties

For a circle of diameter d and radius r:

The *circumference* is πd or $2\pi r$.

The *area* is $\pi d^2/4$ or πr^2.

An *arc* of a circle is part of the circumference.

A *tangent* to a circle is a straight line which meets the circle at one point only. A radius drawn from the point where a tangent meets a circle is at right angles to the tangent.

A *sector* of a circle is the area between an arc of the circle and two radii. The area of a sector is $\frac{1}{2}r^2\theta$, where θ is the angle in radians between the radii.

A *chord* is a straight line joining two points on the circumference of a circle. When two chords intersect, the product of the parts of one chord is equal to the products of the parts of the other chord. In the following figure, $AE \times BE = CE \times ED$.

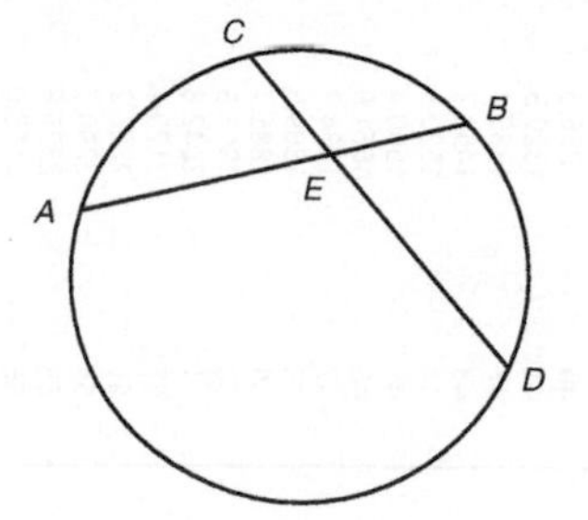

A *segment* of a circle is the area bounded by a chord and an arc. Angles in the same segment of a circle are equal: in the following figure, angle A = angle B.

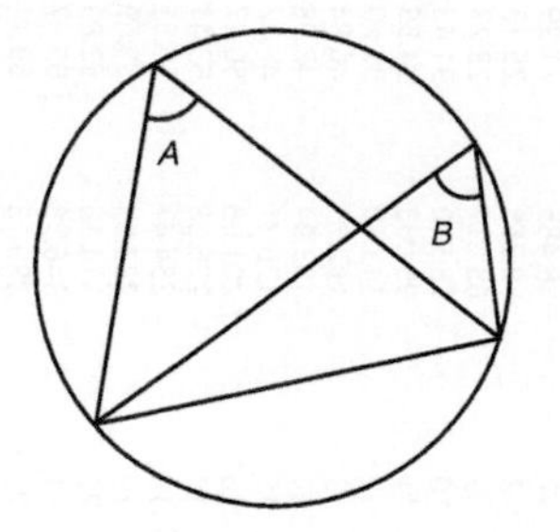

1.10.1 Circles: areas and circumferences

Diameter	Area	Circumference	Diameter	Area	Circumference	Diameter	Area	Circumference
1	0.7854	3.142	34	907.92	106.8	67	3525.7	210.5
2	3.1416	6.283	35	962.11	110.0	68	3631.7	213.6
3	7.0686	9.425	36	1017.9	113.1	69	3739.3	216.8
4	12.566	12.57	37	1075.2	116.2	70	3848.5	219.9
5	19.635	15.71	38	1134.1	119.4	71	3959.2	223.1
6	28.274	18.85	39	1194.6	122.5	72	4071.5	226.2
7	38.485	21.99	40	1256.6	125.7	73	4185.4	229.3
8	50.265	25.13	41	1320.3	128.8	74	4300.8	232.5
9	63.617	28.27	42	1385.4	131.9	75	4417.9	235.6
10	78.540	31.42	43	1452.2	135.1	76	4536.5	238.8
11	95.033	34.56	44	1520.5	138.2	77	4656.6	241.9
12	113.10	37.70	45	1590.4	141.4	78	4778.4	245.0
13	132.73	40.84	46	1661.9	144.5	79	4901.7	248.2
14	153.94	43.98	47	1734.9	147.7	80	5026.5	251.3
15	176.71	47.12	48	1809.6	150.8	81	5153.0	254.5
16	201.06	50.27	49	1885.7	153.9	82	5381.0	257.6
17	226.98	53.41	50	1963.5	157.1	83	5410.6	260.8
18	254.47	56.55	51	2042.8	160.2	84	5541.8	263.9
19	283.53	59.69	52	2123.7	163.4	85	5674.5	267.0
20	314.16	62.83	53	2206.2	166.5	86	5808.8	270.2
21	346.36	65.97	54	2290.2	169.6	87	5944.7	273.3
22	380.13	69.11	55	2375.8	172.8	88	6082.1	276.5
23	415.48	72.26	56	2463.0	175.9	89	6221.1	279.6
24	452.39	75.40	57	2551.8	179.1	90	6361.7	282.7
25	490.87	78.54	58	2642.1	182.2	91	6503.9	285.9
26	530.93	81.68	59	2734.0	185.4	92	6647.6	289.0
27	572.56	84.82	60	2827.4	188.4	93	6792.9	292.2
28	616.75	87.96	61	2922.5	191.6	94	6939.8	295.3
29	660.52	91.11	62	3019.1	194.8	95	7088.2	298.5
30	706.86	94.25	63	3117.2	197.9	96	7238.2	301.6
31	754.77	97.39	64	3217.0	201.1	97	7389.8	304.7
32	804.25	100.5	65	3318.3	204.2	98	7543.0	307.9
33	855.30	103.7	66	3421.2	207.3	99	7697.7	311.0

1.11 Quadratic equations

The solutions (roots) of a quadratic equation:

$ax^2 + bx + c = 0$

are:

$$x = \frac{-b \pm \sqrt{b^2 - 4ac}}{2a}$$

1.12 Natural logarithms

The natural logarithm of a positive real number x is denoted by $\ln x$. It is defined to be a number such that:

$e^{\ln x} = x$

where:
e = 2.1782 which is the base of natural logarithms.

Natural logarithms have the following properties:

$\ln(xy) = \ln x + \ln y$

$\ln(x/y) = \ln x - \ln y$

$\ln y^x = x \ln y$

1.13 Statistics: an introduction

1.13.1 Basic concepts

To understand the fairly advanced statistics underlying quality control, a certain basic level of statistics is assumed by most texts dealing with this subject. The brief introduction given below should help to lead readers into the various texts dealing with quality control.

The *arithmetic mean*, or *mean*, is the average value of a set of data. Its value can be found by adding together the values of the members of the set and then dividing by the number of members in the set. Mathematically:

$X = (X_1 + X_2 + \dots + X_N)/N$

Thus the mean of the set of numbers 4, 6, 9, 3 and 8 is $(4 + 6 + 9 + 3 + 8)/5 = 6$.

The *median* is either the middle value or the mean of the two middle values of a set of numbers arranged in order of magnitude. Thus the numbers 3, 4, 5, 6, 8, 9, 13 and 15 have a median value of $(6 + 8)/2 = 7$, and the numbers 4, 5, 7, 9, 10, 11, 15, 17 and 19 have a median value of 10.

The *mode* is the value in a set of numbers which occurs most frequently. Thus the set 2, 3, 3, 4, 5, 6, 6, 6, 7, 8, 9 and 9 has a modal value of 6.

The *range* of a set of numbers is the difference between the largest value and the smallest value. Thus the range of the set of numbers 3, 2, 9, 7, 4, 1, 12, 3, 17 and 4 is 17 − 1 = 16.

The *standard deviation*, sometimes called the *root mean square deviation*, is defined by:

$$s = \sqrt{[(X_1 - X)^2 + (X_2 - X)^2 + \ldots + (X_N - X)^2]/N}$$

Thus for the numbers 2, 5 and 11, the mean in (2 + 5 + 11)/3, that is 6. The standard deviation is:

$$s = \sqrt{[(2-6)^2 + (5-6)^2 + (11-6)^2]/3}$$

$$= \sqrt{(16 + 1 + 25)/3} = \sqrt{14} \approx \mathbf{3.74}$$

Usually s is used to denote the standard deviation of a population (the whole set of values) and σ is used to denote the standard deviation of a sample.

1.13.2 Probability

When an event can happen x ways out of a total of n possible and equally likely ways, the *probability* of the occurrence of the event is given by $p = x/n$. The probability of an event occurring is therefore a number between 0 and 1. If q is the probability of an event not occurring it also follows that $p + q = 1$. Thus when a fair six-sided dice is thrown, the probability of getting a particular number, say a three, is 1/6, since there are six sides and the number three only appears on one of the six sides.

1.13.3 Binomial distribution

The *binomial distribution* as applied to quality control may be stated as follows.

The probability of having 0, 1, 2, 3, ..., n defective items in a sample of n items drawn at random from a large population, whose probability of a defective item is p and of a non-defective item is q, is given by the successive terms of the expansion of $(q + p)^n$, taking terms in succession from the right.

Thus if a sample of, say, four items is drawn at random from a machine producing an average of 5% defective items, the probability of having 0, 1, 2, 3 or 4 defective items in the sample can be determined as follows. By repeated multiplication:

$$(q + p)^4 = q^4 + 4q^3p + 6q^2p^2 + 4qp^3 + p^4$$

The values of q and p are $q = 0.95$ and $p = 0.05$. Thus:

$$(0.95 + 0.05)^4 = 0.95^4 + (4 \times 0.95^3 \times 0.05) + (6 \times 0.95^2 \times 0.05^2) + (4 \times 0.95 \times 0.05^3) + 0.05^4$$
$$= 0.8145 + 0.1715 + 0.011\,354 + \ldots$$

This indicates that

(a) 81% of the samples taken are likely to have no defective items in them.
(b) 17% of the samples taken are likely to have one defective item.

(c) 1% of the samples taken are likely to have two defective items.
(d) There will hardly ever be three or four defective items in a sample.

As far as quality control is concerned, if by repeated sampling these percentages are roughly maintained, the inspector is satisfied that the machine is continuing to produce about 5% defective items. However, if the percentages alter then it is likely that the defect rate has also altered. Similarly, a customer receiving a large batch of items can, by random sampling, find the number of defective items in the samples and by using the binomial distribution can predict the probable number of defective items in the whole batch.

1.13.4 Poisson distribution

The calculations involved in a binomial distribution can be very long when the sample number n is larger than about six or seven, and an approximation to them can be obtained by using a *Poisson distribution*. A statement for this is:

When the chance of an event occurring at any instant is constant and the expectation np of the event occurring is λ, then the probabilities of the event occurring 0, 1, 2, 3, 4, ... times are given by:

$e^{-\lambda}, \lambda e^{-\lambda}, \lambda^2 e^{-\lambda}/2!, \lambda^3 e^{-\lambda}/3!, \lambda^4 e^{-\lambda}/4!, \ldots$

where:
e is the constant 2.718 28 ... and $2! = 2 \times 1$, $3! = 3 \times 2 \times 1$, $4! = 4 \times 3 \times 2 \times 1$, and so on (where 4! is read 'four factorial').

Applying the Poisson distribution statement to the machine producing 5% defective items, used above to illustrate a use of the binomial distribution, gives:

expectation $np = 4 \times 0.05 = 0.2$
probability of no defective items is $e^{-\lambda} = e^{-0.2} = 0.8187$
probability of one defective item is $\lambda e^{-\lambda} = 0.2e^{-0.2} = 0.1637$
probability of two defective items is $\lambda^2 e^{-\lambda}/2! = 0.2^2 e^{-0.2}/2 = 0.0164$

It can be seen that these probabilities of approximately 82%, 16% and 2% compare quite well with the results obtained previously.

1.13.5 Normal distribution

Data associated with measured quantities such as mass, length, time and temperature is called *continuous*, that is, the data can have any values between certain limits. Suppose that the lengths of items produced by a certain machine tool were plotted as a graph, as shown in the figure; then it is likely that the resulting shape would be mathematically definable. The shape is given by $y = (1/\sigma)e^z$, where $z = -x^2/2\sigma^2$, σ is the standard deviation of the data, and x is the frequency with which the data occurs. Such a curve is called a *normal probability* or a *normal distribution curve*.

Important properties of this curve to quality control are:

(a) The area enclosed by the curve and vertical lines at ± 1 standard deviation from the mean value is approximately 67% of the total area.

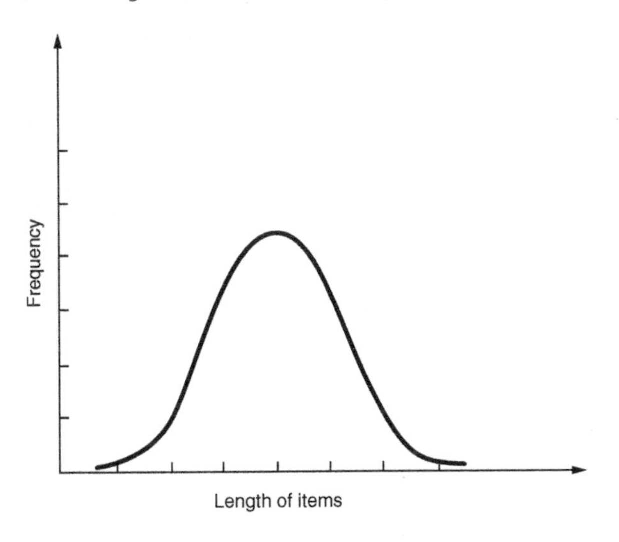

(b) The area enclosed by the curve and vertical lines at ± 2 standard deviations from the mean value is approximately 95% of the total area.
(c) The area enclosed by the curve and vertical lines at ± 3 standard deviations from the mean value is approximately 99.75% of the total area.
(d) The area enclosed by the curve is proportional to the frequency of the population.

To illustrate a use of these properties, consider a sample of 30 round items drawn at random from a batch of 1000 items produced by a machine. By measurement it is established that the mean diameter of the samples is 0.503 cm and that the standard deviation of the samples is 0.0005 cm. The normal distribution curve theory may be used to predict the reject rate if, say, only items having a diameter of 0.502–0.504 cm are acceptable. The range of items accepted is $0.504 - 0.502 = 0.002$ cm. Since the standard deviation is 0.0005 cm, this range corresponds to ± 2 standard deviations. From (b) above, it follows that 95% of the items are acceptable, that is, that the sample is likely to have 28 to 29 acceptable items and the batch is likely to have 95% of 1000, that is, 950 acceptable items.

This example was selected to give exactly ± 2 standard deviations. However, sets of tables are available of *partial areas under the standard normal curve*, which enable any standard deviation to be related to the area under the curve.

1.14 Differential calculus (Derivatives)

If $y = x^n$ then $\dfrac{dy}{dx} = nx^{n-1}$

If $y = ax^n$ then $\dfrac{dy}{dx} = anx^{n-1}$ or

If $f(x) = ax^n$ then $f'(x) = anx^{n-1}$

If $y = \sin\theta$ then $\dfrac{dy}{d\theta} = \cos\theta$

If $y = \cos\theta$ then $\dfrac{dy}{d\theta} = -\sin\theta$

If $y = \tan\theta$ then $\dfrac{dy}{d\theta} = \sec^2\theta$

If $y = \cot\theta$ then $\dfrac{dy}{d\theta} = \text{cosec}^2\theta$

If $y = \sec\theta$ then $\dfrac{dy}{d\theta} = \tan\theta\sec\theta = \dfrac{\sin\theta}{\cos^2\theta}$

If $y = \text{cosec}\,\theta$ then $\dfrac{dy}{d\theta} = -\cot\theta\,\text{cosec}\,\theta = -\dfrac{\cos\theta}{\sin^2\theta}$

If $y = \sin^{-1}\dfrac{x}{a}$ then $\dfrac{dy}{dx} = \dfrac{1}{\sqrt{a^2 - x^2}}$

If $y = \cos^{-1}\dfrac{x}{a}$ then $\dfrac{dy}{dx} = -\dfrac{1}{\sqrt{a^2 - x^2}}$

If $y = \tan^{-1}\dfrac{x}{a}$ then $\dfrac{dy}{dx} = \dfrac{a}{a^2 + x^2}$

If $y = \cot^{-1}\dfrac{x}{a}$ then $\dfrac{dy}{dx} = -\dfrac{a}{a^2 + x^2}$

If $y = \sec^{-1}\dfrac{x}{a}$ then $\dfrac{dy}{dx} = \dfrac{a}{x\sqrt{x^2 - a^2}}$

If $y = \text{cosec}^{-1}\dfrac{x}{a}$ then $\dfrac{dy}{dx} = -\dfrac{a}{x\sqrt{x^2 - a^2}}$

If $y = e^x$ then $\dfrac{dy}{dx} = e^x$

If $y = e^{ax}$ then $\dfrac{dy}{dx} = ae^{ax}$

If $y = a^x$ then $\dfrac{dy}{dx} = a^x \ln a$

If $y = \ln x$ then $\dfrac{dy}{dx} = \dfrac{1}{x}$

Product rule

If $y = uv$ where u and v are both functions of x, then $\dfrac{dy}{dx} = u\dfrac{dv}{dx} + v\dfrac{du}{dx}$

Quotient rule

If $y = \dfrac{u}{v}$ where u and v are both functions of x, then $\dfrac{dy}{dx} = \dfrac{v\dfrac{du}{dx} - u\dfrac{dv}{dx}}{v^2}$

Function of a function

If y is a function of x then $\dfrac{dy}{dx} = \dfrac{dy}{du} \times \dfrac{du}{dx}$

Successive differentiation

If $y = f(x)$ then its first derivative is written $\dfrac{dy}{dx}$ or $f'(x)$.

If this expression is differentiated a second time then the second derivative is obtained and is written $\dfrac{d^2y}{dx^2}$ or $f''(x)$.

1.15 Integral calculus (Standard forms)

$$\int x^n\,dx = \frac{x^{n+1}}{n+1} + c \quad (n \neq -1)$$

Where:
c = the constant of integration.

$$\int ax^n\,dx = \frac{ax^{n-1}}{n+1} + c \quad (n \neq -1)$$

$$\int \cos\theta\,d\theta = \sin\theta + c$$

$$\int \sin\theta\,d\theta = -\cos\theta + c$$

$$\int \sec^2\theta\,d\theta = \tan\theta + c$$

$$\int \mathrm{cosec}^2\theta\,d\theta = -\cot\theta + c$$

$$\int \tan\theta\sec\theta\,d\theta = \sec\theta + c$$

$$\int \cot\theta\,\mathrm{cosec}\,\theta\,d\theta = -\mathrm{cosec}\,\theta + c$$

$$\int \frac{dx}{\sqrt{a^2 - x^2}} = \sin^{-1}\frac{x}{a} + c$$

$$\int \frac{-dx}{\sqrt{a^2 - x^2}} = \cos^{-1}\frac{x}{a} + c$$

$$\int \frac{adx}{a^2 + x^2} = \tan^{-1}\frac{x}{a} + c$$

$$\int \frac{-adx}{a^2 + x^2} = \cot^{-1}\frac{x}{a} + c$$

$$\int \frac{adx}{x\sqrt{x^2 - a^2}} = \sec^{-1}\frac{x}{a} + c$$

$$\int \frac{-adx}{x\sqrt{x^2 - a^2}} = \operatorname{cosec}^{-1}\frac{x}{a} + c$$

$$\int e^x\,dx = e^x + c$$

$$\int e^{ax} = \frac{e^{ax}}{a} + c$$

$$\int a^x\,dx = \frac{ax}{\ln a} + c$$

$$\int \frac{dx}{x} = \ln x + c$$

$$\int \sinh x\,dx = \cosh x + c$$

$$\int \cosh x\,dx = \sinh x + c$$

$$\int \operatorname{sech}^2 x\,dx = \tanh x + c$$

$$\int \frac{dx}{\sqrt{a^2 + x^2}} = \sinh^{-1}\frac{x}{a} + c \quad \text{or} \quad \ln\left[\frac{x + \sqrt{x^2 + a^2}}{a}\right] + c$$

$$\int \frac{dx}{\sqrt{x^2 - a^2}} = \cosh^{-1}\frac{x}{a} + c \quad \text{or} \quad \ln\left[\frac{x\sqrt{x^2 - a^2}}{a}\right] + c$$

$$\int \frac{dx}{a^2 - x^2} = \frac{1}{a}\tanh\frac{x}{a} + c \quad \text{or} \quad \frac{1}{2a}\ln\frac{(a + x)}{(a - x)} + c$$

1.15.1 Integration by parts

$$\int u\,dv = uv - \int v\,du + c$$

1.15.2 Definite integrals

The foregoing integrals contain an arbitrary constant '*c*' and are called *indefinite integrals*.

Definite integrals are those to which limits are applied thus: $[x]_a^b = (b) - (a)$, therefore:

$$y = \int_1^3 x^2\,dx = \left[\frac{x^3}{3} + c\right]_1^3 = \left[\frac{3^3}{3} + c\right] - \left[\frac{1^3}{3} + c\right]$$

$$= (9 + c) - \left(\frac{1}{3} + c\right)$$

$$= \mathbf{8\frac{2}{3}}$$

Note how the constant of integration (*c*) is eliminated in a definite integral.

1.16 Binomial theorem

$$(a + x)^n = a^n + na^{n-1}x + \frac{n(n-1)}{2!}a^{n-2}x^2 + \frac{n(n-1)(n-2)}{3!}a^{n-3}x^3 + \ldots + x^n$$

where 3! is factorial 3 and equals $1 \times 2 \times 3$.

1.17 Maclaurin's theorem

$$f(x) = f(0) + xf'(0) + \frac{x^2}{2!}f''(0) + \ldots\ldots$$

1.18 Taylor's theorem

$$f(x + h) = f(x) + hf'(x) + \frac{h^2}{2!}f''(x) + \ldots\ldots$$

For further information on Engineering Mathematics the reader is referred to the following Pocket Book: Newnes Engineering Mathematics Pocket Book, Third Edition, John Bird, 0750649925

2 Engineering Statics

2.1 Engineering statics

Engineering mechanics can be divided into *statics* and *dynamics*. Basically, *statics* is the study of forces and their effects in equilibrium, whereas *dynamics* is the study of objects in motion.

2.2 Mass, force and weight

2.2.1 Mass

- Mass is the quantity of matter in a body. It is the sum total of the masses of all the sub-atomic particles in that body.
- Matter occupies space and can be solid, liquid or gaseous.
- Unless matter is added to or removed from a body, the mass of that body never varies. It is constant under all conditions. There are as many atoms in a kilogram of, say, metal on the moon as there are in the same kilogram of metal on planet Earth. However, that kilogram of metal will *weigh* less on the moon than on planet Earth (see Section 2.2.4).
- The basic unit of *mass* is the *kilogram* (kg). The most commonly used multiple of this basic unit is the tonne (1000 kg) and the most commonly used sub-multiple is the gram (0.001 kg).

2.2.2 Force

To understand how *mass* and *weight* are related it is necessary to consider the concept of *force*. A force or a system of forces cannot be seen; only the effect of a force or a system of forces can be seen. That is:

- A force can change or try to change the shape of an object.
- A force can move or try to move a body that is at rest. If the magnitude of the force is sufficiently great it will cause the body to move in the direction of the application of the force. If the force is of insufficient magnitude to overcome the resistance to movement it will still *try* to move the body, albeit unsuccessfully.
- A force can change or try to change the motion of a body that is already moving. For example:
  1. The force of a headwind can decrease the speed of an aircraft over the ground, whereas the force of a tailwind can increase the speed of an aircraft over the ground.
  2. The force of a crosswind can cause a car to swerve off course.

- The effect of any force depends on the following:
  1. The magnitude (size) of the force measured in newtons (N).
  2. The direction of the force.
  3. The point of application of the force.
  4. The ability of a body to resist the effects of the force.
- As already stated a force cannot be seen; however, it can be represented by a *vector*.

2.2.3 Vectors

Quantities can be divided into two categories:

- Scalar quantities
- Vector quantities.

Scalar quantities have *magnitude* (size) only. For example:

- The speed of a car can be stated as 50 kilometres per hour (km/h).
- The power of an engine can be rated as 150 kilowatts (kW).
- The current flowing in an electric circuit can be 12 amperes (A).

Vector quantities have both *magnitude* and *direction*. For example, the velocity of a car is 50 km/h due north from London. Velocity is a vector quantity since it has both magnitude (speed) and direction (due north).

Similarly, vectors can be used to represent forces and systems of forces. Figure 2.1 shows how a vector can be used to represent a force (F) acting on the point (P) in a direction at 30° to the horizontal with a magnitude of 150 newtons (N). Vector diagrams will be considered in detail in Section 2.3.

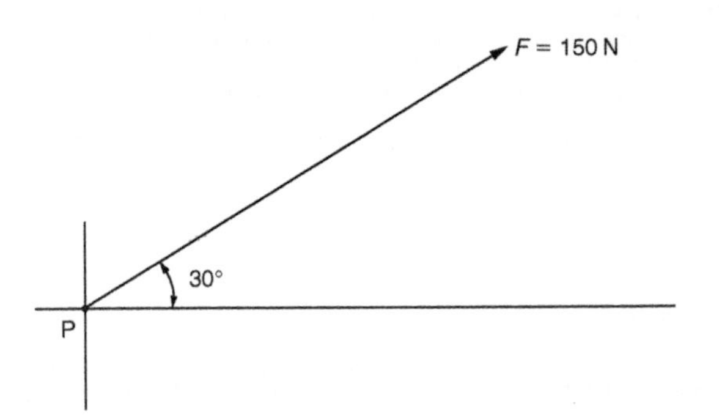

Figure 2.1 Vector representation of a force.

To represent a force by a vector the following information is required:

- The point of application of the force.
- The magnitude (size) of the force.
- The line of action of the force.
- The direction in which the force is acting. The direction in which the force is acting is shown by the arrowhead. This can be at the end of the vector or at the midpoint of the vector, whichever is the more convenient.

2.2.4 Weight

A solid, liquid or gas of mass 1 kg will weigh 9.81 N due to the gravitational attraction of planet Earth. Since planet Earth is not a perfect sphere a mass of 1 kg will weigh slightly more at the poles and slightly less at the equator. However, for all practical purposes, the *average weight* of a mass of 1 kg is taken to be 9.81 N on planet Earth:

- The *force of gravity* attracts the *mass* of any body (solid, liquid or gas) towards the centre of planet Earth.
- The *weight* of a body is the force of gravity acting on that body.
- Therefore weight is a *force* measured in *newtons*.
- The factor 9.81 is the free-fall acceleration of a body due to the force gravity, it is 9.81 metres per second squared (m/s^2). Acceleration due to gravity is discussed further in Section 3.6.7.
- Since the weight of a body of mass 1 kg is approximately 9.81 N (average value) on planet Earth, the weight of that body will change on a planet having a different force of gravity or when travelling through space. For example, the gravitational attraction of the Moon is approximately 1/6 that of planet Earth. Therefore a mass of 1 kg will weigh $1/6 \times 9.81 = 1.64$ N on the Moon.

2.2.5 Mass per unit volume (density)

A solid object has a volume of 60 cm^3. If the mass of this solid is 600 grams (g) then, by proportion,

1 cm^3 has a mass of 600 g/60 cm^3 = 10 gram per cm^3 (10 g/cm^3)

Since 1 cm^3 is unit volume in this instance, then the *mass per unit volume* of the solid shown is 10 g/cm^3. *Mass per unit volume* of any body is also called the *density* of that body. The symbol for density is the Greek letter 'rho' (ρ). The general expression for mass per unit volume (density) is:

$$\rho = m/V$$

where:
ρ = density
m = mass
V = volume

The densities of some common solids, liquids and gases used in engineering are given in Table 2.1.

2.2.6 Weight per unit volume

The weight per unit volume of a substance can be determined by:

(a) direct measurement,
(b) conversion calculation from the density of the substance.

(a) Weight per unit volume by direct measurement:

Weight per unit volume (N/m^3) = weight (N)/volume (m^3) = W/V

Table 2.1 Density of common substances.

Substance	Density	
	kg/m³	g/cm³
Aluminium	2720	2.72
Brass	8480	8.48
Copper	8790	8.79
Cast iron	7200	7.20
Lead	11 350	11.35
Nylon	1120	1.12
PVC	1360	1.36
Rubber	960	0.96
Steel	7820	7.82
Tin	7280	7.28
Zinc	7120	7.12
Alcohol	800	0.80
Mercury	13 590	13.59
Paraffin	800	0.80
Petrol	720	0.72
Water (pure at 4°C)	1000	1.00
Acetylene	1.17	0.001 7
Air (dry)	1.30	0.001 3
Carbon dioxide	1.98	0.001 98
Hydrogen	0.09	0.000 09
Nitrogen	1.25	0.001 25
Oxygen	1.43	0.001 43

(b) Weight per unit volume by conversion calculation:

$$\text{Weight per unit volume} = W/V \quad (1)$$

$$\text{Density } (\rho) = m/V \quad (2)$$

$$\text{Weight} = m \times 9.81 \quad (3)$$

Substituting equation (3) in equation (2):

$$\text{Weight per unit volume} = (m \times 9.81)/V$$

But:

$m/V = \rho$

Therefore:

Weight per unit volume = $\rho \times 9.81$

Note:

- This conversion only applies to planet Earth. Anywhere else in space the local value of gravitational acceleration '*g*' must be used.
- The value for density must be expressed in *kilograms per unit volume*.

2.2.7 Relative density

The relative density of a substance may be defined as the ratio:

(Density of the substance)/(density of pure water at 4°C)

or

(Mass of the substance)/(mass of an equal volume of pure water at 4°C)

Since relative density is a ratio there are no units.

The relative densities of some common substances used in engineering are given in Table 2.2.

Table 2.2 Relative densities of some common substances.

Substance	Relative density (*d*)
Aluminium	2.72
Brass	8.48
Cadmium	8.57
Chromium	7.03
Copper	8.79
Cast iron	7.20
Lead	11.35
Nickel	8.73
Nylon	1.12
PVC	1.36
Rubber	0.96
Steel	7.82
Tin	7.28
Zinc	7.12
Alcohol	0.80
Mercury	13.95
Paraffin	0.80
Petrol	0.72
Water (pure @ 4°C)	1.00
Water (sea)	1.02
Acetylene	0.001 7
Air (dry)	0.001 3
Carbon dioxide	0.001 98
Carbon monoxide	0.001 26
Hydrogen	0.000 09
Nitrogen	0.001 25
Oxygen	0.001 43

2.2.8 Pressure (fluids)

The pressure acting on the surface area of a body is defined as the *normal force per unit area* acting on the surface. Normal force means the force acting at right angles to the surface. The SI unit of fluid pressure is the pascal (Pa), where 1 Pa is equal to 1 Nm^2.

Pressure (p) = F/A

where:

p = pressure in pascals

F = force in newtons acting at right angles (normal) to the surface

A = area of the surface in metres squared

Similarly:

$1\ MPa = 1\ MN/m^2 = 1\ N/mm^2$

Since the pascal is a very small unit a more practical unit is the *bar*, where:

1 bar = 10^5 pascals.

Note:

- *Atmospheric pressure* is usually expressed in *millibars.*
- *Gauge pressure* is the fluid pressure indicated by a bourdon tube pressure gauge, a U-tube manometer or similar device so that the indicated pressure is additional to the atmospheric pressure.
- *Absolute pressure* used in physical science and thermodynamics is the pressure relative to a perfect vacuum, that is:

  Absolute pressure = atmospheric pressure + gauge pressure

2.3 Vector diagrams of forces: graphical solution

It has already been stated that forces cannot be seen so they cannot be drawn; however, the effect of forces and systems of forces can be represented by *vectors*. Vectors were introduced in Section 2.2.2.

2.3.1 Resultant forces

We call a force that can replace two or more other forces and produce the same effect a *resultant force.* For example, the forces F_1 and F_2 acting in the same direction on the point P along the same line of action can be replaced by a single force having the same effect as shown in Fig. 2.2(a).

Since both the forces F_1 and F_2 have a common line of action and act in the same direction, their vector lengths can simply be added together to obtain the magnitude of the vector of the single force that can replace them. This is the force F_R shown in Fig. 2.2(b). It is called the *resultant force.* Note that the vector for a resultant force has a double arrowhead to distinguish it from the other forces acting in the system.

Drawn to a scale of 1 cm = 5 N, the vector representing F_1 will be 4 cm long and the vector representing F_2 will be 5 cm long. When added together to obtain the vector length of the resultant force (F_R), this becomes a single vector of 4 cm + 5 cm = 9 cm long representing a force of 45 N acting at the point P.

Had the forces been acting in *opposite* directions along the same line of action, then the resultant force (F_R) could have been determined by simply subtracting the vector length for F_1 from the vector length for F_2 as shown in Fig. 2.2(c).

2.3.2 Parallelogram of forces

When two or more forces whose lines of action lie at an angle to each other act on a single point P, their resultant force cannot be determined by simple arithmetical addition or subtraction. If only two forces are involved, as shown in Fig. 2.3, the magnitude and direction of the resultant force can be determined by drawing, to scale, a parallelogram of forces.

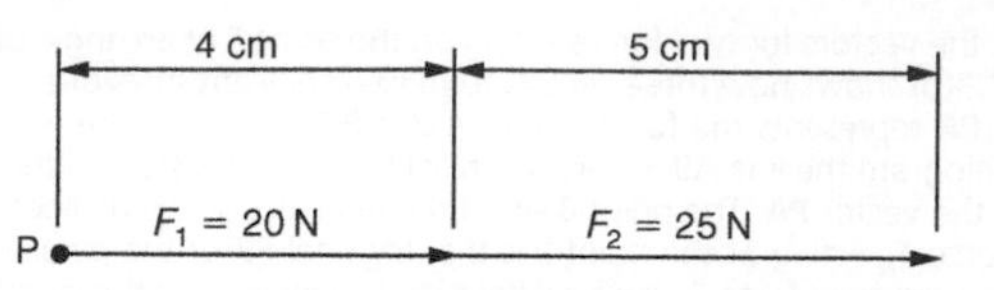

(a) Forces F_1 and F_2 acting on point P

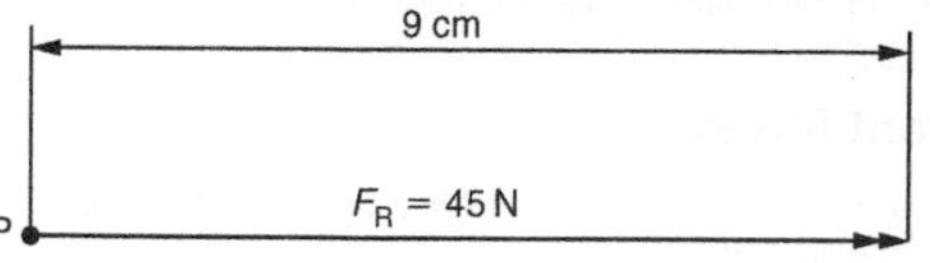

(b) Resultant of forces F_1 and F_2

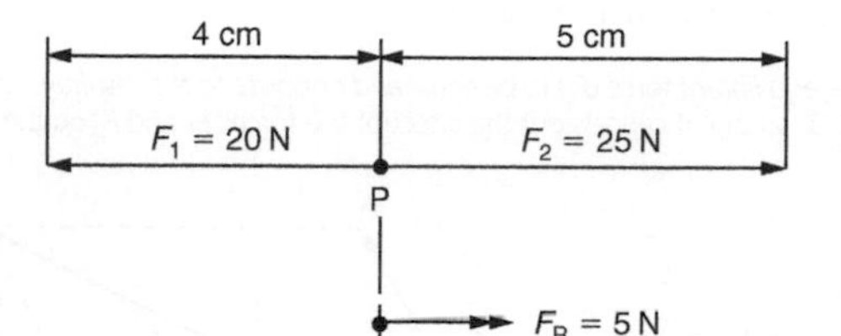

(c) Resultant of forces F_1 and F_2 acting in opposing directions

Scale: 1 cm = 5 N

Figure 2.2 Resultant forces.

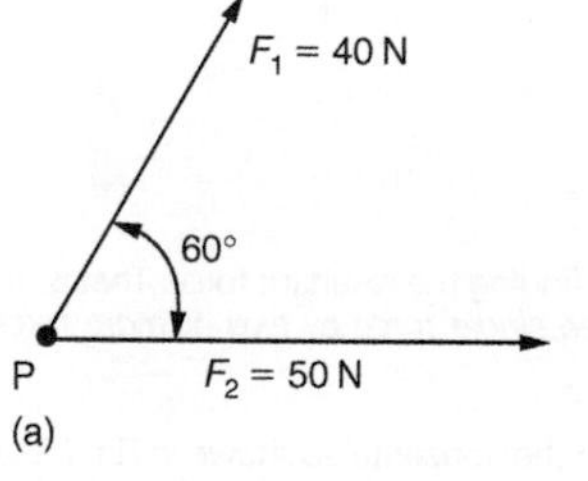

(a)

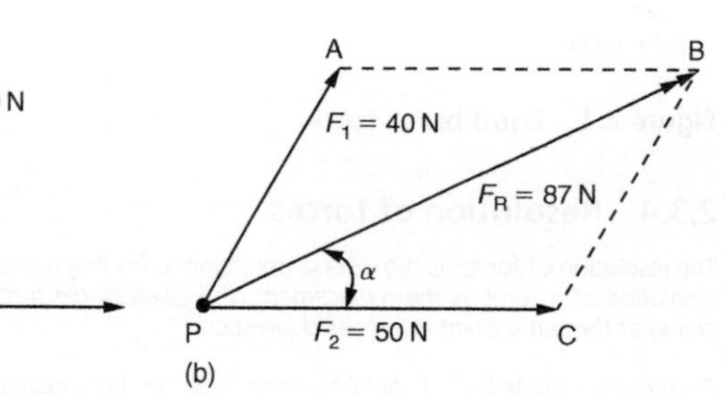

(b)

Scale: 1 cm = 10 N

Figure 2.3 Parallelogram of forces.

Figure 2.3(a) shows the vectors for two forces acting on the point P at an angle of 60° to each other. Figure 2.3(b) shows how these vectors form two adjacent sides of a parallelogram PABC. Vector **PA** represents the force F_1 and vector **PC** represents the force F_2. To complete the parallelogram the line AB is drawn parallel to the vector **PC** and the line CB is drawn parallel to the vector **PA**. The point B lies at the intersection of the lines AB and CB. The resultant force F_R acting at the point P is the diagonal PB of the parallelogram. The magnitude of the resultant force F_R can be determined by measuring the length of the diagonal PB and multiplying it by the scale of the diagram (1 cm = 10 N in this instance). The angle α can be determined by use of a protractor. The values for Fig. 2.3 are: $F_R = 87$ N and $\alpha = 23°$ (to the nearest whole number).

2.3.3 Equilibrant forces

A force which cancels out the effect of another force or system of forces is called the *equilibrant force* (F_E). An equilibrant force:

- has the same magnitude as the resultant force,
- has the same line of action as the resultant force,
- acts in the opposite direction to the resultant force.

Figure 2.4 shows the *equilibrant* force (F_E) to be *equal and opposite to the resultant force* (F_R), determined in Fig. 2.3, so that it cancels out the effect of the forces F_1 and F_2 on the point P.

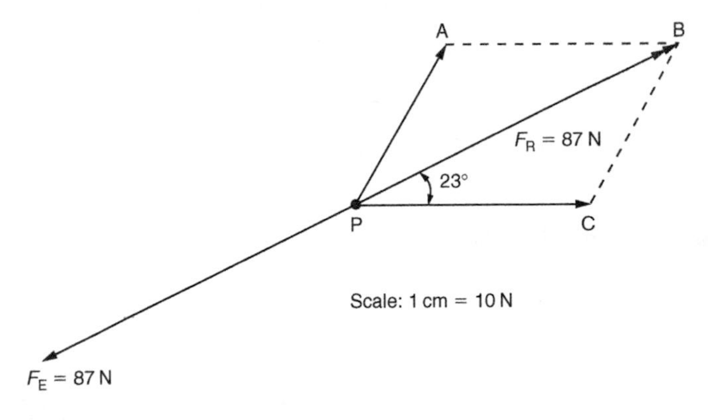

Figure 2.4 Equilibrant force.

2.3.4 Resolution of forces

The resolution of forces is the reverse operation to finding the resultant force. That is, *the resolution of a force is the replacement of a given single force by two or more forces acting at the same point in specified directions.*

Consider the single force F_S of 50 N acting at 30° to the horizontal as shown in Fig. 2.5(a). To resolve the force into its horizontal and vertical component forces first draw the lines of action of these forces from the point P as shown in Fig. 2.5(b). Then complete the parallelogram of forces as shown in Fig. 2.5(b). The magnitude of the forces F_V and F_H can be obtained by scaling the drawing (vector diagram) or by the use of trigonometry.

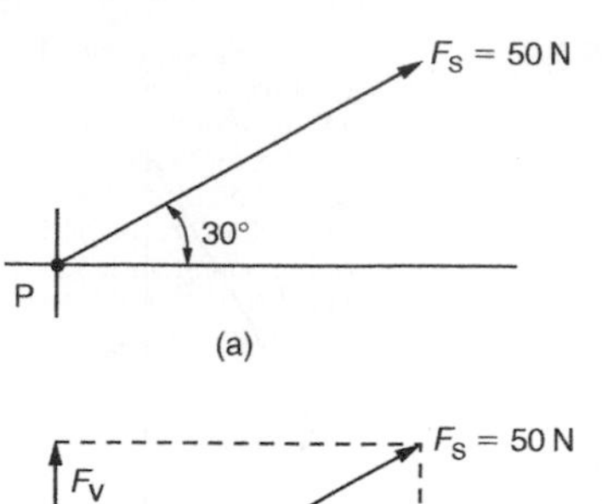

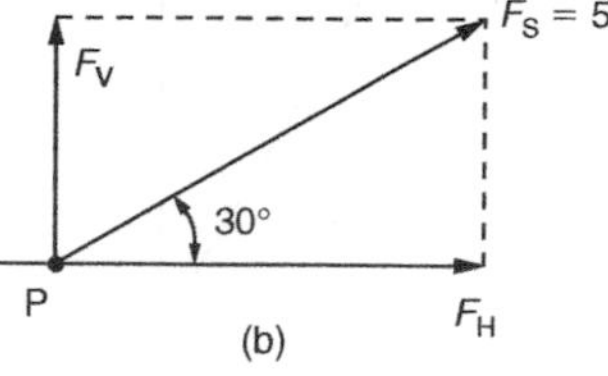

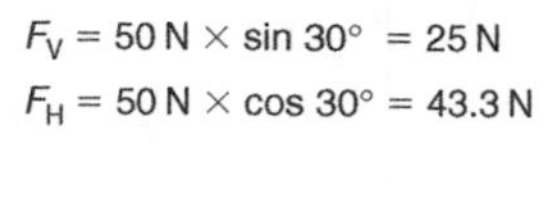

Scale: 1 cm = 10 N

Figure 2.5 Resolution of forces.

2.3.5 Three forces in equilibrium (triangle of forces)

- The vectors of three forces acting in the same plane on a body can be drawn on a flat sheet of paper.
- Forces that act in the same plane are said to be *coplanar*.
- If the three forces acting on a body are in equilibrium, any one force balances out the effects of the other two forces. The three forces may be represented by a *triangle* whose sides represent the vectors of those forces.
- If the lines of action of the three forces act through the same point, they are said to be *concurrent*.

Figure 2.6 shows three coplanar forces acting on a body. Their lines of action are extended backwards to intersect at the point P. The point P is called the *point of concurrency*. If the forces are in equilibrium:

- Their effects will cancel out.
- The body on which they are acting will remain stationary.
- The vectors of the forces will form a closed triangle.
- The vectors follow each other in a closed loop.

Figure 2.7 shows a further worked example. The accuracy of the answers depends on the accuracy of the drawing. This should be as large as possible to aid measurement.

2.3.6 Polygon of forces: Bow's notation

The polygon of forces is used to solve any number of concurrent, coplanar forces acting on a point P. They are concurrent because they act on a single point P. They are coplanar because

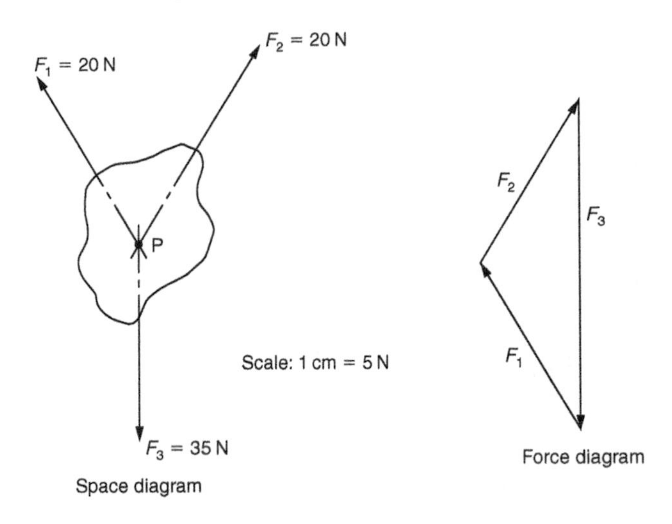

Figure 2.6 Triangle of forces.

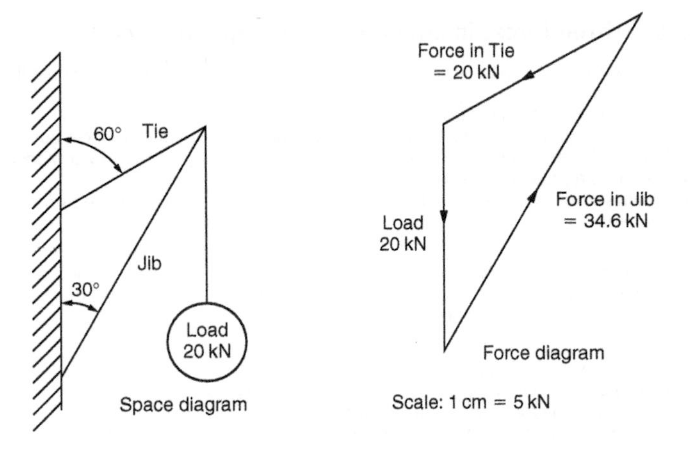

Figure 2.7 Forces in a simple jib crane.

they lie in the same plane. Figure 2.8(a) shows a *space diagram* with the spaces between the forces designated by capital letters. The corresponding *force diagram* is shown in Fig. 2.8(b) with the forces labelled using lower case letters. This is known as *Bow's notation*:

- The forces acting on point P should be represented by vectors whose length and angles must be drawn to scale as accurately as possible. The spaces between the forces are

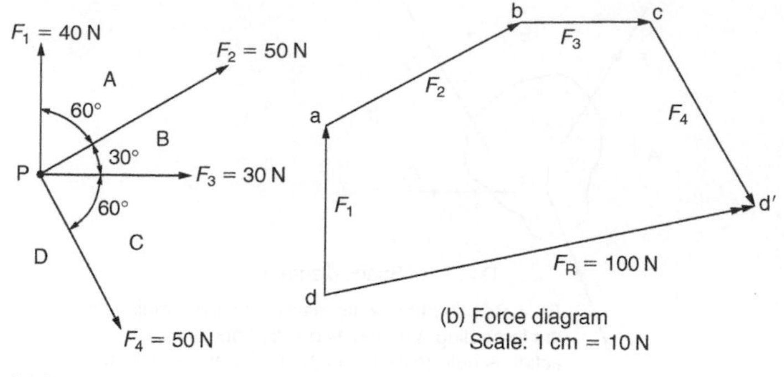

Figure 2.8 Polygon of forces (using Bow's notation).

designated using capital letters. The starting point is not important but the lettering must follow round consecutively, usually in a clockwise direction.

- The force diagram must be drawn strictly to scale, starting with a convenient vector. For example, start with the horizontal force F_3. Since it lies between the spaces D and A this vector is labelled **da**. Remember that conventionally the *spaces* are given *capital* letters whereas the *forces* are given the corresponding *lower case* letters. Add the arrowhead to show the direction in which the force is acting.
- The next force (F_4) lies between the spaces A and B. It is drawn to scale following on from the end of the first vector and labelled **ab**. Again add the arrowhead to indicate the direction in which the force is acting.
- Similarly, add the remaining forces so that they follow on in order labelling them accordingly.
- Finally, the resultant force (F_R) can be determined by joining the points and in the force diagram. Note that the direction of the resultant force is against the general flow of the given forces in the force diagram. If required, the resultant force can now be transferred to the space diagram to show its magnitude and direction relative to the point P.
- Remember that the resultant force can replace all the given forces yet still have the same effect on the point P.
- If the equilibrant had been required then it would have the same magnitude as the resultant force but it would act in the opposite direction. It would follow the general flow of the given forces in the force diagram.

2.3.7 Non-concurrent coplanar forces (funicular link polygon)

As their name implies, non-concurrent coplanar forces lie in the same plane but do not act at a single point (point of concurrency). An example is shown in Fig. 2.9(a). Applying Bow's notation to the space diagram, the resultant force (F_R) can be determined from the force diagram in the usual way as shown in Fig. 2.9(b). However before the position and

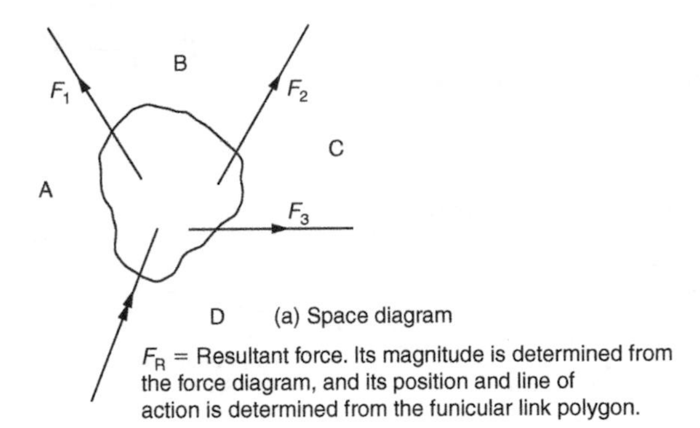

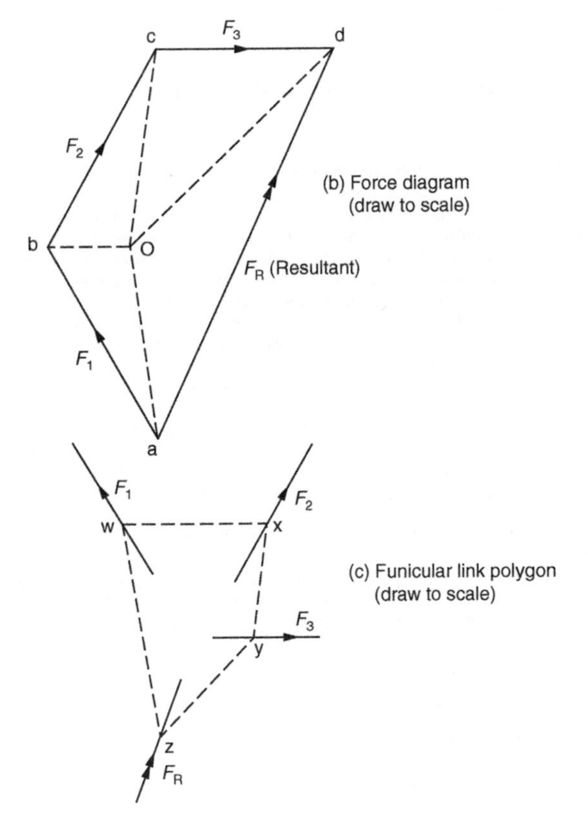

Figure 2.9 The funicular link polygon.

line of action of the resultant force can be determined it is necessary to construct a *funicular link polygon* as shown in Fig. 2.9(c) as follows:

- Draw the force diagram (Fig.2.9(b)) in the usual way, and obtain the magnitude and direction of the resultant force (F_R) as given by the line ad.
- Choose the pole-point O so that it lies inside or outside the force diagram and join the pole-point O to each corner of the force diagram as shown in Fig. 2.9(b). The position of the pole-point O is not critical and can be placed wherever it is most convenient.
- From any point *w* in the line of action of force F_1 draw a line in space A parallel to the line Oa in the force diagram. This line is shown in Fig. 2.9(c).
- From the point w draw the line wx in space B parallel to the line Ob in the space diagram and cutting the force F_2 at the point x. This line can be found in Fig. 2.9(c). It may be necessary to produce the lines of action of the forces to get the necessary intersection.
- Continue with this construction until the lines have been drawn in all the spaces, the space should now resemble Fig. 2.9(c). The lines parallel to Oa and Od can now be produced until they meet at z. Through z draw a line parallel to ad in the force diagram to represent the line of action of the vector of the resultant force FR.

Note, although the resolution of forces and systems of forces by graphical methods is simple and straightforward, the accuracy depends on the scale of the drawing and the quality of the draughtsmanship. Where more accurate results are required they can be obtained by mathematical calculation using trigonometry.

2.4 Moments of forces, centre of gravity and centroids of areas

2.4.1 Moments of forces

Figure 2.10(a) shows how a spanner is used to turn a nut. The force (F) is not acting directly on the nut but a distance (d) from the axis of the nut. This produces a *turning effect* called the *moment of the force* about the axis of the nut. It is also called the *turning moment* or *couple* but all these terms mean the same thing. The distance (d) is called the *moment arm* or the *leverage distance*, both terms being interchangeable. Expressed mathematically:

$$M = F \times d$$

where:
M = moment of the force (Nm)
F = the applied force (N)
d = the moment arm (leverage distance) (m)

Note: It is essential that the moment arm (d) is measured perpendicularly to the line of action of the force. If the force is inclined as shown in Fig. 2.10(b) the effective length of the spanner is shortened. The equation becomes:

$$M = F \times d_e$$

where:
M = moment of the force (Nm)
F = the applied force (N)
d_e = the effective moment arm (leverage distance) (m)

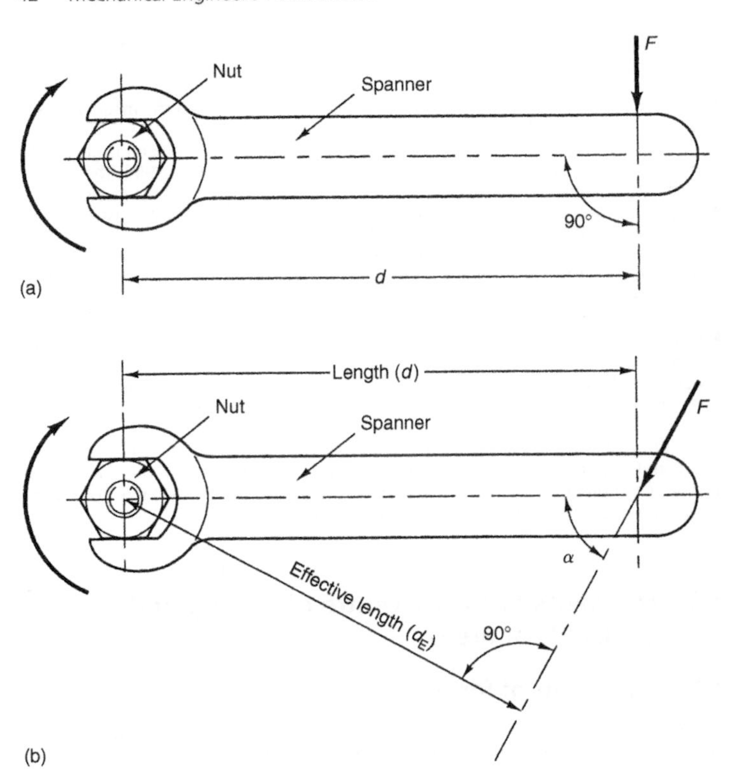

Figure 2.10 Turning moments.

2.4.2 Principle of moments (related terminology)

Fulcrum	The pivot point or axis about which rotation takes place or tends to take place.
Moment arm	The distance between the line of action of an applied force and the fulcrum point measured at right angles to the line of action of the force.
Clockwise moment	Any moment of a force that rotates or tends to rotate a body about its fulcrum in a clockwise direction. Clockwise moments are considered to be positive (+) as shown in Fig. 2.11.
Anticlockwise moment	Any moment of a force that rotates or tends to rotate a body about its fulcrum in an anticlockwise direction. Anticlockwise moments are considered to be negative (−) as shown in Fig. 2.11.
Resultant moment	This is the difference in magnitude between the total clockwise moments and the total anticlockwise moments.

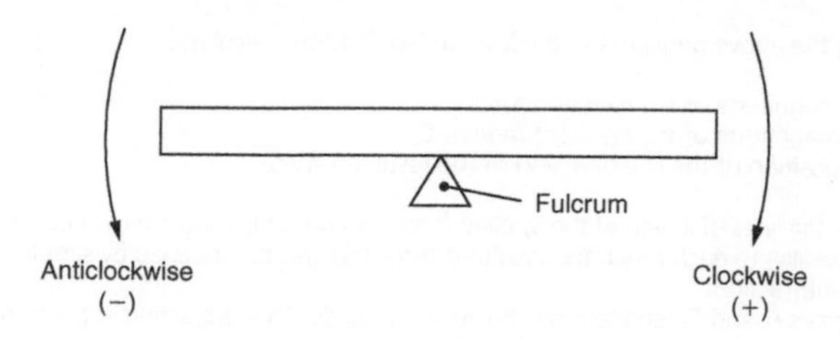

Figure 2.11 Clockwise and anticlockwise moments.

2.4.3 Principle of moments

The algebraic sum of the moments of a number of forces acting about any point is equal to the resultant moment of force about the same point.

Expressed mathematically:

$$M_R = M_1 + M_2 + (-M_3) + \cdots + M_n$$

where:

M_R = the resultant moment of force
M_1 to M_n = any number of moments acting about the same fulcrum point

Remember: The positive moments will be acting in a clockwise direction, whereas the negative moments (such as $-M_3$) will be acting in an anticlockwise direction.

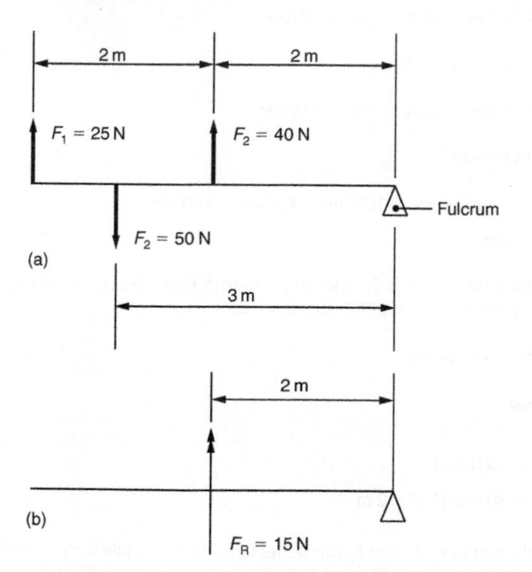

Figure 2.12 Application of principle of moments.

Applying the above principles to the lever in Fig. 2.12(a), calculate:

(a) The magnitude of the resultant force.
(b) The magnitude of the resultant moment.
(c) The position of the line of action of the resultant force.

(a) Since the lines of action of the applied forces acting on the lever shown in Fig. 2.12(a) are parallel to each other, the resultant force (F_R) can be obtained by simple addition and subtraction:

- Forces F_1 and F_2 tend to move the lever upwards. They are acting in the same direction, therefore they can be added together.
- Force F_3 acts in the opposite direction, therefore it is subtracted from the sum of F_1 and F_2.

Therefore:

$$F_R = F_1 + F_2 - F_3$$
$$= 25\,\text{N} + 40\,\text{N} - 50\,\text{N}$$
$$= \mathbf{15\,N}$$

(b) To find the magnitude of the resultant moment (M_R)

Clockwise moments (+):

$$M_1 = F_1 \times (2\,\text{m} + 2\,\text{m}) = 25\,\text{N} \times 4\,\text{m} = 100\,\text{Nm}$$

$$M_2 = F_2 \times 2\,\text{m} = 40\,\text{N} \times 2\,\text{m} = 80\,\text{Nm}$$

Anticlockwise moments (−):

$$M_3 = F_3 \times 3\,\text{m} = 50\,\text{N} \times 3\,\text{m} = 150\,\text{Nm}$$

Resultant moment:

$$M_R = M_1 + M_2 - M_3 = 100\,\text{Nm} + 80\,\text{Nm} - 150\,\text{Nm}$$
$$= \mathbf{30\,Nm}$$

(c) To combine the results of (a) and (b) in order to find the position of the line of action of the resultant force (F_R), refer to Fig. 2.12(b).

$$M_R = F_R \times d = 30\,\text{Nm}$$

Therefore:

$$30\,\text{Nm} = 15\,\text{N} \times d$$

$$d = 30\,\text{Nm}/15\,\text{N} = \mathbf{2\,m}$$

Therefore the forces F_1, F_2 and F_3 can be replaced by the single force F_R acting at a point 2 m from the fulcrum. Since M_R is positive the force F_R will move or tend to move the lever in a clockwise direction so the force F_R will act in an upwards direction as shown.

2.4.4 Equilibrium

In the previous section, the moments have not acted in equal and opposite directions. This has resulted in a residual, *resultant force*, causing or tending to cause a rotation about a pivot point (fulcrum).

Figure 2.13 shows a simple balance scale in which the masses M_1 and M_2 in the scale pans are of equal magnitude so that the scales balance and the beam is horizontal. Since the force of gravity acts equally on M_1 and M_2, the forces F_1 and F_2 are also equal. The reaction force (R) equals the sum of the downward forces.

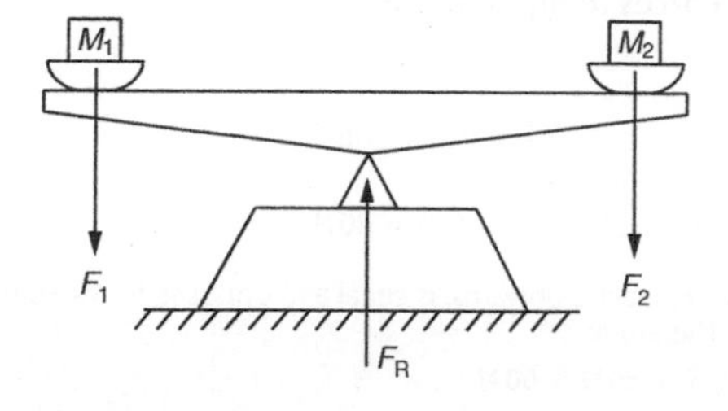

Figure 2.13 Simple balance scale.

The forces are equidistant from the fulcrum, so the turning moments are also equal. Since the turning moments act in opposite directions these moments are also in equilibrium. They balance each other and there is no tendency for the scale-beam to rotate. This can be summarized as follows:

When a body is in a state of equilibrium (balance) under the action of a number of forces, the clockwise moments about any point is equal to the anticlockwise moments about the same point.

Put more elegantly:

The algebraic sum of the moments about any point is zero.

For a body to be in a state of equilibrium the following conditions must exist:

- The algebraic sum of the horizontal forces acting on the body must equal zero, $\Sigma F_H = 0$.
- The algebraic sum of the vertical forces acting on the body must equal zero, $\Sigma F_V = 0$.
- The algebraic sum of the moments acting on the body must equal zero, $\Sigma M = 0$.

Figure 2.14 shows a lever pivoted at its centre. The line of action of a force F_1 is 2 m to the left of the pivot and tends to impart an anticlockwise moment. The line of action of force F_2 is 3 m to the right of the pivot and tends to impart a clockwise moment. If the magnitude of F_2 is 20 N, calculate:

1. The magnitude of the force F_1 required to put the lever into a state of equilibrium.
2. The magnitude of the reaction force (F_R) at the pivot.

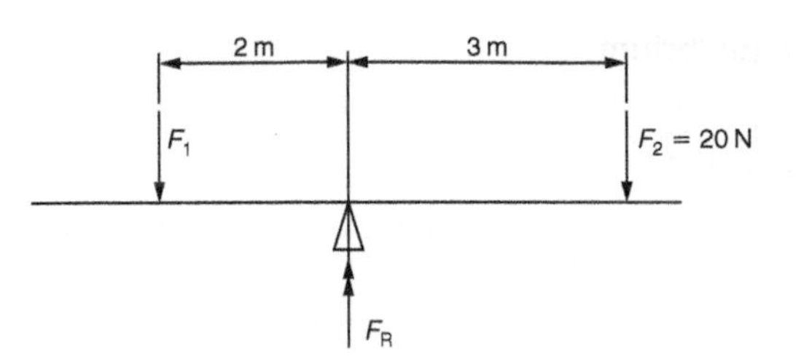

Figure 2.14 Forces in equilibrium.

For equilibrium:

Clockwise moments = anticlockwise moments

$$(20\,\text{N} \times 3\,\text{m}) = (F_1 \times 2\,\text{m})$$

$$F_1 = (20\,\text{N} \times 3\,\text{m})/2\,\text{m} = \mathbf{30\,N}$$

The reaction force F_R, acting upwards, is equal and opposite to the sum of F_1 and F_2 acting downwards. Therefore:

$$R = F_1 + F_2 = 20\,\text{N} + 30\,\text{N} = \mathbf{50\,N}$$

Figure 2.15 shows a simple beam supported at each end. Since it is in equilibrium there is no tendency for it to rotate and the reaction forces R_A and R_B at the supports can be calculated using the principle of moments. This is achieved by taking moments about each support in turn.

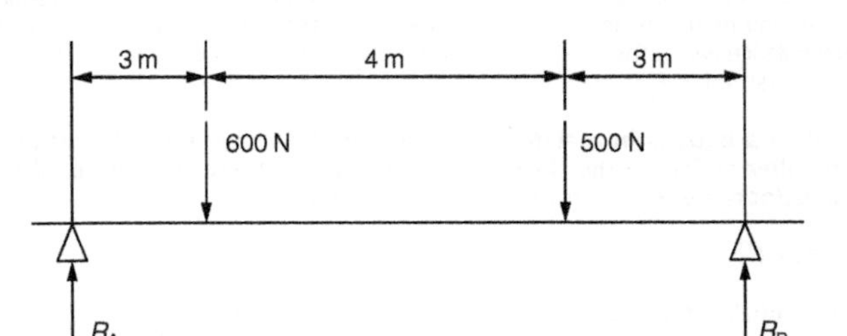

Figure 2.15 Calculation of reaction forces.

To determine the magnitude of R_A take moments about R_B in order to eliminate it from the initial calculation so that only one unknown quantity has to be calculated as follows:

Clockwise moments = anticlockwise moments

$$R_A \times 10\,\text{m} = (600\,\text{N} \times 7\,\text{m}) + (500\,\text{N} \times 3\,\text{m})$$

$$R_A \times 10\,\text{m} = 4200\,\text{Nm} + 1500\,\text{Nm}$$

$$R_A \times 10\,\text{m} = 5700\,\text{Nm}$$

$$R_A = 5700\,\text{Nm}/10\,\text{m}$$

$$R_A = \mathbf{570\,N}$$

Similarly to determine the magnitude of R_B take moments about R_A in order to eliminate it from the initial calculation so that only one unknown quantity has to be calculated as follows:

$$\text{Clockwise moments} = \text{anticlockwise moments}$$

$$(600\,\text{N} \times 3\,\text{m}) + (500\,\text{N} \times 7\,\text{m}) = R_B \times 10\,\text{m}$$

$$1800\,\text{Nm} + 3500\,\text{Nm} = R_B \times 10\,\text{m}$$

$$5300\,\text{Nm} = R_B \times 10\,\text{m}$$

$$R_B = 5300\,\text{Nm}/10\,\text{m}$$

$$R_B = \mathbf{530\,N}$$

For a system of forces in equilibrium the downward forces must equal the upward forces.

Therefore:

$$R_A + R_B = 600\,\text{N} + 500\,\text{N}$$

$$570\,\text{N} + 530\,\text{N} = 600\,\text{N} + 500\,\text{N}$$

$$1100\,\text{N} = 1100\,\text{N}$$

which is true. Therefore the calculation for R_A and R_B are correct. Avoid taking the short-cut of calculating on one of the reaction forces and deducting it from the sum of the downward forces. Any error in the initial calculation will be carried forward and no check will be possible.

The above examples ignore the weight of the beam.

The weight of a uniform beam may be considered to be a concentrated force whose line of action can be assumed to act through the centre of the length of the beam.

2.5 Orders of levers

Figure 2.16 shows the three orders of levers. Applying the principle of moments, the relationship between the load (W), the distance of the load from the fulcrum (d_1) and effort (F), and the distance of the effort from the fulcrum (d_2) is common to all the orders and can be expressed mathematically as:

$$W \times d_1 = F \times d_2$$

Therefore:

$$W = F \times d_2/d_1$$

And:

$$F = W \times d_1/d_2$$

2.5.1 First-order levers

- The fulcrum is positioned between the effort and the load.
- The effort force is smaller than the load.
- The effort moves further than the load.
- This order of levers can be considered as a force magnifier.

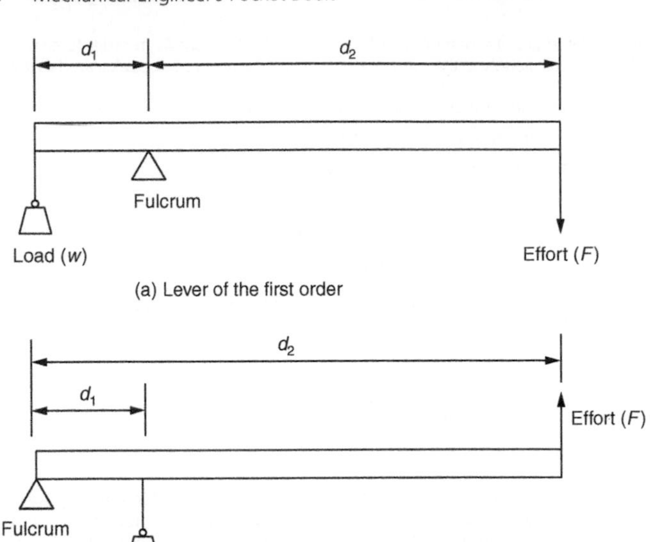

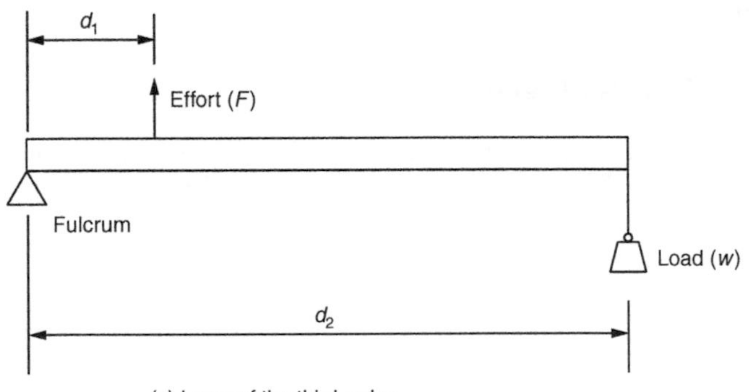

Figure 2.16 Orders of levers.

2.5.2 Second-order levers

- The effort and the load are positioned on the same side of the fulcrum but applied in opposite directions.
- The load lies between the effort and the fulcrum.

- The effort is smaller than the load.
- The effort moves further than the load.
- This order of levers can be considered as a force magnifier.

2.5.3 Third-order levers

- The effort lies between the load, and the load and the fulcrum.
- The effort is greater than the load.
- The load moves further than the effort.
- This order of levers can be considered as a distance magnifier.

2.6 Centre of gravity, centroid of areas and equilibrium

2.6.1 Centre of gravity (solid objects)

Gravitational attraction acts on all the particles within a solid, liquid or gas as shown in Fig. 2.17. It is assumed that the individual particle forces have parallel lines of action. It is further assumed that these individual particle forces can be replaced for calculation purposes by a single, resultant force equal to the sum of the particle forces. The centre of gravity (G) of a solid object is the point through which the resultant gravitational force acts. That is, the point through which the weight of the body acts. For example, the centre of gravity for a cylinder of uniform cross-section lies on the centre line (axis) midway between the two end faces. This and some further examples are shown in Fig. 2.18.

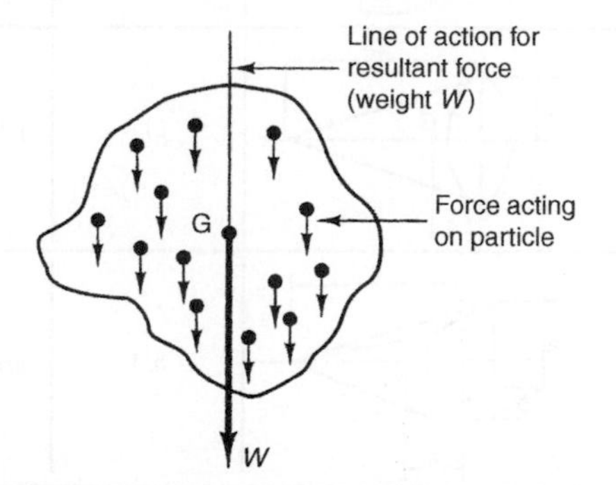

Figure 2.17 Centre of gravity.

2.6.2 Centre of gravity of non-uniform and composite solids

Figure 2.19 shows a cylindrical shaft that has two diameters:

- G_1 is the centre of gravity of the larger diameter and lies on the axis halfway between the end faces of this diameter.

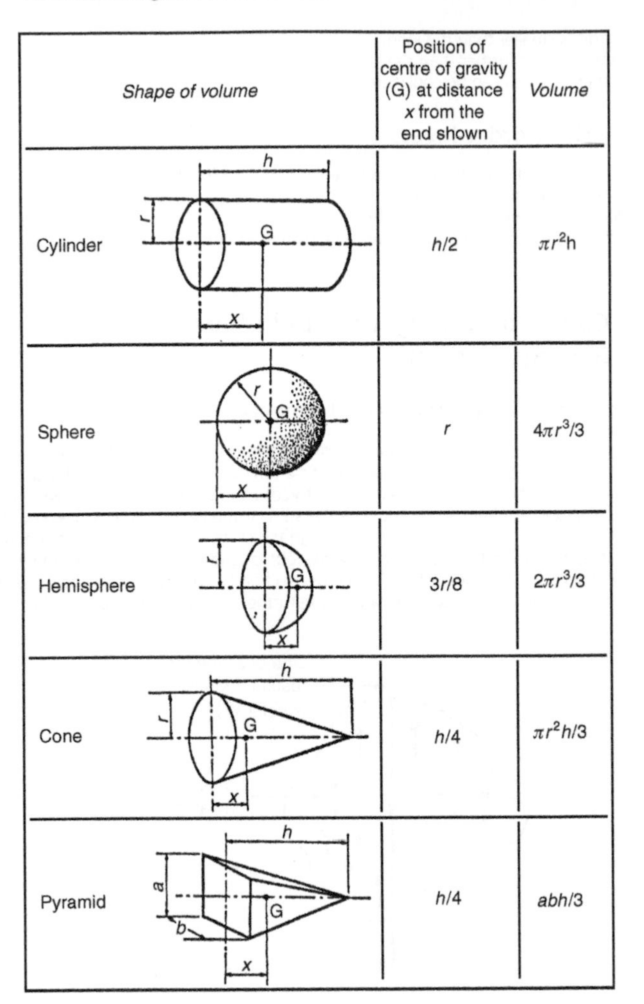

Shape of volume	Position of centre of gravity (G) at distance *x* from the end shown	*Volume*
Cylinder	$h/2$	$\pi r^2 h$
Sphere	r	$4\pi r^3/3$
Hemisphere	$3r/8$	$2\pi r^3/3$
Cone	$h/4$	$\pi r^2 h/3$
Pyramid	$h/4$	$abh/3$

Figure 2.18 Position of centre of gravity (G) for some common solids.

- G_2 is the centre of gravity of the smaller diameter and lies on the axis halfway between the end faces of this diameter.
- G is the centre of gravity for the composite solid as a whole and lies at a distance *d* from the datum end of the shaft. Its position can be determined by the application of the principal of moments.

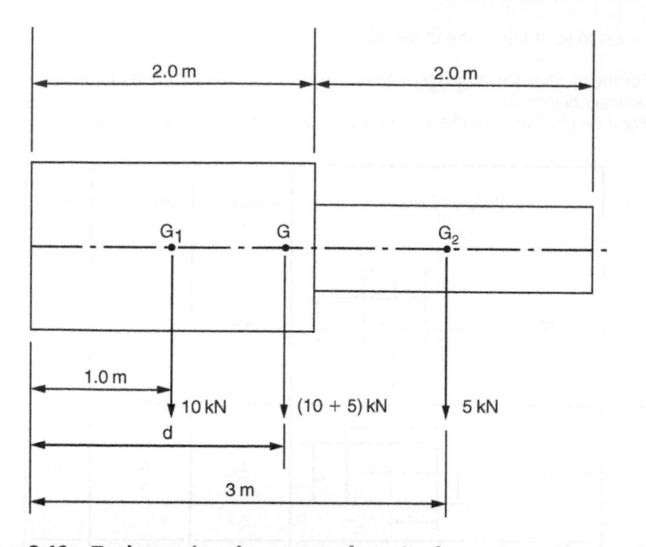

Figure 2.19 To determine the centre of gravity for a non-uniform solid.

1. The overall weight of the shaft is the sum of the weights of the two cylinders. That is, 10 kN + 5 kN = **15 kN**.
2. The total weight of 15 kN will act through the centre of gravity G for the shaft as a whole at some distance *d* from the datum point P about which moments are taken.
3. For this calculation it is conventionally assumed that the moment of G about the point P will be the opposite hand to the moments of G_1 and G_2 about P, despite the fact that they all appear to produce clockwise moments.

$$\text{Clockwise moments} = \text{anticlockwise moments}$$

$$(G_1 \times 1\,\text{m}) + (G_2 \times 4\,\text{m}) = (G \times d)$$

$$(10\,\text{kN} \times 1\,\text{m}) + (5\,\text{kN} \times 3\,\text{m}) = (15\,\text{kN} \times d)$$

$$10\,\text{kN m} + 15\,\text{kN m} = 15\,\text{kN} \times d$$

$$25\,\text{kN m} = 15\,\text{kN} \times d$$

Therefore:

$d = 25\,\text{kN m}/15\,\text{kN} = \mathbf{1.67\,m}$

The same technique can be used for composite solids where the densities of the various elements differ.

2.6.3 Centre of gravity (lamina)

For a thin sheet metal or rigid sheet plastic component of uniform thickness (a *lamina*) where the thickness is negligible compared with its area, the centre of gravity can be assumed to lie at the centre of the area:

- For the rectangular lamina the centre of gravity G lies at the point of intersection of the lamina's diagonals.
- For a circular lamina the centre of gravity G lies at the centre of the circle.

Shape of area	*Distance* x	*Distance* y	*Area*
Square	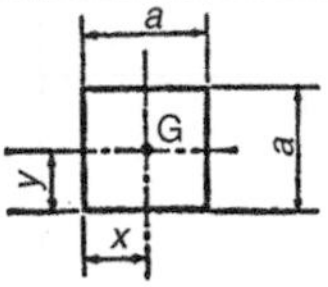 $\frac{a}{2}$	$\frac{a}{2}$	a^2
Rectangle	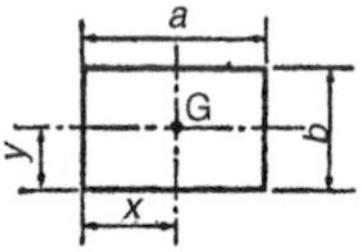 $\frac{a}{2}$	$\frac{b}{2}$	ab
Circle	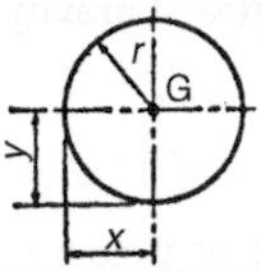 r	r	πr^2
Semi-circle	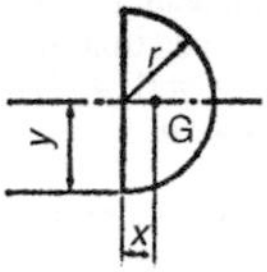 $\frac{4r}{3\pi}$	r	$\frac{\pi r^2}{2}$
Right-angled triangle	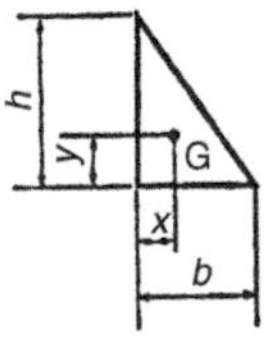 $\frac{b}{3}$	$\frac{h}{3}$	$\frac{bh}{2}$

Figure 2.20 Centroids of areas.

2.6.4 Centroids of areas

The *centroid of an area* refers to a two-dimensional plane figure that has no thickness. It has only area. Since an area has no thickness it can have no mass to be acted upon by the force of gravity. Therefore it has no weight. The term *centroid* is applied to the area of a two-dimensional plane figure in much the way that centre of gravity is applied to a solid lamina. Therefore the centroid of an area is the point on a plane figure where the whole area is considered to be concentrated. Like the centre of gravity, the centroid of an area is also denoted by the letter G. Some examples of centroids are shown in Fig. 2.20.

2.6.5 Equilibrium

Equilibrium was introduced in Section 2.4.4. Consider Fig. 2.21(a) which shows a bar of metal of uniform section pivoted at its centre so that it is free to rotate. Figure 2.21(a) also shows that the turning moments acting on each arm of the bar, due to force of gravity, are equal and opposite. Therefore, being perfectly balanced the bar should remain in any position to which it is turned. The bar is said to be in a state of **neutral equilibrium**.

In Fig. 2.21(b) the same bar is shown pivoted off-centre so that one arm is longer than the other. With the bar upright as shown in Fig. 2.21(b), the line of action of the vector

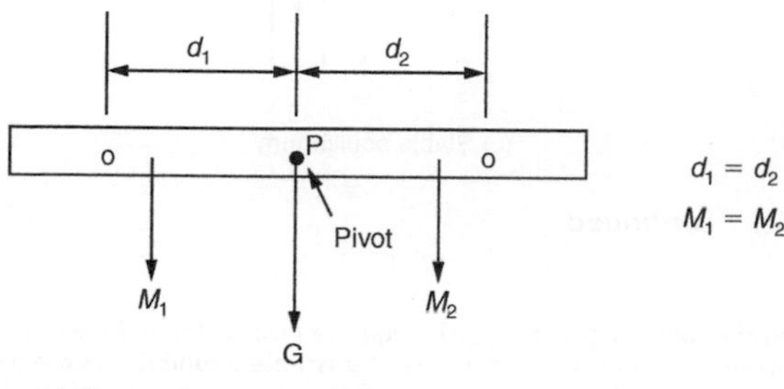

(a) Neutral or stable equilibrium

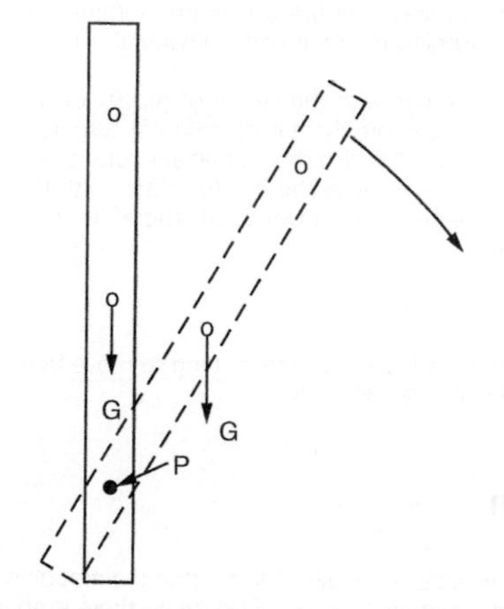

(b) Unstable equilibrium

Figure 2.21 Equilibrium.

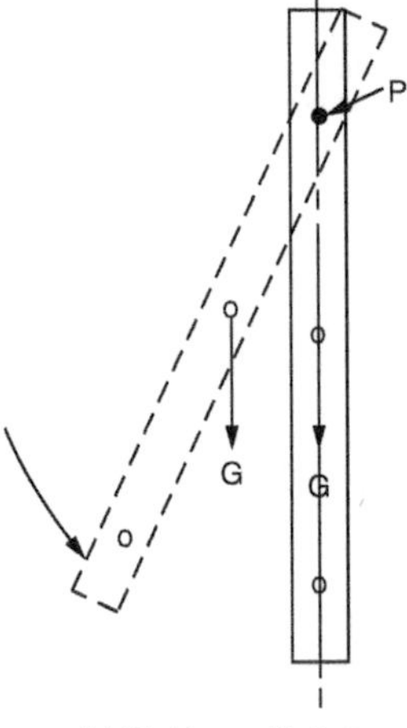

(c) Stable equilibrium

Figure 2.21 ***continued***

representing the force of gravity acts through the pivot point and there is no turning moment. However the bar is now in a state of **unstable equilibrium** since the slightest rotational movement of the bar will cause the centre of gravity to be displaced to one side of the vertical as shown. This will introduce a turning moment that will cause the bar to rotate until the centre of gravity of the longer arm is immediately below the pivot point. There will then be no turning moment and the bar will come to rest.

Figure 2.21(c) shows the bar with the centre of gravity of the longer arm immediately below the pivot point. Therefore there is no moment turning or tending to turn the bar and it will remain stationary. It is also shown that any rotational disturbance of the bar will produce a moment that will cause the bar to rotate until the centre of gravity of the longer arm will again be below the pivot point. Therefore the bar is said to be in a state of **stable equilibrium**.

Note:

When disturbed, a body will always move or tend to move from a state of unstable equilibrium to a state of stable equilibrium.

2.7 Friction

Whenever two bodies slide or roll over one another the motion is opposed by a resistance known as *sliding* (or *kinematic*) *friction*. Even when there is no motion there is dormant resistance known as *static friction*. Static friction is generally greater than sliding friction. More force is needed to overcome the static friction between two given materials and set up motion than maintain that motion at uniform velocity.

2.7.1 Lubrication

Friction and wear between sliding or rolling surfaces can be greatly reduced if the surfaces of the materials are separated by a suitable film of lubricant. Oils are generally used to lubricate rapidly moving surfaces subject to relatively light loads and greases are generally used to lubricate slow moving surfaces subject to heavier loads. If separation of the surfaces is complete, dry friction is eliminated and there is only the *fluid friction* or viscosity of the oil which has to be overcome. As well as reducing wear and friction, lubrication also serves to protect the sliding surfaces against corrosion and reduce the generation of heat which represents wasted energy. Any heat that is generated tends to be dissipated by the lubricant.

2.7.2 Laws of friction

The amount of friction between two surfaces sliding or rolling over one another is governed by the following factors:

- The surface texture of the surfaces involved. Smoothness reduces friction; roughness increases friction.
- The kind of materials in contact (e.g. steel slides more easily on nylon than it does on rubber).
- The magnitude of the normal (at right angles) reaction forces between the surfaces.
- Friction is independent of area.

2.7.3 Coefficient of friction

The coefficient of friction (symbol μ) for a pair of surfaces in contact is given by the expression:

μ = (friction force F_f)/(the normal reaction force F_n)

This is shown in Fig. 2.22. Since F_f and F_n are both expressed in the same units, μ is ratio and has no units.

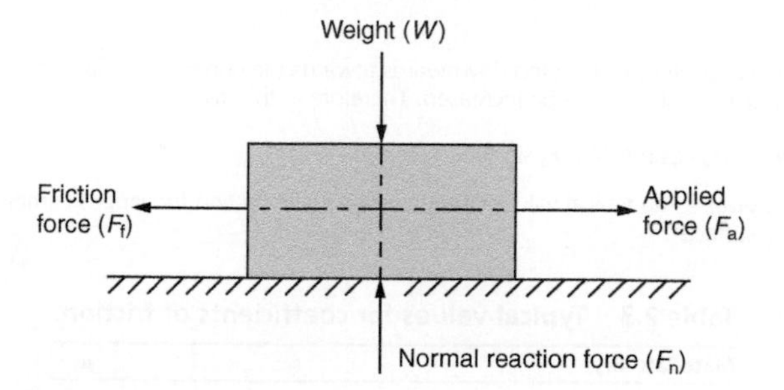

Figure 2.22 Friction.

Case 1
A block of weight W is pulled along a horizontal surface by an applied force of magnitude F_a as shown in Fig. 2.22. Whilst the block moves at a uniform speed or while it is stationary,

a resistance is experienced that is equal and opposite to the applied force F_a. This resistance is the friction force F_f. This friction force always opposes motion. Thus, in this case:

$$\mu = F_f/F_n = F_a/W$$

- If the friction force equals the applied force the block is stationary or in a state of uniform motion and the forces acting on it are in equilibrium.
- If the block tends to accelerate then the applied force is greater than the friction force.
- The friction force can be less than or equal to the applied force. It can never be greater than the applied force. That is, F_f is equal to or less than F_a, and F_n is equal to or less than W.

Case 2

If the applied force (F_a) is inclined at an angle θ to the sliding block as shown in Fig. 2.23 then the friction force F_f will be equal and opposite to the horizontal component force (F_h), where $F_h = F_a \cos\theta$. The vertical component force $F_v = F_a \sin\theta$. And this force is opposing the weight W and, hence, reducing the normal reaction force, F_n. Therefore:

$$\mu = F_f/F_n = (F_a \cos\theta)/(W - F_a \sin\theta)$$

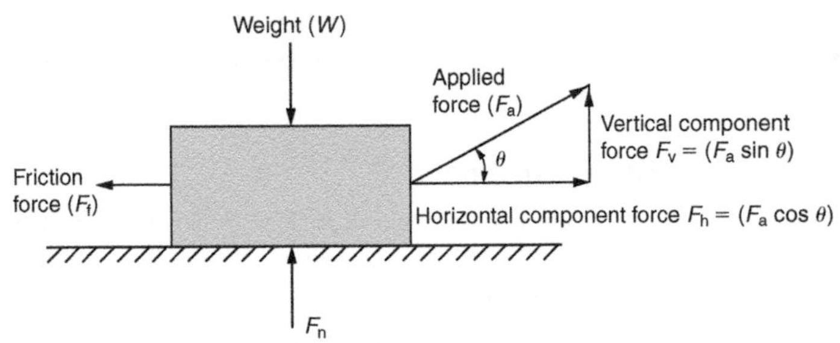

Figure 2.23 Friction – applied force inclined.

Thus if the applied force F_a acted downwards towards the horizontal surface, then F_h the normal reaction force F_n will be increased. Therefore in this instance:

$$\mu = F_f/F_n = (F_a \cos\theta)/(W + F_a \sin\theta)$$

Table 2.3 gives some typical values of the coefficient of friction for various combinations of sliding surfaces.

Table 2.3 Typical values for coefficients of friction.

Materials (dry)	μ
Steel on cast iron	0.20
Steel on brass	0.15
Cast iron on brass	0.15
Leather on cast iron	0.55
Brake lining on cast iron	0.60
Rubber on asphalt	0.65
Rubber on concrete	0.70

2.7.4 Angle of friction

The forces F_f and F_n may be added vectorially as shown in Fig. 2.24 to give the resultant reaction F_r at the surface, since they are both exerted on the sliding block which is in contact with the horizontal surface.

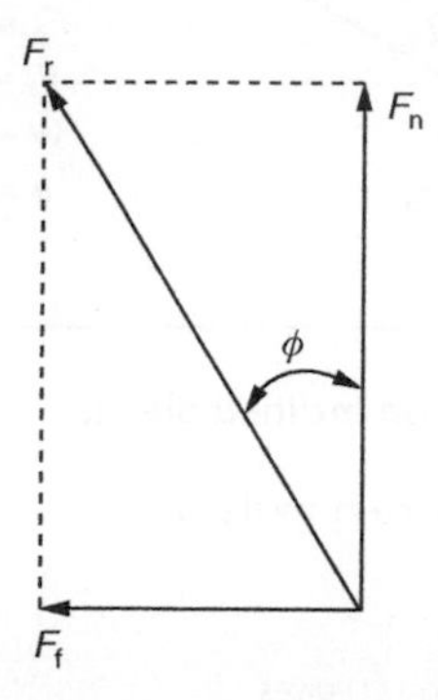

Figure 2.24 Angle of friction.

Since F_f and F_n are at right angles:

$\tan \phi = F_f/F_n$

but $\mu = F_f/F_n$

therefore:

$\mathbf{\tan \phi = \mu}$

The angle ϕ reaches a maximum value when F_f is a maximum. This maximum value of ϕ is known as the **angle of friction**.

2.7.5 Friction on an inclined plane

To determine the applied force F_a to pull a body of weight W up a plane inclined at an angle θ to the horizontal when F_a is acting in at an angle ϕ to the inclined plane as shown in Fig. 2.25.

Force F_a is acting at angle ϕ to the plane, as shown in Fig. 2.25, and the normal reaction between the plane and the body is F_n. Resolving the forces perpendicular to the inclined plane:

$$F_n + F_a \sin \phi = W \cos \theta$$

$$F_n = W \cos \theta - F_a \sin \phi$$

The frictional force opposing the motion of the body is:

$$F_f = \mu F_n$$

$$= \mu(W \cos \theta - F_a \sin \phi) \quad (1)$$

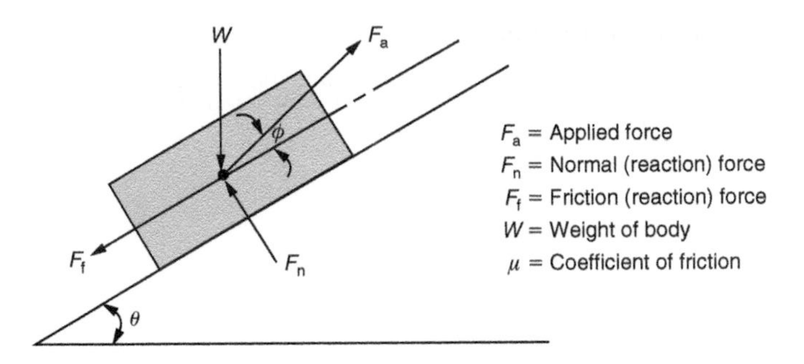

Figure 2.25 Friction on an inclined plane.

Resolving the forces parallel to the inclined plane:

$F_a \cos\phi = F_f + W \sin\theta$

Substituting for F_f from equation (1) gives:

$$F_a \cos\phi = \mu(W\cos\theta - F_a \sin\phi) + W\sin\theta$$

$$F_a \cos\phi = W(\mu\cos\theta + \sin\theta) - \mu F_a \sin\phi$$

$$F_a(\mu\sin\phi + \cos\phi) = W(\mu\cos\theta + \sin\theta)$$

$$F_a = [W(\mu\cos\theta + \sin\theta)]/[(\mu\sin\phi + \cos\phi)] \qquad (2)$$

(i) When F_a is applied parallel to the plane, then $\phi = 0$. Hence $\sin\phi = 0$ and $\cos\phi = 1$. Substituting in equation (2):

$F_a = [W(\mu\cos\theta + \sin\theta)]/(1 + 0)$

$\mathbf{F_a = W(\mu\cos\theta + \sin\theta)}$

(ii) When F_a is applied horizontally, then it assists the motion of the body up the plane, then $\phi = -\theta$. Hence:

$\cos\phi = \cos(-\theta) = \cos\theta$ and

$\sin\phi = \sin(-\theta) = -\sin\theta$

Substituting in equation (2):

$\mathbf{F_a = W(\mu\cos\theta + \sin\theta)/(\cos\alpha + \sin\alpha)}$

2.7.6 Angle of repose

A body resting on a rough plane inclined at an angle θ to the horizontal as shown in Fig. 2.26 is in a state of equilibrium when the gravitational force tending to slide the body

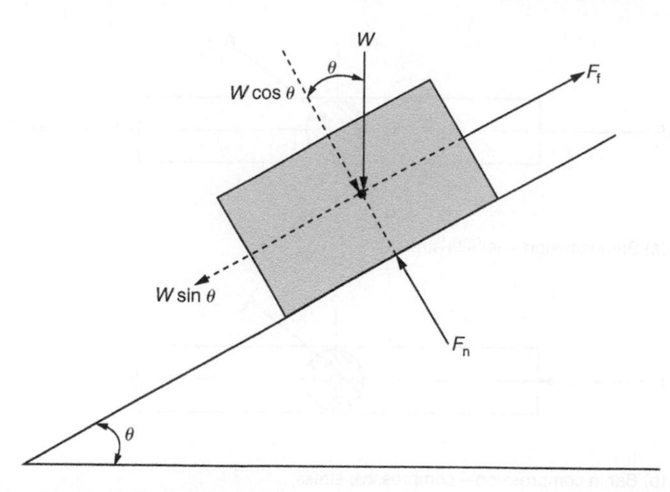

Figure 2.26 Angle of repose.

down the inclined plane is balanced by an equal and opposite frictional force acting up the inclined plane. The angle of inclination of the inclined plane at which the equilibrium conditions apply is called the *angle of repose*.

The gravitational force acting on the body (W) can be resolved into its component forces such that the component force acting perpendicularly to the inclined plane will be $W \cos \theta$ and the component force acting parallel to the inclined plane will be $W \sin \theta$. For equilibrium when the body is just about to slide down the inclined plane:

$$\mu = F_f/F_n = (W \sin \theta)/(W \cos \theta) = \mathbf{\tan \theta}$$

From Section 2.7.4 the angle of friction, $\mu = \tan \phi$; therefore for the body to be in a state of repose, $\theta = \phi$. If $\theta > \phi$ the body will slide down the inclined plane. If angle $\theta < \phi$ the body will remain at rest unless an external force is applied to cause motion.

2.8 Stress and strain

2.8.1 Direct stress

Figure 2.27(a) shows a cylindrical bar of cross-sectional area A in *tension*, whilst Fig. 2.27(b) shows the same bar in *compression*. The applied forces F are in line and are normal (perpendicular) to the cross-sectional area of the bar. Therefore the bar is said to be subject to *direct stress*. Direct stress is given the symbol σ (Greek letter sigma).

Direct stress σ = applied force F/cross-sectional area A

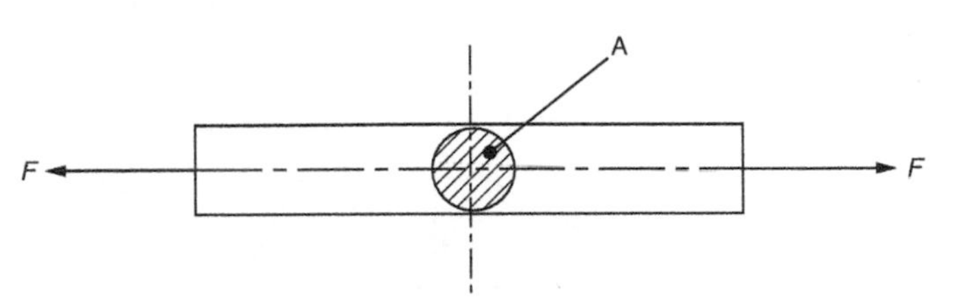

(a) Bar in tension – tensile stress

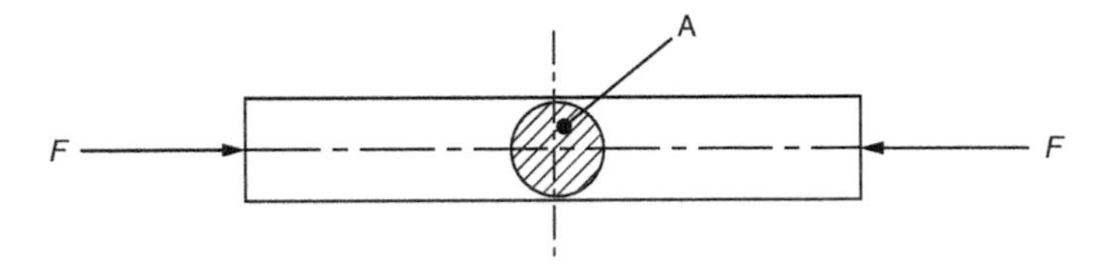

(b) Bar in compression – compressive stress

Figure 2.27 Direct stress.

2.8.2 Shear stress

Figure 2.28 shows a riveted joint. Since the applied forces Q are *offset* (not in line) the rivet is said to be subjected to *shear stress* which is given the symbol τ (Greek letter tau).

Shear stress τ = shear force Q/area in shear A

Direct stress and shear stress are usually of sufficient magnitude to be measured in MN/m^2.

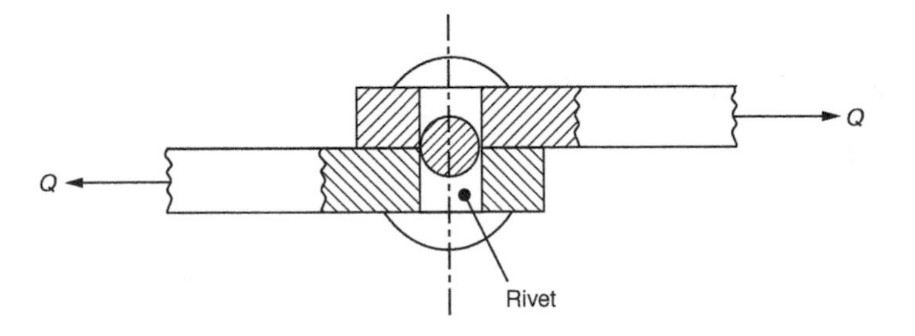

Figure 2.28 Shear (indirect) stress.

2.8.3 Direct strain

The forces acting on a body that produce direct stress also cause a change in the dimensions of that body. For example, a tensile load will produce tensile stress and also cause the body to stretch as shown in Fig. 2.29(a). Similarly, a compressive load causes the body to shorten as shown in Fig. 2.29(b). Direct strain is given the symbol ε (Greek letter epsilon).

Direct strain ε = change in length x/original length l

Since both the change in length and the original length are measured in the same units, *strain* is a *ratio* and has *no units.*

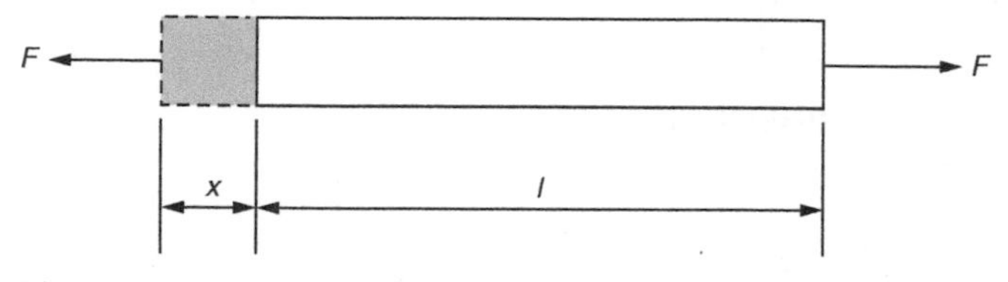

(a) Direct strain resulting from tensile stress

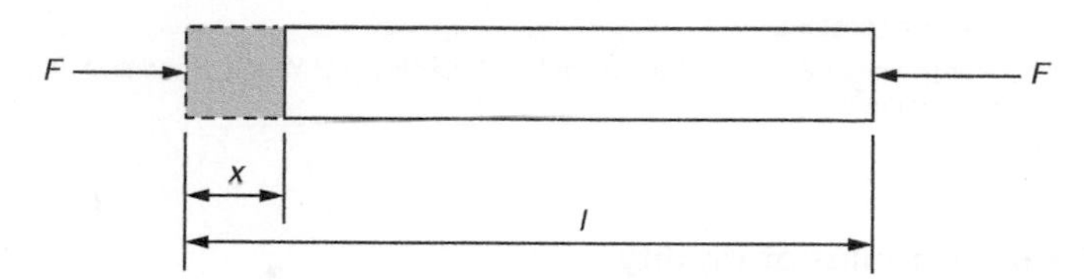

(b) Direct strain resulting from compressive stress

Figure 2.29 Direct strain.

2.8.4 Shear strain

Shear strain is shown in Fig. 2.30. The applied forces Q are offset and cause material to deform. Shear strain is given the symbol γ (Greek letter gamma).

Shear strain γ = deformation x/original dimension l

Again this is a ratio so there are no units.

2.8.5 Modulus of elasticity (Hooke's law)

Hooke's law states that within the elastic range for any given material, the deformation is proportional to the applied force producing it. When this law is applied to a spring:

Stiffness of the spring = direct stress σ/direct strain ε

The modulus of elasticity E can be derived from Hooke's law. Stress is directly proportional to strain whilst the material is stressed within its elastic range, thus:

Stress $\propto$ strain

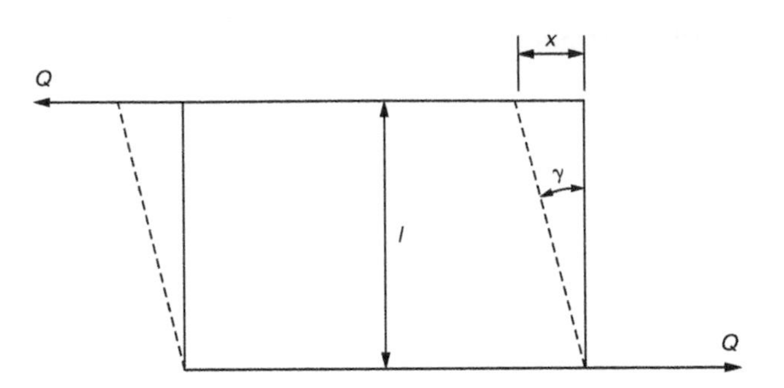

Figure 2.30 Shear (indirect) strain.

Therefore:

Stress = strain × a constant

Therefore:

Stress/strain = a constant

This constant (symbol E) is called the modulus of elasticity (or Young's modulus) and is measured in GN/m^2.

E = stress σ/strain ε

2.8.6 Modulus of rigidity

The modulus of rigidity G relates to shear strain and shear stress such that:

Modulus of rigidity G = shear stress τ/shear strain γ

The modulus of rigidity is also measured in GN/m^2. Table 2.4 shows the modulus of rigidity and the modulus of elasticity for some typical materials.

Table 2.4 Modulus of elasticity and modulus of rigidity for some typical materials.

Material	Modulus of elasticity (E) GN/m^2	Modulus of rigidity (G) GN/m^2
Aluminium	70	26
Brass (70/30)	101	37
Cadmium	50	19
Chromium	279	115
Copper	130	48
Iron (cast)	152	60
Lead	16	6
Steel (mild)	212	82
Tin	50	18
Titanium	116	44
Tungsten	411	161
Zinc	108	43

2.8.7 Torsional stress

With reference to Fig. 2.31(a), it can be seen that when a shaft of circular cross-section acted on by a torque (turning moment) T:

- All sections of the shaft remain circular and of unchanged diameter.
- Plane cross-sections remain plane (circular only) providing the angle of twist is small.

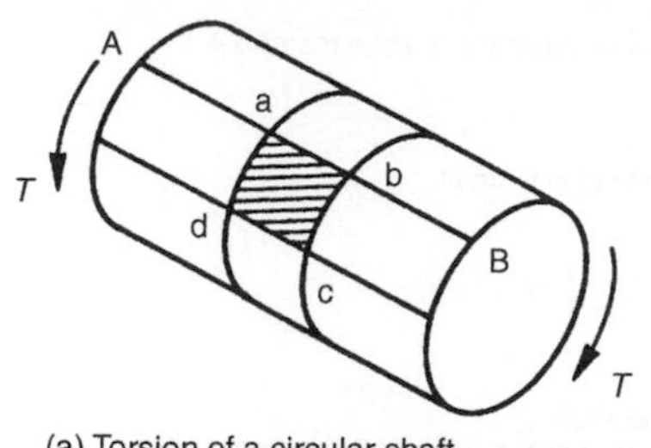

(a) Torsion of a circular shaft

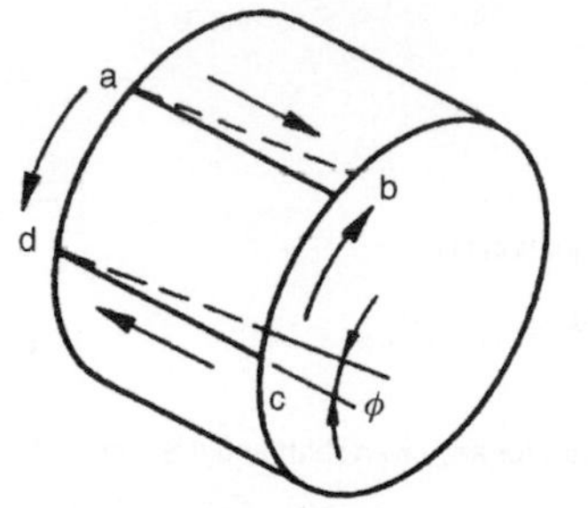

(b) Isolated element abcd

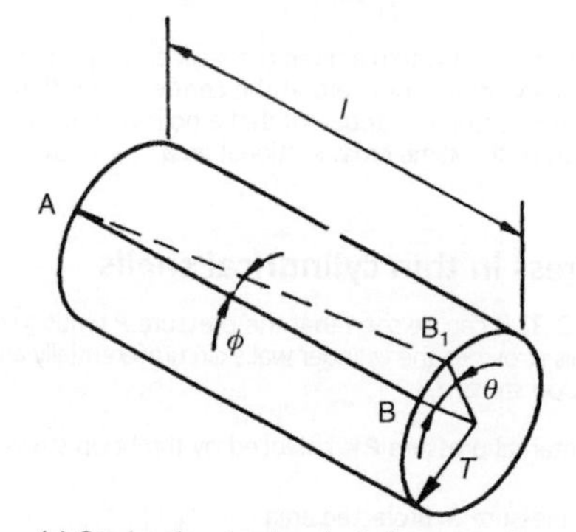

(c) Strain of a circular shaft due to torsion

Figure 2.31 Torsional stress and strain.

For the isolated cylindrical element of the shaft shown in Fig. 2.47(b), the end faces remain plane and any rectangular surface element abcd will be in a state of pure shear. Shear stresses along ad and bc will induce complementary shear stress along ab and cd. Longitudinal twisting of the shaft occurs so that the lengths ab and cd are sheared through angle ϕ relative to the line AB in Fig. 2.47(c).

For pure shear and small angles of twist, angle ϕ represents *shear strain* and τ is the *shear stress* at the surface of the shaft. Since:

Modulus of rigidity G = shear stress τ/shear strain ϕ (1)

and:

Arc $BB_1 = l\phi$ providing ϕ is small

and:

Arc $BB_1 = r\theta$

where:
r = the radius of the shaft
θ = the angle of twist over the full length l of the shaft.

Therefore:

$$l\phi = \text{arc } BB_1 = r\theta$$

$$\phi = r\theta/l$$

Substituting for ϕ in equation (1):

$$G = \tau/(r\theta/l) \text{ or } \tau/r = G\theta/l$$

Note:

Since G and l are constant for any given shaft and θ is constant for all radii at any particular cross-section then:

$$\tau/r = \text{constant}$$

Thus the shear stress at any point within a given cross-section is proportional to the radius, and the stress increases uniformly from zero at the centre of the shaft to a maximum at the outside radius. For this reason it is apparent that a hollow shaft can transmit a greater torque than a solid shaft of the same cross-sectional area.

2.8.8 Hoop stress in thin cylindrical shells

With reference to Fig. 2.32 it can be seen that the pressure P tends to increase the diameter of the cylinder. This stretches the cylinder walls circumferentially and sets up a tensile stress known as the *hoop stress* σ_h.

The force due to the internal pressure P is balanced by the hoop stress σ_h. Hence:

Hoop stress × area = pressure × projected area

$$\sigma_h \times 2lt = Pdl$$

$$\sigma_h = Pd/2t$$

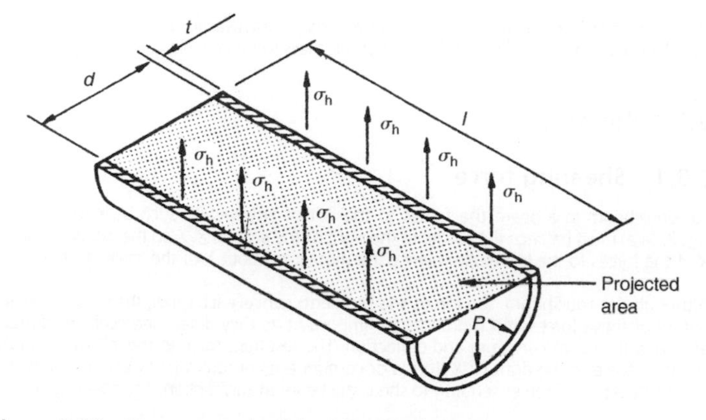

Figure 2.32 Hoop stress in thin cylindrical shells.

2.8.9 Longitudinal stress in thin cylindrical shells

With reference to Fig. 2.33 it can be seen that the internal pressure P is also a tensile stress in the longitudinal direction known as σ_l. Hence the pressure P acting over the area $\pi d^2/4$

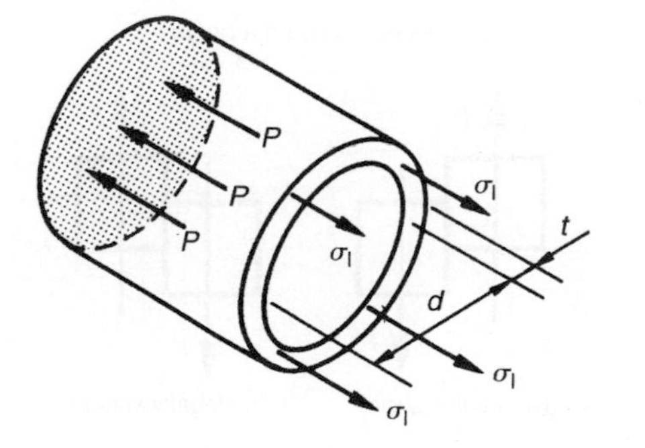

Figure 2.33 Longitudinal stress in thin cylindrical shells.

is balanced by the longitudinal stress σ_l acting over the area πdt. (Strictly, the mean diameter should be used but, as the wall thickness t is small compared with the diameter, this approximation is adequate.) Thus:

$$P \times (\pi d^2/4) = \sigma_l \times \pi dt$$

$$\sigma_l = (\pi P d^2)/(4\pi dt)$$

$$\sigma_l = Pd/4t$$

Note:

The equations for σ_h and σ_l assume stresses are, for all practical purposes, constant over the wall thickness. This is only valid if the ratio of thickness to internal diameter is less than 1:20.

2.9 Beams

2.9.1 Shearing force

For equilibrium in a beam the forces to the left of any section such as X as shown in Fig. 2.34(a) must balance the forces to the right of the section X. Also the moments about X of the forces to the left must balance the moments about X of the forces to the right.

Although, for equilibrium, the forces and moments cancel each other, the magnitude and nature of these forces and moments are important as they determine both the stresses at X, and the beam curvature and deflection. The resultant force to the left of X and the resultant force to the right of X (forces or components of forces transverse to the beam) constitute a pair of forces tending to shear the beam at this section. The *shearing force* is

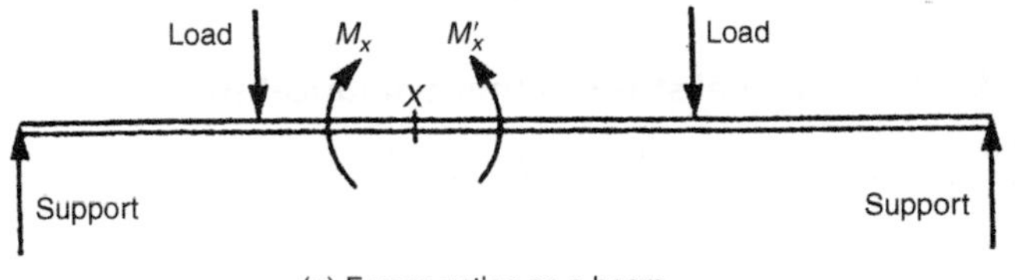

(a) Forces acting on a beam

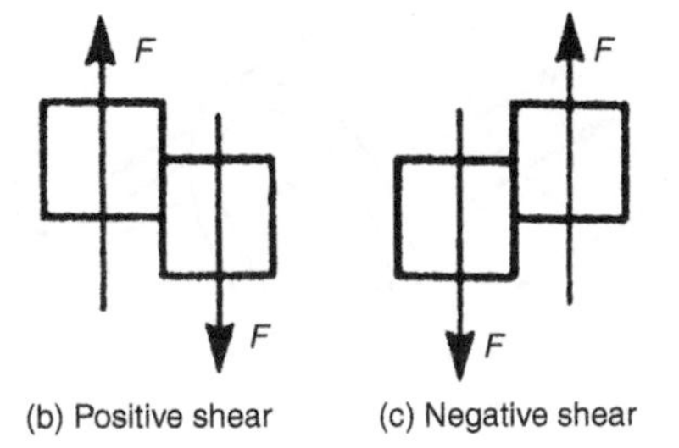

(b) Positive shear (c) Negative shear

Figure 2.34 Forces acting on a beam.

defined as the force transverse to the beam at any given section tending to cause it to shear at that section.

By convention, if the tendency is to shear as shown in Fig. 2.34(b), the shearing force is regarded as positive (i.e. $+F$); if the tendency to shear is as shown in Fig. 2.34(c), it is regarded as negative (i.e. $-F$).

2.9.2 Bending moment

The *bending moment* at a given section of a beam is defined as the resultant moment about that section of either all the forces to the left of the section or of all the forces to the right of the section. In Fig. 2.34(a) the bending moment is either M_x or M'_x. These moments will be clockwise to the left of the section and anticlockwise to the right of the section. They will cause the beam to *sag*. By convention this sagging is regarded as positive bending. That is, positive bending moments produce positive bending (*sagging*). Similarly negative bending moments cause the beam to bend in the opposite direction. That is, negative bending moments produce negative bending (*hogging*). The difference between sagging and hogging is shown in Fig. 2.35.

Contraflexure is present when both hogging and sagging occurs in the same beam as shown in Fig. 2.35(c). For the loading shown in Fig. 2.35(a), the beam will sag between the points B and C, and for the loading shown in Fig. 2.35(b) the beam will hog between the points A and B.

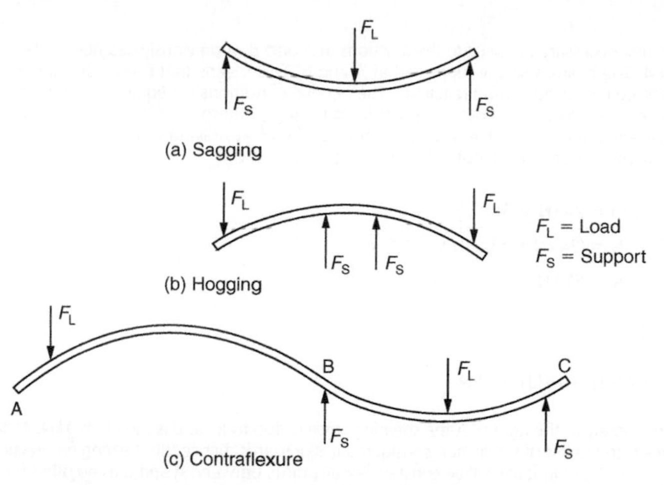

Figure 2.35 Hogging, sagging and contraflexure.

The values of shearing force and bending moment will usually vary along any beam. Diagrams showing the shearing forces and the bending moments for all sections of a beam are called *shearing force diagrams* and *bending moment diagrams*, respectively.

Shearing forces and shearing force diagrams are less important than bending moments and bending moment diagrams; however, they are useful in giving pointers to the more important aspects of a bending moment diagram. For example, wherever the shearing force is zero, the bending moment will be at a maximum or a minimum.

2.9.3 Shearing force and bending moment diagrams

Consider the shearing force and bending moment diagrams for the system of forces acting on the beam in Fig. 2.36. For the moment, only a simple system of three point loads will be considered.

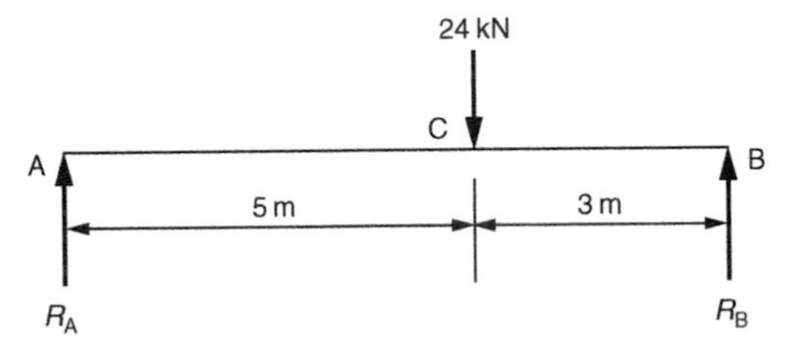

Figure 2.36 Forces acting on a beam.

It is first necessary to calculate the reactions at A and B as previously described in Section 2.4.4. The beam is simply supported at A and B. This means that it rests on supports at these points giving vertical reactions. The general conditions for equilibrium require that the resultant moment about any point must be zero, and the sum of the upward forces must equal the sum of the downward forces. Therefore, taking moments about A, the moment for R_B must balance the moment for the load C:

$$R_B \times 8\,\text{m} = 24\,\text{kN} \times 5\,\text{m}$$

$$R_B = (24\,\text{kN} \times 5\,\text{m})/8\,\text{m}$$

$$R_B = \mathbf{15\,kN}$$

Similarly:

$$R_A = 24\,\text{kN} - 15\,\text{kN} = \mathbf{9\,kN}$$

Immediately to the right of A the shearing force is due to R_A and is therefore 9 kN. As this force is to the left of the section considered it is upwards, hence the shearing force is positive. The shearing force will be constant for all points between A and C as no other forces are applied to the beam between these points.

When a point to the right of C is considered, the load at C as well as the reaction force R_B must be taken into account. Alternatively R_B on its own can be considered. The shearing force will be 15 kN. This is the value previously calculated for R_B or it can be calculated from the load $C - R_A = 24\,\text{kN} - 9\,\text{kN} = 15\,\text{kN}$. For any point between C and B the force to the right is upwards and the shearing force is therefore negative as was shown earlier in Fig. 2.34. It should be noted that the shearing force changes suddenly at point C. This is shown in Fig. 2.37(a).

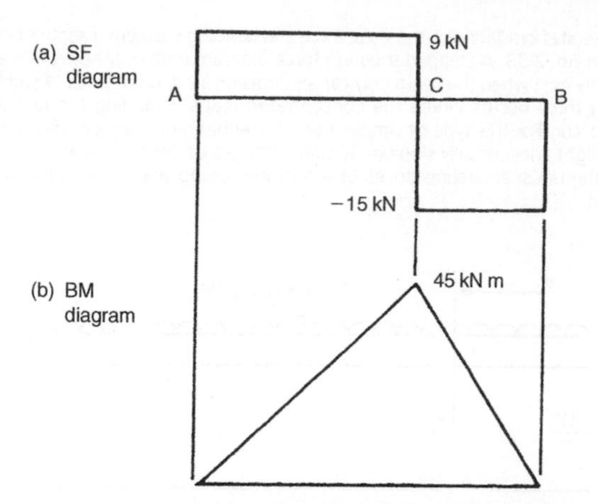

Figure 2.37 Shearing force and bending moment diagrams.

The corresponding bending moments are shown in Fig. 2.37(b). The bending moment at A is zero, since there are no forces to the left of point A. At a point 1 m to the right of point A the moment of the only force R_A to the left of this point is $R_A \times 1\,\text{m} = 9\,\text{kN m}$. As this moment about A is clockwise the moment is positive (+9 kN m). At points 2, 3, 4 and 5 m to the right of A the bending moments are respectively:

$$R_A \times 2\,\text{m} = 9\,\text{kN} \times 2\,\text{m} = 18\,\text{kN m}$$

$$R_A \times 3\,\text{m} = 9\,\text{kN} \times 3\,\text{m} = 27\,\text{kN m}$$

$$R_A \times 4\,\text{m} = 9\,\text{kN} \times 4\,\text{m} = 36\,\text{kN m}$$

$$R_A \times 5\,\text{m} = 9\,\text{kN} \times 5\,\text{m} = 45\,\text{kN m}$$

Since all these are clockwise moments they are all positive bending moments. For points to the right of C, the load at C as well as RA must be considered or, more simply, as previously demonstrated RB alone can be used. At points 5, 6 and 7 m from A, the bending moments are respectively:

$$R_B \times 3\,\text{m} = 15\,\text{kN} \times 3\,\text{m} = 45\,\text{kN m}$$

$$R_B \times 2\,\text{m} = 15\,\text{kN} \times 2\,\text{m} = 30\,\text{kN m}$$

$$R_B \times 1\,\text{m} = 15\,\text{kN} \times 1\,\text{m} = 15\,\text{kN m}$$

As these moments to the right of the points considered are anticlockwise, they are all positive bending moments. At B the bending moment is zero as there is no force to its right. The results are summarized in Table 2.5.

Table 2.5 Shearing force and bending moment values for Figure 2.37.

Distance from A (m)	0	1	2	3	4	5	6	7	8	
Shearing force (kN)	+9	+9	+9	+9	+9	+9	−15	−15	−15	−15
Bending moment (kN m)	0	+9	+18	+27	+36	+45	+30	+15	+15	

Using the results from Table 2.5 the shear force and bending moment diagrams can be drawn as shown in Fig. 2.38. A stepped shearing force diagram with only horizontal and vertical lines can only exist when the beam only carries concentrated, point loads. A sudden change in shearing force occurs where the concentrated loads, including the reactions at the supports, occur. For this type of simple loading the bending moment diagram also consists of straight lines, usually sloping. Sudden changes of bending moment cannot occur except in the unusual circumstances of a moment being applied to a beam as distinct from a load.

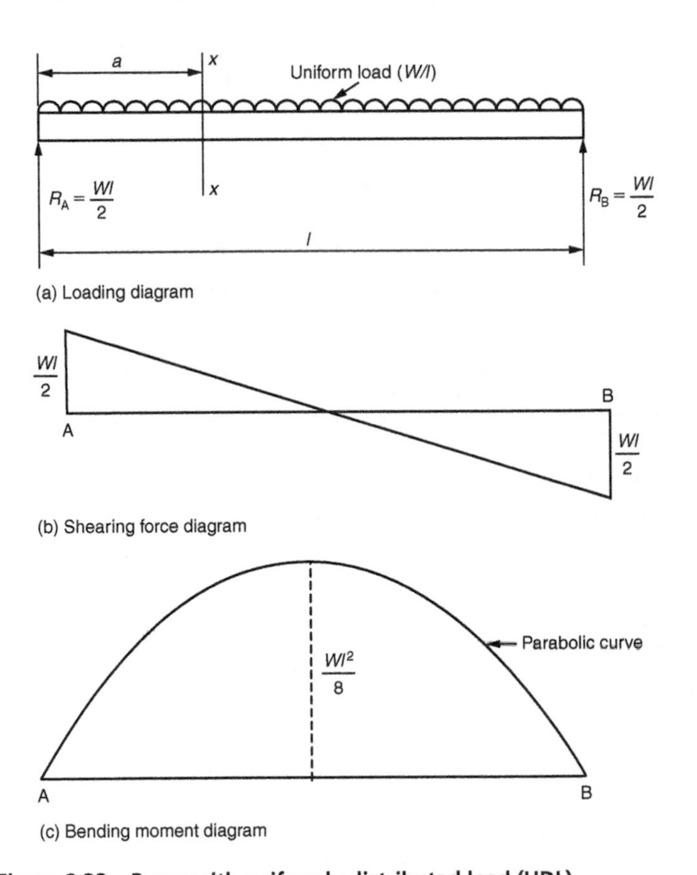

Figure 2.38 Beam with uniformly distributed load (UDL).

In practice a beam loaded with concentrated point loads alone cannot exist. This is because the beam itself has mass and, therefore, weight. In a beam of uniform cross-section this represents a uniform load throughout the length of the beam. The effect of uniform loading will now be considered.

Figure 2.38(a) shows a uniformly loaded beam of length l and weight W. The only point loads being the reactions at the supports R_A and R_B. Since the loading is symmetrical R_A and R_B will both equal $W/2$. The shear force diagram shows maximum values at the points of support and zero shear at the midpoint. This time, however, the line joining the shearing force is a sloping line passing through the midpoint.

Take a section XX at a distance a from the left-hand support R_A:

The shearing force at $XX = (Wl/2) - (W \times a)$

Since this is the equation of a straight line the *shear force diagram* is shown in Fig. 2.38(b). Unlike the previous example, this time the *bending moment diagram* is not made up of straight lines but is a continuous curve with a maximum value at the midpoint.

The bending moment M_x at $XX = [(Wl/2) \times a] - [W \times a \times (a/2)]$

Therefore:

$$M_x = W/2(la - a^2) \qquad (1)$$

This is the equation of a parabola so, its first derivative will be a maximum when:

$$dM_x/da = 0$$

Differentiating equation (1):

$$dM_x/da = W/2(l - 2a) = 0$$

When:

$$a = l/2$$

The maximum bending moment occurs at the centre of the span which is only to be expected with symmetrical loading. Hence substituting $a = l/2$ in equation (1):

$$M_{max} = W/2[(l^2/2) - (l^2/4)] = \mathbf{Wl^2/8}$$

The bending moment diagram is shown in Fig. 2.38(c). Apart from the beam shown in Fig. 2.38(a) where the uniform load resulted from gravity acting on the mass of the beam itself, the only other occasion when a beam is uniformly loaded is when it is carrying a uniform panel of masonry.

More often, there is a combination of point loads and uniform loading as shown in the loading diagram Fig. 2.39(a).

(a) Calculate reactions R_A and R_C by taking moments about A:

$$(45 \times 2) + (12 \times 9) \times 4.5 + (24 \times 12) = 9R_C$$

Thus: $R_C = 96\,kN$ and $R_A = 81\,kN$

(b) Shearing forces (let x be any distance from A):

For $0 \leqslant x \leqslant 2$, shear force $= (81 - 12x)$ kN (1)
For $2 \leqslant x \leqslant 9$, shear force $= (81 - 12x - 45) = (36 - 12x)$ kN
For $9 \leqslant x \leqslant 12$, shear force $= (81 - 12 \times 9 - 45 + 96) = +24\,kN$

These results are depicted by straight lines as shown in Fig. 2.39(b)

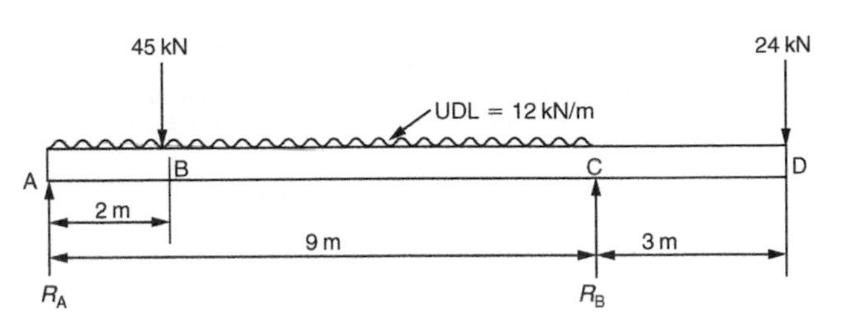

(a) Loading diagram

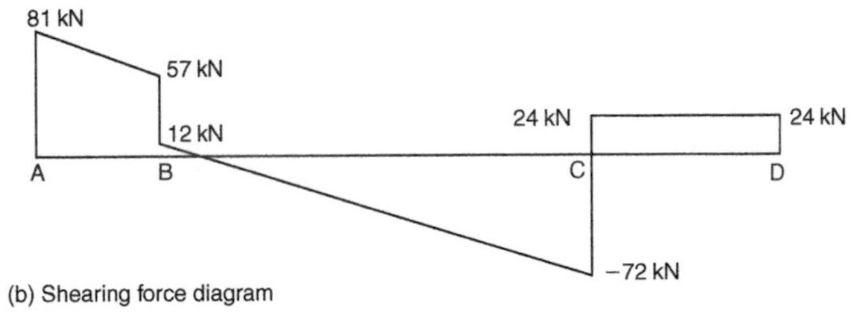

(b) Shearing force diagram

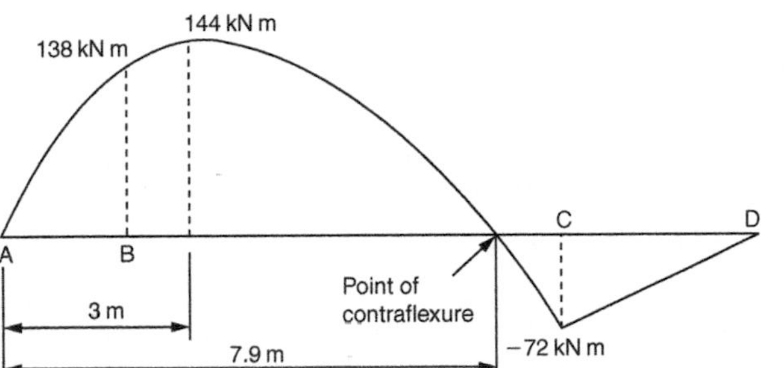

(c) Bending moment diagram

Figure 2.39 Beam with both point and uniform loads. UDL: Uniformly distributed load.

(c) Bending moments (let x be any distance from A):

For $0 \leqslant x \leqslant 2$, $M = 81x - 12(x^2/2) = 81x - 6x^2$ kN m
Therefore: at A, $x = 0$ and $M_A = 0$
at B, $x = 2$ and $M_B = 162 - 24 = 138$ kN m

For $2 \leqslant x \leqslant 9$, $M = 81x - (12x^2/2) - 45(x - 2) = 36x - 6x^2 + 90$ kN m (2)
At $x = 2$ $M_B = 72 - 24 + 90 = 138$ kN m (as above)
At $x = 9$ $M_C = 324 - 486 + 90 = -72$ kN m

The maximum bending moment occurs between the points B and C where $dM/dx = 0$.

Note this is the same point that the shearing force is also zero. From equation (2):

$dM/dx = 36 - 12x = 0$

Therefore:

$x = 3\,\text{m}$

Substituting $x = 3$ in equation (2):

$(36 \times 3) - (6 \times 3^2) + 90 = 144\,\text{kN m}$.

Thus the maximum bending moment is 144 kN m and it occurs at a point 3 m from A.

Contraflexure will occur when the bending moment is zero. That is, the positive root of equation (2) when it is equated to zero.

$36x - 6x^2 + 90 = 0$

Therefore:

$x =$ **7.9 m** (the only positive solution)

Therefore contraflexure occurs at a point 7.9 m from A.

2.9.4 Beams (cantilever)

As well as being simply supported as in the previous examples, beams may also be in the form of a *cantilever*. That is, only one end of the beam is supported and the remote end from the support is unsupported as shown in Fig. 2.40(a). The supported end of the beam may be built into masonry or it may be a projection from a simply supported beam. A typical shearing force diagram and a typical bending moment diagram for a cantilever beam with concentrated, point loads are shown in Fig. 2.40(b) and (c).

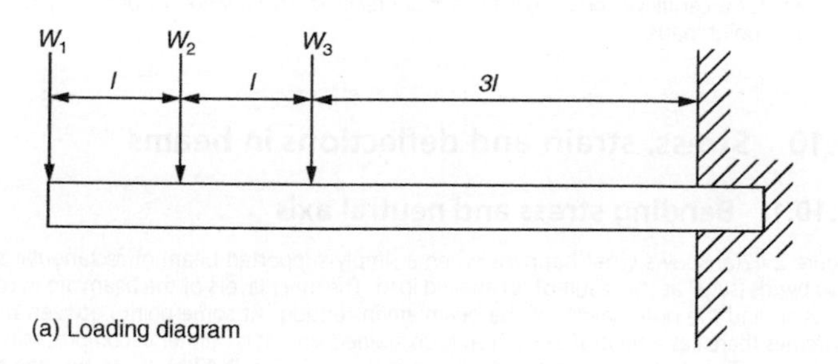

Figure 2.40 Cantilever – point loads only.

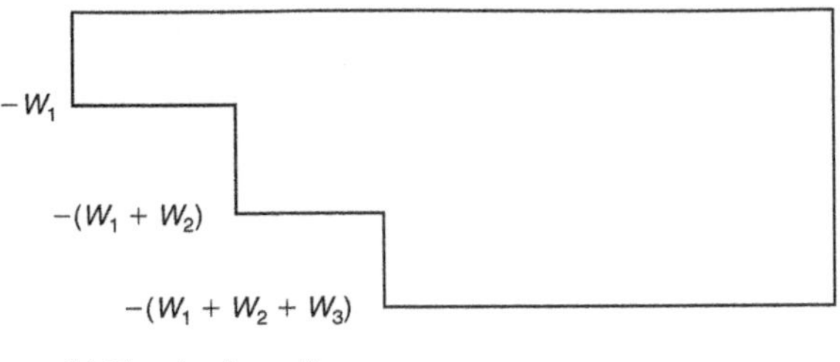

(b) Shearing force diagram

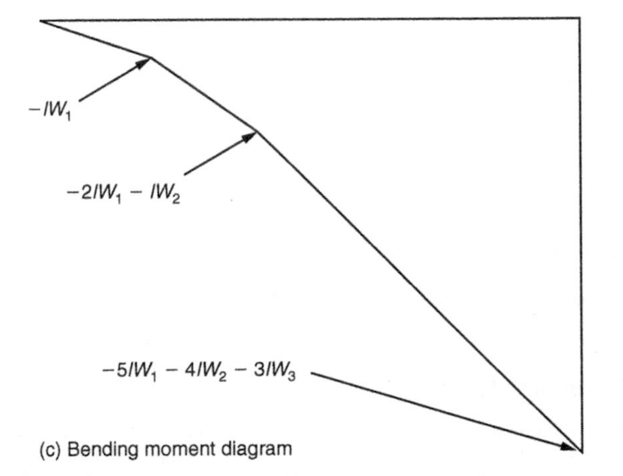

(c) Bending moment diagram

Figure 2.40 ***continued***

Figure 2.41 shows the loading diagram, shearing force diagram and the bending moment diagram for a cantilever beam having a more realistic combination of uniformly distributed and point loads.

2.10 Stress, strain and deflections in beams

2.10.1 Bending stress and neutral axis

Figure 2.42(a) shows what happens when a simply supported beam of rectangular section bends (sags) as the result of an applied load. The inner layers of the beam are in compression and the outer layers of the beam are in tension. At some point between these extremes there lies a *neutral axis* which is unstrained since it is neither in compression nor in tension. The stress diagram for the beam is shown in Fig. 2.42(b). It can be seen that the stress is a maximum at the top and bottom of the beam and zero at the neutral axis.

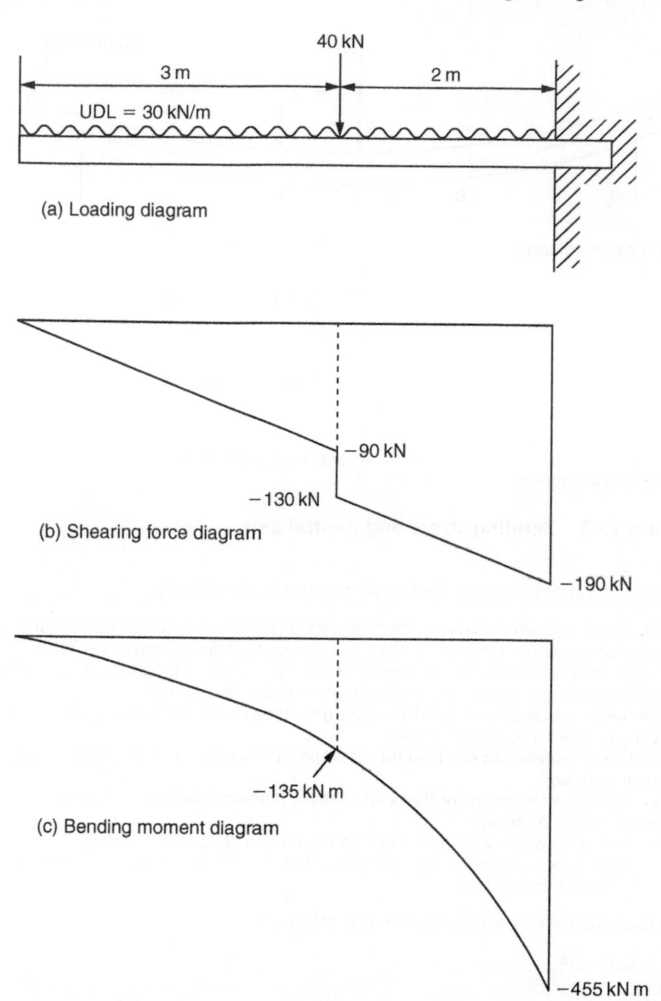

Figure 2.41 Cantilever – combined point and distributed loads.

Therefore a solid rectangular beam is less efficient than a typical **I** section beam which has a much higher strength to weight ratio. In the **I** section beam the maximum stress is carried by the flanges at the top and bottom of the beam, and the lightweight web separating the flanges carries little stress. In fact, to reduce the mass of the beam even further the flange can often be perforated without appreciable loss of strength.

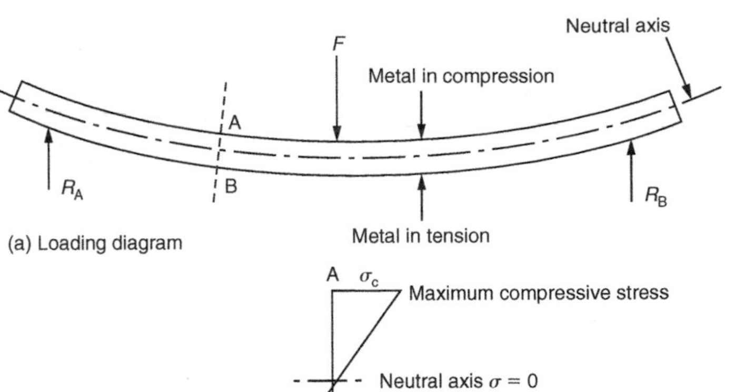

Figure 2.42 Bending stress and neutral axis.

For all practical circumstances the following assumptions are made:

- The deflection is small relative to the cross-sectional dimensions of the beam. That is, the radius of bending is large relative to the cross-sectional dimensions of the beam.
- The beam is assumed to be made up of an infinite number of longitudinal layers which are in contact during bending but are independent of each other.
- The beam is subject to longitudinal stress only. There is no resultant longitudinal force acting on any section of the beam.
- The beam is safely loaded within the elastic limit of its material so that stress is proportional to strain.
- The modulus of elasticity for the beam material remains constant throughout for both tension and compression.
- The transverse plane sections before bending remain plane after bending.
- The neutral axis passes through the centroid of the section irrespective of the shape of the beam cross-section.

The fundamental equation for the bending of beams is:

$\sigma/y = M/I = E/R$

Where:
σ = the stress due to bending at a distance y from the neutral axis
y = the distance from the neutral axis
M = the bending moment
I = the second moment of inertia for the section of the beam.
E = the modulus of elasticity
R = the radius of curvature of the beam

The section formulae including the second moment of inertia, the section modulus and the radius of gyration is shown in Fig. 2.43 together with the cross-sections of the more commonly used beams.

Figure 2.43 Section formulae

Section	Second moment of area I_{xx}	Section modulus $z = I/y$	Radius of gyration $k = \sqrt{(I/A)}$
Rectangle, axis through centroid (b, d, $\frac{d}{2}$, $\frac{d}{2}$, y, x–x)	$\frac{bd^3}{12}$	$\frac{bd^2}{6}$	$\frac{d}{\sqrt{12}}$
Rectangle, axis at base (b, d, y, x–x)	$\frac{bd^3}{3}$	$\frac{bd^2}{3}$	$\frac{d}{\sqrt{3}}$
Triangle (b, d, $d/3$, y_1, y_2, x–x)	$\frac{bd^3}{36}$	$Z_1 = \frac{bd^2}{24}$ $Z_2 = \frac{bd^2}{12}$	$\frac{d}{\sqrt{18}}$
I-section (B, D, d, $\frac{D}{2}$, $\frac{d}{2}$, $\frac{b}{2}$, $\frac{b}{2}$, y, x–x) and channel section (B, b, $\frac{D}{2}$, $\frac{d}{2}$, y, x–x)	$\frac{BD^3 - bd^3}{12}$	$\frac{BD^3 - bd^3}{6D}$	$\sqrt{\left[\frac{BD^3 - bd^3}{12(BD - bd)}\right]}$
Circle (D, y, P)	$\frac{\pi D^4}{32}$	$\frac{\pi D^3}{16}$	$\frac{D}{\sqrt{8}}$

continued

Figure 2.43 ***continued***

Section	*Second moment of area* I_{xx}	*Section modulus* $z = I/y$	*Radius of gyration* $k = \sqrt{(I/A)}$
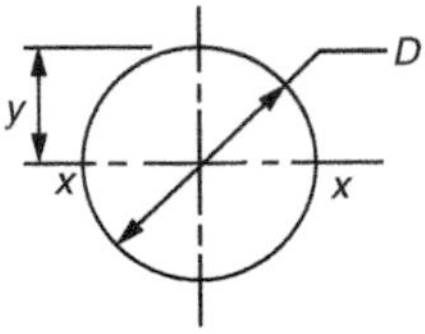	$\frac{\pi D^4}{64}$	$\frac{\pi D^3}{32}$	$\frac{D}{4}$
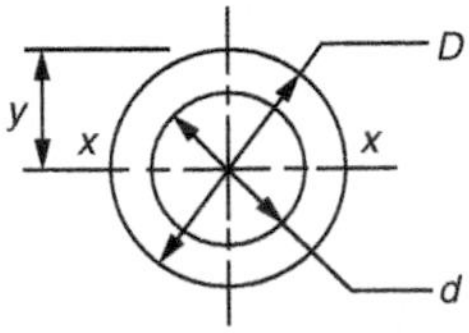	$\frac{\pi(D^4 - d^4)}{64}$	$\frac{\pi(D^4 - d^4)}{32D}$	$\frac{\sqrt{(D^2 - d^2)}}{4}$

2.11 Frameworks

The design of a structure such as a roof truss or a space frame made up from a number of members connected at their ends involves the determination of the forces in each member under a given system of loading. Only 'simple frameworks' will be considered in this section. That is, frameworks in which the members are *pin-jointed*. Figure 2.44(a) shows members rigidly connected, whereas the members shown in Fig. 2.44(b) are pin-jointed and free to rotate about P. In simple frameworks it is assumed that:

- All the connections between the individual members will be pin-jointed.
- Since the members are free to rotate there will be no moments at the joints and no bending of the member.
- External loads are only applied at the pin-joints. Therefore the individual members of the framework are only subjected to tensile or compressive loads
- A *tie* is a member that exerts an inward pull at either end in order to resist a *tensile load* that is trying to stretch the member.
- A *strut* is a member that exerts an outward force at either end in order to resist a *compressive load* that is trying to shorten the member.

Most frameworks are based on a triangle or series of triangles. The reason for this is shown in Fig. 2.44(c) and (d). It can be seen that if a pin-jointed rectangle is subjected to the forces F_1 and F_2 it will be free to distort. However, if a pin-jointed triangle is subjected to the same forces it cannot change shape unless one or more of its members change in length. *A triangular frame is inherently rigid*. Figure 2.44(e) shows a typical trussed girder built up from a series of triangles. This particular example is referred to as being a Warren braced girder.

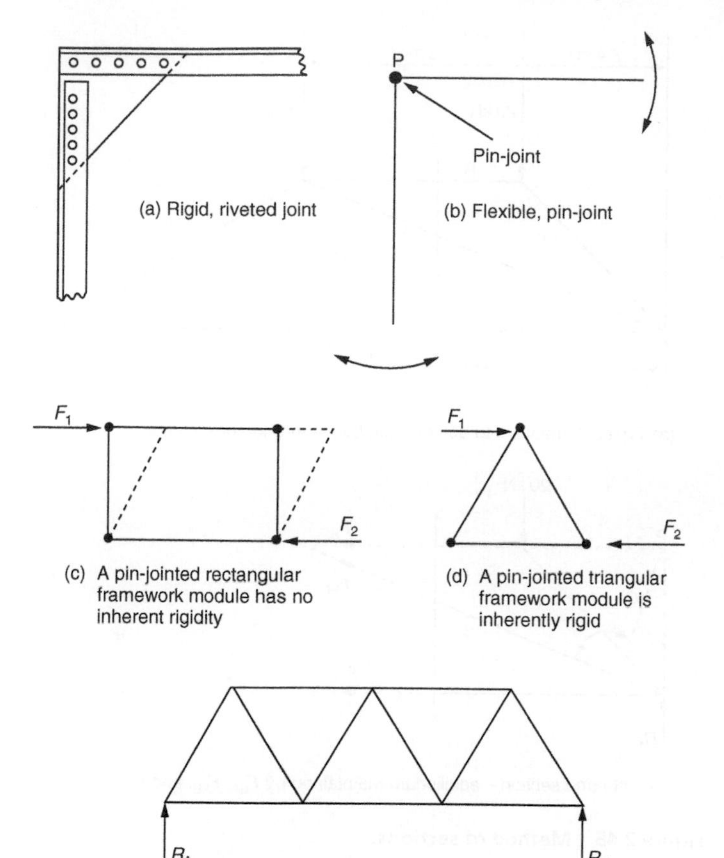

Figure 2.44 Frameworks.

2.11.1 Method of sections

The main advantage of the method of sections is that the force in almost any member can usually be obtained by cutting the frame through that member and applying the principle of moments. For example, consider how the forces in three members cut by the imaginary plane *XX* in Fig. 2.45(a) may be determined by use of the method of sections.

The cutting plane *XX* divides the truss into two parts. Suppose the right-hand part of the framework is removed. This will cause the left hand to collapse about the joint R_1 unless

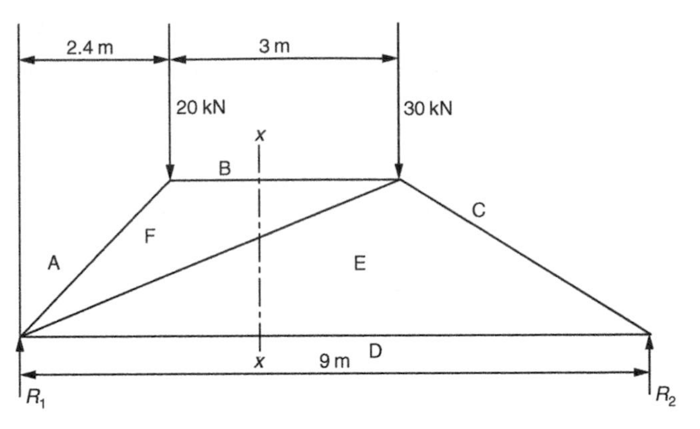

(a) Typical framework to be analysed by the method of sections

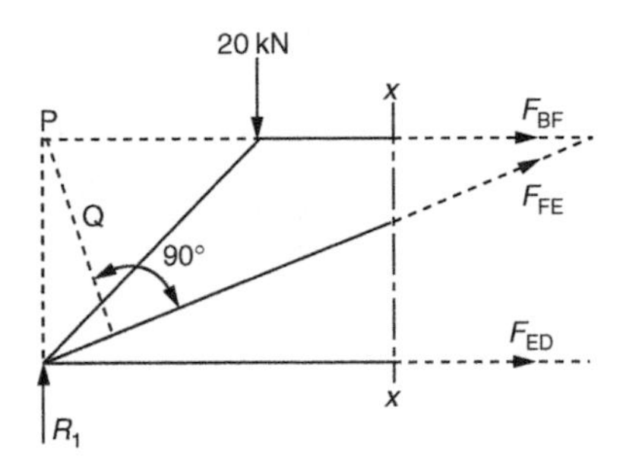

(b) Left-hand section – equilibrium maintained by F_{BF}, F_{FE}, and F_{ED}

Figure 2.45 Method of sections.

the external forces represented by F_{BF}, F_{FE} and F_{ED} are applied to the cut members as indicated in Fig. 2.45(b). Therefore these external forces applied in Fig. 2.45(b) actually represent the internal forces that are to be determined in this example.

Consider the equilibrium conditions of the left-hand section of the truss, the external forces applied to the cut section must either push or pull on the joints to the left of section *XX*. It is good practice to assume that *the cut members are all in tension* as a positive final answer then confirms tension (*tie*) whilst a negative final answer implies a compression (*strut*).

The left-hand section of the frame is now in equilibrium under the reaction R_1, the load of 20 kN and the three unknown forces F_{BF}, F_{FE} and F_{ED}. By taking moments about the left-hand reaction (R_1) – the intersection of two of the unknown forces F_{FE} and F_{ED} – the value of F_{BF} can be easily determined. Similarly the magnitude of F_{ED} can be determined by taking moments about position (1). Finally moments about a point such as P will give the magnitude of the force in FE.

Before commencing the determination of the above forces, it is necessary to calculate the magnitude of R_1 and R_2 using the principle of moments discussed earlier.

Taking moments about R_1:

$$(20 \times 2.4) + (30 \times 5.4) = 9R_2$$

Therefore:

$R_2 =$ **23.33 kN**
$R_1 =$ **26.67 kN**

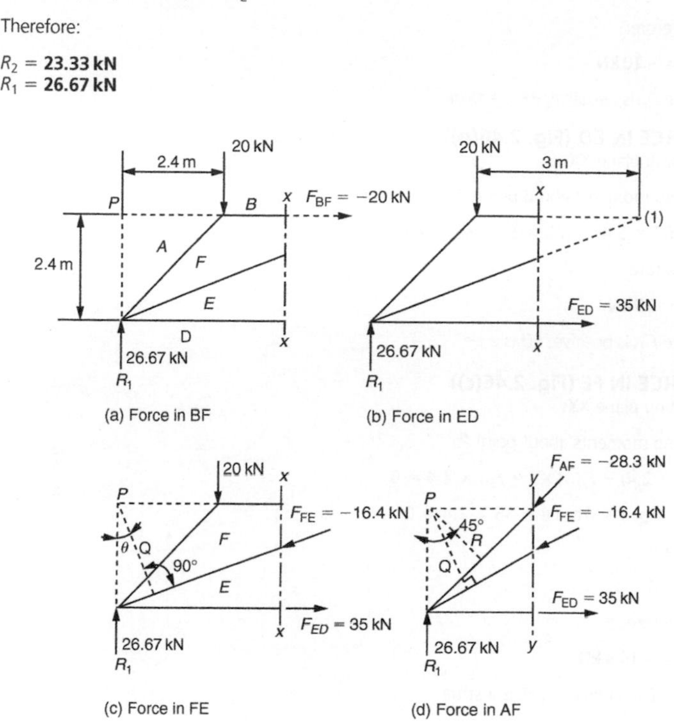

Figure 2.46 Forces in framework members.

FORCE IN BF (Fig. 2.46(a))
Cutting plane XX:

Taking moments about R_1:

$20 \times 2.4 + F_{BF} \times 2.4 = 0$

Therefore:

$F_{BF} = \mathbf{-20\,kN}$

Since F_{BF} is negative, BF is a *strut*.

FORCE IN ED (Fig. 2.46(b))
Cutting plane XX:

Taking moments about point (1):

$(26.67 \times 5.4) - (20 \times 3) - F_{ED} \times 2.4 = 0$

Therefore:

$F_{ED} = \mathbf{35\,kN}$

Since F_{ED} is positive, ED is a *tie*.

FORCE IN FE (Fig. 2.46(c))
Cutting plane XX:

Taking moments about point P:

$(20 \times 2.4) - F_{FE} \times XX - F_{ED} \times 2.4 = 0$

$48 - F_{FE} \times 2.4 \cos\theta - 35 \times 2.4 = 0$

where:

$\theta = 24°$

Therefore:

$F_{FE} = \mathbf{-16.4\,kN}$

Since F_{FE} is negative, FE is a *strut*.

FORCE IN AF (Fig. 2.46(d))
Cutting plane YY:

Taking moments about P:

$16.4\,xx - F_{AF}\,xy - 35 \times 2.4 = 0$

Therefore:

$F_{AF} = \mathbf{-28.3\,kN}$

Since F_{AF} is negative, AF is a *strut*.

FORCE IN EC (Fig. 2.46(e))
Cutting plane ZZ:

Taking moments about point (2):

$23.33 \times 3.6 + F_{EC} \times z = 0$

Therefore:

$F_{EC} = \mathbf{-42\ kN}$

Since F_{EC} is negative, EC is a *strut*.

Note that in all the above instances the force in the unknown member has been assumed initially to be in tension.

2.12 Hydrostatic pressure

The liquid in the container shown in Fig. 2.47(a) exerts a pressure on the sides and base of the container. The pressure exerted by the liquid depends on:

- the density of the liquid (ρ);
- the depth at which the liquid is acting on the body (h);
- the gravitational constant (g).

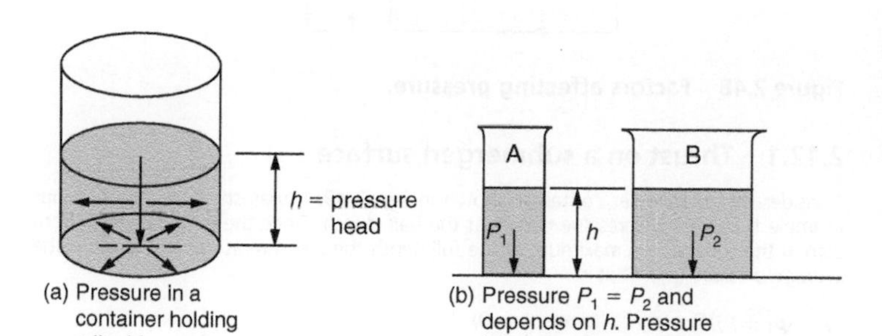

Figure 2.47 Hydrostatic pressure.

Figure 2.47(b) shows two containers A and B in which the height of the liquid (h) is the same. Providing the density of the liquid is the same in each case and since the gravitational constant will be the same in each case, then the pressures p_1 and p_2 will be the same despite the fact that the base areas of the containers are different. The height of the liquid h above a given point is also known as:

- the pressure head
- the head.

The pressure (p) in a liquid is determined using the following expression:

$$p = h \times \rho \times g$$

where:
p = pressure (Pa)
h = depth at which the pressure is measured (m)
ρ = density of the liquid (kg/m^3)
g = the gravitational constant (9.81 m/s^2)

Example

Figure 2.48 shows a container filled with paraffin to a depth of 2 m. Determine the pressures p_1 and p_2, given the density of paraffin = 800 kg/m^3 and $g = 9.81$ m/s^2.

$p_1 = h_1 \times \rho \times g = 2\,\text{m} \times 800\,\text{kg/m}^3 \times 9.81\,\text{m/s}^2 =$ **15696 Pa** or **15.696 kPa**

$p_2 = h_2 \times \rho \times g = 1\,\text{m} \times 800\,\text{kg/m}^3 \times 9.81\,\text{m/s}^2 =$ **7848 Pa** or **7.848 kPa**

At the half depth point the pressure is half that at the bottom of the container. At the surface the pressure is zero since there is no depth. Therefore the pressure is proportional to the depth at which it is measured.

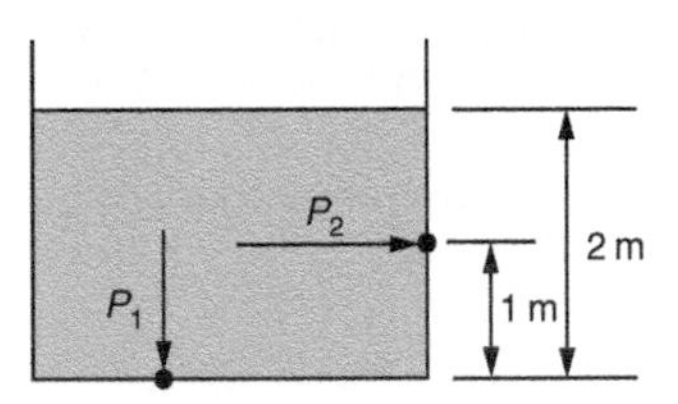

Figure 2.48 Factors affecting pressure.

2.12.1 Thrust on a submerged surface

Consider the rectangular container shown in Fig. 2.49. It was shown in the previous example that the half pressure occurs at the half depth. Since the pressure varies from zero at the surface to a maximum at the full depth the pressure at the half depth is the *average pressure* (p_A). That is:

$p_A = (p_1 + p_2)/2 = p_2/2$ (since $p_1 = 0$)

Therefore the force (thrust) acting on the side of the container will be the submerged area of the side (A_S) multiplied by the average pressure (p_A). That is:

$F_S = p_A \times A_S$

where:
F_S = force (thrust) acting on the side of the container (N)
p_A = average pressure (Pa)
A_S = submerged area of the side of the container (m^2)

Note:

1. The force is assumed to act at the centre of area of the side of the container.
2. The force acting on the bottom of the container is calculated using the maximum pressure and *not* the average pressure.

Example

Figure 2.49 shows a rectangular tank with a base area of 9 m^2. It is filled with a water to a depth of 2 m. Taking $g = 9.81$ m/s^2, calculate:

(i) The pressure acting on the base of the tank.
(ii) The force acting on the base of the tank.
(iii) The average pressure acting on the sides of the tank.

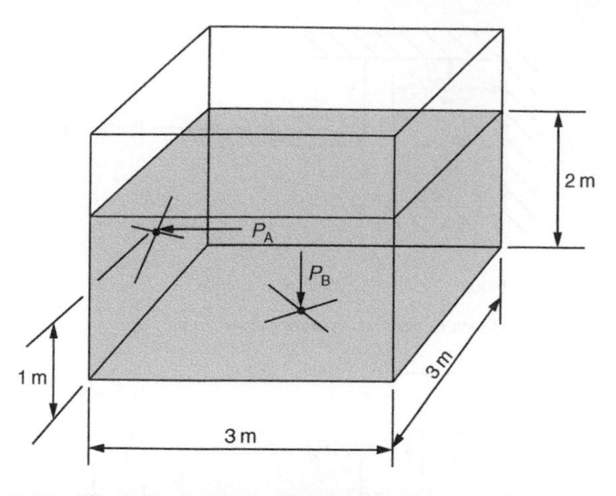

Figure 2.49 Thrust on a submerged surface.

(iv) The force (F_S) acting on the side of the tank.
(v) The height (h) at which the force (F_S) acts above the base of the tank.

(i) $p_B = h \times \rho \times g = 2\,m \times 1000\,kg/m^2 \times 9.81\,m/s^2 = 19220\,Pa =$ **19.22 kPa**
(ii) $F_B = p_B \times A_B = 19220\ Pa \times 9\,m^2 = 172980\,N =$ **17.298 kN**
(iii) $p_A = (p_S + p_B)/2 = (0\,Pa + 19220\,Pa)/2 = 9600\,Pa =$ **9.6 kPa**
(iv) $F_S = p_A \times A_S = 9600\,Pa \times 6\,m^2 = 57600\,N =$ **57.6 kN** (where: A_S = submerged area)
(v) The force F_S acts at the centre of the submerged area of the side of the tank; that is, ½ × 2 m = 1 m up from the base of the tank.

2.12.2 Pascal's law

Pascal's law states that:

If a pressure is exerted on a liquid in a container, the liquid will disperse the pressure uniformly in all directions.

This is shown in Fig. 2.50(a). The pressure (p) which is exerted uniformly within the enclosed space can be calculated as follows:

Pressure (p) = force (F)/area (A)

$= 100\,N/0.5\,m^2$

$= 200\,N/m^2$

= **200 Pa**

In accordance with Pascal's law the pressure of 200 Pa is acting uniformly on every surface. It is acting on the piston, on the cylinder wall and on the cylinder head. Since the mass of the liquid is small, any hydrostatic pressures and forces that may be present have been ignored in this example.

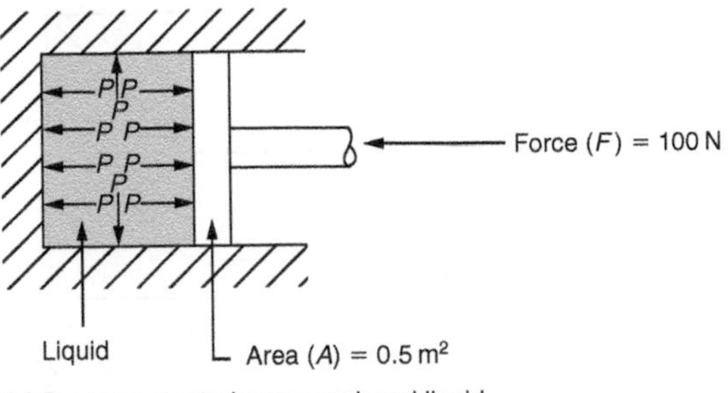

(a) Pressure exerted on an enclosed liquid

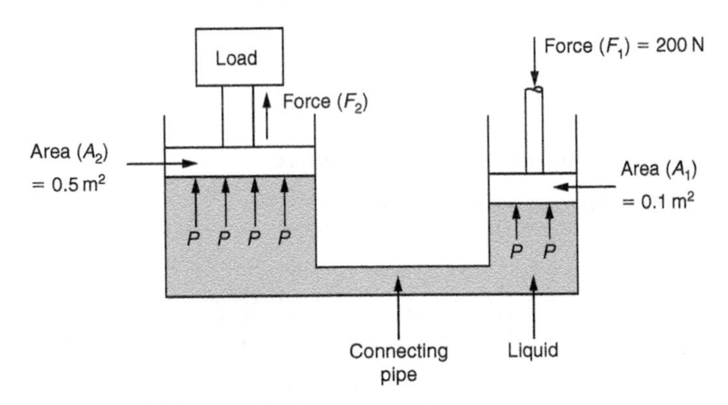

(b) Connected system

Figure 2.50 Pascal's law.

Figure 2.50(b) shows two cylinders of unequal area connected into a single system and filled with water or with hydraulic oil. The pressure in the system is calculated using the previous formula $p = F/A$. For the smaller cylinder the pressure is:

$p = F_1/A_1 = 200\,\text{N}/0.1\,\text{m} = 2000\,\text{N/m}^2 =$ **2000 Pa** or **2 kPa**

But the two cylinders are connected together to form a single system, so according to Pascal's law there is a uniform pressure of 2 kPa acting throughout the system. Therefore the force exerted by the larger piston on the load will be:

$F_2 = p \times A_2 = 2\,\text{kPa} \times 0.5\,\text{m}^2 =$ **1000 N** or **1 kN**

The force F_2 is five times larger than force F_1 because the piston area A_2 is five times larger than the piston area A_1. This ability to magnify forces by the use of hydrostatic processes is used in such devices as hydraulic presses, hydraulic jacks and the hydraulic brakes of vehicles. However it must be noted that although the larger piston exerts a force that is five times greater than the force acting on the smaller piston, the larger piston *only travels one fifth the distance* of the smaller piston.

3

Engineering Dynamics

3.1 Engineering dynamics

As stated at the beginning of Chapter 2, engineering mechanics can be divided into *statics* and *dynamics*. Chapter 3 is concerned with *dynamics* which is the study of *objects in motion*.

3.2 Work

When a force is applied to a body so that it causes the body to move, the force is said to be doing *work*. The work done is the product of the distance moved in metres (m) and the applied force in newtons (N) measured in the direction of motion. Therefore the work done is measured in newton metres and the unit of work is joule (J) where **1 Nm = 1 J.**

Figure 3.1(a) shows a body is being moved by a force F along a horizontal plane. Since the force is acting on the body parallel to the horizontal plane, the work done is product of

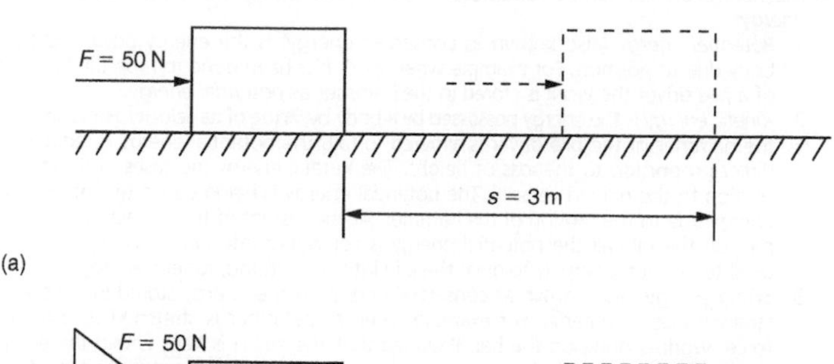

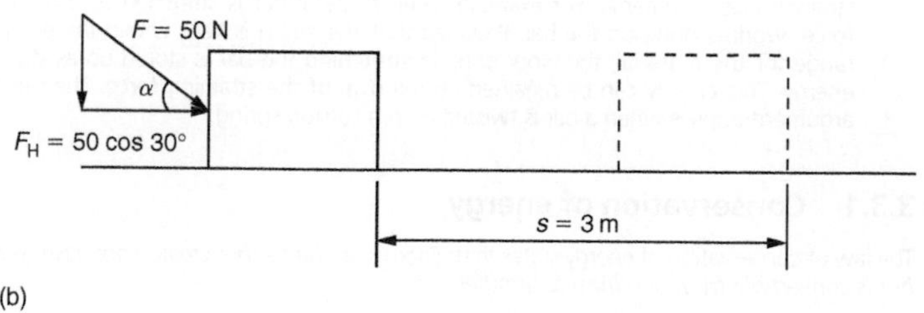

Figure 3.1 Work. (a) Work done by a force F. (b) Work done by the force F_H.

the force F (N) and the distance s (m). If the force is 50 N and the distance is 3 m, then the work done is:

$50\,N \times 3\,m = 150\,Nm =$ **150 J**

In Fig. 3.1(b) the force is inclined through the angle α to the direction of motion. In this case the force vector has to be resolved into its vertical and horizontal components as shown. The component force that is parallel to the direction of motion of the body is $F \cos \alpha$. Therefore the work done is $s \times F \cos \alpha$ joules. If the force (F) is again 50 N, the distance (s) is again 3 m and the angle α is 30° the work done will be:

$W = 3\,m \times 50\,N \times 0.866 = 129.9\,Nm =$ **129.9 J**

3.3 Energy

Anything capable of doing work is said to posses *energy*. Energy exists in various forms and may be considered as 'stored work' awaiting to be used. For example:

- *Chemical energy* is absorbed or released when a reaction occurs (e.g. the combustion of fossil fuels).
- *Atomic energy* is liberated by splitting or combining atomic particles. Splitting atomic particles is called *fission* and combining atomic particles is called *fusion*. These reactions produce large amounts of heat energy.
- *Heat energy* can be subdivided into *sensible heat energy* and *latent heat energy*. Sensible heat energy is the energy taken in or given out when a solid, liquid or gas rises in temperature or falls in temperature. Latent heat energy is the energy required to change the state of a solid, liquid or gas at a constant temperature.
- *Mechanical energy* can be subdivided into *potential energy*, *kinetic energy* and *strain energy*:
  1. *Potential energy* (also known as conserved energy) is the energy possessed by a body due to position. For example when work has been done to raise the hammer of a pile driver the work is stored in the hammer as *potential energy*.
  2. *Kinetic energy* is the energy possessed by a body by virtue of its velocity. For example if the hammer of the pile driver is allowed to fall, the potential energy is reduced in direct proportion to the loss of height. The kinetic energy increases in direct proportion to the gain in velocity. The potential energy is being converted into kinetic energy due to the motion of the hammer. At the instant of the impact of the hammer on the pile all the potential energy is converted into *kinetic energy* which is used to do work, namely to drive the pile into the ground. Kinetic energy $= \frac{1}{2}mv^2$.
  3. *Strain energy* (also known as conserved energy) is the energy stored in a piece of strained elastic material. For example when a metal bar is stretched by a tensile force, work is done on the bar. Provided that the strain is kept within the elastic range for the material, the work done in stretching the bar is stored up as strain energy. This energy can be regained on removal of the straining force. The same argument applies when a bar is twisted as in a torsion spring.

3.3.1 Conservation of energy

The law of conservation of energy states that: *Energy can be neither created nor destroyed but is convertible from one form to another.*

For example, in a power station, the chemical energy of the fuel is released by combustion and is converted into thermal energy which, in turn, produces steam in a boiler. The steam drives a turbine which converts the energy in the steam into mechanical energy. This mechanical energy is converted into electrical energy by the alternator driven by the turbine. During these conversion processes there will be loss of energy due to the inefficiency of the equipment used. This 'lost' energy cannot be destroyed (law of conservation of energy) but is converted into other forms of energy that do not contribute to useful work. For example work done in overcoming friction by heating the bearing surfaces and the lubricating oil.

3.4 Power

Power is the rate of doing work and can be expressed mathematically as:

Power (W) = work done (J)/time taken (s) = force (N) × velocity (m/s)

The unit of power is the watt (W) where 1 watt equals work done at the rate of 1 J/s.

Example

A body of weight 5 kN is to be raised through a height of 15 m in 10 s. Calculate the power required.

Work done = force (weight) × distance moved = 5 kN × 15 m = 75 kNm = 75 kJ

Power required = work done/time = 75 kJ/10 s = **7.5 kW** (assuming 100% efficiency).

3.5 Efficiency

Efficiency is the measure of the 'usefulness' of a machine, process or operation.

Efficiency (η) = [work (or power) output]/[work (or power) input]

Efficiency can be expressed as a percentage (%) or as a decimal fraction of 1 (per-unit efficiency). In practice the efficiency of a device can never achieve 100% or 1.

Example

If the efficiency of the hoist in the previous example is 80%, calculate the actual power required.

Efficiency (%) = (power output × 100)/power input

Therefore

80% = 7.5 kW/power input

Hence

Power input = (7.5 × 100)/80

= **9.375 kW**

3.6 Velocity and acceleration

- Speed is a *scalar quantity* since it has magnitude only (e.g. 30 km/h).
- Velocity is a *vector quantity* since it has magnitude and direction (e.g. 30 km/h due north).
- Acceleration is the *rate of change of velocity*.

3.6.1 Speed

Speed is the rate of change of distance. That is: **speed = distance/time**. Typical units of speed are *kilometres per hour* (km/h) and *metres per second* (m/s).

Note: s (roman type) stands for seconds; *s* (italic type) stands for distance. They may both appear in the same calculation.

3.6.2 Velocity

Velocity is the rate of change of distance in a particular direction. A body moving in a straight line is said to have *uniform velocity in a straight line*. Uniform velocity occurs when a body moving in a straight line covers equal distances in equal interval of time.

Velocity = d*s*/d*t* in a particular direction

3.6.3 Acceleration

Acceleration is the rate of change of velocity. That is:

- *Uniform acceleration* occurs when a body experiences equal changes in velocity in equal time intervals. The basic unit of acceleration = m/s^2.
- *Retardation* or *deceleration* is the reverse process to acceleration. It is frequently referred to as negative *acceleration*.

3.6.4 Equations relating to velocity and acceleration

The symbols that will be used in the following equations are:
v_1 = initial velocity in metres per second (m/s);
v_2 = the final velocity in metres per second (m/s);
a = acceleration in metres per second squared (m/s^2);
s = the distance travelled in metres (m);
t = the time taken in seconds (s).

1. Average velocity $= (v_1 + v_2)/2$
2. Final velocity $v_2 = v_1 + at$
3. Distance travelled $s = (v_1 + v_2)t/2$
4. Distance travelled $s = v_1 t + \frac{1}{2}at^2$
5. Acceleration $a = (v_2 - v_1)/t$
6. Acceleration $a = (v_2^2 - v_1^2)/2s$

3.6.5 Momentum

The linear momentum of a body is defined by:

Momentum = mass × velocity = mv

The basic unit of momentum is kg m/s.

3.6.6 Newton's laws of motion

1. Newton's first law states that: *a body continues in its state of rest, or uniform motion in a straight line, unless acted upon by a resultant force*. That is, a body will not move, change its direction of motion, or accelerate unless it is compelled to by some external

force. Therefore a force can be defined as that which changes or tends to change the motion of a body upon which it acts.

2. Newton's second law states that: *the rate of change of momentum of a body is proportional to the resultant force applied to that body and takes place in the direction in which the force is applied*
3. Newton's third law states that: *for every acting force there is an equal and opposite reaction force* (see Section 3.6.8).

From Newton's second law of motion it is apparent that:

The applied force ∝ rate of change of momentum
The applied force ∝ change in momentum/time

Thus if a force F causes a mass m to experience a change in velocity from v_1 to v_2 in time t, then:

$F \propto (mv_2 - mv_1)/t$

$F \propto m(v_2 - v_1)/t$

$F \propto ma$ where $a = (v_2 - v_1)/t$

Therefore

$\boldsymbol{F = ma \times constant}$

By choosing unit force (N), unit mass (kg) and unit acceleration (m/s^2), the constant becomes 1 and:

$\boldsymbol{F = ma}$

Example

A motor car of mass 4000 kg accelerates from 10 to 60 km/h in 10 s. Determine the applied force required to produce the acceleration.

$v_2 = v_1 + at$ (equation (2) Section 3.6.4)

Where:
$v_2 = 60\,\text{km/h} = 16.67\,\text{m/s}$
$v_1 = 10\,\text{km/h} = 2.78\,\text{m/s}$
$t = 10\,\text{s}$

Then:
$a = (v_2 - v_1)/t$
$a = (16.67 - 2.78)/10$
$a = 1.39\,\text{m/s}^2$

Since:
$F = ma$
$F = 4000\,\text{kg} \times 1.39\,\text{m/s}^2$
$\boldsymbol{F = 5.56\,\textbf{kN}}$

3.6.7 Gravity

Weight is a force. It is the effect of the *force of gravity* acting on the *mass* of a body producing, or trying to produce, an *acceleration*. Note the acceleration of 9.81 m/s^2 is an average value on planet earth. It varies slightly from place to place but 9.81 m/s^2 is acceptable for all practical purposes. The acceleration due to gravity is given the symbol $\boldsymbol{g}$, therefore the expression:

$F = ma$ becomes $F = mg$ in calculations involving gravity.

[On planet earth, the force F exerted by a mass of 5 kg = 5 × 9.81 = 49.05 kN. Since the gravitational attraction of the moon is approximately 1/6 that of the gravitational effect of planet earth, the force F exerted by a mass of 5 kg = 5 × (9.81/6) = 8.175 kN on the moon.]

3.6.8 Conservation of momentum

The Principle of the Conservation of Linear Momentum is derived from Newton's third law of motion. That is, *for every acting force there is an equal and opposite reaction force.* The Principle of Conservation of Linear Momentum states that: the total momentum of as system, in a given direction, remains constant unless an external force is applied to the system in that given direction. This principle enables impact problems involving only internal forces to be easily resolved.

Consider the firing of a projectile from a gun. Equal and opposite forces are acting on the projectile and the gun will cause the projectile to move forward and the gun to move backwards. Since there are no external forces involved, if both gun and projectile had equal masses they would move with equal velocity. Fortunately, in practice, the gun has a much greater mass than the projectile so that the recoil is much less than the forward movement of the projectile.

Example

A body P of mass 15 kg moving with a velocity of 30 m/s when it collides with a second body Q of mass 5 kg which is moving in the same direction at 5 m/s. Determine the common velocity of the two bodies after impact.

By the Principle of the Conservation of Linear Momentum:

Momentum before impact = momentum after impact

Mass × velocity before impact = mass × velocity after impact

$15\,\text{kg} \times 30\,\text{m/s} + 5\,\text{kg} \times 5\,\text{m/s} = (15\,\text{kg} + 5\,\text{kg})v_l$ where v_l = velocity after impact

$\mathbf{v_l = 23.75\,m/s}$

Figure 3.10(a) shows that had the two bodies been moving in opposite directions before impact then the velocity of the second body Q would be considered as a negative quantity and the calculation would be as follows:

Momentum before impact = momentum after impact

$15\,\text{kg}(+30\,\text{m/s}) + 5\,\text{kg}(-5\,\text{m/s}) = (15\,\text{kg} + 5\,\text{kg})v_l$

$\mathbf{v_l = 21.25\,m/s}$

3.6.9 Impact of a fluid jet on a fixed body

- If the jet is deflected from its line of motion then, from Newton's first law, a force must be acting on it.
- If the force that is acting on the jet is proportional to the rate of change of momentum and acts in the direction of the change of momentum, then Newton's second law is applicable.
- If the force exerted by the body on the jet to change its momentum then this force must be equal and opposite to the force exerted by the jet on the body.

Since this example involves the impact of a fluid jet, then it is concerned with jet flow rate and Newton's second law of motion can be written as:

Force = rate of change of momentum
= (mass × change of velocity)/time taken
= mass flow rate × change of velocity in the direction of the force

Example

A water jet of 50 mm diameter impacts perpendicularly on a fixed body with a velocity of 75 m/s. Determine the force acting up the body.

Assume that the jet of water will spray out radially when it impacts on the body; that is, it is deflected through an angle of 90°. Therefore the jet will lose all its momentum at right angles to the body as shown in Fig. 3.2.

$$\text{Area of jet} = (\pi/4) \times (50/1000)^2\,\text{m}^2 = 6.25\pi \times 10^{-4}$$

$$\text{Volume flow rate} = 6.25\pi \times 10^{-4}\,\text{m}^2 \times 75\,\text{m/s} = 0.047\pi\,\text{m}^3/\text{s}$$

$$\text{Mass flow rate (water)} = \text{volume flow rate} \times \text{density of water}$$
$$= 0.047\pi\,\text{m}^3/\text{s} \times 10^3\,\text{kg/m}^3 = 47\pi\,\text{kg/s}$$

Force acting on the body = mass flow rate × change in velocity

Note: *the change in velocity is from 75 m/s to 0 m/s immediately the jet impacts the body.*

Therefore:

Force acting on the body = 47π kg/s × 75 m/s = 3525 N = **3.525 kN**

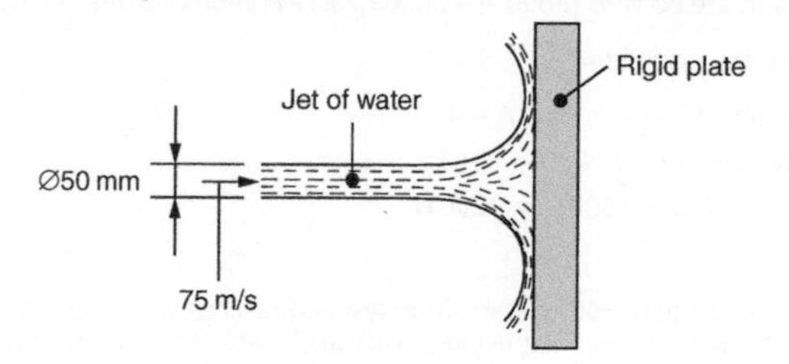

Figure 3.2 Impact of a fluid on a fixed body.

3.6.10 Inertia

Newton's first law of motion states that a body remains in a state of rest or uniform motion in a straight line unless acted on by an external force. This resistance to change is referred to as the inertia of a body and the amount of *inertia* depends on the mass of the body. Therefore it takes a great deal more force to accelerate a fully laden heavy goods vehicle from rest than the force required to produce the same acceleration in a bicycle from rest.

3.6.11 Resisted motion

The following points should be carefully noted when using the expression $F = ma$:

- F can refer to both an *accelerating force* or a *retarding force*.
- It is often necessary to determine the value of an acceleration or a retardation, using the equations of motion considered earlier, when working out engineering calculations.

- When a force causing the acceleration of a body is resisted by a force trying to retard (decelerate) the body, the body is said to be subject to resisted motion.

Figure 3.3(a) shows a force P acting on a mass m against a resistance R and producing an acceleration a. The accelerating force $F_a = (P - R)$ and $F_a = ma$ is written $P - R = ma$.

Figure 3.3(b) shows a suspended body of mass m. Let P be the tension in the rope suspending the body:

1. For motion up or down at constant speed, or when the body is at rest, the acceleration equals zero therefore the accelerating force (F) also equals zero.
2. For motion downwards with acceleration a, W is greater than P where $W = mg$. Then the accelerating force $F = (W - P)$ so that $(W - P) = ma$.
3. For motion upwards with acceleration a, $F = (P - W)$ so that $(P - W) = ma$.

Figure 3.3(c) shows a body on a smooth inclined plane. For this example any friction will be ignored. The component force of weight W acting down the plane $= W \sin \theta$, where $W = mg$ (N):

1. If P is greater than $W \sin\theta$, the body moves up the plane with acceleration a. The accelerating force $F = P - W \sin\theta$ so that $P - W \sin\theta = ma$.
2. If $W \sin\theta$ is greater than P, the body moves down the plane with acceleration a. Then the accelerating force $F = W \sin\theta - P$ so that $W \sin\theta - P = ma$.

Example
A body of mass of 600 kg has a *resistance to motion* of 400 N. Determine the total force to be applied to the body to produce a uniform acceleration of 3 m/s.

Let the total force required $= P$.

Then: The accelerating force $F = (P - 400)$

Since $F = ma$ Then: $(P - 400\,\text{N}) = 600\,\text{kg} \times 3\,\text{m/s}$

Therefore: $P = 400\,\text{N} + 1800\,\text{N} =$ **2200 N**

Example
A passenger lift of mass 1500 kg starts from rest and reaches an upward velocity of 5 m/s in a distance of 10 m. Determine the tension in the lift cable if the acceleration is constant. Take $g = 9.81\,\text{m/s}^2$.

To find the acceleration use the expression $(v_2)^2 = (v_1)^2 + 2as$

Therefore:

$5^2\,\text{m/s} = 0 + 2\,\text{m/s} \times a \times 10\,\text{m}$

So:

$a = 25/(2 \times 10) = 1.25\,\text{m/s}^2$

Whilst the lift is stationary, the tension in the cable is only due to the force of gravity acting on the lift (mg) which is equal to $1500\,\text{kg} \times 9.81\,\text{m/s}^2 = 14715\,\text{N}$. Since the lift is accelerating upwards the tension P is greater than 14715 N.

Let $P =$ the tension in the lift cable.

Then, the accelerating force $F = P - 14715 = ma = 1500 \times 1.25\,\text{m/s}^2$.

Therefore: $P = 14715 + 1500 \times 1.25 = 16590\,\text{N} =$ **16.59 kN**.

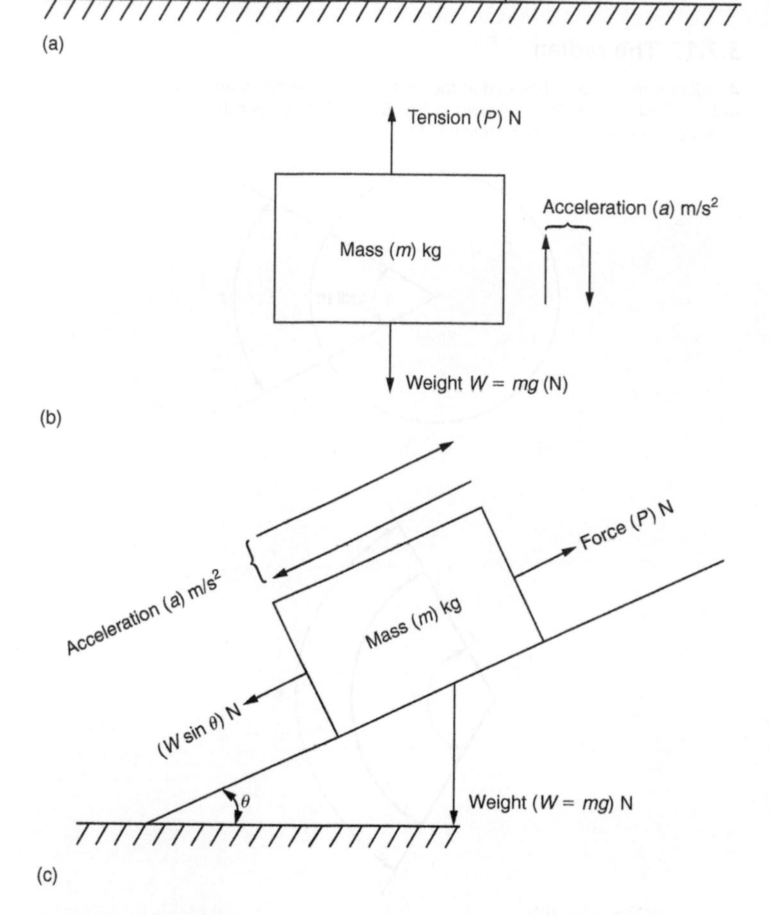

Figure 3.3 Resisted motion. (a) Resisted motion in a horizontal plane, (b) Resisted motion of a suspended body, (c) Resisted motion on an inclined plane.

3.7 Angular motion

When a rigid body rotates about a fixed axis all points on the body are constrained to move in a circular path. Therefore in any given period of time all points on the body will complete the same number of revolutions about the axis of rotation; that is, the speed of rotation is referred to in terms of revolutions per minute (rev/min). However, for many engineering problems it is often necessary to express the speed of rotation in terms of the angle in radians turned through in unit time.

3.7.1 The radian

A radian is the angle subtended at the centre of a circle by an arc equal in length to the radius of that circle. This is shown in Fig. 3.4(a). The relationship between arc length, radius and angle in radians is shown in Fig. 3.4(b).

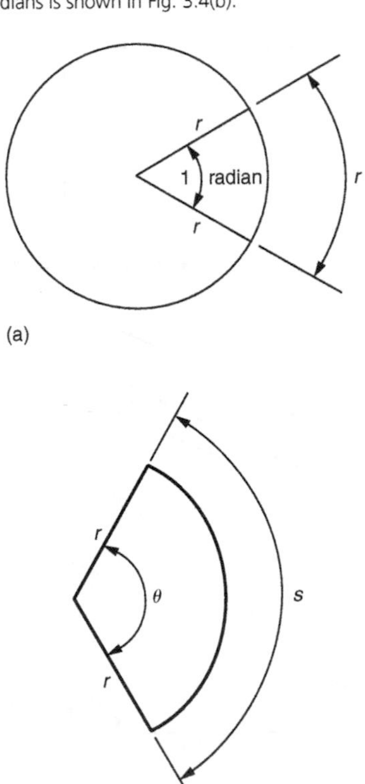

Figure 3.4 (a) The radian (r = radius of the circle), (b) Relationship between arc length (s), radius (r) and angle (θ) in radians.

Where:

s = length of arc
θ = angle in radians
r = radius of the circle

From the definition of a radian:

$s/r = \theta$ therefore $s = r\theta$

In the expression $s = r\theta$ if s is equal to the circumference of a circle then:

$2\pi r = r\theta$ therefore $2\pi = \theta$
That is, 1 revolution = 2π radians (=360°)
Similarly 1 radian = $360/2\pi$ = 57.3° (approximately).

3.7.2 Angular displacement

In *linear motion* the symbol s represents the distance travelled in a straight line; that is, the linear displacement. In *angular motion*, the Greek letter θ is the corresponding symbol for the displacement measured in *radians*.

3.7.3 Angular velocity

A rigid body rotating about a fixed axis O at a uniform speed of n rev/s turns through 2π radians (rad) in each revolution. Therefore the angular velocity ω (Greek letter *omega*) is given by the expression:

$\omega = 2\pi n$ rad/s

Further, if the body rotates through an angle of θ radians in time t seconds then, if that motion is uniform,

The angular velocity $\omega = \theta/t$ rad/s

Note that angular velocity has no linear dimensions.

3.7.4 The relationship between angular and linear velocity

Consider the point A on the outer rim of the solid body shown in Fig. 3.5.

If the body rotates through θ (rad) in t (seconds), and using length of arc = angle in radians × radius

$s = r\theta$

then to obtain the average velocity divide both sides of the expression by t so that:

$s/t = r(\theta/t)$

but

$s/t = v$ and $\theta/t = \omega$

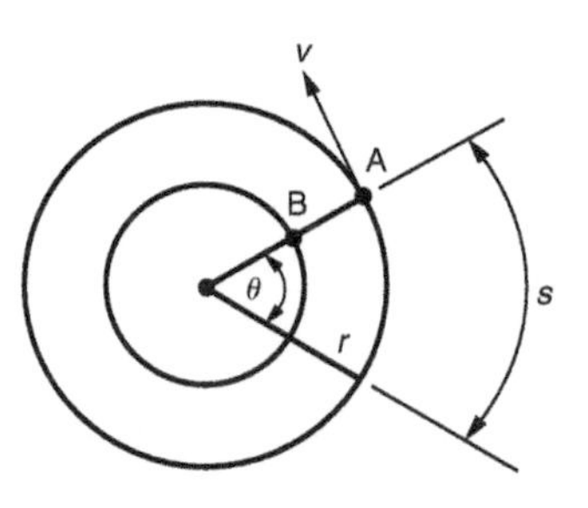

Figure 3.5 Relationship between angular and linear velocity.

Therefore:

$v = r\omega$ or linear velocity = angular velocity × radius.

Note:

1. The point B on the body has the same angular velocity as A but has a smaller linear velocity since it has a smaller radius.
2. If the radius r is in metres and the angular velocity ω is expressed in rad/s, then the linear velocity v is expressed in m/s. Sometimes the angular velocity is expressed in rev/min. It is necessary to convert rev/min to rad/s before substituting in expressions such as $v = r\omega$:

   N (rev/min) = $2\pi N$ (rad/min) = $2\pi N/60$ (rad/s)

3.7.5 Angular acceleration

A similar relationship to that of linear and angular velocity exist between linear and angular acceleration. That is:

$a = r\alpha$

where:
a = linear acceleration, r = radius, α (Greek letter alpha) = angular acceleration.

Section 3.6.4 listed the expressions for the relationships between linear velocity and acceleration. Similarly there is a set of expression for the relationships between angular velocity and acceleration. These are:

$\omega_2 = \omega_1 + \alpha t$	ω_1 = initial angular velocity
$\theta = [(\omega_1 + \omega_2)/2] \times t$	ω_2 = final angular velocity
$\theta = \omega_1 t + ½\alpha t^2$	α = angular acceleration
$\omega_2^2 = \omega_1^2 + 2\alpha\theta$	θ = angle of rotation
$\theta = \omega_2 t - ½\alpha t^2$	t = time taken

3.7.6 Torque

Torque is the *turning moment* of a force. When a force produces the rotation of a body without translation motion, there must be a second force acting upon the body equal and

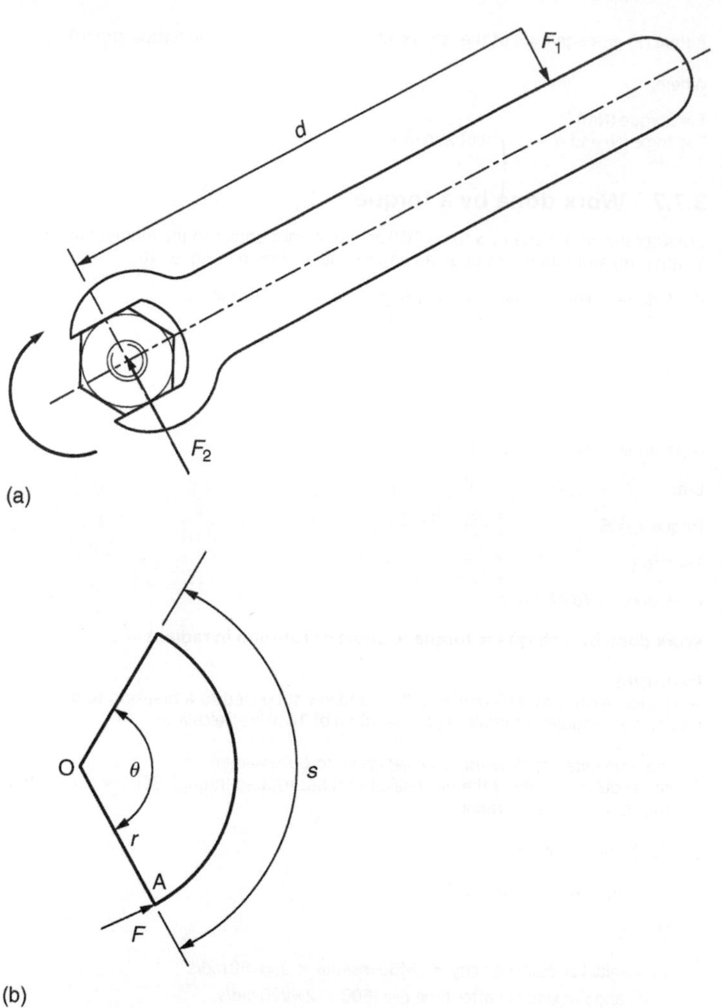

Figure 3.6 (a) Torque (turning moment). (b) Work done by a torque.

opposite to the first so that the resultant force is zero. That is, the forces are in equilibrium. Consider the spanner and nut shown in Fig. 3.6(a). When a force F_1 is applied to the end of the spanner, an equal and opposite parallel force F_2 occurs at the nut. If this were not so, there would be motion of translation in the direction of force F_1. *When two equal and opposite forces act on a body they are said to form a couple.*

Torque (T) = magnitude of the applied force (F_1) × length of the torque arm (d)

Where:

T = torque (Nm)
F = force (N) and d = moment arm (m)

3.7.7 Work done by a torque

Consider the work done by a force F (N) acting at right angles to the moment arm OA of length r (m) and rotating through an angle θ rad as shown in Fig. 3.6(b).

Work done = linear distance moved by a force = $s \times F$ (Nm or J)

But:

$s = r\theta$ (m)

Therefore:

work done = $r\theta \times F = Fr \times \theta$

But:

torque $T = Fr$

Therefore:

work done = $T\theta$ (Nm or J)

Work done by a torque = torque × angle of rotation in radians

Example

A flywheel rotates at 2400 rev/min. It is suddenly subjected to a breaking torque which results in an angular retardation (deceleration) of 10 m/s^2. Calculate:

(a) the time taken to slow the flywheel down to 600 rev/min;
(b) the angular velocity of the flywheel after it has rotated through 300 revolutions from the start of its deceleration.

(a) Apply the equation:

$\omega_2 = \omega_1 + \alpha t$ (Section 3.7.5)

Where:

ω_1 = initial angular velocity = (2400 rev/min × 2π) /60 rad/s
ω_2 = angular velocity after time t = (600 × 2π)/60 rad/s
α = angular acceleration = −10 m/s^2 (the minus sign indicates a deceleration)

Therefore:

(600 × 2π)/60 = (2400 rev/min × 2π)/60 + (−10)t
10t = (1800 × 2π)/60

Therefore:

t = 18.8 s

(b) Apply equation:

$(\omega_2)^2 = (\omega_1)^2 + 2\alpha\theta$ (Section 3.7.5)

Where:

ω_2 = angular velocity after rotating through 300 rev
$\omega_1 = (2400\text{ rev/min} \times 2\pi)/60\text{ rad/s} = 251.33\text{ rad/s}$
$\alpha = -10\text{ rad/s}^2$
$\theta = 300 \times 2\pi\text{ rad}$

Then:

$\omega_2^2 = 251.33^2 + 2 \times (-10) \times 600\pi$
$= 63166.8 - 37699.1$
$=$ **159.6 rad/s** or
$= (159.6 \times 60)/2\pi$
$=$ **1524 rev/min**

3.7.8 Centripetal acceleration and centripetal force

The dictionary definition of *centripetal acceleration* states that it is the acceleration directed towards the centre of curvature of the path of a body which is moving along that path with constant speed. Its value is v^2/r where v is the speed and r is the radius of curvature or its value can be expressed as $\omega^2 r$ where ω is the angular velocity.

Figure 3.7(a) shows a body of mass m moving along a circular path of radius r and centre O with a constant angular velocity ω. When the body is a point P it possesses an instantaneous linear velocity v tangential to the circumference of the circle along which it is travelling. During an incremental period of time, δt, the body moves along the circumference of the circle from P to Q. The angle subtended by this translation from P to Q is $\delta\theta$. Although the body is moving at constant speed, it is not moving along the same line of action. Therefore, there is a change in the velocity of the body. Since there is a change of velocity with respect to time the body is, by definition, subjected to an acceleration.

Figure 3.7(b) shows the velocity vector diagram where:

- Vector *Op* represents the velocity (speed and direction) at point P.
- Vector *Oq* represents the velocity at point Q.
- Vector *pq* represents the radial change in velocity during the translation of the body from P to Q.

The radial change in velocity $= pq = v \times \delta\theta$

However:

$\theta = \omega t$

Therefore:

$\delta\theta = \omega \times \delta t = (v/r) \times \delta t$

Thus:

The vector change in velocity $= (v^2 \times \delta t)/(r \times \delta t) = v^2/r = \omega^2 r$

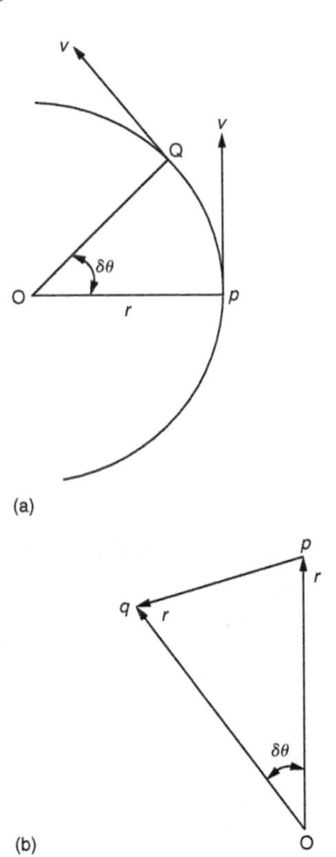

Figure 3.7 (a) Centripetal acceleration, (b) Centripetal velocity diagram.

In this discussion the centripetal acceleration refers to a body of mass *m*. From Newton's second law of motion a body cannot be accelerated unless an external force acts on that body. Therefore the *centripetal force* producing the acceleration = mass × acceleration. Expressed mathematically:

Centripetal force $= m \times (v^2/r) = m\omega^2 r$

3.7.9 Centrifugal force

Newton's third law states that: *for every acting force there is an equal and opposite reaction force*. Therefore there must be an opposing reaction force equal and opposite to the centripetal force. This opposing force acts radially outwards and is known as the *centrifugal force*. Consider a railway train going round a curve. The train wheel flanges push outwards on the rails (centrifugal force) whilst the rails exert a force of equal magnitude on the wheels of the train pushing inwards (centripetal force). Similarly, when a mass restrained by a cord is whirled about an axis, the centrifugal force tries to pull the mass outwards whilst the centripetal force tries to pull the mass inwards, thus the cord is kept in tension.

3.8 Balancing rotating masses

For machines to run smoothly without vibration it is necessary for all the rotating masses to be in a state of balance.

3.8.1 Balancing co-planar masses (static balance)

Co-planar masses refer to masses that lie in the same plane. For a body to be in static equilibrium there must be no resultant force or moment acting on the body. The algebraic sum of the moments must equal zero. Figure 3.8(a) shows three co-planar masses positioned about a point O. For all practical purposes on planet earth the gravitational forces acting on the masses will be 9.81 × mass in each instance.

For the system of masses shown to be balanced it must be in equilibrium and, therefore, the algebraic sum of the moments about the centre of rotation O must be zero. This can be expressed mathematically as:

$$9.81\ m_1r_1\cos\theta_1 + 9.81\ m_2r_2\cos\theta_2 + 9.81\ m_3r_3\cos\theta_3 = 0$$

Dividing through by 9.81 it becomes:

$$m_1r_1\cos\theta_1 + m_2r_2\cos\theta_2 + m_3r_3\cos\theta_3 = 0$$

However this single expression is insufficient for the system of masses to be balanced for all angular positions. The system now has to be rotated through 90° as shown in Fig. 3.12(b) and the algebraic sum of the moments about O must be again be zero.

$$9.81\ m_1r_1\sin\theta_1 + 9.81\ m_2r_2\sin\theta_2 + 9.81\ m_3r_3\sin\theta_3 = 0$$

Again dividing through by 9.81 it becomes:

$$m_1r_1\sin\theta_1 + m_2r_2\sin\theta_2 + m_3r_3\sin\theta_3 = 0$$

If the system is in balance for any two positions at right angles, it will also be in balance for any intermediate positions. Therefore the foregoing equations enable the balance of a system to be checked. If the system is out of balance then the same equations enable the magnitude and position of a balancing mass to be determined.

3.8.2 Balancing co-planar masses (dynamic balance)

The subject of dynamic balancing is highly complex and an in depth treatment will not be attempted in this pocket book as there is insufficient space. However, a simple system of co-planar forces rotating about a point O will be considered, Fig. 3.9(a) shows such a system. When they are rotating, each of the masses exerts a centrifugal force of $mr\omega^2$ on the point O (an axis or shaft). The rotational speed ω is expressed in rad/s. As previously, the algebraic sum of the moments about the point O must equal zero in two directions at right angles (x and y) as shown in Fig. 3.9(a).

Therefore the two equations applying to dynamic balance are:

$$m_1r_1\omega^2\cos\theta_1 + m_2r_2\omega^2\cos\theta_2 + m_3r_3\omega^2\cos\theta_3 = 0 \quad \text{and}$$
$$m_1r_1\omega^2\sin\theta_1 + m_2r_2\omega^2\sin\theta_2 + m_3r_3\omega^2\sin\theta_3 = 0$$

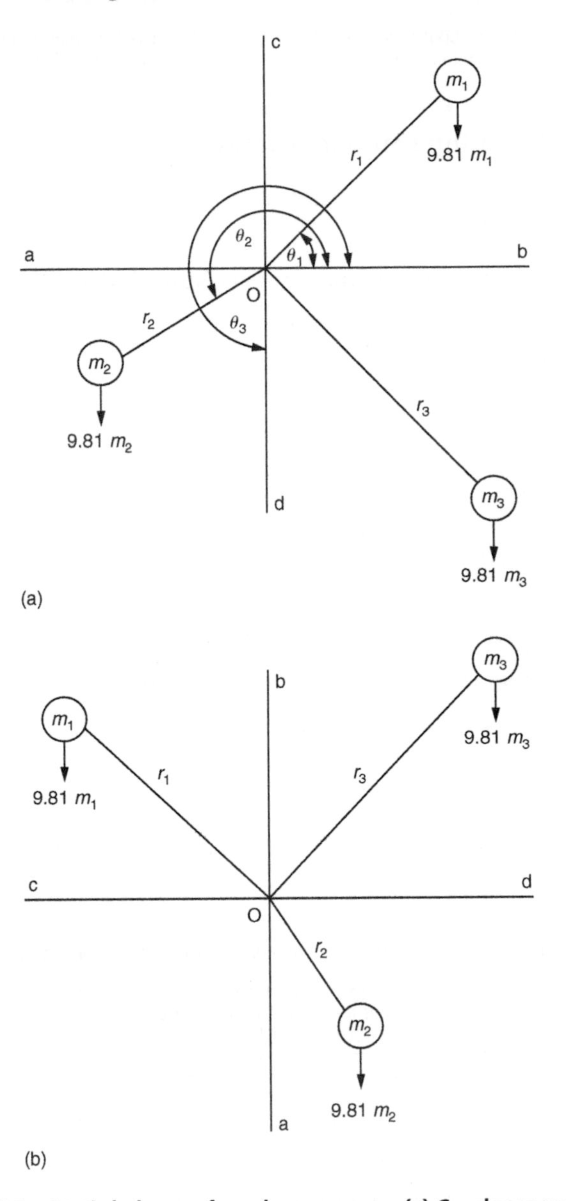

Figure 3.8 Static balance of co-planar masses. (a) Co-planar masses in first position. (b) Co-planar masses rotated through 90° to second position.

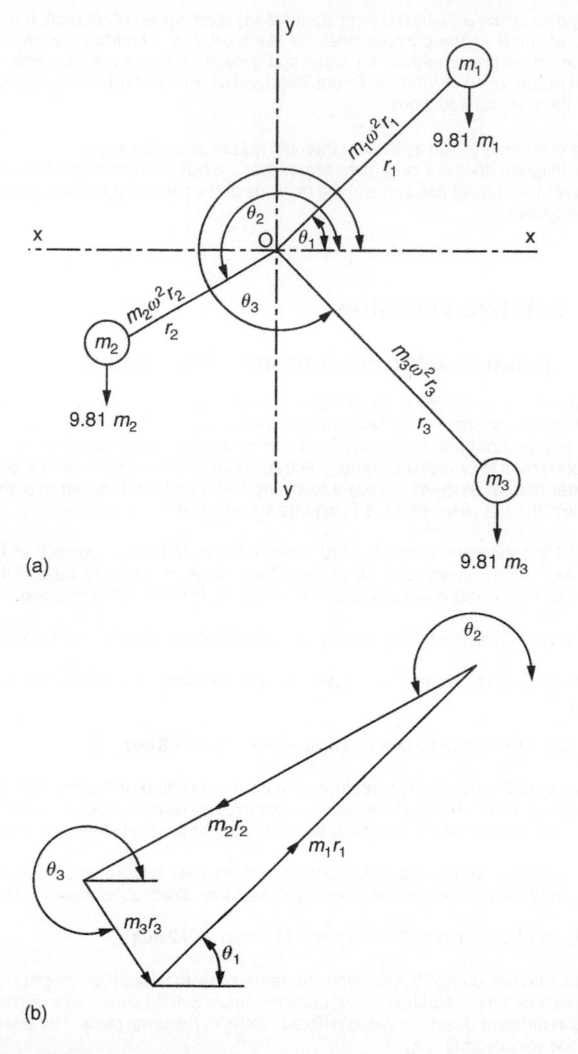

Figure 3.9 (a) Dynamic (rotational) balancing of co-planar masses. (b) Force diagram of co-planar masses.

Dividing through by ω^2, which is a constant for any given speed of rotation, the equations become identical to the previous ones for static balance. Therefore for any system of co-planar forces the conditions for static and dynamic balance are the same and independent of the speed of rotation. Figure 3.9(b) shows the vector diagram of forces representing the previous equations.

- If the diagram is closed as shown, then the masses are in balance.
- If the diagram does not close then the masses are not in balance and the closing vector gives the angular position and the *mr* value of the mass required to achieve balance in the system.

3.9 Relative velocities

3.9.1 Relative velocities (unconnected bodies)

The statement that a body is travelling at 50 km/h means that the body is travelling at 50 km/h relative to some other body such as the surface of planet earth. For example, an aircraft whose indicated airspeed is 300 km/h means that its airspeed is relative to the wind speed through which it is flying. The true speed of the aircraft over the ground may be greater than its airspeed if it has a following (tail) wind, or its speed over the ground will be less than its airspeed if it is flying into a headwind.

Figure 3.10(a) shows two unconnected bodies represented by the points P and Q. These bodies are travelling in opposite directions at 3 and 5 m/s, respectively. Assume that movement from left to right is positive and the movement from right to left is negative.

The velocity of Q relative to P = velocity Q − velocity P = (−5 m/s) − (+3 m/s) = **−8 m/s**

The velocity of Q relative to P is negative because the direction of Q is from right to left. Similarly:

The velocity of P relative to Q = (+3 m/s) − (−5 m/s) = **+8 m/s**

The velocity of P relative to Q is positive because the direction of P is from left to right. In this explanation the choice of direction is arbitrary. The relative directions and the allocation of positive or negative values must be established at the start of each specific problem.

Since the bodies are moving in opposite directions their relative velocity is the *closing velocity*. Had the two bodies been moving in the same direction as shown in Fig. 3.10(b).

The velocity of Q relative to P = +5 m/s − (+3 m/s) = **+2 m/s**

If the two bodies do not travel along the same path their relative velocities cannot be determined by simple addition or subtraction. Figure 3.10(c) shows two bodies P and Q moving in different directions along different paths in the same plane. The velocity of P is v_P and the velocity of Q is v_Q.

To find the velocity of Q with respect to P it is necessary to find the apparent velocity of Q if P is stationary. To do this it is necessary to apply a velocity of $-v_P$ to both bodies as shown in Fig. 3.10(d) in order to bring P to rest. From the vector diagram 3.10(e) it can be seen that the components of the velocity of Q are its true velocity v_Q and its superimposed velocity $(-v_P)$. The resultant velocity of body B is represented by the vector **QR** which represents the velocity of Q relative to P. That is, the apparent movement of body Q as seen

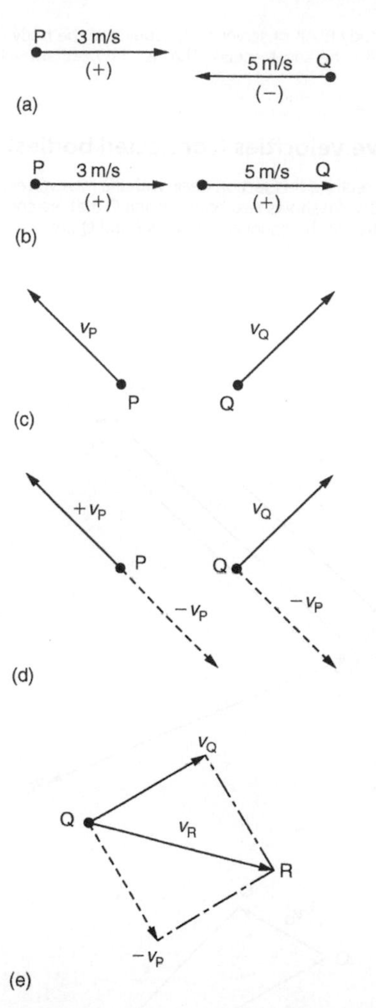

Figure 3.10 Relative velocities of unconnected bodies (a) Relative velocities of two unconnected bodies moving in opposing directions. (b) Relative velocities of two unconnected bodies moving in the same direction. (c) Relative velocities of two unconnected bodies moving along different paths. (d) Method of bringing P to rest and the effect on Q. (e) Velocity of Q relative to P (resultant velocity QR).

by an observer on body P. An observer on Q would see the body P moving in the opposite direction but with the same speed. That is, the resultant velocity vector would be reversed.

3.9.2 Relative velocities (connected bodies)

Unlike the previous section, this section deals with the relative velocities of bodies that are unconnected, Fig. 3.11(a) shows two bodies P and Q that are connected by a rigid link of length l. The velocities of the connected bodies P and Q are v_P and v_Q, respectively in the directions shown.

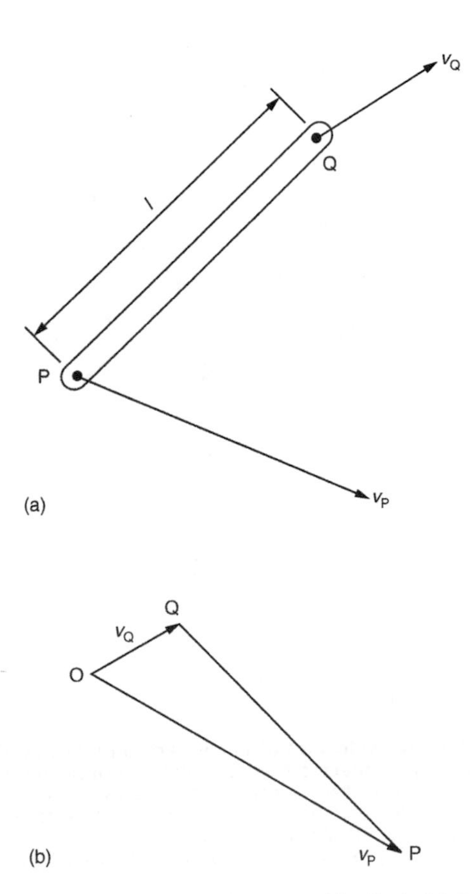

Figure 3.11 Relative velocities (connected bodies). (a) Two bodies (P and Q) connected by a rigid link. (b) Velocity vector diagram.

Since the distance l between the bodies P and Q is fixed, there can be no component velocity along the link connecting P and Q. Therefore any relative velocity will be perpendicular to the link connecting P and Q since this is the only direction that will not produce a component along the link connecting P and Q. The velocity vector diagram is shown in Fig. 3.11(b) from which it can be seen that the vector **pq** is perpendicular to the link connecting the bodies P and Q. Therefore P is apparently rotating with a tangential velocity *pq* with respect to Q. Therefore:

Angular velocity of the link = (tangential velocity)/(radius)

ω = (vector *pq*)/l

The velocity of Q with respect to P would give a similar result since the link connecting the bodies P and Q can have only one angular velocity.

3.10 Kinematics

Kinematics is the study of the motion of bodies without considering the forces necessary to produce the motion. *Linear velocity* is the rate of change of displacement with respect to time and can be expressed as:

Velocity = ds/dt

Remember that the velocity of a body is its speed in a prescribed direction. The numerical values of speed and velocity are identical. Displacement is the area under a velocity–time graph. Since:

$v = ds/dt$ then $\int \mathbf{ds} = \int v\,\mathbf{dt}$

Similarly, *linear acceleration* equals the rate of change of linear velocity:

Acceleration = rate of change of velocity/time = dv/dt or d^2s/dt^2

Note how acceleration is also the second derivative of distance and time. Velocity is the area under an acceleration–time graph. Since:

$a = dv/dt$ then $\int \mathbf{da} = \int a\,\mathbf{dt}$

Similarly *angular acceleration* equals the rate of change of angular velocity.

$\alpha = d\omega/dt$ or $\alpha = d_2\theta/dt^2$

3.10.1 Ballistics

Figure 3.12 shows the trajectory of a projectile fired from a gun with an elevation of θ and an initial velocity of v_1 m/s. The vertical component of the force $v_V = v_1 \sin\theta$ and the horizontal component of the force $v_H = v_1 \cos\theta$. Ignoring the air resistance, the only force acting on the projectile will be the gravitational acceleration (9.81 m/s^2) acting on the mass of the projectile in a downwards direction. Since there is no horizontal force acting on the projectile the projection velocity remains constant throughout its flight. (In reality it would be progressively slowed up by the air resistance.)

From the relationships given in Section 3.6.4, the total time of flight can be obtained by using the expression $s = v_1t + \frac{1}{2}at^2$. Since the projectile is assumed to start from ground

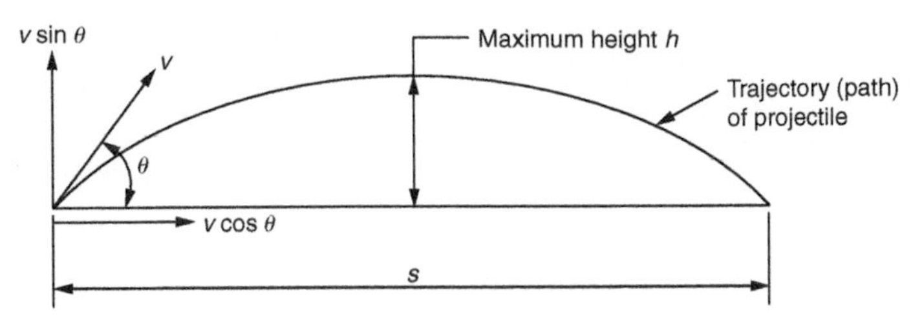

Figure 3.12 Trajectory of a projectile.

level, attain a maximum height of h as shown and return to ground level so that the net vertical distance travelled will be zero. Therefore:

$s = 0,\ v_V = v_1 \sin\theta \quad \text{and} \quad a = -g.$

Since:

$0 = (v_1 \sin\theta)\,t - gt^2$ then $\boldsymbol{t\,(2v_1 \sin\theta)/g}$

Since the horizontal component of the velocity of the projectile remains constant, the horizontal distance s travelled (the range) by the projectile can be calculated as follows:

The distance travelled or range(s) = horizontal component of velocity × time of flight.

$s = (v_1 \cos\theta) \times (2v_1 \sin\theta)/g = \boldsymbol{[(v_1)^2 \sin 2\theta]/g}$ (since: $\sin 2\theta = 2\sin\theta\cos\theta$)

The maximum height (h) reached during the flight of the projectile can be obtained from the expression:

$\boldsymbol{h = [(v_1)^2 \sin^2 2\theta]/2g}$

3.11 Kinetics

Is the study of forces that cause motion as distinct from kinetics which is the study of the geometry of motion. Therefore kinematics is the application on Newton's laws of motion.

3.11.1 Universal gravitation

The law governing the mutual attraction between bodies possessing mass is known as the *law of universal gravitation*. This law was postulated by Sir Isaac Newton and it states that every body in the universe is attracted to every other body by a force that is directly proportional to the product of their masses and inversely proportional to the square of the distance separating them. Stated mathematically:

$F = (Gm_1m_2)/d^2$

Where:

F = mutual force of attraction between masses m_1 and m_2
d = distance between the masses
G = the gravitational constant (independent of the masses involved)

For any two masses on any planet, including planet earth, the mutually attractive force F between them is negligible compared with the force between each mass and the planet on which they are resting. Since planet earth exerts a force on all other bodies it follows from Newton's second law of motion that this force will produce an acceleration g m/s^2. Since:

$$(GM_0M)/r = F = Mg$$

Therefore:

$$g = (GM_0)/r = 9.81 \text{ m/s}^2$$

Where:

G = gravitational constant (6.668×10^{-11} m^3/kg s^2)
M_0 = mass of planet earth (5.98×10^{24} kg) assumed to be concentrated at the centre of the planet.
r = mean radius of planet earth (6.375×10^6 m)
g = mean gravitational acceleration of planet earth
F = mutual force of attraction

3.11.2 Linear translation

Any body acted upon by a number of external forces so that the resultant of these forces passes through the centre of the mass of the body then the resulting motion will be translatory. Further, since the resultant force acts through the centre of the mass it will have no moment and the body will not rotate. Therefore any such motion will be translatory and will move or tend to move the body from one place to another in a straight line.

3.11.3 Translation in a curved path

When a body moves in a curved path of radius r and at a uniform speed v it is subject to an inward radial acceleration of v^2/r as discussed in Section 3.7.8. This is known as the centripetal acceleration. Newton's second law of motion states that a body cannot be accelerated unless it is subjected to an external force. Thus:

Centripetal force = mass $\times$ centripetal acceleration = $(mv^2)/r$
Since $v = r\omega$ for uniform circular motion, the centripetal force = $m\omega^2 r$

The *centripetal force* must be applied externally to the body for example, in the case of a car, by the friction of the tyres in contact with the road or, in the case of a railway rain by the side-thrust of the rails on the flanges of the wheels of the rolling stock. Newton's third law states that there is a reactive force equal and opposite to every active force. Therefore there must be an equal and opposite force to the centripetal force. This is the *centrifugal force* and it acts radially outwards. In the case of the car the car pushes outwards against the road surface (*centrifugal force*) and the road pushes back against the tyres of the car (*centripetal force*) unless the friction force between the tyres and the road breaks down, then the car will skid.

3.11.4 Conic pendulum

If a mass m is suspended from a pivot P by an arm of length l so that it rotates around a circular path of radius r at a constant velocity ω, as shown in Fig. 3.13, then for any

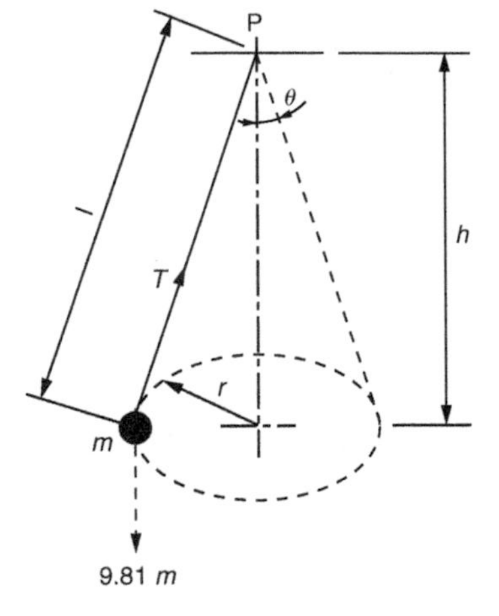

Figure 3.13 A conic pendulum.

specific speed of rotation the inclination of the arm θ and the radius of rotation r will have definite and inter-related values. This is the principle of the *conical pendulum* and also the principle of the *governor* used to control the speed of engines. In the latter case if the load on the engine decreased so that the engine speed started to rise, then r and θ would tend to increase and the linkage to which the mass m was attached would close the throttle valve so that the engine speed would be reduced to the normal (safe) value. Conversely if the load increased so that the engine started to slow down then r and θ would reduce and the linkage to which the mass m was attached would open the throttle and bring the engine back to its correct operating speed. Thus the engine would tend to run at a constant speed.

Let T be the tension in the link supporting the mass m, then the centripetal force to maintain circular motion is provided by horizontal component force of T.

The horizontal component force of T is:

$T \sin\theta = mr\omega^2$

The vertical component force of T for equilibrium is:

$T \cos\theta = 9.81\,m$

Since:

$\sin\theta/\cos\theta = \tan\theta$

Then:

$\mathbf{\tan\theta = (r\omega^2)/9.81}$

Since:

$\tan\theta = (r\omega^2)/9.81$ and $\tan\theta = r/h$ then $r/h = (r\omega^2)/9.81$

Therefore:

$\boldsymbol{h = 9.81/\omega^2}$
$\boldsymbol{r = l \sin \theta}$
$\boldsymbol{l = r/\sin \theta}$
$\boldsymbol{T = m/\omega^2}$

3.11.5 Rotation of a body about a fixed axis

When a body only possesses linear translatory motion, the resultant of the external forces acting on that body will pass through the centre of mass and rotation cannot occur (see Section 3.11.4). However when the resultant force does not pass through the centre of mass of the body a turning moment will be present and the body will rotate. Figure 3.14 shows a body subjected to a torque T so that it rotates about the pivot point P. The instantaneous angular velocity of the body is ω and its angular acceleration is α.

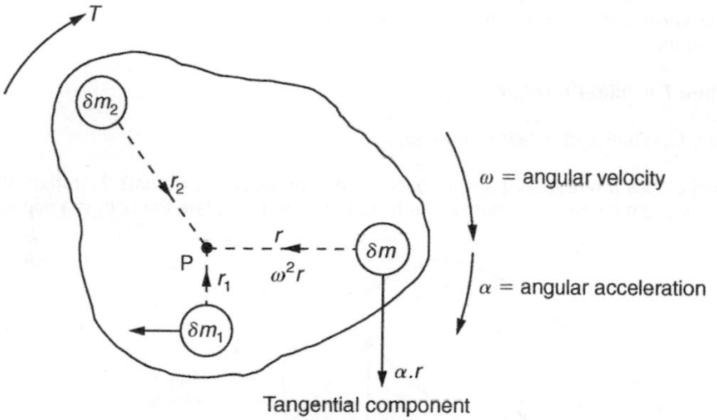

Figure 3.14 Rotation of a body about a fixed point.

The elemental masses δm, δm_1 and δm_2 are each subjected to two acceleration components, a tangential component αr and a centripetal component $\omega^2 r$ which passes through the pivot point P about which the body rotates. Since the centripetal component for each elemental mass passes through the pivot point P there can be no moment about this point. However the component acting tangentially on each elemental mass ($\delta m.\alpha r^2$) about P does provide a turning moment. Therefore the total turning moment about the pivot point P is:

$\Sigma(\delta m.\alpha r^2) = \alpha\Sigma(\delta m.r^2)$ since all the elements rotate about P at the same speed.

$\Sigma(\delta m.r^2)$ being the sum of all the elemental terms (mass $\times$ radius^2) about P is called the ***moment of inertia***. Thus:

The moment of inertia $I_P = \Sigma(\delta m.r^2)$ and the turning moment T about P $= I_P\alpha$.

The units are: T (Nm), I_P (kg m^2) and α (rad/s^2).

3.11.6 Radius of gyration

Since the sum of the elemental masses equals the total mass of a body ($m = \Sigma\delta m$) and if k is the distance from the chosen axis at which the total mass is assumed to be concentrated, then $mk^2 = \Sigma(\delta m.r^2)$ then the moment of inertia can be expressed as:

$I_P = mk^2$ where k is the radius of gyration about the specified axis of rotation.

3.11.7 Centre of percussion

It is often necessary, in rotational problems, to identify the line of action of the resultant force F (as shown in Fig. 3.15) where F equals the mass of the body multiplied by the acceleration of the centre of mass. It is assumed that this resultant force F passes through some point P on the line RGP where R is the axis of rotation and G is the centre of mass. As explained in Section 3.11.6, only the tangential component of the force F can produce a turning moment about R. P is at a position in a body where the moments of all the external forces balance out and equal zero. The point P is known as the ***centre of percussion***. In cases where the body rotates about its centre of mass (R and G are coincident) then:

Torque $T = m(k_G)^2\alpha = I_G\alpha$

where k_G is radius of gyration about G.

When a body rotates about an axis passing through its centre of mass the radius of gyration is at a minimum and, hence, the torque required to rotate the body is a minimum.

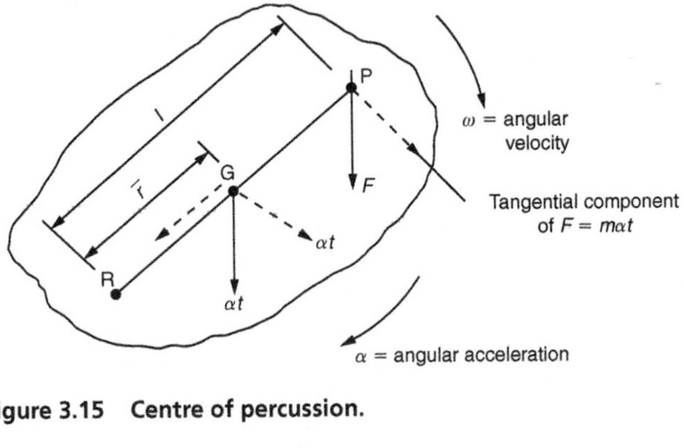

Figure 3.15 Centre of percussion.

3.11.8 Angular momentum

It was stated in Section 3.6.8 that the linear momentum of a body was its mass multiplied by its linear velocity. The angular momentum of a rotating body is the moment of its linear momentum about the point of rotation. Figure 3.16 shows a body rotating about the pivot point P with an angular velocity ω rad/s. Consider the elemental mass, δm, at a distance r from the pivot point P and whose velocity is v m/s.

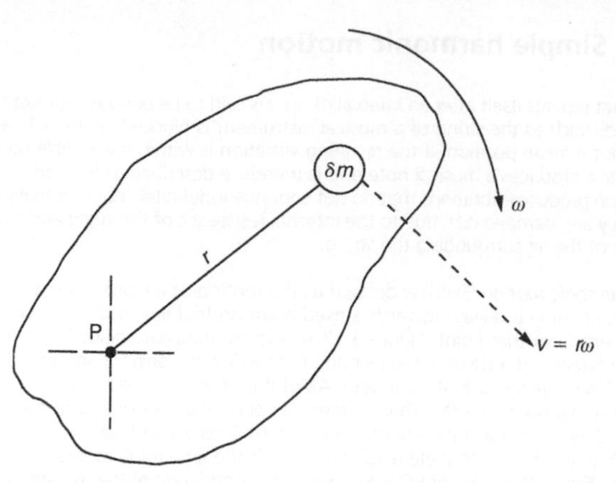

Figure 3.16 Angular momentum. v = linear velocity; ω = angular velocity.

Linear momentum of the elemental mass = $\delta m.v = \delta m.r\omega$
The moment of the linear momentum about P = $r(\delta m.r\omega) = \delta m.r^2\omega$
The moment of momentum about P for the complete body = $\Sigma\delta m.r^2\omega$

But:

$\Sigma\delta m.r^2 = I$

Where:

I = the moment of inertia (see Section 3.11.5)

Thus:

The angular momentum about P = $I\omega$

Where a change of angular velocity occurs the above expression becomes:

Change in angular momentum = $I(\omega_2 - \omega_1)$

Therefore: **angular impulse $Tt = I(\omega_2 - \omega_1)$**

Where:

T = torque (Nm)
t = time (duration of impulse in seconds)

The kinetic energy stored in a rotating body = $I_P\omega^2$

Where: I_P = moment of inertia about the pivot point P.

The kinetic energy stored in a body possessing both rotation and translation
= the kinetic energy of rotation + the kinetic energy of translation
= **$\frac{1}{2}I\omega^2 + \frac{1}{2}mv^2$**

3.12 Simple harmonic motion

Motion that repeats itself after an interval of time is said to be *periodic*. For example, if an elastic body such as the string of a musical instrument is plucked or struck it will vibrate freely about a mean position. If the resulting vibration is within the audible range of the human ear it produces a musical note and is therefore described as *harmonic*. Such periodic motion produces vibrations that do not continue indefinitely but eventually die away. That is they are '*damped out*' due to the internal resistance of the string and the external resistance of the air surrounding the string.

Simple harmonic motion (SHM) is defined as the motion of a point along a straight line, whose acceleration is always towards a fixed point on that line and is proportional to its displacement from that point. Figure 3.17 shows an imaginary point P rotating around the circumference of a circle of radius r and centre O with constant angular velocity ω. If the point P is projected onto the diameter AB of the circle then the point Q will move back and forth across AB with SHM. That is when the point P makes one complete revolution, the point Q will make a complete oscillation from A to B and back to A. Therefore the time for P to make one complete revolution equals the time for Q to make one complete oscillation. The distance AO or BO is known as the *amplitude* of the oscillation:

1. The frequency of oscillation in hertz (Hz) can be found from the expression:

   $$\boldsymbol{f = 1/T = \omega/2\pi}$$

   Where: f = frequency of oscillation, T = periodic time, ω = angular velocity.
2. The maximum velocity of Q occurs at the midpoint when P is immediately above or below the centre O and the acceleration of X is zero. That is: $\mathbf{v_{max}} = \boldsymbol{\omega . r}$
3. The maximum acceleration of Q equals the acceleration of P and occurs at the extreme points A and B where the velocity of Q is zero. That is: $\mathbf{a_{max}} = \boldsymbol{\omega . r}$

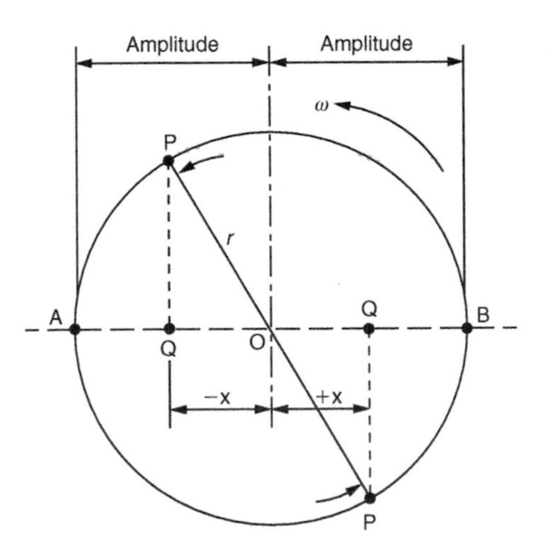

Figure 3.17 Simple harmonic motion.

Example

A body moves with SHM with an amplitude of 50 mm and a frequency of 2 Hz
Since the frequency $f = \omega/2\pi$ then $\omega = f \times 2\pi = 2 \times 2\pi = 12.566$ rad/s.

(i) The maximum velocity $= \omega.r = 12.566 \times 50 =$ **0.628 m/s** at the mean point.
(ii) The maximum acceleration $= \omega^2.r = (12.566)^2 \times 50 =$ **7.895 m/s²** at the limits of its travel.

3.12.1 Simple pendulum

A simple pendulum consists of a mass (the *bob*) suspended on the lower end of a light rod. The upper end of the rod being pivoted so that the rod and the mass can oscillate in a vertical plane as shown in Fig. 3.18.

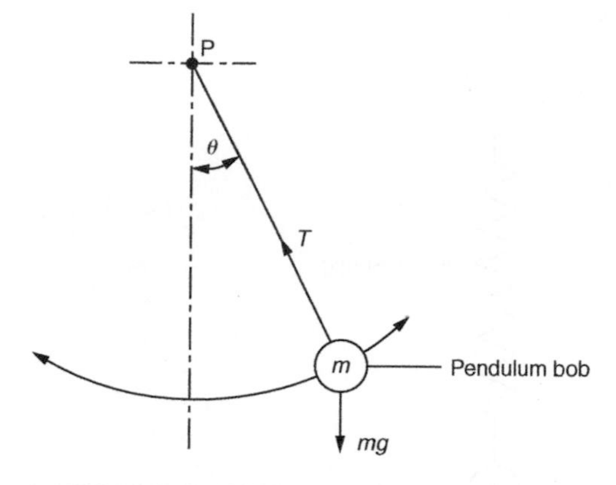

Figure 3.18 Simple pendulum.

Consider a simple pendulum of length l with a bob of mass m which has been displaced through the angle θ. As the mass m of the bob moves in a circular path of radius l, the only forces acting on it are the gravitational force mg and the tension in the rod T. For the displacement shown in the figure, the gravitational force produces a torque about the pivot P tending to restore the mass to its equilibrium position. That is, the position when the pendulum is hanging vertically downwards and the mass of the bob is immediately below the pivot point (P). Therefore:

The restoring torque $= -mgl \sin\theta = -mgl\,\theta$ (for small values of θ)
For the mass m, the moment of inertia about P is $I_P = ml^2$.

Applying Newton's second law of motion to the rotation of the pendulum about P:

$T = I_P\alpha$ where T is the periodic time

Thus:

$-mgl\,\theta = ml^2\alpha$ where $\alpha = (-g\theta)/l$

Therefore acceleration = the displacement × constant and, provided the displacement is small, the pendulum will swing with SHM. Therefore:

1. Periodic time $T = 2\pi/\omega = 2\pi\sqrt{(l/g)}$ seconds.
2. Frequency $f = 1/T = (1/2\pi)\sqrt{(g/l)}$ Hz.

Since the periodic time of the swing of a pendulum (and the related frequency of its oscillation) is dependent only on its length and local gravitational acceleration, it follows that a pendulum is an ideal means of controlling a mechanical clock providing the rod supporting the bob is made from a material that has a very low coefficient of expansion with change of temperature.

3.12.2 Natural vibration

Natural (free) vibrations approximate to SHM when occurring in bodies made from elastic materials that obey Hooke's law. Figure 3.19 shows tension spring supporting a mass *m*.

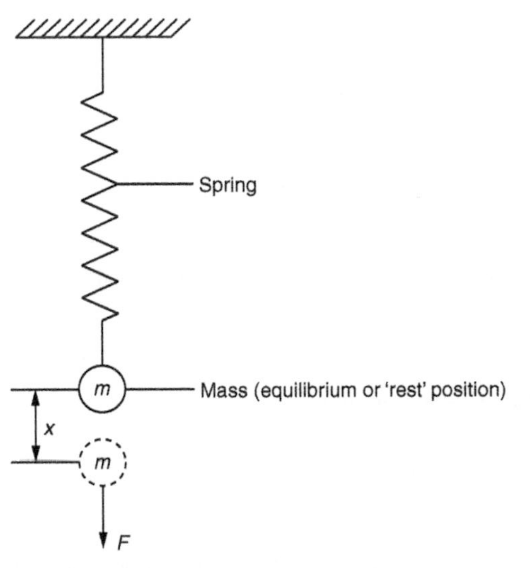

Figure 3.19 Natural vibration.

At rest, the position of the mass will depend on the stiffness of the spring and the force of gravity acting on the mass. That is, these two forces will exactly oppose each other and will be in a state of equilibrium. If the mass is displaced downwards through a distance x and released the mass will oscillate with a frequency (f) under the influence of the spring in a vertical plane.

Let s be the stiffness of the spring (force per unit change of length) and x be the displacement from the rest position, then the change of force in the spring is sx. This is the accelerating force F acting on the mass m so that $F = sx$. Therefore:

Acceleration $a = f/m = sx/m$

Therefore the acceleration is directly proportional to the displacement form the rest position of the mass m and the movement is SHM. The periodic time T of the system in seconds is given by the expression:

$$T = 2\pi\sqrt{(m/s)}$$

The frequency f of vibration in hertz = $1/T$

Example

A load of 5 kg is hung from a tension spring so that it produces an extension of 12 mm when at rest. The load is displaced downwards by a further 15 mm and then release. Ignoring losses due to air resistance and hysteresis losses in the spring material determine: (i) the frequency of vibration, (ii) the maximum velocity of the load, (iii) the maximum acceleration of the load.

(i) The weight of the load is $mg = 5\,\text{kg} \times 9.81\,\text{m/s}^2 = 49.05\,\text{N}$
The stiffness of the spring s = force/extension = 49.05 N/15 mm = 3.27 N/mm
Note: 3.27 N/mm = 3.27 kN/m
The frequency of vibration $f = 1/T = 1/2\pi\sqrt{(s/m)}$

Therefore:

$$f = 1/2\pi\sqrt{[(3.27 \times 10^3)/5]} = \mathbf{4.07\,Hz}$$

(ii) The maximum velocity of the load $= \omega x = 2\pi f x = 2\pi \times 4.07 \times 15 =$ **383.59 mm/s**
(iii) The maximum acceleration of the load $= \omega^2 x = (2\pi f)^2 x = (2\pi \times 4.07)^2 \times 15$
$=$ **9809.3 mm/s²** or **9.81 m/s²**

3.13 Fluid dynamics

Hydrostatics were dealt with in Section 2.13. This section deals with fluids in motion (fluid dynamics). Fundamentally, *steady uniform flow* occurs when the molecules of the liquid are assumed to travel in parallel paths and the pressure and velocity of the fluid do not vary with time. Such conditions usually apply to low rates of flow in smooth-walled parallel pipes of ample cross-section.

Steady non-uniform flow occurs when the fluid is flowing through a tapered pipe. As the cross-section of the pipe becomes smaller the velocity of the fluid molecules increases so that the same volume of fluid can pass through the reduced cross-section in a given time. However, since the velocity and pressure are constant at any particular section of pipe this case is still considered as steady flow, despite the fact that the flow is non-uniform along the length of the tapered length of pipe.

3.13.1 Rate of flow

The rate of flow or discharge of a liquid passing through a pipe can be determined from the following expression:

$$Q = Av$$

Where:

Q = rate of discharge (m^3/s)
A = cross-section area of pipe
v = velocity of flow.

Mass flow rate can be determined as follows:

$$\rho Q = \rho A v$$

Where:

ρ = density of the fluid
ρQ = mass flow rate (kg/s)

3.13.2 Continuity of flow

The above expressions refer to steady flow in pipes of uniform cross-section. For steady flow in a tapered (converging) pipe the mass flow rate passing any given cross-section is constant, Therefore:

$$\rho Q = \rho_1 A_1 v_1 = \rho_2 A_2 v_2$$

Where:

ρQ = mass flow rate
A_1 and A_2 = maximum and minimum cross-sectional areas (m^2), respectively
v_1 = velocity of fluid flow at cross-sectional area A_1 (m/s)
v_2 = velocity of fluid flow at cross-sectional area A_2 (m/s)
ρ_1 = density of the fluid at cross-sectional area A_1
ρ_2 = density of the fluid at cross-sectional area A_2

This expression is known as the *equation of continuity* and is applicable for all fluids be they liquids or gasses.

For incompressible fluids (liquids) ρ is constant for all cross-sectional areas, therefore:

$$Q = A_1 v_1 = A_2 v_2$$

3.13.3 Energy of a fluid in motion (Bernoulli's equation)

Incompressible fluids (liquids) have three forms of energy, namely:

- *Potential energy:* mgZ where m = mass (kg), g = 9.81 m/s^2 and Z = the height of the liquid above a given datum. The potential energy per unit mass = gZ (J/kg).
- *Kinetic energy:* $\frac{1}{2}mv^2$ where: m = mass (kg) and v = uniform velocity of the liquid (m/s). The kinetic energy per unit mass = $\frac{1}{2}v^2$ (J/kg)
- *Pressure energy:* When a liquid flows through a pipe under pressure, work is done in moving the liquid through each section of pipe against the hydrostatic pressure present at that section. Therefore:

  Work done per unit = $pQ/\rho Q = p/\rho$ (J/kg)

Where: p = is the pressure, ρ = density of the liquid, Q = flow rate and p/ρ is the pressure energy of unit mass of the liquid.

Therefore the total energy possessed by unit mass of the liquid will be the sum of its potential energy, its kinetic energy and its pressure energy.

Total energy per unit mass (J/kg) = $gZ + v^2/2 + p/\rho$

For steady flow of an incompressible fluid (liquid) the potential, kinetic and pressure energies may vary at different positions along the flow but, according to the principle of the conservation of energy *the total energy* will remain *constant*. Therefore according to Bernoulli's equation:

$gZ + v^2/2 + p/\rho$ = constant (K) or $Z + v^2/2g + p/w$ = constant (K)

where: w = the weight per unit volume (specific weight).

3.13.4 Flow through orifices

Water flowing from the side of a tank as a horizontal jet has only, that is, a kinetic head or a velocity head – different names for the same thing. Water stored in the tank has an initial potential head only depending on its height h above the datum level or the mean level of the orifice. Therefore:

$v^2/2g = h \quad v = \sqrt{(2gh)}$

Since some energy is lost through friction, the jet of liquid issuing from the orifice will have a velocity slightly less than the theoretical maximum:

Actual velocity of the jet/theoretical velocity of jet = coefficient of velocity (C_v)

The diameter of the jet of liquid issuing from the sharp-edged orifice will, in practice, be less than the actual diameter of the orifice as shown in Fig. 3.20. This is also known as the *vena contracta*:

Cross-section area of jet/cross-section area of orifice = coefficient of contraction (C_c)

Similarly: The actual rate of discharge/actual discharge = coefficient of discharge (C_d). Therefore:

$$C_d = C_v \times C_c$$

3.13.5 Viscosity

Fluids in contact with a surface will tend to adhere to that surface. It is convenient to consider a fluid as having a series of layers so that the layer in contact with the surface of a pipe as being at rest and the remaining layers slide over each other with gradually increasing velocity until the layer most remote from the surface of the pipe is moving with maximum velocity. This is *streamline flow* and this condition applies when the fluid is subject to a shear force. For example the lubricant in a bearing separating the rotating shaft from the bearing shell. The opposing force is the *viscosity* of the fluid which results in the *drag* experienced when starting a vehicle engine in cold weather when the lubricant increases in viscosity. Should the velocity increase beyond a critical value, the *streamline flow* breaks down and becomes *turbulent*. The 'layer' analogy then no longer applies.

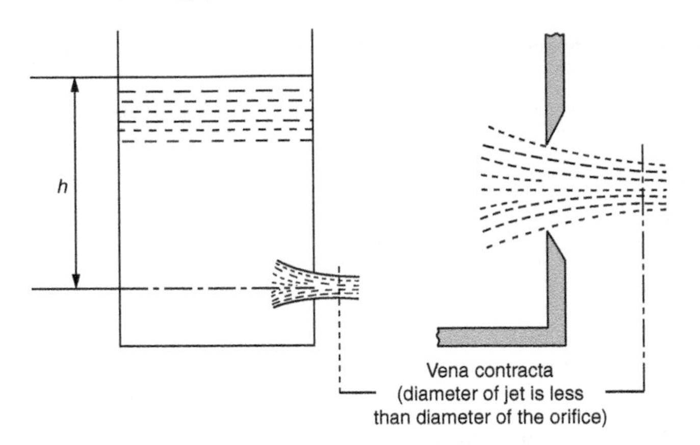

Figure 3.20 Flow through an orifice.

For streamline flow the tangential stress between any two adjacent layers is F/A where A is the contact area between the layers and $\Delta v/\Delta y$ is the velocity gradient. Therefore, since the tangential stress varies directly as the velocity gradient, $F/A = (\Delta v/\Delta y)\eta$ where η is a constant called the *coefficient of viscosity*. Thus: $\boldsymbol{\eta = (F/A)/(\Delta v/\Delta y)}$ where the units of viscosity are $\mathrm{N\,s\,m^{-2}}$ or $\mathrm{kg\,m^{-1}\,s^{-1}}$ since $1\,\mathrm{N} = 1\,\mathrm{kg\,m\,s^{-2}}$.

3.13.6 Poiseulle's formula

For streamline flow through a smooth-walled circular pipe, the volume of liquid (V) being discharged per second is given by Poiseulle's formula which states:

$$\boldsymbol{V = (\pi p r^4)/(8\eta l)}$$

Where:

p = pressure difference between the ends of the pipe
r = internal radius of the pipe (m)
l = length of the pipe (m)
η = viscosity of the fluid flowing through the pipe.

3.13.7 Stoke's formula

Stoke's formula is used to determine the force F resulting from the viscosity of a liquid acting on a smooth sphere with stream line motion. This applies when either the sphere moves through the liquid or when the liquid flows past the sphere.

Force $F = 6\pi\eta r v$ where r = the radius of the sphere (m)

4

Fastenings

4.1 Screwed fastenings

4.1.1 Drawing proportions

Bolts and screws

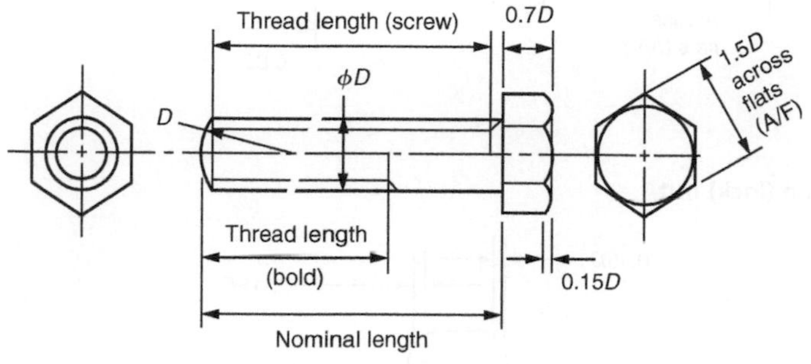

Studs

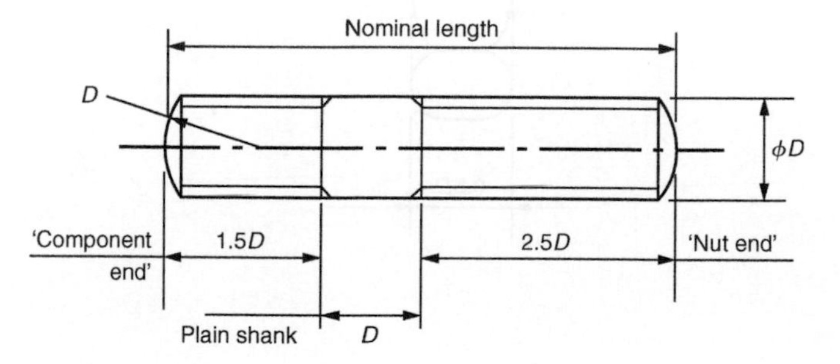

Standard nut

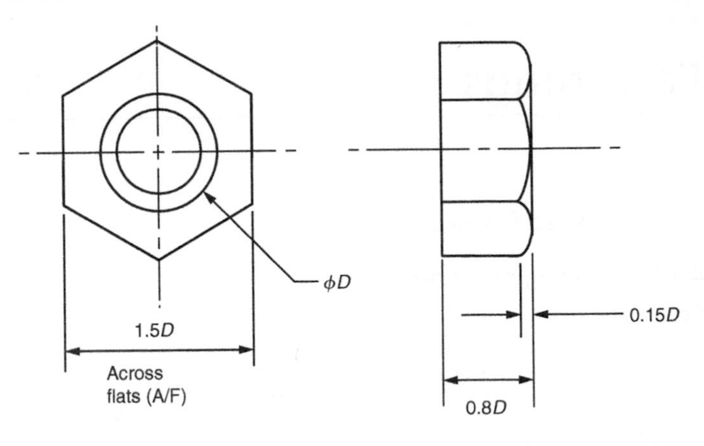

Thin (lock) nut

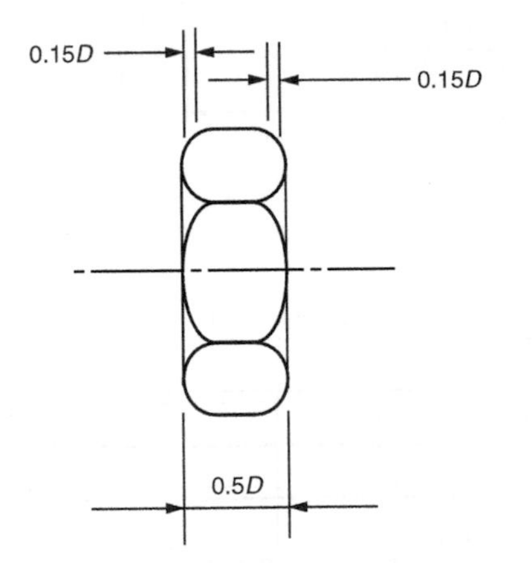

Plain washer

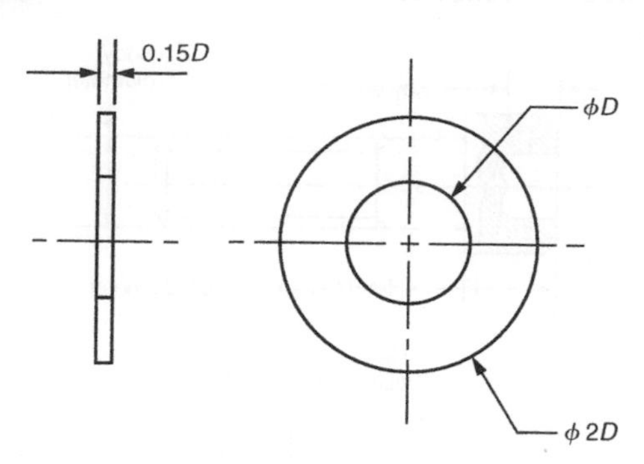

4.1.2 Alternative screw heads

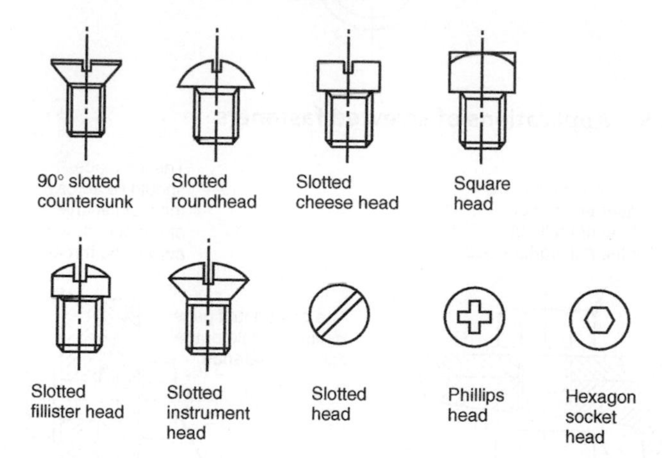

4.1.3 Alternative screw points

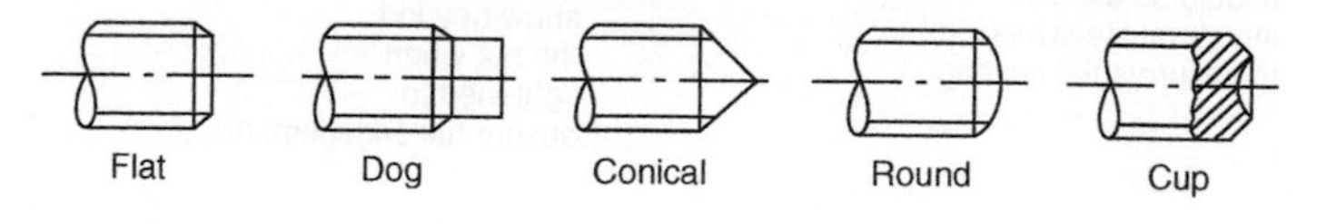

4.1.4 Hexagon socket cap head screw

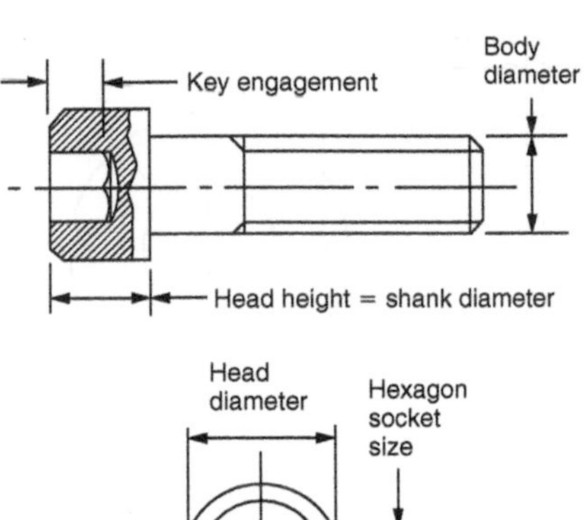

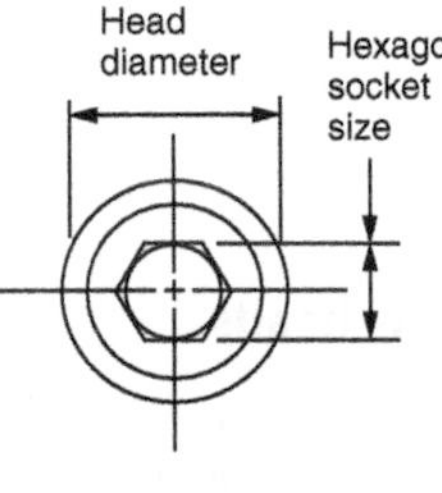

4.1.5 Applications of screwed fasteners

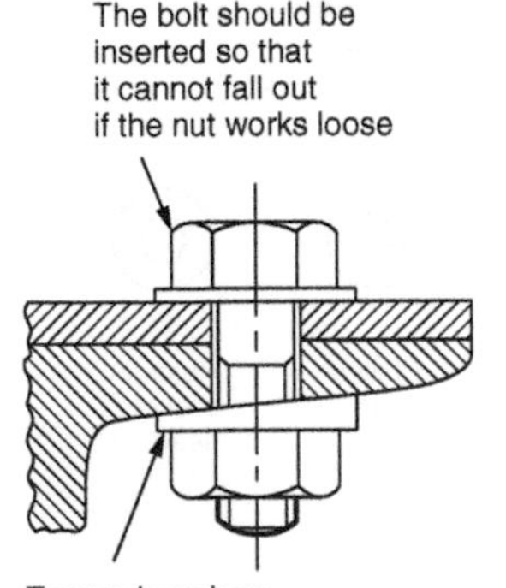

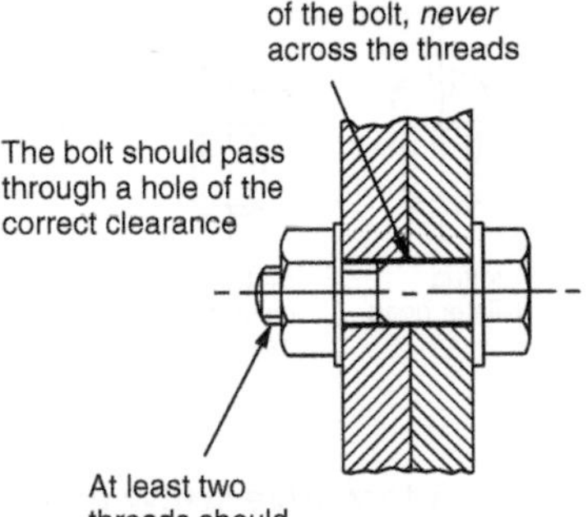

4.1.6 Acme thread form

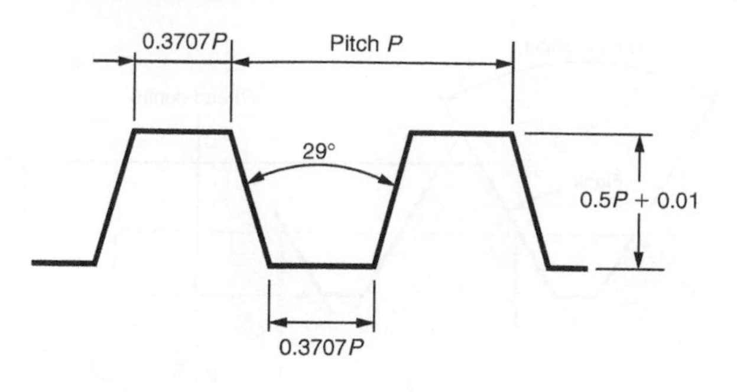

4.1.7 Square thread form

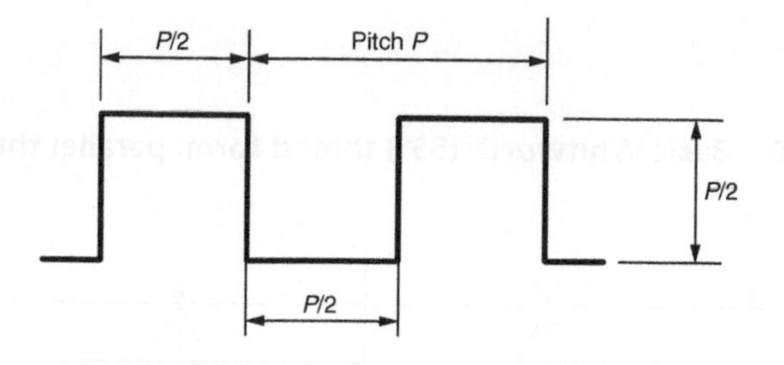

4.1.8 Buttress thread form

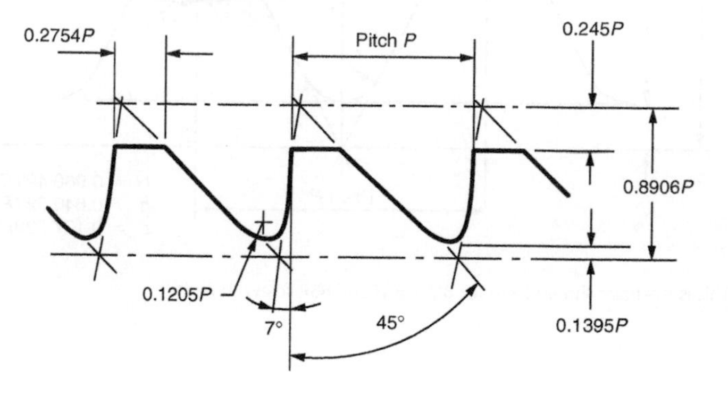

4.1.9 V-thread form

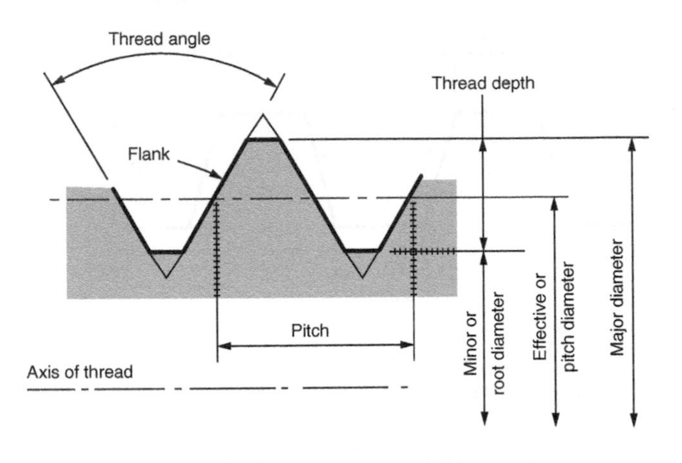

4.1.10 Basic Whitworth (55°) thread form: parallel threads

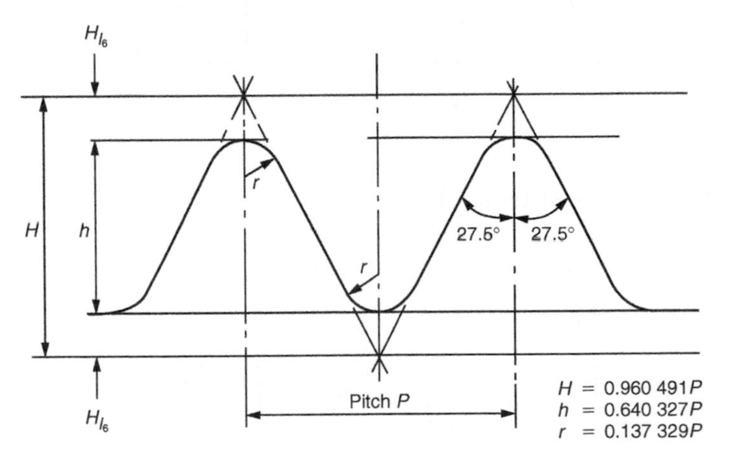

This is the basic thread form for BSW, BSF and BSP screw threads.

4.1.11 ISO metric and ISO 60° unified thread forms

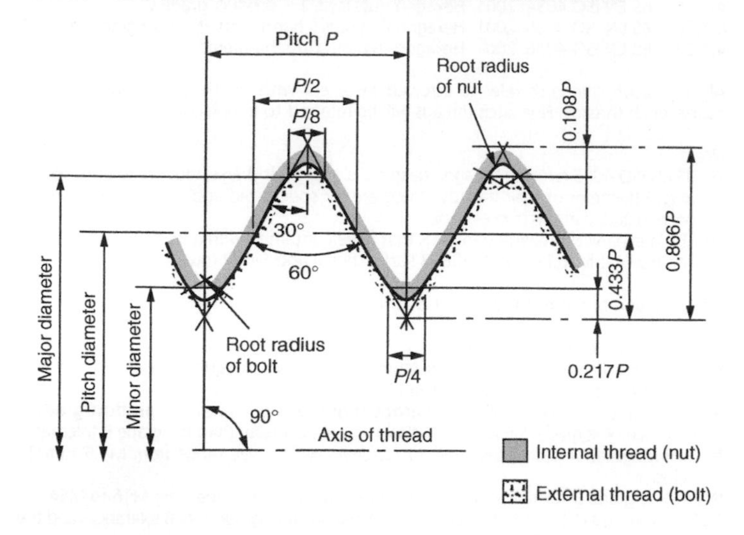

4.1.12 Introduction to screwed fasteners

Although dimensioned in 'inch' units the following screw thread tables (Sections 4.1.38 to 4.1.43) have been retained for maintenance data and similar applications. Screwed fasteners to these specifications are still manufactured:

4.1.38 British Standard Whitworth (BSW) bolts and nuts
4.1.39 British Standard Fine (BSF) bolts and nuts
4.1.40 ISO unified precision internal screw threads: coarse series (UNC)
4.1.41 ISO unified precision external screw threads: coarse series (UNC)
4.1.42 ISO unified precision internal screw threads: fine series (UNF)
4.1.43 ISO unified precision external screw threads: fine series (UNF)

Although obsolescent the following screw thread table (Section 4.1.45) has been retained for the reasons given above:

4.1.45 British Association (BA) internal and external screw threads

The tables based on abstracts from the BS EN 24 000 series have been replaced by the current BS EN ISO 4000 series for screwed fasteners with metric dimensions. These new standards (Sections 4.1.13 to 4.1.21) are as follows:

4.1.13 BS EN ISO 4014: 2001 Hexagon head bolts – product grades A and B
4.1.14 BS EN ISO 4016: 2001 Hexagon head bolts – product grade C
4.1.15 BS EN ISO 4017: 2001 Hexagon head screws – product grades A and B
4.1.16 BS EN ISO 4018: 2001 Hexagon head screws – product grade C
4.1.17 BS EN ISO 4032: 2001 Hexagon nuts style 1 – product grades A and B

4.1.18	BS EN ISO 4033: 2001	Hexagon nuts style 2 – product grades A and B
4.1.19	BS EN ISO 4034: 2001	Hexagon nuts style 1 – product grade C
4.1.20	BS EN ISO 4035: 2001	Hexagon thin nuts (chamfered) – product grades A and B
4.1.21	BS EN ISO 4036: 2001	Hexagon thin nuts (unchamfered) – product grade B

All the above standards refer to screwed fasteners with metric dimensions that have coarse pitch threads. Fine pitch threads will be referred to in due course.

Note:
(i) BS EN ISO 4015 refers to hexagon head bolts with their shanks reduced to the effective (pitch) diameter of their threads. These are for specialized applications and have not been included in this pocket book.
(ii) The mechanical property standards that were contained within BS 4190 and BS 3692 can now be found in BS EN 20898 Part 1 (bolts) and Part 2 (nuts).

Interpretation of the **product grade** is as follows:

- Examine the table in Section 4.1.13.
- The shank diameter (d_s) has a maximum diameter which equals 80 mm, the nominal diameter and also the minimum diameter.
- The minimum diameter can have a **product grade A** tolerance or a **product grade B** tolerance. The grade A tolerance is closer (more accurate) than the grade B tolerance.
- Product grade A tolerances apply to fasteners with a size range from M1.6 to M24 inclusive.
- Product grade B tolerances apply to fasteners with a size range from M16 to M64.
- Therefore sizes M16 to M24 inclusive can have product grade A or B tolerances and the required tolerance must be specified.
- Note that the product grade is not only influenced by the diameter but also by the length as well.

Example 1

An M5 hexagon head bolt will lie within product grade A tolerances and will have a shank diameter lying between 5.00 and 4.82 mm inclusive.

Example 2

An M36 hexagon head bolt will lie within product grade B tolerances and will have a shank diameter lying between 36.00 and 35.38 mm inclusive.

Example 3

An M16 hexagon head bolt will have a shank diameter lying between 16 and 15.73 mm inclusive if it is to product grade A tolerances. If it is to product grade B tolerances, it will have a shank diameter lying between 16 and 15.57 mm inclusive.

Note: The above system of tolerancing applies to all the other dimensions for the fasteners in these tables:

- Examine the table in Section 4.1.14. All the dimensions in this table refer to screwed fasteners with product grade C tolerances. Comparing this table with the previous examples shows that the fasteners made to the product grade C specifications have much coarser tolerances than those manufactured to product grades A and B tolerances.
- An M12 bolt to product grade A has a shank diameter (d_s) lying between 12 and 11.73 mm (a tolerance of 0.27 mm), whereas an M12 bolt made product grade C will have a shank diameter (d_s) lying between 12.77 and 11.3 mm (a tolerance of 1.4 mm). Product grade B does not apply to this size of bolt.

Note: The old terminology of 'black' (hot forged) and 'bright/precision' (cold headed or machined from hexagon bar) bolts and nuts no longer applies. However hot forged (black) bolts and nuts made to special order would only have been made to product grade C.

All the previous notes refer to hexagon head bolts, screws and nuts with **coarse pitch** threads. Bolts have a plain shank between the thread and the head, whereas a screw is threaded virtually all the way up to the head. The following tables and standards (Sections 4.1.22 to 4.1.26) refer to a corresponding **fine pitch** series of screwed fasteners. These fine pitch series of screwed fasteners are only available in product grades A and B.

4.1.22 BS EN ISO 8765: 2001 Hexagon head bolts with metric fine pitch threads – product grades A and B
4.1.23 BS EN ISO 8676: 2001 Hexagon head screws with metric fine pitch threads – product grades A and B
4.1.24 BS EN ISO 8673: 2001 Hexagon nuts style 1 with metric fine pitch threads – product grades A and B
4.1.25 BS EN ISO 8674: 2001 Hexagon nuts style 2 with metric fine pitch threads – product grades A and B
4.1.26 BS EN ISO 8675: 2001 Hexagon thin nuts with metric fine pitch threads – product grades A and B

4.1.13 BS EN ISO 4014: 2001 Hexagon head bolts – product grades A and B

Dimensions

Note: Symbols and descriptions of dimensions are specified in ISO 225.

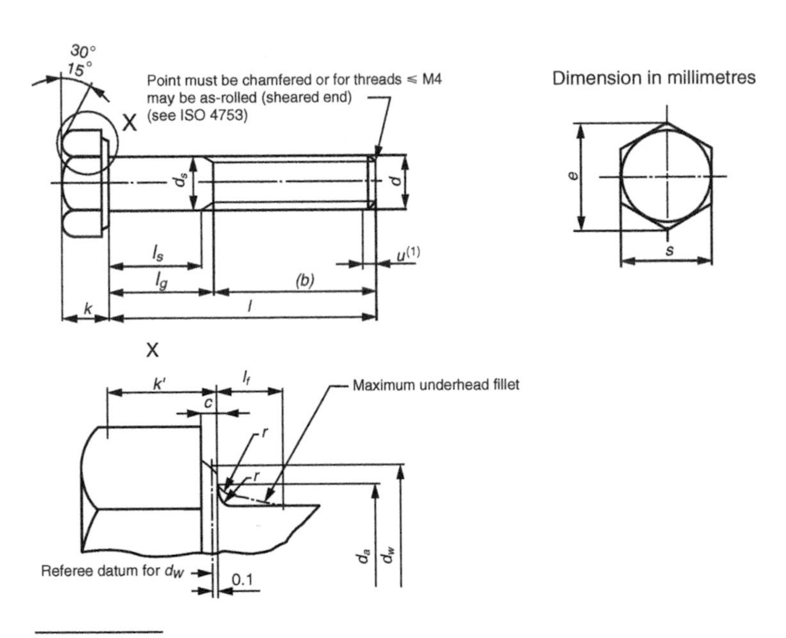

(1) Incomplete thread $u \leqslant 2P$

Example for the designation of a hexagon head bolt with thread M12, nominal length l = 80 mm and property class 8.8:

Hexagon head bolt ISO 4014 – M12 × 80 – 8.8

BS EN ISO 4014: 2001 Table 1 Preferred threads

Dimensions in mm

Thread size, d		M1.6	M2	M2.5	M3	M4	M5	M6	M8	M10	M12	M16	M20	M24	M30	M36	M42	M48	M56	M64
P[a]		0.35	0.4	0.45	0.5	0.7	0.8	1	1.25	1.5	1.75	2	2.5	3	3.5	4	4.5	5	5.5	6
b ref.	_[b]	9	10		12	14	16	18	22	26	30	38	46	54	66	–	–	–	–	–
	_[c]	–	–	–	–	–	–	–	–	–	–	44	52	60	72	84	96	108	–	–
	_[d]	–	–	–	–	–	–	–	–	–	–	–	–	73	85	97	109	121	137	153
c	min.	0.1	0.1	0.1	0.15	0.15	0.15	0.15	0.15	0.15	0.15	0.2	0.2	0.2	0.2	0.2	0.3	0.3	0.3	0.3
	max.	0.25	0.25	0.25	0.4	0.4	0.5	0.5	0.6	0.6	0.6	0.8	0.8	0.8	0.8	0.8	1	1	1	1
d_a	max.	2	2.6	3.1	3.6	4.7	5.7	6.8	9.2	11.2	13.7	17.7	22.4	26.4	33.4	39.4	45.6	52.6	63	71
d_s	nom. = max.	1.6	2	2.5	3	4	5	6	8	10	12	16	20	24	30	36	42	48	56	64
	Product grade A min.	1.46	1.86	2.36	2.86	3.82	4.82	5.82	7.78	9.78	11.73	15.73	19.67	23.67	–	–	–	–	–	–
	Product grade B min.	–	–	–	–	–	–	–	–	–	–	15.57	19.48	23.48	29.48	35.38	41.38	47.38	55.26	63.26
d_w	Product grade A min.	2.27	3.07	4.07	4.57	5.88	6.88	8.88	11.63	14.63	16.63	22.49	28.19	33.61	–	–	–	–	–	–
	Product grade B min.	–	–	–	–	–	–	–	–	–	–	22	27.7	33.25	42.75	51.11	59.95	69.45	78.66	88.16
e	Product grade A min.	3.41	4.32	5.45	6.01	7.66	8.79	11.05	14.38	17.77	20.03	26.75	33.53	39.98	–	–	–	–	–	–
	Product grade B min.	–	–	–	–	–	–	–	–	–	–	26.17	32.95	39.55	50.85	60.79	71.3	82.6	93.56	104.86
l_f	max.	0.6	0.8	1	1	1.2	1.2	1.4	2	2	3	3	4	4	6	6	8	10	12	13
k	nom.	1.1	1.4	1.7	2	2.8	3.5	4	5.3	6.4	7.5	10	12.5	15	18.7	22.5	26	30	35	40
	Product grade A min.	0.975	1.275	1.575	1.875	2.675	3.35	3.85	5.15	6.22	7.32	9.82	12.285	14.785	–	–	–	–	–	–
	Product grade A max.	1.225	1.525	1.825	2.125	2.925	3.65	4.15	5.45	6.58	7.68	10.18	12.715	15.215	–	–	–	–	–	–
	Product grade B min.	–	–	–	–	–	–	–	–	–	–	9.71	12.15	14.65	18.28	22.08	25.58	29.58	34.5	39.5
	Product grade B max.	–	–	–	–	–	–	–	–	–	–	10.29	12.85	15.35	19.12	22.92	26.42	30.42	35.5	40.5
k'[e]	Product grade A min.	0.68	0.89	1.1	1.31	1.87	2.35	2.7	3.61	4.35	5.12	6.87	8.6	10.35	–	–	–	–	–	–
	Product grade B min.	–	–	–	–	–	–	–	–	–	–	6.8	8.51	10.26	12.8	15.46	17.91	20.71	24.15	27.65
r	min.	0.1	0.1	0.1	0.1	0.2	0.2	0.25	0.4	0.4	0.6	0.6	0.8	0.8	1	1	1.2	1.6	2	2
s	nom. = max.	3.2	4	5	5.5	7	8	10	13	16	18	24	30	36	46	55	65	75	85	95
	Product grade A min.	3.02	3.82	4.82	5.32	6.78	7.78	9.78	12.73	15.73	17.73	23.67	29.67	35.38	–	–	–	–	–	–
	Product grade B min.	–	–	–	–	–	–	–	–	–	–	23.16	29.16	35	45	53.8	63.1	73.1	82.8	92.8

continued

BS EN ISO 4014: 2001 Table 1 Preferred threads (*continued*)

Dimensions in mm

Thread size, *d*					M1.6		M2		M2.5		M3		M4		M5		M6		M8		M10		M12		M16		M20		M24		M30		M36		M42		M48		M56		M64	
l	Product grade A		Product grade B		l_s and l_g [f, g]																																					
nom.	min.	max.	min.	max.	l_s min.	l_g max.	l_s min.	l_g max.	l_s min.	l_g max.	l_s min.	l_g max.	l_s min.	l_g max.	l_s min.	l_g max.	l_s min.	l_g max.	l_s min.	l_g max.	l_s min.	l_g max.	l_s min.	l_g max.	l_s min.	l_g max.	l_s min.	l_g max.	l_s min.	l_g max.	l_s min.	l_g max.	l_s min.	l_g max.	l_s min.	l_g max.	l_s min.	l_g max.	l_s min.	l_g max.	l_s min.	l_g max.
12	11.65	12.35	–	–	1.2	3																																				
16	15.65	16.35	–	–	5.2	7	4	6	2.75	5																																
20	19.58	20.42	–	–			8	10	6.75	9	5.5	8																														
25	24.58	25.42	–	–					11.75	14	10.5	13	7.5	11	5	9																										
30	29.58	30.42	–	–							15.5	18	12.5	16	10	14	7	12																								
35	34.5	35.5	–	–									17.5	21	15	19	12	17																								
40	39.5	40.5	–	–									22.5	26	20	24	17	22	11.75	18																						
45	44.5	45.5	–	–											25	29	22	27	16.75	23	11.5	19																				
50	49.5	50.5	–	–											30	34	27	32	21.75	28	16.5	24	11.25	20																		
55	54.4	55.6	–	–													32	37	26.75	33	21.5	29	16.25	25																		
60	59.4	60.6	–	–													37	42	31.75	38	26.5	34	21.25	30																		
65	64.4	65.6	–	–															36.75	43	31.5	39	26.25	35	17	27																
70	69.4	70.0	–	–															41.75	48	36.5	44	31.25	40	22	32																
80	79.4	80.6	–	–															51.75	58	46.5	54	41.25	50	32	42	21.5	34														
90	89.3	90.7	–	–																	56.5	64	51.25	60	42	52	31.5	44	21	36												
100	99.3	100.7	–	–																	66.5	74	61.25	70	52	62	41.5	54	31	46												
110	109.3	110.7	108.25	111.75																			71.25	80	62	72	51.5	64	41	56	26.5	44										
120	119.3	120.7	118.25	121.75																			81.25	90	72	82	61.5	74	51	66	36.5	54										
130	129.2	130.8	128	132																					76	86	65.5	78	55	70	40.5	58										
140	139.2	140.8	138	142																					86	96	75.5	88	65	80	50.5	68	36	56								
150	149.2	150.8	148	152																					96	106	85.5	98	75	90	60.5	78	46	66								
160	–	–	158	162																					106	116	95.5	108	85	100	70.5	88	56	76	41.5	64						
180	–	–	178	182																							115.5	128	105	120	90.5	108	76	96	61.5	84	47	72				
200	–	–	197.7	202.3																							135.5	148	125	140	110.5	128	96	116	81.5	104	67	92				

For sizes above the stepped line, marked thus ____, ISO 4017 is recommended.

220	–	–	217.7	222.3																									132	147	117.5	135	103	123	88.5	111	74	99	55.5	83		
240	–	–	237.7	242.3																									152	167	137.5	155	123	143	108.5	131	94	119	75.5	103		
260	–	–	257.4	262.6																											157.5	175	143	163	128.5	151	114	139	95.5	123	77	107
280	–	–	277.4	292.6																											177.5	195	163	183	148.5	171	134	159	115.5	143	97	127
300	–	–	297.4	302.6																											197.5	215	183	203	168.5	191	154	179	135.5	163	117	147
320	–	–	317.15	322.85																													203	223	188.5	211	174	199	155.5	183	137	167
340	–	–	337.15	342.85																													223	243	208.5	231	194	219	175.5	203	157	187
360	–	–	357.15	362.85																													243	263	228.5	251	214	239	195.5	223	177	207
380	–	–	377.15	382.85																															248.5	271	234	259	215.5	243	197	227
400	–	–	397.15	402.85																															268.5	291	254	279	235.5	263	217	247
420	–	–	416.85	423.15																															288.5	311	274	299	255.5	283	237	267
440	–	–	436.85	443.15																															308.5	331	294	319	275.5	303	257	287
460	–	–	456.85	463.15																																	314	339	295.5	323	277	307
480	–	–	476.85	483.15																																	334	359	315.5	343	297	327
500	–	–	496.85	503.15																																			335.5	363	317	347

[a] P = pitch of thread.
[b] For lengths $l_{nom} \leqslant 125$ mm.
[c] For lengths 125 mm $< l_{nom} \leqslant 200$ mm.
[d] For lengths $l_{nom} > 200$ mm.
[e] $k'_{min} = 0.7\,k_{min}$.
[f] $l_{g\,max} = l_{nom} - b$; $l_{s\,min} = l_{g\,max} - 5P$.
[g] l_g = minimum grip length.

Note: The popular lengths are defined in terms of lengths l_s and l_g:
- product grade A above the stepped line, marked thus - - - -.
- product grade B below this stepped line.

BS EN ISO 4014: 2001 Table 2 Non-preferred threads

Dimensions in mm

Thread size, d		M3.5	M14	M18	M22	M27	M33	M39	M45	M52	M60
P^a		0.6	2	2.5	2.5	3	3.5	4	4.5	5	5.5
b ref.	_b	13	34	42	50	60	–	–	–	–	–
	_c	–	40	48	56	66	78	90	102	116	–
	_d	–	–	–	69	79	91	103	115	129	145
c	min.	0.15	0.15	0.2	0.2	0.2	0.2	0.3	0.3	0.3	0.3
	max.	0.4	0.6	0.8	0.8	0.8	0.8	1	1	1	1
d_a	max.	4.1	15.7	20.2	24.4	30.4	36.4	42.4	48.6	56.6	67
d_s	nom. = max.	3.5	14	18	22	27	33	39	46	52	60
	Product grade $\frac{A}{B}$ min.	3.32	13.73	17.73	21.67	–	–	–	–	–	–
		–	–	17.57	21.48	26.48	32.38	38.38	44.38	51.26	59.26
d_w	Product grade $\frac{A}{B}$ min.	5.07	19.37	25.34	31.71	–	–	–	–	–	–
		–	–	24.85	31.35	38	46.55	55.86	64.7	74.2	83.41
e	Product grade $\frac{A}{B}$ min.	6.58	23.36	30.14	37.72	–	–	–	–	–	–
		–	–	29.56	37.29	45.2	56.37	66.44	76.95	88.25	99.21
l_f	max.	1	3	3	4	6	6	6	8	10	12
k	nom.	2.4	8.8	11.5	14	17	21	25	28	33	38
	Product grade A min.	2.275	8.62	11.285	13.785	–	–	–	–	–	–
	Product grade A max.	2.525	8.98	11.175	14.215	–	–	–	–	–	–
	Product grade A min.	–	–	11.15	13.65	16.65	20.58	24.58	27.58	32.5	37.5
	Product grade A max.	–	–	11.85	14.35	17.35	21.42	25.42	28.42	33.5	38.5
k'^e	Product grade $\frac{A}{B}$ min.	1.59	6.03	7.9	9.65	–	–	–	–	–	–
		–	–	7.81	9.56	11.66	14.41	17.21	19.31	22.75	26.25
r	min.	0.1	0.6	0.6	0.8	1	1	1	1.2	1.6	2
s	nom. = max.	6	21	27	34	41	50	60	70	80	90
	Product grade $\frac{A}{B}$ min.	5.82	20.67	26.67	33.38	–	–	–	–	–	–
		–	–	26.16	33	40	49	58.8	68.1	78.1	87.8

	Product grade A		Product grade B		l_s and l_g[f,g]																			
	l																							
nom.	min.	max.	min.	max.	l_s min.	l_g max.	l_s min.	l_g max.	l_s min.	l_g max.	l_s min.	l_g max.	l_s min.	l_g max.	l_s min.	l_g max.	l_s min.	l_g max.	l_s min.	l_g max.	l_s min.	l_g max.	l_s min.	l_g max.
20	19.58	20.42	–	–	4	7																		
25	24.58	25.42	–	–	9	12																		
30	29.58	30.42	–	–	14	17																		
35	34.5	35.5	–	–	19	22																		
40	39.5	40.5	–	–																				
45	44.5	45.5	–	–																				
50	49.5	50.5	–	–																				
55	54.4	55.6	–	–																				
60	59.4	60.6	–	–			16	26																
65	64.4	65.6	–	–			21	31																
70	69.4	70.6	–	–			26	36	15.5	28														
80	79.4	80.6	–	–			36	46	25.5	38														
90	89.3	90.7	–	–			46	56	35.5	48	27.5	40												
100	99.3	100.7	98.25	101.75			56	66	45.5	58	37.5	50	25	40										
110	109.3	110.7	108.25	111.75			66	76	55.5	68	47.5	60	35	50										
120	119.3	120.7	118.25	121.75			76	86	65.5	78	57.5	70	45	60										
130	129.2	130.8	128	132			80	90	69.5	82	61.5	74	49	64	34.5	52								
140	139.2	140.8	138	142			90	100	79.5	92	71.5	84	59	74	44.5	62								
150	149.2	150.8	148	152					89.5	102	81.5	94	69	84	54.5	72	40	60						
160	–	–	158	162					99.5	112	91.5	104	79	94	64.5	82	50	70						
180	–	–	178	182					119.5	132	111.5	124	99	114	84.5	102	70	90	55.5	78				
200	–	–	197.7	202.3							131.5	144	119	134	104.5	122	90	110	75.6	98	59	84		
220	–	–	217.7	222.3							138.5	151	126	141	111.5	129	97	117	82.5	105	66	91		
240	–	–	237.7	242.3									146	161	131.5	149	117	137	102.5	125	86	111	67.5	95

For sizes above the stepped line. marked thus ________, ISO 4017 is recommended

continued

BS EN ISO 4014: 2001 Table 2 Non-preferred threads (*continued*)

Dimensions in mm

Thread size, *d*					M3.5		M14		M18		M22		M27		M33		M39		M45		M52		M60	
	l Product grade A		l Product grade B		l_s and l_g [f,g]																			
nom.	min.	max.	min.	max.	l_s min.	l_g max.	l_s min.	l_g max.	l_s min.	l_g max.	l_s min.	l_g max.	l_s min.	l_g max.	l_s min.	l_g max.	l_s min.	l_g max.	l_s min.	l_g max.	l_s min.	l_g max.	l_s min.	l_g max.
260	–	–	257.4	262.6									166	181	151.5	169	137	157	122.5	145	106	131	87.5	115
280	–	–	277.4	282.6											171.5	189	157	177	142.5	165	126	151	107.5	135
300	–	–	297.4	302.6											191.5	209	177	197	162.5	185	146	171	127.5	155
320	–	–	317.15	322.85											211.5	229	197	217	182.5	205	166	191	147.5	175
340	–	–	337.15	342.85													217	237	202.5	225	186	211	167.5	195
360	–	–	357.15	362.85													237	257	222.5	245	206	231	187.5	215
380	–	–	377.15	382.85													257	277	242.5	265	226	251	207.5	235
400	–	–	397.15	402.85															262.5	285	246	271	227.5	255
420	–	–	416.85	423.15															282.5	305	266	291	247.5	275
440	–	–	436.85	443.15															302.5	325	286	311	267.5	295
460	–	–	456.85	463.15																	306	331	287.5	315
480	–	–	476.85	483.15																	326	351	307.5	335
500	–	–	496.85	503.15																			327.5	355

[a] P = pitch of thread.
[b] For lengths $l_{nom} \leqslant 125$ mm.
[c] For lengths $125 \text{ mm} < l_{nom} \leqslant 200$ mm.
[d] For lengths $l_{nom} > 200$ mm.
[e] $k'_{min} = 0.7\ k_{min}$.
[f] $l_{g\,max} = l_{nom} - b$; $l_{s\,min} = l_{g\,max} - 5P$.
[g] l_g = minimum grip length.

Note: The popular lengths are defined in terms of lengths l_s and l_g:

- product grade A above the stepped line, marked thus – – – –.
- product grade B below this stepped line.

4.1.14 BS EN ISO 4016: 2001 Hexagon head bolts – product grade C

Dimensions

Note: Symbols and descriptions of dimensions are specified in ISO 225.

Dimensions in millimetres

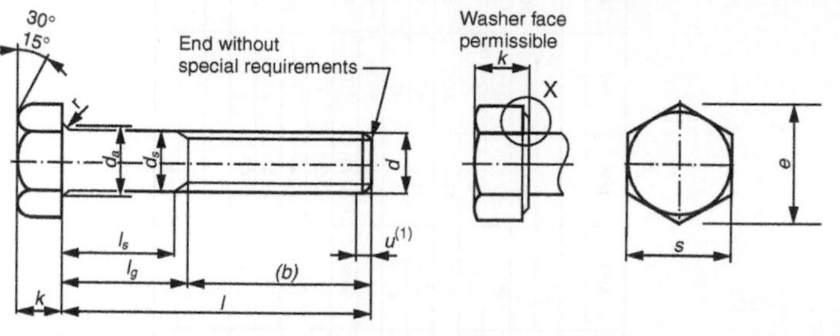

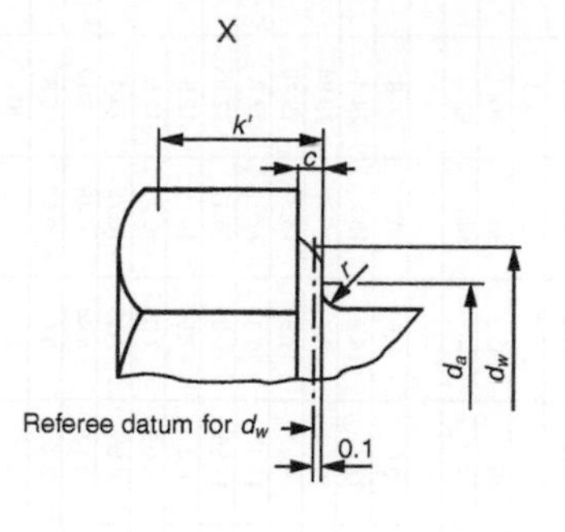

(1) Incomplete thread $u \leqslant 2P$.

Example for the designation of a hexagon head bolt with thread M12, nominal length l = 80 mm and property class 4.6:

Hexagon head bolt ISO 4016 – M12 × 80 – 4.6

BS EN ISO 4016: 2001 Table 1 Preferred threads

Dimensions in mm

Thread size, *d*		M5	M6	M8	M10	M12	M16	M20	M24	M30	M36	M42	M48	M56	M64
P [a]		0.8	1	1.25	1.5	1.75	2	2.5	3	3.5	4	4.5	5	5.5	6
b ref.	_[b]	16	18	22	26	30	38	46	54	66	–	–	–	–	–
	_[c]	–	–	–	–	–	44	52	60	72	84	96	108	–	–
	_[d]	–	–	–	–	–	–	–	73	85	97	109	121	137	153
c	max.	0.5	0.5	0.6	0.6	0.6	0.8	0.8	0.8	0.8	0.8	1	1	1	1
d_a	max.	6	7.2	10.2	12.2	14.7	18.7	24.4	28.4	35.4	42.4	48.6	56.6	67	75
d_s	max.	5.48	6.48	8.58	10.58	12.7	16.7	20.84	24.84	30.84	37	43	49	57.2	65.2
	min.	4.52	5.52	7.42	9.42	11.3	15.3	19.16	23.16	29.16	35	41	47	54.8	62.8
d_w	min.	6.74	8.74	11.47	14.47	16.47	22	27.7	33.25	42.75	51.11	59.95	69.45	78.66	88.16
e	min.	8.63	10.89	14.2	17.59	19.85	26.17	32.95	39.55	50.85	60.79	71.3	82.6	93.56	104.86
k	nom.	3.5	4	5.3	6.4	7.5	10	12.5	15	18.7	22.5	26	30	35	40
	min.	3.125	3.625	4.925	5.95	7.05	9.25	11.6	14.1	17.65	21.45	24.95	28.95	33.75	38.75
	max.	3.875	4.375	5.675	6.85	7.95	10.75	13.4	15.9	19.75	23.55	27.05	31.05	36.25	41.25
k' [e]	min.	2.19	2.54	3.45	4.17	4.94	6.48	8.12	9.87	12.36	15.02	17.47	20.27	23.63	27.13
r	min.	0.2	0.25	0.4	0.4	0.6	0.6	0.8	0.8	1	1	1.2	1.6	2	2
s	nom. = max.	8	10	13	16	18	24	30	36	46	55	65	75	85	95
	min.	7.64	9.64	12.57	15.57	17.57	23.16	29.16	35	45	53.8	63.1	73.1	82.8	92.8

l			l_s and l_g [f, g]																											
nom.	min.	max.	l_s min.	l_g max.	l_s min.	l_g max.	l_s min.	l_g max.	l_s min.	l_g max.	l_s min.	l_g max.	l_s min.	l_g max.	l_s min.	l_g max.	l_s min.	l_g max.	l_s min.	l_g max.	l_s min.	l_g max.	l_s min.	l_g max.	l_s min.	l_g max.	l_s min.	l_g max.	l_s min.	l_g max.
25	23.95	26.05	5	9																										
30	28.95	31.05	10	14	7	12																								
35	33.75	36.25	15	19	12	17																								
40	38.75	41.25	20	24	17	22	11.75	18																						
45	43.75	46.25	25	29	22	27	16.75	23	11.5	19																				
50	48.75	51.25	30	34	27	32	21.75	28	16.5	24																				
55	53.5	56.5			32	37	26.75	33	21.5	29	16.25	25																		
60	58.5	61.5			37	42	31.75	38	26.5	34	21.25	30																		
65	63.5	66.5					36.75	43	31.5	39	26.25	35	17	27																
70	68.5	71.5					41.75	48	36.5	44	31.25	40	22	32																
80	78.5	81.5					51.75	58	46.5	54	41.25	50	32	42	21.5	34														
90	88.25	91.75							56.5	64	51.25	60	42	52	31.5	44														
100	98.25	101.75							66.5	74	61.25	70	52	62	41.5	54	31	46												
110	108.25	111.75									71.25	80	62	72	51.5	64	41	56												
120	118.25	121.75									81.25	90	72	82	61.5	74	51	66	36.5	54										
130	128	132											76	86	65.5	78	55	70	40.5	58										
140	138	142											86	96	75.5	88	65	80	50.5	68	36	56								
150	148	152											96	106	85.5	98	75	90	60.5	78	46	66								
160	156	164											106	116	95.5	108	85	100	70.5	88	56	76								
180	176	184													115.5	128	105	120	90.5	108	76	96	61.5	84						
200	195.4	204.6													135.5	148	125	140	110.5	128	96	116	81.5	104	67	92				

For sizes above the stepped line, marked thus ____, ISO 4018 is recommended.

continued

BS EN ISO 4016: 2001 Table 1 Preferred threads (*continued*)

Dimensions in mm

Thread size, *d*			M5		M6		M8		M10		M12		M16		M20		M24		M30		M36		M42		M48		M56		M64	
l			l_s and l_g[f, g]																											
nom.	min.	max.	l_s min.	l_g max.	l_s min.	l_g max.	l_s min.	l_g max.	l_s min.	l_g max.	l_s min.	l_g max.	l_s min.	l_g max.	l_s min.	l_g max.	l_s min.	l_g max.	l_s min.	l_g max.	l_s min.	l_g max.	l_s min.	l_g max.	l_s min.	l_g max.	l_s min.	l_g max.	l_s min.	l_g max.
220	215.4	224.6															132	147	117.5	135	103	123	88.5	111	74	99				
240	235.4	244.6															152	167	137.5	155	123	143	108.5	131	94	119	75.5	103		
260	254.8	265.2																	157.5	175	143	163	128.5	151	114	139	95.5	123	77	107
280	274.8	285.2																	177.5	195	163	183	148.5	171	134	159	115.5	143	97	127
300	294.8	305.2																	197.5	215	183	203	168.5	191	154	179	135.5	163	117	147
320	314.3	325.7																			203	223	188.5	211	174	199	155.5	183	137	167
340	334.3	345.7																			223	243	208.5	231	194	219	175.5	203	157	187
360	354.3	365.7																			243	263	228.5	251	214	239	195.5	223	177	207
380	374.3	385.7																					248.5	271	234	259	215.5	243	197	227
400	394.3	405.7																					268.5	291	254	279	235.5	263	217	247
420	413.7	426.3																					288.5	311	274	299	255.5	283	237	267
440	433.7	446.3																							294	319	275.5	303	257	287
460	453.7	466.3																							314	339	295.5	323	277	307
480	473.7	486.3																							334	359	315.5	343	297	327
500	493.7	506.3																									335.5	363	317	347

[a] P = pitch of the thread.
[b] For lengths $l_{nom} \leqslant 125$ mm.
[c] For lengths 125 mm $< l_{nom} \leqslant 200$ mm.
[d] For lengths $l_{nom} > 200$ mm.
[e] $k'_{min} = 0.7\ k_{min}$.
[f] $l_{g\,max} = l_{nom} - b$; $l_{s\,min} = l_{g\,max} - 5P$.
[g] l_g = minimum grip length.
Note: The popular lengths are marked by the shank lengths.

BS EN ISO 4016: 2001 Table 2 Non-preferred threads

Dimensions in mm

Thread size, d		M14	M18	M22	M27	M33	M39	M45	M52	M60
p[a]		2	2.5	2.5	3	3.5	4	4.5	5	5.5
b ref	_[b]	34	42	50	60	–	–	–	–	–
	_[c]	40	48	56	66	78	90	102	116	–
	_[d]	–	–	69	79	91	103	115	129	145
c	max.	0.6	0.8	0.8	0.8	0.8	1	1	1	1
d_a	max.	16.7	21.2	26.4	32.4	38.4	45.4	52.6	62.6	71
d_s	max.	14.7	18.7	22.84	27.84	34	40	46	53.2	61.2
	min.	13.3	17.3	21.16	26.16	32	38	44	50.8	58.8
d_w	min.	19.15	24.85	31.35	38	46.55	55.86	64.7	74.2	83.41
e	min.	22.78	29.56	37.29	45.2	55.37	66.44	76.95	88.25	99.21
k	nom.	8.8	11.5	14	17	21	25	28	33	38
	min.	8.35	10.6	13.1	16.1	19.95	23.95	26.95	31.75	36.75
	max.	9.25	12.4	14.9	17.9	22.05	26.05	29.05	34.25	39.25
k'[e]	min.	5.85	7.42	9.17	11.27	13.97	16.77	18.87	22.23	25.73
r	min.	0.6	0.6	0.8	1	1	1	1.2	1.6	2
s	nom. = max.	21	27	34	41	50	60	70	80	90
	min.	20.16	26.16	33	40	49	58.8	68.1	78.1	87.8

l			l_s and l_g[f,g]																	
nom.	min.	max.	l_s min.	l_g max.	l_s min.	l_g max.	l_s min.	l_g max.	l_s min.	l_g max.	l_s min.	l_g max.	l_s min.	l_g max.	l_s min.	l_g max.	l_s min.	l_g max.	l_s min.	l_g max.
60	58.5	61.5	16	26																
65	63.5	66.5	21	31																
70	68.5	71.5	26	36																
80	78.5	81.5	36	46	25.5	38														
90	88.25	91.75	46	56	35.5	48	27.5	40												
100	98.25	101.75	56	66	45.5	58	37.5	50												

For sizes above the stepped line, marked thus ______, ISO 4018 is recommended.

continued

BS EN ISO 4016: 2001 Table 2 Non-preferred threads (*continued*)

Dimensions in mm

Thread size, d			M14		M18		M22		M27		M33		M39		M45		M52		M60	
l			l_s and l_g [f,g]																	
nom.	min.	max.	l_s min.	l_g max.	l_s min.	l_g max.	l_s min.	l_g max.	l_s min.	l_g max.	l_s min.	l_g max.	l_s min.	l_g max.	l_s min.	l_g max.	l_s min.	l_g max.	l_s min.	l_g max.
110	108.25	111.75	66	76	55.5	68	47.5	60	35	50										
120	118.25	121.75	76	86	65.5	78	57.5	70	45	60										
130	128	132	80	90	69.5	82	61.5	74	49	64	34.5	52								
140	138	142	90	100	79.5	92	71.5	84	59	74	44.5	62								
150	148	152			89.5	102	81.5	94	69	84	54.5	72	40	60						
160	156	164			99.5	112	91.5	104	79	94	64.5	82	50	70						
180	176	184			119.5	132	111.5	124	99	114	84.5	102	70	90	55.5	78				
200	195.4	204.6					131.5	144	119	134	104.5	122	90	110	75.5	98	59	84		
220	215.4	224.6					138.5	151	126	141	111.5	129	97	117	82.5	105	66	91		
240	235.4	244.6							146	161	131.5	149	117	137	102.5	125	86	111	67.5	95
260	254.8	265.2							166	181	151.5	167	137	157	122.5	145	106	131	87.5	115
280	274.8	285.2									171.5	189	157	177	142.5	165	126	151	107.5	135
300	294.8	305.2									191.5	209	177	197	162.5	185	146	171	127.5	155
320	314.3	325.7									211.5	229	197	217	182.5	205	166	191	147.5	175
340	334.3	345.7											217	237	202.5	225	186	211	167.5	195
360	354.3	365.7											237	257	222.5	245	206	231	187.5	215
380	374.3	385.7											257	277	242.5	265	226	251	207.5	235
400	394.3	405.7											277	297	262.5	285	246	271	227.5	255
420	413.7	426.3													282.5	305	266	291	247.5	275
440	433.7	446.3													302.5	325	286	311	267.5	295
460	453.7	466.3															306	331	287.5	315
480	473.7	486.3															326	351	307.5	335
500	493.7	506.3															346	371	327.5	355

[a] P = pitch of the thread.
[b] For lengths $l_{nom} \leq 125$ mm.
[c] For lengths $125\text{ mm} < l_{nom} \leq 200$ mm.
[d] For lengths $l_{nom} > 200$ mm.
[e] $k'_{min} = 0.7\, k_{min}$.
[f] $l_{g\,max} = l_{nom} - b$; $l_{s\,min} = l_{g\,max} - 5P$.
[g] l_g is the minimum grip length.
Note: The popular length are marked by the shank lengths.

4.1.15 BS EN ISO 4017: 2001 Hexagon head screws – product grades A and B

Dimensions

Note: Symbols and descriptions of dimensions are specified in ISO 225.

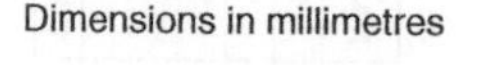

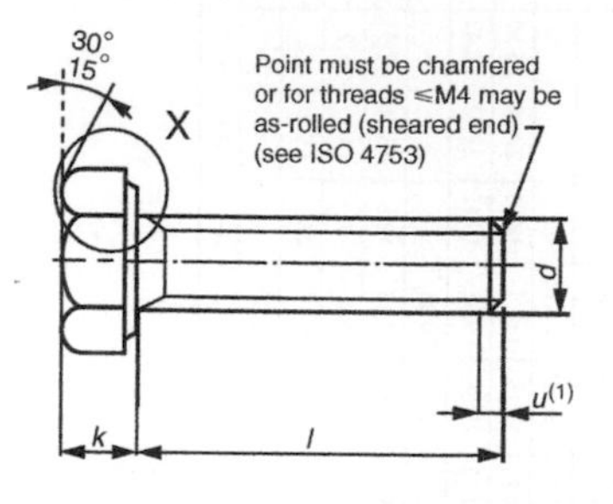

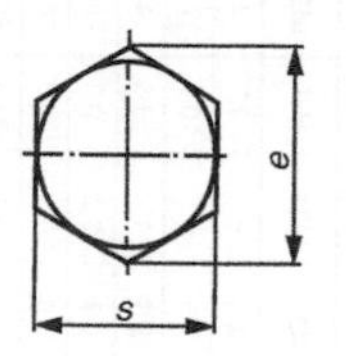

X

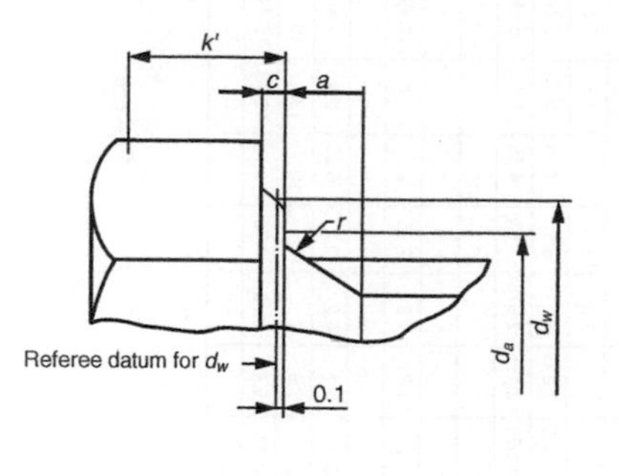

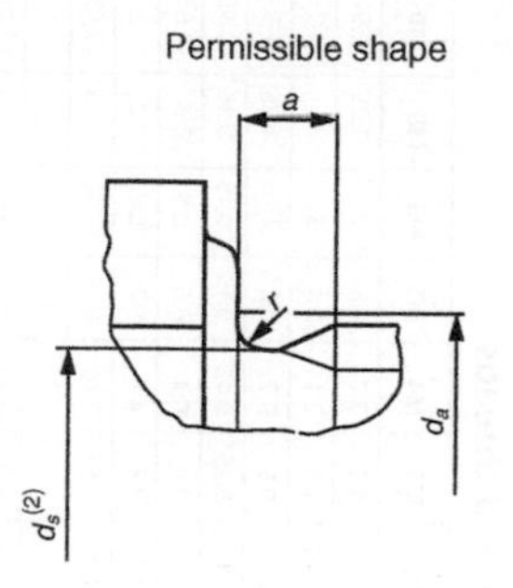

(1) Incomplete thread $u \leqslant 2P$.
(2) d_s = pitch diameter.

Example for the designation of a hexagon head screw with thread size M12, nominal length l = 80 mm and property class 8.8:

Hexagon head screw ISO 4017 – M12 × 80 – 8.8

BS EN ISO 4017: 2001 Table 1 Preferred threads

Dimensions in mm

Thread size, d		M1.6	M2	M2.5	M3	M4	M5	M6	M8	M10	M12	M16	M20	M24	M30	M36	M42	M48	M56	M64
p^a		0.35	0.4	0.45	0.5	0.7	0.8	1	1.25	1.5	1.75	2	2.5	3	3.5	4	4.5	5	5.5	6
a	max.[b]	1.05	1.2	1.35	1.5	2.1	2.4	3	4	4.5	5.3	6	7.5	9	10.5	12	13.5	15	16.5	18
	min.	0.35	0.4	0.45	0.5	0.7	0.8	1	1.25	1.5	1.75	2	2.5	3	3.5	4	4.5	5	5.5	6
c	min.	0.1	0.1	0.1	0.15	0.15	0.15	0.15	0.15	0.15	0.15	0.2	0.2	0.2	0.2	0.2	0.3	0.3	0.3	0.3
	max.	0.25	0.25	0.25	0.4	0.4	0.5	0.5	0.6	0.6	0.6	0.8	0.8	0.8	0.8	0.8	1	1	1	1
d_a	max.	2	2.6	3.1	3.6	4.7	5.7	6.8	9.2	11.2	13.7	17.7	22.4	26.4	33.4	39.4	45.6	52.6	63	71
d_w	Product grade A min.	2.27	3.07	4.07	4.57	6.03	6.88	8.88	11.63	14.63	16.63	22.49	28.19	33.61	–	–	–	–	–	–
	Product grade B min.	–	–	–	–	–	–	–	–	–	–	22	27.7	33.25	42.75	51.11	59.95	69.45	78.66	88.16
e	Product grade A min.	3.41	4.32	5.45	6.01	7.66	8.79	11.05	14.38	17.77	20.03	26.75	33.53	39.98	–	–	–	–	–	–
	Product grade B min.	–	–	–	–	–	–	–	–	–	–	26.17	32.95	39.55	50.85	60.79	71.3	82.6	93.56	104.86
k	nom.	1.1	1.4	1.7	2	2.8	3.5	4	5.3	6.4	7.5	10	12.5	15	18.7	22.5	26	30	35	40
	Product grade A min.	0.975	1.275	1.575	1.875	2.675	3.35	3.85	5.15	6.22	7.32	9.82	12.285	14.785	–	–	–	–	–	–
	Product grade A max.	1.225	1.525	1.825	2.125	2.925	3.65	4.15	5.45	6.58	7.68	10.18	12.715	15.215	–	–	–	–	–	–
	Product grade B min.	–	–	–	–	–	–	–	–	–	–	9.71	12.15	14.65	18.28	22.08	25.58	29.58	34.5	39.5
	Product grade B max.	–	–	–	–	–	–	–	–	–	–	10.29	12.85	15.35	19.12	22.92	26.42	30.42	35.5	40.5
k'^c	Product grade A min.	0.68	0.89	1.1	1.31	1.87	2.35	2.7	3.61	4.35	5.12	6.87	8.6	10.35	–	–	–	–	–	–
	Product grade B min.	–	–	–	–	–	–	–	–	–	–	6.8	8.51	10.26	12.8	15.46	17.91	20.71	24.15	27.65
r	min.	0.1	0.1	0.1	0.1	0.2	0.2	0.25	0.4	0.4	0.6	0.6	0.8	0.8	1	1	1.2	1.6	2	2
s	nom. = max.	3.2	4	5	5.5	7	8	10	13	16	18	24	30	36	46	55	65	75	85	95
	Product grade A min	3.02	3.82	4.82	5.32	6.78	7.78	9.78	12.73	15.73	17.73	23.67	29.67	35.38	–	–	–	–	–	–
	Product grade B min	–	–	–	–	–	–	–	–	–	–	23.16	29.16	35	45	53.8	63.1	73.1	82.8	92.8

	Product grade A		Product grade B	
	l^d			
nom.	min.	max.	min.	max.
2	1.8	2.2	–	–
3	2.8	3.2	–	–
4	3.76	4.24	–	–
5	4.76	5.24	–	–
6	5.76	6.24	–	–
8	7.71	8.29	–	–
10	9.71	10.29	–	–
12	11.65	12.35	–	–
16	15.65	16.35	–	–
20	19.58	20.42	–	–
25	24.58	25.42	–	–
30	29.58	30.42	–	–
35	34.5	35.5	–	–
40	39.5	40.5	–	–
45	44.5	45.5	–	–
50	49.5	50.5	–	–
55	54.4	55.6	–	–
60	59.4	60.6	58.5	61.5
65	64.4	65.6	63.5	66.5
70	69.4	70.6	68.5	71.5
80	79.4	80.6	78.5	81.5
90	89.3	90.7	88.25	91.75
100	99.3	100.7	98.25	101.75
110	109.3	110.7	108.25	111.75

continued

BS EN ISO 4017: 2001 Table 1 Preferred threads (*continued*)

Dimensions in mm

Thread size, d					M1.6	M2	M2.5	M3	M4	M5	M6	M8	M10	M12	M16	M20	M24	M30	M36	M42	M48	M56	M64
	Product grade A		Product grade B																				
	l^{d}																						
nom.	min.	max.	min.	max.																			
120	119.3	120.7	118.25	121.75																			
130	129.2	130.8	128	132																			
140	139.2	140.8	138	142																			
150	149.2	150.8	148	152																			
160	–	–	158	162																			
180	–	–	178	182																			
200	–	–	197.7	202.3																			

[a] P = pitch of the thread.
[b] Values in accordance with a max., normal series, in ISO 3508.
[c] $k'_{min} = 0.7\ k_{min}$.
[d] Range of popular lengths between the stepped lines:

- Product grade A above the stepped line, marked thus – – – –.
- Product grade B below this stepped line.

BS EN ISO 4017: 2001 Table 2 Non-preferred threads

Dimensions in mm

Thread size, d			M3.5	M14	M18	M22	M27	M33	M39	M45	M52	M60
P[a]			0.6	2	2.5	2.5	3	3.5	4	4.5	5	5.5
a		max.[b]	1.8	6	7.5	7.5	9	10.5	12	13.5	15	16.5
		min.	0.6	2	2.5	2.5	3	3.5	4	4.5	5	5.5
c		min.	0.15	0.15	0.2	0.2	0.2	0.2	0.3	0.3	0.3	0.3
		max.	0.4	0.6	0.8	0.8	0.8	0.8	1	1	1	1
d_a		max.	4.1	15.7	20.2	24.4	30.4	36.4	42.4	48.6	56.6	67
d_w	Product grade A	min.	5.07	19.37	25.34	31.71	–	–	–	–	–	–
	Product grade B	min.	–	–	24.85	31.35	38	46.55	55.86	64.7	74.2	83.41
e	Product grade A	min.	6.58	23.36	30.14	37.72	–	–	–	–	–	–
	Product grade B	min.	–	–	29.56	37.29	45.2	55.37	66.44	76.95	88.25	99.21
k		nom.	2.4	8.8	11.5	14	17	21	25	28	33	38
	Product grade A	min.	2.275	8.62	11.285	13.785	–	–	–	–	–	–
		max.	2.525	8.98	11.715	14.215	–	–	–	–	–	–
	Product grade B	min.	–	–	11.15	13.65	16.65	20.58	24.58	27.58	32.5	37.5
		max.	–	–	11.85	14.35	17.35	21.42	25.42	28.42	33.5	38.5
k'[c]	Product grade A	min.	1.59	6.03	7.9	9.65	–	–	–	–	–	–
	Product grade B	min.	–	–	7.81	9.56	11.66	14.41	17.21	19.31	22.75	26.25
r		min.	0.1	0.6	0.6	0.8	1	1	1	1.2	1.6	2
s		nom. = max.	6	21	27	34	41	50	60	70	80	90
	Product grade A	min.	5.82	20.67	26.67	33.38	–	–	–	–	–	–
	Product grade B	min.	–	–	26.16	33	40	49	58.8	68.1	78.1	87.8

BS EN ISO 4017: 2001 Table 2 Non-preferred threads (*continued*)

Dimensions in mm

Thread size, *d*					M3.5	M14	M18	M22	M27	M33	M39	M45	M52	M60
	Product grade A		Product grade B											
	l^d													
nom.	min.	max.	min.	max.										
8	7.71	8.29	–	–										
10	9.71	10.29	–	–										
12	11.65	12.35	–	–										
16	15.65	16.35	–	–										
20	19.58	20.42	–	–										
25	24.58	25.42	–	–										
30	29.58	30.42	–	–										
35	34.5	35.5	–	–										
40	39.5	40.5	–	–										
45	44.5	45.5	–	–										
50	49.5	50.5	–	–										
55	54.4	55.6	53.5	56.5										
60	59.4	60.6	58.5	61.5										
65	64.4	65.6	63.5	66.5										
70	69.4	70.6	68.5	71.5										
80	79.4	80.6	78.5	81.5										
90	89.3	90.7	88.25	91.75										
100	99.3	100.7	98.25	101.75										
110	109.3	110.7	108.25	111.75										
120	119.3	120.7	118.25	121.75										
130	129.2	130.8	128	132										
140	139.2	140.8	138	142										
150	149.2	150.8	148	152										
160	–	–	158	162										
180	–	–	178	182										
200	–	–	197.7	202.3										

[a] P = pitch of the thread.
[b] Values in accordance with a max., normal series, in ISO 3508.
[c] $k'_{min} = 0.7\ k_{min}$.

[d] Range of popular lengths between the stepped lines:
- Product grade A above the stopped line, marked thus – – – –.
- Product grade B below this stepped line.

4.1.16 BS EN ISO 4018: 2001 Hexagon head screws – product grade C

Dimensions

Note: Symbols and descriptions of dimensions are specified in ISO 225.

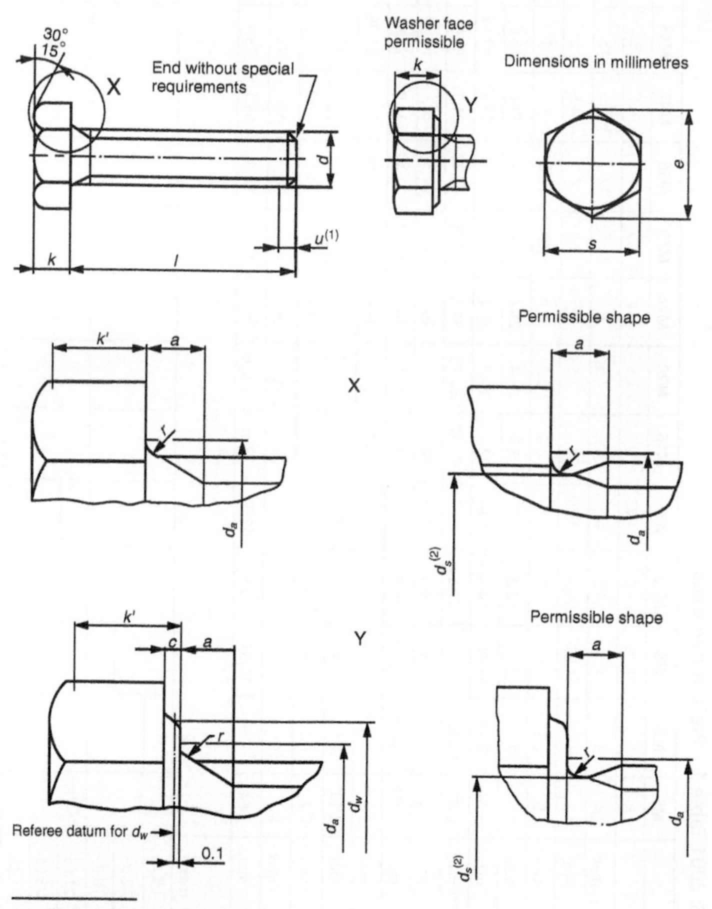

(1) Incomplete thread $u \leqslant 2P$.
(2) d_s = pitch diameter.

Example for the designation of a hexagon head screw with thread M12, nominal length l = 80 mm and property class 4.6:

Hexagon head screw ISO 4018 – M12 × 80 – 4.6

BS EN ISO 4018: 2001 Table 1 Preferred threads

Dimensions in mm

Thread size, *d*			M5	M6	M8	M10	M12	M16	M20	M24	M30	M36	M42	M48	M56	M64
P^a			0.8	1	1.25	1.5	1.75	2	2.5	3	3.5	4	4.5	5	5.5	6
a		max.	2.4	3	4	4.5	5.3	6	7.5	9	10.5	12	13.5	15	16.5	18
		min.	0.8	1	1.25	1.5	1.75	2	2.5	3	3.5	4	4.5	5	5.5	6
c		max.	0.5	0.5	0.6	0.6	0.6	0.8	0.8	0.8	0.8	0.8	1	1	1	1
d_a		max.	6	7.2	10.2	12.2	14.7	18.7	24.4	28.4	35.4	42.4	48.6	56.6	67	75
d_w		min.	6.74	8.74	11.47	14.47	16.47	22	27.7	33.25	42.75	51.11	59.95	69.45	78.66	88.16
e		min.	8.63	10.89	14.2	17.59	19.85	26.17	32.95	39.55	50.85	60.79	71.3	82.6	93.56	104.86
k		nom.	3.5	4	5.3	6.4	7.5	10	12.5	15	18.7	22.5	26	30	35	40
		min.	3.125	3.625	4.925	5.95	7.05	9.25	11.6	14.1	17.65	21.45	24.95	28.95	33.75	38.75
		max.	3.875	4.375	5.675	6.85	7.95	10.75	13.4	15.9	19.75	23.55	27.05	31.05	36.25	41.25
k'^b		min.	2.19	2.54	3.45	4.17	4.94	6.48	8.12	9.87	12.36	15.02	17.47	20.27	23.63	27.13
r		min.	0.2	0.25	0.4	0.4	0.6	0.6	0.8	0.8	1	1	1.2	1.6	2	2
s		nom. = max.	8	10	13	16	18	24	30	36	46	55	65	75	85	95
		min.	7.64	9.64	12.57	15.57	17.57	23.16	29.16	35	45	53.8	63.1	73.1	82.8	92.8
l^c																
nom.	min.	max.														
10	9.25	10.75														
12	11.1	12.9														
16	15.1	16.9														
20	18.95	21.05														
25	23.95	26.05														
30	28.95	31.05														

35	33.75	36.25														
40	38.75	41.25														
45	43.75	46.25														
50	48.75	51.25														
55	53.5	56.5														
60	58.5	61.5														
65	53.5	66.5														
70	68.5	71.5														
80	78.5	81.5														
90	88.25	91.75														
100	98.25	101.75														
110	108.25	111.75														
120	118.25	121.75														
130	128	132														
140	138	142														
150	148	152														
160	156	164														
180	176	184														
200	195.4	204.6														
220	215.4	224.6														
240	235.4	244.6														

continued

BS EN ISO 4018: 2001 Table 1 Preferred threads (*continued*)

Dimensions in mm

Thread size, *d*			M5	M6	M8	M10	M12	M16	M20	M24	M30	M36	M42	M48	M56	M64
260	254.8	265.2														
280	274.8	285.2														
300	294.8	305.2														
320	314.3	325.7														
340	334.3	345.7														
360	354.3	365.7														
380	374.3	385.7														
400	394.3	405.7														
420	413.7	426.3														
440	433.7	446.3														
460	453.7	466.3														
480	473.7	486.3														
500	493.7	506.3														

[a] P = pitch of the thread.
[b] $k'_{min} = 0.7\ k_{min}$.
[c] Range of popular lengths between the stepped lines, marked thus ––.

BS EN ISO 4018: 2001 Table 2 Non-preferred threads

Dimensions in mm

Thread size, d			M14	M18	M22	M27	M33	M39	M45	M52	M60
P^a			2	2.5	2.5	3	3.5	4	4.5	5	5.5
a		max.	6	7.5	7.5	9	10.5	12	13.5	15	16.5
		min.	2	2.5	2.5	3	3.5	4	4.5	5	5.5
c		max.	0.6	0.8	0.8	0.8	0.8	1	1	1	1
d_a		max.	16.7	21.2	26.4	32.4	38.4	45.4	52.6	62.6	71
d_w		min.	19.15	24.85	31.35	38	46.55	55.86	64.7	74.2	83.41
e		min.	22.78	29.56	37.29	45.2	55.37	66.44	76.95	88.25	99.21
k		nom.	8.8	11.5	14	17	21	25	28	33	38
		min.	8.35	10.6	13.1	16.1	19.95	23.95	26.95	31.75	36.75
		max.	9.25	12.4	14.9	17.9	22.05	26.05	29.05	34.25	39.25
k'^b		min.	5.85	7.42	9.17	11.27	13.97	16.77	18.87	22.23	25.73
r		min.	0.6	0.6	0.8	1	1	1	1.2	1.6	2
s		nom. = max.	21	27	34	41	50	60	70	80	90
		min.	20.16	26.16	33	40	49	58.8	68.1	78.1	87.8
l^c nom.	min.	max.									
30	28.95	31.05									
35	33.75	36.25									
40	38.75	41.25									
45	43.75	46.25									
50	48.75	51.25									
55	53.5	56.5									

continued

BS EN ISO 4018: 2001 Table 2 Non-preferred threads (*continued*)

Dimensions in mm

Thread size, *d*			M14	M18	M22	M27	M33	M39	M45	M52	M60
60	58.5	61.5									
65	63.5	66.5									
70	68.5	71.5									
80	78.5	81.5									
90	88.25	91.75									
100	98.25	101.75									
110	108.25	111.75									
120	118.25	121.75									
130	128	132									
140	138	142									
150	148	152									
160	156	164									
180	176	184									
200	195.4	204.6									
220	215.4	224.6									
240	235.4	244.6									
260	254.8	265.2									
280	274.8	285.2									
300	294.8	305.2									
320	314.3	325.7									
340	334.3	345.7									

360	354.3	365.7									
380	374.3	385.7									
400	394.3	405.7									
420	413.7	426.3									
440	433.7	446.3									
460	453.7	466.3									
480	473.7	486.3									
500	493.7	506.3									

[a] P = pitch of the thread.
[b] $k'_{min} = 0.7\ k_{min}$.
[c] Range of popular lengths between the stepped lines, marked thus —— .

4.1.17 BS EN ISO 4032: 2001 Hexagon nuts style 1 – product grades A and B

Dimensions

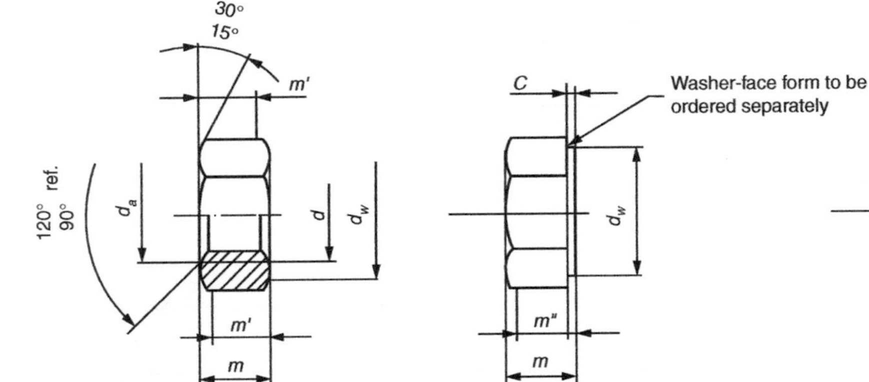

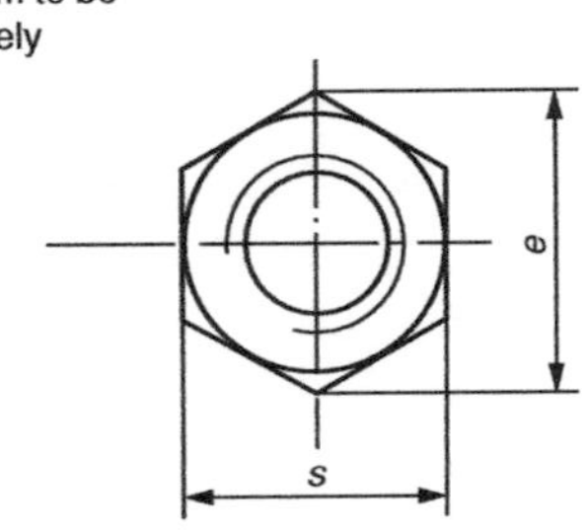

Example for the designation of a hexagon nut, style 1, with thread size d = M12 and property class 8:

Hexagon nut ISO 4032 – M12 – 8

BS EN ISO 4032: 2001 Table 1 Preferred sizes

Dimensions in mm

Thread size, d		M1.6	M2	M2.5	M3	M4	M5	M6	M8	M10	M12
P^{a}		0.35	0.4	0.45	0.5	0.7	0.8	1	1.25	1.5	1.75
c	max.	0.2	0.2	0.3	0.4	0.4	0.5	0.5	0.6	0.6	0.6
	min.	0.1	0.1	0.1	0.15	0.15	0.15	0.15	0.15	0.15	0.15
d_a	min.	1.6	2	2.5	3	4	5	6	8.	10	12
	max.	1.84	2.3	2.9	3.45	4.6	5.75	6.75	8.75	10.8	13
d_w	min.	2.4	3.1	4.1	4.6	5.9	6.9	8.9	11.6	14.6	16.6
e	min.	3.41	4.32	5.45	6.01	7.66	8.79	11.05	14.38	17.77	20.03
m	max.	1.3	1.6	2	2.4	3.2	4.7	5.2	6.8	8.4	10.8
	min.	1.05	1.35	1.75	2.15	2.9	4.4	4.9	6.44	8.04	10.37
m'	min.	0.8	1.1	1.4	1.7	2.3	3.5	3.9	5.2	6.4	8.3
m''	min.	0.7	1	1.2	1.5	2	3.1	3.4	4.5	5.6	7.3
s	nom. = max.	3.2	4	5	5.5	7	8	10	13	16	18
	min.	3.02	3.82	4.82	5.32	6.78	7.78	9.78	12.73	15.73	17.73

continued

BS EN ISO 4032: 2001 Table 1 Preferred sizes (*continued*)

Dimensions in mm

Thread size, *d*		M16	M20	M24	M30	M36	M42	M48	M56	M64
P[a]		2	2.5	3	3.5	4	4.5	5	5.5	6
c	max.	0.8	0.8	0.8	0.8	0.8	1	1	1	1
	min.	0.2	0.2	0.2	0.2	0.2	0.3	0.3	0.3	0.3
d_a	min.	16	20	24	30	36	42	48	56	64
	max.	17.3	21.6	25.9	32.4	38.9	45.4	51.8	60.5	69.1
d_w	min.	22.5	27.7	33.3	42.8	51.1	60	69.5	78.7	88.2
e	min.	26.75	32.95	39.55	50.85	60.79	71.3	82.6	93.56	104.86
m	max.	14.8	18	21.5	25.6	31	34	38	45	51
	min.	14.1	16.9	20.2	24.3	29.4	32.4	36.4	43.4	49.1
m'	min.	11.3	13.5	16.2	19.4	23.5	25.9	29.1	34.7	39.3
m''	min.	9.9	11.8	14.1	17	20.6	22.7	25.5	30.4	34.4
s	nom. = max.	24	30	36	46	55	65	75	85	95
	min.	23.67	29.16	35	45	53.8	63.1	73.1	82.8	92.8

[a] P = pitch of the thread.

BS EN ISO 4032: 2001 Table 2 Non-preferred sizes

Dimensions in mm

Thread size, *d*		M3.5	M14	M18	M22	M27	M33	M39	M45	M52	M60
P[a]		0.6	2	2.5	2.5	3	3.5	4	4.5	5	5.5
c	max.	0.4	0.6	0.8	0.8	0.8	0.8	1	1	1	1
	min.	0.15	0.15	0.2	0.2	0.2	0.2	0.3	0.3	0.3	0.3
d_a	min.	3.5	14	18	22	27	33	39	45	52	60
	max.	4	15.1	19.5	23.7	29.1	35.6	42.1	48.6	56.2	64.8
d_w	min.	5	19.6	24.9	31.4	38	46.6	55.9	64.7	74.2	83.4
e	min.	6.58	23.35	29.56	37.29	45.2	55.37	66.44	76.95	88.25	99.21
m	max.	2.8	12.8	15.8	19.4	23.8	28.7	33.4	36	42	48
	min.	2.55	12.1	15.1	18.1	22.5	27.4	31.8	34.4	40.4	46.4
m'	min.	2	9.7	12.1	14.5	18	21.9	25.4	27.5	32.3	37.1
m''	min.	1.8	8.5	10.6	12.7	15.8	19.2	22.3	24.1	28.3	32.5
s	nom. = max.	6	21	27	34	41	50	60	70	80	90
	min.	5.82	20.67	26.16	33	40	49	58.8	68.1	78.1	87.8

[a] P = pitch of the thread.

4.1.18 BS EN ISO 4033: 2001 Hexagon nuts style 2 – product grades A and B

Dimensions

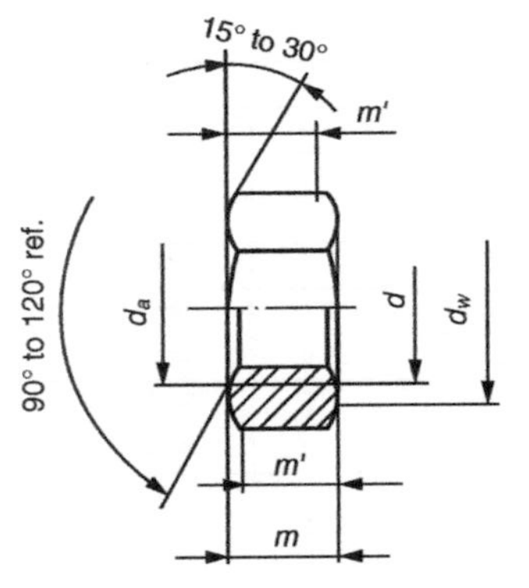

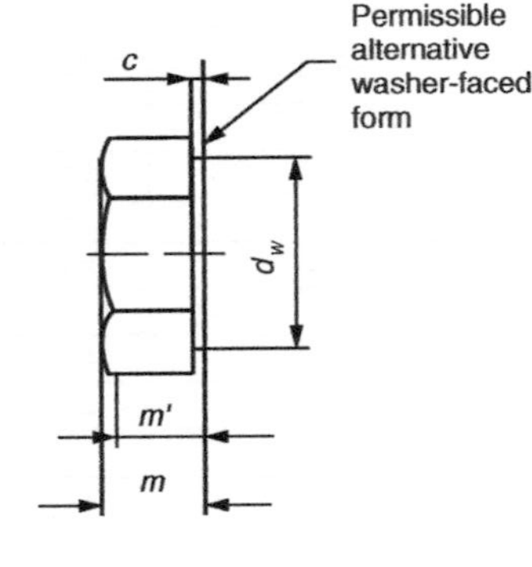

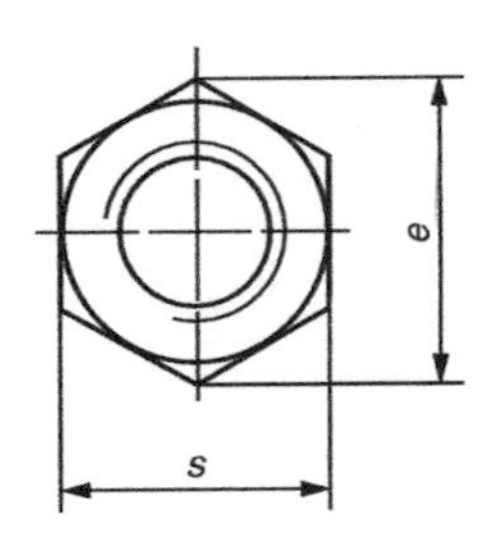

Example for the designation of a hexagon nut with metric thread size d = M12 and property class 9:

Hexagon nut ISO 4033 – M12 – 9

Dimensions in mm

Thread size, d		M5	M6	M8	M10	M12	(M14)	M16	M20	M24	M30	M36
P^a		0.8	1	1.25	1.5	1.75	2	2	2.5	3	3.5	4
c	max.	0.5	0.5	0.6	0.6	0.6	0.6	0.8	0.8	0.8	0.8	0.8
d_a	min.	5	6	8	10	12	14	16	20	24	30	36
	max.	5.75	6.75	8.75	10.8	13	15.1	17.3	21.6	25.9	32.4	38.9
d_w	min.	6.9	8.9	11.6	14.6	16.6	19.6	22.5	27.7	33.2	42.7	51.1
e	min.	8.79	11.05	14.38	17.77	20.03	23.35	26.75	32.95	39.55	50.85	60.79
m	max.	5.1	5.7	7.5	9.3	12	14.1	16.4	20.3	23.9	28.6	34.7
	min.	4.8	5.4	7.14	8.94	11.57	13.4	15.7	19	22.6	27.3	33.1
m'	min.	3.84	4.32	5.71	7.15	9.26	10.7	12.6	15.2	18.1	21.8	26.5
s	max.	8	10	13	16	18	21	24	30	36	46	55
	min.	7.78	9.78	12.73	15.73	17.73	20.67	23.67	29.16	35	45	53.8

Sizes in brackets should be avoided if possible.
[a] P = pitch of the thread.

BS EN ISO 4033: 2001 Table 1

Nominal thread diameter (mm)	Width across flats (mm)	Annular bearing area / Thread stress area *
5	8	1.08
6	10	1.44
8	13	1.23
10	15	0.90
	16	1.30
	17	1.73
12	18	0.91
	19	1.16
14	21	0.96
	22	1.24
16	24	1.02
20	30	0.95
24	36	0.86
30	46	1.02
36	55	1.04

* Calculation based on clearance holes ISO 273 (revised), medium series.

BS EN ISO 4033: 2001 Table 2

Thread size, d		M10		M12	M14
P[a]		1.5		1.75	2
d_w	min.	13.6	15.6	17.4	20.4
e	min.	16.64	18.90	21.10	24.49
m	max.	10	8.8	11.3	13.13
	min.	9.64	8.44	10.87	12.7
m'	min.	7.7	6.75	8.7	10.2
s	max.	15	17	19	22
	min.	14.73	16.73	18.67	21.67

[a] P = pitch of the thread.

4.1.19 BS EN ISO 4034: 2001 Hexagon nuts style 1 – product grade C

Dimensions

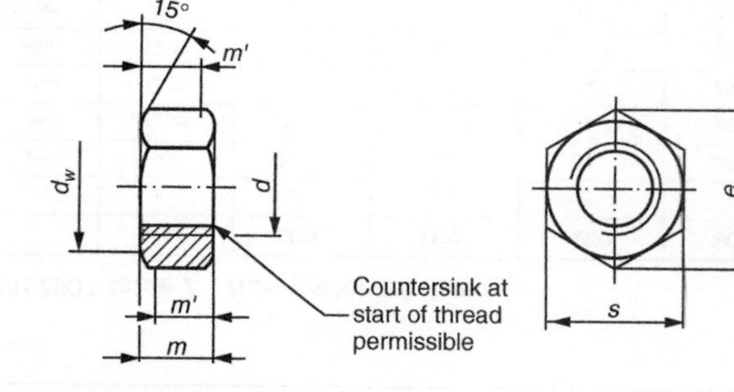

Example for the designation of a hexagon nut with thread size d = M12 and property class 5:

Hexagon nut ISO 4034 – M12 – 5

BS EN ISO 4034: 2001 Table 1 Preferred sizes

Dimensions in mm

Thread size, d		M5	M6	M8	M10	M12	M16	M20
P^a		0.8	1	1.25	1.5	1.75	2	2.5
d_w	min.	6.7	8.7	11.5	14.5	16.5	22	27.7
e	min.	8.63	10.89	14.20	17.59	19.85	26.17	32.95
m	max.	5.6	6.1	7.9	9.5	12.2	15.9	19
	min.	4.4	4.6	6.4	8	10.4	14.1	16.9
m'	min.	3.5	3.7	5.1	6.4	8.3	11.3	13.5
s	nom. = max.	8	10	13	16	18	24	30
	min.	7.64	9.64	12.57	15.57	17.57	23.16	29.16

continued

BS EN ISO 4034: 2001 Table 1 Preferred sizes (*continued*)

Dimensions in mm

Thread size, *d*		M24	M30	M36	M42	M48	M56	M64
P^a		3	3.5	4	4.5	5	5.5	6
d_w	min.	33.3	42.8	51.1	60	69.5	78.7	88.2
e	min.	39.55	50.85	60.79	72.02	82.6	93.56	104.86
m	max.	22.3	26.4	31.5	34.9	38.9	45.9	52.4
	min.	20.2	24.3	28	32.4	36.4	43.4	49.4
m'	min.	16.2	19.5	22.4	25.9	29.1	34.7	39.5
s	nom. = max.	36	46	55	65	75	85	95
	min.	35	45	53.8	63.1	73.1	82.8	92.8

[a] P = pitch of the thread.

BS EN ISO 4034: 2001 Table 2 Non-preferred sizes

Dimensions in mm

Thread size, *d*		M14	M18	M22	M27	M33	M39	M45	M52	M60
P^a		2	2.5	2.5	3	3.5	4	4.5	5	5.5
d_w	min.	19.2	24.9	31.4	38	46.6	55.9	64.7	74.2	83.4
e	min.	22.78	29.56	37.29	45.2	55.37	66.44	76.95	88.25	99.21
m	max.	13.9	16.9	20.2	24.7	29.5	34.3	36.9	42.9	48.9
	min.	12.1	15.1	18.1	22.6	27.4	31.8	34.4	40.4	46.4
m'	min.	9.7	12.1	14.5	18.1	21.9	25.4	27.5	32.3	37.1
s	nom. = max.	21	27	34	41	50	60	70	80	90
	min.	20.16	26.16	33	40	49	58.8	68.1	78.1	87.8

[a] P = pitch of the thread.

4.1.20 BS EN ISO 4035: 2001 Hexagon thin nuts (chamfered) – product grades A and B

Dimensions

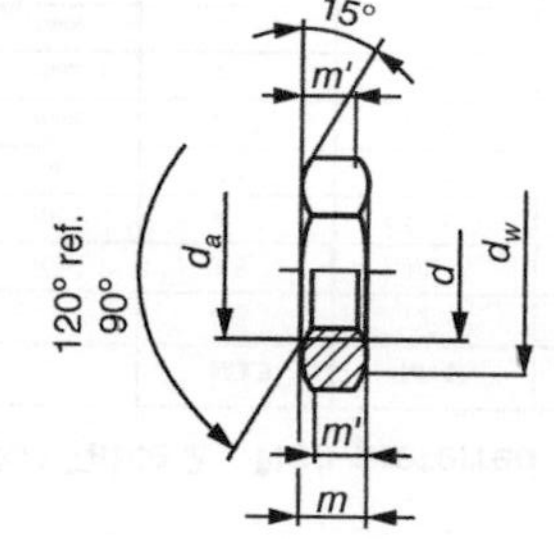

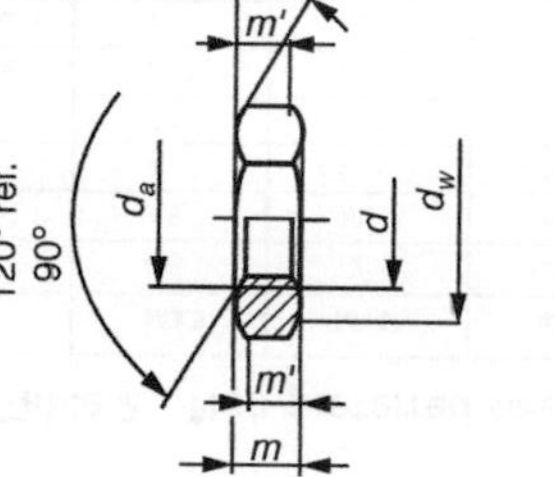

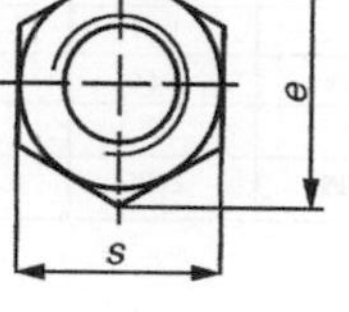

Example for the designation of a chamfered thin nut with thread size d = M12 and property class 05:

Hexagon thin nut ISO 4035 – M12 – 05

BS EN ISO 4035: 2001 Table 1 Preferred sizes

Dimensions in mm

Thread size, *d*		M1.6	M2	M2.5	M3	M4	M5	M6	M8	M10	M12
P^a		0.35	0.4	0.45	0.5	0.7	0.8	1	1.25	1.5	1.75
d_a	min.	1.6	2	2.5	3	4	5	6	8	10	12
	max.	1.84	2.3	2.9	3.45	4.6	5.75	6.75	8.75	10.8	13
d_w	min.	2.4	3.1	4.1	4.6	5.9	6.9	8.9	11.6	14.6	16.6
e	min.	3.41	4.32	5.45	6.01	7.66	8.79	11.05	14.38	17.77	20.03
m	max.	1	1.2	1.6	1.8	2.2	2.7	3.2	4	5	6
	min.	0.75	0.95	1.35	1.55	1.95	2.45	2.9	3.7	4.7	5.7
m'	min.	0.6	0.8	1.1	1.2	1.6	2	2.3	3	3.8	4.6
s	nom. = max.	3.2	4	5	5.5	7	8	10	13	16	18
	min.	3.02	3.82	4.82	5.32	6.78	7.78	9.78	12.73	15.73	17.73

continued

BS EN ISO 4035: 2001 Table 1 Preferred sizes (*continued*)

Dimensions in mm

Thread size, *d*		M16	M20	M24	M30	M36	M42	M48	M56	M64
P[a]		2	2.5	3	3.5	4	4.5	5	5.5	6
d_a	min.	16	20	24	30	36	42	48	56	64
	max.	17.3	21.6	25.9	32.4	38.9	45.4	51.8	60.5	69.1
d_w	min.	22.5	27.7	33.2	42.8	51.1	60	69.5	78.7	88.2
e	min.	26.75	32.95	39.55	50.85	60.79	71.3	82.6	93.56	104.86
m	max.	8	10	12	15	18	21	24	28	32
	min.	7.42	9.10	10.9	13.9	16.9	19.7	22.7	26.7	30.4
m'	min.	5.9	7.3	8.7	11.1	13.5	15.8	18.2	21.4	24.3
s	nom. = max.	24	30	36	46	55	65	75	85	95
	min.	23.67	29.16	35	45	53.8	63.1	73.1	82.8	92.8

[a] P = pitch of the thread.

BS EN ISO 4035: 2001 Table 2 Non-preferred sizes

Dimensions in mm

Thread size, *d*		M3.5	M14	M18	M22	M27	M33	M39	M45	M52	M60
P[a]		0.6	2	2.5	2.5	3	3.5	4	4.5	5	5.5
d_a	min.	3.5	14	18	22	27	33	39	45	52	60
	max.	4	15.1	19.5	23.7	29.1	35.6	42.1	48.6	56.2	64.8
d_w	min.	5.1	19.6	24.9	31.4	38	46.6	55.9	64.7	74.2	83.4
e	min.	6.58	23.35	29.56	37.29	45.2	55.37	66.44	76.95	88.25	99.21
m	max.	2	7	9	11	13.5	16.5	19.5	22.5	26	30
	min.	1.75	6.42	8.42	9.9	12.4	15.4	18.2	21.2	24.7	28.7
m'	min.	1.4	5.1	6.7	7.9	9.9	12.3	14.6	17	19.8	23
s	nom. = max.	6	21	27	34	41	50	60	70	80	90
	min.	5.82	20.67	26.16	33	40	49	58.8	68.1	78.1	87.8

[a] P = pitch of the thread.

4.1.21 BS EN ISO 4036: 2001 Hexagon thin nuts (unchamfered) – product grade B

Dimensions

Dimensions in millimetres

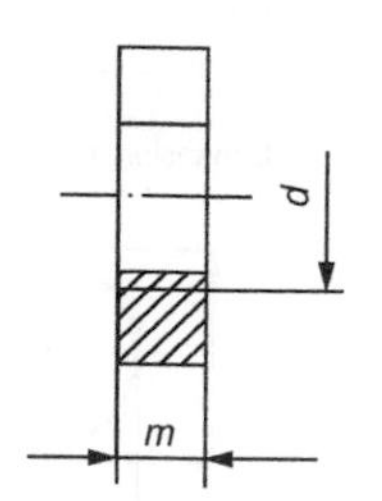

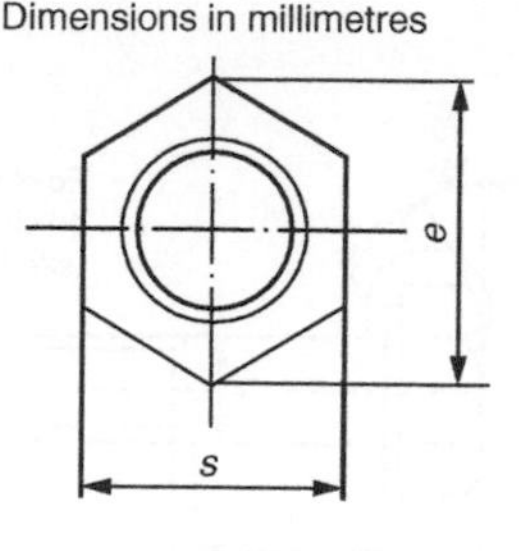

Example for the designation of a hexagon thin nut with metric thread d = M6 made from steel with HV 110 min. (st):

Hexagon nut ISO 4036 – M6-st

BS EN ISO 4036: 2001 Table 1

Nominal thread diameter (mm)	Width across flats (mm)	Annular bearing area / Thread stress area *
5	8	1.08
6	10	1.44
8	13	1.23
10	15	0.90
	16	1.30
	17	1.73

* Calculation based on clearance holes ISO 273 (revised), medium series.

BS EN ISO 4036: 2001 Table 2

Thread size, *d*		M10	
P^a		1.5	
e	min.	16.46	18.73
m	max.	5	5
	min.	4.52	4.52
s	max.	15	17
	min.	14.57	16.57

[a] P = pitch of the thread.

4.1.22 BS EN ISO 8765: 2001 Hexagon head bolts with metric fine pitch threads – product grades A and B

Dimensions

Note: Symbols and descriptions of dimensions are specified in ISO 225.

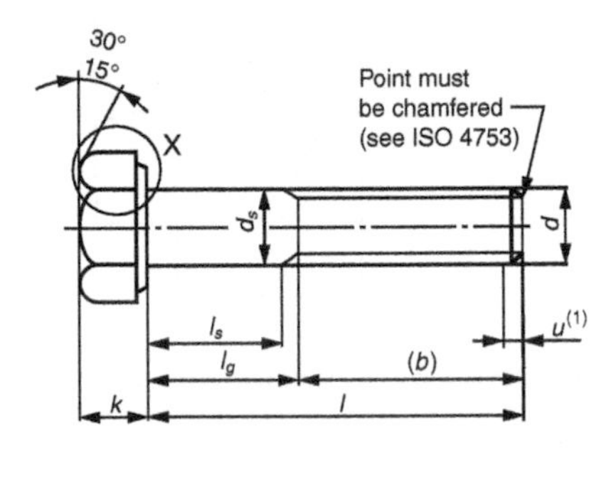

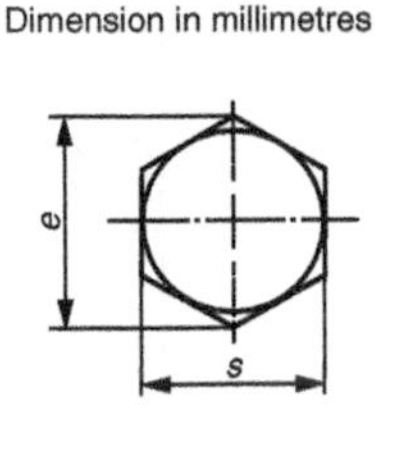

X

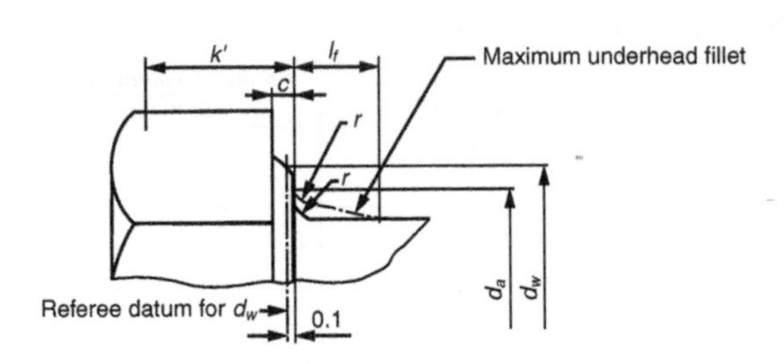

(1) Incomplete thread $u \leq 2P$.

BS EN ISO 8765: 2001 Table 1 Preferred threads

Dimensions in mm

Thread size, $d \times P$			M8 × 1	M10 × 1	M12 × 1.5	M16 × 1.5	M20 × 1.5	M24 × 2	M30 × 2	M36 × 3	M42 × 3	M48 × 3	M56 × 4	M64 × 4
b ref.		–[a]	22	26	30	38	46	54	66	–	–	–	–	–
		–[b]	–	–	–	44	52	60	72	84	96	108	–	–
		–[c]	–	–	–	–	–	73	85	97	109	121	137	153
c		min.	0.15	0.15	0.15	0.2	0.2	0.2	0.2	0.2	0.3	0.3	0.3	0.3
		max.	0.6	0.6	0.6	0.8	0.8	0.8	0.8	0.8	1	1	1	1
d_a		max.	9.2	11.2	13.7	17.7	22.4	26.4	33.4	39.4	45.6	52.6	63	71
d_s		nom. = max.	8	10	12	16	20	24	30	36	42	48	56	64
	Product grade A	min.	7.78	9.78	11.73	15.73	19.67	23.67	–	–	–	–	–	–
	Product grade B	min.	–	–	–	15.57	19.48	23.48	29.48	35.8	41.38	47.38	55.26	63.26
d_w	Product grade A	min.	11.63	14.63	16.63	22.49	28.19	33.61	–	–	–	–	–	–
	Product grade B	min.	–	–	–	22	27.7	33.25	42.75	51.11	59.95	69.45	78.66	88.16
e	Product grade A	min.	14.38	17.77	20.03	26.75	33.53	39.98	–	–	–	–	–	–
	Product grade B	min.	–	–	–	26.17	32.95	39.55	50.85	60.79	71.3	82.6	93.56	104.86
l_f		max.	2	2	3	3	4	4	6	6	8	10	12	13
k		nom.	5.3	6.4	7.5	10	12.5	15	18.7	22.5	26	30	35	40
	Product grade A	min.	5.15	6.22	7.32	9.82	12.285	14.785	–	–	–	–	–	–
		max.	5.45	6.58	7.68	10.18	12.715	15.215	–	–	–	–	–	–
	Product grade B	min.	–	–	–	9.71	12.15	14.65	18.28	22.08	25.58	29.58	34.5	39.5
		max.	–	–	–	10.29	12.85	15.35	19.12	22.92	26.42	30.42	35.5	40.5
k'[d]	Product grade A	min.	3.61	4.35	5.12	6.87	8.6	10.35	–	–	–	–	–	–
	Product grade B	min.	–	–	–	6.8	8.51	10.26	12.8	15.46	17.91	20.71	24.15	27.65
r		min.	0.4	0.4	0.6	0.6	0.8	0.8	1	1	1.2	1.6	2	2
s		nom. = max.	13	16	18	24	30	36	46	55	65	75	85	95
	Product grade A	min.	12.73	15.73	17.73	23.67	29.67	35.38	–	–	–	–	–	–
	Product grade B	min.	–	–	–	23.16	29.16	35	45	53.8	63.1	73.1	82.8	92.8

continued

BS EN ISO 8765: 2001 Table 1 Preferred threads (*continued*)

Dimensions in mm

Thread size, $d \times P$					M8 × 1		M10 × 1		M12 × 1.5		M16 × 1.5		M20 × 1.5		M24 × 2		M30 × 2		M36 × 3		M42 × 3		M48 × 3		M56 × 4		M64 × 4	
	Product grade				l_s and l_g[e,f]																							
	A		B																									
l					l_s	l_g	l_s	l_g	l_s	l_g	l_s	l_g	l_s	l_g	l_s	l_g	l_s	l_g	l_s	l_g	l_s	l_g	l_s	l_g	l_s	l_g	l_s	l_g
nom.	min.	max.	min.	max.	min.	max.	min.	max.	min.	max.	min.	max.	min.	max.	min.	max.	min.	max.	min.	max.	min.	max.	min.	max.	min.	max.	min.	max.
35	34.5	35.5	–	–																								
40	39.5	40.5	–	–	11.75	18																						
45	44.5	45.5	–	–	16.75	23	11.5	19																				
50	49.5	50.5	–	–	21.75	28	16.5	24	11.25	20																		
55	54.4	55.6	–	–	26.75	33	21.5	29	16.25	25																		
60	59.4	60.6	–	–	31.75	38	26.5	34	21.25	30																		
65	64.4	65.6	–	–	36.75	43	31.5	39	26.25	35	17	27																
70	69.4	70.6	–	–	41.75	48	36.5	44	31.25	40	22	32																
80	79.4	80.6	–	–	51.75	58	46.5	54	41.25	50	32	42	21.5	34														
90	89.3	90.7	–	–			56.5	64	51.25	60	42	52	31.5	44														
100	99.3	100.7	–	–			66.5	74	61.25	70	52	62	41.5	54	31	46												
110	109.3	110.7	–	–					71.25	80	62	72	51.5	64	41	56												
120	119.3	120.7	118.25	121.75					81.25	90	72	82	61.5	74	51	66	36.5	54										
130	129.2	130.8	128	132							76	86	65.5	78	55	70	40.5	58										
140	139.2	140.8	138	142							86	96	75.5	88	65	80	50.5	68	36	56								
150	149.2	150.8	148	152							96	106	85.5	98	75	90	60.5	78	46	66								
160	–	–	158	162							106	116	95.5	108	85	100	70.5	88	56	76	41.5	64						
180	–	–	178	182									115.5	128	105	120	90.5	108	76	96	61.5	84						
200	–	–	197.7	202.3									135.5	148	125	140	110.5	128	96	116	81.5	104	67	92				
220	–	–	217.7	222.3											132	147	117.5	135	103	123	88.5	111	74	99	55.5	83		
240	–	–	237.7	242.3											152	167	137.5	155	123	143	108.5	131	94	119	75.5	103		

For sizes above the stepped line marked thus ———, ISO 8676 is recommended.

260	–	–	257.4	262.6													157.5	175	143	163	128.5	151	114	139	95.5	123	77	107
280	–	–	277.4	282.6													177.5	195	163	183	148.5	171	134	159	115.5	143	97	127
300	–	–	297.4	302.6													197.5	215	183	203	168.5	191	154	179	135.5	163	117	147
320	–	–	317.15	322.85															203	223	188.5	211	174	199	155.5	183	137	167
340	–	–	337.15	342.85															223	243	208.5	231	194	219	175.5	203	157	187
360	–	–	357.15	362.85															243	263	228.5	251	214	239	195.9	223	177	207
380	–	–	377.15	382.85																	248.5	271	234	259	215.5	243	197	227
400	–	–	397.15	402.85																	268.5	291	254	279	235.5	263	217	247
420	–	–	416.85	423.15																	288.5	311	274	299	255.5	283	237	267
440	–	–	436.85	443.15																	308.5	331	294	319	275.5	303	257	287
460	–	–	456.85	463.15																			314	339	295.5	323	277	307
480	–	–	476.85	483.15																			334	359	315.5	343	297	327
500	–	–	496.85	503.15																					335.5	363	317	347

[a] For lengths $l_{nom} < 125$ mm.
[b] For lengths 125 mm $< l_{nom} < 200$ mm.
[c] For lengths $l_{nom} > 200$ mm.
[d] $k'_{min} = 0.7\ k_{min}$.
[e] $l_{g\,max} = l_{nom} - b$; $l_{s\,min} = l_{g\,max} - 5P$.
P = pitch of the coarse thread, specified in ISO 261.
[f] l_g = the minimum grip length.

Notes:

- The popular lengths are defined in terms of lengths l_s and l_g:
  - product grade A above the stepped line, marked thus – – – – –.
  - product grade B below this stepped line.
- Thread sizes M10 ×1 and M12 ×1.5 are popular sizes but not included in ISO 262.

BS EN ISO 8765: 2001 Table 2 Non-preferred threads

Dimensions in mm

Thread size, $d \times P$		M10 × 1.25	M12 × 1.25	M14 × 1.5	M18 × 1.5	M20 × 2	M22 × 1.5	M27 × 2	M33 × 2	M39 × 3	M45 × 3	M52 × 4	M60 × 4
b ref.	_[a]	26	30	34	42	46	50	60	–	–	–	–	–
	_[b]	–	–	40	48	52	56	66	78	90	102	116	–
	_[c]	–	–	–	–	–	69	79	91	103	115	129	145
c	min.	0.15	0.15	0.15	0.2	0.2	0.2	0.2	0.2	0.3	0.3	0.3	0.3
	max.	0.6	0.6	0.6	0.8	0.8	0.8	0.8	0.8	1	1	1	1
d_a	max.	11.2	13.7	15.7	20.2	22.4	24.4	30.4	36.4	42.4	48.6	56.6	67
d_s	nom. = max.	10	12	14	18	20	22	27	33	39	45	52	60
	Product grade A min.	9.78	11.73	13.73	17.73	19.67	21.67	–	–	–	–	–	–
	Product grade B min.	–	–	–	17.57	19.48	21.48	26.48	32.38	38.38	44.38	51.26	59.26
d_w	Product grade A min.	14.63	16.63	19.37	25.34	28.19	31.71	–	–	–	–	–	–
	Product grade B min.	–	–	–	24.85	27.7	31.35	38	46.55	55.86	64.7	74.2	83.41
e	Product grade A min.	17.77	20.03	23.36	30.14	33.53	37.72	–	–	–	–	–	–
	Product grade B min.	–	–	–	29.56	32.95	37.29	45.2	55.37	66.44	76.95	88.25	99.21
l_f	max.	2	3	3	3	4	4	6	6	6	8	10	12
k	nom.	6.4	7.5	8.8	11.5	12.5	14	17	21	25	28	33	38
	Product grade A min.	6.22	7.32	8.62	11.285	12.285	13.785	–	–	–	–	–	
	Product grade A max.	6.58	7.68	8.98	11.715	12.715	14.215	–	–	–	–	–	–
	Product grade B min.	–	–	–	11.15	12.15	13.65	16.65	20.58	24.58	27.58	32.5	37.5
	Product grade B max.	–	–	–	11.85	12.85	14.35	17.35	21.42	25.42	28.42	33.5	38.5
k'^{d}	Product grade A min.	4.35	5.12	6.03	7.9	8.6	9.65	–	–	–	–	–	–
	Product grade B min.	–	–	–	7.81	8.51	9.56	11.66	14.41	17.21	1 9.31	22.75	26.25
r	min.	0.4	0.6	0.6	0.6	0.8	0.8	1	1	1.	1.2	1.6	2
s	nom. = max.	16	18	21	27	30	34	41	50	60	70	80	90
	Product grade A min.	15.73	17.73	20.67	26.67	29.67	33.38	–	–	–	–	–	–
	Product grade B min.	–	–	–	26.16	29.16	33	40	49	58.8	68.1	78.1	87.8

	Product grade				l_s and l_g [e,f]																							
	A		B																									
	l				l_s	l_g	l_s	l_g	l_s	l_g	l_s	l_g	l_s	l_g	l_s	l_g	l_s	l_g	l_s	l_g	l_s	l_g	l_s	l_g	l_s	l_g	l_s	l_g
nom.	min.	max.	min.	max.	min.	max.	min.	max.	min.	max.	min.	max.	min.	max.	min.	max.	min.	max.	min.	max.	min.	max.	min.	max.	min.	max.	min.	max.
45	44.5	45.5	–	–	11.5	19																						
50	49.5	50.5	–	–	16.5	24	11.25	20																				
55	54.4	55.6	–	–	21.5	29	16.25	25																				
60	59.4	60.6	–	–	26.5	34	21.25	30	16	26																		
65	64.4	65.6	–	–	31.5	39	26.25	35	21	31																		
70	69.4	70.6	–	–	36.5	44	31.25	40	26	36	15.5	28																
80	79.4	80.6	–	–	46.5	54	41.25	50	36	46	25.5	38	21.5	34														
90	89.3	90.7	–	–	56.5	64	51.25	60	46	56	35.5	48	31.5	44	27.5	40												
100	99.3	100.7	–	–	66.5	74	61.25	70	56	66	45.5	58	41.5	54	37.5	50												
110	109.3	110.7	108.25	111.75			71.25	80	66	76	55.5	68	51.5	64	47.5	60	35	50										
120	119.3	120.7	118.25	121.75			81.25	90	76	86	65.5	78	61.5	74	57.5	70	45	60										
130	129.2	130.8	128	132					80	90	69.5	82	65.5	78	61.5	74	49	64	34.5	52								
140	139.2	140.8	138	142					90	100	79.5	92	75.5	88	71.5	84	59	74	44.5	62								
150	149.2	150.8	148	152							89.5	102	85.5	98	81.5	94	69	84	54.5	72	40	60						
160	–	–	158	162							99.5	122	95.5	108	91.5	104	79	94	64.5	82	50	70						
180	–	–	178	182							119.5	132	115.5	128	111.5	124	99	114	84.5	102	70	90	55.5	78				
200	–	–	197.7	202.3									135.5	148	131.5	144	119	134	104.5	122	90	110	75.5	98	59	84		
220	–	–	217.7	222.3											138.5	151	126	141	111.5	129	97	117	82.5	105	66	91		
240	–	–	237.7	242.3													146	161	131.5	149	117	137	102.5	125	86	111	67.5	95
260	–	–	257.4	262.6													166	181	151.5	169	137	157	122.5	145	106	131	87.5	115
280	–	–	277.4	282.6															171.5	189	157	177	142.5	165	126	151	107.5	135

For sizes above the stepped line marked thus ———, ISO 8676 is recommended.

continued

BS EN ISO 8765: 2001 Table 2 Non-preferred threads (*continued*)

Dimensions in mm

Thread size, $d \times P$					M10 × 1.25		M12 × 1.25		M14 × 1.5		M18 × 1.5		M20 × 2		M22 × 1.5		M27 × 2		M33 × 2		M39 × 3		M45 × 3		M52 × 4		M60 × 4	
	l Product grade A		l Product grade B		l_s and l_g [e,f]																							
nom.	min.	max.	min.	max.	l_s min.	l_g max.	l_s min.	l_g max.	l_s min.	l_g max.	l_s min.	l_g max.	l_s min.	l_g max.	l_s min.	l_g max.	l_s min.	l_g max.	l_s min.	l_g max.	l_s min.	l_g max.	l_s min.	l_g max.	l_s min.	l_g max.	l_s min.	l_g max.
300	–	–	297.4	302.6															191.5	209	177	197	162.5	185	146	171	127.5	155
320	–	–	317.15	322.85															211.5	229	197	217	182.5	205	166	191	147.5	175
340	–	–	337.15	342.85																	217	237	202.5	225	186	211	167.5	195
360	–	–	357.15	362.85																	237	257	222.5	245	206	231	187.5	215
380	–	–	377.15	382.85																	257	277	242.5	265	226	251	207.5	235
400	–	–	397.15	402.85																			262.5	285	246	271	227.5	255
420	–	–	416.85	423.15																			282.5	305	266	291	247.5	275
440	–	–	436.85	443.15																			302.5	325	286	311	267.5	295
460	–	–	456.85	463.15																					306	331	287.5	315
480	–	–	476.85	483.15																					326	351	307.5	335
500	–	–	496.85	503.15																							327.5	355

[a] For lengths $l_{nom} < 125$ mm.
[b] For lengths $125 \text{ mm} < l_{nom} < 200$ mm.
[c] For lengths $l_{nom} > 200$ mm.
[d] $k'_{min} = 0.7\ k_{min}$.
[e] $l_{g\,max} = l_{nom} - b$; $l_{s\,min} = l_{g\,max} - 5P$.
P = pitch of the coarse thread, specified in ISO 261.
[f] l_g is the minimum grip length.

Notes: The popular lengths are defined in terms of lengths l_s and l_g:

- product grade A above the stepped line, marked thus – – – – –.
- product grade B below this stepped line.

4.1.23 BS EN ISO 8676: 2001 Hexagon head screws with metric fine pitch threads – product grades A and B

Dimensions

Note: Symbols and descriptions of dimensions are specified in ISO 225.

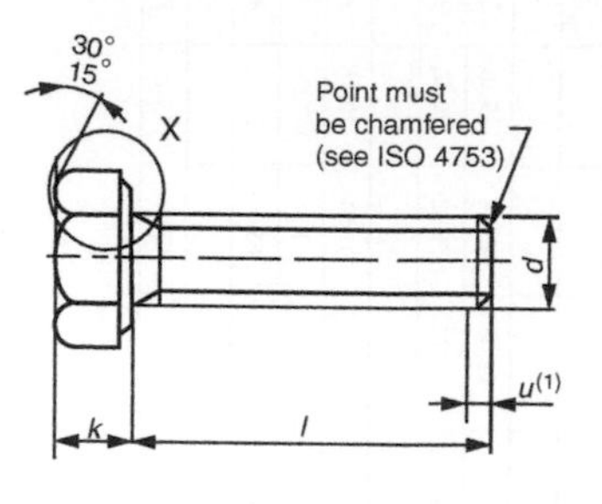

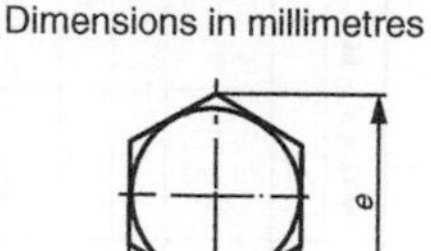

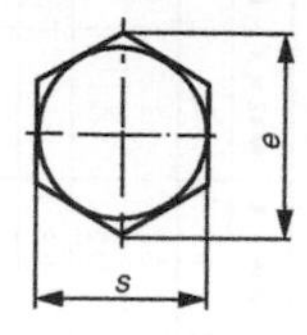

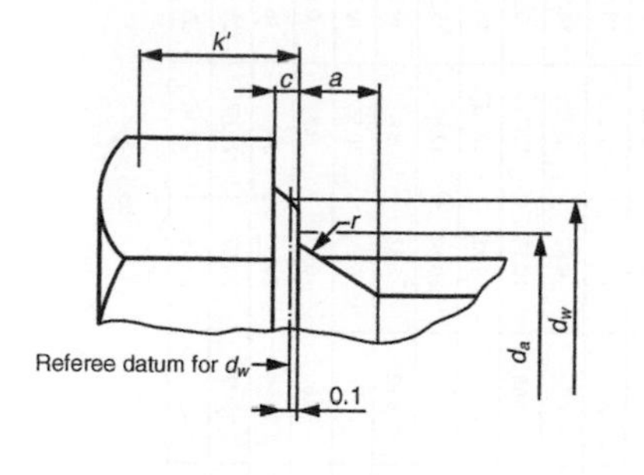

Permissible shape

a

r

$d_s^{(2)}$

d_a

(1) Incomplete thread $u \leqslant 2P$.

(2) d_s = pitch diameter.

BS EN ISO 8676: 2001 Table 1 Preferred threads

Dimensions in mm

Thread size, $d \times P$			M8 × 1	M10 × 1	M12 × 1.5	M16 × 1.5	M20 × 1.5	M24 × 2	M30 × 2	M36 × 3	M42 × 3	M48 × 3	M56 × 4	M64 × 4
a		max.	3	3	4.5	4.5	6	6	6	9	9	9	12	12
		min.	1	1	1.5	1.5	2	2	2	3	3	3	4	4
c		min.	0.15	0.15	0.15	0.2	0.2	0.2	0.2	0.2	0.3	0.3	0.3	0.3
		max.	0.6	0.6	0.6	0.8	0.8	0.8	0.8	0.8	1	1	1	1
d_a		max.	9.2	11.2	13.7	17.7	22.4	26.4	33.4	39.4	45.6	52.6	63	71
d_w	Product grade A	min.	11.63	14.63	16.63	22.49	28.19	33.61	–	–	–	–	–	–
	Product grade B	min.	–	–	–	22	27.7	33.25	42.75	51.11	59.95	69.45	78.66	88.16
e	Product grade A	min.	14.38	17.77	20.03	26.75	33.53	39.98	–	–	–	–	–	–
	Product grade B	min.	–	–	–	26.17	32.95	39.55	50.85	60.79	71.3	82.6	93.56	104.86
k		nom.	5.3	6.4	7.5	10	12.5	15	18.7	22.5	26	30	35	40
	Product grade A	min.	5.15	6.22	7.32	9.82	12.285	14.785	–	–	–	–	–	–
		max.	5.45	6.58	7.68	10.18	12.715	15.215	–	–	–	–	–	–
	Product grade B	min.	–	–	–	9.71	12.15	14.65	18.28	22.08	25.58	29.58	34.5	39.5
		max.	–	–	–	10.29	12.85	15.35	19.12	22.92	26.42	30.42	35.5	40.5
k'[a]	Product grade A	min.	3.61	4.35	5.12	6.87	8.6	10.35	–	–	–	–	–	–
	Product grade B	min.	–	–	–	6.8	8.51	10.26	12.8	15.46	17.91	20.71	24.15	27.65
r		min.	0.4	0.4	0.6	0.6	0.8	0.8	1	1	1.2	1.6	2	2
s		nom. = max.	13	16	18	24	30	36	46	55	65	75	85	95
	Product grade A	min.	12.73	15.73	17.73	23.67	29.67	35.38	–	–	–	–	–	–
	Product grade B	min.	–	–	–	23.16	29.16	35	45	53.8	63.1	73.1	82.8	92.8

l^b	Product grade A		Product grade B	
nom.	min.	max.	min.	max.
16	15.65	16.35	–	–
20	19.58	20.42	–	–
25	24.58	25.42	–	–
30	29.58	30.42	–	–
35	34.5	35.5	–	–
40	39.5	40.5	38.75	41.25
45	44.5	45.5	43.75	46.25
50	49.5	50.5	48.75	51.25
55	54.4	55.6	53.5	56.5
60	59.4	60.6	58.5	61.5
65	64.4	65.6	63.5	66.5
70	69.4	70.6	68.5	71.5
80	79.4	80.6	78.5	81.5
90	89.3	90.7	88.25	91.75
100	99.3	100.7	98.25	101.75
110	109.3	110.7	108.25	111.75
120	119.3	120.7	118.25	121.75
130	129.2	130.8	128	132
140	139.2	140.8	138	142
150	149.2	150.8	148	152
160	–	–	158	162

continued

BS EN ISO 8676: 2001 Table 1 Preferred threads (*continued*)

Dimensions in mm

	Product grade			
	A		B	
l^b				
nom.	min.	max.	min.	max.
180	–	–	178	182
200	–	–	197.7	202.3
220	–	–	217.7	222.3
240	–	–	237.7	242.3
260	–	–	257.4	262.6
280	–	–	277.4	282.6
300	–	–	297.4	302.6
320	–	–	317.15	322.85
340	–	–	337.15	342.85
360	–	–	357.15	362.85
380	–	–	377.15	382.85
400	–	–	397.15	402.85
420	–	–	416.85	423.15
440	–	–	436.85	443.15
460	–	–	456.85	463.15
480	–	–	476.85	483.15
500	–	–	496.85	503.15

[a] $k'_{min} = 0.7\ k_{min}$.

[b] Range of popular lengths between the stepped line, marked thus ——.
- Product grade A above the stepped line, marked thus – – – – –.
- Product grade B below this stepped line.

Note: The threads M10 × 1 and M12 × 1.5 are popular ones but are not included in ISO 262.

BS EN ISO 8676: 2001 Table 2 Non-preferred threads

Dimensions in mm

Thread size, $d \times P$		M10 × 1.25	M12 × 1.25	M14 × 1.5	M18 × 1.5	M20 × 2	M22 × 1.5	M27 × 2	M33 × 2	M39 × 3	M45 × 3	M52 × 4	M60 × 4
a	max.	4	4	4.5	4.5	4.5	4.5	6	6	9	9	12	12
	min.	1.25	1.25	1.5	1.5	1.5	1.5	2	2	3	3	4	4
c	min.	0.15	0.15	0.15	0.2	0.2	0.2	0.2	0.2	0.3	0.3	0.3	0.3
	max.	0.6	0.6	0.6	0.8	0.8	0.8	0.8	0.8	1	1	1	1
d_a	max.	11.2	13.7	15.7	20.2	22.4	24.4	30.4	36.4	42.4	48.6	56.6	67
d_w	Product grade A min.	14.63	16.63	19.37	25.34	28.19	31.71	–	–	–	–	–	–
	Product grade B min.	–	–	–	24.85	27.7	31.35	38	46.55	55.86	64.7	74.2	83.41
e	Product grade A min.	17.77	20.03	23.36	30.14	33.53	37.72	–	–	–	–	–	–
	Product grade B min.	–	–										
– 29.56 32.95 37.29		45.2	55.37	66.44	76.95	88.25	99.21						
k	nom.	6.4	7.5	8.8	11.5	12.5	14	17	21	25	28	33	38
	Product grade A min.	6.22	7.32	8.62	11.285	12.285	13.785	–	–	–	–	–	–
	Product grade A max.	6.58	7.68	8.98	11.715	12.715	14.215	–	–	–	–	–	–
	Product grade B min.	–	–	–	11.15	12.15	13.65	16.65	20.58	24.58	27.58	32.5	37.5
	Product grade B max.	–	–	–	11.85	12.85	14.35	17.35	21.42	25.42	28.42	33.5	38.5
k'^{a}	Product grade A min.	4.35	5.12	6.03	7.9	8.6	9.65	–	–	–	–	–	–
	Product grade B min.	–	–	–	7.81	8.51	9.56	11.66	14.41	17.21	19.31	22.75	26.25
r	min.	0.4	0.6	0.6	0.6	0.8	0.8	1	1	1	1.2	1.6	2
s	nom. = max.	16	18	21	27	30	34	41	50	60	70	80	90
	Product grade A min.	15.73	17.73	20.67	26.67	29.67	33.38	–	–	–	–	–	–
	Product grade B min.	–	–	–	26.16	29.16	33	40	49	58.8	68.1	78.1	87.8

continued

BS EN ISO 8676: 2001 Table 2 Non-preferred threads (*continued*)

Dimensions in mm

Thread size, $d \times P$					M10 × 1.25	M12 × 1.25	M14 × 1.5	M18 × 1.5	M20 × 2	M22 × 1.5	M27 × 2	M33 × 2	M39 × 3	M45 × 3	M52 × 4	M60 × 4
	Product grade															
	A		B													
	l^b															
nom.	min.	max.	min.	max.												
20	19.58	20.42	–	–												
25	24.58	25.42	–	–												
30	29.58	30.42	–	–												
35	34.5	35.5	–	–												
40	39.5	40.5	–	–												
45	44.5	45.5	–	–												
50	49.5	50.5	–	–												
55	54.4	55.6	53.5	56.5												
60	59.4	60.6	58.5	61.5												
65	64.4	65.6	63.5	66.5												
70	69.4	70.6	68.5	71.5												
80	79.4	80.6	78.5	81.5												
90	89.3	90.7	88.25	91.75												
100	99.3	100.7	98.25	101.75												
110	109.3	110.7	108.25	111.75												
120	119.3	120.7	118.25	121.75												
130	129.2	130.8	128	132												
140	139.2	140.8	138	142												
150	149.2	150.8	148	152												
160	–	–	158	162												
180	–	–	178	182												
200	–	–	197.7	202.3												
220	–	–	217.7	222.3												
240	–	–	237.7	242.3												

260	–	–	257.4	262.6												
280	–	–	277.4	282.6												
300	–	–	297.4	302.6												
320	–	–	317.15	322.85												
340	–	–	337.15	342.85												
360	–	–	357.15	362.85												
380	–	–	377.15	382.85												
400	–	–	397.15	402.85												
420	–	–	416.85	423.15												
440	–	–	436.85	443.15												
460	–	–	456.85	463.15												
480	–	–	476.85	483.15												
500	–	–	496.85	503.15												

[a] $k'_{min} = 0.7\,k_{min}$.

[b] Range of popular lengths between the stepped line, marked thus ——:

- Product grade A above the stepped line, marked thus – – – – .
- Product grade B below this stepped line.

4.1.24 BS EN ISO 8673: 2001 Hexagon nuts style 1 with metric fine pitch threads – product grades A and B

Dimensions

Note: Symbols and designations of dimensions are specified in ISO 225.

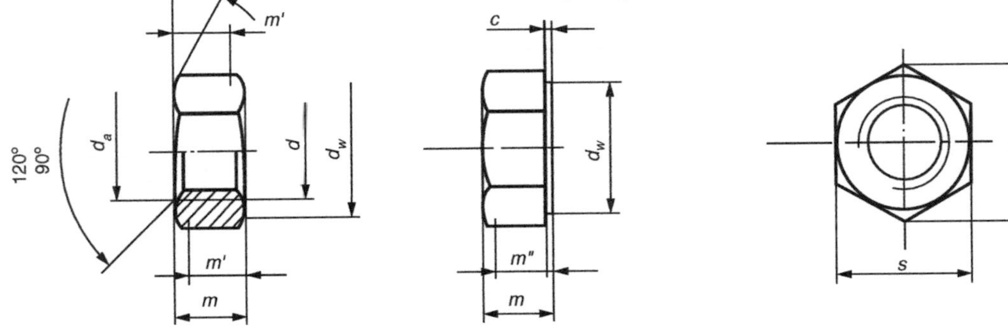

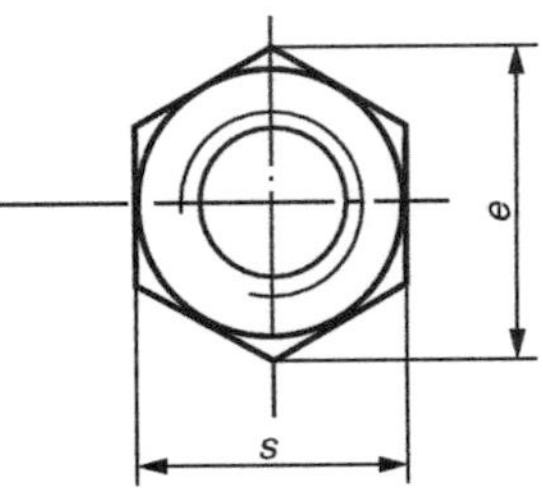

BS EN ISO 8673: 2001 Table 1 Preferred threads

Dimensions in mm

Thread size, $d \times P$		M8 × 1	M10 × 1	M12 × 1.5	M16 × 1.5	M20 × 1.5	M24 × 2	M30 × 2	M36 × 3	M42 × 3	M48 × 3	M56 × 4	M64 × 4
c	max.	0.6	0.6	0.6	0.8	0.8	0.8	0.8	0.8	1	1	1	1
	min.	0.15	0.15	0.15	0.2	0.2	0.2	0.2	0.2	0.3	0.3	0.3	0.3
d_a	min.	8	10	12	16	20	24	30	36	42	48	56	64
	max.	8.75	10.8	13	17.3	21.6	25.9	32.4	38.9	45.4	51.8	60.5	69.1
d_w	min.	11.63	14.63	16.63	22.49	27.7	33.25	42.75	51.11	59.95	69.45	78.66	88.16
e	min.	14.38	17.77	20.03	26.75	32.95	39.55	50.85	60.79	71.3	82.6	93.56	104.86
m	max.	6.8	8.4	10.8	14.8	18	21.5	25.6	31	34	38	45	51
	min.	6.44	8.04	10.37	14.1	16.9	20.2	24.3	29.4	32.4	36.4	43.4	49.1
m'	min.	5.15	6.43	8.3	11.28	13.52	16.16	19.44	23.52	25.92	29.12	34.72	39.28
m''	min.	4.51	5.63	7.26	9.87	11.83	14.14	17.01	20.58	22.68	25.48	30.38	34.37
s	nom. = max.	13	16	18	24	30	36	46	55	65	75	85	95
	min.	12.73	15.73	17.73	23.67	29.16	35	45	53.8	63.1	73.1	82.8	92.8

BS EN ISO 8673: 2001 Table 2 Non-preferred threads

Dimensions in mm

Thread size, $d \times P$		M10 × 1.25	M12 × 1.25	M14 × 1.5	M18 × 1.5	M20 × 2	M22 × 1.5	M27 × 2	M33 × 2	M39 × 3	M45 × 3	M52 × 4	M60 × 4
c	max.	0.6	0.6	0.6	0.8	0.8	0.8	0.8	0.8	1	1	1	1
	min.	0.15	0.15	0.15	0.2	0.2	0.2	0.2	0.2	0.3	0.3	0.3	0.3
d_a	min.	10	12	14	18	20	22	27	33	39	45	52	60
	max.	10.8	13	15.1	19.5	21.6	23.7	29.1	35.6	42.1	48.6	56.2	64.8
d_w	min.	14.63	16.63	19.64	24.85	27.7	31.35	38	46.55	55.86	64.7	74.2	83.41
e	min.	17.77	20.03	23.36	29.56	32.95	37.29	45.2	55.37	66.44	76.95	88.25	99.21
m	max.	8.4	10.8	12.8	15.8	18	19.4	23.8	28.7	33.4	36	42	48
	min.	8.04	10.37	12.1	15.1	16.9	18.1	22.5	27.4	31.8	34.4	40.4	46.4
m'	min.	6.43	8.3	9.68	12.08	13.52	14.48	18	21.92	25.44	27.52	32.32	37.12
m''	min.	5.63	7.26	8.47	10.57	11.83	12.67	15.75	19.18	22.26	24.08	28.28	32.48
s	nom. = max.	16	18	21	27	30	34	41	50	60	70	80	90
	min.	15.73	17.73	20.67	26.16	29.16	33	40	49	58.8	68.1	78.1	87.8

4.1.25 BS EN ISO 8674: 2001 Hexagon nuts style 2 with metric fine pitch threads – product grades A and B

Dimensions

Note: Symbols and descriptions of dimensions are specified in ISO 225.

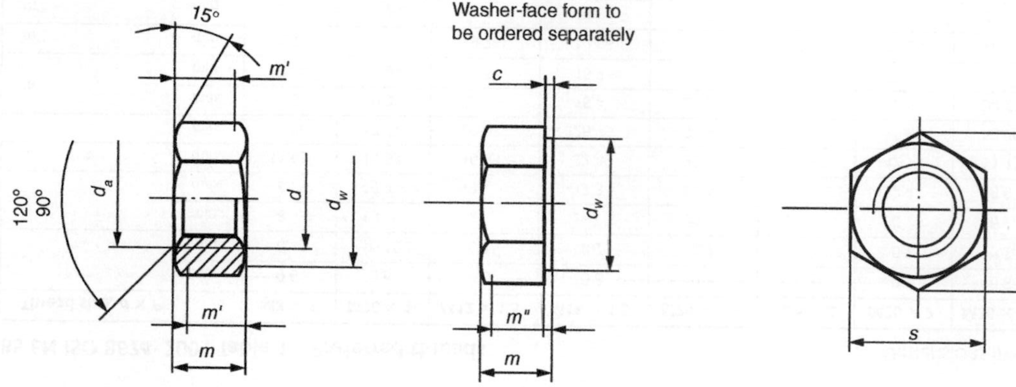

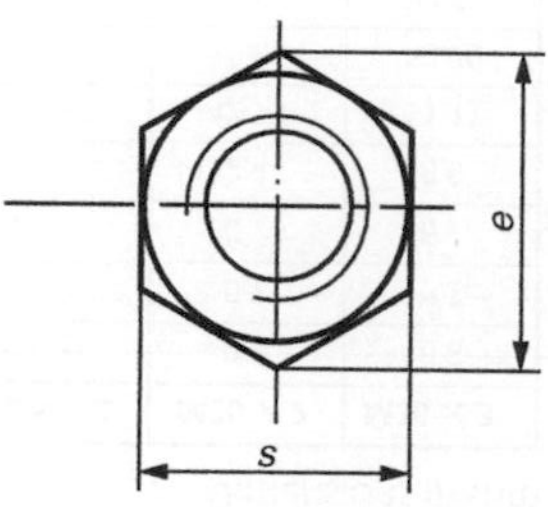

BS EN ISO 8674: 2001 Table 1 Preferred threads

Dimensions in mm

Thread size, $d \times P$		M8 × 1	M10 × 1	M12 × 1.5	M16 × 1.5	M20 × 1.5	M24 × 2	M30 × 2	M36 × 3
c	max.	0.6	0.6	0.6	0.8	0.8	0.8	0.8	0.8
	min.	0.15	0.15	0.15	0.2	0.2	0.2	0.2	0.2
d_a	min.	8	10	12	16	20	24	30	36
	max.	8.75	10.8	13	17.3	21.6	25.9	32.4	38.9
d_w	min.	11.63	14.63	16.63	22.49	27.7	33.25	42.75	51.11
e	min.	14.38	17.77	20.03	26.75	32.95	39.55	50.85	60.79
m	max.	7.5	9.3	12	16.4	20.3	23.9	28.6	34.7
	min.	7.14	8.94	11.57	15.7	19	22.6	27.3	33.1
m'	min.	5.71	7.15	9.26	12.56	15.2	18.08	21.84	26.48
m''	min.	5	6.26	8.1	10.99	13.3	15.82	19.11	23.17
s	nom. = max.	13	16	18	24	30	36	46	55
	min.	12.73	15.73	17.73	23.67	29.16	35	45	53.8

BS EN ISO 8674: 2001 Table 2 Non-preferred threads

Dimensions in mm

Thread size, $d \times P$		M10 × 1.25	M12 × 1.25	M14 × 1.5	M18 × 1.5	M20 × 2	M22 × 1.5	M27 × 2	M33 × 2
c	max.	0.6	0.6	0.6	0.8	0.8	0.8	0.8	0.8
	min.	0.15	0.15	0.15	0.2	0.2	0.2	0.2	0.2
d_a	min.	10	12	14	18	20	22	27	33
	max.	10.8	13	15.1	19.5	21.6	23.7	29.1	35.6
d_w	min.	14.63	16.63	19.64	24.85	27.7	31.35	38	46.55
e	min.	17.77	20.03	23.36	29.56	32.95	37.29	45.2	55.37
m	max.	9.3	12	14.1	17.6	20.3	21.8	26.7	32.5
	min.	8.94	11.57	13.4	16.9	19	20.5	25.4	30.9
m'	min.	7.15	9.26	10.72	13.52	15.2	16.4	20.32	24.72
m''	min.	6.26	8.1	9.38	11.83	13.3	14.35	17.78	21.63
s	nom. = max.	16	18	21	27	30	34	41	50
	min.	15.73	17.73	20.67	26.16	29.16	33	40	49

4.1.26 BS EN ISO 8675: 2001 Hexagon thin nuts with metric fine pitch threads – product grades A and B

Dimensions

Note: Symbols and descriptions of dimensions are specified in ISO 225.

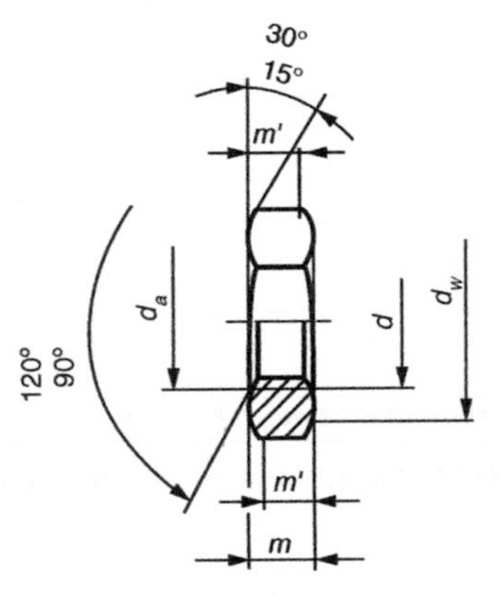

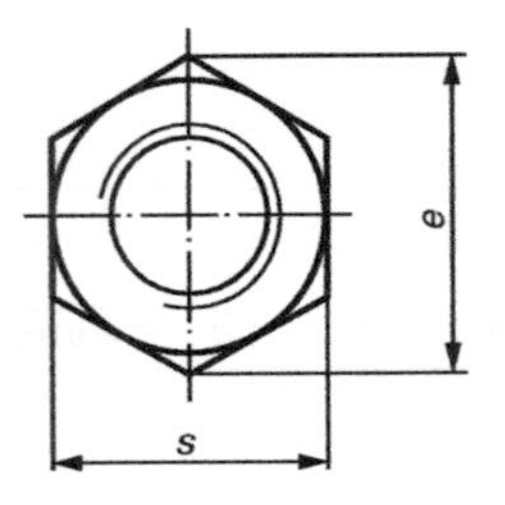

BS EN ISO 8675: 2001 Table 1 Preferred threads

Dimensions in mm

Thread size, $d \times P$		M8 × 1	M10 × 1	M12 × 1.5	M16 × 1.5	M20 × 1.5	M24 × 2	M30 × 2	M36 × 3	M42 × 3	M48 × 3	M56 × 4	M64 × 4
d_a	min.	8	10	12	16	20	24	30	36	42	48	56	64
	max.	8.75	10.8	13	17.3	21.6	25.9	32.4	38.9	45.4	51.8	60.5	69.1
d_w	min.	11.63	14.63	16.63	22.49	27.7	33.25	42.75	51.11	59.95	69.45	78.66	88.16
e	min.	14.38	17.77	20.03	26.75	32.95	39.55	50.85	60.79	71.3	82.6	93.56	104.86
m	max.	4	5	6	8	10	12	15	18	21	24	28	32
	min.	3.7	4.7	5.7	7.42	9.1	10.9	13.9	16.9	19.7	22.7	26.7	30.4
m'	min.	2.96	3.76	4.56	5.94	7.28	8.72	11.12	13.52	15.76	18.16	21.36	24.32
s	nom. = max.	13	16	18	24	30	36	46	55	65	75	85	95
	min.	12.73	15.73	17.73	23.67	29.16	35	45	53.8	63.1	73.1	82.8	92.8

BS EN ISO 8675: 2001 Table 2 Non-preferred threads

Dimensions in mm

Thread size, $d \times P$		M10 × 1.25	M12 × 1.25	M14 × 1.5	M18 × 1.5	M20 × 2	M22 × 1.5	M27 × 2	M33 × 2	M39 × 3	M45 × 3	M52 × 4	M60 × 4
d_a	min.	10	12	14	18	20	22	27	33	39	45	52	60
	max.	10.8	13	15.1	19.5	21.6	23.7	29.1	35.6	42.1	48.6	56.2	64.8
d_w	min.	14.63	16.63	19.64	24.85	27.7	31.35	38	46.55	55.86	64.7	74.2	83.41
e	min.	17.77	20.03	23.36	29.56	32.95	37.29	45.2	55.37	66.44	76.95	88.25	99.21
m	max.	5	6	7	9	10	11	13.5	16.5	19.5	22.5	26	30
	min.	4.7	5.7	6.42	8.42	9.1	9.9	12.4	15.4	18.2	21.2	24.7	28.7
m'	min.	3.76	4.56	5.14	6.74	7.28	7.92	9.92	12.32	14.56	16.96	19.76	22.96
s	nom. = max.	16	18	21	27	30	34	41	50	60	70	80	90
	min.	15.73	17.73	20.67	26.16	29.16	33	40	49	58.8	68.1	78.1	87.8

4.1.27 BS 7764: 1994 Hexagon slotted nuts and castle nuts

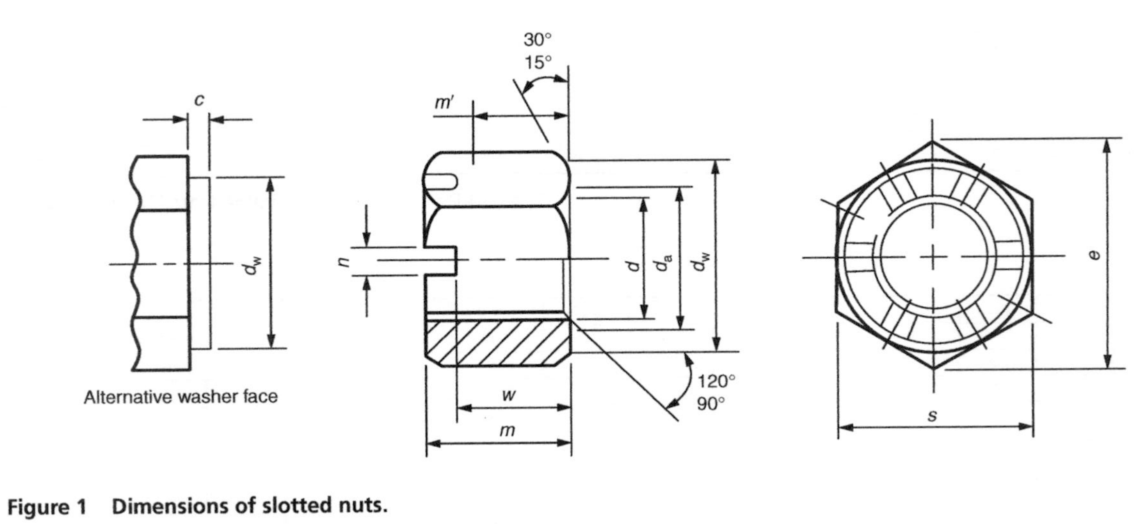

Figure 1 Dimensions of slotted nuts.

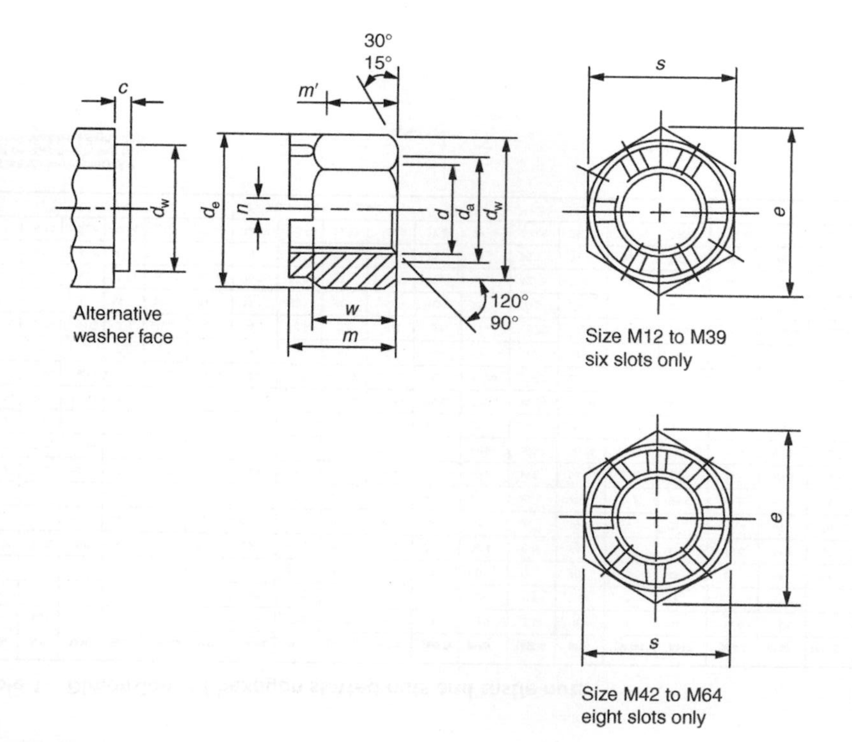

Figure 2 Dimensions of castle nuts.

BS 7764: 1994 Table 1 Dimensions of hexagon slotted nuts and castle nuts

Dimensions in mm

d		M4	M5	M6	M8	M10	M12	(M14)	M16	(M18)	M20	(M22)	M24	(M27)	M30	(M33)	M36	(M39)	M42	(M45)	M48	(M52)	M56	(M60)	M64
p		0.7	0.8	1	1.25	1.5	1.75	2	2	2.5	2.5	2.5	3	3	3.5	3.5	4	4	4.5	4.5	5	5	5.5	5.5	6
c	max.	0.4	0.5	0.5	0.6	0.6	0.6	0.6	0.8	0.8	0.8	0.8	0.8	0.8	0.8	0.8	0.8	1	1	1	1	1	1	1	1
d_a	min.	4	5	6	8	10	12	14	16	18	20	22	24	27	30	33	36	39	42	45	48	52	56	60	64
	max.	4.6	5.75	6.75	8.75	10.8	13	15.1	17.3	19.5	21.6	23.7	25.9	29.1	32.4	35.6	38.9	42.1	45.4	48.6	51.8	56.2	60.5	64.8	69.1
d_e	max.	–	–	–	–	–	16	19	22	25	28	30	34	38	42	46	50	55	58	62	65	70	75	80	85
	min.	–	–	–	–	–	15.57	18.48	21.48	24.16	27.16	29.16	33	37	41	45	49	53.8	56.8	60.8	63.8	68.8	73.8	78.8	83.8
d_w	min.	5.9	6.9	8.9	11.6	14.6	16.6	19.6	22.5	24.8	27.7	31.4	33.2	38	42.7	46.6	51.1	55.9	59.9	64.7	69.4	74.2	78.7	83.4	88.2
e	min.	7.66	8.79	11.05	14.38	17.77	20.03	23.36	26.75	29.58	32.95	37.29	39.55	45.2	50.85	55.37	60.79	66.44	71.3	76.95	82.6	88.25	93.56	99.21	104.86
m	max.	5	6	7.5	9.5	12	15	16	19	21	22	26	27	30	33	35	38	40	46	48	50	54	57	63	66
	min.	4.7	5.7	7.14	9.14	11.57	14.57	15.57	18.48	21.48	21.48	25.48	25.48	29.48	32.38	34.38	37.38	39.38	45.38	47.38	49.38	53.26	56.26	62.26	65.26
m'	min.	2.32	3.52	3.92	5.15	6.43	8.3	9.68	11.28	12.08	13.52	14.48	16.16	18	19.44	21.92	23.52	25.44	25.92	27.52	29.12	32.32	34.72	37.12	39.3
n	min.	1.2	1.4	2	2.5	2.8	3.5	3.5	4.5	4.5	4.5	5.5	5.5	5.5	7	7	7	7	9	9	9	9	9	11	11
	max.	1.45	1.65	2.25	2.75	3.05	3.8	3.8	4.8	4.8	4.8	5.8	5.8	5.8	7.36	7.36	7.36	7.36	9.36	9.36	9.36	9.36	9.36	11.43	11.43
s	max.	7	8	10	13	16	18	21	24	27	30	34	36	41	46	50	55	60	65	70	75	80	85	90	95
	min.	6.78	7.78	9.78	12.73	15.73	17.73	20.67	23.67	26.16	29.16	33	35	40	45	49	53.8	58.8	63.1	68.1	73.1	78.1	82.8	87.8	92.8
w	max.	2.9	4	5	6.5	8	10	11	13	15	16	18	19	22	24	26	29	31	34	36	38	42	45	48	51
	min.	3.2	3.7	4.7	6.14	7.64	9.64	10.57	12.59	14.57	15.57	17.57	18.48	21.48	23.48	25.48	28.48	30.38	33.38	35.38	37.38	41.38	44.38	47.38	50.26
For product grade A see Table 2										For product grade B see Table 2															

Notes: 1. Non-preferred sizes are shown in brackets.
2. Castle nuts shall not be specified below M12.
3. Castle nuts above M39 shall have eight slots.

BS 7764: 1994 Table 2 Characteristics of hexagon slotted nuts and castle nuts

Characteristic	Requirement		
Material	**Steel**	**Stainless steel**	**Non-ferrous metal**
Thread tolerance	Conforming to class 6H of BS 3643: Part 1: 1981 and BS 3643: Part 2: 1981		
Mechanical properties[a]	Conforming to material property class 6, 8 or 10 of BS EN 20898-2: 1992	Conforming to material property class A2-70 of BS 6105: 1981	Conforming to BS EN 28839: 1992
Permissible dimensional deviations and deviations of form	Conforming to product grade A of BS 6322: Part 1: 1982 for thread sizes $\leq$ M16 Conforming to product grade B of BS 6322: Part 1: 1982 for thread sizes $>$ M16		
Finish	As processed with a lubricant, unless otherwise agreed between the purchaser and supplier[b,c]	As processed, unless otherwise agreed between the purchaser and supplier[c]	As processed, unless otherwise agreed between the purchaser and supplier[c]
Acceptability	In accordance with BS 6587: 1985		

[a] Including the marking requirement.
[b] The lubricant shall contain a rust inhibitor unless otherwise agreed between the purchaser and supplier.
[c] Alternative surface finishes (e.g. electroplating) may be applied, if agreed between the purchaser and supplier.

4.1.28 BS EN ISO 898-1: 1999 Mechanical properties of fasteners: bolts, screws and studs

Mechanical properties

When tested by the methods described in clause 8, the bolts, screws and studs shall, at room temperature, have the mechanical properties set out in BS EN ISO 898-1: 1999 Table 3.

BS EN ISO 898-1: 1999 Table 3 Mechanical properties of bolts, screws and studs

Sub-clause No.	Mechanical property		Property class										
			3.6	4.6	4.8	5.6	5.8	6.8	8.8[a]		9.8[c]	10.9	12.9
									$d \leqslant$ 16 mm	$d >$ 16 mm[b]			
5.1 and 5.2	Tensile strength, R_m[d,e], N/mm^2	nom.	300	400		500		600	800	800	900	1 000	1 200
		min.	330	400	420	500	520	600	800	830	900	1 040	1 220
5.3	Vickers hardness, HV, $F > 98$ N	min.	95	120	130	155	160	190	250	255	290	320	385
		max.	250						320	335	360	380	435
5.4	Brinell hardness, HB, $F = 30D^2$	min.	90	114	124	147	152	181	238	242	276	304	366
		max.	238						304	318	342	361	414
5.5	Rockwell hardness, HR	min. HRB	52	67	71	79	82	89	–	–	–	–	–
		min. HRC	–	–	–	–	–	–	22	23	28	32	39
		max. HRB	99.5						–	–	–	–	–
		max. HRC	–						32	34	37	39	44
5.6	Surface hardness, HV 0.3	max.	–						–[f]				
5.7	Lower yield stress, R_{eL}[g], N/mm^2	nom.	180	240	320	300	400	480	–	–	–	–	–
		min.	190	240	340	300	420	480	–	–	–	–	–

5.8	Proof stress, $R_{p0.2}$, N/mm^2	nom.	–						640	640	720	900	1080
		min.	–						640	660	720	940	1100
5.9	Stress under proofing load, S_p	S_p/R_{eL} or $S_p/R_{p0.2}$	0.94	0.94	0.91	0.93	0.90	0.92	0.91	0.91	0.90	0.88	0.88
		N/mm^2	180	225	310	280	380	440	580	600	650	830	970
5.10	Elongation after fracture, A	min.	25	22	14	20	10	8	12	12	10	9	8
5.11	Strength under wedge loading[e]		The values for full size bolts and screws (not studs) shall not be smaller than the minimum values for tensile strength shown in 5.2										
5.12	Impact strength, J	min.	–			25	–		30	30	25	20	15
5.13	Head soundness		no fracture										
5.14	Minimum height of non-decarburized thread zone, E		–						$\frac{1}{2}H_1$			$\frac{2}{3}H_1$	$\frac{3}{4}H_1$
	Maximum depth of complete decarburization, G	mm	–						0.015				

[a] For bolts of property class 8.8 in diameters $d \leqslant 16$ mm, there is an increased risk of nut stripping in the case of inadvertent over-tightening inducting a load in excess of proofing load. Reference to ISO 898.2 is recommended.

[b] For structural bolting the limit is 12 mm.

[c] Applies only to nominal thread diameters $d \leqslant 16$ mm.

[d] Minimum tensile properties apply to products of nominal length $l \leqslant 2.5d$. Minimum hardness applies to products of length $l < 2.5d$ and other products which cannot be tensile tested (e.g. due to head configuration).

[e] For testing of full-size bolts, screws and studs, the loads given in BS EN ISO 898-1: 1999 Tables 6 to 9 (of the full standard) shall be applied.

[f] Surface hardness shall not be more than 30 Vickers points above the measured core hardness on the product when readings of both surface and core are carried out at HV 0.3. For property class 10.9, any increase in hardness at the surface which indicates that the surface hardness exceeds 390 HV is not acceptable.

[g] In cases where the lower yield stress R_{eL} cannot be determined, it is permissible to measure the proof stress $R_{p0.2}$.

Mechanical properties to be determined

Two test programmes, A and B, for mechanical properties of bolts, screws and studs, using the methods described in clause 8, are set out in BS EN ISO 898-1: 1999 Table 5.

The application of programme B is always desirable, but is mandatory for products with breaking loads less than 500 kN. Programme A is suitable for machined test pieces and for bolts with a shank area less than the stress area.

BS EN ISO 898-1: 1999 Table 4 Key to test programmes (see BS EN ISO 898-1: 1999 Table 5)

Size	Bolts and screws with thread diameter $d \leqslant 4$ mm or length $l < 2.5d$[a]	Bolts and screws with thread diameter $d > 4$ mm and length $l \geqslant 2.5d$
Test decisive for acceptance	○	●

[a] Also bolts and screws with special head or shank configurations which are weaker than the threaded section.

BS EN ISO 898-1: 1999 Table 5 Test programmes A and B for acceptance purposes

Test group	Property		Test programme A				Test programme B			
			Test method		Property class		Test method		Property class	
					3.6, 4.6, 5.6	8.8, 9.8, 10.9, 12.9			3.6, 4.6, 4.8, 5.6, 5.8, 6.8	8.8, 9.8, 10.9, 12.9
I	5.1 and 5.2	Minimum tensile strength, R_m	8.1	Tensile test	●	●	8.2	Tensile test[a]	●	●
	5.3	Minimum hardness[b]	8.3	Hardness test[c]	○	○	8.3	Hardness test[c]	○	○
	5.4 and 5.5	Maximum hardness			● ○	● ○			● ○	● ○
	5.6	Maximum surface hardness				● ○				● ○
II	5.7	Minimum lower yield stress, R_{eL}	8.1	Tensile test	●					
	5.8	Proof stress, $R_{p0.2}$	8.1	Tensile test		●				
	5.9	Stress under proofing load, S_p					8.4	Proofing load test	●	●

continued

BS EN ISO 898-1: 1999 Table 5 Test programmes A and B for acceptance purposes (*continued*)

Test group	Property		Test programme A				Test programme B			
			Test method		Property class		Test method		Property class	
					3.6, 4.6, 5.6	8.8, 9.8, 10.9, 12.9			3.6, 4.6, 4.8, 5.6, 5.8, 6.8	8.8, 9.8, 10.9, 12.9
III	5.10	Minimum elongation after fracture, A_{min}.	8.1	Tensile test	●	●				
	5.11	Strength under wedge loading[d]					8.5	Wedge loading test[a]	●	●
IV	5.12	Minimum impact strength	8.6	Impact test[e]	●[f]	●	8.6			
	5.13	Head soundness[g]					8.7	Head soundness test	○	○
V	5.14	Maximum decarburized zone	8.8	Decarburization test		● ○	8.8	Decarburization test		● ○
	5.15	Minimum tempering temperature	8.9	Retempering test		● ○	8.9	Retempering test		● ○
	5.16	Surface integrity	8.10	Surface integrity test	● ○	● ○	8.10	Surface integrity test	● ○	● ○

[a] If the wedge loading test is satisfactory, the axial tensile test is not required.
[b] Minimum hardness applies only to products of nominal length $l < 2.5d$ and other products which cannot be tensile tested (e.g. due to head configuration).
[c] Hardness may be Vickers, Brinell or Rockwell. In case of doubt, the Vickers hardness test is decisive for acceptance.
[d] Special head bolts and screws with configurations which are weaker than the threaded section are excluded from wedge tensile testing requirements.
[e] Only for bolts, screws and studs with thread diameters $d \geqslant 16$ mm and only if required by the purchaser.
[f] Only for property class 5.6.
[g] Only for bolts and screws with thread diameters $d \leqslant 16$ mm and lengths too short to permit wedge load testing.
Note: These procedures apply to mechanical but not chemical properties.

Minimum ultimate tensile loads and proofing loads

See BS EN ISO 898-1: 1999 Tables 6 to 9.

BS EN ISO 898-1: 1999 Table 6 Minimum ultimate tensile loads – ISO metric coarse pitch thread

Thread[a]	Nominal stress area $A_{s,\,nom}$ (mm^2)	Property class 3.6	4.6	4.8	5.6	5.8	6.8	8.8	9.8	10.9	12.9
		Minimum ultimate tensile load ($A_s \times R_m$), N									
M3	5.03	1 660	2 010	2 110	2 510	2 620	3 020	4 020	4 530	5 230	6 140
M3.5	6.78	2 240	2 710	2 850	3 390	3 530	4 070	5 420	6 100	7 050	8 270
M4	8.78	2 900	3 510	3 690	4 390	4 570	5 270	7 020	7 900	9 130	10 700
M5	14.2	4 690	5 680	5 960	7 100	7 380	8 520	11 350	12 800	14 800	17 300
M6	20.1	6 630	8 040	8 440	10 000	10 400	12 100	16 100	18 100	20 900	24 500
M7	28.9	9 540	11 600	12 100	14 400	15 000	17 300	23 100	26 000	30 100	35 300
M8	36.6	12 100	14 600	15 400	18 300	19 000	22 000	29 200	32 900	38 100	44 600
M10	58	19 100	23 200	24 400	29 000	30 200	34 800	46 400	52 200	60 300	70 800
M12	84.3	27 800	33 700	35 400	42 200	43 800	50 600	67 400[b]	75 900	87 700	103 000
M14	115	38 000	46 000	48 300	57 500	59 800	69 000	92 000[b]	104 000	120 000	140 000
M16	157	51 800	62 800	65 900	78 500	81 600	94 000	125 000[b]	141 000	163 000	192 000
M18	192	63 400	76 800	80 600	96 000	99 800	115 000	159 000	–	200 000	234 000
M20	245	80 800	98 000	103 000	122 000	127 000	147 000	203 000	–	255 000	299 000
M22	303	100 000	121 000	127 000	152 000	158 000	182 000	252 000	–	315 000	370 000
M24	353	116 000	141 000	148 000	176 000	184 000	212 000	293 000	–	367 000	431 000
M27	459	152 000	184 000	193 000	230 000	239 000	275 000	381 000	–	477 000	560 000
M30	561	185 000	224 000	236 000	280 000	292 000	337 000	466 000	–	583 000	684 000
M33	694	229 000	278 000	292 000	347 000	361 000	416 000	576 000	–	722 000	847 000
M36	817	270 000	327 000	343 000	408 000	425 000	490 000	678 000	–	850 000	997 000
M39	976	322 000	390 000	410 000	488 000	508 000	586 000	810 000	–	1 020 000	1 200 000

[a] Where no thread pitch is indicated in a thread designation, coarse pitch is specified. This is given in ISO 261 and ISO 262.
[b] For structural bolting 70 000, 95 500 and 130 000 N, respectively.

BS EN ISO 898-1: 1999 Table 7 Proofing loads ISO metric coarse pitch thread

Thread[a]	Nominal stress area $A_{s,nom}$ (mm^2)	Property class									
		3.6	4.6	4.8	5.6	5.8	6.8	8.8	9.8	10.9	12.9
		Proofing load ($A_s \times S_p$), N									
M3	5.03	910	1 130	1 560	1 410	1 910	2 210	2 920	3 270	4 180	4 880
M3.5	6.78	1 220	1 530	2 100	1 900	2 580	2 980	3 940	4 410	5 630	6 580
M4	8.78	1 580	1 980	2 720	2 460	3 340	3 860	5 100	5 710	7 290	8 520
M5	14.2	2 560	3 200	4 400	3 980	5 400	6 250	8 230	9 230	11 800	13 800
M6	20.1	3 620	4 520	6 230	5 630	7 640	8 840	11 600	13 100	16 700	19 500
M7	28.9	5 200	6 500	8 960	8 090	11 000	12 700	16 800	18 800	24 000	28 000
M8	36.6	6 590	8 240	11 400	10 200	13 900	16 100	21 200	23 800	30 400	35 500
M10	58	10 400	13 000	18 000	16 200	22 000	25 500	33 700	37 700	48 100	56 300
M12	84.3	15 200	19 000	26 100	23 600	32 000	37 100	48 900[b]	54 800	70 000	81 800
M14	115	20 700	25 900	35 600	32 200	43 700	50 600	66 700[b]	74 800	95 500	112 000
M16	157	28 300	35 300	48 700	44 000	59 700	69 100	91 000[b]	102 000	130 000	152 000
M18	192	34 600	43 200	59 500	53 800	73 000	84 500	115 000	–	159 000	186 000
M20	245	44 100	55 100	76 000	68 600	93 100	108 000	147 000	–	203 000	238 000
M22	303	54 500	68 200	93 900	84 800	115 000	133 000	182 000	–	252 000	294 000
M24	353	63 500	79 400	109 000	98 800	134 000	155 000	212 000	–	293 000	342 000
M27	459	82 600	103 000	142 000	128 000	174 000	202 000	275 000	–	381 000	445 000
M30	561	101 000	126 000	174 000	157 000	213 000	247 000	337 000	–	466 000	544 000
M33	694	125 000	156 000	215 000	194 000	264 000	305 000	416 000	–	570 000	673 000
M36	817	147 000	184 000	253 000	229 000	310 000	359 000	490 000	–	678 000	792 000
M39	976	176 000	220 000	303 000	273 000	371 000	429 000	586 000	–	810 000	947 000

[a] Where no thread pitch is indicated in a thread designation, coarse pitch is specified. This is given in ISO 261 and ISO 262.
[b] For structural bolting 50 700, 68 800 and 94 500 N, respectively.

BS EN ISO 898-1: 1999 Table 8 Minimum ultimate tensile loads – ISO metric fine pitch thread

Thread	Nominal stress area $A_{s,nom}$ (mm^2)	Property class									
		3.6	4.6	4.8	5.6	5.8	6.8	8.8	9.8	10.9	12.9
		Minimum ultimate tensile load ($A_s \times R_m$), N									
M8 × 1	39.2	12 900	15 700	16 500	19 600	20 400	23 500	31 360	35 300	40 800	47 800
M10 × 1	64.5	21 300	25 800	27 100	32 300	33 500	38 700	51 600	58 100	67 100	78 700
M12 × 1.5	88.1	29 100	35 200	37 000	44 100	45 800	52 900	70 500	79 300	91 600	107 500
M14 × 1.5	125	41 200	50 000	52 500	62 500	65 000	75 000	100 000	112 000	130 000	152 000
M16 × 1.5	167	55 100	66 800	70 100	83 500	86 800	100 000	134 000	150 000	174 000	204 000
M18 × 1.5	216	71 300	86 400	90 700	108 000	112 000	130 000	179 000	–	225 000	264 000
M20 × 1.5	272	89 800	109 000	114 000	136 000	141 000	163 000	226 000	–	283 000	332 000
M22 × 1.5	333	110 000	133 000	140 000	166 000	173 000	200 000	276 000	–	346 000	406 000
M24 × 2	384	127 000	154 000	161 000	192 000	200 000	230 000	319 000	–	399 000	469 000
M27 × 2	496	164 000	194 000	208 000	248 000	258 000	298 000	412 000	–	516 000	605 000
M30 × 2	621	205 000	248 000	261 000	310 000	323 000	373 000	515 000	–	646 000	758 000
M33 × 2	761	251 000	304 000	320 000	380 000	396 000	457 000	632 000	–	791 000	928 000
M36 × 3	865	285 000	346 000	363 000	432 000	450 000	519 000	718 000	–	900 000	1 055 000
M39 × 3	1 030	340 000	412 000	433 000	515 000	536 000	618 000	855 000	–	1 070 000	1 260 000

BS EN ISO 898-1: 1999 Table 9 Proofing loads – ISO metric fine pitch thread

Thread	Nominal stress area $A_{s,nom}$ (mm^2)	Property class									
		3.6	4.6	4.8	5.6	5.8	6.8	8.8	9.8	10.9	12.9
		Proofing load ($A_s \times S_p$), N									
M8 × 1	39.2	7 060	8 820	12 200	11 000	14 900	17 200	22 700	25 500	32 500	38 000
M10 × 1	64.5	11 600	14 500	20 000	18 100	24 500	28 400	37 400	41 900	53 500	62 700
M12 × 1.5	88.1	15 900	19 800	27 300	24 700	33 500	38 800	51 100	57 300	73 100	85 500
M14 × 1.5	125	22 500	28 100	38 800	35 000	47 500	55 000	72 500	81 200	104 000	121 000
M16 × 1.5	167	30 100	37 600	51 800	46 800	63 500	73 500	96 900	109 000	139 000	162 000
M18 × 1.5	216	38 900	48 600	67 000	60 500	82 100	95 000	130 000	–	179 000	210 000
M20 × 1.5	272	49 000	61 200	84 300	76 200	103 000	120 000	163 000	–	226 000	264 000
M22 × 1.5	333	59 900	74 900	103 000	93 200	126 000	146 000	200 000	–	276 000	323 000
M24 × 2	384	69 100	86 400	119 000	108 000	146 000	169 000	230 000	–	319 000	372 000
M27 × 2	496	89 300	112 000	154 000	139 000	188 000	218 000	298 000	–	412 000	481 000
M30 × 2	621	112 000	140 000	192 000	174 000	236 000	273 000	373 000	–	515 000	602 000
M33 × 2	761	137 000	171 000	236 000	213 000	289 000	335 000	457 000	–	632 000	738 000
M36 × 3	865	156 000	195 000	268 000	242 000	329 000	381 000	519 000	–	718 000	839 000
M39 × 3	1 030	185 000	232 000	319 000	288 000	391 000	453 000	618 000	–	855 000	999 000

4.1.29 BS EN ISO 898-1: 1999 Marking

Symbols

Marking symbols are shown in BS EN ISO 898-1: 1999 Table 1.

Identification

Hexagon bolts and screws

Hexagon bolts and screws shall be marked with the designation symbol of the property class described in clause 3.

The marking is obligatory for all property classes, preferably on the top of the head by indenting or embossing or on the side of the head by indenting (see Fig. 1).

Marking is required for hexagon bolts and screws with nominal diameters $d \geqslant 5\,\text{mm}$ where the shape of the product allows it, preferably on the head.

Hexagon socket head cap screws

Hexagon socket head cap screws shall be marked with the designation symbol of the property class described in clause 3.

The marking is obligatory for property classes equal to or higher than 8.8, preferably on the side of the head by indenting or on the top of the head by indenting or embossing (see Fig. 2).

Marking is required for hexagon socket head cap screws with nominal diameters $d \geqslant 5\,\text{mm}$ where the shape of the product allows it, preferably on the head.

The clock-face marking system as given for nuts in ISO 898-2 may be used as an alternative method on small hexagon socket head cap screws.

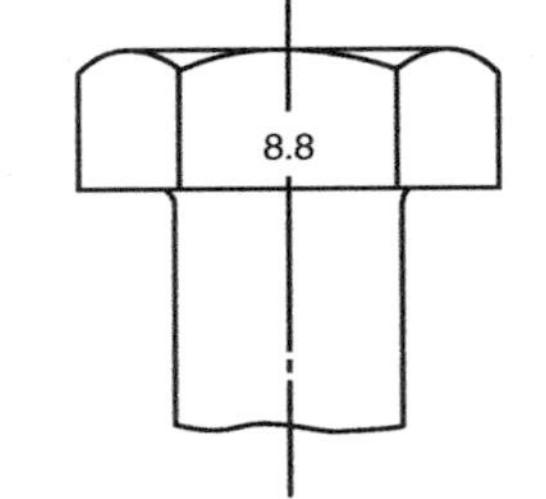

Figure 1 Examples of marking on hexagon bolts and screws.

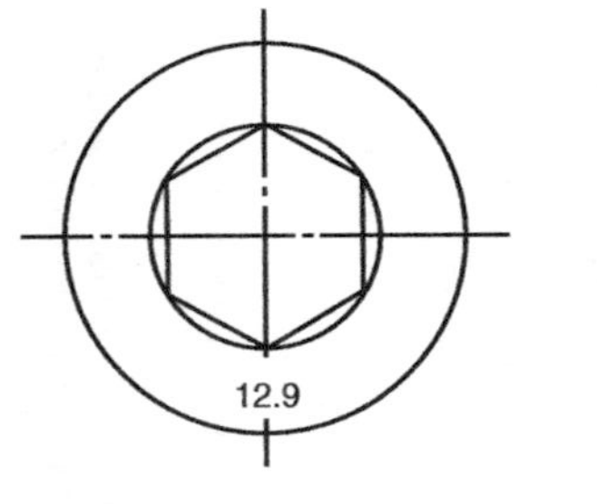

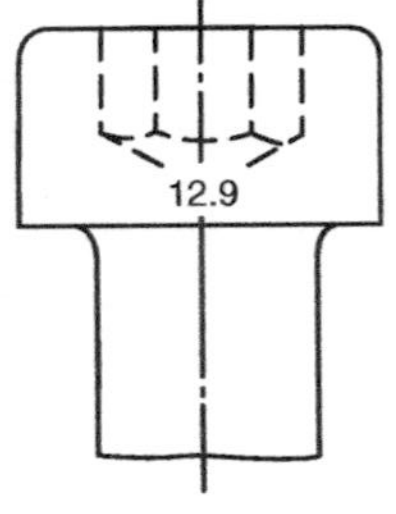

Figure 2 Examples of marking on hexagon socket head cap screws.

BS EN ISO 898-1: 1999 Table 1 Marking symbols

Property class	3.6	4.6	4.8	5.6	5.8	6.8	8.8	9.8	10.9	12.9
Marking symbol[a,b]	3.6	4.6	4.8	5.6	5.8	6.8	8.8	9.8	10.9	12.9

[a] The full-stop in the marking symbol may be omitted.
[b] When low carbon martensitic steels are used for property class 10.9 (see BS EN ISO 898-1: 1999 Table 2), the symbol 10.9 shall be underlined: 10.9.

BS EN ISO 898-1: 1999 Table 2 Identification marks for studs

Property class	8.8	9.8	10.9	12.9
Identification mark	○	+	□	△

Studs

Studs shall be marked with the designation symbol of the property class described in clause 3.

The marking is obligatory for property classes equal to or higher than 8.8, preferably on the extreme end of the threaded portion by indenting (see Fig. 3). For studs with interference fit, the marking shall be at the nut end.

Marking is required for studs with nominal diameters equal to or greater than 5 mm.

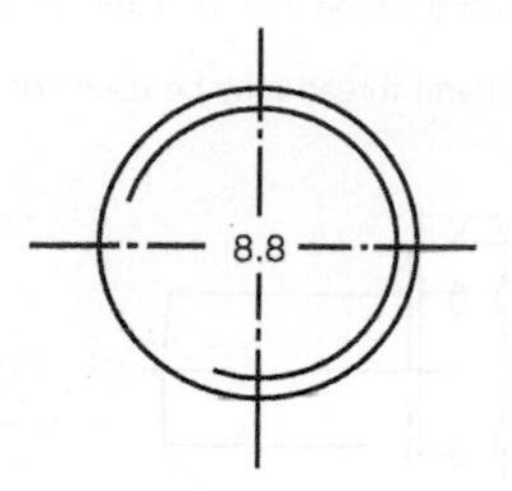

Figure 3 Marking of stud.

The symbols in BS EN ISO 898-1: 1999 Table 2 are permissible as an alternative identification method.

Other types of bolts and screws

The same marking system as described in the sections Hexagon bolts and screws, and Hexagon socket head cap screws shall be used for other types of bolts and screws of property classes 4.6, 5.6 and all classes equal to or higher than 8.8, as described in the appropriate International Standards or, for special components, as agreed between the interested parties.

Marking of left-hand thread

Bolts and screws with left-hand thread shall be marked with the symbol shown in Fig. 4, either on the top of the head or the point.

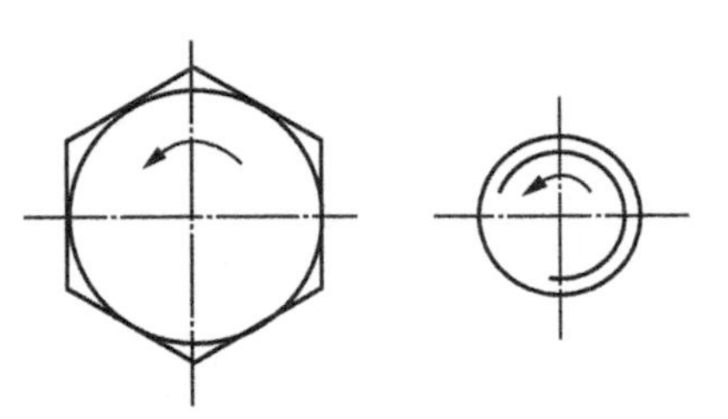

Figure 4 Left-hand thread marking.

Marking is required for bolts and screws with nominal thread diameters $d \geqslant 5\,\text{mm}$.

Alternative marking for left-hand thread may be used for hexagon bolts and screws as shown in Fig. 5.

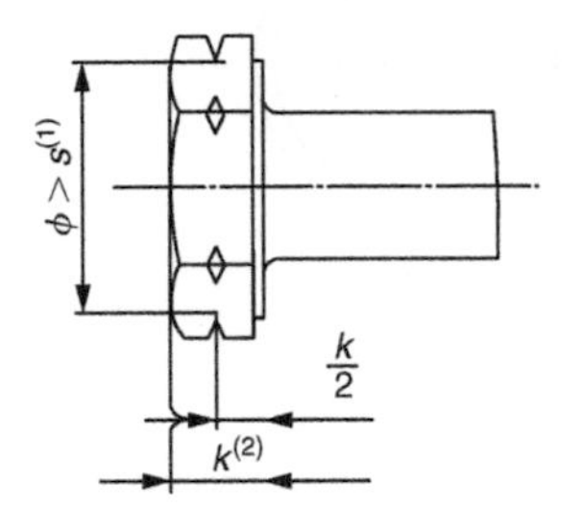

(1) s = width across flats.
(2) k = height of the head.

Figure 5 Alternative left-hand thread marking.

Alternative marking

Alternative or optional permitted marking as stated in the sections Symbols, Identification and Marking of left-hand thread should be left to the choice of the manufacturer.

Trade (identification) marking

The trade (identification) marking of the manufacturer is mandatory on all products which are marked with property classes.

4.1.30 BS EN 20898-2: 1994 Mechanical properties of fasteners: nuts with specified proof load values – coarse thread

BS EN 20898-2: 1994 Table 1 Mechanical properties

Thread		Property class														
		04					05									
		Stress under proof load S_p (N/mm²)	Vickers hardness HV		Nut		Stress under proof load S_p (N/mm²)	Vickers hardness HV		Nut		Stress under proof load S_p (N/mm²)	Vickers hardness HV		Nut	
greater than	less than or equal to		min.	max.	state	style		min.	max.	state	style		min.	max.	state	style
–	M4															
M4	M7															
M7	M10	380	188	302	NQT[a]	thin	500	272	353	QT[b]	thin	–	–	–	–	–
M10	M16															
M16	M39											510	117	302	NQT[a]	1

continued

BS EN 20898-2: 1994 Table 1 Mechanical properties (*continued*)

Thread		Property class																			
		5[c]					6					8									
		Stress under proof load S_p (N/mm^2)	Vickers hardness HV		Nut		Stress under proof load S_p (N/mm^2)	Vickers hardness HV		Nut		Stress under proof load S_p (N/mm^2)	Vickers hardness HV		Nut		Stress under proof load S_p (N/mm^2)	Vickers hardness HV		Nut	
greater than	less than or equal to		min.	max.	state	style		min.	max.	state	style		min.	max.	state	style		min.	max.	state	style
–	M4	520					600					800	180								
M4	M7	580	130				670	150				855		302	NQT[a]		–	–	–	–	–
M7	M10	590		302	NQT[a]	1	680		302	NQT[a]	1	870	200			1					
M10	M16	610					700					880									
M16	M38	630	146				720	170				920	233	353	QT[b]		990	180	302	NQT[a]	2

Thread		Property class																			
		9					10					12									
		Stress under proof load S_p (N/mm^2)	Vickers hardness HV		Nut		Stress under proof load S_p (N/mm^2)	Vickers hardness HV		Nut		Stress under proof load S_p (N/mm^2)	Vickers hardness HV		Nut		Stress under proof load S_p (N/mm^2)	Vickers hardness HV		Nut	
greater than	less than or equal to		min.	max.	state	style		min.	max.	state	style		min.	max.	state	style		min.	max.	state	style
–	M4	900	170				1 040					1 140					1 150				
M4	M7	915					1 040					1 140					1 150				
M7	M10	940	188	302	NQT[a]	2	1 040	272	353	QT[b]	1	1 140	295	353	QT[b]	1	1 160	272	353	QT[b]	2
M10	M16	950					1 050					1 170					1 190				
M16	M38	920					1 060					–	–	–	–	–	1 200				

[a] NQT = not quenched or tempered.
[b] QT = quenched and tempered.
[c] The maximum bolt hardness of property classes 5.6 and 5.8 will be changed to be 220 HV in the next revision of ISO 896-1: 1988. This is the maximum bolt hardness in the thread engagement areas whereas only the thread end or the head may have a maximum hardness of 250 HV. Therefore the values of stress under proof load are based on a maximum bolt hardness of 220 HV.
Note: Minimum hardness is mandatory only for heat-treated nuts and nuts too large to be proof-load tested. For all other nuts, minimum hardness is not mandatory but is provided for guidance only.
For nuts which are not hardened and tempered, and which satisfy the proof-load test, minimum hardness shall not be cause for rejection.

Proof load values

Proof load values are given in BS EN 20898-2: 1994 Table 2.
The nominal stress area A_s is calculated as follows:

$$A_s = \frac{\pi}{4}\left(\frac{d_2 + d_3}{2}\right)^2$$

where:

d_2 = basic pitch diameter of the external thread (see ISO 724);
d_3 = minor diameter of the external thread;

$$d_3 = d_1 - \frac{H}{6}$$

where:

d_1 = basic minor diameter of the external thread;
H = height of the fundamental triangle of the thread.

BS EN 20898-2: 1994 Table 2 Proof load values – Coarse thread

Thread	Thread pitch (mm)	Nominal stress area of the mandrel A_s (mm^2)	Property class										
			04	05	4	5	6	8		9	10	12	
			Proof load ($A_s \times S_p$) (N)										
					style 1	style 1	style 1	style 1	style 2	style 2	style 1	style 1	style 2
M3	0.5	5.03	1 910	2 500	–	2 600	3 000	4 000	–	4 500	5 200	5 700	5 800
M3.5	0.6	6.78	2 580	3 400	–	3 550	4 050	5 400	–	6 100	7 050	7 700	7 800
M4	0.7	8.78	3 340	4 400	–	4 550	5 250	7 000	–	7 900	9 150	10 000	10 100
M5	0.8	14.2	5 400	7 100	–	8 250	9 500	12 140	–	13 000	14 800	16 200	16 300
M6	1	20.1	7 640	10 000	–	11 700	13 500	17 200	–	18 400	20 900	22 900	23 100
M7	1	28.9	11 000	14 500	–	16 800	19 400	24 700	–	26 400	30 100	32 900	33 200
M8	1.25	36.6	13 900	18 300	–	21 600	24 900	31 800	–	34 400	38 100	41 700	42 500
M10	1.5	58	22 000	29 000	–	34 200	39 400	50 500	–	54 500	60 300	66 100	67 300
M12	1.75	84.3	32 000	42 200	–	51 400	59 000	74 200	–	80 100	88 500	98 600	100 300
M14	2	115	43 700	57 500	–	70 200	80 500	101 200	–	109 300	120 800	134 600	136 900
M16	2	157	59 700	78 500	–	95 800	109 900	138 200	–	149 200	164 900	183 700	186 800
M18	2.5	192	73 000	96 000	97 900	121 000	138 200	176 600	170 900	176 600	203 500	–	230 400
M20	2.5	245	93 100	122 500	125 000	154 400	176 400	225 400	218 100	225 400	259 700	–	294 000
M22	2.5	303	115 100	151 500	154 500	190 900	218 200	278 800	269 700	278 800	321 200	–	363 600
M24	3	353	134 100	176 500	180 000	222 400	254 200	324 800	314 200	324 800	374 200	–	423 600
M27	3	459	174 400	229 500	234 100	289 200	330 500	422 300	408 500	422 300	486 500	–	550 800
M30	3.5	561	213 200	280 500	286 100	353 400	403 900	516 100	499 300	516 100	594 700	–	673 200
M33	3.5	694	263 700	347 000	353 900	437 200	499 700	638 500	617 700	638 500	735 600	–	832 800
M36	4	817	310 500	408 500	416 700	514 700	588 200	751 600	727 100	751 600	866 000	–	980 400
M39	4	976	370 900	488 000	497 800	614 900	702 700	897 900	868 600	897 900	1 035 000	–	1 171 000

BS EN 20898-2: 1994 Table 3 Marking symbols for nuts with property classes in accordance with BS EN 20898-2 clause 3.1

	Property class	4	5	6	8	9	10	12[a]
Alternative marking	either designation symbol	4	5	6	8	9	10	12
	or code symbol (clock-face system)	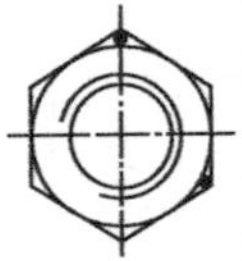	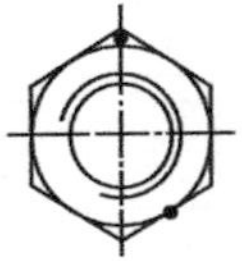			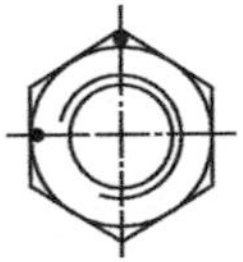	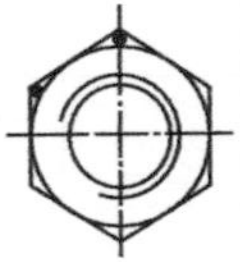	

[a]The marking dot cannot be replaced by the manufacturer's mark.

BS EN 20898-2: 1994 Table 4 Marking symbols for nuts with property classes in accordance with BS EN 20898-2 clause 3.2

Property class	04	05
Marking	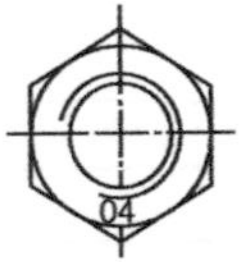 04	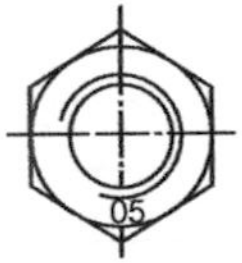 05

Marking of left-hand thread

Nuts with left-hand thread shall be marked as shown in Fig. 1 on one bearing surface of the nut by indenting.

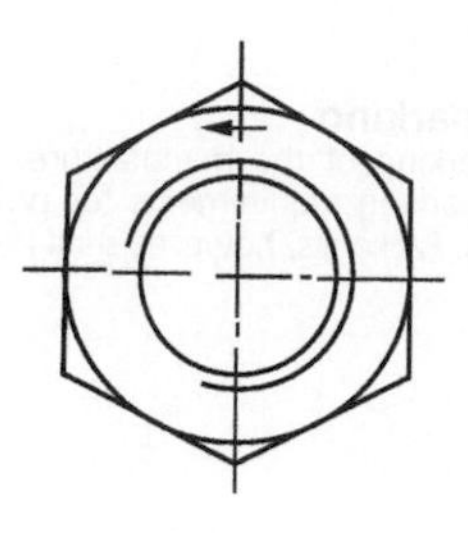

Figure 1 Left-hand thread marking.

Marking is required for nuts with threads ⩾M5.

The alternative marking for left-hand thread shown in Fig. 2 may also be used.

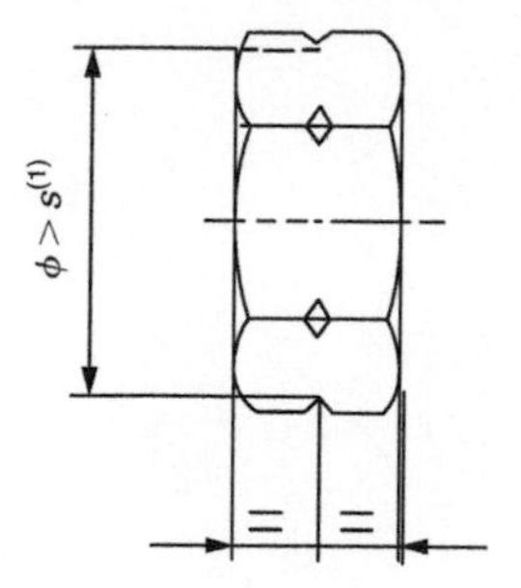

(1) s = width across flats.

Figure 2 Alternative left-hand thread marking.

Alternative marking

Alternative or optional permitted marking as stated in the sections Symbols, Identification and Marking of left-hand thread under Section 4.1.29 is left to the choice of the manufacturer.

Trade (identification) marking

The trade (identification) marking of the manufacturer is mandatory on all products covered by the obligatory marking requirements for property classes, provided this is possible for technical reasons. Packages, however, shall be marked in all cases.

4.1.31 BS EN ISO 898-6: 1996 Mechanical properties of fasteners: nuts with specified proof load values – fine pitch thread

BS EN ISO 898-6: 1996 Table 1 Designation system for nuts with nominal heights ⩾0.8*d*

Property class of nut	Mating bolts		Nuts	
			Style 1	Style 2
	Property class	Nominal thread diameter range (mm)	Nominal thread diameter range (mm)	
5	3.6, 4.6, 4.8	$d \leqslant 39$	$d \leqslant 39$	–
	5.6, 5.8			
6	6.8	$d \leqslant 39$	$d \leqslant 39$	–
8	8.8	$d \leqslant 39$	$d \leqslant 39$	$d \leqslant 16$
10	10.9	$d \leqslant 39$	$d \leqslant 16$	$d \leqslant 39$
12	12.9	$d \leqslant 16$	–	$d \leqslant 16$

Note: In general, nuts of a higher property class can replace nuts of a lower property class. This is advisable for a bolt/nut assembly going into a stress higher than the yield stress or the stress under proof load of the bolt.

However, should tightening beyond bolt proof load take place, the net design is intended to ensure at least 10% of the over-tightened assemblies fail through bolt breakage in order to warn the user that the installation practice is not appropriate.

Note: For more detailed information on the strength of screw thread assemblies and for the styles of nuts, see ISO 898-2: 1992, Annex A.

Nuts with nominal heights $\geqslant 0.5d$ and $<0.8d$ (effective heights of thread $\geqslant 0.4d$ and $<0.6d$)

Nuts with nominal heights $\geqslant 0.5d$ and $<0.8d$ (effective height of thread $\geqslant 0.4d$ and $<0.6d$) are designated by a combination of two numbers: the second indicates the nominal stress under proof load on a hardened test mandrel, while the first indicates that the load ability of a bolt-nut assembly is reduced in comparison with the loadability on a hardened test mandrel and also in comparison with a bolt-nut assembly described in Section 3.1. The effective loading capacity is not only determined by the hardness of the nut and the effective height of thread but also by the tensile strength of the bolt with which the nut is assembled. BS EN ISO 898-6: 1996 Table 2 gives the designation system and the stresses under proof load of the nuts. Proof loads are shown in BS EN ISO 898-6: 1996 Table 3. A guide for minimum expected stripping strengths of the joints when these nuts are assembled with bolts of various property classes is shown in BS EN ISO 898-6: 1996 Table 4.

BS EN ISO 898-6: 1996 Table 2 Designation system and stresses under proof load for nuts with nominal heights $\geqslant 0.5d$ and $<0.8d$

Property class of nut	Nominal stress under proof load (N/mm^2)	Actual stress under proof load (N/mm^2)
04	400	380
05	500	500

BS EN ISO 898-6: 1996 Table 3 Proof load values

Thread size, $d \times P$	Nominal stress area of mandrel A_s (mm^2)	Property class								
		04	05	5	6	8		10		12
		Proof load ($A_s \times S_p$) (N)								
				style 1	style 1	style 1	style 2	style 1	style 2	style 2
M8 ×1	39.2	14 900	19 600	27 000	30 200	37 400	34 900	43 100	41 400	47 000
M10 ×1	64.5	24 500	32 200	44 500	49 700	61 600	57 400	71 000	68 000	77 400
M10 ×1.25	61.2	23 300	30 600	44 200	47 100	58 400	54 500	67 300	64 600	73 400
M12 ×1.25	92.1	35 000	46 000	63 500	71 800	88 000	82 000	102 200	97 200	110 500
M12 ×1.5	88.1	33 500	44 000	60 800	68 700	84 100	78 400	97 800	92 900	105 700
M14 ×1.5	125	47 500	62 500	86 300	97 500	119 400	111 200	138 800	131 900	150 000
M16 ×1.5	167	63 500	83 500	115 200	130 300	159 500	148 600	185 400	176 200	200 400
M18 ×1.5	215	81 700	107 500	154 800	187 000	221 500	–	–	232 200	–
M18 ×2	204	77 500	102 000	146 900	177 500	210 100	–	–	220 300	–
M20 ×1.5	272	103 400	136 000	195 800	236 600	280 200	–	–	293 800	–
M20 × 2	258	98 000	129 000	185 800	224 500	265 700	–	–	278 600	–
M22 × 1.5	333	126 500	166 500	239 800	289 700	343 000	–	–	359 600	–
M22 × 2	318	120 800	159 000	229 000	276 700	327 500	–	–	343 400	–
M24 × 2	384	145 900	192 000	276 500	334 100	395 500	–	–	414 700	–
M27 × 2	496	188 500	248 000	351 100	431 500	510 900	–	–	535 700	–
M30 × 2	621	236 000	310 500	447 100	540 300	639 600	–	–	670 700	–
M33 × 2	761	289 200	380 500	547 900	662 100	783 800	–	–	821 900	–
M36 × 3	865	328 700	432 500	622 800	804 400	942 800	–	–	934 200	–
M39 × 3	1 030	391 400	515 000	741 600	957 900	1 123 000	–	–	1 112 000	–

BS EN ISO 898-6: 1996 Table 4 Minimum stripping strength of nuts as a percentage of the proof load of bolts

Property class of the nut	Minimum stripping strength of nuts as a percentage of the proof load of bolts with property classes			
	6.8	**8.8**	**10.9**	**12.9**
04	85	65	45	40
05	100	85	60	50

Materials

Nuts of property classes 05, 8 (style 1), 10 and 12 shall be hardened and tempered.

Nuts shall be made of steel conforming to the chemical composition limits specified in BS EN ISO 898-6: 1996 Table 5. The chemical composition shall be analysed in accordance with relevant International Standards.

BS EN ISO 898-6: 1996 Table 5 Limits of chemical composition

Property class		Chemical composition limits (check analysis), %			
		C max.	Mn min.	P max.	S max.
5[a]; 6	–	0.50	–	0.060	0.150
8[b]	**04[a]**	0.58	0.25	0.060	0.150
10[b]	**05[b]**	0.58	0.30	0.048	0.058
12[b]	–	0.58	0.45	0.048	0.058

[a] Nuts of this property class may be manufactured from free-cutting steel unless otherwise agreed between the purchaser and the manufacturer. In such cases, the following maximum sulphur, phosphorus and lead contents are permissible:

Sulphur 0.34%; phosphorus 0.11%; lead 0.35%

[b] Alloying elements may be added, if necessary to develop the mechanical properties of the nuts.

Mechanical properties

When tested by the methods described in clause 8, the nuts shall have the mechanical properties set out in BS EN ISO 898-6: 1996 Table 1 under Section 4.1.30.

Proof load values

Proof load values are given in BS EN ISO 898-6: 1996 Table 3.

The nominal stress area A_s is calculated as follows:

$$A_s = \frac{\pi}{4}\left(\frac{d_2 + d_3}{2}\right)^2$$

where:

d_2 = basic pitch diameter of the external thread (see ISO 724);
d_3 = minor diameter of the external thread;

$$d_3 = d_1 - \frac{H}{6}$$

where:

d_1 = basic minor diameter of the external thread;
H = height of the fundamental triangle of the thread.

4.1.32 BS EN 20898-7: 1995 Mechanical properties of fasteners: torsional test and minimum torques for bolts and screws with nominal diameters 1–10 mm

BS EN 20898-7: 1995 Table 1 Strength ratio *X*

Property class	8.8	9.8	10.9	12.9
Ratio *X*	0.84	0.815	0.79	0.75

BS EN 20898-7: 1995 Table 2 Minimum breaking torques

Thread	**Pitch (mm)**	**Minimum breaking torque[a] $M_{B\,min}$ (N·m)**			
		Property class			
		8.8	**9.8**	**10.9**	**12.9**
M1	0.25	0.033	0.036	0.040	0.045
M1.2	0.25	0.075	0.082	0.092	0.10
M1.4	0.3	0.12	0.13	0.14	0.16
M1.6	0.35	0.16	0.18	0.20	0.22
M2	0.4	0.37	0.40	0.45	0.50
M2.5	0.45	0.82	0.90	1.0	1.1
M3	0.5	1.5	1.7	1.9	2.1
M3.5	0.6	2.4	2.7	3.0	3.3
M4	0.7	3.6	3.9	4.4	4.9
M5	0.8	7.6	8.3	9.3	10
M6	1	13	14	16	17
M7	1	23	25	28	31
M8	1.25	33	36	40	44
M8 ×1	–	38	42	46	52
M10	1.5	66	72	81	90
M10 ×1	–	84	92	102	114
M10 ×1.25	–	75	82	91	102

[a] These minimum breaking torques are valid for bolts and screws with the thread tolerances 6g, 6f and 6e.

4.1.33 BS EN ISO 4762: 2004 Metric hexagon socket head screws

ISO 10683, *Fasteners – Non-electrolytically applied zinc flake coatings.*
ISO 23429, *Gauging of hexagon sockets.*
See Fig. 1 and BS EN ISO 4762: 2004 Table 1.

Dimensions

Note: Symbols and designations of dimensions are defined in ISO 225.

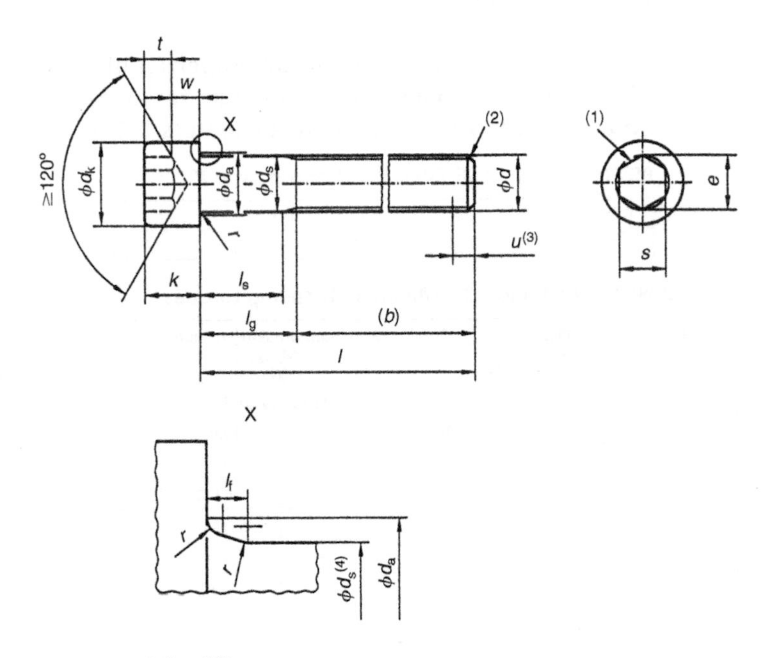

Maximum underhead fillet

$l_{f,max} = 1.7\ r_{max}$

$$r_{max} = \frac{d_{a,max} - d_{s,max}}{2}$$

r_{min}, see BS EN ISO 4762: 2004 Table 1

Figure 1

Permissible alternative form of socket

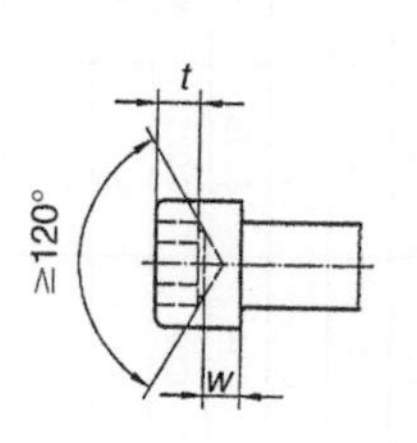

Top and bottom edge of the head

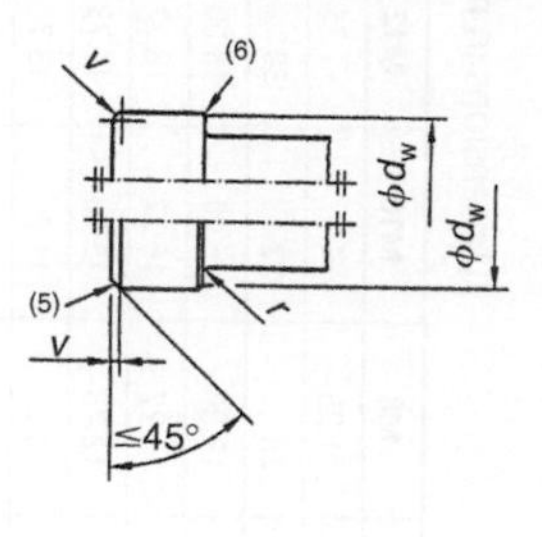

For broached sockets which are at the maximum limit of size the overcut resulting from drilling shall not exceed 1/3 of the length of any flat of the socket which is $e/2$.

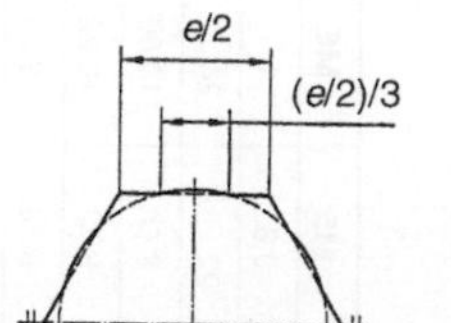

(1) A slight rounding or countersink at the mouth of the socket is permissible.
(2) Point chamfered or for sizes M4 and below 'as rolled' according to ISO 4753.
(3) Incomplete thread $u \leq 2P$.
(4) d_s applies if values of $l_{s,min}$ are specified.
(5) Top edge of head may be rounded or chamfered as shown at the option of the manufacturer.
(6) Bottom edge of head may be rounded or chamfered to d_w but in every case shall be free from burrs.

Figure 1 ***continued***

BS EN ISO 4762: 2004 Table 1

Dimensions in mm

Thread size, d		M1.6	M2	M2.5	M3	M4	M5	M6	M8	M10	M12
P^a		0.35	0.4	0.45	0.5	0.7	0.8	1	1.25	1.5	1.75
b^b	ref.	15	16	17	18	20	22	24	28	32	36
d_k	max.[c]	3.00	3.80	4.50	5.50	7.00	8.50	10.00	13.00	16.00	18.00
	max.[d]	3.14	3.98	4.68	5.68	7.22	8.72	10.22	13.27	16.27	18.27
	min.	2.86	3.62	4.32	5.32	6.78	8.28	9.78	12.73	15.73	17.73
d_a	max.	2	2.6	3.1	3.6	4.7	5.7	6.8	9.2	11.2	13.7
d_s	max.	1.60	2.00	2.50	3.00	4.00	5.00	6.00	8.00	10.00	12.00
	min.	1.46	1.86	2.36	2.86	3.82	4.82	5.82	7.78	9.78	11.73
$e^{e,f}$	min.	1.733	1.733	2.303	2.873	3.443	4.583	5.723	6.863	9.149	11.429
l_f	max.	0.34	0.51	0.51	0.51	0.6	0.6	0.68	1.02	1.02	1.45
k	max.	1.60	2.00	2.50	3.00	4.00	5.00	6.0	8.00	10.00	12.00
	min.	1.46	1.86	2.36	2.86	3.82	4.82	5.7	7.64	9.64	11.57
r	min.	0.1	0.1	0.1	0.1	0.2	0.2	0.25	0.4	0.4	0.6
s^f	nom.	1.5	1.5	2	2.5	3	4	5	6	8	10
	max.	1.58	1.58	2.08	2.58	3.08	4.095	5.14	6.14	8.175	10.175
	min.	1.52	1.52	2.02	2.52	3.02	4.020	5.02	6.02	8.025	10.025
t	min.	0.7	1	1.1	1.3	2	2.5	3	4	5	6
v	max.	0.16	0.2	0.25	0.3	0.4	0.5	0.6	0.8	1	1.2
d_w	min.	2.72	3.48	4.18	5.07	6.53	8.03	9.38	12.33	15.33	17.23
w	min.	0.55	0.55	0.85	1.15	1.4	1.9	2.3	3.3	4	4.8

l_g			Shank length l_s and grip length l_g																			
			l_s	l_g	l_s	l_g	l_s	l_g	l_s	l_g	l_s	l_g	l_s	l_g	l_s	l_g	l_s	l_g	l_s	l_g	l_s	l_g
nom.	min.	max.	min.	max.	min.	max.	min.	max.	min.	max.	min.	max.	min.	max.	min.	max.	min.	max.	min.	max.	min.	max.
2.5	2.3	2.7																				
3	2.8	3.2																				
4	3.76	4.24																				
5	4.76	5.24																				
6	5.76	6.24																				
8	7.71	8.29																				
10	9.71	10.29																				
12	11.65	12.35																				
16	15.65	16.35																				
20	19.58	20.42		2		4																
25	24.58	25.42					5.75	8	4.5	7												
30	29.58	30.42							9.5	12	6.5	10	4	8								
35	34.5	35.5									11.5	15	9	13	6	11						
40	39.5	40.5									16.5	20	14	18	11	16	5.75	12				
45	44.5	45.5											19	23	16	21	10.75	17	5.5	13		
50	49.5	50.5											24	28	21	26	15.75	22	10.5	18		
55	54.4	55.6													26	31	20.75	27	15.5	23	10.25	19
60	59.4	60.6													31	36	25.75	32	20.5	28	15.25	24
65	64.4	65.6															30.75	37	25.5	33	20.25	29
70	69.4	70.6															35.75	42	30.5	38	25.25	34
80	79.4	80.6															45.75	52	40.5	48	35.25	44

continued

BS EN ISO 4762: 2004 Table 1 *(continued)*

l_g			Shank length l_s and grip length l_g																			
			l_s	l_g	l_s	l_g	l_s	l_g	l_s	l_g	l_s	l_g	l_s	l_g	l_s	l_g	l_s	l_g	l_s	l_g	l_s	l_g
nom.	min.	max.	min.	max.	min.	max.	min.	max.	min.	max.	min.	max.	min.	max.	min.	max.	min.	max.	min.	max.	min.	max.
90	89.3	90.7																	50.5	58	45.25	54
100	99.3	100.7																	60.5	68	55.25	64
110	109.3	110.7																			65.25	74
120	119.3	120.7																			75.25	84
130	129.2	130.8																				
140	139.2	140.8																				
150	149.2	150.8																				
160	159.2	160.8																				
180	179.2	180.8																				
200	199.075	200.925																				
220	219.075	220.925																				
240	239.075	240.925																				
260	258.95	261.05																				
280	278.95	281.05																				
300	298.95	301.05																				

Thread size, d		(M14)[h]	M16	M20	M24	M30	M36	M42	M48	M56	M64
P[a]		2	2	2.5	3	3.5	4	4.5	5	5.5	6
b[b]	ref.	40	44	52	60	72	84	96	108	124	140
d_k	max.[c]	21.00	24.00	30.00	36.00	45.00	54.00	63.00	72.00	84.00	96.00
	max.[d]	21.33	24.33	30.33	36.39	45.39	54.46	63.46	72.46	84.54	96.54
	min.	20.67	23.67	29.67	35.61	44.61	53.54	62.54	71.54	83.46	95.46
d_a	max.	15.7	17.7	22.4	26.4	33.4	39.4	45.6	52.6	63	71
d_s	max.	14.00	16.00	20.00	24.00	30.00	36.00	42.00	48.00	56.00	64.00
	min.	13.73	15.73	19.67	23.67	29.67	35.61	41.61	47.61	55.54	63.54
e[e,f]	min.	13.716	15.996	19.437	21.734	25.154	30.854	36.571	41.131	46.831	52.531
l_f	max.	1.45	1.45	2.04	2.04	2.89	2.89	3.06	3.91	5.95	5.95
k	max.	14.00	16.00	20.00	24.00	30.00	36.00	42.00	48.00	56.00	64.00
	min.	13.57	15.57	19.48	23.48	29.48	35.38	41.38	47.38	55.26	63.26
r	min.	0.6	0.6	0.8	0.8	1	1	1.2	1.6	2	2
s[f]	nom.	12	14	17	19	22	27	32	36	41	46
	max.	12.212	14.212	17.23	19.275	22.275	27.275	32.33	36.33	41.33	46.33
	min.	12.032	14.032	17.05	19.065	22.065	27.065	32.08	36.08	41.08	46.08
t	min.	7	8	10	12	15.5	19	24	28	34	38
v	max.	1.4	1.6	2	2.4	3	3.6	4.2	4.8	5.6	6.4
d_w	min.	20.17	23.17	28.87	34.81	43.61	52.54	61.34	70.34	82.26	94.26
w	min.	5.8	6.8	8.6	10.4	13.1	15.3	16.3	17.5	19	22

continued

BS EN ISO 4762: 2004 Table 1 *(continued)*

l_g			Shank length l_s and grip length l_g																	
			l_s	l_g	l_s	l_g	l_s	l_g	l_s	l_g	l_s	l_g	l_s	l_g	l_s	l_g	l_s	l_g	l_s	l_g	l_s	l_g
nom.	min.	max.	min.	max.	min.	max.	min.	max.	min.	max.	min.	max.	min.	max.	min.	max.	min.	max.	min.	max.	min.	max.
2.5	2.3	2.7																				
3	2.8	3.2																				
4	3.76	4.24																				
5	4.76	5.24																				
6	5.76	6.24																				
8	7.71	8.29																				
10	9.71	10.29																				
12	11.65	12.35																				
16	15.65	16.35																				
20	19.58	20.42																				
25	24.58	25.42																				
30	29.58	30.42																				
35	34.5	35.5																				
40	39.5	40.5																				
45	44.5	45.5																				
50	49.5	50.5																				
55	54.4	55.6																				
60	59.4	60.6	10	20																		
65	64.4	65.6	15	25	11	21																
70	69.4	70.6	20	30	16	26																
80	79.4	80.6	30	40	26	36	15.5	28														

90	89.3	90.7	40	50	36	46	25.5	38	15	30												
100	99.3	100.7	50	60	46	56	35.5	48	25	40												
110	109.3	110.7	60	70	56	66	45.5	58	35	50	20.5	38										
120	119.3	120.7	70	80	66	76	55.5	68	45	60	30.5	48	16	36								
130	129.2	130.8	80	90	76	86	65.5	78	55	70	40.5	58	26	46								
140	139.2	140.8	90	100	86	96	75.5	88	65	80	50.5	68	36	56	21.5	44						
150	149.2	150.8			96	106	85.5	98	75	90	60.5	78	46	66	31.5	54						
160	159.2	160.8			106	116	95.5	108	85	100	70.5	88	56	76	41.5	64	27	52				
180	179.2	180.8					115.5	128	105	120	90.5	108	76	96	61.5	84	47	72	28.5	56		
200	199.075	200.925					135.5	148	125	140	110.5	128	96	116	81.5	104	67	92	48.5	76	30	60
220	219.075	220.925													101.5	124	87	112	68.5	96	50	80
240	239.075	240.925													121.5	155	107	132	88.5	116	70	100
260	258.95	261.05													141.5	164	127	152	108.5	136	90	120
280	278.95	281.05													161.5	184	147	172	128.5	156	110	140
300	298.95	301.05													181.5	204	167	192	148.5	176	130	160

[a] P = pitch of the thread.
[b] For lengths between the bold stepped lines in the unshaded area.
[c] For plain heads.
[d] For knurled heads.
[e] $e_{min} = 1{,}14\ s_{min}$.
[f] Combined gauging of socket dimensions e and s, see ISO 23429.
[g] The range of commercial lengths is between the bold stepped lines. Lengths in the shaded area are threaded to the head within $3P$. Lengths below the shaded area have values of l_g and l_s in accordance with the following formulae:
$l_{g,max} = l_{nom} - b$; $l_{s,min} = l_{g,max} - 5P$.
[h] The size in brackets should be avoided if possible.

4.1.34 BS EN ISO 10642: 2004 Hexagon socket countersunk head screws

Dimensions and gauging of head

See Figs 1 and 2, and BS EN ISO 10642: 2004 Table 1.

Dimensions

Note: Symbols and designations of dimensions are defined in ISO 225.

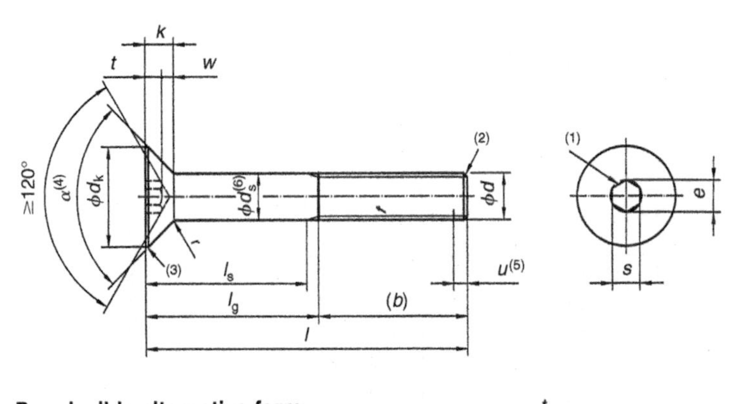

Figure 1 Hexagon socket countersunk head screws.

For broached sockets which are at the maximum limit of size the overcut resulting from drilling shall not exceed 1/3 of the length of any flat of the socket which is $e/2$.

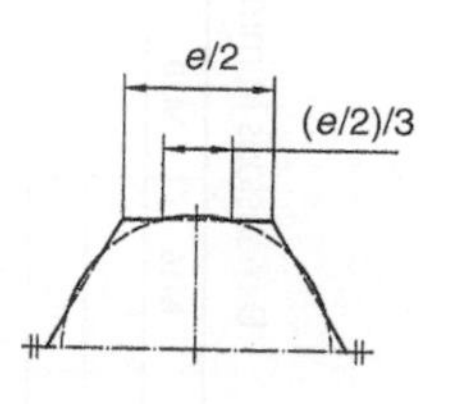

(1) A slight rounding or countersink at the mouth of the socket is permissible.
(2) Point to be chamfered or for sizes M4 and below 'as rolled' in accordance to ISO 4753.
(3) Edge of the head to be truncated or rounded.
(4) $\alpha = 90°-92°$
(5) Incomplete thread $u \leq 2P$.
(6) d_s applies if values of $l_{s,min}$ are specified.

Figure 1 ***continued***

Gauging of head

The top surface of the screw shall be located between the gauge surfaces A and B.

Tolerances in mm

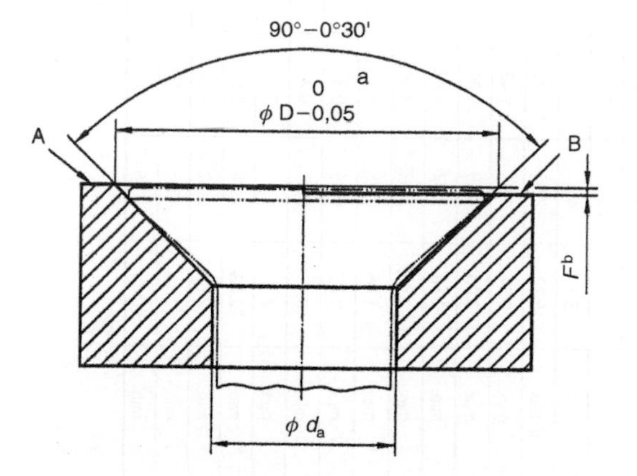

[a] $D = d_{k,\ theor.,\ max}$ (see BS EN ISO 10642: 2004 Table 1).
[b] F = flushness tolerance of the head (see BS EN ISO 10642: 2004 Table 1).

Figure 2 Flushness gauge.

BS EN ISO 10642: 2004 Table 1

Dimensions in mm

Thread size, d			M3	M4	M5	M6	M8	M10	M12	(M14)[g]	M16	M20
P^a			0.5	0.7	0.8	1	1.25	1.5	1.75	2	2	2.5
b^b	ref.		18	20	22	24	28	32	36	40	44	52
d_a	max.		3.3	4.4	5.5	6.6	8.54	10.62	13.5	15.5	17.5	22
d_k	theoretical max.		6.72	8.96	11.20	13.44	17.92	22.40	26.88	30.8	33.60	40.32
	actual	min.	5.54	7.53	9.43	11.34	15.24	19.22	23.12	26.52	29.01	36.05
d_s		max.	3.00	4.00	5.00	6.00	8.00	10.00	12.00	14.00	16.00	20.00
		min.	2.86	3.82	4.82	5.82	7.78	9.78	11.73	13.73	15.73	19.67
$e^{c,d}$		min.	2.303	2.873	3.443	4.583	5.723	6.863	9.149	11.429	11.429	13.716
k		max.	1.86	2.48	3.1	3.72	4.96	6.2	7.44	8.4	8.8	10.16
F^e		max.	0.25	0.25	0.3	0.35	0.4	0.4	0.45	0.5	0.6	0.75
r		min.	0.1	0.2	0.2	0.25	0.4	0.4	0.6	0.6	0.6	0.8
		nom.	2	2.5	3	4	5	6	8	10	10	12
s^d		max.	2.08	2.58	3.08	4.095	5.14	6.140	8.175	10.175	10.175	12.212
		min.	2.02	2.52	3.02	4.020	5.02	6.020	8.025	10.025	10.025	12.032
t		min.	1.1	1.5	1.9	2.2	3	3.6	4.3	4.5	4.8	5.6
w		min.	0.25	0.45	0.66	0.7	1.16	1.62	1.8	1.62	2.2	2.2

l^{f}			Shank length l_s and grip length l_g																			
			l_s	l_g	l_s	l_g	l_s	l_g	l_s	l_g	l_s	l_g	l_s	l_g	l_s	l_g	l_s	l_g	l_s	l_g	l_s	l_g
nom.	min.	max.	min.	max.	min.	max.	min.	max.	min.	max.	min.	max.	min.	max.	min.	max.	min.	max.	min.	max.	min.	max.
8	7.71	8.29																				
10	9.71	10.29																				
12	11.65	12.35																				
16	15.65	16.35																				
20	19.58	20.42																				
25	24.58	25.42																				
30	29.58	30.42	9.5	12	6.5	10																
35	34.5	35.5			11.5	15	9	13														
40	39.5	40.5			16.5	20	14	18	11	16												
45	44.5	45.5					19	23	16	21												
50	49.5	50.5					24	28	21	26	15.75	22										
55	54.4	55.6							26	31	20.75	27	15.5	23								
60	59.4	60.6							31	36	25.75	32	20.5	28								
65	64.4	65.6									30.75	37	25.5	33	20.25	29						
70	69.4	70.6									35.75	42	30.5	38	25.25	34	20	30				
80	79.4	80.6									45.75	52	40.5	48	35.25	44	30	40	26	36		
90	89.3	90.7											50.5	58	45.25	54	40	50	36	46		
100	99.3	100.7											60.5	68	55.25	64	50	60	46	56	35.5	48

[a] P = pitch of the thread.
[b] For lengths between the bold stepped lines in the unshaded area.
[c] $e_{min} = 1{,}14\ s_{min}$.
[d] Combined gauging of socket dimensions e and s, see ISO 23429.
[e] F is the flushness of the head, see Fig. 2. The guage dimension F has the tolerance $^{\ 0}_{-0.01}$.

[f] The range of commercial lengths is between the bold stepped lines. Lengths in the shaded area are threaded to the head within $3P$. Lengths below the shaded area have values of l_g and l_s in accordance with the following formulae:
$l_{g,max} = l_{nom} - b$; $l_{s,min} = l_{g,max} - 5P$.
[g] The size in brackets should be avoided if possible.

Requirements and reference International Standards

See BS EN ISO 10642: 2004 Tables 2 and 3.

BS EN ISO 10642: 2004 Table 2 Requirements and reference International Standards

Material		Steel
General requirements	International Standard	ISO 8992
Thread	Tolerance	6g for property classes 8.8 and 10.9; 5g6g for property class 12.9
	International Standards	ISO 261, ISO 965-2, ISO 965-3
Mechanical properties	Property class[a]	8.8, 10.9, 12.9
	International Standard	ISO 898-1
Tolerances	Product grade	A
	International Standard	ISO 4759-1
Finish		As processed Requirements for electroplating are covered in ISO 4042 Requirements for non-electrolytically applied zinc flake coatings are covered in ISO 10683
Surface discontinuities		Limits for surface discontinuities are given in ISO 6157-1 and ISO 6157-3 for property class 12.9
Acceptability		Acceptance procedure is covered in ISO 3269

[a] Because of their head configurations, these screws may not meet the minimum ultimate tensile load for property classes 8.8, 10.9 and 12.9, specified in ISO 898-1, when tested in accordance with test programme B. They shall nevertheless meet the other material and property requirements for property classes 8.8, 10.9 and 12.9 specified in ISO 898-1. In addition, when full-size screws are loaded with the head supported on a suitable collar (conical bearing surface) using the type of testing fixture illustrated in ISO 898-1, they shall withstand, without fracture, the minimum ultimate tensile loads given in BS EN ISO 10642: 2004 Table 3. If tested to failure, the fracture may occur in the threaded section, the head, the shank or at the head/shank junction.

BS EN ISO 10642: 2004 Table 3 Minimum ultimate tensile loads for hexagon socket countersunk head screws

Thread size, d	Property class		
	8.8	**10.9**	**12.9**
	Minimum ultimate tensile load (N)		
M3	3 220	4 180	4 190
M4	5 620	7 300	8 560
M5	9 080	11 800	13 800
M6	12 900	16 700	19 600
M8	23 400	30 500	35 700
M10	37 100	48 200	56 600
M12	53 900	70 200	82 400
M14	73 600	96 000	112 000
M16	100 000	130 000	154 000
M20	162 000	204 000	239 000

Eighty per cent of the values specified in ISO 898-1.

Designation

Example. A hexagon socket countersunk head screw with thread M12 nominal length l = 40 mm and property class 12.9 is designated as follows:

Hexagon socket countersunk head screw ISO 10642-M12 × 40-12.9

For information concerning hexagon socket countersunk head screws in inch dimensions see BS2470.

4.1.35 BS4827 ISO metric screw threads, miniature series

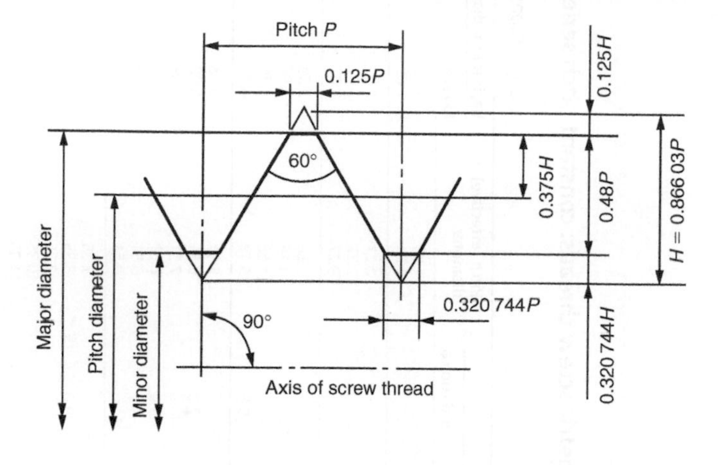

Dimensions in mm

Nominal size		Pitch of thread *P*	Major diameter	Pitch (effective) diameter	Minor diameter
1st choice	2nd choice				
S-0.3		0.080	0.300 000	0.248 038	0.223 200
	S-0.35	0.090	0.350 000	0.291 543	0.263 600
S-0.4		0.100	0.400 000	0.335 048	0.304 000
	S-0.45	0.100	0.450 000	0.385 048	0.354 000
S-0.5		0.125	0.500 000	0.418 810	0.380 000
	S-0.55	0.125	0.550 000	0.468 810	0.430 000
S-0.6		0.150	0.600 000	0.502 572	0.456 000
	S-0.7	0.175	0.700 000	0.586 334	0.532 000
S-0.8		0.200	0.800 000	0.670 096	0.608 000
	S-0.9	0.225	0.900 000	0.753 858	0.684 000
	S-1	0.250	1.000 000	0.837 620	0.760 000
	S-1.1	0.250	1.100 000	0.937 620	0.860 000
	S-1.2	0.250	1.200 000	1.037 620	0.960 000
	S-1.4	0.300	1.400 000	1.205 144	1.112 000

For full range and further information see BS 4827.

4.1.36 BS3643-1/2 ISO metric screw threads: constant pitch series

Dimensions in mm

Pitch of thread	Basic major diameter			Pitch (effective) diameter	Basic minor diameter	
	1st choice	2nd choice	3rd choice		External	Internal
0.25	2.0	–	–	1.84	1.69	1.73
0.25	–	2.2	–	2.04	1.89	1.93
0.35	2.5	–	–	2.27	2.07	2.12
0.35	3.0	–	–	2.77	2.57	2.62
0.35	–	3.5	–	3.27	3.07	3.12
0.50	4.0	–	–	3.68	3.39	3.46
0.50	–	4.5	–	4.18	3.86	3.96
0.50	5.0	–	–	4.68	4.39	4.46
0.50	–	–	5.5	5.18	4.86	4.96
0.75	6.0	–	–	5.51	5.08	5.19
0.75	–	–	7.0	6.51	6.08	6.19
0.75	8.0	–	–	7.51	7.08	7.19
0.75	–	–	9.0	8.51	8.08	8.19
0.75	10.0	–	–	9.51	9.08	9.19
0.75	–	–	11.0	10.51	10.08	10.19
1.0	8.0	–	–	7.35	6.77	6.92
1.0	–	–	9.0	8.35	7.77	7.92
1.0	10.0	–	–	9.35	8.77	8.92
1.0	–	–	11.0	10.35	9.77	9.92
1.0	12.0	–	–	11.35	10.77	10.92

1.0	–	14.0	–	13.35	12.77	12.92
1.0	–	–	15.0	14.35	13.77	13.92
1.0	16.0	–	–	15.35	14.77	14.92
1.0	–	–	17.0	16.35	15.77	15.92
1.0	–	18.0	–	17.35	16.77	16.92
1.0	20.0	–	–	19.35	18.77	18.92
1.0	–	22.0	–	21.35	21.77	21.92
1.0	24.0	–	–	23.35	22.77	22.92
1.0	–	–	25.0	24.35	23.77	23.92
1.0	–	27.0	–	26.35	25.77	25.92
1.0	–	–	28.0	27.35	26.77	26.92
1.0	30.0	–	–	29.35	28.77	28.92
1.25	10.0	–	–	9.19	8.47	8.65
1.25	12.0	–	–	11.19	10.47	10.65
1.25[a]	–	14.0[a]	–	13.19	12.47	12.65
1.5	12.0	–	–	11.03	10.16	10.38
1.5	–	14.0	–	13.03	12.16	12.38
1.5	–	–	15.0	14.03	13.16	13.38
1.5	16.0	–	–	15.03	14.16	14.38
1.5	–	–	17.0	16.03	15.16	15.38
1.5	–	18.0	–	17.03	16.16	16.38
1.5	20.0	–	–	19.03	18.16	18.38
1.5	–	22.0	–	21.03	20.16	20.38
1.5	24.0	–	–	23.03	22.16	22.38
1.5	–	–	25.0	24.03	23.16	23.38
1.5	–	–	26.0	25.03	24.16	24.38
1.5	–	27.0	–	26.03	25.16	25.38
1.5	–	–	28.0	27.03	26.16	26.38
1.5	30.0	–	–	29.03	28.16	28.38

continued

Section 4.1.36 (*continued*)

Dimensions in mm

Pitch of thread	Basic major diameter			Pitch (effective) diameter	Basic minor diameter	
	1st choice	2nd choice	3rd choice		External	Internal
1.5	–	–	32.0	31.03	30.16	30.38
1.5	–	33.0	–	32.03	31.16	31.38
1.5	–	–	35.0	34.03	33.16	33.38
The 1.5 mm pitch series continues to a maximum diameter of 80 mm						
2.0	–	18.0	–	16.70	15.55	15.84
2.0	20.0	–	–	18.70	17.55	17.84
2.0	–	22.0	–	20.70	19.55	19.84
2.0	24.0	–	–	22.70	21.55	21.84
2.0	–	–	25.0	23.70	22.55	22.84
2.0	–	–	26.0	24.70	23.55	23.84
2.0	–	27.0	–	25.70	24.55	24.84
2.0	–	–	28.0	26.70	25.55	25.84
2.0	30.0	–	–	28.70	27.55	27.84
2.0	–	–	32.0	30.70	29.55	29.84
2.0	–	33.0	–	31.70	30.55	30.84
2.0	–	–	35.0	33.70	32.55	32.84
The 2.0 mm pitch series continues to a maximum diameter of 150 mm						
3.0	30.0	–	–	28.05	26.32	26.75
3.0	–	33.0	–	31.05	29.32	29.75
3.0	36.0	–	–	34.05	32.32	32.75
3.0	–	–	38.0	36.05	34.32	34.75
3.0	–	39.0	–	37.05	35.32	35.75
3.0	–	–	40.0	38.05	36.32	36.75
3.0	42.0	–	–	40.05	38.32	38.75
3.0	–	45.0	–	43.05	41.32	41.75
3.0	48.0	–	–	46.05	44.32	44.75

3.0	–	–	50.0	48.05	46.32	46.75
3.0	–	52.0	–	50.05	48.32	48.75
3.0	–	–	55.0	53.05	51.32	51.75
The 3.0 mm pitch series continues to a maximum diameter of 250 mm						
4.0	42.0	–	–	39.40	37.09	37.67
4.0	–	45.0	–	42.40	40.09	40.67
4.0	48.0	–	–	45.40	43.09	43.67
4.0	–	–	50.0	47.40	45.09	45.67
4.0	–	52.0	–	49.40	47.09	47.67
4.0	–	–	55.0	52.40	50.09	50.67
4.0	56.0	–	–	53.40	51.09	51.67
4.0	–	–	58.0	55.40	53.09	53.67
4.0	–	60.0	–	57.40	55.09	55.67
4.0	–	–	62.0	59.40	57.09	57.67
4.0	64.0	–	–	61.40	59.09	59.67
4.0	–	–	65.0	62.40	60.09	60.67
The 4.0 mm pitch series continues to a maximum diameter of 300 mm						
6.0	–	–	70.0	66.10	62.64	63.50
6.0	72.0	–	–	68.10	64.64	65.50
6.0	–	76.0	–	72.10	68.64	69.50
6.0	80.0	–	–	76.10	72.64	73.50
6.0	–	85.0	–	81.10	77.64	78.50
6.0	90.0	–	–	86.10	82.64	83.50
6.0	–	95.0	–	91.10	87.64	88.50
6.0	100.0	–	–	96.10	92.64	93.50
6.0	–	105.0	–	101.10	97.64	98.50
6.0	110.0	–	–	106.10	102.64	103.50
6.0	–	115.0	–	111.10	107.64	108.50
6.0	–	120.0	–	116.10	112.64	113.50
6.0	125.0	–	–	121.10	117.64	118.50
The 6.0 mm pitch series continues to a maximum diameter of 300 mm						

[a] This size sparking plugs only.
For further information see BS 3643.

4.1.37 BS EN ISO 228-1: 2003 Pipe threads where pressure-tight joints are not made on the threads

Part 1: Dimensions, tolerances and designation

Scope

This part of ISO 228 specifies the requirements for thread form, dimensions, tolerances and designation for fastening pipe threads, thread sizes 1/16 to 6 inclusive. Both internal and external threads are parallel threads, intended for the mechanical assembly of the component parts of fittings, cocks and valves, accessories, etc.

These threads are not suitable as jointing threads where a pressure-tight joint is made on the thread. If assemblies with such threads must be made pressure-tight, this should be effected by compressing two tightening surfaces outside the threads, and by interposing and appropriate seal.

Notes:

- For pipe threads where pressure-tight joints are made on the threads, see ISO 7-1.
- ISO 228-2 gives details of methods for verification of fastening thread dimensions and form, and recommended gauging systems.

Normative reference

The following normative document contains provisions which, through reference in this text, constitute provisions of this part of ISO 228. For dated references, subsequent amendments to, or revisions of, any of these publications do not apply. However, parties to agreements based on this part of ISO 228 are encouraged to investigate the possibility of applying the most recent editions of the normative document indicated below. For undated references, the latest edition of the normative document referred to applies. Members of ISO and IEC maintain registers of currently valid International Standards.

ISO 7-1:1994, *Pipe threads where pressure-tight joints are made on the threads – Part 1: Dimensions, tolerances and designation.*

Symbols

For the purposes of this part of ISO 228, the following symbols apply:

A Tighter class of tolerance of external pipe threads where pressure-tight joints are not made on the threads.
B Wider class of tolerance of external pipe threads where pressure-tight joints are not made on the threads.
D $= d$; major diameter of the internal thread.
D_1 $= D - 1{,}280\ 654\ P = d_1$; minor diameter of the internal thread.
D_2 $= D - 0{,}640\ 327\ P = d_2$; pitch diameter of the internal thread.
d Major diameter of the external thread.
d_1 $= d - 1{,}280\ 654\ P$; minor diameter of the external thread.
d_2 $= d - 0{,}640\ 327\ P$; pitch diameter of the external thread.
G Pipe thread where pressure-tight joints are not made on the threads.
H Height of the fundamental triangle of the thread.
h Height of the thread profile with rounded crests and roots.
P Pitch.
r Radius of rounded crests and roots.
T_{D1} Tolerance on the minor diameter of the internal thread.
T_{D2} Tolerance on the pitch diameter of the internal thread.
T_d Tolerance on the major diameter of the external thread.
T_{d2} Tolerance on the pitch diameter of the external thread.

Dimensions

The profile of these threads is identical with that of the parallel thread specified in ISO 7-1. The internal and external threads covered by this part of ISO 228 are both parallel.

Unless otherwise specified, the thread in accordance with this part of ISO 228 is a right-hand thread. (See also clause 5.)

Threads are normally of the truncated form, with crests truncated to the limits of tolerance as given in columns 14 and 15 of BS EN ISO 228-1: 2003 Table 1. The exception to this is on internal threads, where they are likely to be assembled with external threads in accordance with ISO 7-1, and in which case the thread length shall be equal to or greater than that specified in ISO 7-1.

The tolerances on the pitch diameter of the internal threads correspond to the positive deviation of the diameter tolerances in ISO 7-1, with the exception of those for thread sizes 1/16, 1/8, 1/4 and 3/8, for which slightly higher values are specified.

For external threads, two classes of tolerances on the pitch diameter are specified (see BS EN ISO 228-1: 2003 Table 1).

- Class A (column 10) consists of entirely negative tolerances, each equivalent in value to the tolerance for the respective internal thread.
- Class B (column 11) consists of entirely negative tolerances, each with a value of twice that of the respective internal thread.

The choice between class A and class B depends on the conditions of application and shall be made in product standards where threads in accordance with this part of ISO 228 are specified.

Pipe thread dimensions, in millimetres, are given in BS EN ISO 228-1: 2003 Table 1.

Fig. 1 shows fastening threads with full-form profiles and their tolerance zones.

Fig. 2 shows fastening threads with truncated profiles and their tolerance zones.

BS EN ISO 228-1: 2003 Table 1 Thread dimensions

Dimensions in mm

Designation of thread	Number of threads in 25.4 mm	Pitch P	Height of thread h	Diameters			Tolerances on pitch diameter[a]					Tolerance on minor diameter		Tolerance on major diameter	
							Internal thread T_{D2}		External thread T_{d2}			Internal thread T_{D1}		External thread T_d	
				major $d = D$	pitch $d_2 = D_2$	minor $d_1 = D_1$	Lower deviation	Upper deviation	Lower deviation Class A	Lower deviation Class B	Upper deviation	Lower deviation	Upper deviation	Lower deviation	Upper deviation
1	2	3	4	5	6	7	8	9	10	11	12	13	14	15	16
1/16	28	0.907	0.581	7.723	7.142	6.561	0	+0.107	−0.107	−0.214	0	0	+0.282	−0.214	0
1/8	28	0.907	0.581	9.728	9.147	8.566	0	+0.107	−0.107	−0.214	0	0	+0.282	−0.214	0
1/4	19	1.337	0.856	13.157	12.301	11.445	0	+0.125	−0.125	−0.250	0	0	+0.445	−0.250	0
3/8	19	1.337	0.856	16.662	15.806	14.950	0	+0.125	−0.125	−0.250	0	0	+0.445	−0.250	0
1/2	14	1.814	1.162	20.955	19.793	18.631	0	+0.142	−0.142	−0.284	0	0	+0.541	−0.284	0
5/8	14	1.814	1.162	22.911	21.749	20.587	0	+0.142	−0.142	−0.284	0	0	+0.541	−0.284	0
3/4	14	1.814	1.162	26.441	25.279	24.117	0	+0.142	−0.142	−0.284	0	0	+0.541	−0.284	0
7/8	14	1.814	1.162	30.201	29.039	27.877	0	+0.142	−0.142	−0.284	0	0	+0.541	−0.284	0
1	11	2.309	1.479	33.249	31.770	30.291	0	+0.180	−0.180	−0.360	0	0	+0.640	−0.360	0
1 1/8	11	2.309	1.479	37.897	36.418	34.939	0	+0.180	−0.180	−0.360	0	0	+0.640	−0.360	0
1 1/4	11	2.309	1.479	41.910	40.431	38.952	0	+0.180	−0.180	−0.360	0	0	+0.640	−0.360	0
1 1/2	11	2.309	1.479	47.803	46.324	44.845	0	+0.180	−0.180	−0.360	0	0	+0.640	−0.360	0
1 3/4	11	2.309	1.479	53.746	52.267	50.788	0	+0.180	−0.180	−0.360	0	0	+0.640	−0.360	0
2	11	2.309	1.479	59.614	58.135	56.656	0	+0.180	−0.180	−0.360	0	0	+0.640	−0.360	0
2 1/4	11	2.309	1.479	65.710	64.231	62.752	0	+0.217	−0.217	−0.434	0	0	+0.640	−0.434	0
2 1/2	11	2.309	1.479	75.184	73.705	72.226	0	+0.217	−0.217	−0.434	0	0	+0.640	−0.434	0
2 3/4	11	2.309	1.479	81.534	80.055	78.576	0	+0.217	−0.217	−0.434	0	0	+0.640	−0.434	0
3	11	2.309	1.479	87.884	86.405	84.926	0	+0.217	−0.217	−0.434	0	0	+0.640	−0.434	0
3 1/2	11	2.309	1.479	100.330	98.851	97.372	0	+0.217	−0.217	−0.434	0	0	+0.640	−0.434	0
4	11	2.309	1.479	113.030	111.551	110.072	0	+0.217	−0.217	−0.434	0	0	+0.640	−0.434	0
4 1/2	11	2.309	1.479	125.730	124.251	122.772	0	+0.217	−0.217	−0.434	0	0	+0.640	−0.434	0
5	11	2.309	1.479	138.430	139.951	135.472	0	+0.217	−0.217	−0.434	0	0	+0.640	−0.434	0
5 1/2	11	2.309	1.479	151.130	149.651	148.172	0	+0.217	−0.217	−0.434	0	0	+0.640	−0.434	0
6	11	2.309	1.479	163.830	162.351	160.872	0	+0.217	−0.217	−0.434	0	0	+0.640	−0.434	0

[a]For thin-walled parts, the tolerances apply to the mean pitch diameter, which is the arithmetical mean of two diameters measured at right angles to each other.

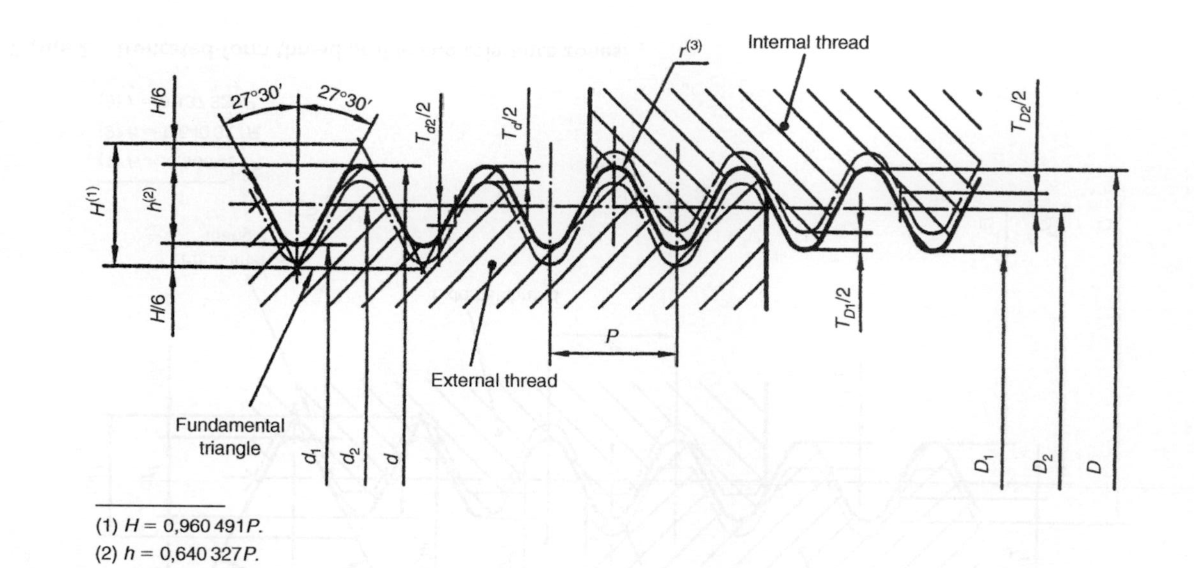

(1) $H = 0{,}960\,491P$.

(2) $h = 0{,}640\,327P$.

(3) $r = 0{,}137\,329P$.

Figure 1 Full-form thread profile and tolerance zones.

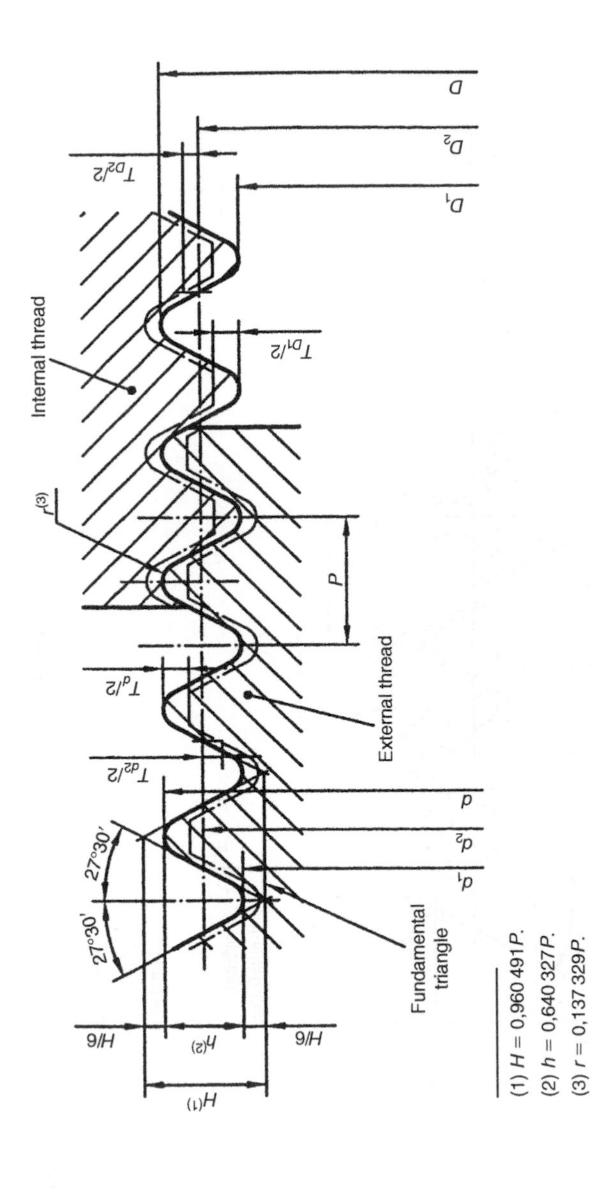

(1) $H = 0{,}960\,491P$.

(2) $h = 0{,}640\,327P$.

(3) $r = 0{,}137\,329P$.

Figure 2 Truncated-form thread profile and tolerance zones.

Designation

The designation of threads shall consist of the following elements in the given sequence:

(a) Description block: pipe thread.
(b) International Standard number block: ISO 228.
(c) Individual item block (one of the following, as applicable):
  - the letter G followed by the designation of the thread size from column 1 of BS EN ISO 228-1: 2003 Table 1 for internal threads (one class of tolerance only); or
  - the letter G followed by the designation of the thread size from column 1 of BS EN ISO 228-1: 2003 Table 1 and the letter A for class A external threads; or
  - the letter G followed by the designation of the thread size from column 1 of BS EN ISO 228-1: 2003 Table 1 and the letter B for class B external threads.

(d) For left-hand threads, the letters LH shall be added to the designation. Right-hand threads require no special designation.

Examples. Complete designation for a right-hand thread size 1 1/2:

- Internal thread (one tolerance class only) **Pipe thread ISO 228 – G 1 1/2**
- External thread { tolerance class A **Pipe thread ISO 228 – G 1 1/2 A**
  tolerance class B **Pipe thread ISO 228 – G 1 1/2 B**

Combination with jointing thread

Combining an external parallel thread G, tolerance class A or B, in accordance with ISO 228-1, with an internal parallel thread Rp, in accordance with ISO 7-1, needs special consideration.

When this combination is necessary, the tolerance of the internal thread in accordance with ISO 7-1 shall be considered in the relevant product standards, where external parallel threads G are used.

Note:
Such a combination of threads does not necessarily achieve a leaktight joint.

4.1.38 ISO Pipe threads, tapered: basic sizes

Terms relating to taper pipe threads

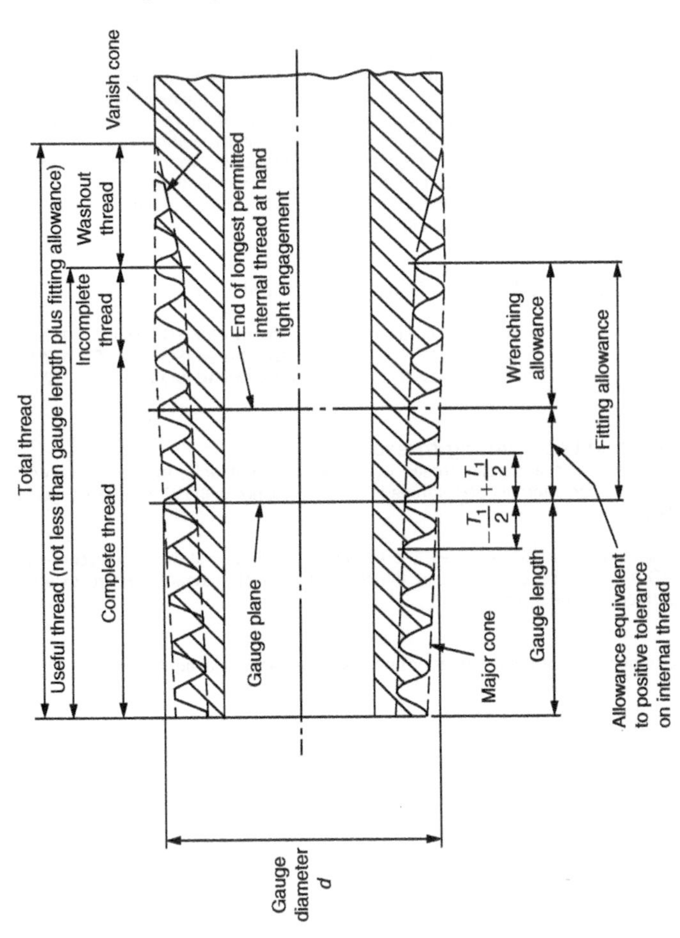

ISO pipe threads, tapered: basic sizes (see Tables 1 and 2).

Table 1

Nominal (bore) size of pipe[a]		Number of threads per inch	Pitch of thread (mm)	Depth of thread (mm)	Basic diameters at gauge plane		
in.	mm				Major (gauge) diameter (mm)	Pitch (effective) diameter (mm)	Minor diameter (mm)
$\frac{1}{8}$	6	28	0.907	0.581	9.728	9.147	8.566
$\frac{1}{4}$	8	19	1.337	0.856	13.157	12.301	11.445
$\frac{3}{8}$	10	19	1.337	0.856	16.662	15.806	14.950
$\frac{1}{2}$	15	14	1.814	1.162	20.955	19.793	18.631
$\frac{3}{4}$	20	14	1.814	1.162	26.441	25.279	24.117
1	25	11	2.309	1.479	33.249	31.770	30.291
$1\frac{1}{8}$	32	11	2.309	1.479	41.910	40.431	38.952
$1\frac{1}{8}$	40	11	2.309	1.479	47.803	46.324	44.845
2	50	11	2.309	1.479	59.614	58.135	56.656
$2\frac{1}{8}$	65	11	2.309	1.479	75.184	73.705	72.226
3	80	11	2.309	1.479	87.884	86.405	84.926
4	100	11	2.309	1.479	113.030	111.551	110.072
5	125	11	2.309	1.479	138.430	136.951	135.472
6	150	11	2.309	1.479	163.830	162.351	160.872

[a]Nominal pipe size equivalents, *not* conversions.

Table 2

Nominal (bore) size of pipe[a]		Gauge length[b]				Useful thread (min.)			Fitting allowance	Wrenching allowance	Position of gauge plane tolerance[c] ±	Diametral tolerance[d] ±
in.	mm	Basic	Tolerance ±	max.	min.	Basic	max.	min.				
$\frac{1}{8}$	6	$4\frac{3}{8}$ (4.0)	1 (0.9)	$5\frac{3}{8}$ (4.9)	$3\frac{3}{8}$ (3.1)	$7\frac{1}{8}$ (6.5)	$8\frac{1}{8}$ (7.4)	$6\frac{1}{8}$ (5.6)	$2\frac{3}{4}$ (2.5)	$1\frac{1}{2}$ (1.4)	$1\frac{1}{4}$ (1.1)	0.071
$\frac{1}{4}$	8	$4\frac{1}{2}$ (5.0)	1 (1.3)	$5\frac{1}{2}$ (7.3)	$3\frac{1}{2}$ (4.7)	$7\frac{1}{4}$ (9.7)	$8\frac{1}{4}$ (11.0)	$6\frac{1}{4}$ (8.4)	$2\frac{3}{4}$ (3.7)	$1\frac{1}{2}$ (2.0)	$1\frac{1}{4}$ (1.7)	0.104
$\frac{3}{8}$	10	$4\frac{3}{4}$ (6.4)	1 (1.3)	$5\frac{3}{4}$ (7.7)	$3\frac{3}{4}$ (5.1)	$7\frac{1}{2}$ (10.1)	$8\frac{1}{2}$ (11.4)	$6\frac{1}{2}$ (8.8)	$2\frac{3}{4}$ (3.7)	$1\frac{1}{2}$ (2.0)	$1\frac{1}{4}$ (1.7)	0.104
$\frac{1}{2}$	15	$4\frac{1}{2}$ (8.2)	1 (1.8)	$5\frac{1}{2}$ (10.0)	$3\frac{1}{2}$ (6.4)	$7\frac{1}{4}$ (13.2)	$8\frac{1}{4}$ (15.0)	$6\frac{1}{4}$ (11.4)	$2\frac{3}{4}$ (5.0)	$1\frac{1}{2}$ (2.7)	$1\frac{1}{4}$ (2.3)	0.142
$\frac{3}{4}$	20	$5\frac{1}{4}$ (9.5)	1 (1.8)	$6\frac{1}{4}$ (11.3)	$4\frac{1}{4}$ (7.7)	8 (14.5)	9 (16.3)	7 (12.7)	$2\frac{1}{4}$ (5.0)	$1\frac{1}{2}$ (2.7)	$1\frac{1}{4}$ (2.3)	0.142
1	25	$4\frac{1}{2}$ (10.4)	1 (2.3)	$5\frac{1}{2}$ (12.7)	$3\frac{1}{2}$ (8.1)	$7\frac{1}{4}$ (16.8)	$8\frac{3}{4}$ (19.1)	$6\frac{1}{4}$ (14.5)	$2\frac{3}{4}$ (6.4)	$1\frac{1}{2}$ (3.5)	$1\frac{1}{4}$ (2.9)	0.180
$1\frac{1}{4}$	32	$5\frac{1}{2}$ (12.7)	1 (2.3)	$6\frac{1}{2}$ (15.0)	$4\frac{1}{2}$ (10.4)	$8\frac{1}{4}$ (19.1)	$9\frac{1}{4}$ (21.4)	$7\frac{1}{4}$ (16.8)	$2\frac{3}{4}$ (6.4)	$1\frac{1}{2}$ (3.5)	$1\frac{1}{4}$ (2.9)	0.180
$1\frac{1}{2}$	40	$5\frac{1}{2}$ (12.7)	1 (2.3)	$6\frac{1}{2}$ (15.0)	$4\frac{1}{2}$ (10.4)	$8\frac{1}{4}$ (19.1)	$9\frac{1}{4}$ (21.4)	$7\frac{1}{4}$ (16.8)	$2\frac{3}{4}$ (6.4)	$1\frac{1}{2}$ (3.5)	$1\frac{1}{4}$ (2.9)	0.180

2	50	$6\frac{7}{8}$ (15.9)	1 (2.3)	$7\frac{7}{8}$ (18.2)	$5\frac{7}{8}$ (15.6)	$10\frac{1}{8}$ (23.4)	$11\frac{1}{8}$ (25.7)	$9\frac{1}{8}$ (21.1)	$3\frac{1}{4}$ (7.5)	2 (4.6)	$1\frac{1}{4}$ (2.9)	0.180
$2\frac{1}{2}$	65	$7\frac{9}{16}$ (17.5)	$1\frac{1}{2}$ (3.5)	$9\frac{1}{16}$ (21.0)	$6\frac{1}{16}$ (14.0)	$11\frac{9}{16}$ (26.7)	$13\frac{1}{16}$ (30.2)	$10\frac{1}{16}$ (23.2)	4 (9.2)	$2\frac{1}{2}$ (5.8)	$1\frac{1}{2}$ (3.5)	0.216
3	80	$8\frac{15}{16}$ (20.6)	$1\frac{1}{2}$ (3.5)	$10\frac{7}{16}$ (24.1)	$7\frac{7}{16}$ (17.1)	$12\frac{15}{16}$ (29.8)	$14\frac{7}{16}$ (33.3)	$11\frac{7}{16}$ (26.3)	4 (9.2)	$2\frac{1}{2}$ (5.8)	$1\frac{1}{2}$ (3.5)	0.216
4	100	11 (25.4)	$1\frac{1}{2}$ (3.5)	$12\frac{1}{2}$ (28.9)	$9\frac{1}{2}$ (21.9)	$15\frac{1}{2}$ (35.8)	17 (19.3)	14 (32.3)	$4\frac{1}{2}$ (10.4)	3 (6.9)	$1\frac{1}{2}$ (3.5)	0.216
5	125	$12\frac{3}{8}$ (28.6)	$1\frac{1}{2}$ (3.5)	$13\frac{7}{8}$ (32.1)	$10\frac{7}{8}$ (25.1)	$17\frac{3}{8}$ (40.1)	$18\frac{7}{8}$ (43.6)	$15\frac{7}{8}$ (36.6)	5 (11.5)	$3\frac{1}{2}$ (8.1)	$1\frac{1}{2}$ (3.5)	0.216
6	150	$12\frac{3}{8}$ (28.6)	$1\frac{1}{2}$ (3.5)	$13\frac{7}{8}$ (32.1)	$10\frac{7}{8}$ (25.1)	$17\frac{7}{8}$ (40.1)	$18\frac{7}{8}$ (43.6)	$15\frac{7}{8}$ (36.6)	5 (11.5)	$3\frac{1}{2}$ (8.1)	$1\frac{1}{2}$ (3.5)	0.216

[a] Nominal pipe size equivalents, *not* conversions.
[b] Gauge length in number of turns of thread {() = linear equivalent to nearest 0.1 mm}.
[c] Tolerance on position of gauge plane relative to face of internally taper threaded parts.
[d] Diametral tolerance on parallel internal threads (millimetres).
For further information see BS 2779.

Note: The screwed fasteners to be described in the following tables (Sections 4.1.39 to 4.1.46) are now obsolete or obsolescent. Therefore they are not recommended for use in new product design or manufacture. However they are still manufactured and still in widespread use. For this reason they have been retained in this edition.

4.1.39 British Standard Whitworth (BSW) bolts and nuts

Dimensions in inches

Nominal size	Threads per inch (TPI)	Pitch	Depth	Diameters			Hexagon (Bolt heads)					Hexagon (nuts)				
				Major	Effective	Minor	Across flats (A/F)		Across corners	Head thickness		Across flats (A/F)		Across corners	Nut thickness	
							max.	min.	max.	max.	min.	max.	min.	max.	max.	min.
$\frac{1}{4}$	20	0.0500	0.0320	0.250	0.2180	0.1860	0.455	0.438	0.51	0.19	0.18	0.455	0.438	0.51	0.200	0.190
$\frac{5}{16}$	18	0.05536	0.0356	0.3125	0.2769	0.2413	0.525	0.518	0.61	0.22	0.21	0.525	0.518	0.61	0.250	0.240
$\frac{3}{8}$	16	0.06250	0.0400	0.3750	0.3350	0.2950	0.600	0.592	0.69	0.27	0.26	0.600	0.592	0.69	0.312	0.302
$\frac{7}{16}$	14	0.07141	0.0457	0.4375	0.3981	0.3461	0.710	0.702	0.82	0.33	0.32	0.710	0.702	0.82	0.375	0.365
$\frac{1}{2}$	12	0.08333	0.0534	0.5000	0.4466	0.3932	0.820	0.812	0.95	0.38	0.37	0.820	0.812	0.95	0.437	0.427
$\frac{7}{16}$	12	0.08333	0.0534	0.5625	0.5091	0.4557	0.920	0.912	1.06	0.44	0.43	0.920	0.912	1.06	0.500	0.490
$\frac{5}{8}$	11	0.09091	0.0542	0.6250	0.5668	0.5086	1.010	1.000	1.17	0.49	0.48	0.010	1.000	1.17	0.562	0.552
$\frac{3}{4}$	10	0.10000	0.0640	0.7500	0.6860	0.3039	1.200	1.190	1.39	0.60	0.59	1.200	1.190	1.39	0.687	0.677
$\frac{7}{8}$	8	0.11111	0.0711	0.8750	0.8039	0.7328	1.300	1.288	1.50	0.66	0.65	1.300	1.288	1.50	0.750	0.740
1	8	0.12500	0.0800	1.0000	0.9200	0.8400	1.480	1.468	1.71	0.77	0.76	1.480	1.468	1.71	0.875	0.865
$1\frac{1}{8}$	7	0.14286	0.0915	1.1250	1.0335	0.9402	1.670	1.658	1.93	0.88	0.87	1.670	1.658	1.93	1.000	0.990
$1\frac{1}{4}$	7	0.14286	0.0915	1.2500	1.1585	1.0670	1.860	1.845	2.15	0.98	0.96	1.860	1.845	2.15	1.125	1.105
$1\frac{1}{2}$	6	0.16667	0.1067	1.5000	1.3933	1.2866	2.220	2.200	2.56	1.20	1.18	2.220	2.200	2.56	1.375	1.355
$1\frac{3}{4}$	5	0.20000	0.1281	1.7500	1.6219	1.4938	2.580	2.555	2.98	1.42	1.40	2.580	2.555	2.98	1.625	1.605
2	4.5	0.20222	0.1423	2.0000	1.8577	1.7154	2.760	2.735	3.19	1.53	1.51	2.760	2.735	3.19	1.750	1.730
$2\frac{1}{4}$	4	0.25000	0.1601	2.2500	2.0899	1.9298	–	–	–	–	–	–	–	–	–	–
$2\frac{1}{2}$	4	0.25000	0.1601	2.5000	2.3399	2.1798	–	–	–	–	–	–	–	–	–	–
$2\frac{3}{4}$	3.5	0.28571	0.1830	2.7500	2.5670	2.3840	–	–	–	–	–	–	–	–	–	–
3	3.5	0.28571	0.1830	3.0000	2.8170	2.6340	–	–	–	–	–	–	–	–	–	–
$3\frac{1}{2}$	3.25	0.30769	0.1970	3.5000	3.3030	3.1060	–	–	–	–	–	–	–	–	–	–
4	3	0.33333	0.2134	4.0000	3.7866	3.5732	–	–	–	–	–	–	–	–	–	–
$4\frac{1}{2}$	2.875	0.34783	0.2227	4.5000	4.2773	4.0546	–	–	–	–	–	–	–	–	–	–
5	2.75	0.36364	0.2328	5.0000	4.7672	4.5344	–	–	–	–	–	–	–	–	–	–

4.1.40 British Standard Fine (BSF) bolts and nuts

Dimensions in inches

Nominal size	Threads per inch (TPI)	Pitch	Depth	Diameters			Hexagon (Bolt heads)					Hexagon (nuts)				
				Major	Effective	Minor	Across flats (A/F)		Across corners	Head thickness		Across flats (A/F)		Across corners	Nut thickness	
							max.	min.	max.	max.	min.	max.	min.	max.	max.	min.
$\frac{1}{4}$	26	0.03846	0.0246	0.2500	0.2254	0.2008	0.455	0.438	0.51	0.19	0.18	0.455	0.438	0.51	0.200	0.190
$\frac{5}{16}$	22	0.04545	0.0291	0.3125	0.2834	0.2543	0.525	0.518	0.61	0.22	0.21	0.525	0.518	0.61	0.250	0.240
$\frac{3}{8}$	20	0.05000	0.0320	0.3750	0.3430	0.3110	0.600	0.592	0.69	0.27	0.26	0.600	0.592	0.69	0.312	0.302
$\frac{7}{16}$	18	0.05556	0.0356	0.4375	0.4019	0.3663	0.710	0.708	0.82	0.33	0.32	0.710	0.706	0.82	0.375	0.365
$\frac{1}{2}$	16	0.06250	0.0400	0.5000	0.4600	0.4200	0.820	0.812	0.95	0.38	0.37	0.820	0.812	0.95	0.437	0.427
$\frac{9}{16}$	16	0.06250	0.0400	0.5625	0.5225	0.4825	0.920	0.912	1.06	0.44	0.43	0.920	0.912	1.06	0.500	0.490
$\frac{5}{8}$	14	0.07143	0.0457	0.6250	0.5793	0.5335	1.010	1.000	1.17	0.49	0.48	1.010	1.000	1.17	0.562	0.552
$\frac{3}{4}$	12	0.08333	0.0534	0.7500	0.6966	0.6432	1.200	1.190	1.39	0.60	0.59	1.200	1.190	1.39	0.687	0.677
$\frac{7}{8}$	11	0.09091	0.0582	0.8750	0.8168	0.7586	1.300	1.288	1.50	0.66	0.65	1.300	1.288	1.50	0.750	0.740
1	10	0.10000	0.0640	1.0000	0.9360	0.8720	1.480	1.468	1.71	0.77	0.76	1.480	1.468	1.71	0.875	0.865
$1\frac{1}{8}$	9	0.11111	0.0711	1.1250	1.0539	0.9828	1.670	1.658	1.93	0.88	0.87	1.670	1.658	1.93	1.000	0.990
$1\frac{1}{4}$	9	0.11111	0.0711	1.2500	1.1789	1.1078	1.860	1.845	2.15	0.98	0.96	1.860	1.845	2.15	1.125	1.105
$1\frac{3}{8}$	8	0.12500	0.0800	1.1370	1.2950	1.2150	2.050	2.035	2.37	1.09	1.07	2.050	2.035	2.37	1.250	1.230
$1\frac{1}{2}$	8	0.12500	0.0800	1.5000	1.4200	1.3400	2.220	2.200	2.56	1.20	1.18	2.220	2.200	2.56	1.375	1.355
$1\frac{5}{8}$	8	0.12500	0.0800	1.6250	1.5450	1.4650	–	–	–	–	–	–	–	–	–	–
$1\frac{3}{4}$	7	0.14826	0.0915	1.7500	1.6585	1.5670	2.580	2.555	2.98	1.42	1.40	2.580	2.555	2.98	1.625	1.605
2	7	0.14826	0.0915	2.0000	1.9085	1.8170	2.760	2.735	3.19	1.53	1.51	2.760	2.735	3.19	1.750	1.730
$2\frac{1}{4}$	6	0.16667	0.1067	2.2500	2.1433	2.0366	–	–	–	–	–	–	–	–	–	–

4.1.41 ISO unified precision internal screw threads: coarse series (UNC)

Dimensions in inches

Designation	Major diameter min.	Pitch (effective) diameter		Minor diameter		Hexagon (nut)				
						Max. width across flats (A/F)	Max. width across corners (A/C)	Nut thickness		
		max.	min.	max.	min.			Thick	Normal	Thin
$\frac{1}{4}$-20 UNC-2B	0.2500	0.2223	0.2175	0.2074	0.1959	0.4375	0.505	0.286	0.224	0.161
$\frac{5}{16}$-18 UNC-2B	0.3125	0.2817	0.2764	0.2651	0.2524	0.5000	0.577	0.333	0.271	0.192
$\frac{3}{8}$-16 UNC-2B	0.3750	0.3401	0.3344	0.3214	0.3073	0.5625	0.560	0.411	0.333	0.224
$\frac{7}{16}$-14 UNC-2B	0.4375	0.3972	0.3911	0.3760	0.3602	0.6875	0.794	0.458	0.380	0.255
$\frac{1}{2}$-13 UNC-2B	0.5000	0.4565	0.4500	0.4336	0.4167	0.7500	0.866	0.567	0.442	0.317
$\frac{9}{16}$-12 UNC-2B[a]	0.5625	0.5152	0.5084	0.4904	0.4723	0.8750	1.010	0.614	0.489	0.349
$\frac{5}{8}$-11 UNC-2B	0.6250	0.5732	0.5660	0.5460	0.5266	0.9375	1.083	0.724	0.552	0.380
$\frac{3}{4}$-10 UNC-2B	0.7500	0.6927	0.6850	0.6627	0.6417	1.1250	1.300	0.822	0.651	0.432
$\frac{7}{8}$-9 UNC-2B	0.8750	0.8110	0.8028	0.7775	0.7547	1.3125	1.515	0.916	0.760	0.494
1-8 UNC-2B	1.0000	0.9276	0.9188	0.8897	0.8647	1.5000	1.732	1.015	0.874	0.562
$1\frac{1}{8}$-7 UNC-2B	0.1250	1.0416	1.0322	0.9980	0.9704	1.6875	1.948	1.176	0.989	0.629
$1\frac{1}{4}$-7 UNC-2B	1.2500	1.1668	1.1572	1.1230	1.0954	1.8750	2.165	1.275	1.087	0.744

$1\frac{3}{8}$-6 UNC-2B[a]	1.3750	1.2771	1.2667	1.2252	1.1946	2.0625	2.382	1.400	1.197	0.806
$1\frac{3}{8}$-6 UNC-2B	1.5000	1.4022	1.3917	1.3502	1.3196	2.2500	2.598	1.530	1.311	0.874
$1\frac{3}{4}$-5 UNC-2B	1.7500	1.6317	1.6201	1.5675	1.5335	2.6250	3.031	–	1.530	0.999
2-$4\frac{1}{2}$ UNC-2B	2.0000	1.8681	1.8557	1.7952	1.7594	3.0000	3.464	–	1.754	1.129
$2\frac{1}{4}$-$4\frac{1}{2}$ UNC-2B	2.2500	2.1183	2.1057	2.0452	2.0094					
$2\frac{1}{2}$-4 UNC-2B	2.5000	2.3511	2.3376	2.2669	2.2294					
$2\frac{3}{4}$-4 UNC-2B	2.7500	2.6013	2.5876	2.5169	2.4794					
3-4 UNC-2B	3.0000	2.8515	2.8376	2.7669	2.7294					
$3\frac{1}{4}$-4 UNC-2B	3.2500	3.1017	3.0876	3.0169	2.9794					
$3\frac{1}{2}$-4 UNC-2B	3.5000	3.3519	3.3376	3.2669	3.2294					
$3\frac{3}{4}$-4 UNC-2B	3.7500	3.6021	3.5876	3.5169	3.4794					
4-4 UNC-2B	4.0000	3.8523	3.8376	3.7669	3.7294					

[a] To be dispensed with wherever possible.
For full range and further information see BS 1768.

Example

The interpretation of designation $\frac{1}{2}$-13 UNC-2B is as follows: nominal diameter $\frac{1}{2}$ inch; threads per inch 13; ISO unified thread, coarse series; thread tolerance classification 2B.

4.1.42 ISO unified precision external screw threads: coarse series (UNC)

Dimensions in inches

Designation	Major diameter		Pitch (effective) diameter		Minor diameter		Shank diameter		Hexagon head (bolt)		
	max.	min.	max.	min.	max.	min.	max.	min.	Max. width across flats (A/F)	Max. width across corners (A/C)	Max. height
$\frac{1}{4}$-20 UNC-2A	0.2489	0.2408	0.2164	0.2127	0.1876	0.1803	0.2500	0.2465	0.4375	0.505	0.163
$\frac{5}{16}$-18 UNC-2A	0.3113	0.3026	0.2752	0.2712	0.2431	0.2351	0.3125	0.3090	0.5000	0.577	0.211
$\frac{3}{4}$-16 UNC-2A	0.3737	0.3643	0.3331	0.3287	0.2970	0.2881	0.3750	0.3715	0.5625	0.650	0.243
$\frac{7}{16}$-14 UNC-2A	0.4361	0.4258	0.3897	0.3850	0.3485	0.3387	0.4375	0.4335	0.6250	0.722	0.291
$\frac{1}{2}$-13 UNC-2A	0.4985	0.4876	0.4485	0.4435	0.4041	0.3936	0.5000	0.4960	0.7500	0.866	0.323
$\frac{9}{16}$-12 UNC-2A[a]	0.5609	0.5495	0.5068	0.5016	0.4587	0.4475	0.5625	0.5585	0.8125	0.938	0.371
$\frac{5}{8}$-11 UNC-2A	0.6234	0.6113	0.5644	0.5589	0.5119	0.4999	0.6250	0.6190	0.9375	1.083	0.403
$\frac{3}{4}$-10 UNC-2A	0.7482	0.7353	0.6832	0.6773	0.6255	0.6124	0.7500	0.7440	1.1250	1.300	0.483
$\frac{7}{8}$-9 UNC-2A	0.8731	0.8592	0.8009	0.7946	0.7368	0.7225	0.8750	0.8670	1.3125	1.515	0.563
1-8 UNC-2A	0.9980	0.9830	0.9168	0.9100	0.8446	0.8288	1.0000	0.9920	1.5000	1.732	0.627
$1\frac{1}{8}$-7 UNC-2A	1.1228	1.1064	1.0300	1.0228	0.9475	0.9300	1.1250	1.1170	1.6875	1.948	0.718
$1\frac{1}{4}$-7 UNC-2A	1.2478	1.2314	1.1550	0.1476	1.0725	1.0548	1.2500	1.2420	1.8750	2.165	0.813
$1\frac{3}{8}$-6 UNC-2A[a]	1.3726	1.3544	1.2643	1.2563	1.1681	1.1481	1.3750	1.3650	2.0625	2.382	0.878

$1\frac{1}{2}$-6 UNC-2A	1.4976	1.4794	1.3893	1.3812	1.2931	1.2730	1.5000	1.4900	2.2500	2.598	0.974
$1\frac{3}{4}$-5 UNC-2A	1.7473	1.7268	1.6174	1.6085	1.5019	1.4786	1.7500	1.7400	2.6250	3.031	1.134
2-$4\frac{1}{2}$ UNC-2A	1.9971	1.9751	1.8528	1.8433	1.7245	1.6990	2.000	1.9900	3.000	3.464	1.263
$2\frac{1}{4}$-$4\frac{1}{2}$ UNC-2A	2.2471	2.2251	2.1028	2.0931	1.9745	1.9488					
$2\frac{1}{2}$-4 UNC-2A	2.4969	2.4731	2.3345	2.3241	2.1902	2.1618					
$2\frac{3}{4}$-4 UNC-2A	2.7468	2.7230	2.5844	2.5739	2.4401	2.4116					
3-4 UNC-2A	2.9968	2.9730	2.8344	2.8237	2.6901	2.6614					
$3\frac{1}{4}$-4 UNC-2A	3.2467	3.2229	3.0843	3.0734	2.9400	2.9111					
$3\frac{1}{2}$-4 UNC-2A	3.4967	3.4729	3.3343	3.3233	3.1900	3.1610					
$3\frac{3}{4}$-4 UNC-2A	3.7466	3.7228	3.5842	3.5730	3.4399	3.4107					
4-4 UNC-2A	3.9966	3.9728	3.8342	3.8229	3.6899	3.6606					

[a] To be dispensed with wherever possible.
For full range and further information see BS 1768.

Example

The interpretation of designation $\frac{1}{2}$-13 UNC-2A is as follows: nominal diameter $\frac{1}{2}$ inch; threads per inch 13; ISO unified thread, coarse series, thread tolerance classification 2a.

4.1.43 ISO unified precision internal screw threads: fine series (UNF)

Dimensions in inches

Designation	Major diameter min.	Pitch (effective) diameter		Minor diameter		Hexagon (nut)				
						Max. width across flats (A/F)	Max. width across corners (A/C)	Nut thickness		
		max.	min.	max.	min.			Thick	Normal	Thin
$\frac{1}{4}$-28 UNF-2B	0.2500	0.2311	0.2268	0.2197	0.2113	0.4375	0.505	0.286	0.224	0.161
$\frac{5}{16}$-24 UNF-2B	0.3125	0.2902	0.2854	0.2771	0.2674	0.5000	0.577	0.333	0.271	0.192
$\frac{3}{8}$-24 UNF-2B	0.3750	0.3528	0.3479	0.3396	0.3299	0.5625	0.650	0.411	0.333	0.224
$\frac{7}{16}$-20 UNF-2B	0.4375	0.4104	0.4050	0.3949	0.3834	0.6875	0.794	0.458	0.380	0.255
$\frac{1}{2}$-20 UNF-2B	0.5000	0.4731	0.4675	0.4574	0.4459	0.7500	0.866	0.567	0.442	0.317
$\frac{9}{16}$-18 UNF-2B[a]	0.5625	0.5323	0.5264	0.5151	0.5024	0.8750	1.010	0.614	0.489	0.349
$\frac{5}{8}$-18 UNF-2B	0.6250	0.5949	05889	0.5776	0.5649	0.9375	1.083	0.724	0.552	0.380
$\frac{3}{4}$-16 UNF-2B	0.7500	0.7159	0.7094	0.6964	0.6823	1.1250	1.300	0.822	0.651	0.432
$\frac{7}{8}$-14 UNF-2B	0.8750	0.8356	0.8286	0.8135	0.7977	1.3125	1.515	0.916	0.760	0.494
1-12 UNF-2B	1.000	0.9535	0.9459	0.9279	0.9098	1.5000	1.732	1.015	0.874	0.562
$1\frac{1}{8}$-12 UNF-2B	1.1250	1.0787	1.0709	1.0529	1.0348	1.6875	1.948	1.176	0.984	0.629
$1\frac{1}{4}$-12 UNF-2B	1.2500	1.2039	1.1959	1.1779	1.1598	1.8750	2.165	1.275	1.087	0.744
$1\frac{3}{8}$-12 UNF-2B[a]	1.3750	1.3291	1.3209	1.3029	1.2848	2.0625	2.382	1.400	1.197	0.806
$1\frac{1}{2}$-12 UNF-2B	1.500	1.4542	1.4459	1.4279	1.4098	2.2500	2.598	1.530	1.311	0.874

[a] To be dispensed with wherever possible
For full range and further information see BS 1768.

Example
The interpretation of designation $\frac{1}{2}$-20 UNF-2B is as follows: nominal diameter $\frac{1}{2}$ inch; threads per inch 20; ISO unified thread, fine series: thread tolerance classification 2B.

4.1.44 ISO unified precision external screw threads: fine series (UNF)

Dimensions in inches

Designation	Major diameter		Pitch (effective) diameter		Minor diameter		Shank diameter		Hexagon head (bolt)		
	max.	min.	max.	min.	max.	min.	max.	min.	Max. width across flats (A/F)	Max. width across corners (A/C)	Max. height
$\frac{1}{4}$-28 UNF-2A	0.2490	0.2425	0.2258	0.2225	0.2052	0.1993	0.2500	0.2465	0.4375	0.505	0.163
$\frac{5}{16}$-24 UNF-2A	0.3114	0.3042	0.2843	0.2806	0.2603	0.2536	0.3125	0.3090	0.5000	0.577	0.211
$\frac{3}{8}$-24 UNF-2A	0.3739	0.3667	0.3468	0.3430	0.3228	0.3160	0.3750	0.3715	0.5625	0.650	0.243
$\frac{7}{16}$-20 UNF-2A	0.4362	0.4281	0.4037	0.3995	0.3749	0.3671	0.4375	0.4335	0.6250	0.722	0.291
$\frac{1}{2}$-20 UNF-2A	0.4987	0.4906	0.4662	0.4615	0.4374	0.4295	0.5000	0.4960	0.7500	0.866	0.323
$\frac{9}{16}$-18 UNF-2A[a]	0.5611	0.5524	0.5250	0.5205	0.4929	0.4844	0.5625	0.5585	0.8125	0.938	0.371
$\frac{5}{8}$-18 UNF-2A	0.6236	0.6149	0.5875	0.5828	0.5554	0.5467	0.6250	0.6190	0.9375	1.083	0.403
$\frac{3}{4}$-16 UNF-2A	0.7485	0.7391	0.7079	0.7029	0.6718	0.6623	0.7500	0.7440	1.1250	1.300	0.483
$\frac{7}{8}$-14 UNF-2A	0.8734	0.8631	0.8270	0.8216	0.7858	0.7753	0.8750	0.8670	1.3125	1.515	0.563
1-12 UNF-2A	0.9982	0.9868	0.9441	0.9382	0.8960	0.8841	1.0000	0.9920	1.5000	1.732	0.627
$1\frac{1}{8}$-12 UNF-2A	1.1232	1.1118	1.0691	1.0631	0.0210	1.0090	1.1250	1.1170	1.6875	1.948	0.718
$1\frac{1}{4}$-12 UNF-2A	1.2482	1.2368	1.1941	1.1879	1.1460	1.1338	1.2500	1.2420	1.8750	2.165	0.813
$1\frac{3}{8}$-12 UNF-2A[a]	1.3731	1.3617	1.3190	1.3127	1.2709	1.2586	1.3750	1.3650	2.0625	2.382	0.878
$1\frac{1}{2}$-12 UNF-2A	1.4981	1.4867	1.4440	1.4376	1.3959	1.3835	1.500	1.4900	2.2500	2.598	0.974

[a] To be dispensed with wherever possible.
For full range and further information see BS 1768.

Example

The interpretation of designation $\frac{1}{2}$-20 UNF-2A is as follows: nominal diameter $\frac{1}{2}$ inch; threads per inch 20; ISO unified thread, fine series: thread tolerance classification 2A.

4.1.45 British Association thread form

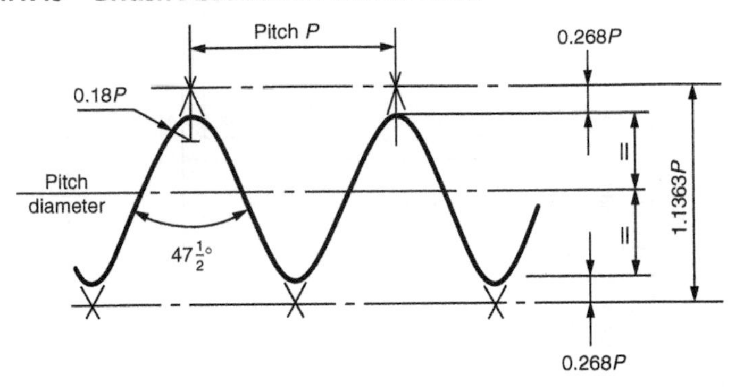

British Association (BA) thread forms are obsolete but are still used in repairs and maintenance.

4.1.46 BA internal and external screw threads

Dimensions in mm

Designation number	Pitch	Depth of thread	Major diameter	Pitch (effective) diameter	Minor diameter	Crest radius	Root radius
0	1.0000	0.600	6.00	5.400	4.80	0.1808	0.1808
1	0.9000	0.540	5.30	4.760	4.22	0.1627	0.1627
2	0.8100	0.485	4.70	4.215	3.73	0.1465	0.1465
3	0.7300	0.440	4.10	3.660	3.22	0.1320	0.1320
4	0.6600	0.395	3.60	3.205	2.81	0.1193	0.1193
5	0.5900	0.355	3.20	2.845	2.49	0.1067	0.1067
6	0.5300	0.320	2.80	2.480	2.16	0.0958	0.0958
7	0.4800	0.290	2.50	2.210	1.92	0.0868	0.0868
8	0.4300	0.260	2.20	1.940	1.68	0.0778	0.0778
9	0.3900	0.235	1.90	1.665	1.43	0.0705	0.0705
10	0.3500	0.210	1.70	1.490	1.28	0.0633	0.0633
11	0.3100	0.185	1.50	1.315	1.13	0.0561	0.0561
12	0.2800	0.170	1.30	1.130	0.96	0.0506	0.0506
13	0.2500	0.150	1.20	1.050	0.90	0.0452	0.0452
14	0.2300	0.140	1.00	0.860	0.72	0.0416	0.0416
15	0.2100	0.125	0.90	0.775	0.65	0.0380	0.0380
16	0.1900	0.115	0.79	0.675	0.56	0.0344	0.0344
17	0.1700	0.100	0.70	0.600	0.50	0.0307	0.0307
18	0.1500	0.090	0.62	0.530	0.44	0.0271	0.0271
19	0.1400	0.085	0.54	0.455	0.37	0.0253	0.0253
20	0.1200	0.070	0.48	0.410	0.34	0.0217	0.0217
21	0.1100	0.065	0.42	0.355	0.29	0.0199	0.0199
22	0.1000	0.060	0.37	0.310	0.25	0.0181	0.0181
23	0.0900	0.055	0.33	0.275	0.22	0.0163	0.0163
24	0.0800	0.050	0.29	0.240	0.19	0.0145	0.0145
25	0.0700	0.040	0.25	0.210	0.17	0.0127	0.0127

For further information see BS 57 and BS 93.
BA thread forms are obsolete but are still used in repairs and maintenance.

4.1.47 BA threads: tapping and clearance drills

BA no.	Tapping size drill		Clearance size drill	
	mm	Number or fraction size	mm	Number, letter or fraction size
0	5.10	8	6.10	D
1	4.50	16	5.50	2
2	4.00	22	4.85	10
3	3.40	29	4.25	18
4	3.00	32	3.75	24
5	2.65	37	3.30	29
6	2.30	43	2.90	32
7	2.05	45/46	2.60	36
8	1.80	50	2.25	41
9	1.55	53	1.95	45
10	1.40	54	1.75	49
11	1.20	56	1.60	52
12	1.05	59	1.40	54
13	0.98	62	1.30	55
14	0.78	68	1.10	57
15	0.70	70	0.98	60
16	0.60	73	0.88	65

4.1.48 ISO metric tapping and clearance drills, coarse thread series

Nominal size	Tapping drill size (mm)		Clearance drill size (mm)		
	Recommended 80% engagement	Alternative 70% engagement	Close fit	Medium fit	Free fit
M1.6	1.25	1.30	1.7	1.8	2.0
M2	1.60	1.65	2.2	2.4	2.6
M2.5	2.05	2.10	2.7	2.9	3.1
M3	2.50	2.55	3.2	3.4	3.6
M4	3.30	3.40	4.3	4.5	4.8
M5	4.20	4.30	5.3	5.5	5.8
M6	5.00	5.10	6.4	6.6	7.0
M8	6.80	6.90	8.4	9.0	10.0
M10	8.50	8.60	10.5	11.0	12.0
M12	10.20	10.40	13.0	14.0	15.0
M14	12.00	12.20	15.0	16.0	17.0
M16	14.00	14.25	17.0	18.0	19.0
M18	15.50	15.75	19.0	20.0	21.0
M20	17.50	17.75	21.0	22.0	24.0
M22	19.50	19.75	23.0	24.0	26.0
M24	21.00	21.25	25.0	26.0	28.0
M27	24.00	24.25	28.0	30.0	32.0
M30	26.50	26.75	31.0	33.0	35.0
M33	29.50	29.75	34.0	36.0	38.0
M36	32.00	–	37.0	39.0	42.0
M39	35.00	–	40.0	42.0	45.0
M42	37.50	–	43.0	45.0	48.0
M45	40.50	–	46.0	48.0	52.0
M48	43.00	–	50.0	52.0	56.0
M52	47.00	–	54.0	56.0	62.0

4.1.49 ISO metric tapping and clearance drills, fine thread series

Nominal size	Tapping drill size (mm)		Clearance drill size (mm)		
	Recommended 80% engagement	Alternative 70% engagement	Close fit	Medium fit	Free fit
M6	5.20	5.30	6.4	6.6	7.0
M8	7.00	7.10	8.4	9.0	10.0
M10	8.80	8.90	10.5	11.0	12.0
M12	10.80	10.90	13.0	14.0	15.0
M14	12.50	12.70	15.0	16.0	17.0
M16	14.50	14.75	17.0	18.0	19.0
M18	16.50	16.75	19.0	20.0	21.0
M20	18.50	18.75	21.0	22.0	24.0
M22	20.50	20.75	23.0	24.0	26.0
M24	22.00	22.25	25.0	26.0	28.0
M27	25.00	25.25	28.0	30.0	32.0
M30	28.00	28.25	31.0	33.0	35.0
M33	31.00	31.25	34.0	36.0	38.0
M36	33.00	–	37.0	39.0	42.0
M39	36.00	–	40.0	42.0	45.0
M42	39.00	–	43.0	45.0	48.0

4.1.50 ISO unified tapping and clearance drills, coarse thread series

Nominal size in.	Tapping drill size		Clearance drill size	
	mm	in.	mm	Letter or in.
$\frac{1}{4} \times 20$	5.20	$\frac{13}{64}$	6.50	F or $\frac{17}{64}$
$\frac{5}{16} \times 18$	6.60	$\frac{17}{64}$	8.00	O or $\frac{21}{64}$
$\frac{3}{4} \times 16$	8.00	$\frac{5}{16}$	9.80	W or $\frac{25}{64}$
$\frac{7}{16} \times 14$	9.40	$\frac{3}{8}$	11.30	$\frac{33}{64}$
$\frac{1}{2} \times 13$	10.80	$\frac{27}{64}$	13.00	$\frac{33}{64}$
$\frac{9}{16} \times 12$	12.20	$\frac{31}{64}$	14.75	$\frac{37}{64}$
$\frac{5}{8} \times 11$	13.50	$\frac{17}{32}$	16.25	$\frac{41}{64}$
$\frac{3}{4} \times 10$	16.50	$\frac{21}{32}$	19.50	$\frac{47}{64}$
$\frac{7}{8} \times 9$	19.25	$\frac{49}{64}$	20.25	$\frac{51}{64}$
1×8	22.25	$\frac{7}{8}$	25.75	$1\frac{1}{64}$
$1\frac{1}{8} \times 7$	25.00	$\frac{63}{64}$	26.00	$1\frac{9}{64}$
$1\frac{1}{4} \times 7$	28.25[a]	$1\frac{7}{64}$	28.25	$1\frac{17}{64}$
$1\frac{3}{8} \times 6$	30.50[a]	$1\frac{13}{64}$	30.75	$1\frac{25}{64}$
$1\frac{1}{2} \times 6$	34.00[a]	$1\frac{21}{64}$	34.00	$1\frac{33}{64}$
$1\frac{3}{4} \times 5$	39.50[a]	$1\frac{35}{64}$	45.00	$1\frac{49}{64}$
$2 \times 4\frac{1}{2}$	45.50[a]	$1\frac{25}{64}$	52.00	$2\frac{1}{64}$

[a] Nearest standard metric size: approximately. 0.25 mm over recommended inch size.

4.1.51 ISO unified tapping and clearance drills, fine thread series

Nominal size in.	Tapping drill size		Clearance drill size	
	mm	Letter or in.	mm	Letter or in.
$\frac{1}{4} \times 28$	5.50	$\frac{7}{32}$	6.50	F
$\frac{5}{16} \times 24$	6.90	I	8.00	O
$\frac{3}{8} \times 24$	8.50	R	9.80	W
$\frac{7}{16} \times 20$	9.90	$\frac{25}{64}$	11.30	$\frac{29}{64}$
$\frac{1}{2} \times 20$	11.50	$\frac{29}{64}$	13.00	$\frac{33}{64}$
$\frac{9}{16} \times 18$	12.90	$\frac{33}{64}$	14.75	$\frac{37}{64}$
$\frac{5}{8} \times 18$	14.50	$\frac{37}{64}$	16.50	$\frac{41}{64}$
$\frac{3}{4} \times 16$	17.50	$\frac{11}{16}$	19.50	$\frac{49}{64}$
$\frac{7}{8} \times 14$	20.50	$\frac{13}{16}$	22.75	$\frac{57}{64}$
1×12	23.25	$\frac{59}{64}$	25.80	$1\frac{1}{64}$
$1\frac{1}{8} \times 12$	26.50	$1\frac{3}{64}$	29.00	$1\frac{9}{64}$
$1\frac{1}{4} \times 12$	29.50	$1\frac{11}{64}$	32.50	$1\frac{17}{64}$
$1\frac{3}{8} \times 12$	33.00	$1\frac{19}{64}$	35.50	$1\frac{25}{64}$
$1\frac{1}{2} \times 12$	36.00	$1\frac{27}{64}$	38.50	$1\frac{33}{64}$

4.1.52 ISO metric tapping and clearance drills, miniature series

Nominal size		Pitch mm	Threads per inch	Tapping drill size		Clearance drill size	
ISO mm	ASA B1.10 mm			mm	Number or fraction	mm	Number or fraction
S-0.3	0.30 unm	0.080	318	0.25	–	0.32	–
(S-0.35)	(0.35 unm)	0.090	282	0.28	–	0.38	79
S-0.4	0.40 unm	0.100	254	0.35	80	0.45	77
(S-0.45)	(0.45 unm)	0.100	254	0.38	79	0.50	76
S-0.5	0.50 unm	0.125	203	0.42	78	0.55	75, 74
(S-0.55)	(0.55 unm)	0.125	203	0.45	77	0.60	73
S-0.6	0.60 unm	0.150	169	0.50	76	0.65	72
(S-0.7)	(0.70 unm)	0.175	145	0.58	74	0.78	$\frac{1}{32}$in.
S-0.8	0.80 unm	0.200	127	0.65	72	0.88	66, 65
(S-0.9)	(0.90 unm)	0.225	113	0.72	70	0.98	62
S-1.0	1.00 unm	0.250	102	0.80	$\frac{1}{32}$in.	1.10	57
(S-1.1)	(1.10 unm)	0.250	102	0.90	65	1.20	$\frac{3}{64}$in.
S-1.2	1.20 unm	0.250	102	1.00	61	1.30	55
(S-1.4)	(1.40 unm)	0.300	85	1.15	$\frac{3}{64}$in.	1.50	53

4.1.53 BSW threads, tapping and clearance drills

Size	TPI	Tapping	Clearing
$\frac{1}{16}$	60	58	49
$\frac{3}{32}$	48	49	36
$\frac{1}{8}$	40	38	29
$\frac{5}{32}$	32	31	19
$\frac{3}{16}$	24	26	9
$\frac{7}{32}$	24	15	1
$\frac{1}{4}$	20	7	G
$\frac{5}{16}$	18	F	P
$\frac{3}{8}$	16	O	W
$\frac{7}{16}$	14	U	$\frac{29}{64}$
$\frac{1}{2}$	12	$\frac{27}{64}$	$\frac{33}{64}$

4.1.54 BSF threads, tapping and clearance drills

Size	T.P.I	Tapping	Clearing
$\frac{7}{32}$	28	14	1
$\frac{1}{4}$	26	3	G
$\frac{9}{32}$	26	C	M
$\frac{5}{16}$	22	H	P
$\frac{3}{8}$	20	$\frac{21}{64}$	W
$\frac{7}{16}$	18	W	$\frac{29}{64}$
$\frac{1}{2}$	16	$\frac{7}{16}$	$\frac{33}{64}$

4.1.55 Plain washers, bright: metric series

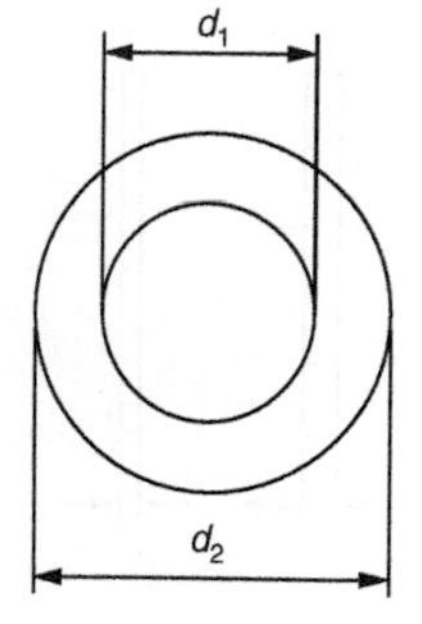

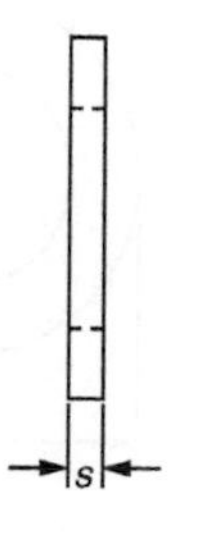

Dimensions in mm

Designation (thread diameter)[a]	Internal diameter d_1		External diameter d_2		Thickness s			
					Thick (normal)		Thin	
	max.	min.	max.	min.	max.	min.	max.	min.
M1.0	1.25	1.1	2.5	2.3	0.4	0.2	–	–
M1.2	1.45	1.3	3.0	2.8	0.4	0.2	–	–
(M1.4)	1.65	1.5	3.0	2.8	0.4	0.2	–	–
M1.6	1.85	1.7	4.0	3.7	0.4	0.2	–	–
M2.0	2.35	2.2	5.0	4.7	0.4	0.2	–	–
(M2.2)	2.55	2.4	5.0	4.7	0.6	0.4	–	–
M2.5	2.85	2.7	6.5	6.2	0.6	0.4	–	–
M3	3.4	3.2	7.0	6.7	0.6	0.4	–	–
(M3.5)	3.9	3.7	7.0	6.7	0.6	0.4	–	–
M4	4.5	4.3	9.0	8.7	0.9	0.7	–	–
(M4.5)	5.0	4.8	9.0	8.7	0.9	0.7	–	–
M5	5.5	5.3	10.0	9.7	1.1	0.9	–	–
M6	6.7	6.4	12.5	12.1	1.8	1.4	0.9	0.7
(M7)	7.7	7.4	14.0	13.6	1.8	1.4	0.9	0.7
M8	8.7	8.4	17.0	16.6	1.8	1.4	1.1	0.9
M10	10.9	10.5	21.0	20.5	2.2	1.8	1.45	1.05
M12	13.4	13.0	24.0	23.5	2.7	2.3	1.8	1.4
(M14)	15.4	15.0	28.0	27.5	2.7	2.3	1.8	1.4
M16	17.4	17.0	30.0	29.5	3.3	2.7	2.2	1.8
(M18)	19.5	19.0	34.0	33.2	3.3	2.7	2.2	1.8
M20	21.5	21.0	37.0	36.2	3.3	2.7	2.2	1.8
(M22)	23.5	23.0	39.0	38.2	3.3	2.7	2.2	1.8
M24	25.5	25.0	44.0	43.2	4.3	3.7	2.7	2.3
(M27)	28.5	28.0	50.0	49.2	4.3	3.7	2.7	2.3
M30	31.6	31.0	56.0	55.0	4.3	3.7	2.7	2.3
(M33)	34.6	34.0	60.0	59.0	5.6	4.4	3.3	2.7
M36	37.6	37.0	66.0	65.0	5.6	4.4	3.3	2.7
(M39)	40.6	40.0	72.0	71.0	6.6	5.4	3.3	2.7

[a] Non-preferred sizes in brackets ( ).
For full information see BS 4320.

4.1.56 Plain washers, black: metric series

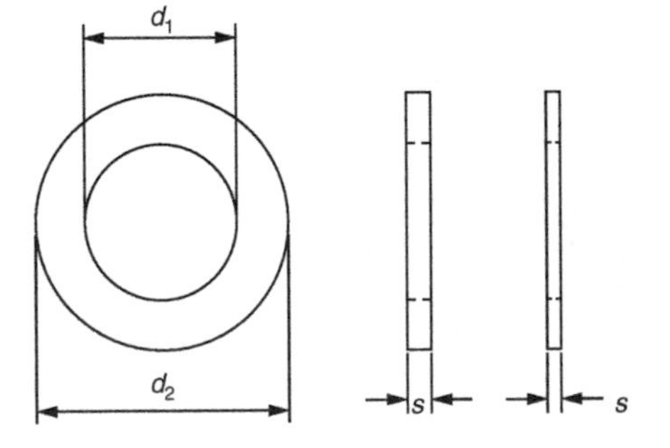

Dimensions in mm

Designation (thread diameter)[a]	Internal diameter d_1		External diameter d_2		Thickness s	
	max.	min.	max.	min.	max.	min.
M5	5.8	5.5	10.0	9.2	1.2	0.8
M6	7.0	6.6	12.5	11.7	1.9	1.3
(M7)	8.0	7.6	14.0	13.2	1.9	1.3
M8	9.4	9.0	17.0	16.2	1.9	1.3
M10	11.5	11.0	21.0	20.2	2.3	1.7
M12	14.5	14.0	24.0	23.2	2.8	2.2
(M14)	16.5	16.0	28.0	27.2	2.8	2.2
M16	18.5	18.0	30.0	29.2	3.6	2.4
(M18)	20.6	20.0	34.0	32.8	3.6	2.4
M20	22.6	22.0	37.0	35.8	3.6	2.4
(M22)	24.6	24.0	39.0	37.8	3.6	2.4
M24	26.6	26.0	44.0	42.8	4.6	3.4
(M27)	30.6	30.0	50.0	48.8	4.6	3.4
M30	33.8	33.0	56.0	54.5	4.6	3.4
(M33)	36.8	36.0	60.0	58.5	6.0	4.0
M36	39.8	39.0	66.0	64.5	6.0	4.0
(M39)	42.8	42.0	72.0	70.5	7.0	5.0
M42	45.8	45.0	78.0	76.5	8.2	5.8
(M45)	48.8	48.0	85.0	83.0	8.2	5.8
M48	53.0	52.0	92.0	90.0	9.2	6.8
(M52)	57.0	56.0	98.0	96.0	9.2	6.8
M56	63.0	62.0	105.0	103.0	10.2	7.8
(M60)	67.0	66.0	110.0	108.0	10.2	7.8
M64	71.0	70.0	115.0	113.0	10.2	7.8
(M68)	75.0	74.0	120.0	118.0	11.2	8.8

[a]Non-preferred sizes in brackets ( ).
For full information see BS 4320.

4.1.57 Friction locking devices

Lock nut

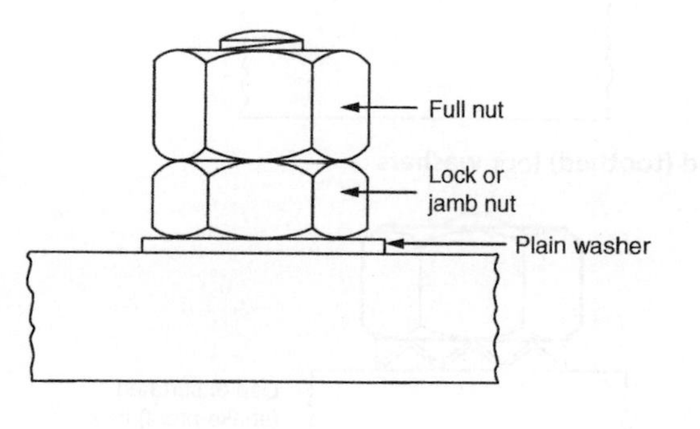

Stiff nut (insert)

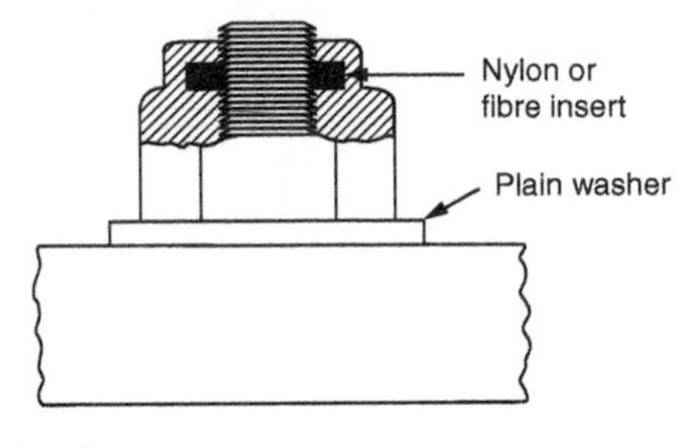

Stiff nut (slit head)

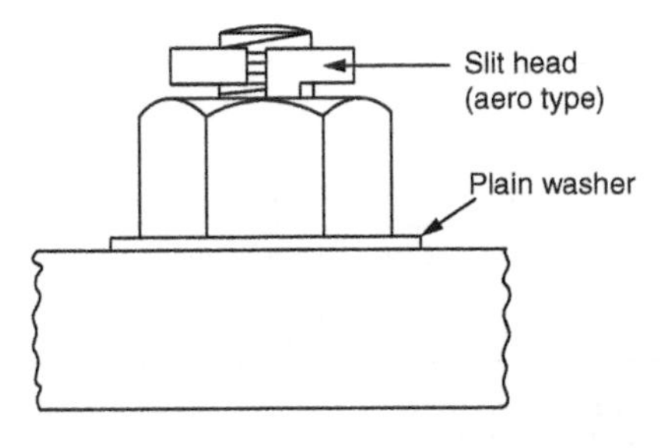

Stiff nut (slit head)

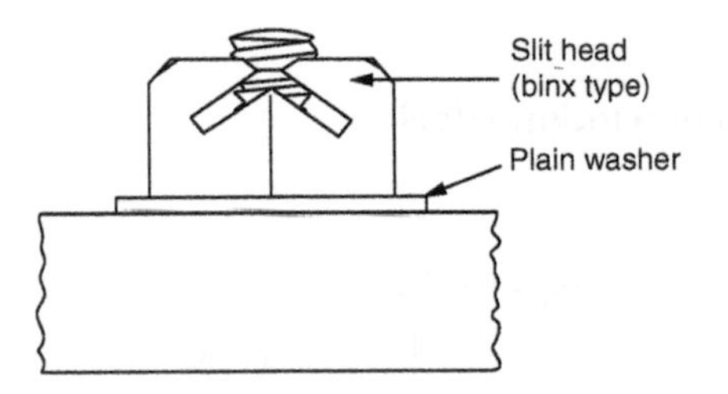

Serrated (toothed) lock washers

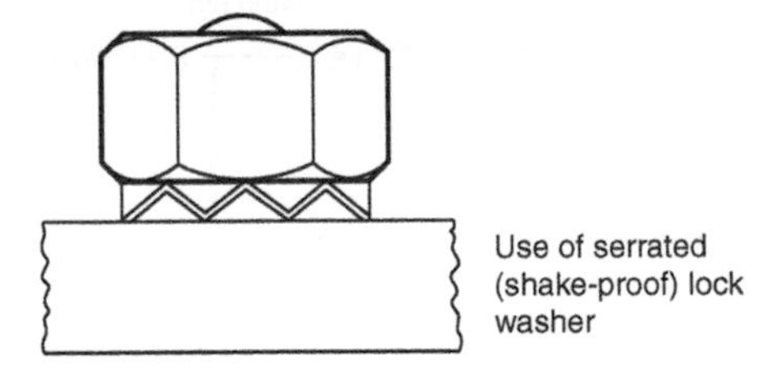

Lock washers, see:
Spring washers 4.1.59, 4.1.60, 4.1.61
Toothed lock washers 4.1.62
Serrated lock washers 4.1.63
Crinkle washers 4.1.64

4.1.58 Positive locking devices

Slotted nut

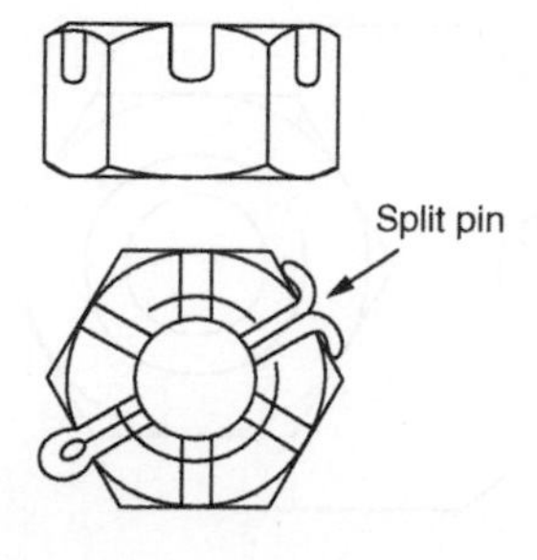

Castle nut

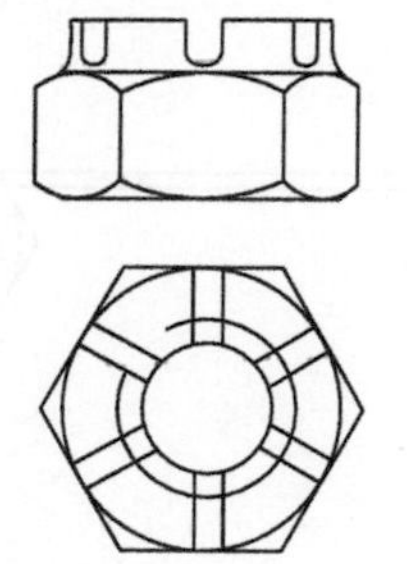

Tab washer

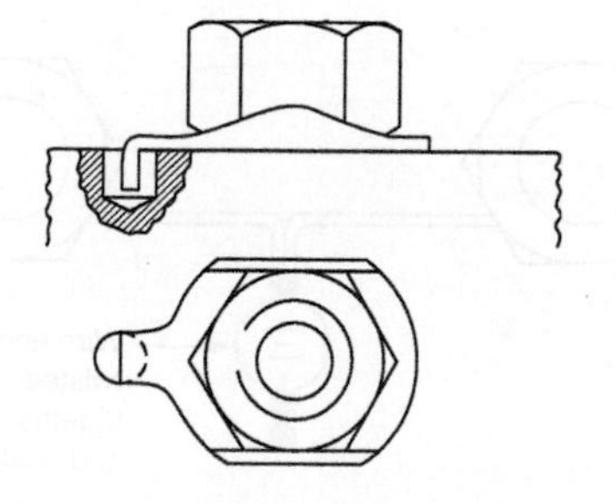

Lock plate

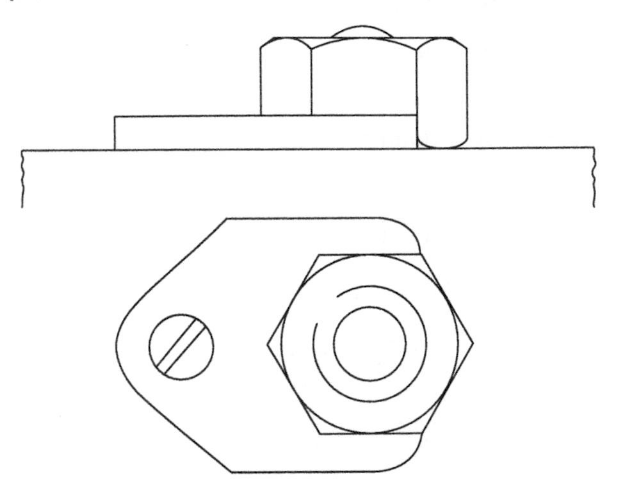

Wiring

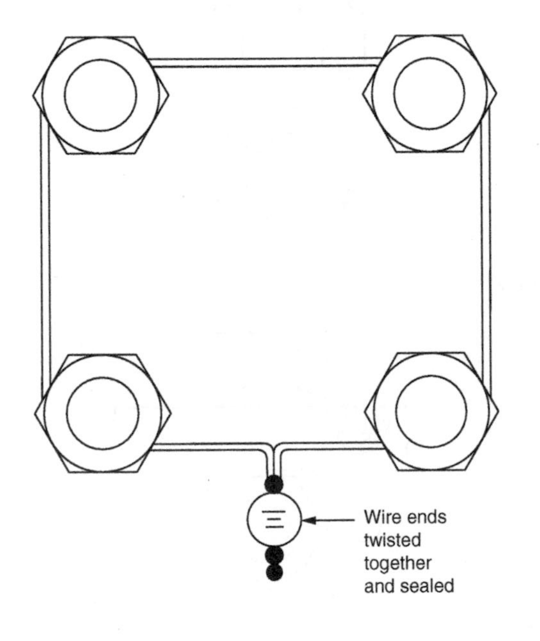

4.1.59 Single coil square section spring washers: metric series, type A

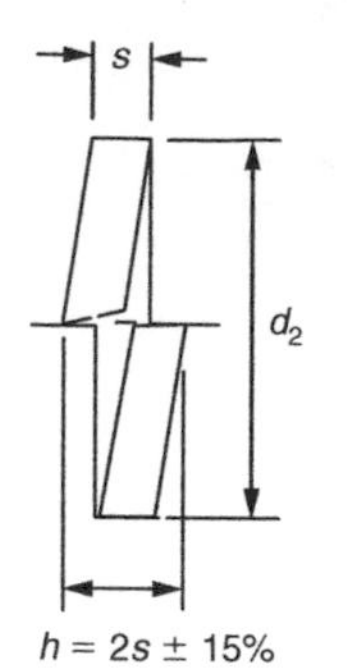

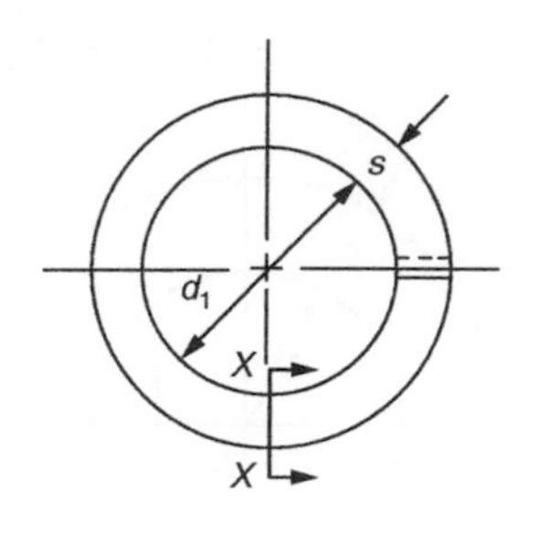

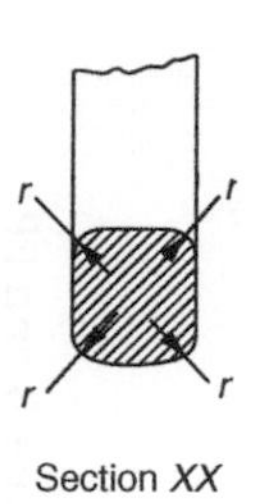

Dimensions in mm

Nominal size and thread diameter[a] d	Inside diameter d_1		Thickness and width s	Outside diameter d_2 max.	Radius r max.
	max.	min.			
M3	3.3	3.1	1 ± 0.1	5.5	0.3
(M3.5)	3.8	3.6	1 ± 0.1	6.0	0.3
M4	4.35	4.1	1.2 ± 0.1	6.95	0.4
M5	5.35	5.1	1.5 ± 0.1	8.55	0.5
M6	6.4	6.1	1.5 ± 0.1	9.6	0.5
M8	8.55	8.2	2 ± 0.1	12.75	0.65
M10	10.6	10.2	2.5 ± 0.15	15.9	0.8
M12	12.6	12.2	2.5 ± 0.15	17.9	0.8
(M14)	14.7	14.2	3 ± 0.2	21.1	1.0
M16	16.9	16.3	3.5 ± 0.2	24.3	1.15
(M18)	19.0	18.3	3.5 ± 0.2	26.4	1.15
M20	21.1	20.3	4.5 ± 0.2	30.5	1.5
(M22)	23.3	22.4	4.5 ± 0.2	32.7	1.5
M24	25.3	24.4	5 ± 0.2	35.7	1.65
(M27)	28.5	27.5	5 ± 0.2	38.9	1.65
M30	31.5	30.5	6 ± 0.2	43.9	2.0
(M33)	34.6	33.5	6 ± 0.2	47.0	2.0
M36	37.6	36.5	7 ± 0.25	52.1	2.3
(M39)	40.8	39.6	7 ± 0.25	55.3	2.3
M42	43.8	42.6	8 ± 0.25	60.3	2.65
(M45)	46.8	45.6	8 ± 0.25	63.3	2.65
M48	50.0	48.8	8 ± 0.25	66.5	2.65

[a] Sizes shown in brackets are non-preferred and are not usually stock sizes.
For further information see BS 4464.

4.1.60 Single coil rectangular section spring washers: metric series, types B and BP

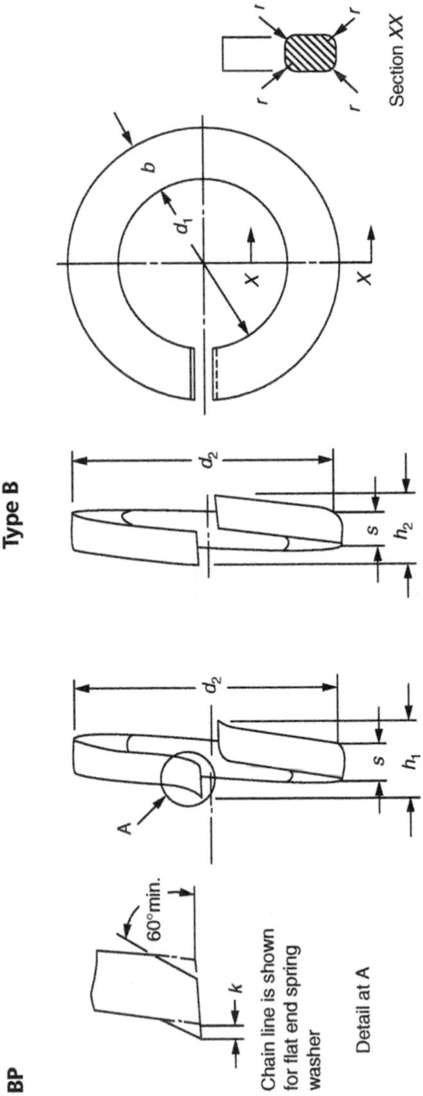

Type BP

60°min.

A

d_2

k

Chain line is shown for flat end spring washer

Detail at A

s

h_1

$h_1 = (2s + 2k) \pm 15\%$

Type B

d_2

s

h_2

$h_2 = 2s \pm 15\%$

b

d_1

X

X

r r r r

Section XX

Dimensions in mm

Nominal size and thread diameter[a] d	Inside diameter d_1		Width b	Thickness s	Outside diameter d_2 max.	Radius r max.	k (type BP only)
	max.	min.					
M1.6	1.9	1.7	0.7 ± 0.1	0.4 ± 0.1	3.5	0.15	–
M2	2.3	2.1	0.9 ± 0.1	0.5 ± 0.1	4.3	0.15	–
(M2.2)	2.5	2.3	1.0 ± 0.1	0.6 ± 0.1	4.7	0.2	–
M2.5	2.8	2.6	1.0 ± 0.1	0.6 ± 0.1	5.0	0.2	–
M3	3.3	3.1	1.3 ± 0.1	0.8 ± 0.1	6.1	0.25	–
(M3.5)	3.8	3.6	1.3 ± 0.1	0.8 ± 0.1	6.6	0.25	0.15
M4	4.35	4.1	1.5 ± 0.1	0.9 ± 0.1	7.55	0.3	0.15
M5	5.35	5.1	1.8 ± 0.1	1.2 ± 0.1	9.15	0.4	0.15
M6	6.4	6.1	2.5 ± 0.15	1.6 ± 0.1	11.7	0.5	0.2
M8	8.55	8.2	3 ± 0.15	2 ± 0.1	14.85	0.65	0.3
M10	10.6	10.2	3.5 ± 0.2	2.2 ± 0.15	18.0	0.7	0.3
M12	12.6	12.2	4 ± 0.2	2.5 ± 0.15	21.0	0.8	0.4
(M14)	14.7	14.2	4.5 ± 0.2	3 ± 0.15	24.1	1.0	0.4
M16	16.9	16.3	5 ± 0.2	3.5 ± 0.2	27.3	1.15	0.4
(M18)	19.0	18.3	5 ± 0.2	3.5 ± 0.2	29.4	1.15	0.4
M20	21.1	20.3	6 ± 0.2	4 ± 0.2	33.5	1.3	0.4
(M22)	23.3	22.4	6 ± 0.2	4 ± 0.2	35.7	1.3	0.4
M24	25.3	24.4	7 ± 0.25	5 ± 0.2	39.8	1.65	0.5
(M27)	28.5	27.5	7 ± 0.25	5 ± 0.2	43.0	1.65	0.5
M30	31.5	30.5	8 ± 0.25	6 ± 0.25	48.0	2.0	0.8
(M33)	34.6	33.5	10 ± 0.25	6 ± 0.25	55.1	2.0	0.8
M36	37.6	36.5	10 ± 0.25	6 ± 0.25	58.1	2.0	0.8
(M39)	40.8	39.6	10 ± 0.25	6 ± 0.25	61.3	2.0	0.8
M42	43.8	42.6	12 ± 0.25	7 ± 0.25	68.3	2.3	0.8
(M45)	46.8	45.6	12 ± 0.25	7 ± 0.25	71.3	2.3	0.8
M48	50.0	48.8	12 ± 0.25	7 ± 0.25	74.5	2.3	0.8
(M52)	54.1	52.8	14 ± 0.25	8 ± 0.25	82.6	2.65	1.0
M56	58.1	56.8	14 ± 0.25	8 ± 0.25	86.6	2.65	1.0
(M60)	62.3	60.9	14 ± 0.25	8 ± 0.25	90.8	2.65	1.0
M64	66.3	64.9	14 ± 0.25	8 ± 0.25	93.8	2.65	1.0
(M68)	70.5	69.0	14 ± 0.25	8 ± 0.25	99.0	2.65	1.0

[a] Sizes shown in brackets are non-preferred, and are not usually stock sizes.
For further information see BS 4464.

4.1.61 Double coil rectangular section spring washers: metric series, type D

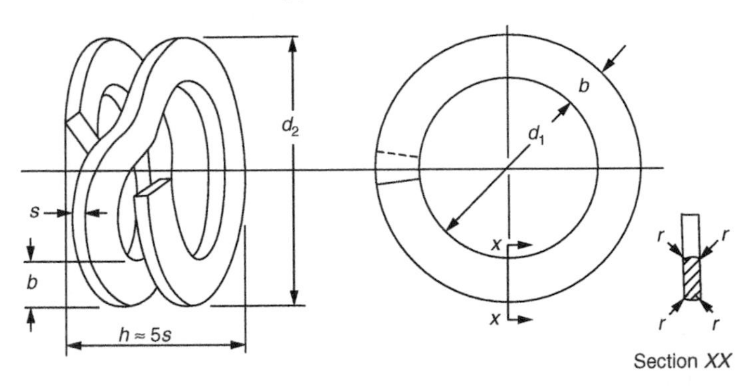

Dimensions in mm

Nominal size and thread diameter[a] d	Inside diameter d_1 max.	Inside diameter d_1 min.	Width b	Thickness s	Outside diameter d_2 max.	Radius r max.
M2	2.4	2.1	0.9 ± 0.1	0.5 ± 0.05	4.4	0.15
(M2.2)	2.6	2.3	1.0 ± 0.1	0.6 ± 0.05	4.8	0.2
M2.5	2.9	2.6	1.2 ± 0.1	0.7 ± 0.1	5.5	0.23
M3.0	3.6	3.3	1.2 ± 0.1	0.8 ± 0.1	6.2	0.25
(M3.5)	4.1	3.8	1.6 ± 0.1	0.8 ± 0.1	7.5	0.25
M4	4.6	4.3	1.6 ± 0.1	0.8 ± 0.1	8.0	0.25
M5	5.6	5.3	2 ± 0.1	0.9 ± 0.1	9.8	0.3
M6	6.6	6.3	3 ± 0.15	1 ± 0.1	12.9	0.33
M8	8.8	8.4	3 ± 0.15	1.2 ± 0.1	15.1	0.4
M10	10.8	10.4	3.5 ± 0.20	1.2 ± 0.1	18.2	0.4
M12	12.8	12.4	3.5 ± 0.2	1.6 ± 0.1	20.2	0.5
(M14)	15.0	14.5	5 ± 0.2	1.6 ± 0.1	25.4	0.5
M16	17.0	16.5	5 ± 0.2	2 ± 0.1	27.4	0.65
(M18)	19.0	18.5	5 ± 0.2	2 ± 0.1	29.4	0.65
M20	21.5	20.8	5 ± 0.2	2 ± 0.1	31.9	0.65
(M22)	23.5	22.8	6 ± 0.2	2.5 ± 0.15	35.9	0.8
M24	26.0	25.0	6.5 ± 0.2	3.25 ± 0.15	39.4	1.1
(M27)	29.5	28.0	7 ± 0.25	3.25 ± 0.15	44.0	1.1
M30	33.0	31.5	8 ± 0.25	3.25 ± 0.15	49.5	1.1
(M33)	36.0	34.5	8 ± 0.25	3.25 ± 0.15	52.5	1.1
M36	40.0	38.0	10 ± 0.25	3.25 ± 0.15	60.5	1.1
(M39)	43.0	41.0	10 ± 0.25	3.25 ± 0.15	63.5	1.1
M42	46.0	44.0	10 ± 0.25	4.5 ± 0.2	66.5	1.5
M48	52.0	50.0	10 ± 0.25	4.5 ± 0.2	72.5	1.5
M56	60.0	58.0	12 ± 0.25	4.5 ± 0.2	84.5	1.5
M64	70.0	67.0	12 ± 0.25	4.5 ± 0.2	94.5	1.5

[a] Sizes shown in brackets are non-preferred, and are not usually stock sizes.

Note: The free height of double coil washers before compression is normally approximately five times the thickness but, if required, washers with other free heights may be obtained by arrangement between the purchaser and the manufacturer.

For further information see BS 4464.

4.1.62 Toothed lock washers, metric

Type A externally toothed

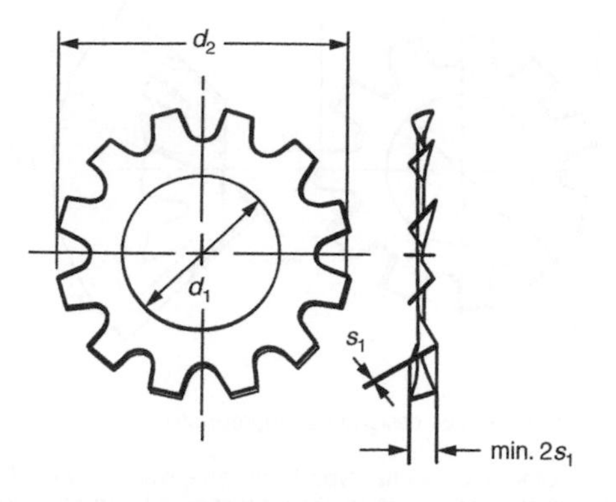

Type J internally toothed

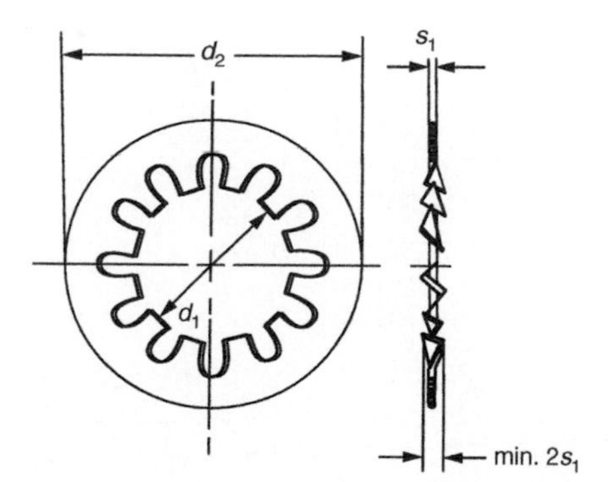

Type V countersunk

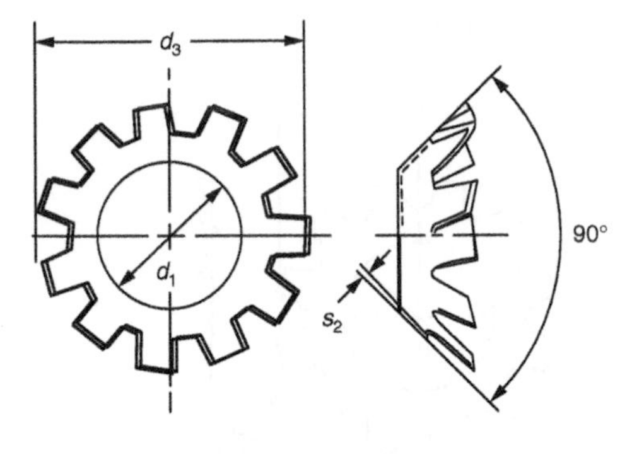

Details left unspecified are to be designed as appropriate.

Designation of a toothed lock washer type J with hole diameter d_1 = 6.4 mm of spring steel, surface phosphated for protection against rusting (phr): toothed lock washer J 6.4 DIN 6797 – phr.

If toothed lock washers are required for left hand threaded bolts, the designation reads: toothed lock washer J 6.4 left DIN 6797 – phr.

Dimensions in mm

d_1 (H13)	d_2 (h14)	$d_3 \approx$	s_1	s_2	Number of teeth min.		Weight (7.85 kg/dm^3) kg/1000 pieces ≈			For thread diameter
					A and J	V	A	J	V	
1.7	3.6	–	0.3	–	6	–	0.01	–	–	1.6
1.8	3.8	–	0.3	–	6	–	0.015	–	–	1.7
1.9	4	–	0.3	–	6	–	0.02	0.03	–	1.8
2.2	4.5	4.2	0.3	0.2	6	6	0.025	0.04	0.02	2
2.5	5	–	0.4	0.2	6	6	0.03	0.025	–	2.3
2.7	5.5	5.1	0.4	0.2	6	6	0.04	0.045	0.025	2.5
2.8	5.5	–	0.4	0.2	6	6	0.04	0.045	–	2.6
3.2	6	6	0.4	0.2	6	6	0.045	0.045	0.025	3
3.7	7	7	0.5	0.25	6	6	0.075	0.085	0.04	3.5
4.3	8	8	0.5	0.25	8	8	0.095	0.1	0.05	4
5.1[a]	9	–	0.5	–	8	–	0.14	0.15	–	5
5.3	10	9.8	0.6	0.3	8	8	0.18	0.2	0.12	5
6.4	11	11.8	0.7	0.4	8	10	0.22	0.25	0.2	6
7.4	12.5	–	0.8	–	8	–	0.3	0.35	–	7
8.2[a]	14	–	0.8	–	8	–	0.4	0.45	–	8
8.4	15	15.3	0.8	0.4	8	10	0.45	0.55	0.4	8
10.5	18	19	0.9	0.5	9	10	0.8	0.9	0.7	10
12.5	20.5	23	1	0.5	10	10	1.1	1.3	1.2	12
14.5	24	26.2	1	0.6	10	12	1.7	2	1.4	14
16.5	26	30.2	1.2	0.6	12	12	2.1	2.5	1.4	16
19	30	–	1.4	–	12	–	3.5	3.7	–	18
21	33	–	1.4	–	12	–	3.8	4.1	–	20
23	36	–	1.5	–	14	–	5	6	–	22
25	38	–	1.5	–	14	–	6	6.5	–	24
28	44	–	1.6	–	14	–	8	8.5	–	27
31	48	–	1.6	–	14	–	9	9.5	–	30

[a] Only for hexagon head bolts.
For further details see DIN 6797.

4.1.63 Serrated lock washers, metric

Type A serrated externally

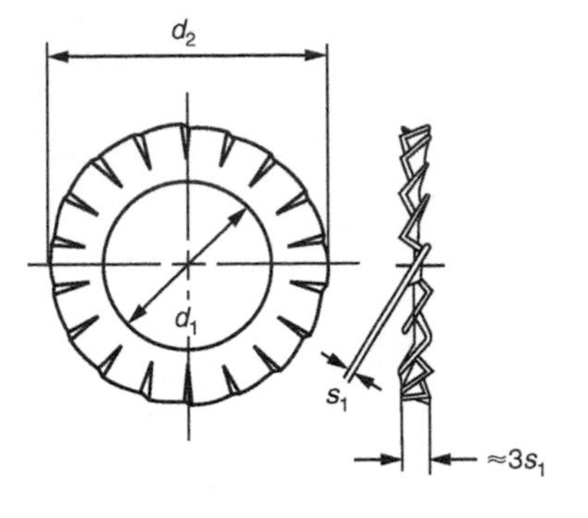

Type J serrated internally

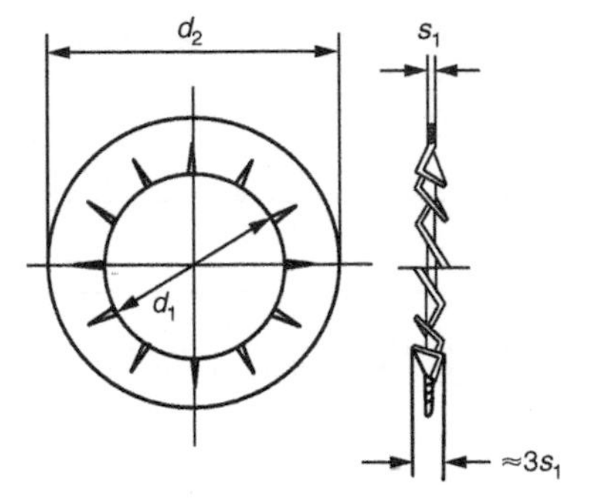

Type V countersunk

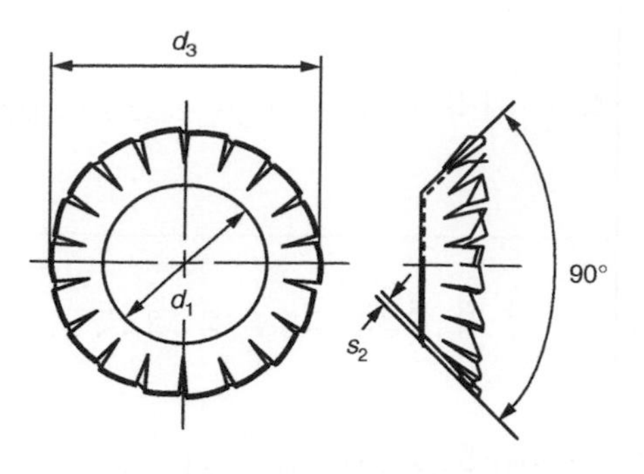

Details left unspecified are to be designed as appropriate.

Designation of a serrated lock washer type J with hole diameter $d_1 = 6.4$ mm in spring steel, surface phosphated for protection against rusting (phr): serrated lock washer J 6.4 DIN 6798 – phr.

If serrated lock washers are required for left hand threaded bolts, the designation reads: serrated lock washer J 6.4 left DIN 6798 – phr.

Dimensions in mm

d_1 (H13)	d_2 (h14)	$d_3 \approx$	s_1	s_2	Number of teeth min.			Weight (7.85 kg/dm³) kg/1000 pieces ≈		For thread diameter
					A	J	V	A and J	V	
1.7	3.6		0.3	–	9	7	–	0.02	–	1.6
1.8	3.8		0.3	–	9	7	–	0.02	–	1.7
1.9	4		0.3	–	9	7	–	0.025	–	–
2.2	4.5	4.2	0.3	0.2	9	7	10	0.03	0.025	2
2.5	5		0.4	0.2	9	7	10	0.04	–	2.3
2.7	5.5	5.1	0.4	0.2	9	7	10	0.045	0.03	2.5
2.8	5.5		0.4	0.2	9	7	10	0.05	–	2.6
3.2	6	6	0.4	0.2	9	7	12	0.06	0.04	3
3.7	7	7	0.5	0.25	10	8	12	0.11	0.075	3.5
4.3	8	8	0.5	0.25	11	8	14	0.14	0.1	4
5.1[a]	9		0.5	–	11	8	–	0.22	–	5
5.3	10	9.8	0.6	0.3	11	8	14	0.28	0.2	5
6.4	11	11.8	0.7	0.4	12	9	16	0.36	0.3	6
7.4	12.5		0.8	–	14	10	–	0.5	–	7
8.2[a]	14		0.8	–	14	10	–	0.75	–	8
8.4	15	15.3	0.8	0.4	14	10	18	0.8	0.5	8
10.5	18	19	0.9	0.5	16	12	20	1.25	1	10
12.5	20.5	23	1	0.5	16	12	26	1.7	1.5	12
14.5	24	26.2	1	0.6	18	14	28	2.4	2	14
16.5	26	30.2	1.2	0.6	18	14	30	3	2.4	16
19	30	–	1.4	–	18	14	–	5	–	18
21	33	–	1.4	–	20	16	–	6	–	20
23	36	–	1.5	–	20	16	–	7.5	–	22
25	38	–	1.5	–	20	16	–	8	–	24
28	44	–	1.6	–	22	18	–	12	–	27
31	48	–	1.6	–	22	18	–	14	–	30

[a] Only for hexagon head bolts.
For further details see DIN 6797.

4.1.64 ISO metric crinkle washers: general engineering

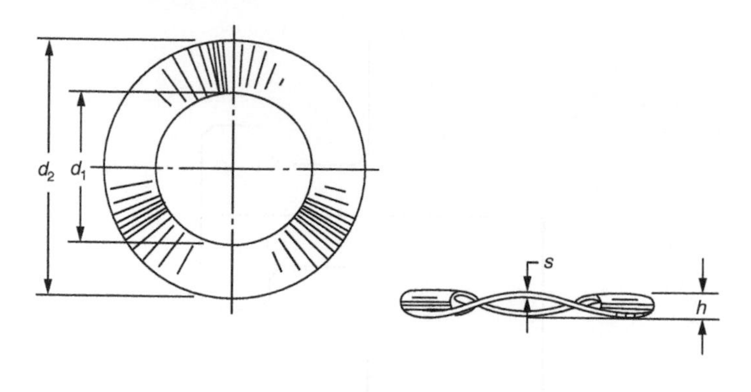

Dimensions in mm

Nominal (thread.) diameter[a]	Inside diameter d_1		Outside diameter d_2		*Height* h		Thickness s
	max.	min.	max.	min.	max.	min.	
M1.6	1.8	1.7	3.7	3.52	0.51	0.36	0.16
M2	2.3	2.2	4.6	4.42	0.53	0.38	0.16
M2.5	2.8	2.7	5.8	5.62	0.53	0.38	0.16
M3	3.32	3.2	6.4	6.18	0.61	0.46	0.16
M4	4.42	4.3	8.1	7.88	0.84	0.69	0.28
M5	5.42	5.3	9.2	8.98	0.89	0.74	0.30
M6	6.55	6.4	11.5	11.23	1.14	0.99	0.40
M8	8.55	8.4	15.0	14.73	1.40	1.25	0.40
M10	10.68	10.5	19.6	19.27	1.70	1.55	0.55
M12	13.18	13.0	22.0	21.67	1.90	1.65	0.55
(M14)	15.18	15.0	25.5	25.17	2.06	1.80	0.55
M16	17.18	17.0	27.8	27.47	2.41	2.16	0.70
(M18)	19.21	19.0	31.3	30.91	2.41	2.16	0.70
M20	21.21	21.0	34.7	34.31	2.66	2.16	0.70

[a] Second choice sizes in brackets ( ).
For full range and further information see BS 4463.

4.1.65 T-slot profiles

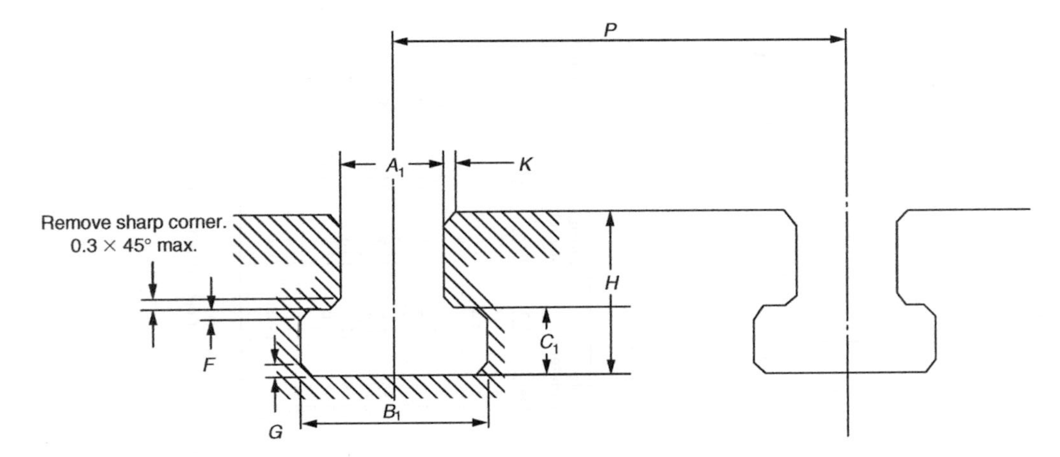

Dimensions in mm

Designations of T-slot	Width of throat A_1			Width of recess B_1		Depth of recess C_1		Overall depth of T-slot H		Chamfer × 45° or radius			Pitch P (avoid pitch values in brackets as they lead to weakness)			
	Nominal	Ordinary (H12)	For use as tenon (HB)	min.	max.	min.	max.	min.	max.	K max.	F max.	G max.				
M4	5	+0.12	+0.018	10	11	3.5	4.5	8	10					20	25	32
M5	6	0	0	11	12.5	5	6	11	13					25	32	40
M6	8	+0.15	+0.022	14.5	16	7	8	15	18	1.0		1.0		32	40	50
M8	10	0	0	16	18	7	8	17	21		0.6			40	50	63
M10	12	+0.18	+0.027	19	21	8	9	20	25				(40)	50	63	80
M12	14	0	0	23	25	9	11	23	28				(50)	63	80	100
M16	18			30	32	12	14	30	36			1.6	(63)	80	100	125
M20	22	+0.21	+0.033	37	40	16	18	38	45	1.6			(80)	100	125	160
M24	28	0		46	50	20	22	48	56		1.0	2.5	100	125	160	200
M30	36	+0.25	+0.039	56	60	25	28	61	71				125	160	200	250
M36	42	0	0	68	72	32	35	74	85		1.6	4.0	160	200	250	320
M42	48			80	85	36	40	84	95	2.5			200	250	320	400
M48	54	+0.30 0	+0.046 0	90	95	40	44	94	106		2.0	6.0	250	320	400	500

Tolerance on pitches *p* of T-slots

Pitch (mm)	Tolerance (mm)
20 to 25	±0.2
32 to 100	±0.3
125 to 250	±0.5
320 to 500	±0.8

For further information see BS 2485.

4.1.66 Dimensions of T-bolts and T-nuts

T-nut

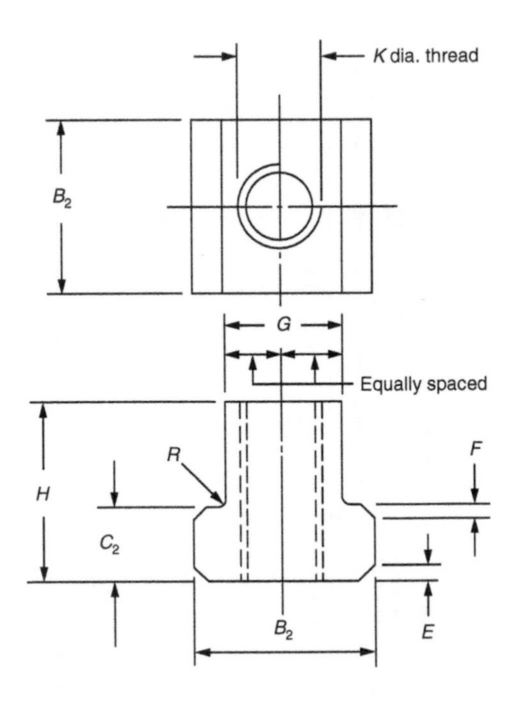

T-bolt

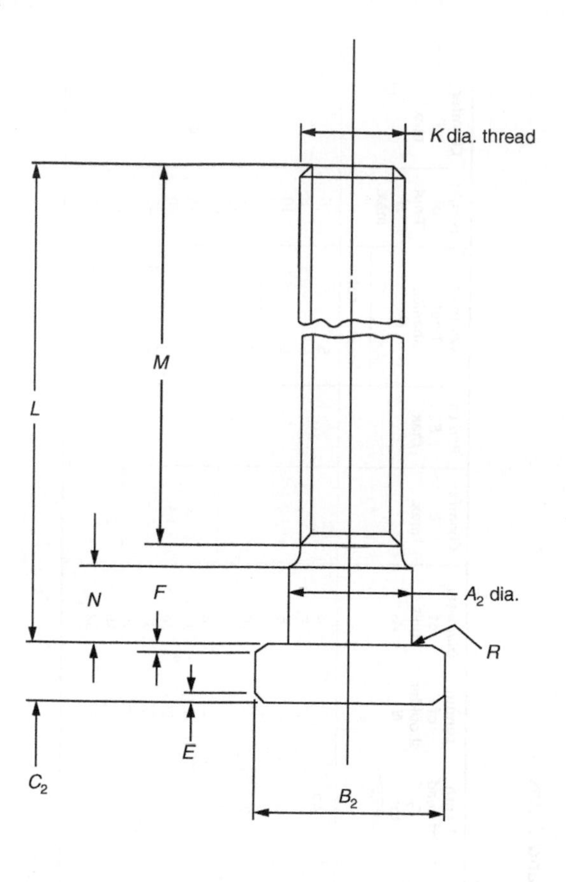

Dimensions of T-bolts and T-nuts

Table 1

Designation of T-bolt and diameter of thread K	Width of head (square) B_2		Depth of head C_2 tol. ±0.25	Length of shoulder N	Diameter of bolt A_2	Chamfer E max.	Radius R max.	Width of T-nut shank G		Height of T-nut H max.	Chamfer F max.
	nom.	tol.						nom.	tol.		
M4	9	0 −0.5	2.5	–	4	1.0	1.0	5	−0.3 −0.5	6	0.3 × 45°
M5	10		4	–	5			6		8	
M6	13		6	–	6			8		10	
M8	15		6	–	8			10		12	
M10	18		7	–	10			12	−0.3 −0.6	16	
M12	22		8	–	12			14		19	
M16	28		10	–	16	1.5		18		25	
M20	34		14	–	20			22		30	
M24	43		18	20	26			28		40	
M30	53		23	20	33			36	−0.4 −0.7	45	
M36	64		28	25	39	2.5		42		55	
M42	75		32	25	46			48		65	
M48	85		36	30	52			54	−0.4 −0.8	75	

For further information See BS 2485.

Table 2 Dimensions in mm

Diameter of thread K	Recommended length of bolt stem L													Length of threaded portion of bolt stem M
M4	30	40	50	60	70	80	100							For $L \leqslant 100$ $M = 0.5L$
M5	30	40	50	60	70	80	100							
M6				60	70	80	100							
M8				60	70	80	100							
M10				60	70	80	100	125	160	180				
M12					70	80	100	125	160	180				For $L > 100$ $M = 0.3L$
M16					70	80	100	125	160	180	200	250	300	
M20					70	80	100	125	160	180	200	250	300	
M24, M30, M36, M42, M48							100	125	160	180	200	250	300	

For further information see BS 2485.

4.1.67 Dimensions of tenons for T-slots

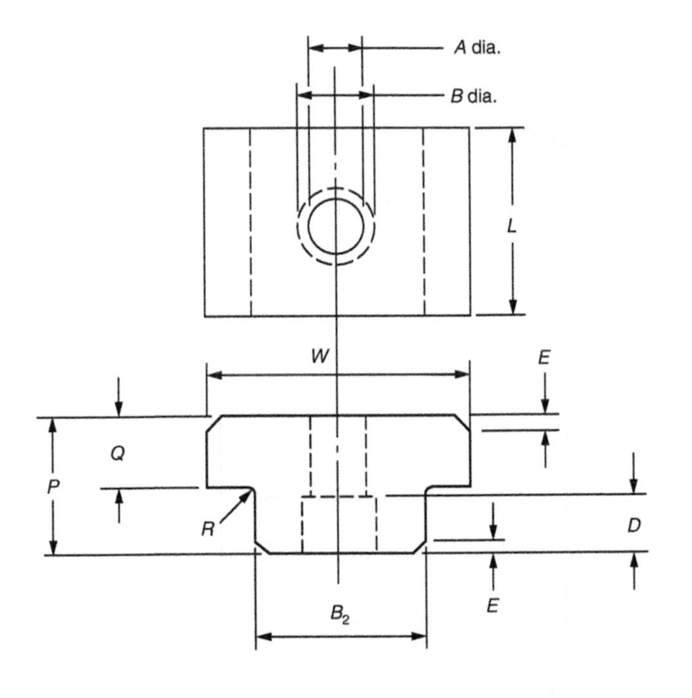

Dimensions in mm

Designation of T-slot	Width of tenon shank B_2		Overall width of tenon W		Depth of head of tenon Q	Overall height of tenon P	Length of tenon L	Radius R max.	Fixing hole			
	nom.	tol. (h7)	nom.	tol. (h7)					To suit socket head cap screw to BS 4168: Pt 1	Clearance hole diameter A to BS 4186 medium fit	Counterbores diameter B tolerance H13	Counterbore depth D tolerance +0.2
M4	5	0 −0.012	16	0 −0.018	5	10	25	0.6	M2	2.4	4.3	2.5
M5	6	0 −0.015										
M6	8								M3	3.4	6.0	3.5
M8	10	0 −0.018	30	0 −0.021	5.5	12	30					
M10	12											
M12	14											
M16	18							1.0	M6	6.6	11.0	6.5
M20	22	0 −0.021										
M24	28		50	0 −0.025	15	30	40					
M30	36	0 −0.025										
M36	42											
M42	48	0 −0.030	70	0 −0.030	25	40	60	1.5	M8	9.0	14.0	8.5
M48	54											

For further information see BS 2485: 1987.

4.2 Riveted joints

4.2.1 Typical rivet heads and shanks

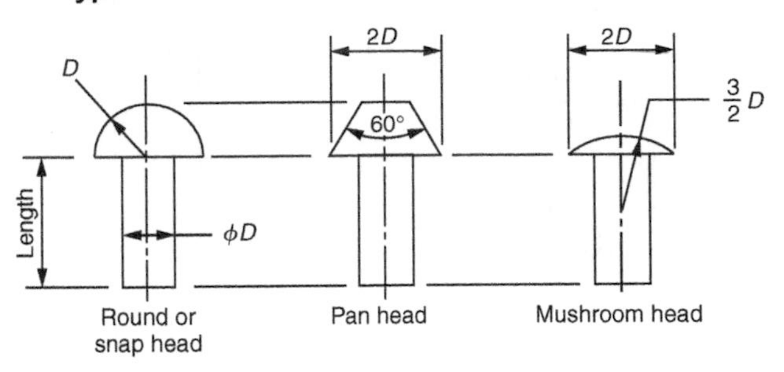

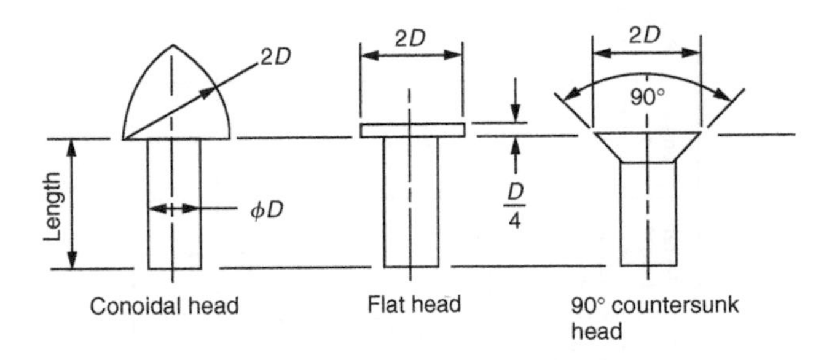

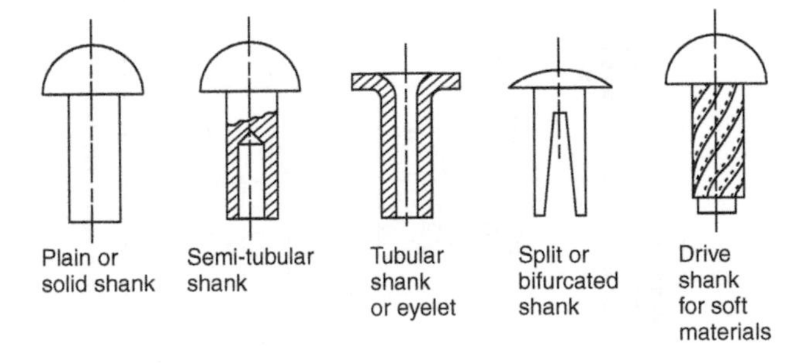

4.2.2 Typical riveted lap joints

Single row lap joint

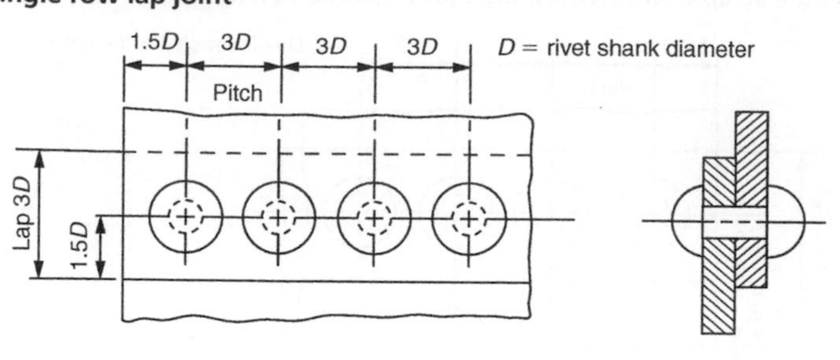

Double row (chain) lap joint

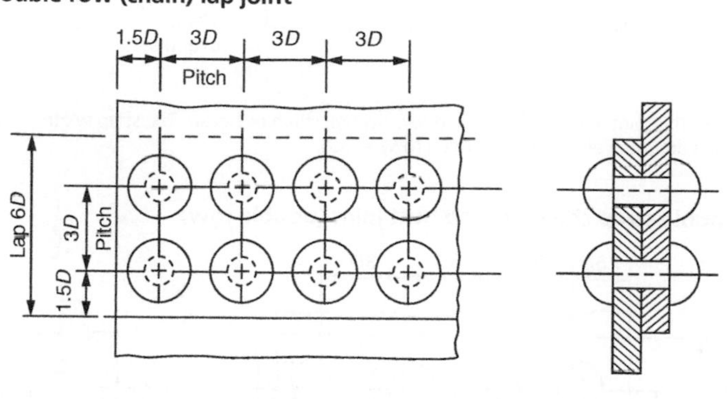

Double row (zigzag) lap joint

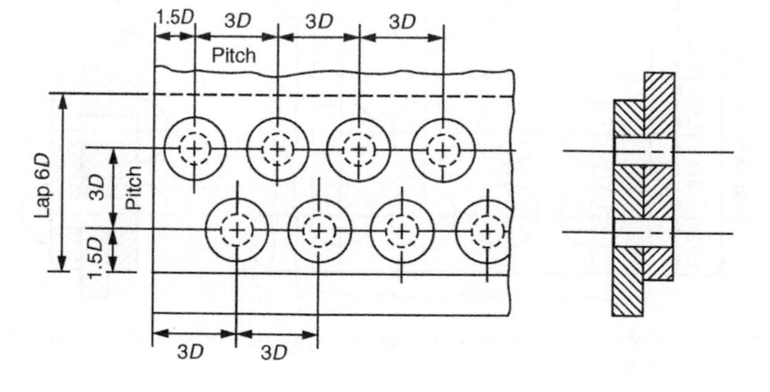

4.2.3 Typical riveted butt joints

Single strap chain riveted butt joint (single row)

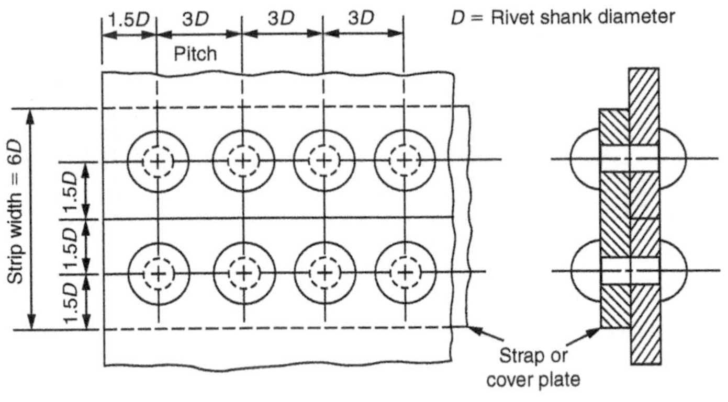

Note: This joint may also be double row riveted, chain or zigzag. The strap width = 12D when double riveted (pitch between rows = 3D).

Double strap chain riveted butt joint (double row)

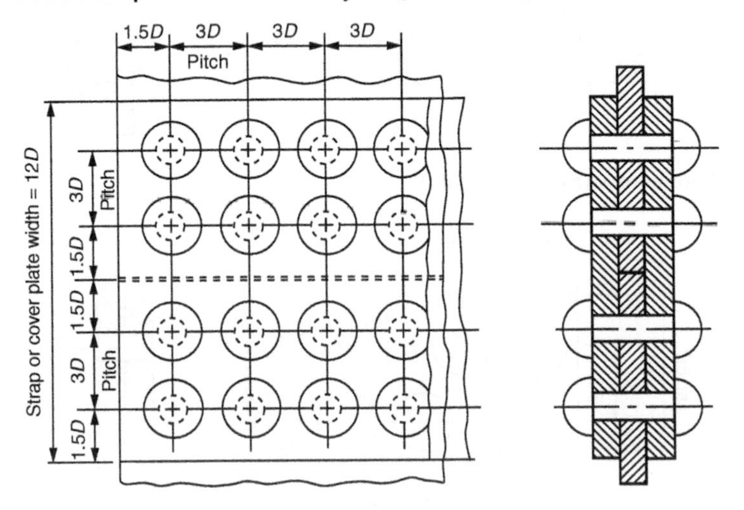

Note: This joint may also be double row zigzag riveted (see Section 4.2.2) or it may be single riveted as above.

4.2.4 Proportions for hole diameter and rivet length

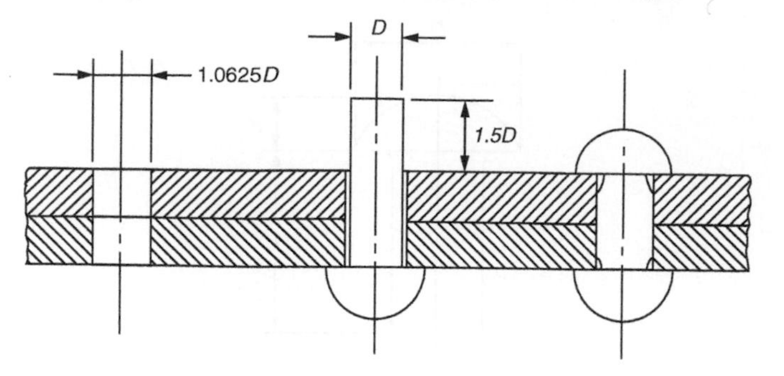

4.2.5 Cold forged snap head rivets

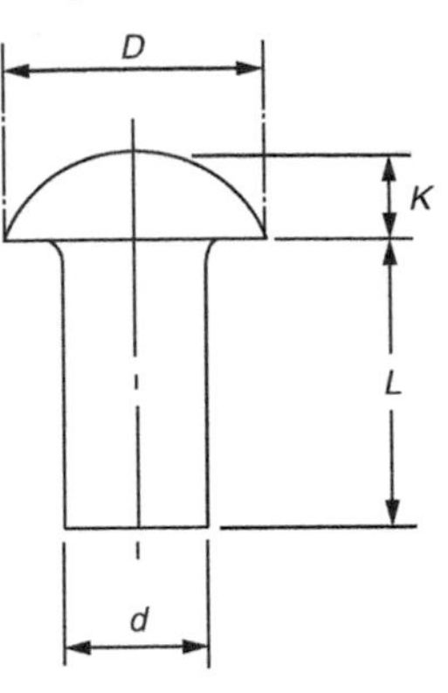

With d = 16 mm or smaller
D = 1.75d
K = 0.6d
L = length

Dimensions in mm

Nominal shank diameter[a] d	Tolerance on diameter d	Nominal head diameter D	Tolerance on diameter D	Nominal head depth K	Tolerance on head depth K	Tolerance on length L
1	±0.07	1.8	±0.2	0.6	+0.2 −0.0	+0.5 −0.0
1.2		2.1		0.7		
1.6		2.8		1.0		
2.0		3.5	±0.24	1.2	+0.24 −0.0	
2.5		4.4		1.5		
3.0		5.3		1.8		
(3.5)	±0.09	6.1	±0.29	2.1	+0.29 −0.0	
4		7.0		2.4		
5		8.8		3.0		
6		10.5	±0.35	3.6	+0.35 −0.0	
(7)	±0.11	12.3		4.2		+0.8 −0.0
8		14.0		4.8		
10		18.0	±0.42	6.0	+0.42 −0.0	
12	±0.14	21.0		7.2		+1.0 −0.0
(14)		25.0		8.4		
16		28.0		9.6		

[a]Rivet sizes shown in brackets are non-preferred.
For further information see BS 4620: 1970.

4.2.6 Hot forged snap head rivets

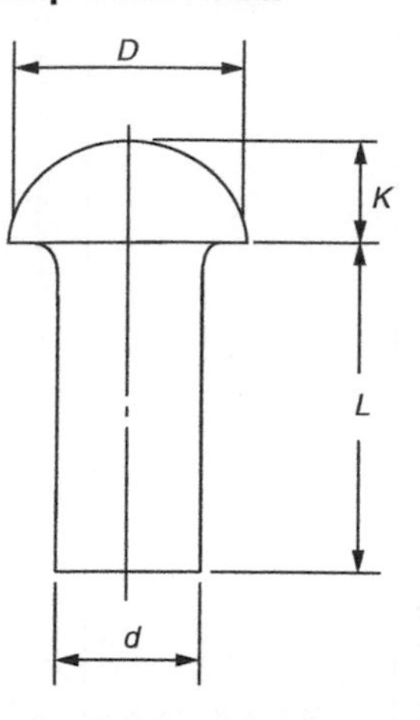

With d = 14 mm or larger
$D = 1.6d$
$K = 0.65d$
L = length

Dimensions in mm

Nominal shank diameter[a] d	Tolerance on diameter d	Nominal head diameter D	Tolerance on diameter D	Nominal head depth K	Tolerance on head depth K	Tolerance on length L
(14)		22		9		+1.0
16	±0.43	25	±1.25	10	+1.00	−0.0
18		28		11.5	−0.0	
20		32		13	+1.5	+1.6
(22)	±0.52	36	+1.8	14	−0.0	−0.0
24		40		16		
(27)		43		17	+2.0	
30		48	±2.5	19	−0.0	
(33)	±0.62	53		21		+3.0 −0.0
36		58	±3.0	23	+2.5	
39		62		25	−0.0	

[a] Rivet sizes shown in brackets are non-preferred.
For further information see BS 4620: 1970.

4.2.7 Tentative range of nominal lengths associated with shank diameters

Dimensions in mm

Nominal shank diameter[a] d	Nominal length* L																					
	3	4	5	6	8	10	12	14	16	(18)	20	(22)	25	(28)	30	(32)	35	(38)	40	45	50	55
1.0	×	×	×	×	×	×	×	×	×		×											
1.2	×	×	×	×	×	×	×	×	×		×											
1.6	×	×	×	×	×	×	×	×	×	×		×	×									
2.0	×	×	×	×	×	×	×	×	×		×	×	×									
2.5	×	×	×	×	×	×	×	×	×	×		×	×									
3.0		×	×	×	×	×	×	×	×	×	×	×	×									
(3.5)																						
4.0				×	×	×	×	×	×	×		×	×									
5.0				×	×	×	×	×	×	×	×	×	×	×	×		×	×				
6.0				×	×	×	×	×	×	×	×	×	×	×	×		×	×		×		

[a] Sizes and lengths shown in brackets are non-preferred and should be avoided if possible. The inclusion of dimensional data is not intended to imply that all the products described are stock production sizes. The purchaser should consult the manufacturer concerning lists of stock production sizes.

For the full range of head types and sizes up to and including 39 mm diameter by 160 mm shank length see BS 4620: 1970.

4.2.8 POP® rivets

POP® or blind riveting is a technique which enables a mechanical fastening to be made when access is limited to only one side of the parts to be assembled, although reliability, predictability, reduction of assembly costs and simplicity in operation, mean that blind rivets are also widely used where access is available to both sides of an assembly. POP® and other brands of blind riveting systems have two elements, the blind rivet and the chosen setting tool.

The blind rivet is a two-part mechanical fastener. It comprises of a headed tubular body mounted on a mandrel which has self-contained features that create (when pulled into the body during setting) both an upset of the blind end and an expansion of the tubular body, thus joining together the component parts of the assembly. The setting tool basically comprises an anvil which supports the head of the rivet and jaws which grip and pull the mandrel to cause it to set the rivet before the mandrel breaks at a predetermined load.

Many different styles of blind rivet are available but the most widely used is the Open End Rivet Body type defined in BS EN ISO 14588: 2000 as: 'A blind rivet having a body hollow throughout its length and able to use an standard mandrel'. The principle of blind riveting using open style rivets is shown in the following figure.

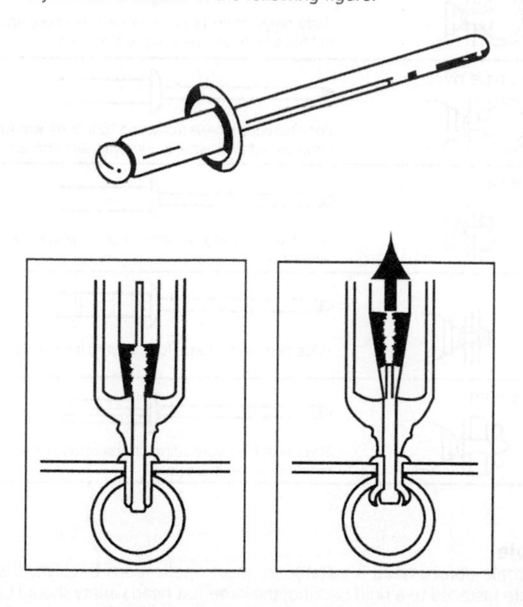

With the mandrel held in the setting tool, the rivet body is inserted into the pre-punched or pre-drilled component. Operation of the setting tool pulls the mandrel head into the rivet body causing it to expand on the blind side of the assembly, whilst drawing the components together and eliminating any gaps between them as it does so. At a predetermined point, when the blind side head is fully formed, continued operation of the setting tool causes the mandrel to break, the spent portion of the mandrel is pulled clear and the installation of the rivet is complete.

The Closed End Rivet Body type is defined in BS EN ISO 14588: 2000 as: 'A blind rivet body which is closed and remains closed after setting'. This type is also commonly known as 'sealed'. The closed end rivet prevents ingress of vapour or moisture through the bore of the installed rivet and also ensures mandrel head retention, particularly important in electrical equipment, for example.

POP® blind rivets are available in a variety of materials, body styles and head forms to provide fastening options for a broad spectrum of assembly and environmental requirements from brittle and fragile materials such as acrylic plastics through to stainless steel. A summary is shown in Section 4.2.9. This figure and the following tables are taken from the publications of Emhart Teknologies (Tucker Fasteners Ltd.) from whom further information can be obtained. This company's address is listed in Appendix 3.

4.2.9 POP® range guide

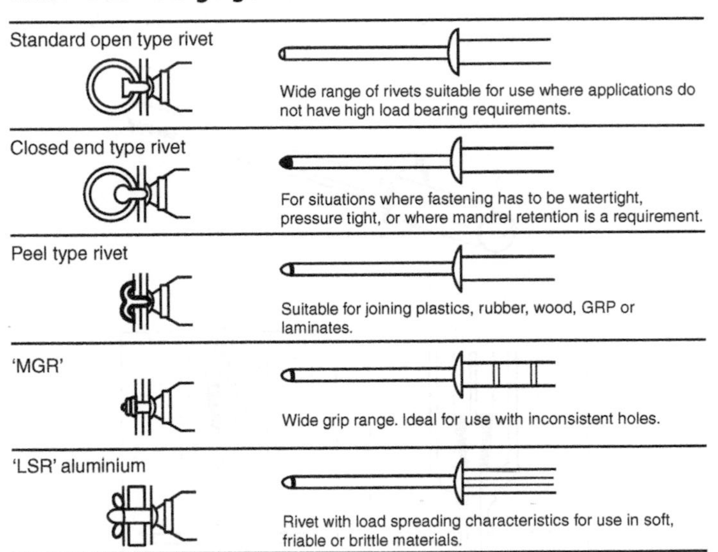

Head style

The low-profile domed head is suitable for most applications but, where soft or brittle materials are fastened to a rigid backing, the large flat head variety should be considered. The 120° countersunk head style should be used wherever a flush surface is required.

Mandrel types

POP® open type rivets are normally supplied with 'Break stem' mandrels (code BS) designed to retain the mandrel head when the rivet is set. 'Break head' (code BH) mandrels, designed to eject the mandrel head from the rivet body, can be supplied for most open rivets and are particularly useful in the pre-assembly of electrical circuit boards.

Finish

- *Rivet body standard finishes*: Steel and nickel-copper rivet bodies are normally supplied zinc plated.
- *Paint and other finishes*: Rivets with differing surface finishes and paint colours can be provided on request. Aluminium alloy rivets are available anodized and dyed, matt or gloss for aesthetic and environmental reasons.

4.2.10 Good fastening practice

Blind riveting is a highly reliable and proven method of fixing material together permanently. To achieve a superior fastening, the following principles should be considered.

Workpiece materials

When materials of different thickness or strengths are being joined, the stronger material – if possible – should be on the blind side. For example, if plastic and metal are to be joined, the plastic sheet should be beneath the rivet head and the metal component should be on the blind side.

Hole size and preparation

Achieving a good joint depends on good hole preparation, preferably punched and, if necessary, de-burred to the sizes recommended in the POP® blind rivet data tables.

Rivet diameter

As a guide for load-bearing joints, the rivet diameter should be at least equal to the thickness of the thickest sheet and not more than three times the thickness of the sheet immediately under the rivet head. Refer to data tables for rivet strength characteristics.

Edge distance

Rivet holes should be drilled or punched at least two diameters away from an edge – but no more than 24 diameters from that edge.

Rivet pitch

As a guide to the distance between the rivets in load-carrying joint situations, this distance should never exceed three rivet diameters. In butt construction it is advisable to include a reinforcing cover strip, fastening it to the underlying sheet by staggered rivets.

Rivet material

Choosing rivets of the correct material normally depends on the strength needed in the riveted joint. When this leads to rivets of material different to the sheets being joined it is important to be aware that electrolytic action may cause corrosion. (See Section 4.2.12.)

Setting and safety

The type of setting tool is usually selected to suit the production environment. The tool must be cleared of spent materials before setting the next rivet and, in the case of power operated tools, **must not** be operated without the mandrel deflector or mandrel collection system being in position. Safety glasses or goggles should always be worn.

4.2.11 Selection of POP® (or blind) rivets

Joint strength

First assess the tensile strength and the shear-load strength required by the joint, both of which can be achieved by the correct number and spacing of fastenings, and by choosing a rivet with a body of the correct material and diameter. The strength columns on the data pages enable a rivet of the correct strength to be chosen.

Joint thickness

The next stage is to work out the combined thickness of the materials to be joined, remembering to allow for any air gaps or intermediate layers such as sealants. Then identify the selected rivet in the size with the necessary grip by consulting the data page. It is important to do this because a rivet with the incorrect grip range cannot satisfactorily grip the back of the workpiece or assembly.

Corrosion acceleration (nature of materials)

Finally, follow the general rule that the rivet chosen should have the same physical and mechanical properties as the workpiece. A marked difference in properties may cause joint failure through metal fatigue or galvanic corrosion. Corrosion is accelerated by certain combinations of materials and environments. Generally, avoid contact between dissimilar metals. The significance of the letters A, B, C and D in the following chart is as follows:

A The corrosion of the metal considered is not accelerated by the contact metal.
B The corrosion of the metal being considered may be slightly accelerated by the contact metal.
C The corrosion of the metal considered may be markedly accelerated by the contact metal.
D When moisture is present, this combination of metal considered and contact metals is inadvisable, even under mild conditions, without adequate protective measures.

Where two symbols are given (for instance **B** or **C**) the acceleration is likely to change with changes in the environmental conditions or the condition of the metal.

Rivet material	Contact metal					
	Nickel Copper Alloy	**Stainless Steel**	**Copper**	**Steel**	**Aluminium and Alloys**	**Zinc**
Nickel Copper Alloy	–	A	A	A	A	A
Stainless Steel	A	–	A	A	A	A
Copper	B or C	B or C	–	A	A	A
Steel	C	C	C	–	B	A
Aluminium and Alloys	C	B or C	D	B or C	–	A
Zinc	C	C	C	C	C	–

4.2.12 Design guidelines

Soft materials to hard

A large flange rivet can be used with the flange on the side of the soft material. Alternatively, POP® LSR type rivets spreads the clamping loads over a wide area so as to avoid damage to soft materials.

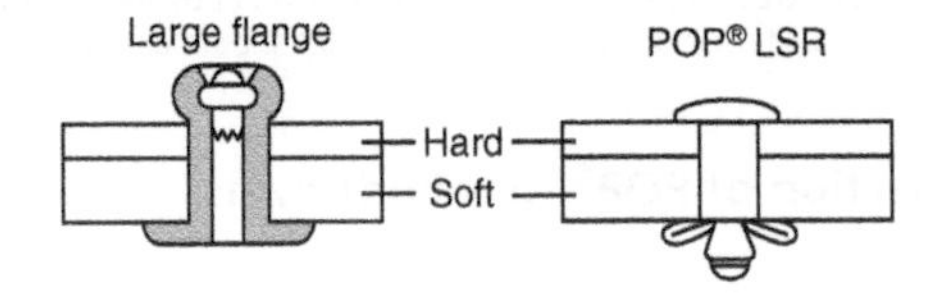

Plastics and brittle materials

For fragile plastics and brittle materials, POP® riveting offers a variety of application solutions. Soft-set/All Aluminium rivets offer low setting loads, whereas both the 'Peel' and

'LSR' ranges afford enhanced support to the materials being joined. For stronger plastics, standard POP® rivets – open or closed end products – may be used.

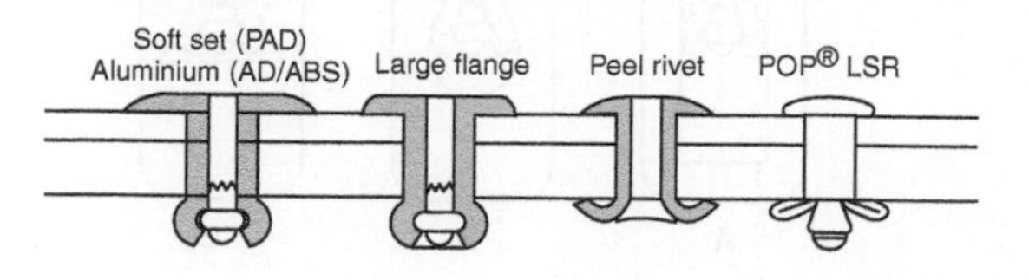

Channel section material

An extended nosepiece can be used to reach to the bottom of a narrow channel section (A) (see figure below). A longer mandrel rivet should be used and the maximum nosepiece diameter should be equal to that of the rivet flange. Alternatively, the rivet can be set from the other side (B) or the channel widened to accept a standard rivet and setting tool (C).

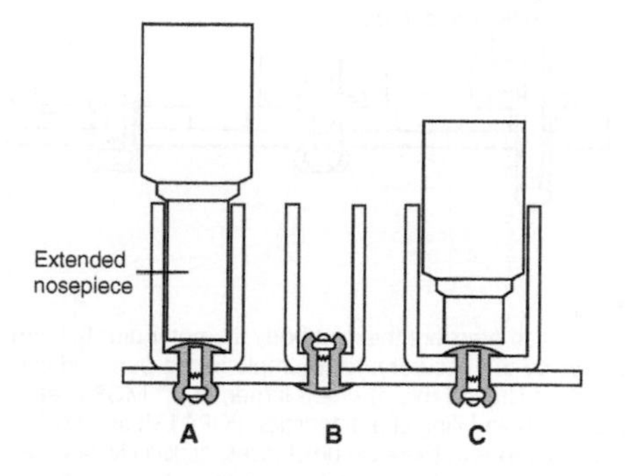

Thick/thin sheet

When materials of different thickness are to be fastened, it is best to locate the thicker plate at the fastened side (A) (see figure below). When the hole diameter in the thinner plate is large, a large flange rivet should be used (B). When the thinner plate is located at the fastened side, either use a backing washer (C) or ensure that the diameter of the hole in the thicker plate is smaller than the one in the thinner plate (D).

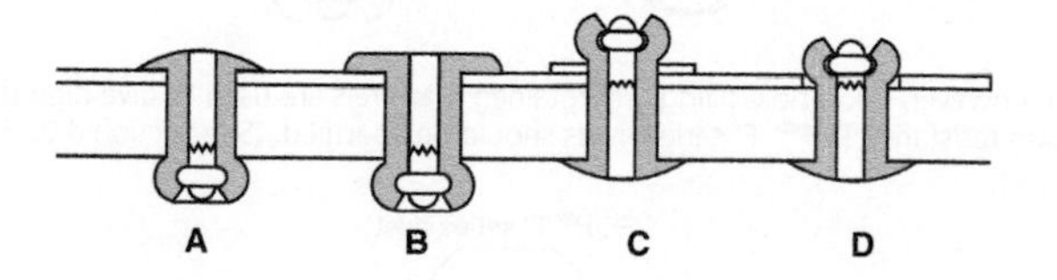

Blind holes and slots

The setting of a POP® or blind rivet against the side of a blind hole, or into and against a milled slot, intersecting hole or internal cavity, is possible because of the expansion of the rivet body on installation.

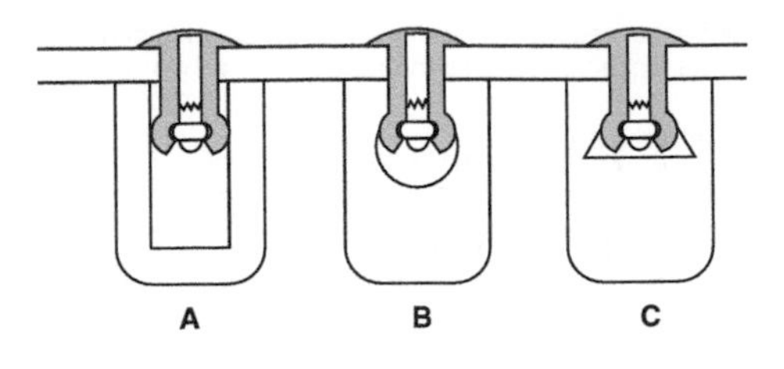

Pivot fastening

Use of a special nosepiece will provide a small gap between the rivet flange and the assembly, so providing for pivot action.

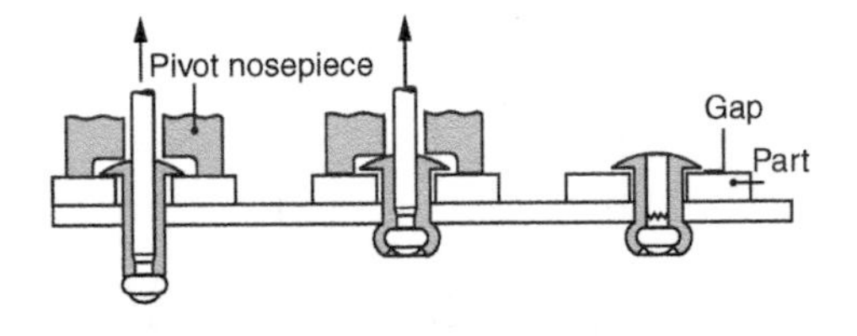

Hole diameters

Whilst standard hole diameters are the rivet body diameter plus 0.1 mm, component hole sizes may not always be this accurate, for example in pre-punched components. In cases when the hole on the fastened (blind) side is larger, POP® MGR rivets should be selected because of its superior hole filling characteristics. POP® LSR and POP® Peel rivets are also possible alternative solutions in these circumstances, especially when working with friable or fragile materials. Alternatively, when then larger hole is on the flange side, a large flange rivet should be chosen.

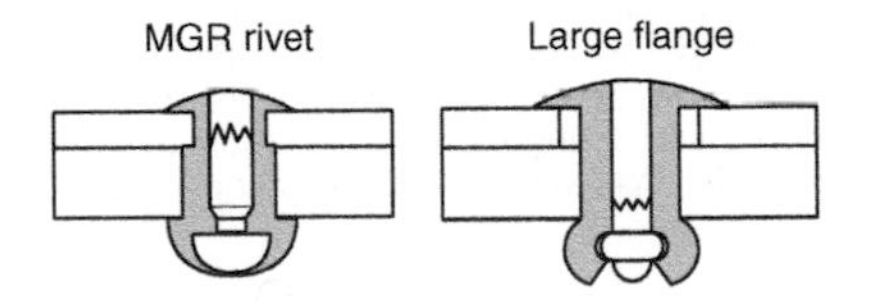

When, however, good hole filling and retained mandrels are used to give high shear and vibration resistance POP® 'F' series rivets should be specified. (See Section 4.2.13.)

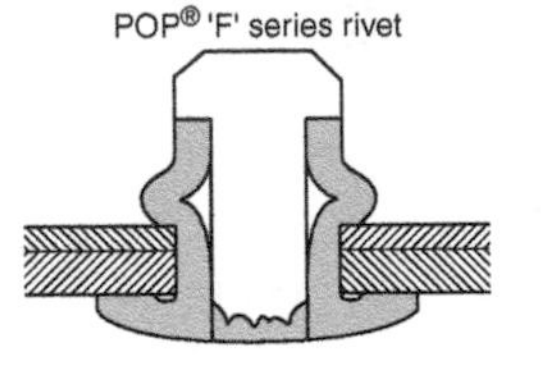

Elevated temperature performance

Elevated temperature strengths will vary from the figures quoted in the data tables. The following curves are offered for guidance.

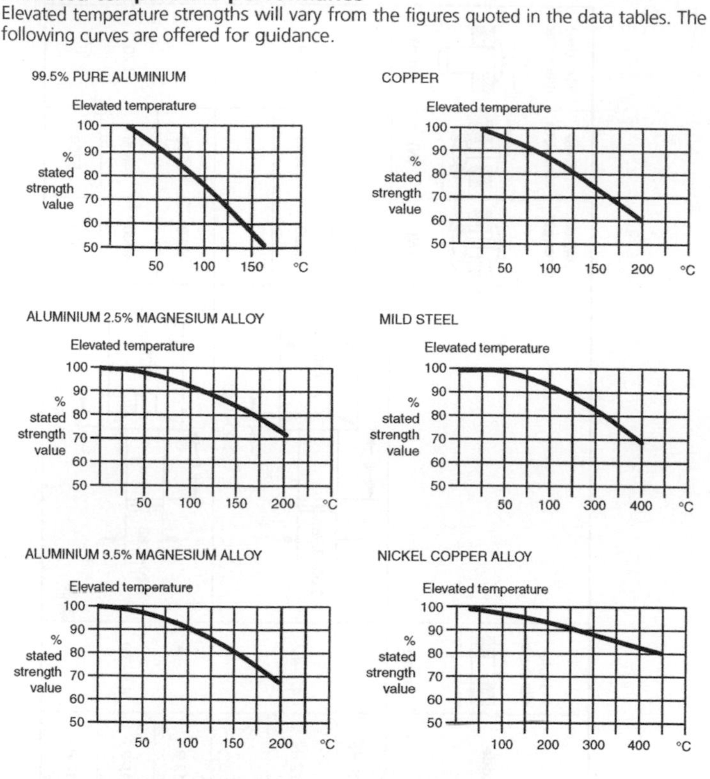

4.2.13 POP® 'F' series

'F' Series: Domed head
Carbon steel (AISI 1006) mandrel: carbon steel

Nominal rivet diameter (mm)	'L' Length (mm)	Grip range (mm)	Rivet code	Hole diameter (mm)		Tensile strength (N)	Shear strength (N)	Mandrel diameter (mm)
4.0	98–10.6	1.5–3.5	**FSD 4010 BS**	4.1–4.2	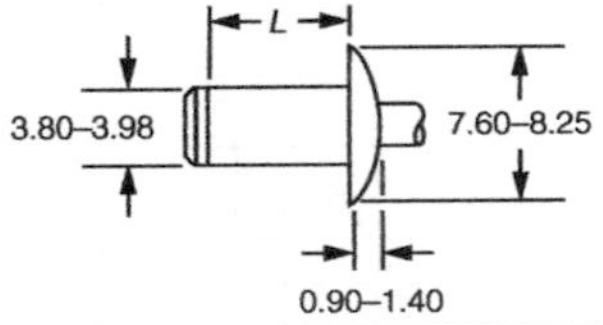 3.80–3.98; *L*; 7.60–8.25; 0.90–1.40	2800	3000	2.64
	12.2–13.0	3.5–6.0	**FSD 4012 BS**			2800	3000	
4.8	11.3–12.1	1.5–3.5	**FSD 4812 BS**	4.9–5.0	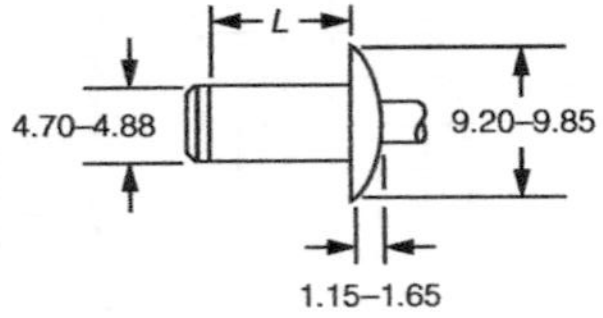 4.70–4.88; *L*; 9.20–9.85; 1.15–1.65	4250	3600	3.20
	14.2–15.0	3.5–6.0	**FSD 4815 BS**			4000	4600	
	16.3–17.1	6.0–8.5	**FSD 4817 BS**			3550	5700	

'F' Series: Domed head
Aluminium 3.5% Magnesium alloy (5052) mandrel: carbon steel

Nominal rivet diameter (mm)	'L' Length (mm)	Grip range (mm)	Rivet code	Hole diameter (mm)		Tensile strength (N)	Shear strength (N)	Mandrel diameter (mm)
4.8	10.3–11.1	1.5–3.5	**FSD 4811 BS**	4.9–5.0	L; 4.70–4.88; 9.20–9.85; 1.15–1.65	2050	1850	3.20
	13.2–14.0	3.5–6.0	**FSD 4814 BS**			2200	2100	
	15.6–16.4	6.0–8.5	**FSD 4816 BS**			2200	3100	

Note: Shear and tensile strengths are typical values. Joint strengths will be dependent upon the following application criteria:
1. Hole size, 2. Materials to be fastened, 3. Application material thicknesses
It is recommended users conduct their own test(s) to determine suitability for their application(s).

The following symbols are used in the tables in Sections 4.2.14–4.2.16.

Maximum riveting thickness	Nominal diameter hole	Tensile/shear performance	Nominal mandrel diameter
	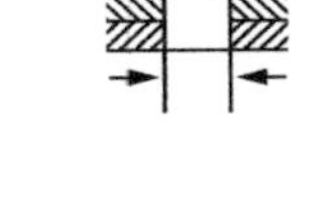	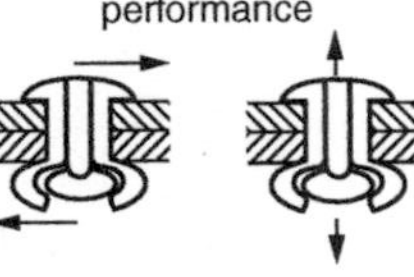 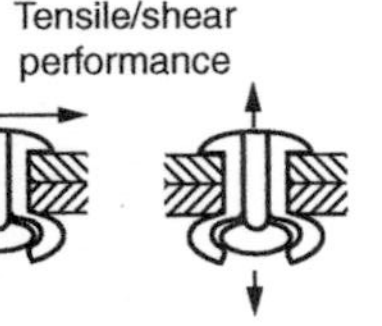	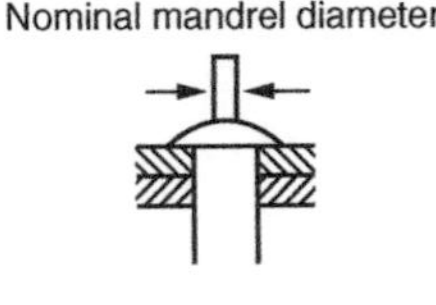
This is the recommended maximum thickness of the materials to be riveted together assuming a hole diameter indicated as nominal in the specification tables. The thickness should include any gap between materials prior to setting.	This is the recommended diameter of the drilled/punched hole which for reliable rivet setting should be burr free. The rivet will in most circumstances set satisfactorily in holes up to 0.1 mm greater than the nominal quoted.	The shear and tensile figures quoted are indicative of the performance of the rivet under standard test conditions. Actual performance will depend on material types and thickness. For safety, testing in the application is recommended for critical assemblies.	This is the nominal diameter of the mandrel and is shown to assist in the selection of the correct 'POP®' rivet tool nosepiece – essential for correct rivet setting performance.

4.2.14 Open type aluminium 3.5% magnesium alloy

Material composition: aluminium 3.5% magnesium alloy; mandrel: carbon steel

Nominal rivet dia. (mm)	'L' Nominal rivet body length (mm)	Max. riveting thickness (mm)	Rivet code carbon steel mandrel	Nominal (mm)	Rivet dimensions and limits (mm)	(N)	(N)	Nominal (mm)
2.4	3.5	0.8	TAPD 31 BS	2.5	2.48–2.30; 5.00–4.00; 0.90 max.	550	400	1.42
	5.0	2.4	TAPD 33 BS					
	7.5	4.8	TAPD 36 BS					
3.0	5.0	2.0	SNAD 3050 BS	3.1	3.08–2.90; 6.3–5.7; 1.10 max.	1000	800	1.83
	6.5	3.5	SNAD 3065 BS					
	8.0	5.0	SNAD 3080 BS					
	10.0	6.0	SNAD 3100 BS					
	12.0	9.0	SNAD 3120 BS					

continued

Section 4.2.14 (*continued*)

Nominal rivet dia. (mm)	'L' Nominal rivet body length (mm)	Max. riveting thickness (mm)	Rivet code carbon steel mandrel	Nominal (mm)	Rivet dimensions and limits (mm)	(N)	(N)	Nominal (mm)
	15.0	12.0	SNAD 3150 BS					
3.2	4.5	1.6	TAPD 42 BS	3.3	3.28–3.10; L; 6.65–6.05; 1.10 max.	1200	800	1.83
	6.0	3.2	TAPD 44 BS					
	8.0	4.8	TAPD 46 BS					
	9.7	6.4	TAPD 48 BS					
	11.5	7.9	TAPD 410 BS					
	13.5	9.5	TAPD 412 BS					
	15.0	11.1	TAPD 414 BS					
	17.0	12.7	TAPD 416 BS					

	18.5	14.3	**TAPD 418 BS**					
	20.0	16.7	**TAPD 421 BS**					
	24.0	20.7	**TAPD 425 BS**					
3.2	6.2	3.2	**TAPD 44 BSLF9.5**	3.3	L 3.28–3.10 10.00–9.00 1.50 max.	1200	800	1.83
	8.0	4.8	**TAPD 46 BSLF9.5**					
	9.7	6.4	**TAPD 48 BSLF9.5**					
	11.5	7.9	**TAPD 410 BSLF9.5**					
	13.5	9.5	**TAPD 412 BSLF9.5**					
	17.0	12.7	**TAPD 416 BSLF9.5**					
4.0	7.0	3.2	**TAPD 54 BS**					
	8.5	4.8	**TAPD 56 BS**					
	10.5	6.4	**TAPD 58 BS**					
	12.2	7.9	**TAPD 510 BS**					

continued

Section 4.2.14 (*continued*)

Nominal rivet dia. (mm)	'L' Nominal rivet body length (mm)	Max. riveting thickness (mm)	Rivet code carbon steel mandrel	Nominal (mm)	Rivet dimensions and limits (mm)	(N)	(N)	Nominal (mm)
4.0	14.0	9.5	TAPD 512 BS	4.1	4.08–3.90; L; 8.22–7.62; 1.35 max.	1910	1330	2.29
	15.7	11.1	TAPD 514 BS					
	17.5	12.7	TAPD 516 BS					
	18.5	13.5	TAPD 517 BS					
	20.2	15.9	TAPD 520 BS					
	22.5	17.4	TAPD 522 BS					
	24.7	19.8	TAPD 525 BS					
	8.5	4.8	TAPD 56 BSLF12					
	10.5	6.4	TAPD 58 BSLF12					

4.0	12.2	7.9	**TAPD 510 BSLF12**	4.1	4.08–3.90; *L*; 12.150–11.50; 1.60 max.	1910	1330	2.29
	14.0	9.5	**TAPD 512 BSLF12**					
	15.7	11.1	**TAPD 514 BSLF12**					
	17.5	12.7	**TAPD 516 BSLF12**					
	19.4	14.3	**TAPD 518 BSLF12**					
4.8	7.5	3.2	**TAPD 64 BS**	4.9	4.88–4.70; *L*; 9.80–9.20; 1.60 max.	2800	2020	2.64
	9.2	4.8	**TAPD 66 BS**					
	11.0	6.4	**TAPD 68 BS**					
	12.7	7.9	**TAPD 610 BS**					
	14.7	9.5	**TAPD 612 BS**					
	16.5	11.1	**TAPD 614 BS**					
	19.0	13.5	**TAPD 617 BS**					
	25.5	19.8	**TAPD 625 BS**					
	32.0	26.2	**TAPD 633 BS**					

continued

Section 4.2.14 (*continued*)

Nominal rivet dia. (mm)	'L' Nominal rivet body length (mm)	Max. riveting thickness (mm)	Rivet code carbon steel mandrel	Nominal (mm)	Rivet dimensions and limits (mm)	(N)	(N)	Nominal (mm)
4.8	38.2	32.8	**TAPD 6150 BS**[a]	4.9	4.88–4.90; L; 8.97–8.37; 1.20 max.	2710	1950	2.64
	44.5	39.1	**TAPD 6175 BS**[a]					
4.8	9.20	4.8	**TAPD 66 BSLF14**	4.9	4.88–4.70; L; 14.30–13.70; 2.00 max.	2800	2020	2.64
	11.0	6.4	**TAPD 68 BSLF14**					
	12.7	7.9	**TAPD 610 BSLF14**					
	14.5	9.5	**TAPD 612 BSLF14**					
	16.5	11.1	**TAPD 614 BSLF14**					

	19.0	13.5	**TAPD 617 BSLF14**					
	25.5	19.8	**TAPD 625 BSLF14**					
5.0	6.5	3.5	**SNAD 5065 BS**	5.1	5.08–4.85; L; 9.30–8.70; 1.60 max.	2600	2200	2.64
	8.0	4.5	**SNAD 5080 BS**					
	10.0	6.0	**SNAD 5100 BS**					
	12.0	7.5	**SNAD 5120 BS**					
	14.0	9.5	**SNAD 5140 BS**					
	16.0	11.5	**SNAD 5160 BS**					
	18.0	13.5	**SNAD 5180 BS**					
	20.0	15.5	**SNAD 5200 BS**					
	22.0	17.5	**SNAD 5220 BS**					
	25.0	20.5	**SNAD 5250 BS**					
6.0	8.0	3.5	**SNAD 6080 BS**					

continued

Section 4.2.14 (*continued*)

Nominal river dia. (mm)	'*L*' Nominal river body length (mm)	Max. riveting thickness 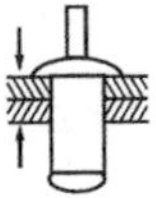 (mm)	Rivet code carbon steel mandrel	Nominal 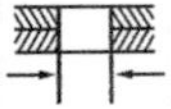 (mm)	Rivet dimensions and limits (mm)	 (N)	 (N)	Nominal 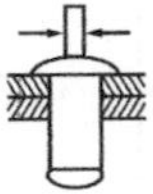 (mm)
6.0	10.0	5.5	SNAD 6100 BS	6.1	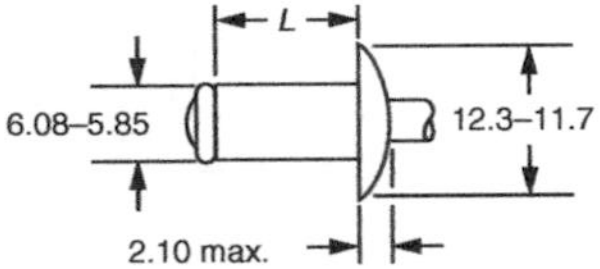 *L*; 6.08–5.85; 12.3–11.7; 2.10 max.	3800	3000	3.20
	12.0	7.5	SNAD 6120 BS					
	14.0	9.5	SNAD 6140 BS					
	16.0	11.5	SNAD 6160 BS					
	18.0	13.5	SNAD 6180 BS					
	20.0	15.5	SNAD 6200 BS					
6.4	12.7	6.4	TAPD 88 BS	6.5	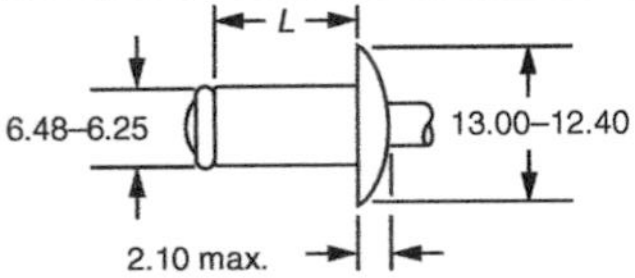 *L*; 6.48–6.25; 13.00–12.40; 2.10 max.	4525	3200	3.66
	19.5	12.7	TAPD 816 BS					
	26.2	19.8	TAPD 824 BS					

4.2.15 Open type carbon steel

Material composition: carbon steel; mandrel: carbon steel

Nominal rivet dia. (mm)	'L' Nominal rivet body length (mm)	Max. riveting thickness (mm)	Rivet code carbon steel mandrel	Nominal (mm)	Rivet dimensions and limits (mm)	(N)	(N)	Nominal (mm)
2.8	5.3	2.9	**TSPD 33 BS**	2.9	2.88–2.90; 5.84–5.24; 1.32 max.; L	930	715	1.83
3.0	5.0	2.0	**SNSD 3050 BS**	3.1	3.08–2.90; 6.3–5.7; 1.1 max.; L	1400	1100	1.83
	6.5	3.5	**SNSD 3065 BS**					
	8.0	5.0	**SNSD 3080 BS**					
	10.0	7.0	**SNSD 3100 BS**					
	12.0	9.0	**SNSD 3120 BS**					

continued

Section 4.2.15 *(continued)*

Nominal rivet dia. (mm)	'L' Nominal rivet body length. (mm)	Max. riveting thickness (mm)	Rivet code carbon steel mandrel	Nominal (mm)	Rivet dimensions and limits (mm)	(N)	(N)	Nominal (mm)
	15.0	12.0	**SNAD 3150 BS***					
3.2	4.5	1.6	**TSPD 42 BS**	3.3	3.28–3.10; L; 6.65–6.05; 1.10 max.	1550	1150	1.93
	6.0	3.2	**TSPD 44 BS**					
	8.0	4.8	**TSPD 46 BS**					
	9.5	6.4	**TSPD 48 BS**					
	11.5	7.9	**TSPD 410 BS**					
	13.5	9.5	**TSPD 412 BS**					
	5.0	1.6	**TSPD 52 BS**					
	7.0	3.2	**TSPD 54 BS**					

4.0	8.5	4.8	**TSPD 56 BS**	4.1	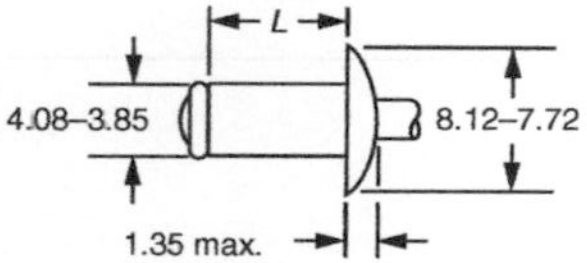 4.08–3.85; 8.12–7.72; L; 1.35 max.	2500	1730	2.29
	10.5	6.4	**TSPD 58 BS**					
	12.2	7.9	**TSPD 510 BS**					
	14.0	9.5	**TSPD 512 BS**					
	15.9	11.1	**TSPD 514 BS**					
	17.6	12.7	**TSPD 516 BS**					
4.8	6.5	2.4	**TSPD 63 BS**	4.9	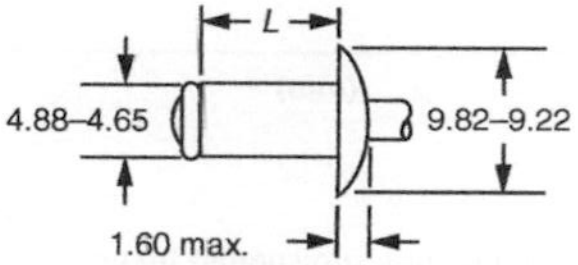 4.88–4.65; 9.82–9.22; L; 1.60 max.	3500	2620	2.9
	7.5	3.2	**TSPD 64 BS**					
	9.0	4.8	**TSPD 66 BS**					
	11.0	6.4	**TSPD 68 BS**					
	12.7	7.9	**TSPD 610 BS**					
	14.5	9.5	**TSPD 612 BS**					
	16.5	11.1	**TSPD 614 BS**					
	18.3	12.7	**TSPD 616 BS**					
	19.0	13.5	**TSPD 617 BS**					

continued

Section 4.2.15 (*continued*)

Nominal rivet dia. (mm)	'L' Nominal rivet body length (mm)	Max. riveting thickness (mm)	Rivet code carbon steel mandrel	Nominal (mm)	Rivet dimensions and limits (mm)	(N)	(N)	Nominal (mm)
5.0	6.5	2.5	SNSD 5065 BS	5.1	5.08–4.85; L; 9.30–8.70; 1.6 max.	3790	2880	2.9
	8.0	4.0	SNSD 5080 BS					
	10.0	6.0	SNSD 5100 BS					
	12.0	8.0	SNSD 5120 BS					
	14.0	10.0	SNSD 5140 BS					
	16.0	12.0	SNSD 5160 BS					
	18.0	14.0	SNSD 5180 BS					
	10.0	4.0	SNSD 6100 BS					

6.0	12.0	6.0	**SNSD 6120 BS**	6.1	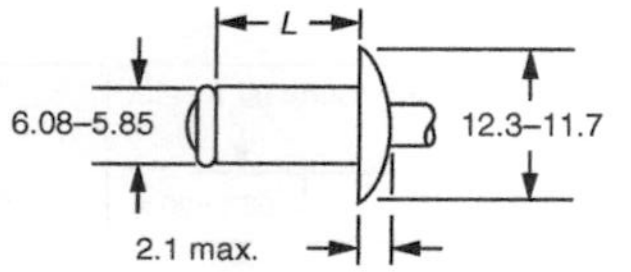 L; 6.08–5.85; 12.3–11.7; 2.1 max.	5500	4200	3.65
	14.0	8.0	**SNSD 6140 BS**					
	16.0	10.0	**SNSD 6160 BS**					
	18.0	12.0	**SNSD 6180 BS**					
6.4	9.5	3.8	**TSPD 8095 BS**	6.5	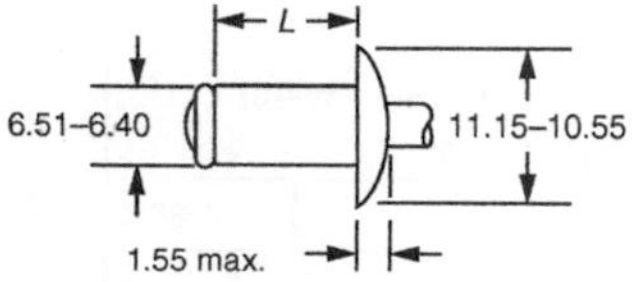 L; 6.51–6.40; 11.15–10.55; 1.55 max.	6900	5000	3.86
	13.0	7.6	**TSPD 8130 BS**					
	18.5	12.7	**TSPD 8185 BS**					

[a] Body – 5% magnesium alloy.

4.2.16 Closed end type aluminium 5% magnesium alloy

Material composition: aluminium 5% magnesium alloy; mandrel: carbon steel or stainless steel

Nominal rivet dia. (mm)	'L' Nominal rivet body length (mm)	Max. riveting thickness (mm)	Rivet code carbon steel mandrel	Rivet code stainless steel mandrel	Nominal (mm)	Rivet dimensions and limits (mm)	(N)	(N)	Nominal (mm)
3.2	6.0	1.6	AD 42 SB	AD 42 SS	3.3	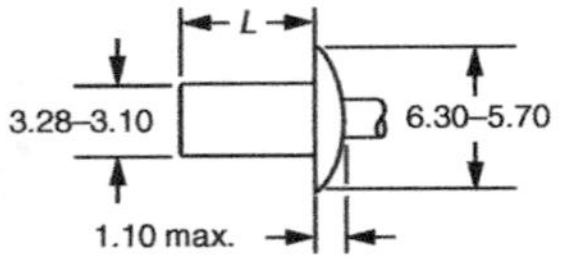 L; 3.28–3.10; 6.30–5.70; 1.10 max.	1400	1110	1.63
	7.5	3.2	AD 44 SB	AD 44 SS					
	9.0	4.8	AD 46 SB	AD 46 SS					
	11.0	6.4	AD 48 SB	AD 48 SS					
	12.0	7.9	AD 410 SB	AD 410 SS					
4.0	8.0	3.2	AD 54 SB	AD 54 SS	4.1	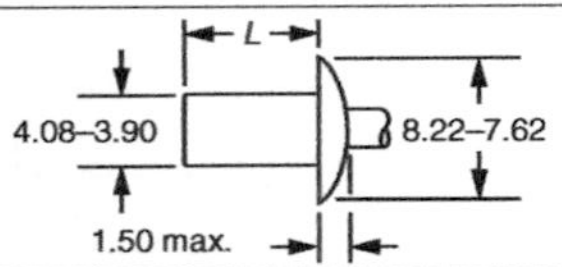 L; 4.08–3.90; 8.22–7.62; 1.50 max.	2220	1640	2.18
	9.5	4.8	AD 56 SB	AD 56 SS					
	11.0	6.4	AD 58 SB	AD 58 SS					

	12.5	7.9	AD 510 SB	AD 510 SS					
4.8	8.30	3.2	AD 64 SB	AD 64 SS	4.9	L; 4.88–4.70; 9.85–9.20; 1.75 max.	3110	2260	2.64
	10.0	4.8	AD 66 SB	AD 66 SS					
	11.5	6.4	AD 68 SB	AD 68 SS					
	13.0	7.9	AD 610 SB	AD 610 SS					
	14.5	9.5	AD 612 SB	AD 612 SS					
	18.0	12.7	AD 616 SB	AD 616 SS					
	22.0	15.9	AD 620 SB	AD 620 SS					
6.4	13.0	6.4	AD 84 H	–	6.5	L; 6.48–6.32; 13.34–12.06; 2.51 max.	4800	4000	3.66
	16.0	9.5	AD 86 H	–					

4.2.17 Blind rivet nuts

Blind rivet nuts are especially designed to offer a means of providing a stronger threaded joint in sheet and other materials and, like blind rivets, only require access to one side of the workpiece for installation. They are not only a form of captive nut but, unlike some other types, they also allow components to be riveted together as well providing a screw thread anchorage. The principle of their use is shown below.

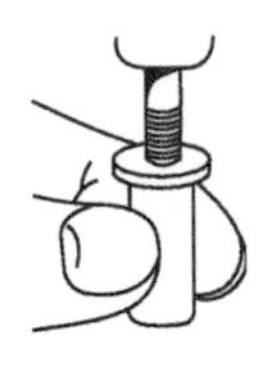

The POP® blind rivet nut is screwed onto the threaded mandrel of the setting tool and is then inserted into the drilled or punched hole.

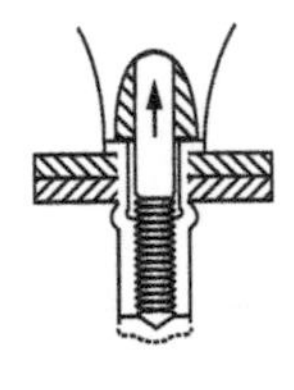

The tool is operated, retracting the mandrel. The unthreaded part of the nut expands on the blind side of the workpiece to form a collar and applies a powerful clenching force that rivets the components firmly together.

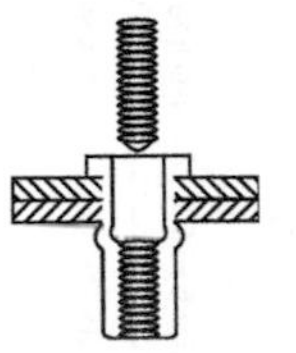

With the POP® nut firmly in position the tool mandrel is simply unscrewed from the nut. The threaded portion is now ready to act as a secure anchorage point.

Blind rivet nuts are generally available in thread sizes M3 to M12 in numerous combinations of head style, body form and material. Three head styles are available, flat head, 90° countersunk and, for thin gauge materials, a small flange which provides a near flush appearance without the need for countersinking.

Standard bodies are round with open or closed ends. The closed end prevents the ingress of moisture or vapour through the bore of the nut. Where a higher torque resistance is required, body forms may be fully or partially hexagonal (set in a hexagonal hole), or splined (set in a round hole).

4.2.18 POP® Nut Threaded Inserts: application

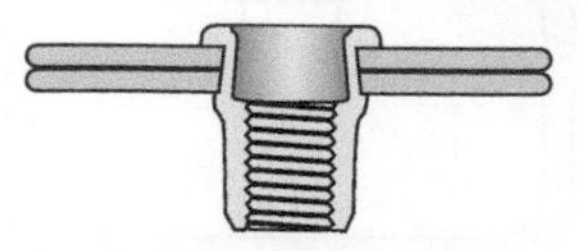

POP® Nut Threaded Inserts provide a simple and effective way to join materials with the benefit of an internal thread, in a variety of applications.

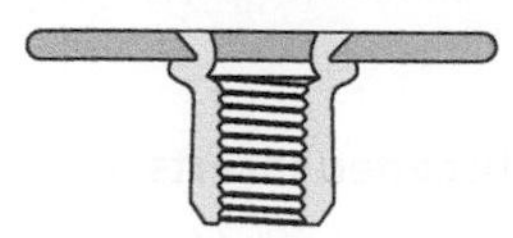

POP® Nut Threaded Inserts are the perfect solution for providing high-quality, load-bearing threads in various materials where alternative methods cannot maintain torque and pull out loads. **POP®** Nut Threaded Inserts are suitable for single sheets down to 0.5 mm.

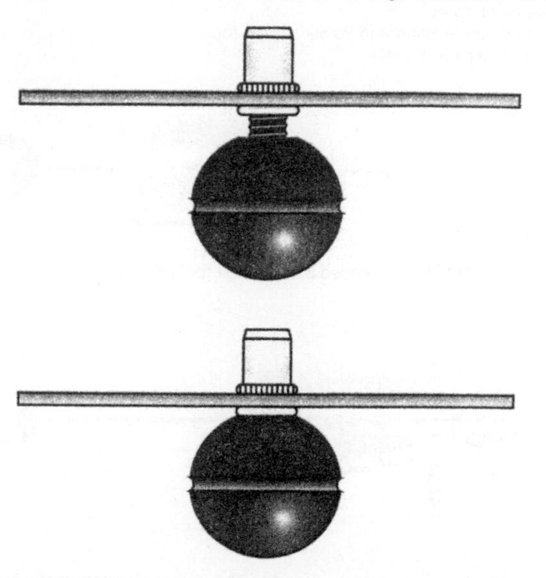

POP® Nut Threaded Inserts enable components, which are assembled later in the production cycle, to be adjusted.

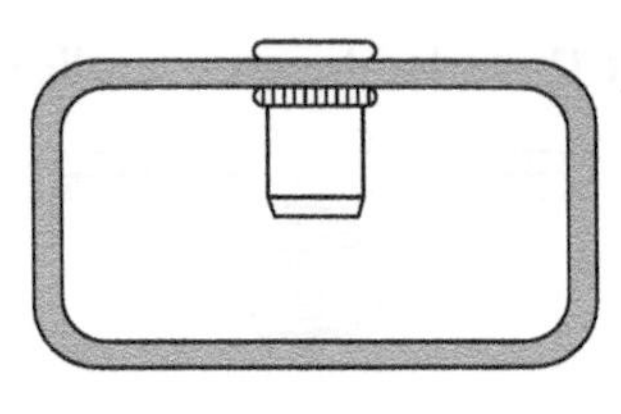

POP® Nut Threaded Inserts are ideally suited to applications where access is only available from one side of the workpiece.

4.2.19 POP® Nut Threaded Inserts: installation

POP® Nut Threaded Inserts are easily installed from one side of the workpiece without damaging surrounding surfaces of previously finished or delicate components and are suitable for use with all materials in today's manufacturing environment:

- Available in a variety of materials
- Wide range of styles
- Complete range of Hand and Power setting tools
- Bulk and small pack available.

Install/screw the **POP**® Nut Threaded Insert to the tool's threaded mandrel.

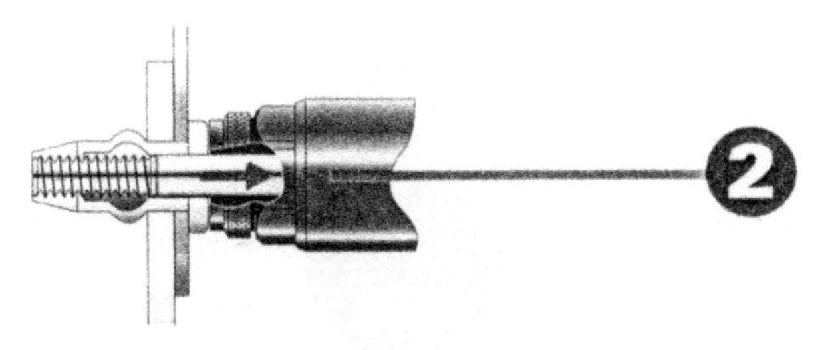

Operating the tool then retracts the mandrel. The unthreaded part of the nut then compresses to form a collar on the blind side of the workpiece, applying a powerful clenching force that firmly joins the components together.

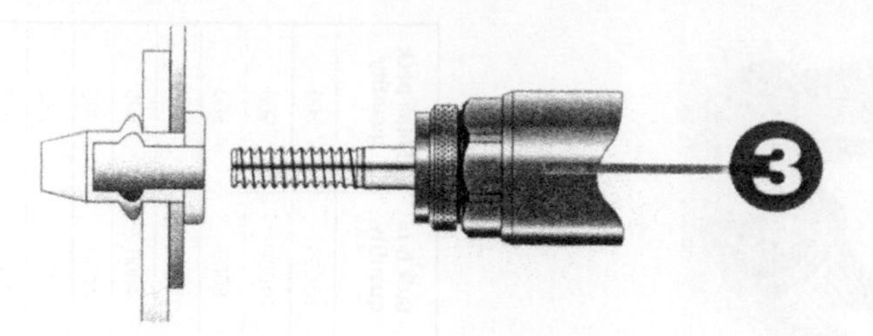

With the **POP®** Nut Threaded Insert firmly in position, the tool mandrel is simply demounted/unscrewed from the insert.

The threaded part of the insert then acts as a secure anchorage point for subsequent assembly work.

4.2.20 POP® Nut: Steel

Flat head open end (with knurls)

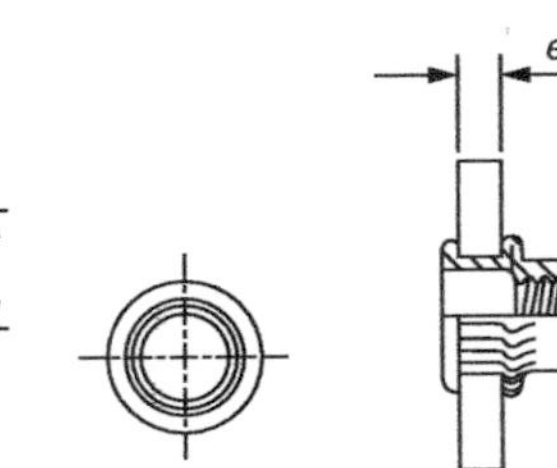

Thread d	Description	Length (mm) L	Grip (mm) e	Hole dia. (mm)	Barrel dia. (mm) D	Flange dia. (mm) B	Flange thickness (mm) S	Bulk box quantity	Small pack quantity
M4	**PSFON430**	10.0	0.3–3.0	6.0	5.9	9.0	1.0	10 000	500
	PSFON440	11.5	3.1–4.0					10 000	500
M5	**PSFON530**	12.0	0.3–3.0	7.0	6.9	10.0	1.0	8000	500
	PSFON540	15.0	3.1–4.0					5000	500
M6	**PSFON630**	14.5	0.5–3.0	9.0	8.9	12.0	1.5	4000	500
	PSFON645	16.0	3.1–4.5					4000	500
M8	**PSFON830**	16.0	0.5–3.0	11.0	10.9	15.0	1.5	2000	250
	PSFON855	18.5	3.1–5.5					2000	250
M10	**PSFON1030**	17.0	0.5–3.0	12.0	11.9	16.0	2.0	1500	200
	PSFON1060	22.0	3.0–6.0					1500	200
M12	**PSFON1240**	23.0	1.0–4.0	16.0	15.9	22.0	2.0	1500	200

Flat head closed end (with knurls)

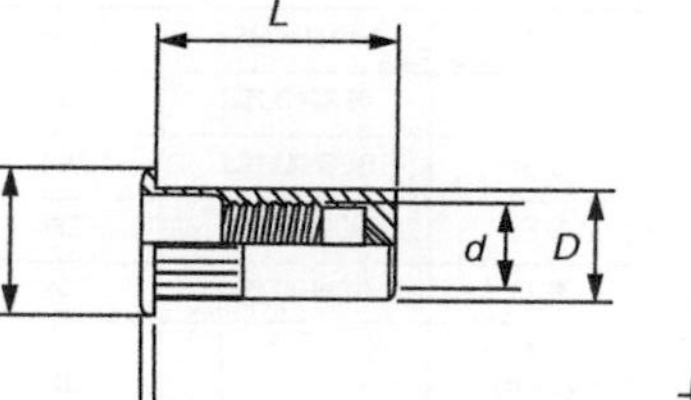

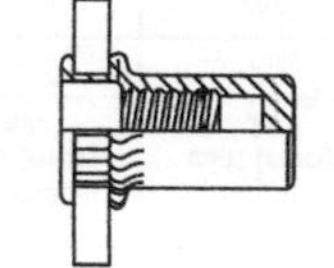

Thread *d*	Description	Length (mm) *L*	Grip (mm) *e*	Hole dia. (mm)	Barrel dia. (mm) *D*	Flange dia. (mm) *B*	Flange thickness (mm) *S*	Bulk box quantity	Small pack quantity
M5	**PSFCN530**	18.0	0.3–3.0	7.0	6.9	10.0	1.0	5000	500
M6	**PSFCN630**	20.5	0.5–3.0	9.0	8.9	12.0	1.5	2000	500
M8	**PSFCN830**	25.0	0.5–3.0	11.0	10.9	15.0	1.5	1500	250

Flat head open end hexagonal

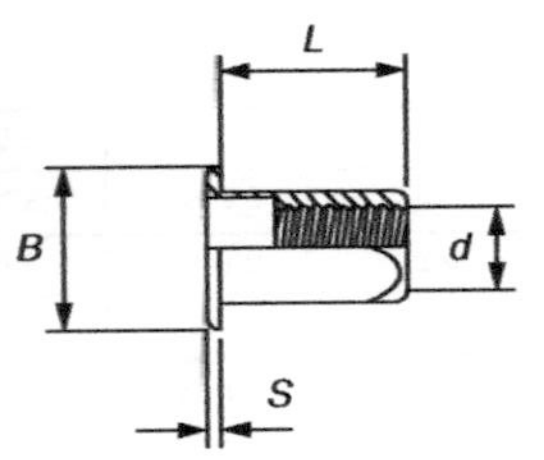

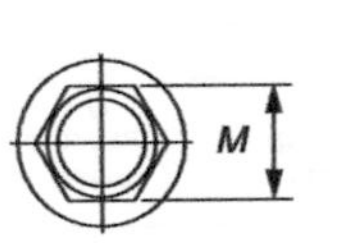

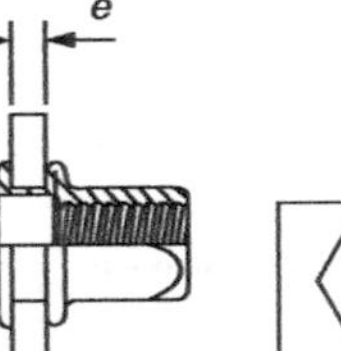

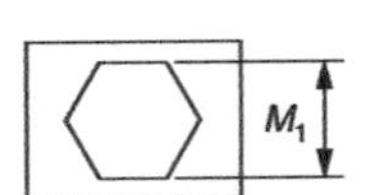

Thread d	Description	Length (mm) L	Grip (mm) e	Hole dia. (mm) M_1	Barrel dia. (mm) M	Flange dia. (mm) B	Flange thickness (mm) S	Bulk box quantity	Small pack quantity
M4	**PSFOH430**	11.5	0.5–3.0	6.1	6.0	9.3	1.0	10 000	500
M5	**PSFOH530**	13.5	0.5–3.0	7.1	7.0	10.3	1.0	8 000	500
M6	**PSFOH630**	15.5	0.5–3.0	9.1	9.0	12.3	1.5	4 000	500
M8	**PSFOH830**	17.5	0.5–3.0	11.1	11.0	14.3	1.5	2 000	250
M10	**PSFOH1040**	22.0	1.0–4.0	13.1	13.0	16.3	2.0	1 500	200

Countersunk head open end (with knurls)

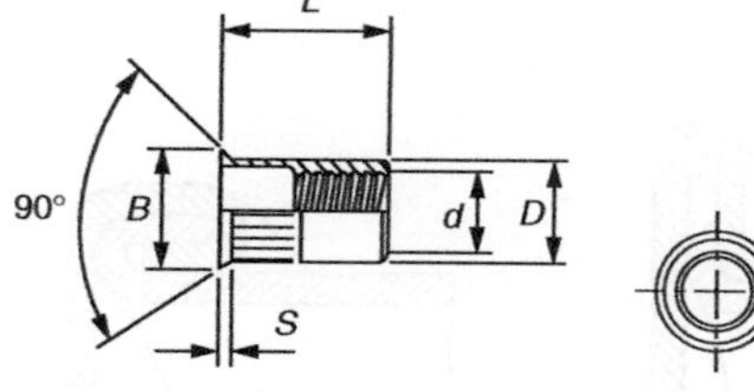

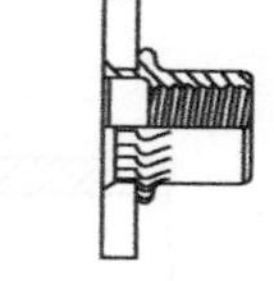

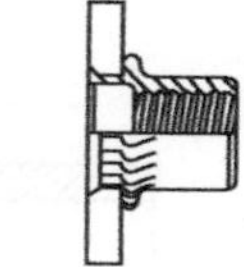

Thread *d*	Description	Length (mm) *L*	Grip (mm) e	Hole dia. (mm)	Barrel dia. (mm) *D*	Flange dia. (mm) *B*	Flange thickness (mm) *S*	Bulk box quantity	Small pack quantity
M4	**PSKON435**	11.5	2.0–3.5	6.0	5.9	9.0	1.5	10 000	500
M5	**PSKON540**	13.5	2.0–4.0	7.0	6.9	10.0	1.5	8 000	500
M6	**PSKON645**	16.0	2.0–4.5	9.0	8.9	12.0	1.5	4 000	500
M8	**PSKON845**	19.0	2.0–4.5	11.0	10.9	14.0	1.5	2 000	250
M10	**PSKON1045**	21.0	2.0–4.5	12.0	11.9	14.7	1.5	1 500	200

For further information see the wide range of publications concerning POP® riveting issued by Emhart Teknologies (Tucker Fasteners Ltd) – See Appendix 3.

4.3 Self-secured joints

4.3.1 Self-secured joints

Grooved seam

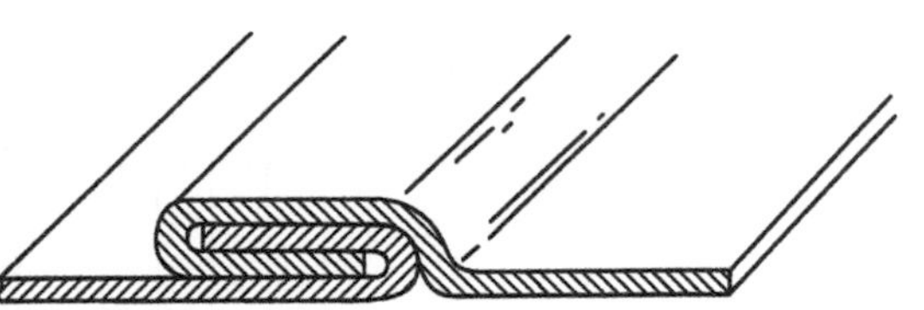
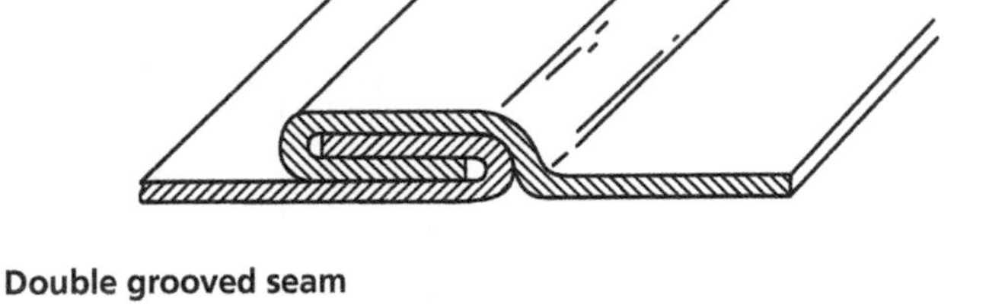

Double grooved seam

Paned down seam

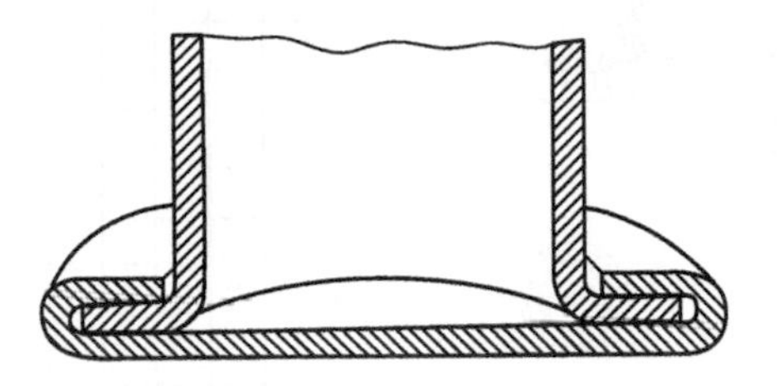

Knocked up seam

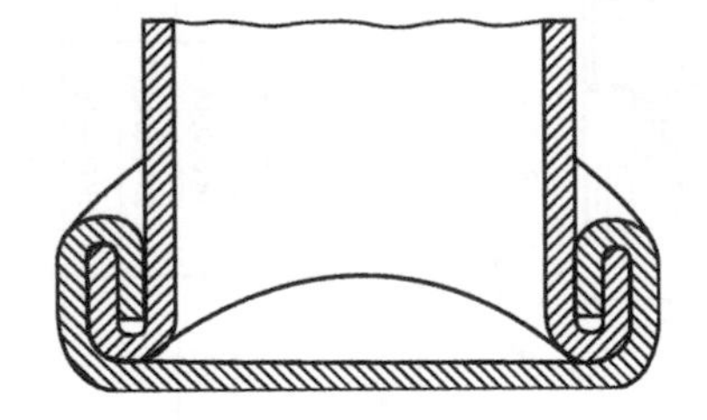

Making a grooved seam

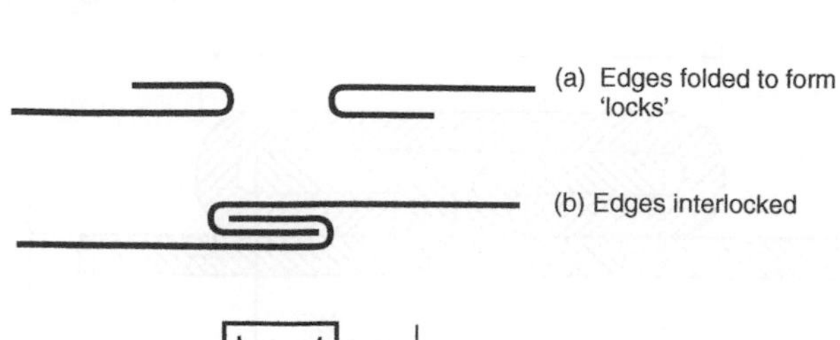

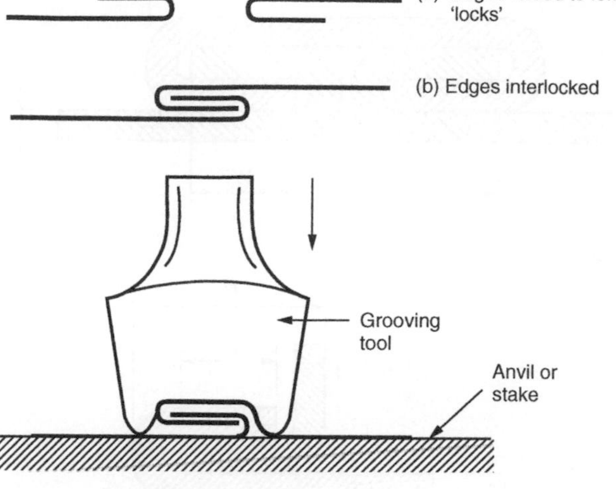

(c) Seam finally closed 'locked' using a grooving tool of the correct width

4.3.2 Allowances for self-secured joints

Grooved seam

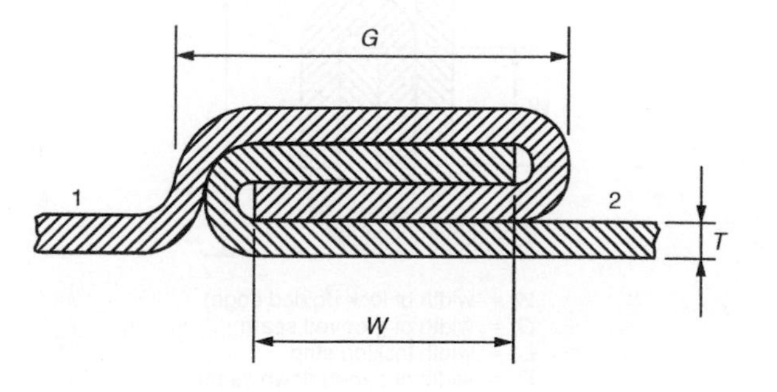

Double grooved seam

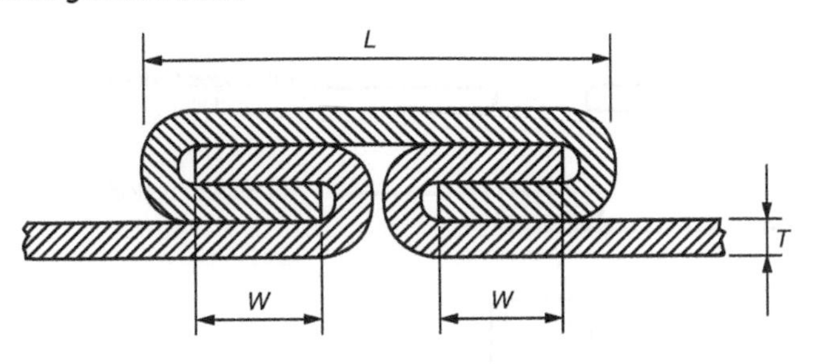

Paned down seam

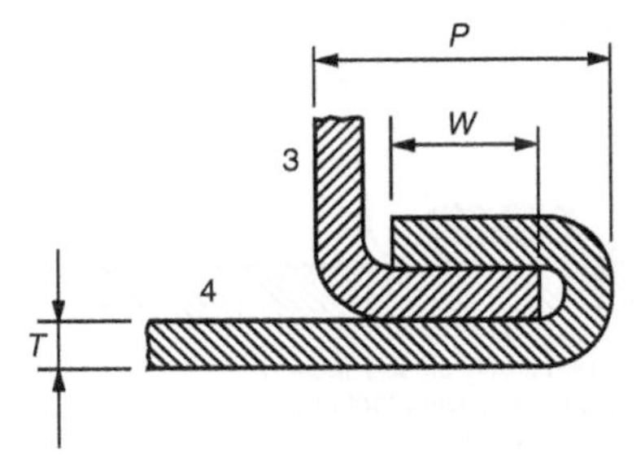

Knocked up seam

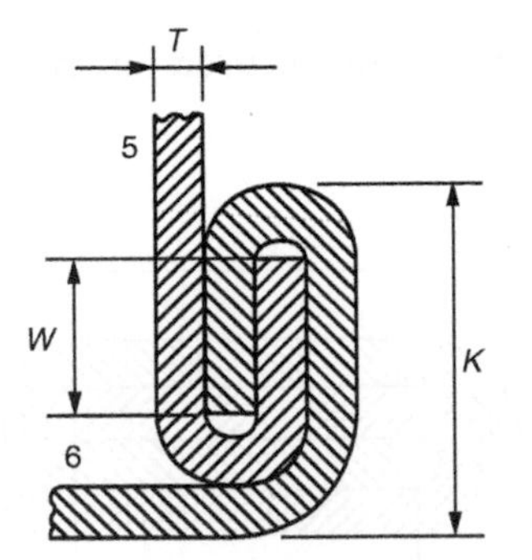

W = width of lock (folded edge)
G = width of grooved seam
L = width locking strip
P = width of paned down seam
K = width of knocked up seam
T = thickness of metal

Type of joint	Approximate allowance
Grooved seam	Total allowance = $3G - 4T$ shared: (a) equally between limbs 1 and 2; *or* (b) two-thirds limb 1 and one-third limb 2 where joint centre position is critical.
Double grooved seam	Add $W - T$ to the edge of each blank to be joined. Allowance for capping strip = $4W + 4T$, where $L = 2W + 4T$.
Paned down seam	Add W to the single edge 3. Add $2W + T$ to the double edge 4. $P = 2W + 2T$.
Knocked up joint	Add W to the single edge 5. Add $2W + T$ to the double edge 6. $K = 2W + 3T$.

4.4 Miscellaneous fasteners

4.4.1 Taper pins, unhardened

Dimensions

Type A (ground pins): Surface finish $R_a = 0.8\,\mu m$

Type B (turned pins): Surface finish $R_a = 3.2\,\mu m$

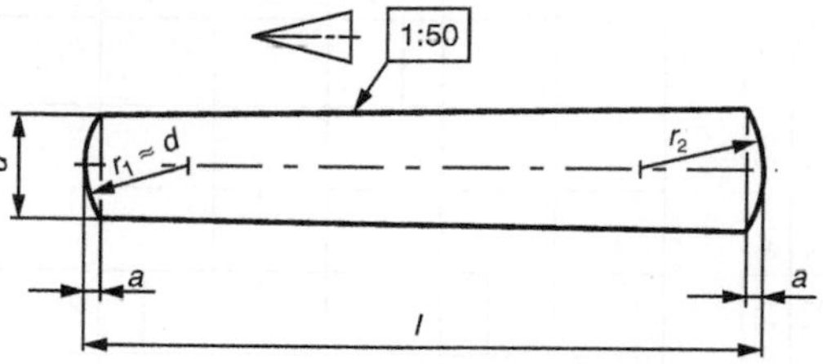

1:50

$r_1 \approx d$, r_2, d, a, l

$$r_2 \approx \frac{a}{2} + d + \frac{(0.02l)^2}{8a}$$

Dimensions in mm

d		h10[a]	0.6	0.8	1	1.2	1.5	2	2.5	3	4	5	6	8	10	12	16	20	25	30	40	50
a		≈	0.08	0.1	0.12	0.16	0.2	0.25	0.3	0.4	0.5	0.63	0.8	1	1.2	1.6	2	2.5	3	4	5	6.3
l[b] nom.	min.	max.																				
2	1.75	2.25																				
3	2.75	3.25																				
4	3.75	4.25																				
5	4.75	5.25																				
6	5.75	6.25																				
8	7.75	8.25																				
10	9.75	10.25																				
12	11.5	12.5			Range																	
14	13.5	14.5																				
16	15.5	16.5																				
18	17.5	18.5							of													
20	19.5	20.5																				
22	21.5	22.5																				
24	23.5	24.5																				
26	25.5	26.5																				
28	27.5	28.5																				
30	29.5	30.5											commercial									
32	31.5	32.5																				

35	34.5	35.5																				
40	39.5	40.5																				
45	44.5	45.5																				
50	49.5	50.5																				
55	54.25	55.75																				
60	59.25	60.75																				
65	64.25	65.75																				
70	69.25	70.75														**lengths**						
75	74.25	75.75																				
80	79.25	80.75																				
85	84.25	85.75																				
90	89.25	90.75																				
95	94.25	95.75																				
100	99.25	100.75																				
120	119.25	120.75																				
140	139.25	140.75																				
160	159.25	160.75																				
180	179.25	180.75																				
200	199.25	200.75																				

[a] Other tolerances, for example, a11, c11, f8, as agreed between customer and supplier.
[b] For nominal lengths above 200 mm, steps of 20 mm.

Specifications and reference International Standards

Material	St = Free-cutting steel, hardness 125 to 245 HV. Other materials as agreed between customer and supplier.
Surface finish	Plain, that is, pins to be supplied in natural finish treated with a rust-preventative lubricant, unless otherwise specified by agreement between customer and supplier.
	Preferred coatings are black oxide, phosphate coating or zinc plating with chromate conversion coating (see ISO 2081 and ISO 4520). Other coatings as agreed between customer and supplier. All tolerances shall apply prior to the application of a plating or coating.
Workmanship	Parts shall be uniform in quality and free of irregularities or detrimental defects. No burrs shall appear on any part of the pin.
Taper	The taper shall be inspected by use of an adequate optical comparator.
Acceptability	The acceptance procedure is covered in ISO 3269.

Designation

Example for the designation of an unhardened steel taper pin, type A, with nominal diameter $d = 6\,\text{mm}$ and nominal length $l = 30\,\text{mm}$:

Taper pin ISO 2339 – A – 6 × 30 – St

4.4.2 Circlips, external: metric series

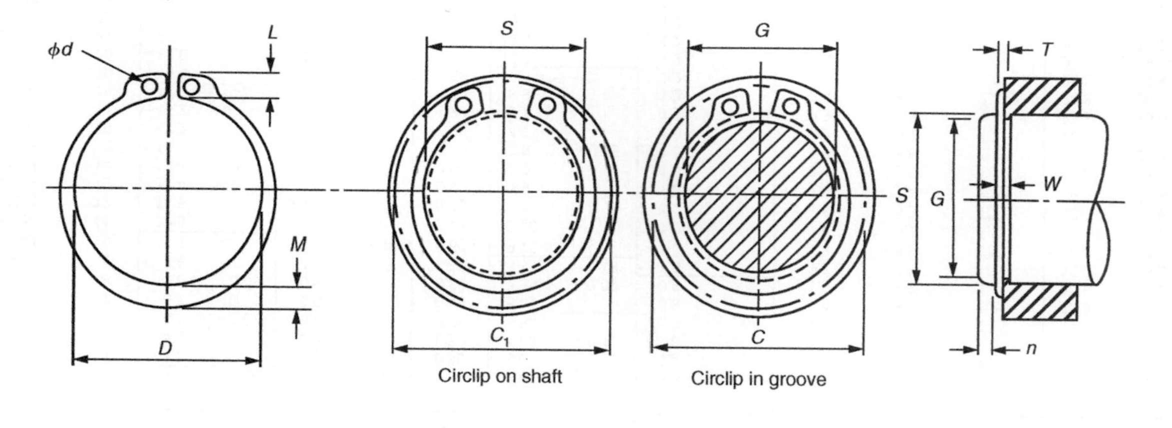

Circlip on shaft

Circlip in groove

Dimensions in mm

Reference number of circlip	Shaft diameter *S*	Groove details					Circlip details							Minimum external clearance	
		Diameter *G*	Tolerance	Width *W*	Tolerance	Edge margin (min.) *n*	Diameter *D*	Tolerance	Thickness *T*	Tolerance	Beam (approx.) *M*	Lug depth (max.) *L*	Lug hole diameter (min.) *d*	Fitted *C*	During fitting (C_1)
S003M	3	2.8	0 −0.06	0.5		0.30	2.66		0.4		0.8	1.9	0.8	6.6	7.2
S004M	4	3.8		0.5		0.30	3.64	+0.06	0.4		0.9	2.2	1.0	8.2	8.8
S005M	5	4.8	0	0.7		0.30	4.64	−0.15	0.6	0	1.1	2.5	1.0	9.8	10.6
S006M	6	5.7	−0.075	0.8		0.45	5.54		0.7	−0.04	1.3	2.7	1.15	11.1	12.1
S007M	7	6.7	0	0.9		0.45	6.45		0.8		1.4	3.1	1.2	12.9	14.0
S008M	8	7.6	−0.09	0.9		0.60	7.35	+0.09	0.8		1.5	3.2	1.2	14.0	15.2
S009M	9	8.6		1.1		0.60	8.35	−0.18	1.0		1.7	3.3	1.2	15.2	16.6
S010M	10	9.6		1.1	+0.14	0.60	9.25		1.0		1.8	3.3	1.5	16.2	17.6
S011M	11	10.5		1.1	0	0.75	10.20		1.0		1.8	3.3	1.5	17.1	18.6
S012M	12	11.5		1.1		0.75	11.0		1.0		1.8	3.3	1.7	18.1	19.6
S013M	13	12.4	0	1.1		0.90	11.9		1.0	0	2.0	3.4	1.7	19.2	20.8
S014M	14	13.4	−0.11	1.1		0.90	12.9	+0.18	1.0	−0.06	2.1	3.5	1.7	20.4	22.0
S015M	15	14.3		1.1		1.10	13.8	−0.36	1.0		2.2	3.6	1.7	21.5	23.2
S016M	16	15.2		1.1		1.20	14.7		1.0		2.2	3.7	1.7	22.6	24.4
S017M	17	16.2		1.1		1.20	15.7		1.0		2.3	3.8	1.7	23.8	25.6
S018M	18	17.0		1.3		1.50	16.5		1.2		2.4	3.9	2.0	24.8	26.8
S019M	19	18.0		1.3		1.50	17.5		1.2		2.5	3.9	2.0	25.8	27.8
S020M	20	19.0		1.3		1.50	18.5		1.2		2.6	4.0	2.0	27.0	29.0
S021M	21	20.0		1.3		1.50	19.5		1.2		2.7	4.1	2.0	28.2	30.2
S022M	22	21.0		1.3		1.50	20.5		1.2		2.8	4.2	2.0	29.4	31.4
S023M	23	22.0		1.3		1.50	21.5		1.2		2.9	4.3	2.0	30.6	32.6
S024M	24	22.9	0	1.3		1.70	22.2	+0.21	1.2		3.0	4.4	2.0	31.7	33.8
S025M	25	23.9	−0.21	1.3		1.70	23.2	−0.42	1.2		3.0	4.4	2.0	32.7	34.8

S026M	26	24.9		1.3		1.70	24.2		1.2		3.1	4.5	2.0	33.9	36.0
S027M	27	25.6		1.3		2.10	24.9		1.2		3.1	4.6	2.0	34.8	37.2
S028M	28	26.6		1.6		2.10	25.9		1.5		3.2	4.7	2.0	36.0	38.4
S029M	29	27.6		1.6		2.10	26.9		1.5		3.4	4.8	2.0	37.2	39.6
S030M	30	28.6		1.6	+0.14	2.10	27.9		1.5		3.5	5.0	2.0	38.6	41.0
S031M	31	29.3		1.6	0	2.60	28.6		1.5		3.5	5.1	2.5	39.5	42.2
S032M	32	30.3		1.6		2.60	29.6		1.5		3.6	5.2	2.5	40.7	43.4
S033M	33	31.3		1.6		2.60	30.5		1.5	0	3.7	5.3	2.5	41.9	44.4
S034M	34	32.3		1.6		2.60	31.5		1.5	−0.06	3.8	5.4	2.5	43.1	45.8
S035M	35	33.0		1.6		3.00	32.2		1.5		3.9	5.6	2.5	44.2	47.2
S036M	36	34.0		1.85		3.00	33.2		1.75		4.0	5.6	2.5	45.2	48.2
S037M	37	35.0		1.85		3.00	34.2		1.75		4.1	5.7	2.5	46.4	49.4
S038M	38	36.0		1.85		3.00	35.2		1.75		4.2	5.8	2.5	47.6	50.6
S039M	39	37.0	0	1.85		3.00	36.0	+0.25	1.75		4.3	5.9	2.5	48.8	51.8
S040M	40	37.5	−0.25	1.85		3.80	36.5	−0.50	1.75		4.4	6.0	2.5	49.5	53.0
S041M	41	38.5		1.85		3.80	37.5		1.75		4.5	6.1	2.5	50.7	54.2
S042M	42	39.5		1.85		3.80	38.5		1.75		4.5	6.5	2.5	52.5	56.0
S043M	43	40.5		1.85		3.80	39.5		1.75		4.6	6.6	2.5	53.7	57.2
S044M	44	41.5		1.85		3.80	40.5		1.75		4.6	6.6	2.5	54.7	58.2
S045M	45	42.5		1.85		3.80	41.5		1.75		4.7	6.7	2.5	55.9	59.4
S046M	46	43.5		1.85		3.80	42.5	+0.39	1.75		4.8	6.7	2.5	56.9	60.4
S047M	47	44.5		1.85		3.80	43.5	−0.78	1.75		4.9	6.8	2.5	58.1	61.6
S048M	48	45.5		1.85		3.80	44.5		1.75		5.0	6.9	2.5	59.3	62.8
S049M	49	46.5		1.85		3.80	44.8		1.75		5.0	6.9	2.5	60.3	63.8
S050M	50	47.0		2.15		4.50	45.8		2.00		5.1	6.9	2.5	60.8	64.8
S052M	52	49.0		2.15		4.50	47.8		2.00		5.2	7.0	2.5	63.0	67.0

For full range of sizes and types and for full information see BS 3673: Part 4: 1977.

4.4.3 Circlips, internal: metric series

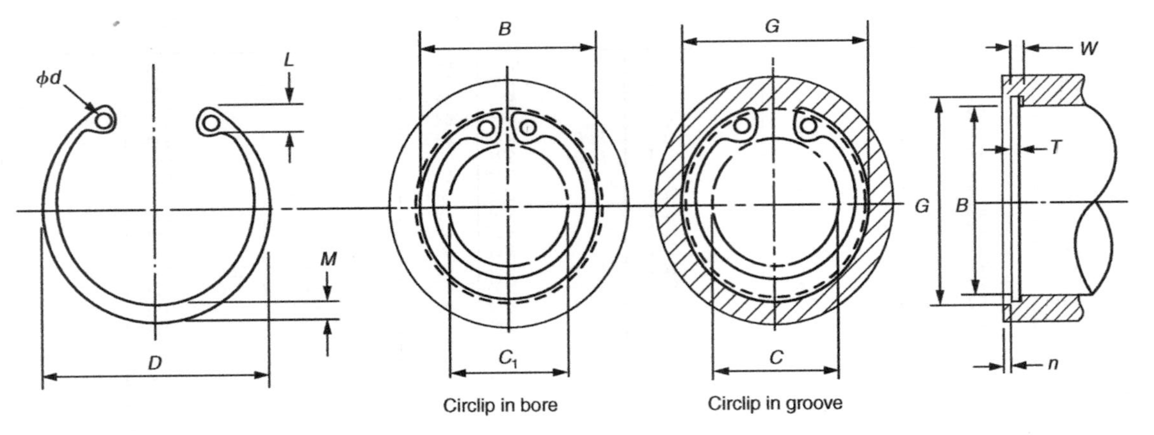

Dimensions in mm

Reference number of circlip	Shaft diameter *B*	Groove details					Circlip details							Minimum internal clearance	
		Diameter *G*	Tolerance	Width *W*	Tolerance	Edge margin (min.) *n*	Diameter *D*	Tolerance	Thickness *T*	Tolerance	Beam (approx.) *M*	Lug depth (max.) *L*	Lug hole diameter (min.) *d*	Fitted *C*	During fitting (C_1)
B008M	8	8.4	+0.09	0.9		0.6	8.7		0.8		1.1	2.4	1.0	3.6	2.8
B009M	9	9.4	0	0.9		0.6	9.8		0.8		1.3	2.5	1.0	4.0	3.1
B010M	10	10.4		1.1		0.6	10.8		1.0		1.4	3.2	1.2	4.4	3.6
B011M	11	11.4		1.1		0.6	11.8	+0.36	1.0		1.5	3.3	1.2	4.8	3.9
B012M	12	12.5		1.1		0.75	13.0	−0.18	1.0		1.7	3.4	1.5	5.7	4.7
B013M	13	13.6	+0.11	1.1		0.9	14.1		1.0		1.8	3.6	1.5	6.4	5.3
B014M	14	14.6	0	1.1		0.9	15.1		1.0		1.9	3.7	1.7	7.2	6.1
B015M	15	15.7		1.1	+0.14	1.1	16.2		1.0		2.0	3.7	1.7	8.3	7.1
B016M	16	16.8		1.1	0	1.2	17.3		1.0	0	2.0	3.8	1.7	9.2	7.9
B017M	17	17.8		1.1		1.2	18.3		1.0	−0.06	2.1	3.9	1.7	10.0	8.7
B018M	18	19.0		1.1		1.5	19.5		1.0		2.2	4.1	2.0	10.8	9.3
B019M	19	20.0		1.1		1.5	20.5		1.0		2.2	4.1	2.0	11.8	9.8
B020M	20	21.0		1.1		1.5	21.5	+0.42	1.0		2.3	4.2	2.0	12.6	10.6
B021M	21	22.0		1.1		1.5	22.5	−0.21	1.0		2.4	4.2	2.0	13.6	11.6
B022M	22	23.0	+0.21	1.1		1.5	23.5		1.0		2.5	4.2	2.0	14.6	12.6
B023M	23	24.1	0	1.1		1.5	24.6		1.0		2.5	4.2	2.0	15.7	13.6
B024M	24	25.2		1.1		1.8	25.9		1.0		2.6	4.4	2.0	16.4	14.2
B025M	25	26.2		1.1		1.8	26.9		1.0		2.7	4.5	2.0	17.2	15.0
B026M	26	27.2		1.1		1.8	27.9		1.0		2.8	4.7	2.0	17.8	15.6
B027M	27	28.4		1.1		2.1	29.1		1.0		2.8	4.7	2.0	19.0	16.6
B028M	28	29.4		1.3		2.1	30.1	+0.92	1.2		2.9	4.8	2.0	19.8	17.4
B029M	29	30.4	+0.25	1.3		2.1	31.1	−0.46	1.2		3.0	4.8	2.0	20.8	18.4
B030M	30	31.4	0	1.3		2.1	32.1		1.2		3.0	4.8	2.0	21.8	19.4

continued

Section 4.4.3 (*continued*)

Reference number of circlip	Shaft diameter *B*	Groove details					Circlip details							Minimum internal clearance	
		Diameter *G*	Tolerance	Width W	Tolerance	Edge margin (min.) *n*	Diameter *D*	Tolerance	Thickness *T*	Tolerance	Beam (approx.) *M*	Lug depth (max.) *L*	Lug hole diameter (min.) *d*	Fitted C	During fitting (C_1)
B031M	31	32.7		1.3		2.6	33.4		1.2		3.2	5.2	2.5	22.3	19.6
B032M	32	33.7		1.3		2.6	34.4	+0.50	1.2		3.2	5.4	2.5	22.9	20.2
B033M	33	34.7		1.3	+0.14	2.6	35.5	−0.25	1.2		3.3	5.4	2.5	23.9	21.2
B034M	34	35.7		1.6	0	2.6	36.5		1.5		3.3	5.4	2.5	24.9	22.2
B035M	35	37.0		1.6		3.0	37.8		1.5		3.4	5.4	2.5	26.2	23.2
B036M	36	38.0	+0.25	1.6		3.0	38.8		1.5		3.5	5.4	2.5	27.2	24.2
B037M	37	39.0	0	1.6		3.0	39.8		1.5	0	3.6	5.5	2.5	28.0	25.0
B038M	38	40.0		1.6		3.0	40.8		1.5	−0.06	3.7	5.5	2.5	29.0	26.0
B039M	39	41.0		1.6		3.0	42.0		1.5		3.8	5.6	2.5	29.8	26.8
B040M	40	42.5		1.85		3.8	43.5		1.75		3.9	5.8	2.5	30.9	27.4
B041M	41	43.5		1.85		3.8	44.5		1.75		4.0	5.9	2.5	31.7	28.2
B042M	42	44.5		1.85		3.8	45.5	+0.78	1.75		4.1	5.9	2.5	32.7	29.2
B043M	43	45.5		1.85		3.8	46.5	−0.39	1.75		4.2	6.0	2.5	33.5	30.0
B044M	44	46.5		1.85		3.8	47.5		1.75		4.3	6.0	2.5	34.5	31.0
B045M	45	47.5		1.85		3.8	48.5		1.75		4.3	6.2	2.5	35.1	31.6
B046M	46	48.5		1.85		3.8	49.5		1.75		4.4	6.3	2.5	35.9	32.4
B047M	47	49.5		1.85		3.8	50.5		1.75		4.4	6.4	2.5	36.7	33.2
B048M	48	50.5		1.85		3.8	51.5	+0.92	1.75		4.5	6.4	2.5	37.7	34.2
B049M	49	51.5	+0.30	1.85		3.8	52.5	−0.46	1.75		4.5	6.4	2.5	38.7	35.2
B050M	50	53.0	0	2.15		4.5	54.2		2.0		4.6	6.5	2.5	40.0	36.0
B051M	51	54.0		2.15		4.5	55.2		2.0		4.7	6.5	2.5	41.0	37.0
B052M	52	555.0		2.15		4.5	56.2		2.0		4.7	6.7	2.5	41.6	37.6

For full range of sizes and types and for full information see BS 3673: Part 4: 1977.

4.5 Adhesive bonding of metals

4.5.1 Anaerobic adhesives

These are high-strength adhesives widely used in the engineering industry. Anaerobic adhesives are single-compound materials that remain inactive when in contact with oxygen. The adhesives require two conditions to be present simultaneously in order to cure. That is, the absence of oxygen and in the presence of metal parts. These adhesives are widely used for threadlocking, threadsealing, gasketing and retaining (or cylindrical part bonding) at room temperature where ideal conditions exist for the adhesive to cure as shown in Fig. 1.

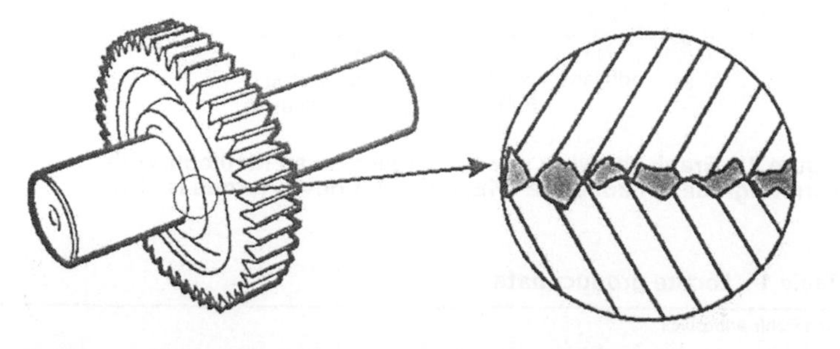

Figure 1 The adhesive flows into the spaces between the gear and the shaft where it has the ideal conditions to cure (absence of oxygen and metal part activity).

Copper, brass and plain carbon steel are 'active' as far as anaerobic adhesives are concerned and result in rapid curing at normal room temperatures. Stainless steel, aluminium and aluminium alloys, and plated products are 'passive' and require the use of an activator to ensure a full cure. As a liquid, the adhesive flows into the interstitial spaces between the male and female components of the joint where, devoid of atmospheric oxygen, it cures and becomes 'keyed' to the surface roughness. The curing process is also stimulated by contact between the adhesive and the metal surface that acts as a catalyst in the case of 'active'. Once cured to provide a tough thermoset plastic, the adhesive provides up to 60% of the torsional strength of the joint as shown in Fig. 2. To ensure a successful joint the mating surfaces must be mechanically and chemically clean before applying the adhesive and any activator necessary – normally to the male component for convenience. Handling strength is achieved within 10–20 min of the joint being achieved and full strength is attained in 4–24 h depending on the gap between the mating components, the materials being joined and the temperature. The cured adhesives will resist temperatures up to 150°C and are resistant to most chemicals, although some higher temperature versions are available. Typical anaerobic adhesives from the Loctite product range and some applications are listed in Table 1.

4.5.2 Adhesives cured by ultraviolet light

These are adhesives that are cured by the application of ultraviolet (UV) light. The cure times depend on the intensity and wavelength of the UV light which initiates polymerization.

Note: Suitable eye-protection should always be used when working with UV light.

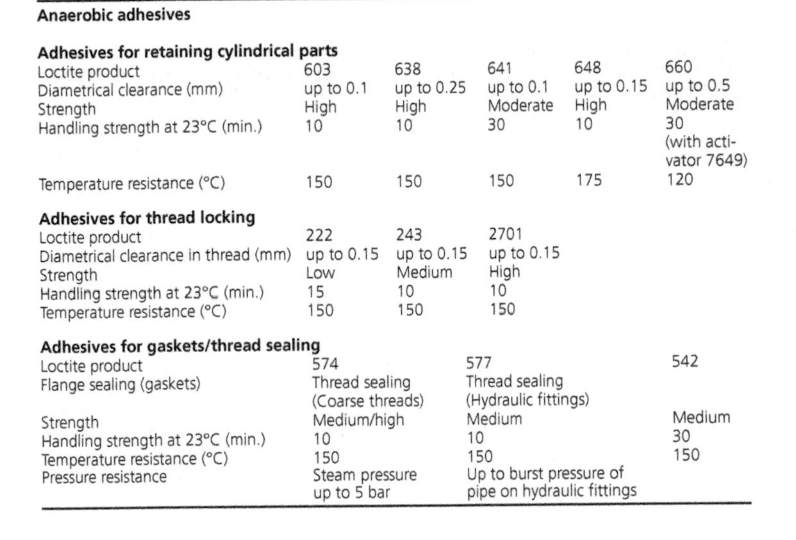

Figure 2 Graph showing release torques on an M10 bolt with various grades of adhesive where the Put down torque is 40 Nm.

Table 1 Loctite product data

Anaerobic adhesives					
Adhesives for retaining cylindrical parts					
Loctite product	603	638	641	648	660
Diametrical clearance (mm)	up to 0.1	up to 0.25	up to 0.1	up to 0.15	up to 0.5
Strength	High	High	Moderate	High	Moderate
Handling strength at 23°C (min.)	10	10	30	10	30 (with activator 7649)
Temperature resistance (°C)	150	150	150	175	120
Adhesives for thread locking					
Loctite product	222	243	2701		
Diametrical clearance in thread (mm)	up to 0.15	up to 0.15	up to 0.15		
Strength	Low	Medium	High		
Handling strength at 23°C (min.)	15	10	10		
Temperature resistance (°C)	150	150	150		
Adhesives for gaskets/thread sealing					
Loctite product	574		577		542
Flange sealing (gaskets)	Thread sealing (Coarse threads)		Thread sealing (Hydraulic fittings)		
Strength	Medium/high		Medium		Medium
Handling strength at 23°C (min.)	10		10		30
Temperature resistance (°C)	150		150		150
Pressure resistance	Steam pressure up to 5 bar		Up to burst pressure of pipe on hydraulic fittings		

Adhesives cured by UV light can be described as having high strength, high gap-filling capacity, very short curing times to handling strength, good to very good environmental resistance and good dispensing capacity with automatic systems as single-component adhesives.

The UV lamp is an essential part of the process and light sources can range from a simple bench-top open unit through to a fully automated conveyor system with several flood lights, incorporating special fixturing to hold the components in place during the curing cycle. UV light guides (or 'wand' systems) are often specified for small components as these units produce high-intensity light over a diameter of about 10 mm. These light guides ensure that the UV light is directed to the precise area for curing, thus minimizing glare and stray UV light. An example of a light guide is shown in the following figure.

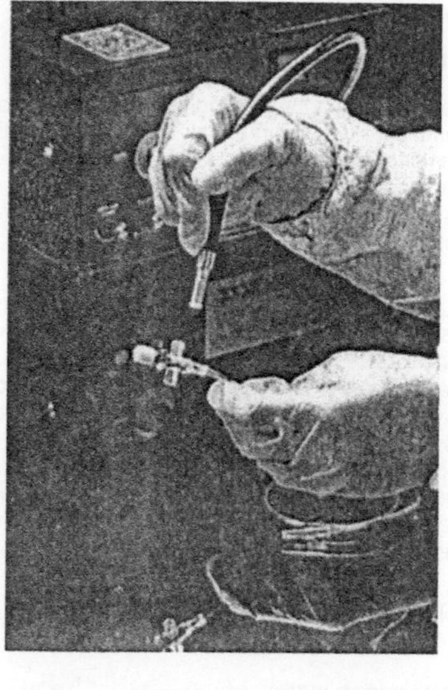

A flexible UV light guide unit for curing UV adhesives.

In some applications, it is not always possible to ensure that the entire adhesive is exposed to the UV light and products are available which will cure after the UV cycle. There are several options for the secondary cure as follows.

Heat

A secondary heat cure will ensure the full cure of many grades of UV adhesives. Cure temperatures are typically in the range 100–150°C.

Anaerobic

UV anaerobic grades are available for applications where a shadow cure is required. These anaerobic adhesives will cure through a depth of 0.2 mm due to contact with metal parts and the absence of oxygen.

Moisture

UV silicone products will immobilize after UV cure and then continue to cure over 24 h due to surrounding atmospheric moisture.

Surface moisture

UV curing cyanoacrylates will cure through several millimetres with UV light but also cure due to surface moisture in shadowed areas.

One of the main reasons why UV adhesives are used is the benefit of 'cure on demand' (i.e. the ability to cure the product exactly when required once, and not until, the components are fully aligned).

4.5.3 Adhesives cured by anionic reaction (cyanoacrylates)

Single-component cyanoacrylate adhesives polymerize on contact with slightly alkaline surfaces. In general, ambient humidity in the air and on the bonding surface is sufficient to initiate curing to handing strength within a few seconds. The best results are achieved when the relative humidity value is 40–60% at the workplace at room temperature. Lower humidity leads to slower curing, higher humidity accelerates the curing but in extreme cases may impair the final strength of the bond.

After adhesive application, the parts must be assembled quickly since polymerization begins in only a few seconds. The open time is dependent on the relative humidity, the humidity of the bonding surfaces and the ambient temperature. Due to their very fast curing times, cyanoacrylate adhesives are particularly suitable for bonding small parts. UV curing grades are also available to improve the cure throughout the depth of the adhesive and to cure any excess adhesive. Cyanoacrylate adhesives should be applied only to one surface and the best bond is achieved if only enough adhesive is applied to fill the joint gap. Activators may be used to speed the curing process and to cure any excess adhesive. A further benefit of cyanoacrylates is that they can be used with primers to bond low-energy plastics such as polypropylene and polyethylene.

Features of cyanoacrylate adhesives are:

- very high shear and tensile strength;
- very fast curing speed (seconds rather than minutes);
- minimum adhesive consumption;
- almost all materials may be bonded;
- simple dispensing by good ageing resistance of single-component adhesives;
- simultaneous sealing effect.

4.5.4 Adhesives cured with activator systems (modified acrylics)

These adhesives cure at room temperature when used with activators which are applied separately to the bonding surfaces. These components of the adhesive system are not pre-mixed, so it is not necessary to be concerned about 'pot life'. The characteristic properties of modified acrylics are:

- very high shear and tensile strengths;
- good impact and peel resistance;
- wide useful temperature (−55°C to +120°C);
- almost all materials can be bonded;
- large gap-filling capacity;
- good environmental resistance.

4.5.5 Adhesives cured by ambient moisture

These adhesives/sealants polymerize (in most cases) through a condensation effect that involves a reaction with ambient moisture. There are three types of adhesive in this category, namely:

- silicones
- polyurethanes
- modified silanes.

Silicones

These materials vulcanize at room temperature by reacting with ambient moisture (RTV). The solid rubber silcone is characterized by the following properties:

- Excellent thermal resistance
- Flexible, tough, low modulus and high elongation
- Effective sealants for a variety of fluid types.

Polyurethanes

Polyurethanes are formed through a mechanism in which water reacts (in most cases) with a formulative additive containing isocyanate groups. These products are characterized by the following properties:

- Excellent toughness and flexibility
- Excellent gap filling (up to 5 mm)
- Paintable once cured
- Can be used with primers to improve adhesion.

Modified silanes

The modified silanes again cure due to reaction with atmospheric moisture and offer the following properties:

- Paintable once cured
- Good adhesion to a wide range of substrates
- Isocyanate free
- Excellent toughness and flexibility.

It has only been possible to give a brief review of the main types of these specialized adhesives and the reader is referred to the technical literature published by the **Henkel Loctite Adhesives Limited** for detailed information on adhesive types and applications.

4.5.6 Epoxy adhesives

These are based on resins derived from epichlorhydrin and bisphenol-A. They are characterized by low shrinkage on polymerization and by good adhesion and high mechanical strength. These are usually 'two-pack' adhesives that are cured by mixing the adhesive with a catalyst (hardener) immediately prior to use. Curing commences as soon as the catalyst is added. However, in some instances the hardener is premixed by the adhesive manufacturer and is activated by heating to (typically) 150°C for 30 min. Typical epoxy adhesives are found in the Loctite® Hysol® range. The substrate selector guide is shown in Fig. 1.

Typical Hysol® epoxy adhesives are as follows:

1. General purpose epoxy adhesives:
   - Multi-purpose adhesives for general bonding applications.
   - Ultra-clear adhesives for transparent bond lines.
   - Flexible adhesives for low stress bonding of plastics and dissimilar substrates.

Loctite® Hysol®	3421 A&B	3423 A&B	9481 A&B	9483 A&B	9484 A&B	9489 A&B	9533	3430 A&B	9455 A&B	3422 A&B	9450 A&B	3450 A&B	3455 A&B	3463 A&B	3425 A&B	9466 A&B	9463 A&B	9461 A&B	9464 A&B	9514	9491 A&B	9492 A&B	9502	9509	9493 A&B	9496 A&B
Steel	++	++	+	+	++	++	++	++	++	++	++	++	++	++	++	++	++	++	++	++	+	+	++	++	+	+
Stainless steel	++	++	+	+	+	+	++	++	++	+	++	++	++	++	++	++	++	+	+	++	+	+	++	++	+	+
Zinc dichromate steel	++	++	+	++	++	++	++	+	+	+	++	++	++	++	+	+	++	+	+	++	+	+	++	++	+	+
Hot dipped galvanised steel	++	++	+	++	++	++	+	+	+	+	++	++	++	++	+	+	++	+	+	++	+	+	++	++	+	+
Aluminium	++	++	+	++	++	++	+	++	++	+	++	++	++	++	++	++	++	++	+	++	+	+	++	++	+	+
Copper	++	++	+	+	+	+	+	++	++	+	++	++	++	++	++	++	++	+	+	++			++	++	+	+
Brass	++	++	+	+	+	+	+	++	++	+	++	++	++	++	++	++	++	+	+	++			++	++	+	+
Thermoplastics	+		+		+	+	+	+	+		+					+	++	++	++		+	+			++	+
Thermoset plastics	+		+	+	++	++	+	+	+		+					+	++	+	+						+	+
GRP/SMC composites	+	+	+	+	+	+	++	+	+		+	+	+			+	++	++	++	++		+	+	+	++	+
Rubbers	+	+	+		+	+		+	+	+					+	+	+	+	+							
Glass	++		+	++	+	+		++	++	+	+					++	+									
Ceramic	++	+		++	+	+	+	+	+	+	+	+	+		+	+	+	++	++	++	++	++	++	++	+	+
Masonry	++	+		+	+	+		+	+	+	+	+	+		++	+	+	+	+		+	+			+	
Wood	++	+	+					+	+	++	+	+	+		++	+	+	+	+						+	

Column groups: 3421 A&B – 9533: General purpose Epoxies; 3430 A&B – 3463 A&B: Five-minute Epoxies; 3425 A&B – 9514: Toughened Epoxies; 9491 A&B – 9496 A&B: High temperature Epoxies

++ Highly recommended
\+ Recommended

- Thermoplastics: ABS, Acrylic, PA, PC, PVC, SAN
- Thermoset plastics: DAP, Epoxy, Phenolic, Polyster
- Rubbers: Butyl, EPDM, Natural, Neoprene, Nitrile, SBR

Figure 1 Substrate selector guide.

2. Five-minute epoxy adhesives:
   - Rapid cure for a fast fixture time.
   - Ultra-clear adhesives for fast transparent bonding.
   - Emergency repair of metal parts like pipes and tanks.
3. Toughened epoxy resins:
   - Highest shear and peel strength.
   - Excellent impact resistance.
   - Adhesives for bonding high performance composites such as glass-reinforced plastics (GRP) and CFRP.
4. High-temperature epoxy resins:
   - Resistance to high operating temperatures.
   - UL94-V0 rating for fire retardant parts.
   - High thermal conductivity.
5. Metal-filled epoxy resins:
   - Rebuild and restore worn and damaged metal parts.
   - Repair metal pipes and castings.
   - Form moulds, tools, fixtures and models.
   - Reduce sliding wear on moving parts.

Typical performance characteristics are shown in Fig. 2. The reader is referred to the comprehensive data published by **Henkel Loctite Adhesives Limited** who may be contacted at the address given in Appendix 3.

	General purpose epoxides			Five-minute epoxides	
	General bonding	Clear bonding	Flexible bonding	Multipurpose	Metal repair
Shear strength (GBMS) Peel strength (GBMS) Operating temperature	17–23 N/mm^2 2.7–2.5 N/mm Up to 100°C	19–23 N/mm^2 1.0–1.5 N/mm Up to 80°C to 100°C	22–25 N/mm^2 1.3–2.0 N/mm Up to 100°C	(Flowable) (Toughened) 16 N/mm^2–20 N/mm^2 <1 N/mm–1.5 N/mm Up to 80°C	(Gap filling) (Keadable stick) 6.8 N/mm^2–23 N/mm^2 (Steel filled) <1.0 N/mm–1.5 N/mm Up to 80°C to 100°C

	Toughened epoxides			High temperature epoxides			
	Extended working life	Medium working life	Heat curing	Room temperature curing	Heat curing	Fire retardant	Thermally conductive
Shear strength (GBMS) Peel strength (GBMS) Operating temperature	25–37 N/mm^2 20–8.0 N/mm Up to 120°C	22–24 N/mm^2 10.0–10.5 N/mm Up to 100°C	46 N/mm^2 9.5 N/mm Up to 180°C	15–20 N/mm^2 4.2–5.1 N/mm Up to 180°C	29–35 N/mm^2 19–25 N/mm Up to 180°C	20 N/mm^2 2.1 N/mm Up to 100°C	17 N/mm^2 6.8 N/mm Up to 150°C

	Metal-filled epoxides					
	Steel filled			Aluminium filled		Metallic parts under friction
	Putty	Pourable	Fast cure	Multipurpose	High temperature resistant	Wear resistant
Shear strength (GBMS) Compressive strength Operating temperature	20 N/mm^2 70 N/mm^2 Up to 120°C	25 N/mm^2 70 N/mm^2 Up to 120°C	20 N/mm^2 60 N/mm^2 Up to 120°C	20 N/mm^2 70 N/mm^2 Up to 120°C	20 N/mm^2 90 N/mm^2 Up to 190°C	20 N/mm^2 70 N/mm^2 Up to 120°C

Figure 2 Typical performance characteristics for epoxy adhesives.

4.5.7 Redux process

A technique developed for bonding primary sheet metal aircraft components with a two component adhesive under closely controlled heat and pressure.

4.5.8 Bonded joints

It has already been stated that correct surface preparation is necessary for optimum bonding. Bond strength is determined to a great extent by the adhesion between the adhesive and the substrate. Therefore joint preparation should remove oil, grease, oxide films and dirt in order to ensure chemical and mechanical cleanliness. Mechanical surface treatment by grit-blasting or wire brushing ensures good adhesion.

In addition it is necessary to design the joint correctly. Adhesives are relatively strong in tension and shear but weak in cleavage and peel. Therefore the joint must be designed with this in mind and also provide an adequate surface area between the mating parts to ensure the required strength. That is, the joint must be designed with adhesive bonding in mind from the start and not just be an adaptation of an existing mechanical/welded joint. Fig. 1 below shows a number of correctly and incorrectly designed joints suitable for adhesive bonding.

The author is indebted to Mr Bob Goss, Senior Technology Specialist at Henkel Loctite Adhesives Limited for his assistance in compiling the above information on adhesive bonding.

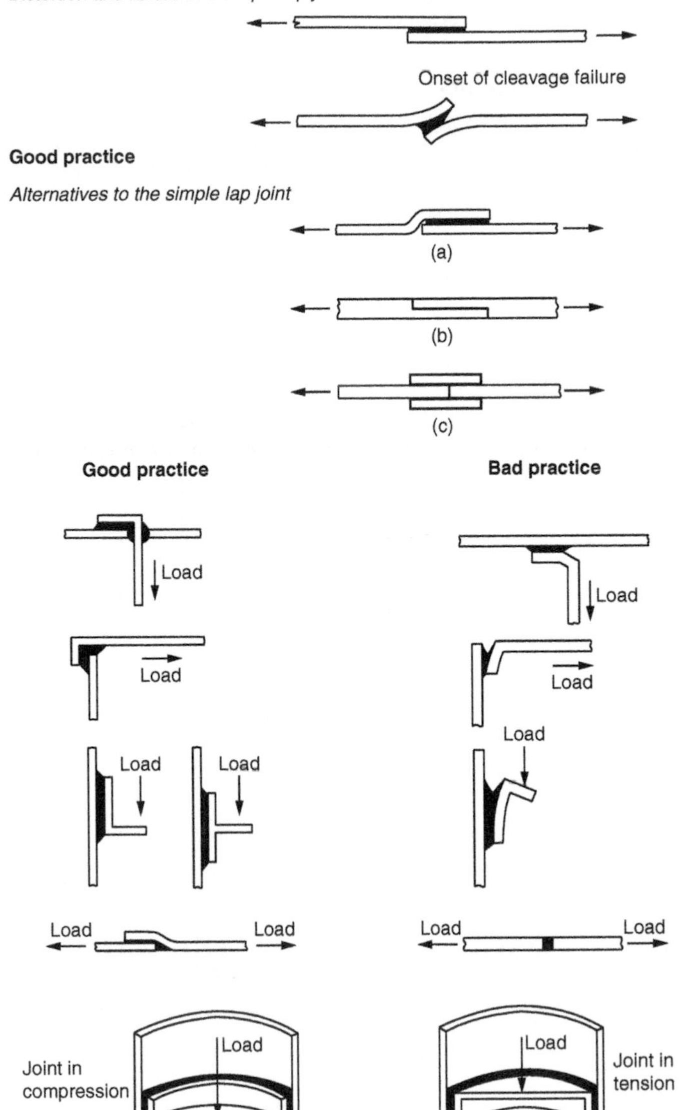

Figure 1 Bonded joint design.

5
Power Transmission

5.1 Power transmission: gear drives

5.1.1 Some typical gear drives

Straight tooth spur gears

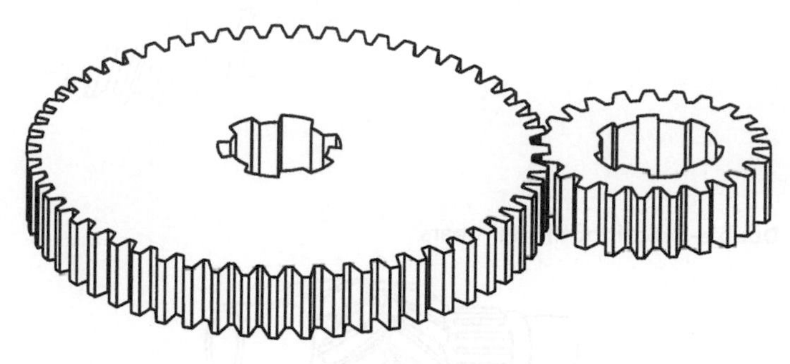

Rack and pinion: converts rotary to linear motion

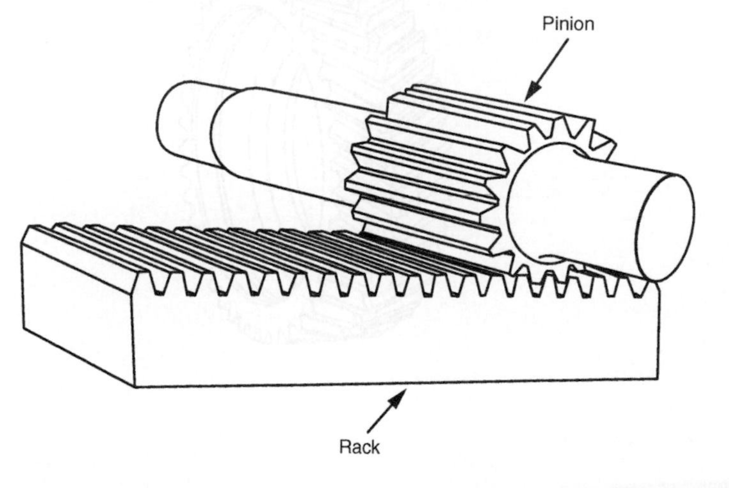

Single-helical tooth spur gears

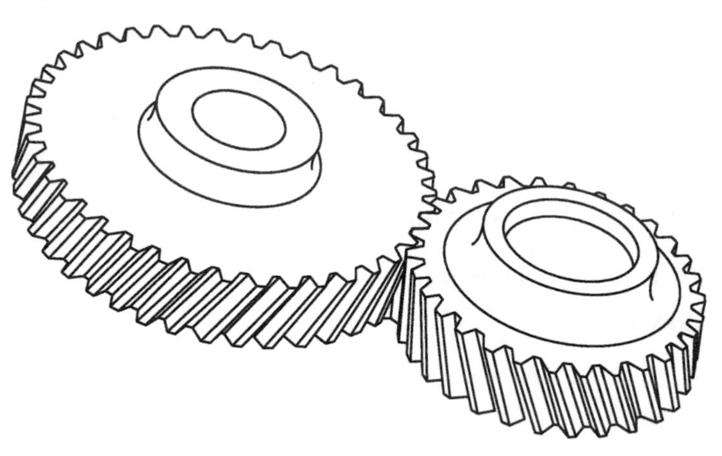

Double-helical tooth spur gears

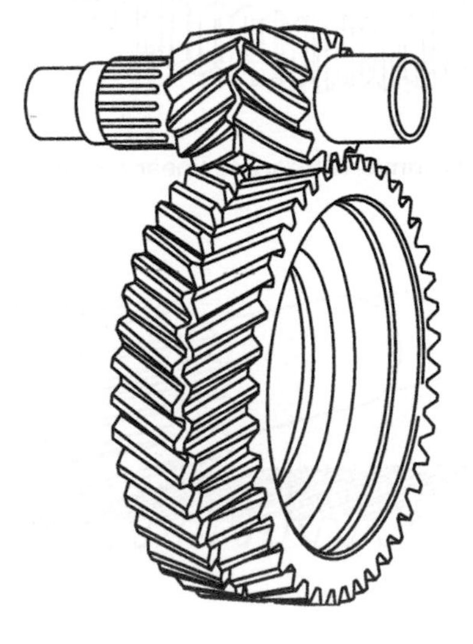

Straight tooth bevel gears

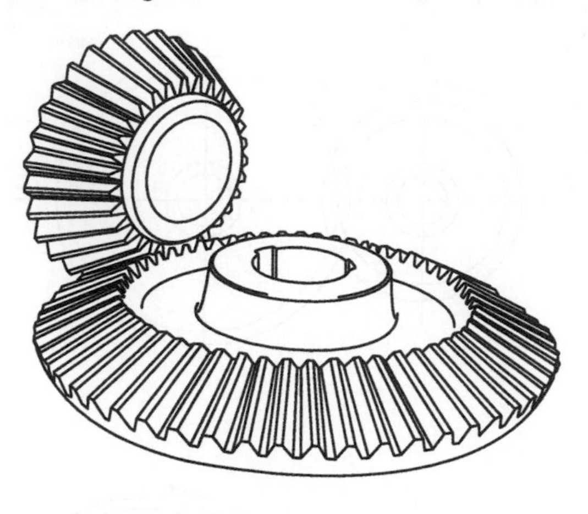

Worm and worm wheel

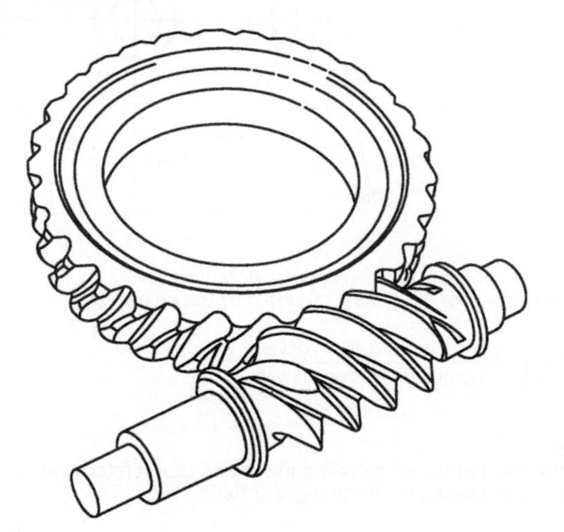

5.1.2 Simple spur gear trains

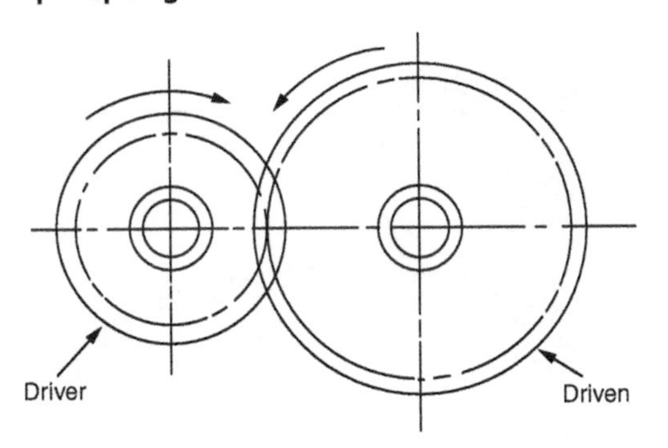

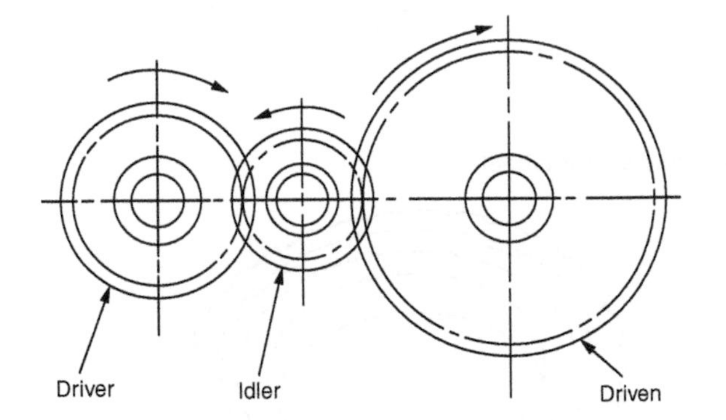

Simple train

(a) Driver and driven gears rotate in opposite directions.
(b) The relative speed of the gears is calculated by the expression:

$$\frac{\text{rev/min driver}}{\text{rev/min driven}} = \frac{\text{number of teeth on driven}}{\text{number of teeth on driver}}$$

Example

Calculate the speed of the driven gear if the driving gear is rotating at 120 rev/min. The driven gear has 150 teeth and the driving gear has 50 teeth:

$$\frac{120}{\text{rev/min driven}} = \frac{150}{50}$$

$$\text{rev/min driven} = \frac{120 \times 50}{150} = \mathbf{40\,rev/min}$$

Simple train with idler gear

(a) Driver and driven gears rotate in the same direction if there is an odd number of idler gears, and in the opposite direction if there is an even number of idler gears.
(b) Idler gears are used to change the direction of rotation and/or to increase the centre distance between the driver and driven gears.
(c) The number of idler gears and the number of teeth on the idler gears do not affect the overall relative speed.
(d) The overall relative speed is again calculated using the expression:

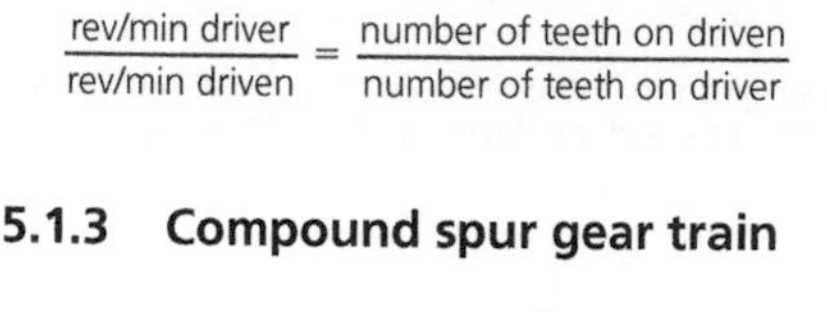

$$\frac{\text{rev/min driver}}{\text{rev/min driven}} = \frac{\text{number of teeth on driven}}{\text{number of teeth on driver}}$$

5.1.3 Compound spur gear train

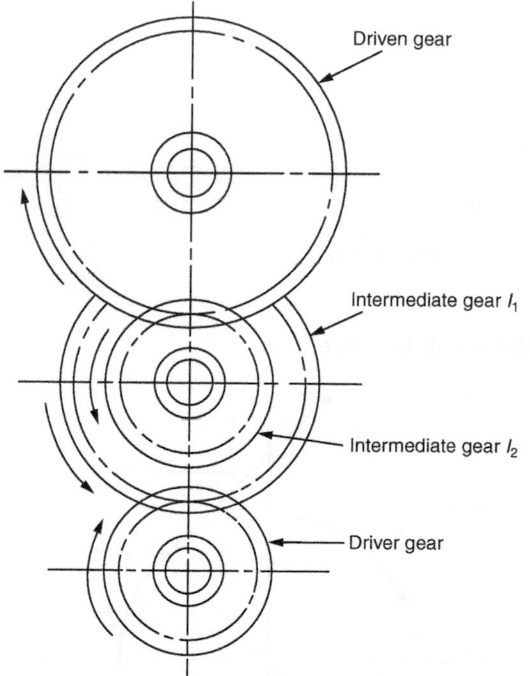

(a) Unlike the idler gear of a simple train, the intermediate gears of a compound train do influence the overall relative speeds of the driver and driven gears.
(b) Both intermediate gears (I_1 and I_2) are keyed to the same shaft and rotate at the same speed.

(c) Driver and driven gears rotate in the same direction. To reverse the direction of rotation an idler gear has to be inserted either between the driver gear and I_1 or between I_2 and the driven gear.
(d) The overall relative speed can be calculated using the expression:

$$\frac{\text{rev/min driver}}{\text{rev/min driven}} = \frac{\text{number of teeth on } I_1}{\text{number of teeth on driver}} \times \frac{\text{number of teeth on driven}}{\text{number of teeth on } I_2}$$

Example
Calculate the speed of the driven gear given that: the driver rotates at 600 rev/min and has 30 teeth; I_1 has 60 teeth; I_2 has 40 teeth; and the driven gear has 80 teeth:

$$\frac{\text{rev/min driver}}{\text{rev/min driven}} = \frac{60 \times 80}{30 \times 40} = \frac{4}{1}$$

but speed of driver = 600 rev/min

Therefore:

$$\frac{600 \text{ rev/min}}{\text{rev/min driven}} = \frac{4}{1}$$

$$\text{Speed driven} = \frac{600 \times 1}{4}$$

$$= \mathbf{150 \text{ rev/min}}$$

5.1.4 The involute curve

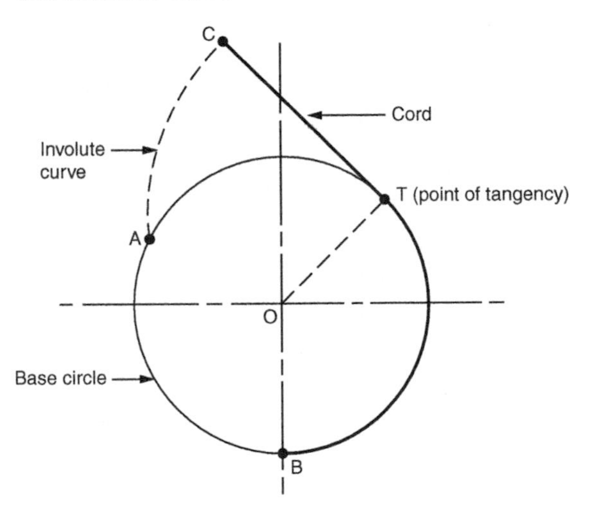

The involute curve is used for the teeth of the majority of industrially produced spur gears. The figure shows that the involute curve can be produced by unwinding an imaginary taut cord BTC from around a base circle, centre O, commencing with the end of the cord at A. The curve AC which is produced this way has an *involute form.*

The curvature of AC depends on the radius of the base circle. As the radius of the base circle increases, the curvature of the involute decreases. For a base circle of infinite radius, the line AC ceases to be a curve and becomes a straight line. Thus a *rack gear* has teeth with straight sides, since a rack can be considered as part of the circumference of a gear of infinite radius. That is, a rack with straight sided teeth will always mesh satisfactorily with a spur gear of involute form.

The advantages of the involute form for spur gear teeth are as follows:

(a) Involute form teeth produce smooth running gears capable of transmitting heavy loads at a constant velocity.
(b) Mating teeth of involute form make mainly rolling contact between their flanks as they move in and out of mesh. Since sliding contact is minimized, scuffing and wear of the flanks of the teeth is also minimized.
(c) Due to the relationship between the involute form and the straight sided rack form, gear teeth can be rapidly and accurately *generated* using rack type cutters with easily produced straight flank cutting teeth. This has obvious advantages for tooling costs.

5.1.5 Basic gear tooth geometry

In order to develop a practical gear tooth form from the basic involute form considered in Section 5.1.4, it is necessary to consider two further geometrical relationships:

(a) A tangent to an involute curve, at any point, is always perpendicular to a tangent to the base circle drawn from the same point. Reference to Section 5.1.4 shows that the line CT is tangential to the base circle and that such a tangent is perpendicular to a tangent drawn to the involute curve at C. It will be shown that the tangent CT is also the line of action of a gear teeth.
(b) The length of the line CT is also equal to the length of the arc AT.

Consider two circles touching at the point P as shown overleaf. The point P is called the *pitch point*, and the circles are called *reference circles.* If the reference circles were friction discs they would drive each other at the required velocity ratio depending on their relative diameters. However, gears which depended solely on friction to transmit power would be very limited. In practice they are provided with interlocking teeth to prevent slip. If involute form teeth are used then two additional circles have to be added to the figure as shown. These additional circles are the *base circles* from which the involutes are drawn (Section 5.1.4). Involutes are now drawn from these base circles so that they touch on the *line of action.* This line of action is the common tangent to the base circles and is an extension of the line CT (Section 5.1.4).

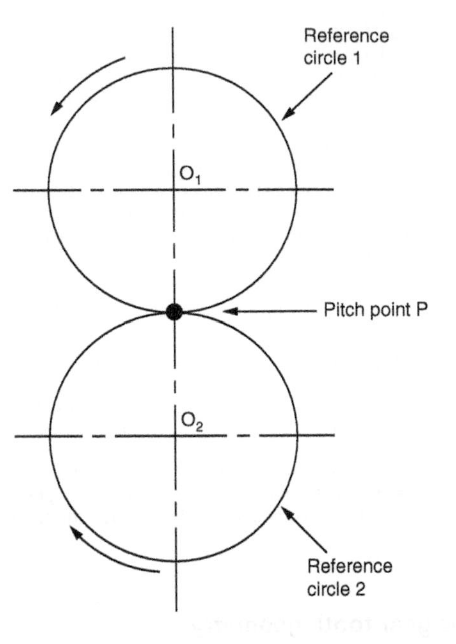

The *pressure angle* ϕ shown opposite is the angle between the line of action and a common tangent to the base circles drawn through the point P. Experience has shown that a value of $\phi = 20°$ gives optimum strength combined with smooth running. This is the value adopted in most modern gears.

To develop a working gear tooth from an abstract involute curve, additional data is required as shown:

Top circle This fixes the height of the tooth and prevents it from interfering with the root of the meshing tooth on the opposite gear.
Root circle This fixes the position of the bottom or root of the tooth and thus fixes the depth to which the tooth form is cut into the gear blank.
Addendum The radial distance between the top circle and the reference circle.
Dedendum The radial distance between the root circle and the reference circle.
Tooth height The radial distance between the top circle and the root circle. That is, tooth height = addendum + dedendum.

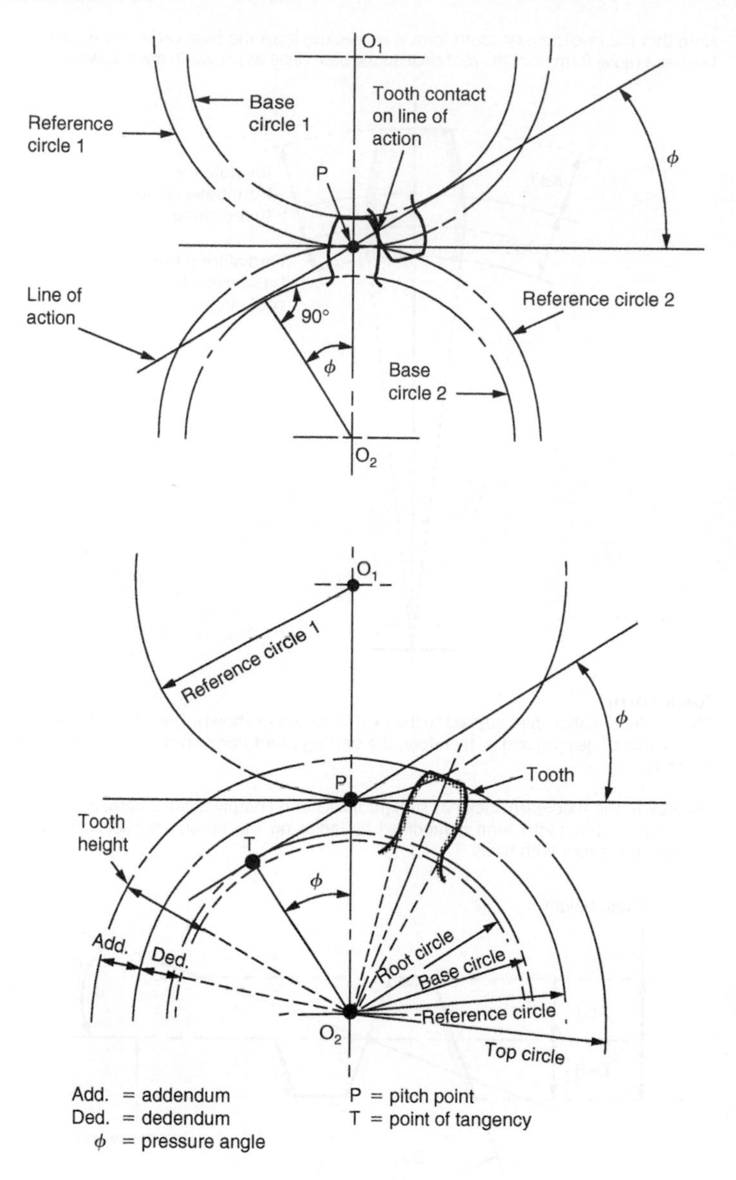
O1
Base
circle 1
Reference
circle 1
Tooth contact
on line of
action
P
ϕ
Line of
action
90°
ϕ
Reference circle 2
Base
circle 2
O2
O1
Reference circle 1
ϕ
Tooth
P
Tooth
height
T
ϕ
Add.
Ded.
Root circle
Base circle
Reference circle
O2
Top circle
Add. = addendum
Ded. = dedendum
ϕ = pressure angle
P = pitch point
T = point of tangency

Note that the involute gear tooth form is an involute from the base circle to the top circle, but has a radial form from the root circle to the base circle as shown in the following figure.

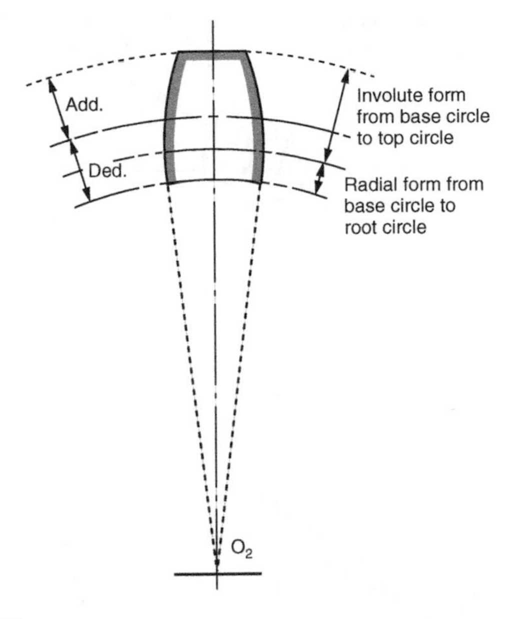

Rack form

The involute relationships applied to the rack tooth are as shown. The rack is the basis for any system of gearing and is, therefore, the starting point from which to build up a series of gears.

As shown, the theoretical rack has straight flanks and sharply defined corners. For practical reasons this tooth form is modified by radiusing the tip and replacing the sharp corners at the root with fillets.

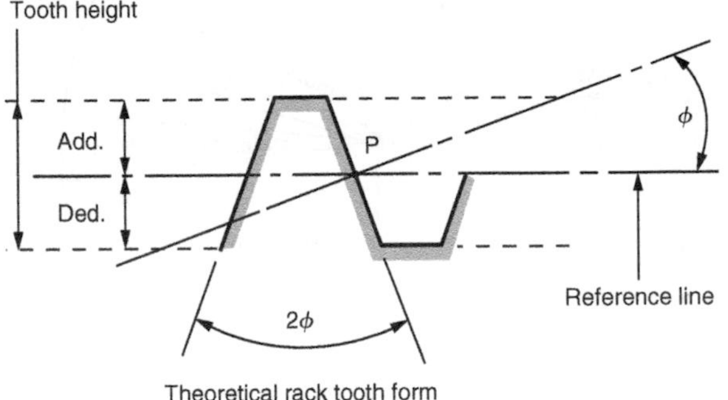

Theoretical rack tooth form

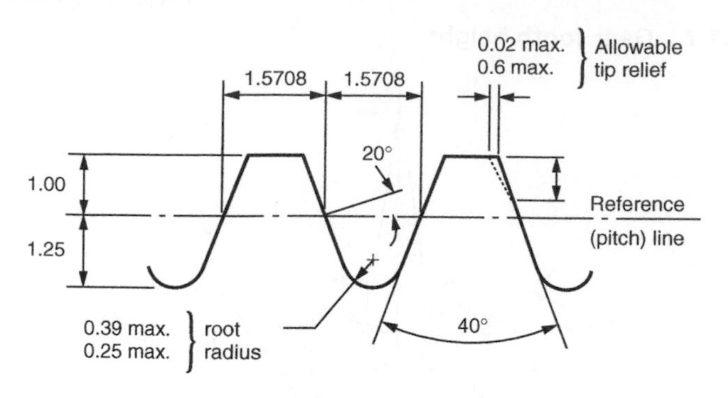

British Standard rack tooth profile
(dimensions for unit modular pitch)

A practical rack tooth form to BS 436 (metric units) is as shown above. This is the British standard basic rack dimensioned for unit modular pitch. For any module other than unity the tooth dimensions are determined by multiplying the figures shown by the module value. For practical reasons this basic rack can be modified further as follows:

(a) Variation of the total depth within the limits of 2.25–2.40 times the module value.
(b) Variation of the root radius within the limits of 0.25–0.39 times the module value.

Gears are graded in 10 grades of accuracy, and these are given in BS 436.

5.1.6 Gear tooth pitch

Spur gear teeth are spaced out around the reference circle at equal intervals (or increments) known as the *pitch*. There are three different systems by which the pitch may be defined.

Circular pitch (p)

This is the centre distance between two adjacent teeth on a gear measured around the circumference of the reference circle. For this reason the reference circle is also known as the *pitch circle*. As both the width of the tooth and the space between adjacent teeth are equal (ignoring clearance), the thickness of each tooth (as measured on the reference circle) is half the circular pitch.

Diametral pitch (P)

This is an 'inch' system for specifying the pitch of gear teeth:

$$\text{Diametral pitch}(\boldsymbol{P}) = \frac{\text{number of teeth}}{\text{diameter of reference circle in inches}}$$

Modular pitch (m)

This is the modern international metric parameter for specifying pitch. It is calculated in a similar manner to diametral pitch:

$$\text{Modular pitch}(m) = \frac{\text{diameter of reference circle in millimetres}}{\text{number of teeth}}$$

5.1.7 Gear tooth height

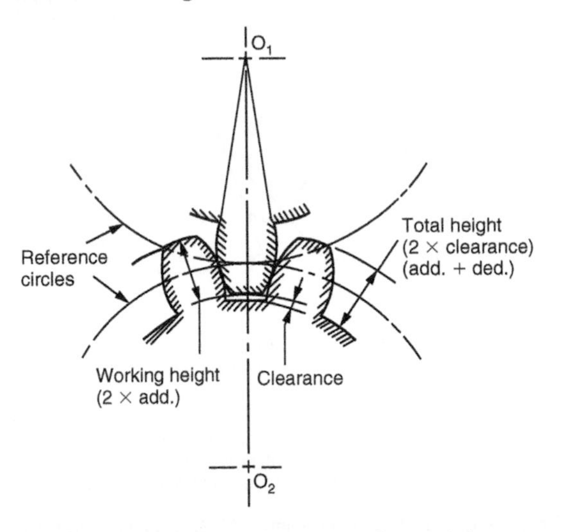

The figure shows that when two gear teeth mesh the reference circles are tangential, and that the addendum of one gear fits into the dedendum space of the mating gear. Since there must be clearance between the top (crest) of one tooth and the root of the space into which it engages, the dedendum is greater than the addendum by the clearance.

Thus the *working height* of the tooth is twice the addendum, whilst the *total height* is twice the addendum plus the clearance (i.e. the addendum plus the dedendum). Reference to the rack gear (Section 5.1.5) shows that the clearance is equal to the root radius. This is, ideally, 0.25–0.39 m for modular pitch teeth. Hence:

Addendum = 1.00 m
Dedendum = 1.25 m
Tooth height (cutting depth) = addendum + dedendum
= 1.00 m + 1.25–1.39 m
= 2.25–2.39 m
Working height = addendum + addendum
= **2.00 m**

where:
m = modular pitch

5.1.8 Standard gear tooth elements (in.)

Normal diametral pitch	Normal circular pitch	Standard circular tooth thickness	Standard addendum	Standard work depth	Minimum clearance*	Minimum whole depth*	Total shaving stock on tooth thickness
2.5	1.256637	0.6283	0.4000	0.8000	0.090	0.890	0.0040/0.0050
3	1.047198	0.5236	0.3333	0.6666	0.083	0.750	0.0040/0.0050
3.5	0.897598	0.4488	0.2857	0.5714	0.074	0.645	0.0040/0.0050
4	0.785398	0.3927	0.2500	0.5000	0.068	0.568	0.0035/0.0045
4.5	0.698132	0.3491	0.2222	0.4444	0.060	0.504	0.0035/0.0045
5	0.628319	0.3142	0.2000	0.4000	0.056	0.456	0.0035/0.0045
6	0.523599	0.2618	0.1667	0.3334	0.050	0.383	0.0030/0.0040
7	0.448799	0.2244	0.1429	0.2858	0.044	0.330	0.0030/0.0040
8	0.392699	0.1963	0.1250	0.2500	0.040	0.290	0.0030/0.0040
9	0.349066	0.1745	0.1111	0.2222	0.038	0.260	0.0025/0.0035
10	0.314159	0.1571	0.1000	0.2000	0.035	0.235	0.0025/0.0035
11	0.285599	0.1428	0.0909	0.1818	0.032	0.214	0.0020/0.0030
12	0.261799	0.1309	0.0833	0.1666	0.030	0.197	0.0020/0.0030
14	0.224399	0.1122	0.0714	0.1428	0.026	0.169	0.0020/0.0030
16	0.196350	0.0982	0.0625	0.1250	0.023	0.148	0.0015/0.0025
18	0.174533	0.0873	0.0556	0.1112	0.021	0.132	0.0015/0.0025

* Based on standard tip radius on preshaving hobs and shaper cutters. Increase whole depth by approximately 0.005–0.010 in. for crown shaving.

5.1.9 Fine pitch gear tooth elements (in.)

Normal diametral pitch	Normal circular pitch	Standard circular tooth thickness	Standard addendum	Standard work depth	Minimum clearance*	Minimum whole depth*	Total shaving stock on tooth thickness
20	0.157080	0.0785	0.0500	0.1000	0.020	0.120	
22	0.142800	0.0714	0.0454	0.0908	0.020	0.111	
24	0.130900	0.0654	0.0417	0.0834	0.020	0.103	
26	0.120830	0.0604	0.0385	0.0770	0.019	0.096	
28	0.112200	0.0561	0.0357	0.0714	0.019	0.090	
30	0.104720	0.0524	0.0333	0.0666	0.018	0.085	0.0005–0.0015
32	0.098175	0.0491	0.0312	0.0624	0.018	0.080	
36	0.087266	0.0436	0.0278	0.0556	0.016	0.072	
40	0.078540	0.0393	0.0250	0.0500	0.015	0.065	
44	0.071400	0.0357	0.0227	0.0454	0.014	0.059	
48	0.065450	0.0327	0.0208	0.0416	0.013	0.055	
52	0.060415	0.0302	0.0192	0.0384	0.012	0.050	
56	0.056100	0.0281	0.0178	0.0356	0.011	0.047	
60	0.052360	0.0262	0.0167	0.0334	0.010	0.043	0.0003–0.0007
64	0.049087	0.0245	0.0156	0.0312	0.010	0.041	
72	0.043633	0.0218	0.0139	0.0278	0.010	0.038	

* Based on standard tip radius on preshaving hobs and shaper cutter. Increase whole depth by approximately 0.005–0.010 in. for crown shaving.

5.1.10 Standard stub gear tooth elements (in.)

Normal diametral pitch	Normal circular pitch	Standard circular tooth thickness	Standard addendum	Standard work depth	Minimum clearance*	Minimum whole depth*	Total shaving stock on tooth thickness
3/4	1.047198	0.5236	0.2500	0.5000	0.083	0.583	0.0040/0.0050
4/5	0.785398	0.3927	0.2000	0.4000	0.068	0.468	0.0035/0.0045
5/7	0.628319	0.3142	0.1429	0.2858	0.056	0.342	0.0035/0.0045
6/8	0.523599	0.2618	0.1250	0.2500	0.050	0.300	0.0030/0.0040
7/9	0.448799	0.2244	0.1111	0.2222	0.044	0.266	0.0030/0.0040
8/10	0.392699	0.1963	0.1000	0.2000	0.040	0.240	0.0030/0.0040
9/11	0.349066	0.1745	0.0909	0.1818	0.038	0.220	0.0025/0.0035
10/12	0.314159	0.1571	0.0833	0.1666	0.035	0.202	0.0025/0.0035
12/14	0.261799	0.1309	0.0714	0.1428	0.030	0.173	0.0020/0.0030
14/18	0.224399	0.1122	0.0556	0.1112	0.026	0.137	0.0020/0.0030
16/21	0.196350	0.0982	0.0476	0.0952	0.023	0.118	0.0015/0.0025
18/24	0.174533	0.0873	0.0417	0.0834	0.021	0.104	0.0015/0.0025
20/26	0.157080	0.0785	0.0385	0.0770	0.020	0.097	0.0010/0.0015
22/29	0.142800	0.0714	0.0345	0.0690	0.020	0.089	0.0010/0.0015
24/32	0.130900	0.0654	0.0313	0.0626	0.020	0.083	0.0010/0.0015
26/35	0.120830	0.0604	0.0286	0.0572	0.019	0.076	0.0010/0.0015
28/37	0.112200	0.0561	0.0270	0.0540	0.019	0.073	0.0010/0.0015
30/40	0.104720	0.0524	0.0250	0.0500	0.018	0.068	0.0010/0.0015
32/42	0.098175	0.0491	0.0238	0.0476	0.018	0.066	0.0010/0.0015
34/45	0.092400	0.0462	0.0222	0.0444	0.017	0.061	0.0010/0.0015
36/48	0.087266	0.0436	0.0208	0.0416	0.016	0.058	0.0010/0.0015
38/50	0.082673	0.0413	0.0200	0.0400	0.016	0.056	0.0010/0.0015
40/54	0.078540	0.0393	0.0185	0.0370	0.015	0.052	0.0010/0.0015

* Based on standard tip radius on preshaving hobs and shaper cutters. Increase whole depth by approximately 0.005–0.010 in. for crown shaving.

5.1.11 Standard gear tooth elements (metric)

Normal module	Normal circular pitch (mm)	Standard circular tooth thickness (mm)	Minimum root clearance (mm)*	Minimum whole depth (mm)*	Total shaving stock on tooth thickness (mm)	Normal diametral pitch	Normal circular pitch (in.)	Standard circular tooth thickness (in.)	Standard addendum (in.)	Minimum root clearance (in.)*	Minimum whole depth (in.)*
1.0	3.142	1.571	0.390	2.390	0.015/0.040	25.400	0.1237	0.0618	0.0394	0.0154	0.0941
1.25	3.927	1.963	0.480	2.980	0.020/0.045	20.320	0.1546	0.0773	0.0492	0.0189	0.1173
1.50	4.712	2.356	0.560	3.560	0.025/0.050	16.933	0.1855	0.0928	0.0591	0.0220	0.1402
1.75	5.498	2.749	0.640	4.140	0.030/0.055	14.514	0.2164	0.1082	0.0689	0.0252	0.1630
2.0	6.283	3.142	0.720	4.720	0.035/0.060	12.700	0.2474	0.1237	0.0787	0.0283	0.1858
2.25	7.069	3.534	0.790	5.290	0.040/0.065	11.289	0.2783	0.1391	0.0886	0.0311	0.2083
2.50	7.854	3.927	0.860	5.860	0.045/0.070	10.160	0.3092	0.1546	0.0984	0.0339	0.2307
2.75	8.639	4.320	0.920	6.420	0.045/0.070	9.2364	0.3401	0.1701	0.1083	0.0362	0.2528
3.0	9.425	4.712	0.990	6.990	0.050/0.075	8.4667	0.3711	0.1855	0.1181	0.0390	0.2752
3.25	10.210	5.105	1.050	7.550	0.050/0.075	7.8154	0.4020	0.2010	0.1280	0.0413	0.2972
3.50	10.996	5.498	1.110	8.110	0.055/0.080	7.2571	0.4329	0.2165	0.1378	0.0437	0.3193
3.75	11.781	5.890	1.160	8.660	0.055/0.080	6.7733	0.4638	0.2319	0.1476	0.0457	0.3409
4.0	12.566	6.283	1.220	9.220	0.060/0.085	6.3500	0.4947	0.2474	0.1575	0.0480	0.3630
4.25	13.352	6.676	1.270	9.770	0.060/0.085	5.9765	0.5256	0.2628	0.1673	0.0500	0.3846

4.50	14.137	7.069	1.320	10.320	0.060/0.085	5.6444	0.5566	0.2783	0.1772	0.0520	0.4063
4.75	14.923	7.461	1.370	10.870	0.065/0.090	5.3474	0.5875	0.2938	0.1870	0.0539	0.4280
5.0	15.708	7.854	1.420	11.420	0.065/0.090	5.0800	0.6184	0.3092	0.1969	0.0559	0.4496
5.25	16.493	8.247	1.470	11.970	0.065/0.090	4.8381	0.6493	0.3247	0.2067	0.0579	0.4713
5.50	17.278	8.639	1.520	12.520	0.070/0.095	4.6182	0.6803	0.3401	0.2165	0.0598	0.4929
5.75	18.064	9.032	1.560	13.060	0.070/0.095	4.4174	0.7112	0.3556	0.2264	0.0614	0.5142
6.0	18.850	9.425	1.610	13.610	0.070/0.095	4.2333	0.7421	0.3711	0.2362	0.0634	0.5358
6.5	20.420	10.210	1.710	14.710	0.070/0.095	3.9077	0.8040	0.4020	0.2559	0.0673	0.5791
7.0	21.991	10.996	1.820	15.820	0.075/0.100	3.6286	0.8658	0.4239	0.2756	0.0717	0.6228
7.5	23.562	11.781	1.920	16.920	0.075/0.100	3.3867	0.9276	0.4638	0.2953	0.0756	0.6661
8.0	25.133	12.566	2.020	18.020	0.075/0.100	3.1750	0.9895	0.4947	0.3150	0.0795	0.7094
9.0	28.274	14.137	2.240	20.240	0.080/0.105	2.8222	1.1132	0.5566	0.3543	0.0882	0.7968
10.0	31.416	15.708	2.440	22.440	0.080/0.105	2.5400	1.2368	0.6184	0.3937	0.0960	0.8835

* Based on standard tip radius on preshaving hobs and shaper cutters. Increase whole depth by approximately 0.130–0.250 mm (0.005–0.010 in.) for crown shaving.

5.1.12 Letter symbols for gear dimensions and calculations

a	addendum
b	dedendum
B	backlash
B_n	normal circular backlash
c	clearance
C	centre distance
d	diameter of measuring pin or ball
D	pitch diameter
D_b	base diameter
D_c	contact diameter
D_f	form diameter
D_F	fillet diameter
D_i	minor diameter (internal gears)
D_m	diameter of circle through centre of measuring pins or balls
D_{Me}	measurement over pins or balls (external)
D_{Mi}	measurement between pins or balls (internal)
D_o	outside diameter (O/D)
D_r	rolling or operating pitch diameter
D_R	root diameter
D_s	shaved diameter
F	face width
h_k	working depth
h_t	whole depth
L	lead
LA	total length of line of action
m	module (metric system)
m_F	face contact ratio (helical overlap)
m_n	normal module
m_p	involute contact ratio (involute overlap)
m_t	transverse module
N	number of teeth in gear (N_G) or pinion (N_p)
p	circular pitch
p_b	base pitch
p_n	normal circular pitch
p_t	transverse circular pitch
p_x	axial pitch
P	diametral pitch
P_n	normal diametral pitch
P_t	transverse diametral pitch
SAP	start of active profile
S	circular tooth space
SRP	start of radius profile
S_n	normal circular tooth space
S_t	transverse circular tooth space
t	circular tooth thickness (t_n, t_t, etc.)
t_c	chordal tooth thickness (t_{nc}, t_{tc}, etc.)
Z	length of contact

Angles

ϕ	pressure angle
ϕ_m	pressure angle to centre of measuring pin or ball
ϕ_n	normal pressure angle
ϕ_r	operating pressure angle (ϕ_{nr}, ϕ_{tr}, etc.)
ϕ_t	transverse pressure angle

ϕ_{to} transverse pressure angle at O/D
ϕ_x axial pressure angle
ψ helix angle
ψ_b base helix angle
ψ_r operating helix angle
ψ_o helix angle at O/D
X crossed axes angle

Notes

(a) The addition of an arc (∩) over the symbol for an angle indicates that the angle is in radians rather than degrees.
(b) Subscripts are used with symbols to differentiate between various diameters and angles and to indicate whether pinion or gear characteristics are involved.

Terminology

Transverse characteristics (subscript t) are taken in the plane of rotation, parallel to the gear face and perpendicular to the axis.
Normal characteristics (subscript n) are taken from a section of the gear teeth which is normal to the helix at a given diameter.
Axial plane characteristics (subscript x) are in a plane through the teeth and axis of the gear, perpendicular to the gear face.
Start of active profile (*SAP*) is the lowest point of mating gear contact as measured along the line or action in inches or degrees of roll from zero (base diameter).
Contact diameter (D_c) is the diameter through the lowest point of mating gear contact.
Form diameter (D_f) is the diameter through the lowest point on the gear profile where the desired involute tooth form is to start.
Start of radius profile (*SRP*) is the height of the generated root fillet on a gear as measured along the line of action from the base diameter.
Fillet diameter (D_F) is the diameter through the start of radius profile.
Shaved diameter (D_s) is the diameter through the lowest point of contact of the shaving cutter; that is, the start of the shaved profile of a gear.
Crossed axes angle (X) is the sum or difference of the gear and shaving cutter helix angles, dependent on the hand of the helix and centre distance. If the hands are opposite, the crossed axes angle will be equal to the difference between the helix angles; if the hands are the same, it will be equal to the sum.

5.1.13 Basic spur gear calculations

To find	in.	mm
Pitch diameter D	N/P	mN
Addendum a	$1/P$	Module m (in mm and parts mm)
Standard outside diameter D_o	$D + 2a$	$D + 2m$
Circular pitch p	π/P	πm
Average backlash per pair B	$0.040/P$	$0.040\,m$
Root diameter D_R	$D_o - 2h_t$	
Base diameter D_b	$D \cos \phi$	
Standard circular tooth thickness t	$p/2$	

Conversions

Diametral pitch $P = 25.4/m$
Module $m = 25.4/P$
millimetres = in./0.039 37 = 25.4 in.
inches = 0.039 37 mm = mm/25.4

5.1.14 Basic helical gear equations

Transverse diametral pitch $P_t = P_n \cos \psi$
Pitch diameter $D = N/P_t$
Addendum a standard $= 1/P_n$
Outside diameter $D_o = D + 2a$
Transverse pressure angle ϕ_t: $\tan \phi_t = \tan \phi_n / \cos \psi$
Base diameter $D_b = D \cos \phi_t$
Lead $L = \pi D \cot \psi = \pi D / \tan \psi$
Normal circular pitch $p_n = \pi / P_n$
Standard normal circular tooth thickness $t_n = p_n/2$
Axial pitch $p_x = \pi / P_n \sin \psi = p_n / \sin \psi = L/N$
Transverse circular pitch $p_t = \pi / P_t = p_n / \cos \psi$

5.1.15 Miscellaneous gear equations

1. Helix angle ψ, when given centre distance is standard:

$$\cos \psi = \frac{N_p + N_G}{2 p_n C}$$

2. Operating pitch diameter (pinion), with non-standard centre distance C:

$$D_{rp} = \frac{2 C N_p}{N_p + N_G}$$

3. Operating pressure angle ϕ_{rt}, with non-standard centre distance C:

$$\cos \phi_{rt} = \frac{D_{bp} + D_b^G}{2C}$$

4. Normal diametral pitch:

$$P_n = P_t \sec \psi$$

5. Helix angle ψ:

$$\cos \psi = \frac{N}{D P_n}$$

$$\sin \psi = \frac{\pi N}{P_n L}$$

At any diameter D_2:

$$\tan \psi_2 = \frac{D_2 \tan \psi_1}{D_1}$$

6. Transverse circular pitch p_{t2} at any diameter D_2:

$$p_{t2} = \frac{\pi D_2}{N}$$

7. Involute function of pressure angle (function tables are available):

$$\text{inv}\,\phi = \tan\phi - \phi$$

8. Normal pressure angle (ϕ_n):

$$\tan\phi_n = \tan\phi_t \cos\psi$$

9. Transverse pressure angle ϕ_t at any diameter D_2:

$$\cos\phi_{t2} = \frac{D_b}{D_2}$$

10. Base helix angle ψ_b:

$$\cos\psi_b = \frac{\cos\psi\cos\phi_n}{\cos\phi_t} = \frac{\sin\phi_n}{\sin\phi_t}$$

$$\sin\psi_b = \sin\psi\cos\phi_n$$

$$\tan\psi_b = \tan\psi\cos\phi_t$$

11. Base pitch:

$$p_b = \frac{\pi D_b}{N} = p\cos\phi$$

5.1.16 Straight bevel gear nomenclature

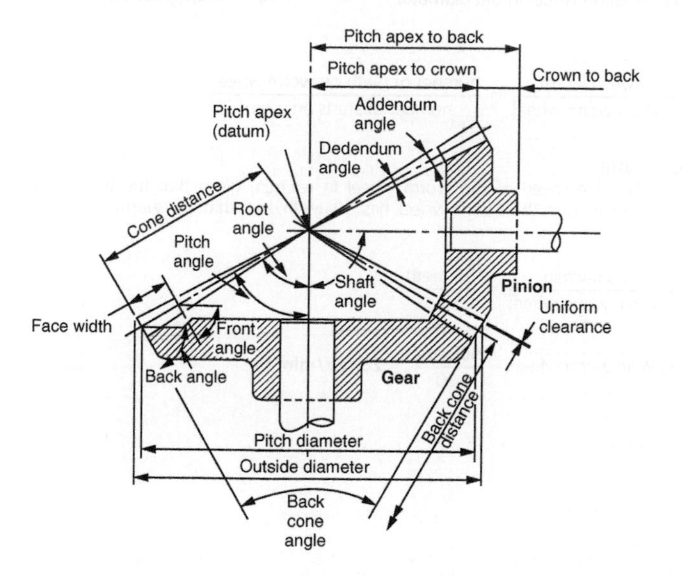

5.1.17 Worm and worm wheel nomenclature

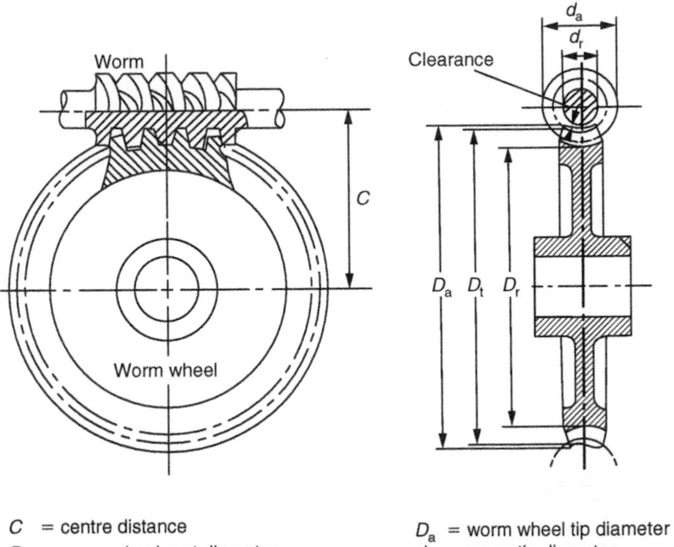

C = centre distance
D_r = worm wheel root diameter
D_t = worm wheel throat diameter
D_a = worm wheel tip diameter
d_a = worm tip diameter
d_r = worm root diameter

$$\frac{\text{rev/min worm}}{\text{rev/min worm wheel}} = \frac{\text{number of teeth on worm wheel}}{\text{number of starts on worm}}$$

Example

Calculate the speed of the worm wheel in rev/min, given that the worm rotates at 500 rev/min, that the worm wheel has 50 teeth and that the worm has a two-start helix.

$$\frac{500 \text{ rev/min}}{\text{rev/min worm wheel}} = \frac{50 \text{ teeth}}{2 \text{ starts}}$$

$$\text{rev/min worm wheel} = \frac{500 \times 2}{50} = \mathbf{20\ rev/min}$$

5.2 Power transmission: belt drives

5.2.1 Simple flat-belt drives

Open-belt drive

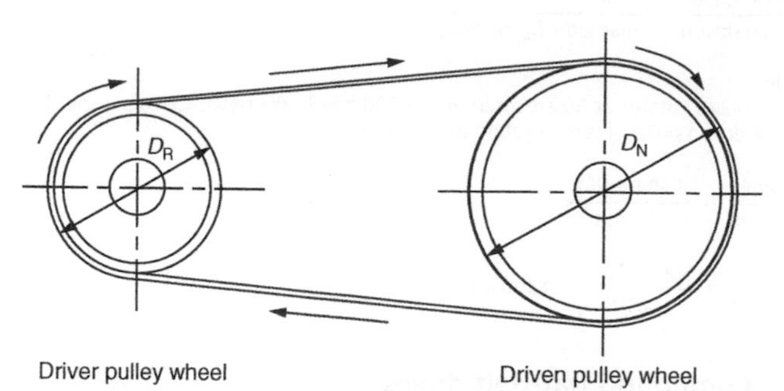

(a) Driver and driven pulley wheels rotate in the same direction.
(b) The relative speed of the pulley wheels is calculated by the expression:

$$\frac{\text{rev/min driver}}{\text{rev/min driven}} = \frac{\text{diameter } D_N \text{ of driven}}{\text{diameter } D_R \text{ of driver}}$$

Example
Calculate the speed in rev/min of the driven pulley if the driver rotates at 200 rev/min. Diameter D_R is 500 mm and diameter D_N is 800 mm.

$$\frac{200 \text{ rev/min}}{\text{rev/min driven}} = \frac{800 \text{ mm}}{500 \text{ mm}}$$

$$\text{rev/min driven} = \frac{200 \times 500}{800} = \mathbf{125 \text{ rev / min}}$$

Crossed-belt drive

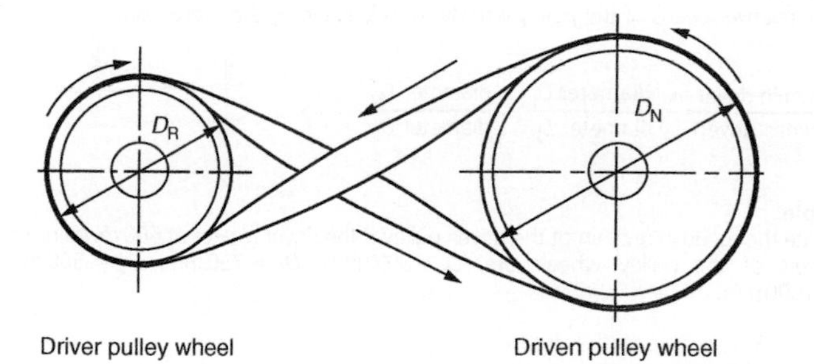

(a) Driver and driven pulley wheels rotate in opposite directions.
(b) Crossed-belt drives can only be used with flat section belts (long centre distances) or circular section belts (short centre distances).
(c) The relative speed of the pulley wheels is again calculated by the expression:

$$\frac{\text{rev/min driver}}{\text{rev/min driven}} = \frac{\text{diameter } D_N \text{ of driven}}{\text{diameter } D_R \text{ of driver}}$$

Example
The driver pulley rotates at 500 rev/min and is 600 mm in diameter. Calculate the diameter of the driven pulley if it is to rotate at 250 rev/min.

$$\frac{500 \text{ rev/min}}{250 \text{ rev/min}} = \frac{\text{diameter } D_N}{600 \text{ mm}}$$

$$\text{diameter } D_N = \frac{500 \times 600}{250} = \mathbf{1200\,mm}$$

5.2.2 Compound flat-belt drives

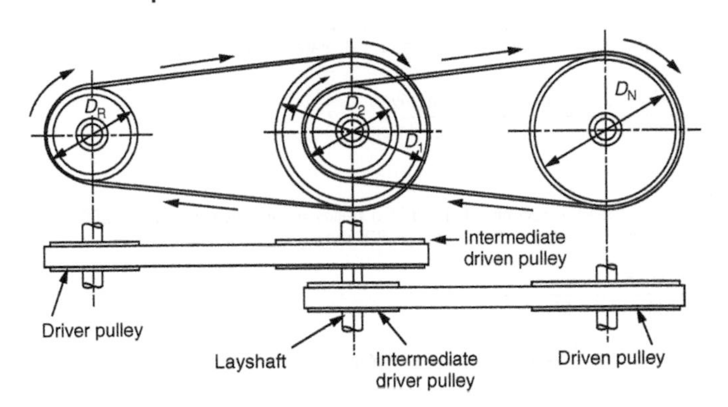

(1) To identify the direction of rotation, the rules for open and crossed-belt drives apply (Section 5.2.1).
(2) The relative speeds of the pulley wheels are calculated by the expression:

$$\frac{\text{rev/min driver}}{\text{rev/min driven}} = \frac{\text{diameter } D_1}{\text{diameter } D_R} \times \frac{\text{diameter } D_N}{\text{diameter } D_2}$$

Example
Calculate the speed in rev/min of the driven pulley if the driver rotates at 600 rev/min. The diameters of the pulley wheels are: $D_R = 250$ mm, $D_1 = 750$ mm, $D_2 = 500$ mm, $D_N = 1000$ mm.

$$\frac{600 \text{ rev/min}}{\text{rev/min driven}} = \frac{750 \text{ mm}}{250 \text{ mm}} \times \frac{1000 \text{ mm}}{500 \text{ mm}}$$

$$\text{rev/min driven} = \frac{600 \times 250 \times 500}{750 \times 1000} = \mathbf{100\ rev/min}$$

5.2.3 Typical belt tensioning devices

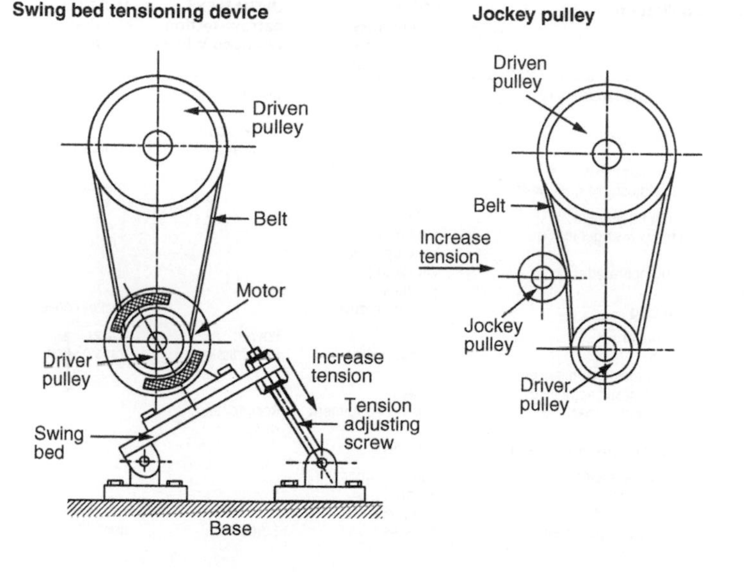

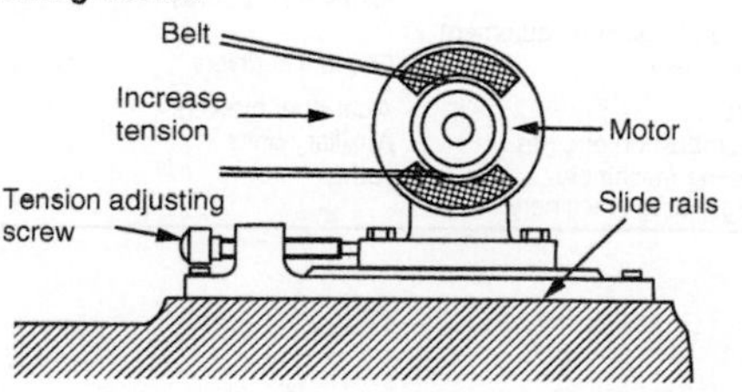

5.2.4 Typical V-belt and synchronous-belt drive applications

Application	FO®-Z heavy-duty cogged raw edge V-belts	ULTRAFLEX® narrow-section wrapped V-belts	MULTIFLEX® classical-section wrapped V-belts
Construction machinery	Soil-compacting equipment	Concrete mixers	Impact crushers
Mining		Underground application	Underground application
Office equipment	Paper shredders		
Speed adjustable gear units			Interlocking pulleys
Printing equipment	Rotary presses		
Electric power generation	Emergency back-up systems		
Electric-operated tools	Oscillating grinders		
Conveying	Conveyor drives		Bucket conveyors
Industrial drives	Power transmissions	Power transmissions	Power transmissions
Large domestic appliances			
Small domestic appliances	Mixers		
Woodworking machinery	Planing equipment	Reciprocating saws	
Rubber and plastics processing	Extruders		
Automotive sector	Auxiliary units		
Compressors	Piston compressors		
Agricultural equipment	Blowers	Spreaders	Beet lifters
Motorcycles			
Paper-making machinery	Drying cylinders		
Pumps	Radial-flow pumps		
Lawn care and cleaning equipment			
Textile machinery	Cylindrical dryers		
Ventilators	Axial-flow blowers		
Internal combustion engines	Auxiliary units		
Metalworking machinery	Lathes		
Crushing/grinding machinery			

CONTI-V MULTIRIB® multiple V-ribbed belts	**VARIDUR®-Z** ***Variable*** **speed cogged raw edge belts**
Franking machines	
Offset machines	Compact units Multicolour offset
Hand-held planes Lift door mechanisms	
	Adjusting pulley sets
Washing machines Floor polishers Millers	
Auxiliary units	
	Threshing cylinders Automatic transmission
	Lawnmower drives Spoolers
Auxiliary units Main spindle drives Shredders	Lathes

continued

Section 5.2.4 (*continued*)

Application	VARIFLEX®-Z variable speed cogged belts
Office equipment Data processing Speed adjustable gear units	
Printing equipment	Letterpress printing machines
Electric-operated tools Film projectors Conveying	
Industrial drives Small domestic appliances Woodworking machinery	Adjusting pulley sets
Automotive sector Sewing machines	
Lawn care and cleaning equipment Robotics Textile machinery	
Ventilators Internal combustion engines Packaging machines	
Metalworking machinery	

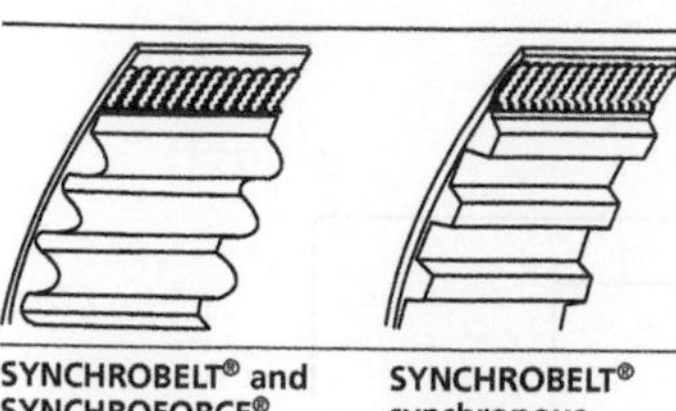

SYNCHROBELT® and SYNCHROFORCE® high torque drive belts	SYNCHROBELT® synchronous drive belts
Photocopiers Hard-copy printers	Typewriters Plotters
Rotary presses	
Hand-held planes	Belt grinders Reel drives Lift control systems
Reduction gears Domestic appliances Adhesive bonding equipment	Reduction gear Turbo-brushers
Camshaft drives	Camshaft drives Looper drives
Cultivators Positioning drives Weaving machines	Positioning drives
Main drives Camshaft drives	Camshaft drives Counters
Power feed drives	

All data is supplied by ContiTech P.T.S. Ltd whose technical publications should be consulted for full product range, applications and technical data. (See Appendix 3 for full details.)
Note: CONTICORD® flat belts and CON-FLEX® driving belts, are no longer available.

5.2.5 ULTRAFLEX® narrow-section wrapped V-belts

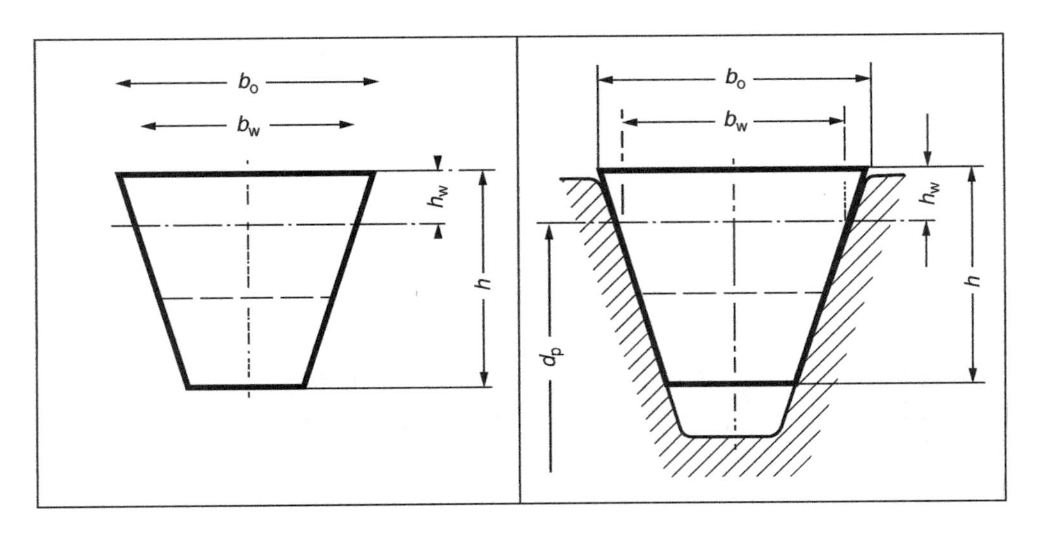

Belt section	DIN symbol	SPZ	SPA	SPB	SPC	19
	BS/ISO symbol	SPZ	SPA	SPB	SPC	–
Top belt width b_o≈	mm	9.7	12.7	16.3	22	18.6
Pitch width b_w	mm	8.5	11	14	19	16
Bottom belt width b_u≈	mm	4	5.6	7.1	9.3	8
Height of belt h≈	mm	8	10	13	18	15
Pitch height h_w≈	mm	2	2.8	3.5	4.8	4
Min. pulley pitch diameter $d_{w\,min}$	mm	63	90	140	224	180
Max. flexing frequency $f_{B\,max}$	per s	100	100	100	100	100
Max. belt speed V_{max}	m/s	40	40	40	40	40
Weight per metre	kg/m	0.073	0.100	0.178	0.380	0.360
Range of pitch length L_w:						
From	mm	512	647	1250	2000	1175
To	mm	3550	4500	8000	12250	5000
Length differential value from L_w:						
$\Delta L = L_a - L_w$	mm	13	18	22	30	25

ULTRAFLEX® narrow-section wrapped V-belts complying with BS 3790 and DIN standard 7753 Part 1 are used in demanding drive systems in all spheres of mechanical engineering. They have high-power transmission capacity and economic efficiency coupled with a long service life.

5.2.6 FO®-Z heavy-duty cogged raw edge V-belts

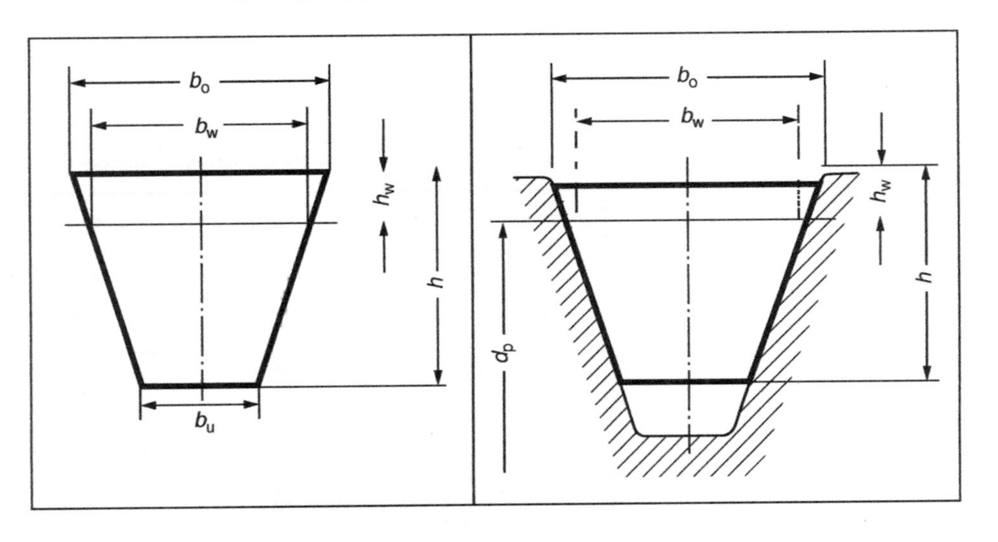

Belt section	DIN symbol	5	6	SPZ	SPA	SPB
	BS/ISO symbol	–	Y	SPZ	SPA	SPB
Top belt width $b_o \approx$	mm	5	6	10	13	16.3
Pitch width b_w	mm	4.2	5.3	8.5	11	14
Height of belt $h \approx$	mm	3	4	8	9	13
Pitch height $h_w \approx$	mm	1.3	1.6	2	2.8	3.5
Min. pulley pitch diameter $d_{w\,min}$	mm	16	20	50	63	100
Max. flexing frequency $f_{B\,max}$	per s	120	120	120	120	120
Max. belt speed v_{max}	m/s	50	50	50	50	50
Weight per metre	kg/m	0.015	0.023	0.073	0.100	0.178
Range of pitch length L_w:						
From	mm	171	295	590	590	1250
To	mm	611	865	3550	3550	3550
Length differential value from L_w:						
$\Delta L = L_w - L_i$	mm	11	15	–	–	–

FO®-Z heavy-duty cogged V-belts are manufactured in a raw edge type and supplement the proven range of CONTI® V-belts. They comply in their dimensions as narrow-section V-belts with DIN standard 7753 Part 1 and BS 3790 and as standard V-belts with DIN standard 2215 and BS 3790.

5.2.7 MULTIFLEX® classical-section wrapped V-belts

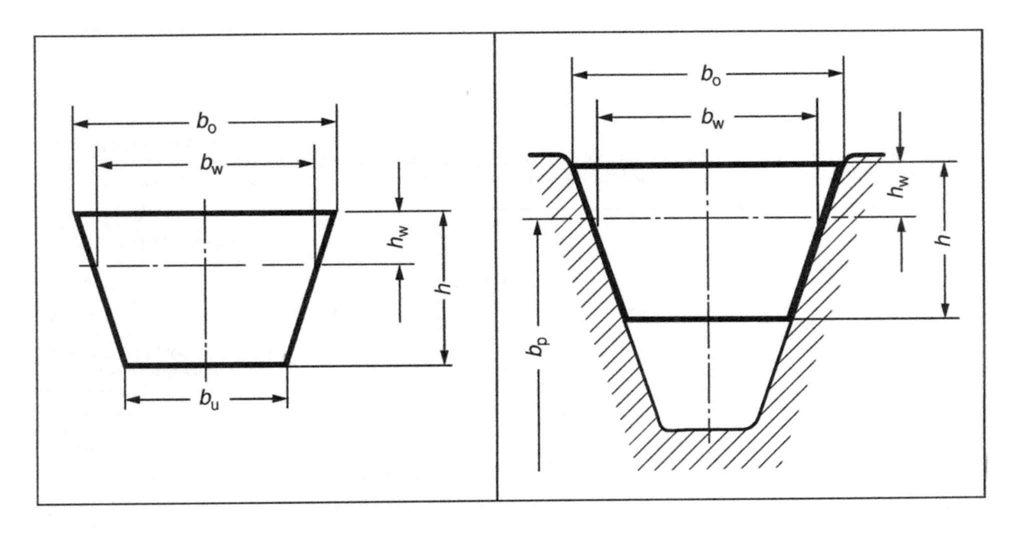

Belt section	DIN symbol	8	10	13	17	20	22	25	32	40
	BS/ISO symbol	–	Z	A	B	–	C	–	D	E
Top belt width $b_o \approx$	mm	8	10	13	17	20	22	25	32	40
Pitch width b_w	mm	6.7	8.5	11	14	17	19	21	27	32
Bottom belt width $b_u \approx$	mm	4.6	5.9	7.5	9.4	11.4	12.4	14	18.3	22.8
Height of belt $h \approx$	mm	5	6	8	11	12.5	14	16	20	25
Pitch height $h_w \approx$	mm	2	2.5	3.3	4.2	4.8	5.7	6.3	8.1	12
Min. pulley pitch diameter $d_{w\ min}$	mm	40	50	80	125	160	200	250	355	500
Max. flexing frequency $f_{B\ max}$	per s	60	60	60	60	60	60	60	60	60
Max. belt speed V_{max}	m/s	30	30	30	30	30	30	30	30	30
Weight per metre	kg/m	0.040	0.060	0.105	0.170	0.240	0.300	0.430	0.630	0.970
Range of pitch length L_w:										
From	mm	160	375	400	615	900	1150	1400	2080	5080
To	mm	1250	2500	5000	8763	8000	8000	9000	13460	11280
Length differential value from L_i:										
$\Delta L = L_w - L_i$	mm	19	22	30	43	48	52	61	75	82

MULTIFLEX® classical-section wrapped V-belts complying with BS 3790 and DIN standard 2215 are designed for all industrial applications from precision engineering to heavy machine construction.

5.2.8 MULTIBELT banded V-belts

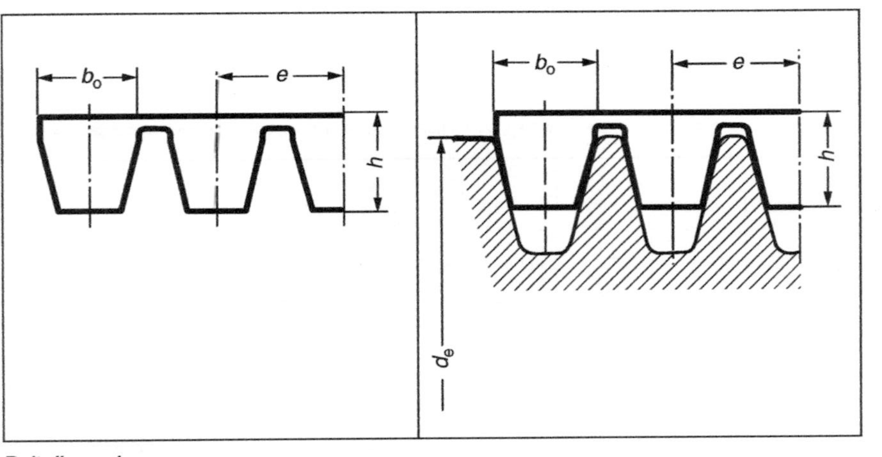

Belt dimensions

MULTIBELT banded V-belts form a drive unit combining the advantage of a single-belt with the increased power transmission capacity of a multi-groove V-belt set. This drive unit helps solve problems relating to drives that formerly involve extensive constructional expenditure.

Dimensional data

Belt section	DIN/ISO symbol	9J	SPA	15J	17/B	19	SPC
	RMA/ASAE symbol	3V	–	5V	B/HB	–	–
Top section width $b_o \approx$	mm	9	12.5	15	16	19	22
Height of belt $h \approx$	mm	11	12	15	13	21	23
Lateral rib pitch e	mm	10.3	15	17.5	19.05	22	25.5
Min. pulley pitch diameter $d_{e\,min}$	mm	67	95.6	180	100	188	265
Max. flexing frequency $f_{B\,max}$	s^{-1}	60	60	60	60	60	60
Max. belt speed v_{max}	m/s	30	30	30	30	30	30
Weight per rib per metre	kg/m	0.115	0.170	0.250	0.240	0.405	0.505
Range-outside length L_a:							
From	mm	750	1350	1350	1350	2500	2500
To	mm	2800	2800	5600	5600	5600	5600
Length differential value from L_a:		*	–	*	–	–	–
$\Delta L = L_a - L_i$	mm	–	–	–	62	–	–
$\Delta L = L_a - L_w$	mm	–	18	–	–	25	30

* The reference length L_a is the designated length for sections 9J and 15J.

5.2.9 V-belt pulleys complying with BS 3790 and DIN standard 2211 for FO®-Z and ULTRAFLEX® belts. R_z xx refers to surface roughness

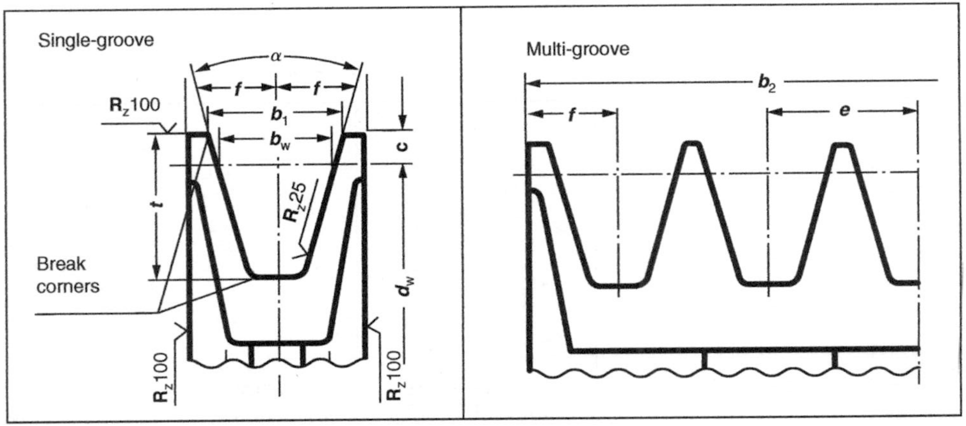

Dimensions in mm

Belt section						
To DIN 7753 Part 1 and BS 3790	**DIN symbol** **BS/ISO symbol**	**SPZ** **SPZ**	**SPA** **SPA**	**SPB** **SPB**	**SPC** **SPC**	**19** **–**
To DIN 2215 and BS 3790	**DIN symbol** **BS/ISO symbol**	**10** **Z**	**13** **A**	**17** **B**	**22** **C**	**–** **–**
Pitch width	b_w	8.5	11	14	19	16
Top groove width	$b_1\approx$	9.7	12.7	16.3	22	18.6
	c	2	2.8	3.5	4.8	4
Groove spacing	e	$12_{\pm 0.3}$	$15_{\pm 0.3}$	$19_{\pm 0.4}$	$25.5_{\pm 0.5}$	$22_{\pm 0.4}$
	f	$8_{\pm 0.6}$	$10_{\pm 0.6}$	$12.5_{\pm 0.8}$	$17_{\pm 1}$	$14.5_{\pm 0.8}$
Groove depth	t	$11^{+0.6}_{0}$	$14^{+0.6}_{0}$	$18^{+0.6}_{0}$	$24^{+0.6}_{0}$	$20^{+0.6}_{0}$
x | 34 | for pitch diameter	d_w	$\leqslant 80$	$\leqslant 118$	$\leqslant 190$	$\leqslant 315$	$\leqslant 250$
x | 38 | for pitch diameter		>80	>118	>190	>315	>250
Tolerance for $\alpha = 34 - 38$		$\pm 1°$	$\pm 1°$	$\pm 1°$	$\pm 0.5°$	$\pm 1°$
Pulley face width b_2 for number of grooves z: $b_2 = (z - 1)e + 2f$	1	16	20	25	34	29
	2	28	35	44	59.5	51
	3	40	50	63	85	73
	4	52	65	82	110.5	95
	5	64	80	101	136	117
	6	76	95	120	161.5	139
	7	88	110	139	187	161
	8	100	125	158	212.5	183
	9	112	140	177	238	205
	10	124	155	196	263.5	227
	11	136	170	215	289	249
	12	148	185	234	314.5	271

5.2.10 V-belt pulleys complying with DIN standard 2217 Part 1 for FO®-Z and MULTIFLEX® belts

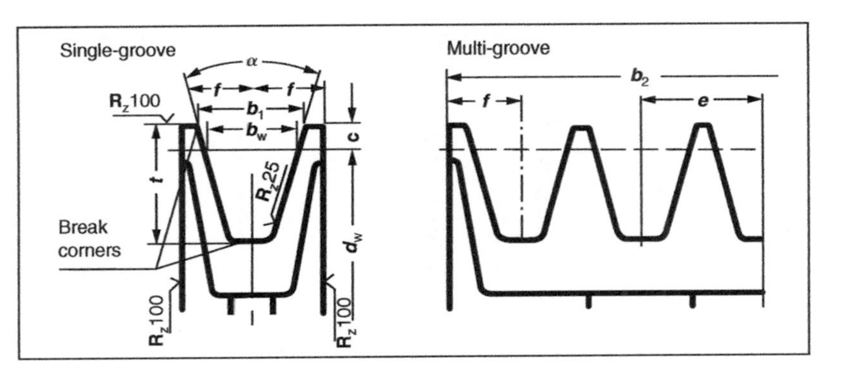

Dimensions in mm

Belt section												
To DIN 2215:*	**DIN symbol BS/ISO symbol**	**5** –	**6** **Y**	**(8)** –	**10** **Z**	**13** **A**	**17** **B**	**(20)** –	**22** **C**	**(25)** –	**32** **D**	**40** **E**
Alternative section (DIN 2211 Part 1) to BS 3790	**BS/ISO symbol**	–	–	–	**SPZ**	**SPA**	**SPB**	–	**SPC**	–	–	–
Pitch width	b_w	4.2	5.3	6.7	8.5	11	14	17	19	21	27	32
Top grove width	$b_1 \approx$	5	6.3	8		Pulleys for narrow-section V-belts (DIN 2211 Part 1 and BS 3790) must be used for drives with these sections		20	As columns SPZ SPA SPB	25	32	40
	c	1.3	1.6	2				5.1		6.3	8.1	12
Groove spacing	e	$6_{\pm 0.3}$	$8_{\pm 0.3}$	$10_{\pm 0.3}$				$23_{\pm 0.4}$		$29_{\pm 0.5}$	$37_{\pm 0.6}$	$44.5_{\pm 0.3}$
	f	$5_{\pm 0.5}$	$6_{\pm 0.5}$	$7_{\pm 0.6}$				$15_{\pm 0.8}$		$19_{\pm 1}$	$24_{\pm 2}$	$29_{\pm 2}$
Groove depth	t	$6^{+0.6}_{0}$	$7^{+0.6}_{0}$	$9^{+0.6}_{0}$				$18^{+0.6}_{0}$		$22^{+0.6}_{0}$	$28^{+0.6}_{0}$	$33^{+0.6}_{0}$
x 32 for pitch diameter	d_w	≤50	≤63	≤75				–		–	–	–
x 34		–	–	–				≤250		≤355	–	–
x 36		>50	>63	≥75				–		–	≤500	≤630
x 38		–	–	–				>250		>355	>500	>630
Tolerance for		±1	±1	±1				±1		≤30′	±30′	±30′
$x = 32 - 38$ Pulley face width b_2 for number of grooves: $b_2 = (Z - 1)e + 2$	1	10	12	14				30		38	48	58
	2	16	20	24				53		67	85	102.5
	3	22	28	34				76		96	122	147
	4	28	36	44				99		125	159	191.5
	5	34	44	54				122		154	196	236
	6	40	52	64				145		183	233	280.5
	7		60	74				168		212	270	325
	8			84				191		241	307	369.5
	9							214		270	344	414
	10							237		299	381	458.5
	11							260		328	418	503
	12							283		357	455	547.5

*Sections in brackets should not be used for new constructions.

5.2.11 Deep-groove pulleys

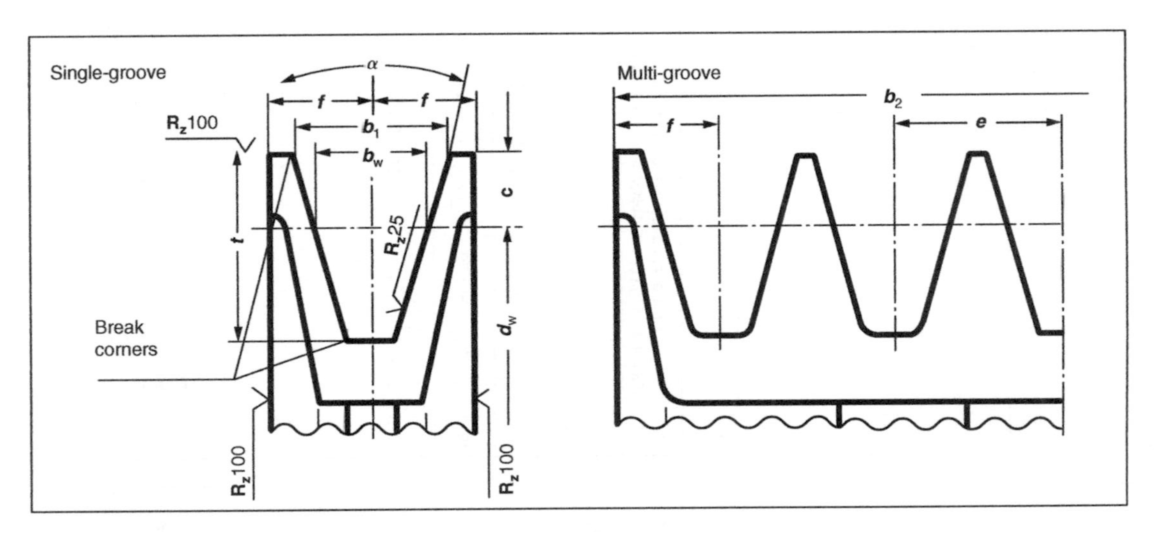

Dimensions in mm

Belt section						
To DIN 7753 Part 1 and BS 3790	DIN symbol BS/ISO symbol	SPZ SPZ	SPA SPA	SPB SPB	SPC SPC	19 –
To DIN 2215 and BS 3790	DIN symbol BS/ISO symbol	10 Z	13 A	17 B	22 C	– –
Pitch width	b_w	8.5	11	14	19	16
Increased groove width $b_1 \approx$	$\alpha = 34°$	11	15	18.9	26.3	22.1
	$\alpha = 38°$	11.3	15.4	19.5	27.3	22.9
	c	4	6.5	8	12	10
Groove spacing	e	$14_{\pm 0.3}$	$18_{\pm 0.3}$	$23_{\pm 0.4}$	$31_{\pm 0.5}$	$27_{\pm 0.5}$
	f	$9_{\pm 0.6}$	$11.5_{\pm 0.6}$	$14.5_{\pm 0.8}$	$20_{\pm 1}$	$17_{\pm 1}$
Increased groove depth	t_{min}	13	18	22.5	31.5	26
α | 34° | for pitch diameter d_w		63–80	90–118	140–190	224–315	180–250
| 38° | with belts to DIN 7753 1 and BS 3790		>80	>118	>190	>315	>250
α | 34° | for pitch diameter d_w		50–80	71–118	112–190	180–315	–
| 38° | with belts to DIN 2215 and BS 3790		>80	>118	>190	>315	>250
Tolerance for $\alpha = 34°–38°$		±1°	±1°	±1°	±0.5°	±1°
Pulley face width b_2 for number of grooves z: $b_2 = (z - 1)e + 2f$	1	18	23	29	40	34
	2	32	41	52	71	61
	3	46	59	75	102	88
	4	60	77	98	133	115
	5	74	95	121	164	142
	6	88	113	144	195	169
	7	102	131	167	226	196
	8	116	149	190	257	223
	9	130	167	213	288	250
	10	144	185	236	319	277
	11	158	203	259	350	304
	12	172	221	282	381	331

Minimum pulley diameter must be adhered to. Not to be used for banded V-belts.

5.2.12 Synchronous belt drives: introduction

Synchronous (toothed) belt drives are now widely used in place of traditional, roller chain drives for many applications. Unlike a flat belt or a V-belt the toothed belt cannot slip, therefore it can be used where the rotation of input (driver) and output (driven) elements of a system must always be synchronized.

The main advantages of synchronous-belt drives compared with traditional roller chain drives are as follows:

1. Substantially lower cost.
2. Quieter running.
3. Ability to operate in environments which would be hostile to a roller chain drive.
4. No need for lubrication of the drive.
5. The elastomer material from which the belt is made tends to absorb vibrations rather than transmit them.

Synchronous-belt drive applications are found where their special properties can be exploited. For example:

1. Office equipment where quiet running and lack of lubrication is important.
2. Food processing machinery where conventional lubrication, necessary with a chain drive, might contaminate the foodstuffs being processed.
3. Motor vehicle camshaft drives where synchronous, trouble-free, quiet, lubrication-free, smooth running is required.
4. The coupling of stepper motors and servo-motors to the feed mechanisms of computer controlled machine tools where synchronous, trouble-free, vibration-free, smooth running is required.

The following tables of synchronous-belt data and associated toothed pulley data have been selected to assist in the design of synchronous-belt drive systems.

Construction

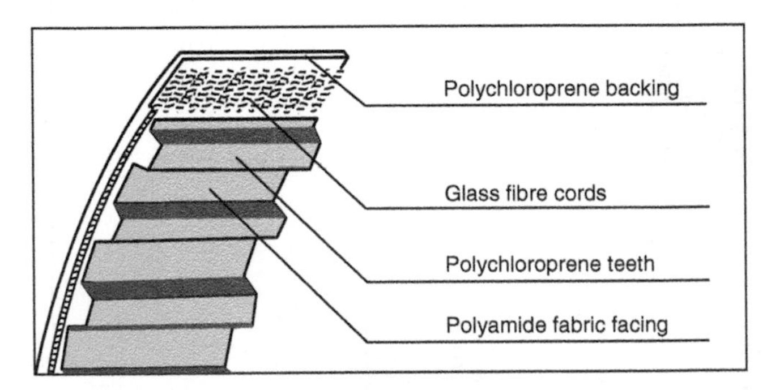

Synchronous-drive belts are composed of:

- polychloroprene teeth and backing,
- polyamide fabric facing,
- glass fibre cords.

Polychloroprene teeth and backing

The teeth and belt backing are made from highly loadable polychloroprene-based elastomer materials, firmly bonded to the tension member and fabric facing.

Polyamide fabric facing
Lasting protection of the teeth is an essential precondition for smooth drive operation and a long belt life. This is guaranteed by using tough, wear-resistant polyamide fabric with a low coefficient of friction.

Glass fibre cords
Synchronous drives call for maximum length stability and tensile strength of the belt. These requirements are optimally met by low-stretch glass fibre cords helically wound over the entire belt width. Lateral mistracking is minimized by using S/Z cords arranged in pairs.

Properties

Constant angular velocity and uniform speed transmission
In much the same manner as the teeth on a gear, the teeth of the belt make positive engagement with the mating tooth spaces on the pulleys. The positive-grip drive principle ensures synchronous operation and constant circumferential speed.

Freedom from high tension
The tooth forming principle requires only a very low belt tensioning and so the load on axles and bearings is kept to a minimum.

High-power transmission
High-power transmission is guaranteed by the combination of the extra-stiff teeth and the wear-resistant fabric facing as well as by the tension member's high resistance to dynamic load.

High-speed ratio
Reliable drive function is achieved through positive engagement, even with small arcs of contact and small pulley diameters.

Minimal space requirements
High dynamic loadability and high-power transmission capacity allow the use of small pulley diameters and short centre distances. This enables designers to design economical drives which are not only compact but also light in weight.

High belt speed
Low inertia forces and outstanding flex life ensure reliable drive systems up to belt speeds of 60 m/s.

Quiet operation
The supple belt design with fabric-faced rubber teeth in combination with metal or plastic pulleys reduces drive noise to a minimum.

Freedom from lubrication and maintenance
SYNCHROBELT® synchronous drive belts are maintenance free. They need no lubrication or retensioning. Constant belt tension is guaranteed by the use of high-strength glass fibre cords as the load-bearing element.

High efficiency
The flexible and supple belt design as well as optimum dimensional match between the belt toothing and pulley tooth spacing allow virtually friction-free drives with an efficiency of 98%.

CONTI SYNCHROBELT® synchronous drive belts
CONTI SYNCHROBELT® synchronous drive belts have the following standard properties. They are resistant to:

- certain oils
- ozone
- temperatures ranging from −40°C to 100°C and are
- unaffected by tropical climates
- insensitive to weathering.

Designation

SYNCHROBELT® synchronous drive belts are fully specified by a coding system based on DIN ISO 5296 and showing the following:

- Pitch length in tenths of an inch.
- The pitch length of a synchronous drive belt is equal to its overall circumference, measured along the pitch line which keeps the same length when the belt is bent. The pitch line lies in the centre of the tension member and can only be precisely located with the aid of suitable measuring fixtures. More details are given Section 5.2.17 under 'Length measurement'.
- Tooth pitch.
- The tooth pitch is the linear distance between two adjacent teeth along the pitch line.
- Belt width in hundredths of an inch.
- Belt width and belt width reference are identical.

Example

SYNCHROBELT® synchronous drive belt 1100 H 100
1100→pitch length 110 in. = 2794.0 mm
H→tooth pitch 0.5 in. = 12.7 mm
100→belt width 1 in. = 25.4 mm

The number of teeth z is the function of pitch length L_W and pitch t:

$$z = \frac{Lw}{t}$$

Available sizes

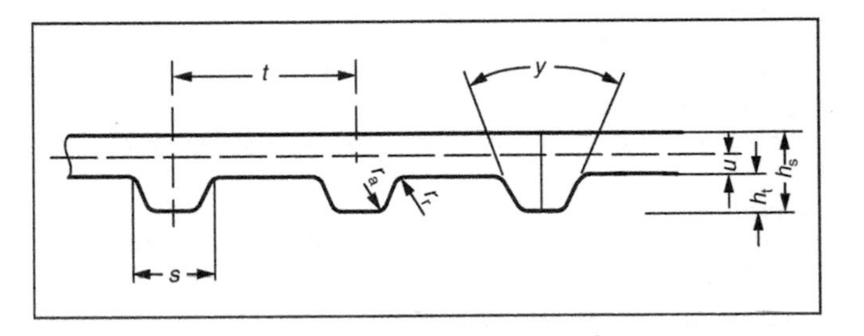

Cross-section of synchronous drive belt.

Pitches

SYNCHROBELT® synchronous drive belts are supplied in six tooth pitch sizes. They comply with DIN ISO 5296 standard and can be used internationally. Stock pitches and their dimensions are shown in the above table.

Lengths

SYNCHROBELT® synchronous drive belts are available in lengths to cover a broad range of applications. In addition, special belt lengths can be furnished on a made-to-order basis. For stock lengths, refer to tables in MXL pitch, XL pitch, L pitch, H pitch, XH pitch and XXH pitch on pages 396–402.

Widths

Stock widths are contained in the tables for stock lengths. Non-stock widths are also available on request.

Parameters

Pitch	DIN ISO code	MXL	XL	L	H	XH	XXH
Tooth pitch t	mm	2.032	5.080	9.525	12.700	22.225	31.750
	in.	0.080	0.200	0.375	0.500	0.875	1.250
Flank angle γ	degrees	40	50	40	40	40	40
Belt thickness h_s	mm	1.14	2.3	3.6	4.3	11.2	15.7
Tooth height h_t	mm	0.51	1.27	1.91	2.29	6.35	9.53
Top width of tooth s	mm	1.14	2.57	4.65	6.12	12.57	19.05
Top radius r_a	mm	0.13	0.38	0.51	1.02	1.19	1.52
Bottom radius r_r	mm	0.13	0.38	0.51	1.02	1.57	2.29
Pitch zone u	mm	0.254	0.254	0.381	0.686	1.397	1.524
Weight (belt width 25.4 mm)	kg/m	0.013	0.016	0.089	0.117	0.235	0.321
Range of pitch lengths L_w:							
From	mm	109.73	152.40	314.33	609.60	1289.05	1778.00
To	mm	920.50	1473.20	1524.00	4318.00	4445.00	4572.00
Stock widths b:							
From	mm	3.18	6.35	12.7	19.05	50.8	50.8
To	mm	6.35	25.4	76.2	127.0	177.8	127.0

Belts with special characteristics

SYNCHROBELT® synchronous drive belts have the standard properties listed on Page 6. Consult our engineers for special types.

5.2.13 Synchronous belt drives: belt types and sizes

MXL pitch

Tooth pitch 2.032 mm (0.080 in.)

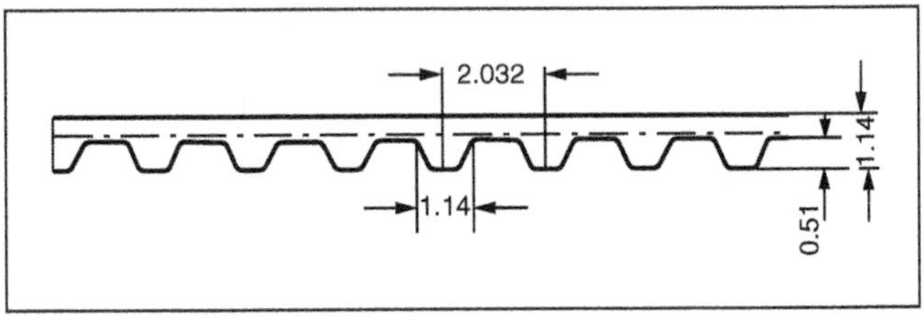

2.032
1.14
1.14
0.51

Stock lengths

Length code	Pitch length L_w (mm)	No. of teeth z	Length code	Pitch length L_w (mm)	No. of teeth z
43.2 MXL	109.73	54	91.2 MXL	231.65	114
44.0 MXL	111.76	55	96.0 MXL*	243.84	120
44.8 MXL*	113.79	56	97.6 MXL	247.90	122
46.4 MXL*	117.86	58	98.4 MXL*	249.94	123
48.0 MXL	121.92	60	100.8 MXL	256.06	126
54.5 MXL	138.18	68	112.0 MXL*	284.48	140
56.0 MXL*	142.24	70	120.0 MXL*	304.80	150
56.8 MXL	144.27	71	132.0 MXL*	335.28	165
57.6 MXL*	146.30	72	136.0 MXL	345.44	170
60.0 MXL*	152.40	75	140.0 MXL	355.60	175
64.0 MXL*	162.56	80	144.0 MXL	365.76	180
65.6 MXL*	166.62	82	147.2 MXL	373.89	184
67.2 MXL	170.69	84	180.0 MXL*	457.20	225
68.0 MXL*	172.72	85	188.8 MXL*	479.55	236
70.4 MXL*	178.81	88	200.8 MXL*	510.03	251
72.0 MXL	182.88	90	204.8 MXL*	520.19	256
75.2 MXL	191.01	94	238.4 MXL*	605.54	298
77.6 MXL*	197.10	97	277.6 MXL	705.10	347
80.0 MXL*	203.20	100	282.4 MXL*	717.30	353
80.8 MXL*	205.23	101	292.0 MXL*	741.68	365
82.4 MXL*	209.30	103	320.0 MXL*	812.80	400
84.0 MXL	213.36	105	347.2 MXL*	881.89	434
84.8 MXL*	215.39	106	362.4 MXL*	920.50	453
88.0 MXL	223.52	110	370.4 MXL*	940.82	463
89.6 MXL*	227.58	112	398.4 MXL*	1011.94	498
90.4 MXL	229.62	113	404.0 MXL*	1026.16	505

* Made-to-order belts.

Stock widths

Width reference	012	019	025
Width (mm)	3.18	4.76	6.35

XL pitch

Tooth pitch 5.080 mm (0.200 in.)

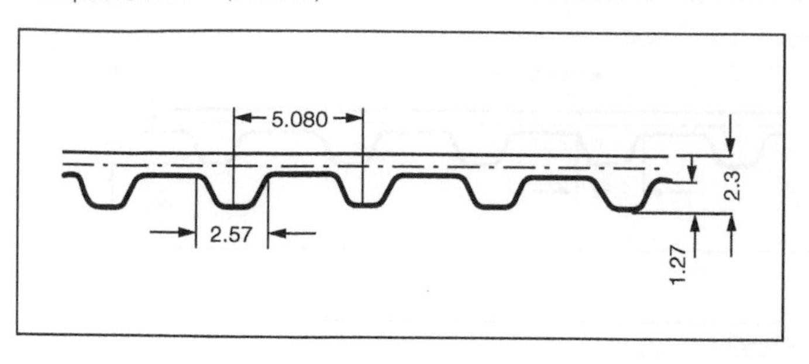

Stock lengths

Length code	Pitch length L_w (mm)	No. of teeth z	Length code	Pitch length L_w (mm)	No. of teeth z
60 XL	152.40	30	136 XL*	345.44	68
70 XL	177.80	35	138 XL*	350.62	69
80 XL	203.20	40	140 XL	355.60	70
86 XL*	218.44	43	148 XL*	376.92	74
90 XL	228.60	45	150 XL	381.00	75
92 XL*	233.68	46	156 XL*	396.24	78
94 XL*	238.76	47	160 XL	406.40	80
96 XL*	243.84	48	162 XL*	411.48	81
100 XL	254.00	50	168 XL*	426.72	84
102 XL*	259.08	51	170 XL	431.80	85
106 XL*	269.24	53	174 XL*	441.96	87
108 XL*	274.32	54	176 XL*	447.04	88
110 XL	279.40	55	178 XL*	452.12	89
112 XL*	284.48	56	180 XL	457.20	90
116 XL*	294.84	58	182 XL*	462.28	91
118 XL*	299.72	59	184 XL*	467.36	92
120 XL	304.80	60	188 XL*	477.52	94
124 XL*	314.96	62	190 XL	482.60	95
126 XL*	320.04	63	196 XL*	497.82	98
130 XL	330.20	65	200 XL	508.00	100
134 XL*	340.36	67			

* Made-to-order belts.

Stock widths

Width reference Width (mm)	025 6.35	031 7.94	037 9.53	050 12.7	075 19.05	100 25.4

XL pitch (continued)
Tooth pitch 5.080 mm (0.200 in.)

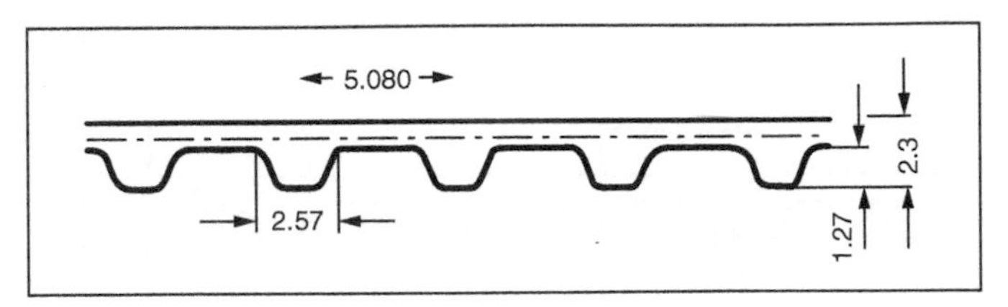

Stock lengths

Length code	Pitch length L_w (mm)	No. of teeth z	Length code	Pitch length L_w (mm)	No. of teeth z
210 XL	533.40	106	322 XL*	817.88	161
220 XL	558.80	110	330 XL*	838.20	165
230 XL	584.20	115	340 XL*	863.60	170
240 XL	609.60	120	344 XL*	873.76	172
244 XL*	619.76	122	350 XL*	889.00	175
248 XL*	629.92	124	380 XL*	965.20	190
250 XL	635.00	125	382 XL*	970.28	191
260 XL	660.40	130	388 XL*	985.52	194
270 XL*	685.80	135	392 XL*	995.68	196
272 XL*	690.88	136	412 XL*	1046.48	206
274 XL*	695.96	137	414 XL*	1051.56	207
280 XL*	711.20	140	438 XL*	1112.52	219
286 XL*	726.44	143	460 XL*	1168.40	230
290 XL*	736.60	145	498 XL*	1264.92	249
296 XL*	751.84	148	506 XL*	1285.24	253
300 XL*	762.00	150	514 XL*	1305.56	257
306 XL*	777.24	153	580 XL*	1473.20	290
316 XL*	802.64	158	630 XL*	1600.20	315

* Made-to-order belts.

Stock widths

Width reference	025	031	037	050	075	100
Width (mm)	6.35	7.94	9.53	12.7	19.05	25.4

L pitch

Tooth pitch 9.525 mm (0.375 in.)

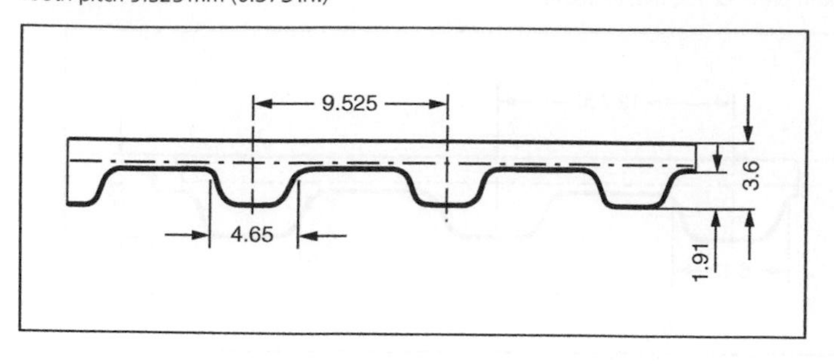

Stock lengths

Length code	Pitch length L_w (mm)	No. of teeth z	Length code	Pitch length L_w (mm)	No. of teeth z
124 L	314.33	33	322 L	819.15	86
150 L	381.00	40	345 L	876.30	92
187 L	476.25	50	367 L	933.45	98
210 L	533.40	56	390 L	990.60	104
225 L	571.50	60	420 L	1066.80	112
236 L*	600.70	63	450 L	1143.00	120
240 L	609.60	64	454 L*	1152.53	121
255 L	647.70	68	480 L	1219.20	128
270 L	685.80	72	510 L	1295.40	136
285 L	723.90	76	540 L	1371.60	144
300 L	762.00	80	600 L	1524.00	160

* Made-to-order belts.

Stock widths

Width reference	050	075	100	150	200	300
Width (mm)	12.7	19.05	25.4	38.1	50.8	76.2

H pitch

Tooth pitch 12.700 mm (0.500 in.)

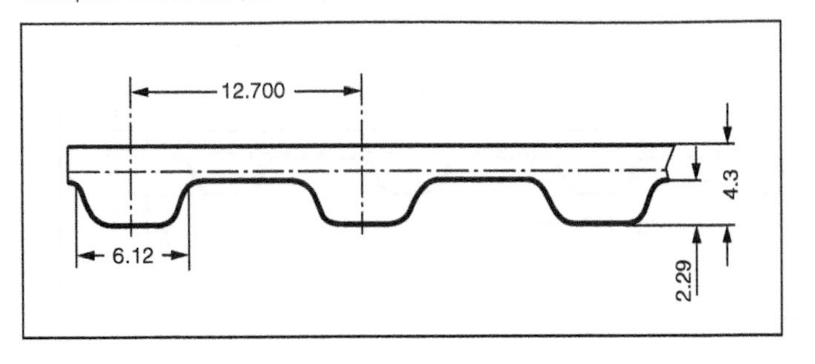

Stock lengths

Length code	Pitch length L_w (mm)	No. of teeth z	Length code	Pitch length L_w (mm)	No. of teeth z
240 H	609.60	48	600 H	1524.00	120
255 H*	647.70	51	630 H	1600.20	126
270 H	685.80	54	660 H	1676.40	132
300 H	782.00	60	700 H	1778.00	140
330 H	838.20	66	730 H*	1854.20	146
335 H*	850.90	67	750 H	1905.00	150
360 H	914.40	72	800 H	2032.00	160
370 H*	939.80	74	850 H	2159.00	170
390 H	990.60	78	900 H	2286.00	180
420 H	1066.80	84	1000 H	2540.00	200
450 H	1143.00	90	1100 H	2794.00	220
480 H	1219.20	96	1250 H	3175.00	250
510 H	1295.40	102	1400 H	3556.00	280
540 H	1371.60	108	1700 H	4318.00	340
570 H	1447.80	114			

* Made-to-order belts.

Stock widths

Width reference	075	100	150	200	300	400	500
Width (mm)	19.05	25.4	38.1	50.8	76.2	101.6	127.0

XH pitch

Tooth pitch 22.225 mm (0.875 in.)

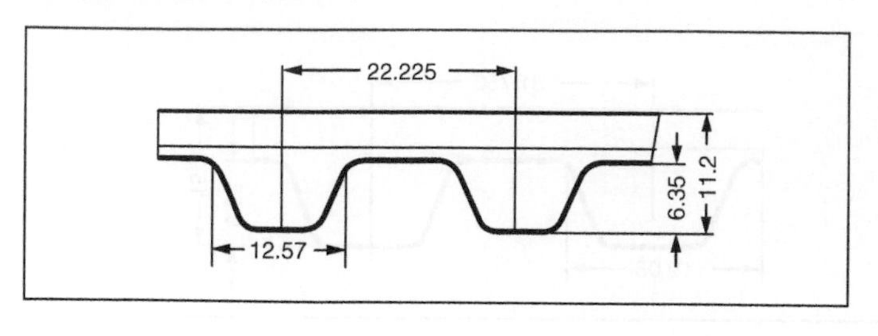

Stock lengths

Length code	Pitch length L_w (mm)	No. of teeth *z*	Length code	Pitch length L_w (mm)	No. of teeth *z*
507 XH	1289.05	58	980 XH	2489.20	112
560 XH	1422.40	64	1120 XH	2844.80	128
630 XH	1600.20	72	1260 XH	3200.40	144
700 XH	1778.00	80	1400 XH	3556.00	160
770 XH	1955.80	88	1540 XH	3911.60	176
840 XH	2133.60	96	1750 XH	4445.00	200

Stock widths

Width reference	200	300	400	500	600	700
Width (mm)	50.8	76.2	101.6	127.0	152.4	177.8

XXH pitch
Tooth pitch 31.750 mm (1.250 in.)

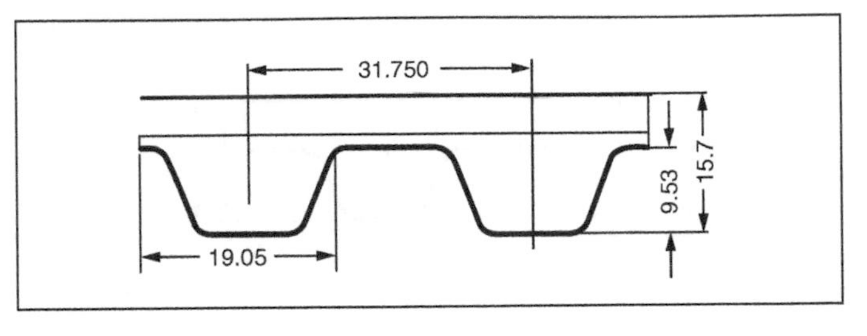

Stock lengths

Length code	Pitch length L_w (mm)	No. of teeth z	Length code	Pitch length L_w (mm)	No. of teeth z
700 XXH	1778.00	56	1200 XXH	3048.00	96
800 XXH	2032.00	64	1400 XXH	3556.00	112
900 XXH	2286.00	72	1600 XXH	4064.00	128
1000 XXH	2540.00	80	1800 XXH	4572.00	144

Stock widths

Width reference	200	300	400	500
Width (mm)	50.8	76.2	101.6	127.0

5.2.14 Synchronous belt drives: pulleys

The right quality and the right choice of pulleys are all-important factors affecting the performance of a synchronous-belt drive. They are precision made and machined by special milling cutters accurately to pitch so as to ensure precise meshing of the teeth.

Pulleys for synchronous-belt drives should be manufactured to DIN ISO 5294 specifications and can be obtained from your nearest pulley stockist.

Some useful pulley data is given below.

Designation
SYNCHROBELT® pulleys bear the following designation:

- Number of teeth
- The number of teeth is calculated from the pitch circumference and the pitch as follows:

$$z = \frac{U_w}{t} = \frac{\pi \cdot d_w}{t}$$

Tooth pitch
Tooth pitch is the distance between two reference points on adjacent teeth at the perimeter of the pitch diameter. The pitch diameter is larger than the outside diameter of the pulley by double the pitch zone of the corresponding belt and is located at the pitch line of the belt.

Pulley width in hundredths of an inch

The width reference indicates the width of the matching belt. Precise pulley widths and their corresponding belt widths are contained in the Table in p. 407.

Example

SYNCHROBELT® pulley 28 H 100
28→28 teeth
H→tooth pitch 0.500 in. = 12.700 mm
100→width reference for a 1 in. = 25.4 mm wide synchronous drive belt

Materials

The choice of material depends on the size of the pulley and the power to be transmitted. The most widely used materials are:

- aluminium alloy: AlCuMgPb F 36 or F 38
- steel: St 9 S20K
- grey cast iron: GG-25
- plastic: PA 6 and 6.6, POM, PBTP, PC

Tooth space measurements

Synchronous-belt drives may be fitted with pulleys with straight tooth flanks or with involute toothing. The type of pulley is determined by the manufacturing process and the intended application.

Table in p. 404 shows the tooth space measurements for pulleys with straight tooth flanks. These are not now generally used.

With an involute tooth profile, tooth space measurements may vary depending on the pulley diameter and so an elaborate table would be required to specify involute toothing dimensions. Table in p. 406 therefore shows the hob cutter measurements for pulleys with involute toothing.

Pulleys with straight tooth flanks

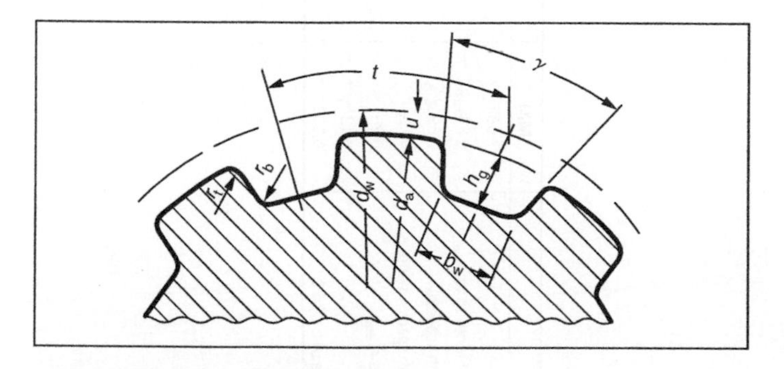

Tooth space with straight flanks.

Dimensions in mm

Pitch	MXL	XL	L	H	XH	XXH
Tooth pitch t	2.032	5.080	9.525	12.700	22.225	31.750
Root width of tooth space b_w	$0.84_{\pm 0.05}$	$1.32_{\pm 0.05}$	$3.05_{\pm 0.10}$	$4.19_{\pm 0.13}$	$7.90_{\pm 0.15}$	$12.17_{\pm 0.18}$
Depth of tooth space h_g	$0.69^{0}_{-0.05}$	$1.65^{0}_{-0.08}$	$2.67^{0}_{-0.10}$	$3.05^{0}_{-0.13}$	$7.14^{0}_{-0.13}$	$10.31^{0}_{-0.13}$
Angle of tooth space γ degrees	$40_{\pm 3}$	$50_{\pm 3}$	$40_{\pm 3}$	$40_{\pm 3}$	$40_{\pm 3}$	$40_{\pm 3}$
Bottom radius $r_{b\ max}$	0.25	0.41	1.19	1.60	1.98	3.96
Top radius r_t	$0.13^{+0.05}_{0}$	$0.64^{+0.05}_{0}$	$1.17^{+0.13}_{0}$	$1.60^{+0.13}_{0}$	$2.39^{+0.13}_{0}$	$3.18^{+0.13}_{0}$
$2 \cdot$ pitch zone u	0.508	0.508	0.762	1.372	2.794	3.048

Pulleys with involute toothing

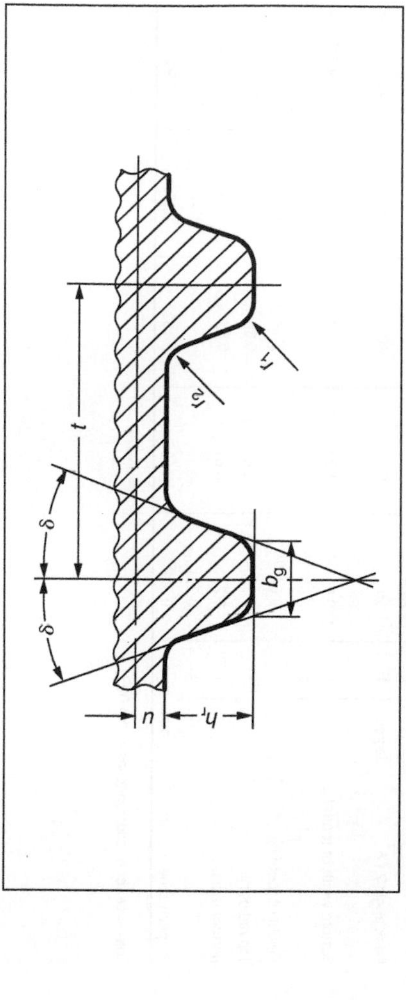

Hob cutter for involute toothing.

Dimensions in mm

Pitch			MXL	XL	L	H	XH	XXH
Tooth pitch t			2.032	5.080	9.525	12.700	22.225	31.750
No. of pulley teeth		1	10–23	⩾10	⩾10	14–19	⩾18	⩾18
		2	⩾24			>19		
Flank angle $2 \cdot \delta$	degrees	1	56	50	40	40	40	40
		2	40			40		
Bottom width of tooth b_g		1	$0.61^{+0.05}_{0}$	$1.27^{+0.05}_{0}$	$3.10^{+0.05}_{0}$	$4.24^{+0.05}_{0}$	$7.59^{+0.05}_{0}$	$11.61^{+0.05}_{0}$
		2	$0.67^{+0.05}_{0}$					
Height of tooth h_r			$0.64^{+0.05}_{0}$	$1.40^{+0.05}_{0}$	$2.13^{+0.05}_{0}$	$2.59^{+0.05}_{0}$	$6.88^{+0.05}_{0}$	$10.29^{+0.05}_{0}$
Top radius r_1			$0.30_{\pm 0.03}$	$0.61_{\pm 0.03}$	$0.86_{\pm 0.03}$	$1.47_{\pm 0.03}$	$2.01_{\pm 0.03}$	$2.69_{\pm 0.03}$
Bottom radius r_2		1	$0.23_{\pm 0.03}$	$0.61_{\pm 0.03}$	$0.53_{\pm 0.03}$	$1.04_{\pm 0.03}$	$1.93_{\pm 0.03}$	$2.82_{\pm 0.03}$
		2	$0.23_{\pm 0.03}$			$1.42_{\pm 0.03}$		
$2 \cdot$ pitch zone u	mm		0.508	0.508	0.762	1.372	2.794	3.048

Tolerance for all tooth pitches = ±0.003

Pulley widths

Width references, nominal widths and minimum toothing widths for pulleys with and without flanges are listed in following Table. When using pulleys with one flange, the minimum toothing width for pulleys with two flanges is recommended.

Pulley widths (all dimensions in mm)

Pitch	Width reference	Nominal width	Minimum toothing width	
			with flanges	without flanges
MXL	012	3.2	3.8	5.6
	018	4.8	5.3	7.1
	025	6.4	7.1	8.9
XL	025	6.4	7.1	8.9
	031	7.9	8.6	10.4
	037	9.5	10.4	12.2
H	075	19.1	20.3	24.8
	100	25.4	26.7	31.2
	150	38.1	39.4	43.9
	200	50.8	52.8	57.3
	300	76.2	79.0	83.5
XH	200	50.8	56.6	62.6
	300	76.2	83.8	89.8
	400	101.6	110.7	116.7
XXH	200	50.8	56.6	64.1
	300	76.2	83.8	91.3
	400	101.6	110.7	118.2
	500	127.0	137.7	145.2

Pulley diameters

Pitch and outside diameters for SYNCHROBELT® pulleys, together with the number of teeth are as follows:

Synchrobelt pulleys (MXL pitch)
Range: 10 teeth: P/D 6.47 mm,
O/D 5.96 mm ***to*** 150 teeth: P/D
97.02 mm,
O/D 96.51 mm in steps of 1 tooth

Synchrobelt pulleys (XL pitch)
Range: 10 teeth: P/D 16.17 mm,
O/D 15.66 mm ***to*** 150 teeth: P/D
242.55 mm,
O/D 242.04 mm in steps of 1 tooth

Synchrobelt pulleys (L pitch)
Range: 12 teeth: P/D 36.38 mm,
O/D 35.62 mm ***to*** 150 teeth: P/D
454.78 mm,
O/D 454.02 mm in steps of 1 tooth

Synchrobelt pulleys (H pitch)
Range: 16 teeth: P/D 64.68 mm,
O/D 63.61 mm ***to*** 150 teeth: P/D
606.38 mm,
O/D 605.01 mm in steps of 1 tooth

Synchrobelt pulleys (XH pitch)
Range: 20 teeth: P/D 141.49 mm, O/D
138.69 mm ***to*** 150 teeth: P/D
1061.17 mm,
O/D 1058.37 mm in steps of 1 tooth

Synchrobelt pulleys (XXH pitch)
Range: 22 teeth: P/D 222.34 mm, O/D
219.29 mm ***to*** 120 teeth: P/D
1212.79 mm,
O/D 1209.71 mm in steps of 1 tooth

(P/D = pitch diameter, O/D = overall diameter)

5.2.15 SYNCHROBELT® HTD

Construction

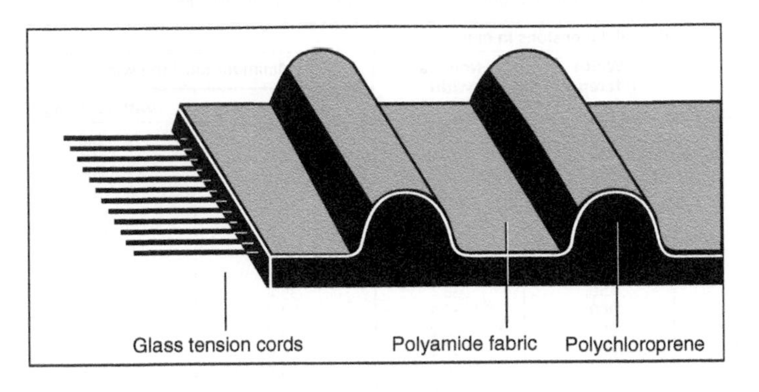

The teeth and the belt top are made from highly loadable polychloroprene-based elastomer compounds. They have excellent adhesion both on the tensile member and on the facing fabric.

A durable protection of the teeth is an essential precondition for a smooth operation and a long service life. This is ensured by the application of particularly abrasion-resistant polyamide fabrics with low friction coefficients.

Synchronous-belt drives call for a high degree of length stability and tensile strength. These requirements are optimally met by low-elongation tensile members of glass cord helically wound over the entire belt width. Any longitudinal off-track running will be largely prevented by the use of S/Z tensile cords arranged in pairs.

The belts are also resistant to fatigue failure, temperature change, ageing, deformation and a wide range of environmental conditions.

Designation

SYNCHROBELT® HTD belts are designated by the following data:

Pitch length (mm) The pitch length of the belt is the overall circumference measured on the neutral pitch line. The pitch length is located in the middle of the tensile member. The precise pitch length can only be ascertained on suitable measuring devices (see Section 5.2.4).

Tooth pitch (mm) The tooth pitch is the linear distance between two adjacent teeth along the pitch line.

Belt width (mm) The belt width and the width designation are identical.

For example, the SYNCHROBELT® HTD 960 8M 50 belt has 960 mm pitch length, 8 mm tooth pitch and 50 mm belt width.

The number of teeth z is a function of pitch length and pitch: $z = L_W/t$

Available belt range

SYNCHROBELT® HTD belts are supplied in four tooth pitch versions:

HTD-3M:	3 mm tooth pitch
HTD-5M:	5 mm tooth pitch
HTD-8M:	8 mm tooth pitch
HTD-14M:	14 mm tooth pitch
HTD-20M:	20 mm tooth pitch

The length and width dimensions that can be supplied are shown in Section 5.2.11. The range of pulleys is illustrated in Section 5.2.16.

5.2.16 SYNCHROBELT® HTD synchronous (toothed) belts: tooth profiles

Tooth pitch 3 mm

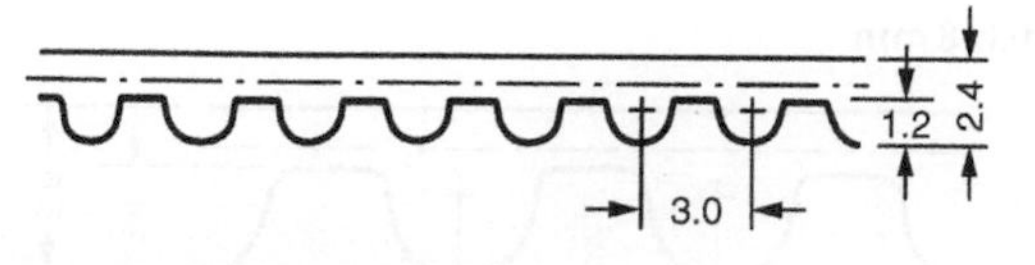

Standard lengths

Designation	Pitch length L_w (mm)	No. of teeth z
144-3M	144	48
177-3M	177	59
225-3M	225	75
255-3M	255	85
300-3M	300	100
339-3M	339	113
384-3M	384	128
420-3M	420	140
474-3M	474	158
513-3M	513	171
564-3M	564	188
633-3M	633	211
711-3M	711	237
1125-3M	1125	375

Standard widths: 6, 9, 15 mm; intermediate widths on request.

Tooth pitch 5 mm

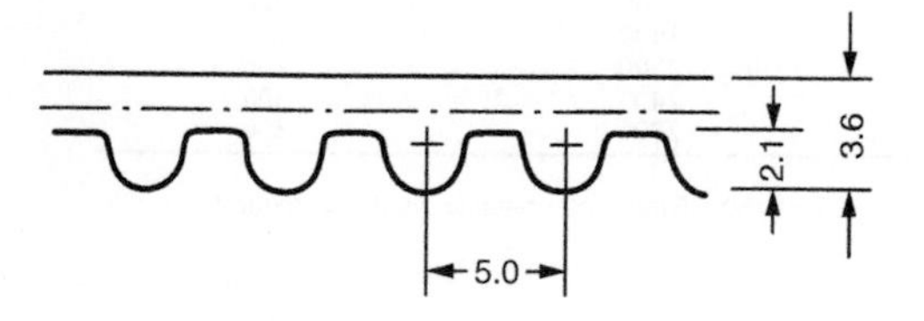

Standard lengths

Designation	Pitch length L_w (mm)	No. of teeth z
350-5M	350	70
400-5M	400	80
450-5M	450	90
500-5M	500	100
600-5M	600	120
710-5M	710	142
800-5M	800	160
890-5M	890	178
1000-5M	1000	200
1125-5M	1125	225
1270-5M	1270	254
1500-5M	1500	300

Standard widths: 9, 15, 25 mm; intermediate widths on request.

Tooth pitch 8 mm

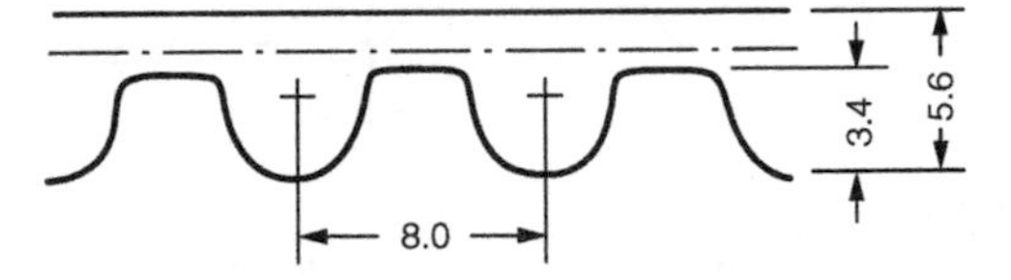

Standard lengths

Designation	Pitch length L_w (mm)	No. of teeth z
480-8M	480	60
560-8M	560	70
600-8M	600	75
640-8M	640	80
656-8M	656	82
720-8M	720	90
800-8M	800	100
880-8M	880	110
960-8M	960	120
1040-8M	1040	130
1120-8M	1120	140
1200-8M	1200	150
1280-8M	1280	160
1440-8M	1440	180
1600-8M	1600	200
1760-8M	1760	220
1800-8M	1800	225
2000-8M	2000	250
2400-8M	2400	300
2800-8M	2800	350

Standard widths: 20, 30, 50, 85 mm; intermediate widths on request.

Tooth pitch 14 mm

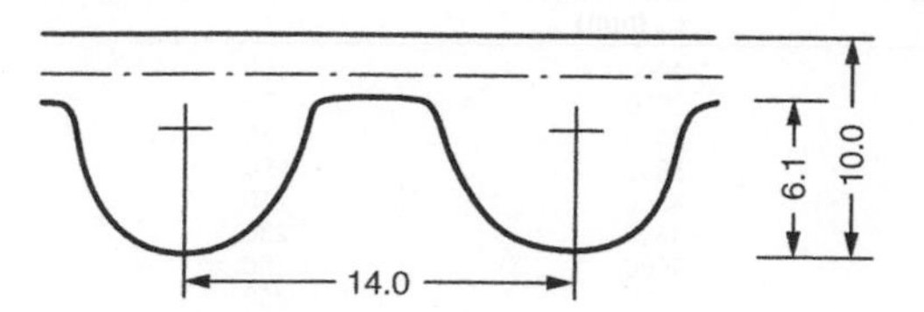

Standard lengths

Designation	**Pitch length L_w (mm)**	**No. of teeth *z***
966-14M	966	69
1190-14M	1190	85
1400-14M	1400	100
1610-14M	1610	115
1778-14M	1778	127
1890-14M	1890	135
2100-14M	2100	150
2310-14M	2310	165
2450-14M	2450	175
2590-14M	2590	185
2800-14M	2800	200
3150-14M	3150	225
3500-14M	3500	250
3850-14M	3850	275
4326-14M	4326	309
4578-14M	4578	327

Standard widths: 40, 55, 85, 115, 170 mm; intermediate widths on request.

Tooth profile HTD-20M

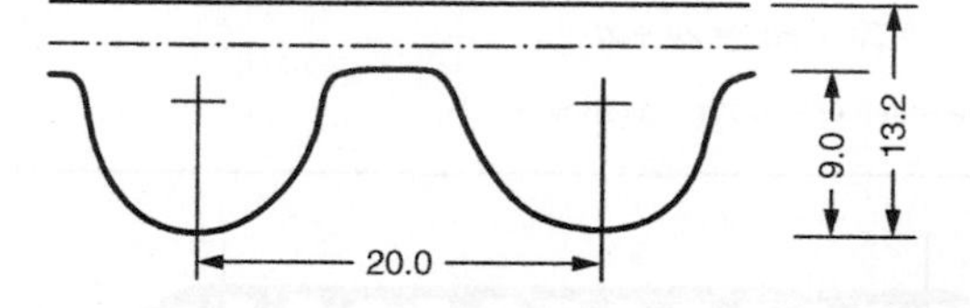

Standard lengths

Designation	Pitch length L_w (mm)	No. of teeth z
2000-20M*	2000	100
2500-20M*	2500	125
3400-20M*	3400	170
3800-20M*	3800	190
4200-20M*	4200	210
4600-20M*	4600	230
5000-20M*	5000	250
5200-20M*	5200	260
5400-20M*	5400	270
5600-20M*	5600	280
5800-20M*	5800	290
6000-20M*	6000	300
6200-20M*	6200	310
6400-20M*	6400	320
6600-20M*	6600	330

* Non-stock items, delivery on request.
Standard widths: 115, 170, 230, 290, 340 mm; intermediate widths on request.
Note: pulley sizes are only available on request to the company.

5.2.17 Synchronous (toothed) belts: length measurement

The pitch length is decisive for the calculation and application of synchronous drive belts. A precise measurement can only be made on suitable measuring equipment.

The belt is placed over two equal size measuring pulleys with the same pitch diameters. The movable measuring pulley is loaded in such a way that the measuring force F will act on the belt. To ensure a correct position of the belt on the pulleys and a uniform tension on both belt sides, the belt must have completed at least two rotations under load. The centre distance a is then measured between the two pulleys.

The pitch length L_w is double the centre distance a plus the pitch circumference U_w of the test pulleys:

$$L_w = 2a + U_w = 2a + \pi d_w = 2a + zt$$

The test measurement layout is as shown.

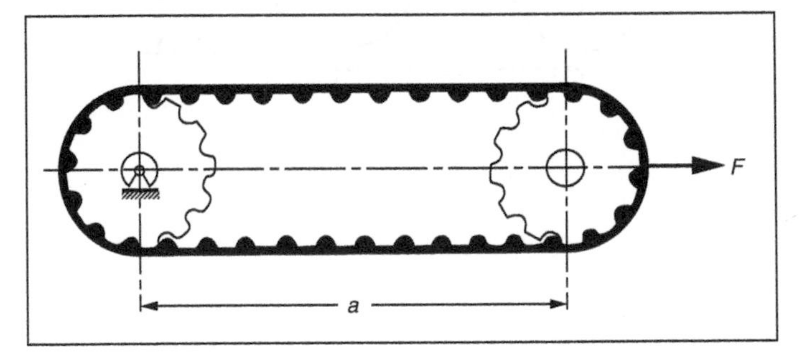

5.2.18 SYNCHROBELT® HTD toothed pulleys: preferred sizes

Tooth pitch 3 mm

Dimensions in mm

No. of teeth z	Pitch diameter d_w	Outside diameter d_a	Flanged pulley diameter d_b	Stock bore diameter d_v	Finished bore diameter $d_{F\,max}$
10	9.55	8.79	12	3	3
12	11.46	10.70	14	3	4
14	13.37	12.61	16	3	6
15	14.32	13.56	16	3	6
16	15.28	14.52	18	3	8
17	16.23	15.47	20	4	8
18	17.19	16.43	20	4	8
19	18.14	17.38	21	4	8
20	19.10	18.34	23	4	10
22	21.01	20.25	25	4	10
24	22.94	22.16	26	4	12
26	24.83	24.07	28	4	14
28	26.74	25.98	30	4	15
30	28.65	27.89	33	6	17
32	30.56	29.80	34	6	19
34	32.47	31.71	36	6	20
36	34.38	33.62	38	6	20
38	36.29	35.53	42	6	25
40	38.20	37.44	42	6	25
44	42.02	41.26	47	8	28
50	47.75	46.99	51	8	32
56	53.48	52.72	59	8	36
62	59.21	58.45	64	8	42
72	68.75	67.99	73	8	50

The relationship between standard belt width and pulley width.

Dimensions in mm

Standard belt width b	Pulley width = toothing width w/o flanged pulleys	Toothed width for flanged pulleys
6	11	≈9
9	14	≈12
15	20	≈18

Tooth pitch 5 mm

Dimensions in mm

No. of teeth z	Pitch diameter d_w	Outside diameter d_a	Flanged pulley diameter d_b	Stock bore diameter d_v	Finished bore diameter $d_{F\ max}$
14	22.28	21.14	26	4	12
15	23.87	22.73	28	4	14
16	25.46	24.32	30	4	14
17	27.06	25.92	32	4	15
18	28.65	27.51	33	6	17
19	30.24	29.10	34	6	18
20	31.83	30.69	36	6	20
22	35.01	33.87	40	6	21
24	38.20	37.06	42	6	25
26	41.38	40.24	45	8	25
28	44.56	43.42	48	8	30
30	47.75	46.61	51	8	32
32	50.93	49.79	55	8	35
34	54.11	52.97	59	8	36
36	57.30	56.16	61	8	38
38	60.48	59.34	64	8	40
40	63.66	62.52	67	8	45
44	70.03	68.89	73	8	50
50	79.58	78.94	85	8	60
56	89.13	87.99	95	8	70
62	98.68	97.54	103	8	75
72	114.59	113.45	118	10	90

Relation between standard belt width and pulley width.

Dimensions in mm

Standard belt width b	Pulley width = toothing width w/o flanged pulleys	Toothed width for flanged pulleys
9	14	≈12
15	20	≈18
25	30	≈28

Tooth pitch 8 mm

Dimensions in mm

No. of teeth z	Pitch diameter d_w	Outside diameter d_a	Flanged pulley diameter d_b	Stock bore diameter d_v	Finished bore diameter $d_{F\,max}$
22	56.02	54.65	61	8	38
24	61.12	59.75	67	8	45
26	66.21	64.84	75	8	50
28	71.30	70.08	80	8	55
30	76.39	75.13	82	8	60
32	81.49	80.16	86	8	60
34	86.58	85.22	95	8	70
36	91.67	90.30	99	8	75
38	96.77	95.39	103	8	75
40	101.86	100.49	107	10	80
44	112.05	110.67	118	10	90
48	122.23	120.86	127	10	98
56	142.60	141.23	150	10	124
64	162.97	161.60	168	16	138
72	183.35	181.97	189	16	155
80	203.72	202.35	210	20	170
90	229.18	227.81	235	20	190
112	285.21	283.83	292	20	250
144	366.69	365.32	–	30	300
168	427.81	426.44	–	30	350
192	488.92	487.55	–	30	400

Relation between standard belt width and pulley width.

Dimensions in mm

Standard belt width b	Pulley width = toothing width w/o flanged pulleys	Toothed width for flanged pulleys
20	26	≈22
30	38	≈34
50	58	≈54
85	94	≈90

Tooth pitch 14 mm

Dimensions in mm

No. of teeth z	Pitch diameter d_w	Outside diameter d_a	Flanged pulley diameter d_b	Stock bore diameter d_v	Finished bore diameter $d_{F\,max}$
28	124.78	122.12	130	10	95
29	129.23	126.57	134	10	100
30	133.69	130.99	138	10	100
32	142.60	139.88	148	10	110
34	151.52	148.79	156	16	120
36	160.43	157.68	166	16	130
40	178.25	175.49	184	16	145
48	213.90	211.11	220	20	180
56	249.55	246.76	254	20	210
64	285.21	282.41	290	20	240
72	320.86	318.06	326	30	260
80	356.51	353.71	362	30	290
90	401.07	398.28	–	30	330
112	499.11	496.32	–	30	420
144	641.71	638.92	–	30	550
192	855.62	852.62	–	30	750

Relation between standard belt width and pulley width.

Dimensions in mm

Standard belt width b	Pulley width = toothing width w/o flanged pulleys	Toothed width for flanged pulleys
40	54	≈47
55	70	≈63
85	102	≈95
115	133	≈126
170	187	≈180

5.3 Power transmission: chain drives

The following information is taken from the Renold Chain Designer Guide with the permission of that company.

5.3.1 Chain performance

Renold chain products that are dimensionally in line with the ISO standard far exceed the stated ISO minimum tensile strength requirements. However Renold does not consider breaking load to be a key indicator of performance because it ignores the principal factors of wear and fatigue. In these areas, Renold products are designed to produce the best possible results and independent testing proves this.

In this text, where the ISO breaking load is quoted, it should be noted that we are stating that the Renold product conforms to the ISO minimum standard. Independent test results show that the minimum (many companies quote averages) breaking loads were far in excess of the ISO minimum.

Where the quoted breaking load is not described as being the ISO minimum, the product has no relevant ISO standard. In this case, the breaking loads quoted are the minimum guaranteed.

The performance of a chain is governed by a number of key factors. The tensile strength is the most obvious since this is the means by which a chain installation is roughly sized. However, since a chain is constructed from steel, the yield strength of which is around 65% of the ultimate tensile strength, any load above this limit will cause some permanent deformation to take place with consequent rapid failure.

Reference to the *s–n* curve below shows that at loads below this 65% line, finite life may be expected and at subsequent reductions in load the expected life increases until the fatigue endurance limit is reached at around 8 000 000 operations.

Loads below the endurance limit will result in infinite fatigue life. The failure mode will then become wear related which is far safer, since a controlled monitor of chain extension can take place at suitable planned intervals. In practice, if a load ratio of tensile strength to maximum working load of 8:1 is chosen, then the endurance limit will not normally be exceeded. Careful consideration of the expected maximum working loads should be given since these are often much higher than the designer may think! It is also a requirement that any passenger lift applications are designed with a safety factor of not less than 10:1.

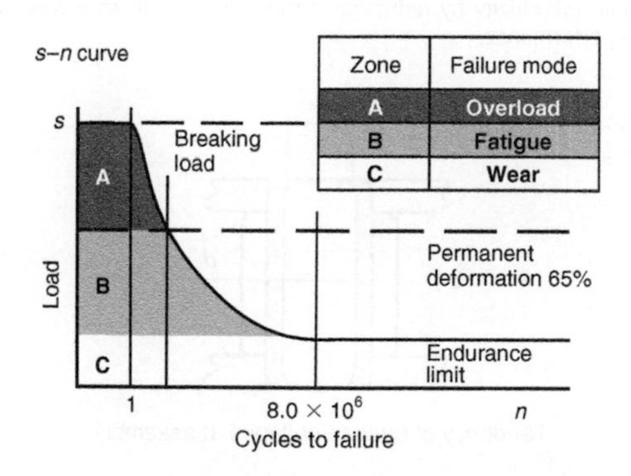

In most applications the failure mode is designed to be wear and therefore some consideration of how a chain behaves in this mode are shown below.

Examination of the wear characteristics graph below shows that chain tends to wear in three distinct phases. The first phase, shown as 'bedding in', is a very rapid change in chain length associated with components adjusting to the loads imposed on them. The degree of this initial movement will depend to a large extent on the quality of chain used. For example, good component fits, chain preloaded at manufacture, plates assembled squarely, etc. Renold chain has many features that minimise the degree of bedding in.

Wear characteristics

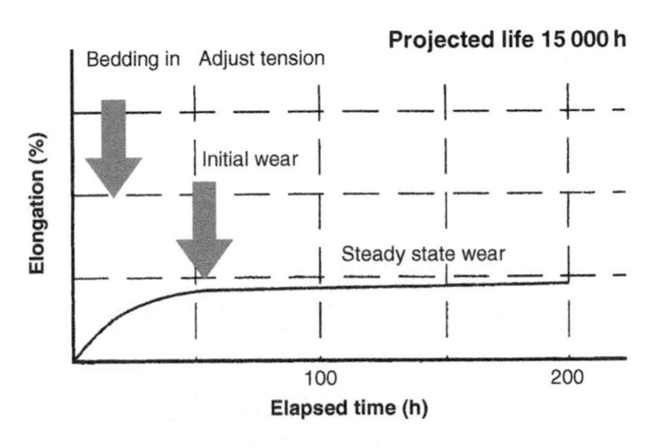

The second phase, shown as 'initial wear', might also be described as secondary 'bedding in'. This is caused firstly by the rapid abrasion of local high spots between the mating surfaces of the pin and bush, and secondly by displacement of material at the bush ends. This is explained more clearly by the inner link assembly diagram shown, where it may be seen that in order to ensure good fatigue life, the bush and plate have a high degree of interference fit resulting in a tendency of the bush ends to collapse inwards slightly. This localised bulge will wear rapidly until the pin bears equally along the length of the bush. Renold limits this effect by introducing special manufacturing techniques. Some manufacturers maintain cylindricity by reducing the interference fit to a very low level. This reduces fatigue performance.

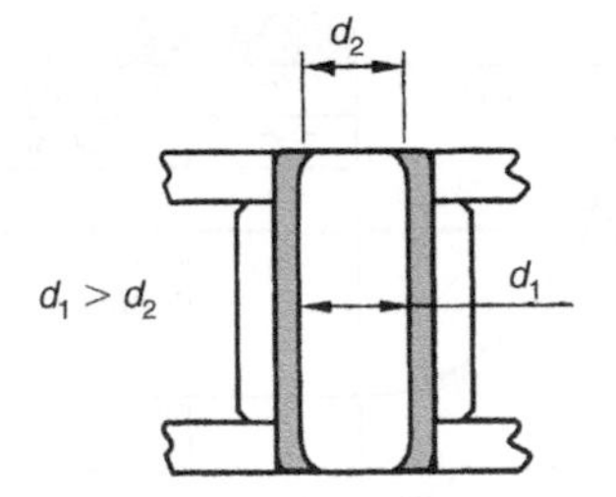

Tendency of bush to collapse at assembly

The final steady state of wear will continue at a very low rate until the chain needs renewal. In a correctly designed and lubricated system, 15 000 h continuous running should be normal.

The reason that wear takes place at all is demonstrated with reference to the Stribeck diagram below. It may be seen from this that where two mating surfaces are in contact, the coefficient of friction is very high at the point of initial movement, known as static friction. The reason for this is that the surface irregularities of the two bodies are interlocked with little or no separating lubrication layer. As the surface speeds increase, lubricant is drawn between the two surfaces and friction takes place with some surface contact. This condition is known as 'mixed friction'. These two conditions result in material loss over time. With a continuing increase in surface speed, hydrodynamic friction takes place, a condition where there is no metal to metal contact.

Stribeck diagram

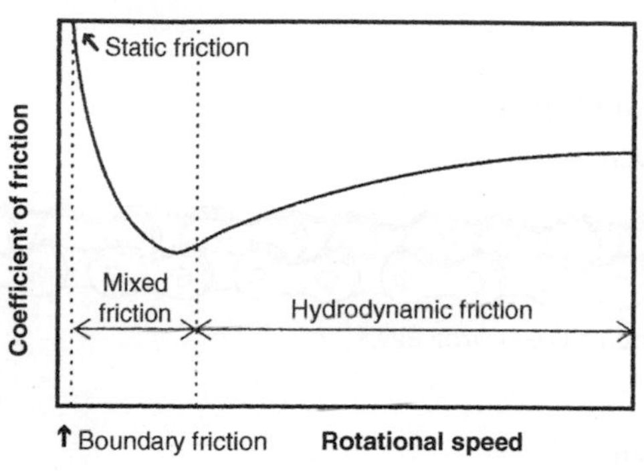

If we consider the action of the mating surfaces of the bush and pin during one cycle of a two sprocket system, it will quickly be realised that these components are stationary with respect to each other during travel from one sprocket to the other, and accelerate rapidly through a very small angle when engaging with the sprocket before coming to rest once more. This means that the pin/bush combination is operating between the static and mixed friction states and that lubrication will therefore be an important aspect of system design.

5.3.2 Wear factors

As already shown, wear takes place from the friction between the mating of the pin and bush. The rate of wear is primarily determined by the bearing area and the specific pressure on these surfaces. The hardened layers of the pin and bush are eroded in such a way that the chain will become elongated.

Elongation may amount to a *Maximum* of 2% of the nominal length of the chain. Above 2% elongation, there can be problems with the chain riding up and jumping the sprocket teeth.

Elongation should be limited to 1% when:

- A sprocket in the system has 90 teeth or more.
- Perfect synchronization is required.
- Centre distances are greater than recommended and not adjustable.

When the demands of the system become even higher, it is necessary to reduce the allowable percentage elongation further.

Wear depends on the following variables in a drive system:

- **Speed** The higher the speed of a system, the higher the frequency of bearing articulations, so accelerating wear.
- **Number of sprockets** The more sprockets used in a drive system, the more frequently the bearings articulate.
- **Number of teeth** The fewer the number of teeth in a sprocket, the greater the degree of articulation, the higher the wear.
- **Chain length** The shorter the length of chain, the more frequently the bearings in the chain will have to operate, the faster wear takes place.
- **Lubrication** As already shown, using the correct lubrication is critical to giving good wear life.

5.3.3 Chain types

Simplex chain

Standard ISO 606, ANSI B29.1.

Duplex chain

Standard ISO 606, ANSI B29.

Triplex chain

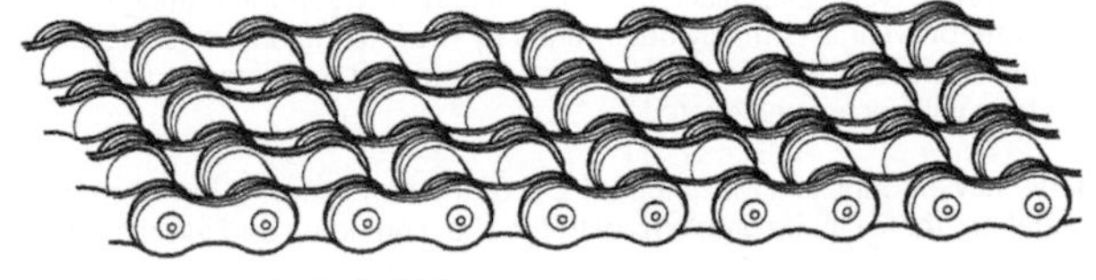

Standard ISO 606, ANSI B29.1.

As with all engineered products, industry demands that chain be produced to a formal standard. The key transmission chain standards are summarised in Sections 5.3.4 and 5.3.5 on pages 423 and 424.

5.3.4 International Standards

European Standard

Chains manufactured to the above standards are covered by ISO 606 and DIN 8187. These standards cover three versions:

- Simplex
- Duplex
- Triplex

The range of pitch sizes can vary between 4 mm, (0.158 in.) to 114.3 mm, (4.500 in.).

They are characterised by a large pin diameter, especially for the larger pitch sizes. This results in better wear resistance due to the greater bearing area.

The ISO Standard has a simple form of part numbering, for example 1/2 in. pitch duplex chain would be 08B-2.

- The first two digits are the pitch size in 1/16's of an inch, therefore 08 = 8/16 or 1/2 in.
- The letter 'B' indicates European Standard.
- The suffix two indicates the number of strands in the chain, in this case a duplex chain.

American Standard

American Standard chains are covered by ISO 606, ANSI B29.1 and DIN 8188 and eight versions are covered.

- Simplex, duplex and triplex as for the European Standard chains
- Quadruplex, 4 strands
- Quintuplex, 5 strands
- Sextuplex, 6 strands
- Octuplex, 8 strands
- Decuplex, 10 strands.

The pitch sizes covered by this standard are 1/4 to 3 in. pitch.

American Standard chains have a smaller pin diameter than their European Standard equivalent. Wear resistance is therefore reduced when compared with European Standard chains with the one exception, 5/8 in. pitch. In this case the pin and bush diameter is larger in an American Standard chain.

American Standard chains are normally referred to under the ANSI Standard numbering system, for example a 1/2 in. pitch duplex chain would be, ANSI 40-2.

The ANSI numbering system works as follows:

- The first number is the pitch size in 1/8 in.; that is, 4/8 = 1/2 in. pitch.
- The second number refers to the chain being a roller chain, 0 = roller chain. A 5 replacing the 0 would indicate a bush chain.
- The suffix, as with European Standard chain, refers to the number of strands in the chain, that is 2 = duplex chain.

ANSI chain is also available in heavy duty options with thicker plates (H) and through hardened pins (V). An ANSI heavy chain would be specified using these suffixes. That is,

ANSI 140-2HV	Duplex, thick plates, through hardened pin
ANSI 80H	Simplex, thick plates

Range of application

The transmission chain market worldwide is divided between these two chain standards, based on the economic and historical influences within their regions.

- American Standard chain is used primarily in the USA, Canada, Australia, Japan and some Asiatic countries.
- European Standard chains dominate in Europe, the British Commonwealth, Africa and Asian countries with a strong British historical involvement.

In Europe around 85% of the total market uses European Standard chain. The remaining 15% is American Standard chains found on:

- Machinery imported from countries where American Standard chain dominates.
- Machinery manufactured in Europe under licence from American dominated markets.

Chain not conforming to ISO Standards

There are also Renold manufacturing standards for special or engineered chain which can be split as follows:

1. **Higher breaking load chain** This chain usually has plates that undergo a special treatment, has thicker side plate material and/or pin diameters that slightly deviate from the standards.
2. **Special dimensions** Some chains can be a mixture of American and European Standard dimensions or the inner width and roller diameters vary, such as in motorcycle chains.
3. **Applicational needs** Special or engineered chain is manufactured for specific applicational use, examples being:
   - Stainless steel chain
   - Zinc or nickel plated chain
   - Chain with plastic lubricating bushes
   - Chains with hollow bearing pins
   - Chain that can bend sideways (SIDEBOW).

5.3.5 Standards reference guide

Transmission chain types

	ISO	ANSI	Other
Short pitch transmission chain and sprockets	606	B29.1M	DIN 8187 DIN 8188
Short pitch bush chains and sprockets	1395		DIN 8154
Double pitch roller chain and sprockets	1275	B29.3M	DIN 8181
Oilfield chain and sprockets	606	B29.1M	API Spec 7F
Cycle chains	9633		
Motorcycle chains	10190		
Cranked link chain and sprockets	3512	B29.1M	DIN 8182

Lifting chain types

	ISO	ANSI	Other
Leaf chain, clevises and sheaves	4347	B29.8M	DIN 8152
Roller load chains for overhead hoists		B29.24M	

5.3.6 Advantages of chain drives

Steel transmission roller chain is made to close tolerances with excellent joint articulation, permitting a smooth efficient flow of power. Any friction between the chain rollers and sprocket teeth is virtually eliminated because the rollers rotate on the outside of the bushes, independent of bearing pin articulation inside the bush. As a result, very little energy is wasted and tests have shown chain to have an efficiency of between 98.4% and 98.9%.

This high level of efficiency, achieved by a standard stock chain drive under the correct conditions of lubrication and installation, is equalled only by gears of the highest standard with teeth ground to very close tolerances.

Roller chain offers a positive, non-slip driving medium. It provides an accurate pitch by pitch positive drive which is essential on synchronised drives such as those to automobile and marine camshafts, packaging and printing machinery. Under conditions of high speed and peak load when efficiency is also required, the roller chain has proved consistently quiet and reliable.

Centre distances between shafts can range from 50 mm up to more than 9 m in a very compact installation envelope. Drives can be engineered so that the sprocket teeth just clear each other or so that a considerable span is traversed by the chain. In this later category, double pitch chain comes into its own.

Roller chain has a certain degree of inherent elasticity and this, plus the cushioning effect of an oil film in the chain joints, provides good shock absorbing properties. In addition, the load distribution between a chain and sprocket takes place over a number of teeth, which assists in reducing wear. When, after lengthy service, it becomes necessary to replace a chain, this is simple and does not normally entail sprocket or bearing removal.

Roller chain minimises loads on the drive motor and driven shaft bearings since no preload is required to tension the chain in the static condition.

One chain can drive several shafts simultaneously and in almost any configuration of centre distance or layout. Its adaptability is not limited to driving one or more shafts from a common drive point. It can be used for an infinite variety of devices including reciprocation, racks, cam motions, internal or external gearing, counterbalancing, hoisting or weight suspension. Segmental tooth or 'necklace' chain sprocket rims can be fitted to large diameter drums.

Since there are no elastomeric components involved, chain is tolerant of a wide variety of environmental conditions, including extremes of temperature. Chain is used successfully in such harsh environments as chemical processing, mining, baking, rock drilling and wood processing. Special coatings can easily be applied for further enhancement.

Roller chain can also be fitted with link plate attachments and extended bearing pins, etc., which allow them to be used for mechanical handling equipment and the operation of mechanisms.

Roller chain drives are available for ratios up to 9:1 and to transmit up to 520 kW at 550 rev/min. Beyond this, four matched strands of triplex chain can achieve 3200 kW at 300 rev/min.

Roller chain does not deteriorate with the passage of time, the only evidence of age being elongation due to wear which normally is gradual and can be accommodated by centre distance adjustment or by an adjustable jockey sprocket. Provided a chain drive is selected correctly, properly installed and maintained, a life of 15 000 h can be expected without chain failure either from fatigue or wear.

Where complete reliability and long life are essential, chains can be selected on their assured performance for applications such as hoists for control rods in nuclear reactors and control systems for aircraft.

Chain is a highly standardised product available in accordance with ISO Standards all over the world. It is also totally recyclable and causes no harmful effects to the environment.

Shown below is a simple table comparing the merits of different transmission/lifting media.

Summary of advantages

Feature	Gears	Rope	Belt	Chain
Efficiency	**A**	X	B	**A**
Positive drive	**A**	X	B	**A**
Centre distance	C	**A**	B	**A**
Elasticity	C	**A**	**A**	B
Wear resistance	**A**	C	B	**A**
No pre-load	**A**	C	C	**A**
Multiple drives	C	X	C	**A**
Heat resistant	B	B	C	**A**
Chemical resistant	B	**A**	C	**A**
Oil resistant	**A**	**A**	C	**A**
Adaptations	C	B	C	**A**
Power range	**A**	X	B	**A**
Ease of maintenance	C	B	B	**A**
Standardized	C	B	B	**A**
Environment	**A**	**A**	C	**A**

A = excellent B = good
C = poor X = not appropriate

Note: To achieve the above ratings, different types of belt would be required.

5.3.7 Chain selection

The notes given below are general recommendations and should be followed in the selection and installation of a chain drive, in order that satisfactory performance and drive life may be ensured.

Chain pitch

The Rating Charts (pages 441 and 442) give the alternative sizes of chains that may be used to transmit the load at a given speed. The smallest pitch of a simplex chain should be used, as this normally results in the most economical drive. If the simplex chain does not satisfy the requirements dictated by space limitations, high speed, quietness or smoothness of running, then consider a smaller pitch of duplex or triplex chain.

When the power requirement at a given speed is beyond the capacity of a single strand of chain, then the use of multistrand drives permits higher powers to be transmitted.

These drives can also be made up from multiples of matched simplex, duplex or triplex ISO chains or in the case of ANSI chain, multiplex chain up to decuplex (10 strands) are available.

Please consult Renold's technical staff for further information (see Appendix 3).

Maximum operating speeds

For normal industrial drives, experience has established a maximum sprocket speed for each pitch of chain. These speeds, which relate to driver sprockets having 17–25 teeth inclusive, are given in the graph below; they are applicable only if the method of lubrication provided is in line with recommendations.

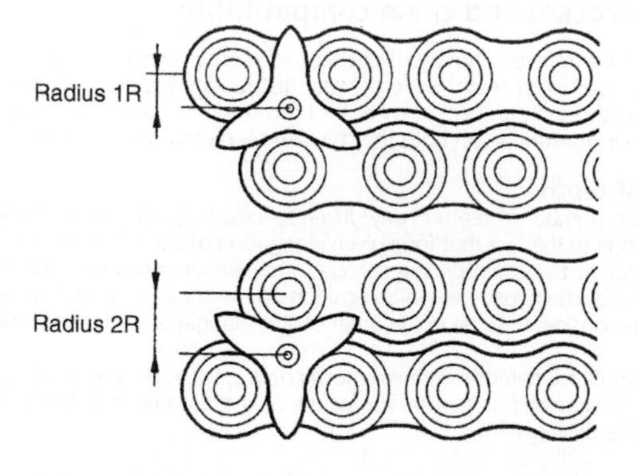

Polygonal effect

Four important advantages of a chain drive are dependent directly upon the number of teeth in the driver sprocket (Z_1).

The advantages are smooth uniform flow of power, quietness of operation, high efficiency and long life, the reason for their dependence being that chain forms a polygon on the sprocket. Thus, when the sprocket speed is constant, the chain speed (due to the many sided shape of its path around the teeth) is subject to a regular cyclic variation. This cyclic variation becomes less marked as the path of the chain tends towards a true circle and in fact, becomes insignificant for most applications as the number of teeth in the driver sprocket exceeds 19.

The effect of this cyclic variation can be shown in the extreme case of a driver sprocket with the absolute minimum number of teeth; that is, three. In this instance, for each revolution of the sprocket the chain is subjected to a three-phase cycle; each phase being associated with the engagement of a single tooth. As the tooth comes into engagement, for a sixth of a revolution the effective distance, or driving radius from the sprocket centre to the chain is gradually doubled; for the remaining sixth of a revolution, it falls back to its original position. Thus, as the linear speed of the chain is directly related to the effective driving radius of the driver sprocket, the chain speed fluctuates by 50% six times during each revolution of the driver sprocket.

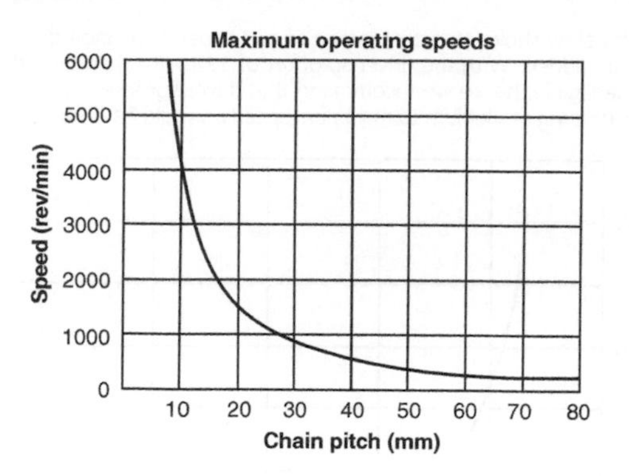

As the graph below shows, the percentage of cyclic speed variation decreases rapidly as more teeth are added. With the driver sprocket of 19 teeth, therefore, this cyclic speed variation is negligible; hence we recommend that driver sprockets used in normal application drives running at medium to maximum speeds, should have not less than 19 teeth.

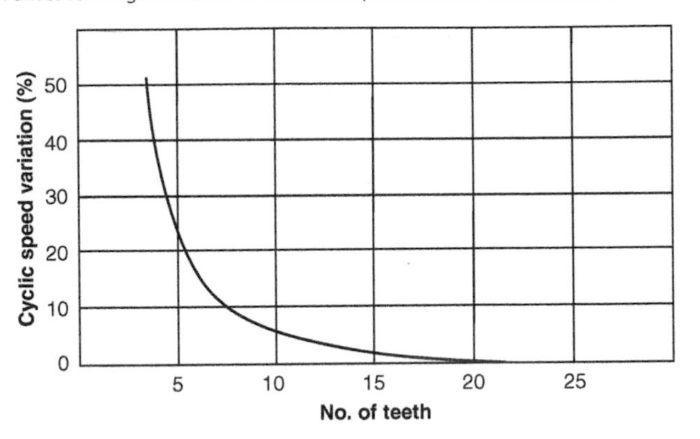

There are, however, applications where space saving is a vital design requirement and the speed/power conditions are such that the smaller numbers of teeth (i.e. below 17) give acceptable performance so that a compact, satisfactory drive is achieved; for example, office machinery, hand operated drives, mechanisms, etc.

The limiting conditions with steady loading for using small numbers of teeth are:

No. of teeth	Percentage of maximum rated speed	Percentage of maximum rated power
11	20	30
13	30	40
15	50	60
17	80	90

5.3.8 Sprocket and chain compatibility

Most drives have an even number of pitches in the chain and by using a driver sprocket with an odd number of teeth, uniform wear distribution over both chain and sprocket teeth is ensured. Even numbers of teeth for both the driver and driven sprockets can be used, but wear distribution on both the sprocket teeth and chain is poor.

Number of teeth

The maximum number of teeth in any driven sprocket (Z_2) should not exceed 114. This limitation is due to the fact that for a given elongation of chain due to wear, the working pitch diameter of the chain on the sprocket increases in relation to the nominal pitch diameter; that is, chain assumes a high position on the sprocket tooth. The allowable safe chain wear is considered to be in the order of 2% elongation over nominal length.

A simple formula for determining how much chain elongation a sprocket can accommodate is 200/N expressed as percentage where N is the number of teeth on the largest sprocket in the drive system.

It is good practice to have the sum of teeth not less than 50 where both the driver and driven sprockets are operated by the same chain, for example, on a 1:1 ratio drive, both sprockets should have 25 teeth each.

Centre distance

For optimum wear life, centre distance between two sprockets should normally be within the range 30–50 times the chain pitch. On drive proposals with centre distances below 30 pitches or greater than 2 m, we would recommend that the drive details are discussed with Renold's technical staff.

The minimum centre distance is sometimes governed by the amount of chain lap on the driver sprocket, our normal recommendation in this circumstance being not less than 6 teeth in engagement with the chain.

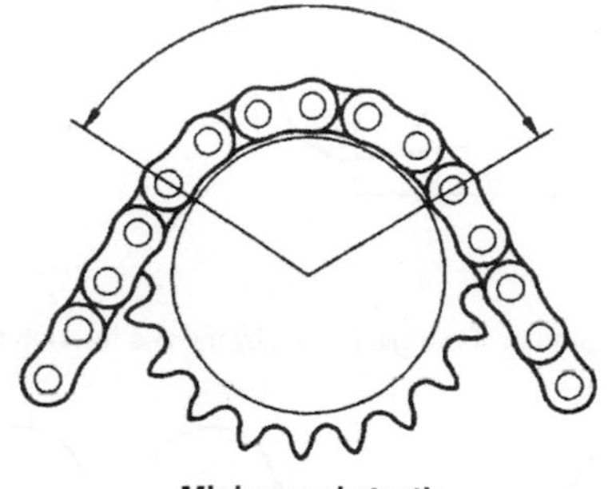

Minimum six teeth

The centre distance is also governed by the desirability of using a chain with an even number of pitches to avoid the use of a cranked link, a practice that is not recommended except in special circumstances.

For a drive in the horizontal plane the shortest centre distance possible should be used consonant with recommended chain lap on the driver sprocket.

Formulae for the calculation of chain length and centre distance for two-point drives are given on page 437.

Recommended centre distances for drives are:

	in.	3/8	1/2	5/8	3/4	1	1 1/4
Pitch	**mm**	**9.525**	**12.70**	**15.87**	**19.05**	**25.40**	**31.75**
Centre distance	mm	450	600	750	900	1000	1200

	in.	1 1/2	1 3/4	2	2 1/2	3
Pitch	**mm**	**38.1**	**44.45**	**50.80**	**63.50**	**76.20**
Centre distance	mm	1350	1500	1700	1800	2000

Lie of drive

Drives may be arranged to run horizontally, inclined or vertically. In general, the loaded strand of the chain may be uppermost or lowermost as desired. Where the lie of the drive is vertical, or nearly so, it is preferable for the driver sprocket (Z_1) to be above the driven sprocket (Z_2); however, even with a drive of vertical lie it is quite feasible for the driver sprocket to be lowermost, provided care is taken that correct chain adjustment is maintained at all times.

Centres

The centre distance between the axis of two shafts or sprockets.

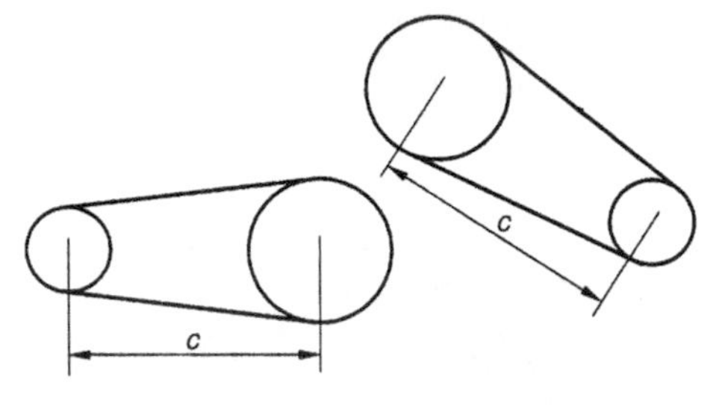

Angle

The lie of the drive is given by the angle formed by the line through the shaft centres and a horizontal line.

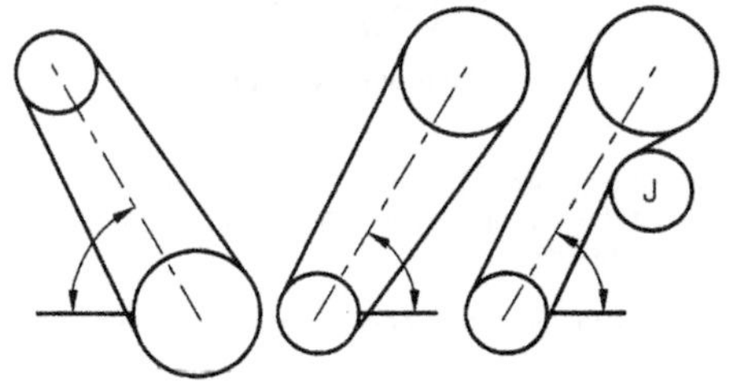

Rotation

Viewed along the axis of the driven shaft the rotation can be clockwise or anti-clockwise.

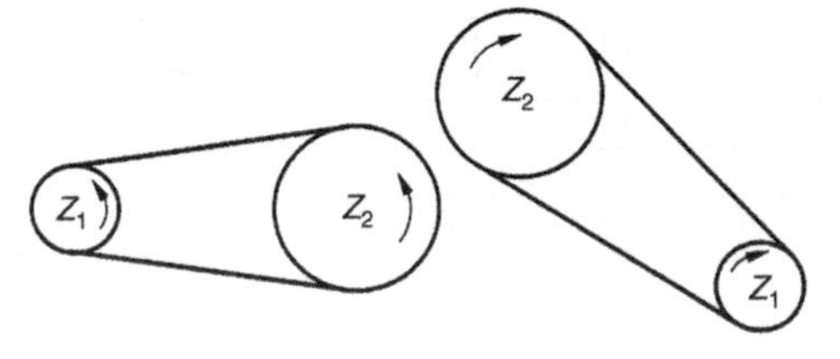

5.3.9 Drive layout

One chain can be used for driving a number of shafts and due to the ability of roller chains to gear on either face, individual shafts in the same drive can be made to rotate in the same or opposite directions by arranging the driven sprockets to gear in different faces of the chain. The number of driven sprockets permissible in any one drive depends on the layout.

A selection of possible drive layouts is shown below.

Drives with variable shaft positions

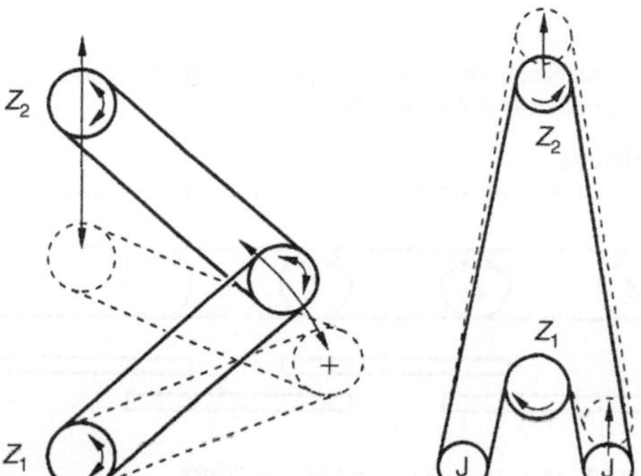

Floating countershaft and floating jockey

Chain lap: Recommended 120°. Minimum of 90° permissible for sprockets of 27 teeth or over.
Centres: Pitch of chain multiplied by 30–50.

Drives with abnormally long centres

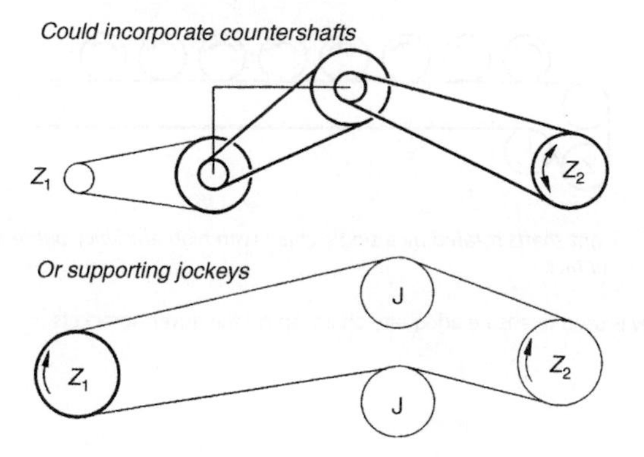

For slow and medium chain speed applications up to 150 m/min.

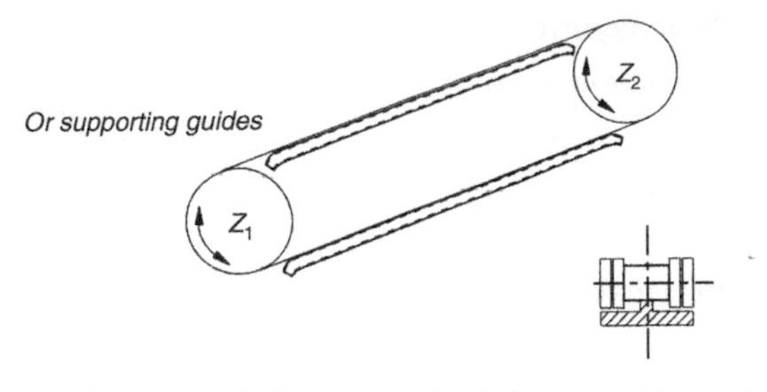

For applications where countershafts or supporting jockeys cannot be employed and where the chain speed does not exceed 60 m/min.

Multi-shaft drives

The permissible number of driven shafts will vary according to drive characteristics.

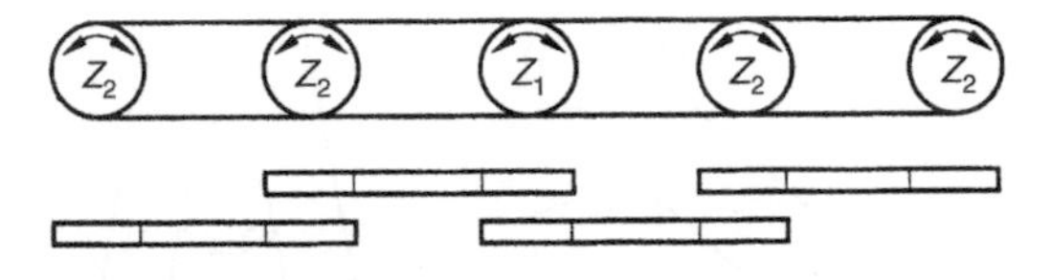

Five sprockets coupled by four simple drives.

Whilst the efficiency of a single stage drive is approximately 98%, where a series of drives are interconnected as in live roller conveyors, the overall efficiency will vary with the number of drives involved. It is necessary in applications of this nature to increase the calculated motor power to allow for this reduced efficiency.

4 drives overall efficiency = 94%
8 drives overall efficiency = 87%
12 drives overall efficiency = 80%

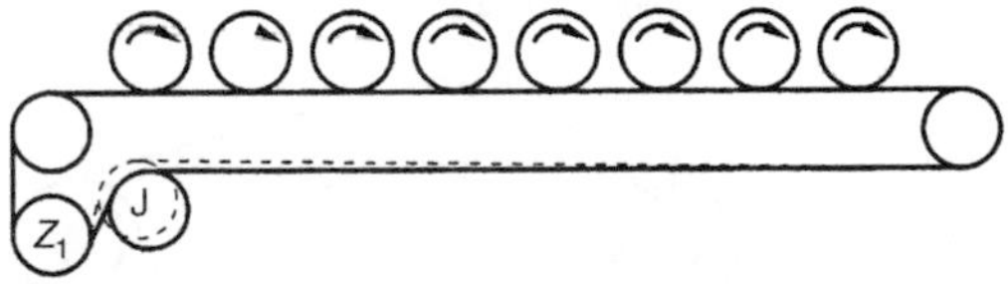

Eight shafts rotated by a single chain with high efficiency but reduced tooth contact

The jockey is used to ensure adequate chain lap on the driven sprockets.

Horizontal drives

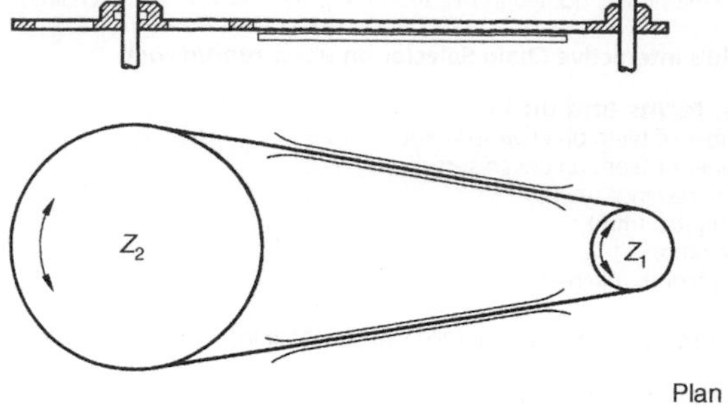

Two shafts vertically mounted

When centres are long, use guide strips to support chain strands with generous 'lead-in' to ensure smooth entry and exit of chain.

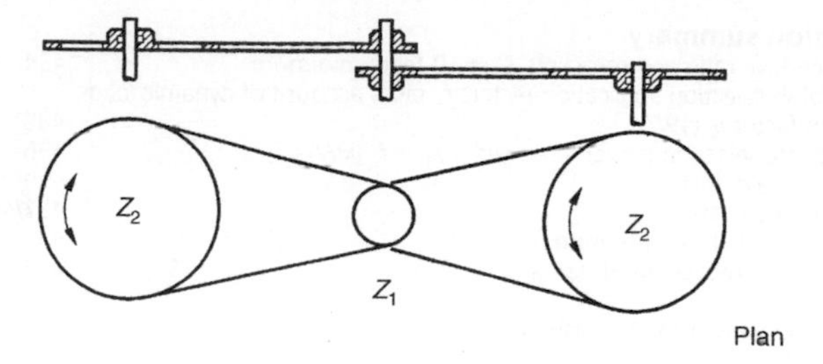

Three shafts vertically mounted

Chain lap: Recommended 120°. Minimum of 90° permissable for sprockets of 27 teeth or over.
Centres: Shortest possible.

5.3.10 Selection method

Introduction

Chain selected using this method will have a minimum life expectancy with proper installation and lubrication of 15 000 h.

Warning

The rating charts pages 441 and 442 exceed the minimum standards and selection of chain using the figures quoted in this section is only valid for Renold chain.

Use Renold's interactive Chain Selector on *www.renold.com.*

Symbols, terms and units

Z_1 = number of teeth on drive sprocket.
Z_2 = number of teeth on driven sprocket.
C = centre distance (mm).
P = chain pitch (mm).
i = drive ratio.
L = chain length (pitches).

In order to select a chain drive the following essential information must be known:

- The power in kilowatts to be transmitted.
- The speed of the driving and driven shafts.
- The characteristics of the drive.
- Centre distance.

From this base information, the selection power to be applied to the ratings chart is derived.

Selection summary

1	Select drive ratio and sprockets Z_1 = 19 teeth minimum	434
2	Establish selection application factors f_1 takes account of dynamic loads	
	Tooth factor f_2 $(19/Z_1)$	435
3	Calculate selection power = power $\times f_1 \times f_2$ (kW)	436
4	Select chain drive.	
	Use rating charts	437/441/442
5	Calculate chain length using formulae	437
6	Calculate exact centre distance	437

Finally, choose lubrication method.

Select drive and ratio

Chart 1 may be used to choose a ratio based on the standard sprocket sizes available. it is best to use an odd number of teeth combined with an even number of chain pitches.

Ideally, chain sprockets with a minimum of 19 teeth should be chosen. If the chain drive operates at high speed or is subjected to impulsive loads, the smaller sprockets should have at least 25 teeth and should be hardened.

It is recommended that chain sprockets should have a maximum of 114 teeth.

Drive ratio can otherwise be calculated using the formula:

$$i = \frac{Z_2}{Z_1}$$

For large ratio drives, check that the angle of lap of Z_1 is not less than 120°.

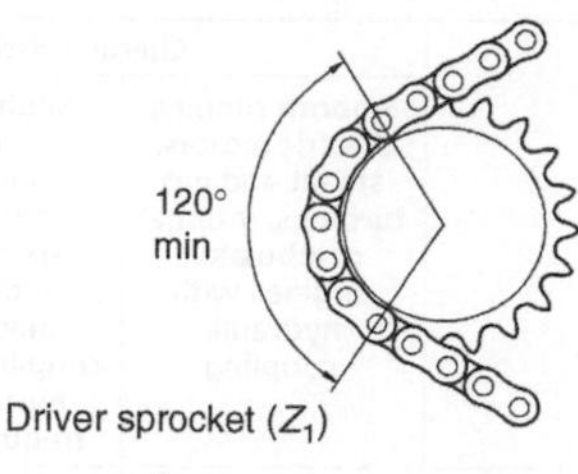

Select drive ratio and sprockets $= \frac{Z_2}{Z_1}$

Chain reduction ratios to one using preferred sprockets

No. of teeth driven sprocket Z_2	No. of teeth drive sprocket Z_1					
	15	17	19	21	23	25
25	–	–	–	–	–	1.00
38	2.53	2.23	2.00	1.80	1.65	1.52
57	3.80	3.35	3.00	2.71	2.48	2.28
76	5.07	4.47	4.00	3.62	3.30	3.04
95	6.33	5.59	5.00	4.52	4.13	3.80
114	7.60	6.70	6.00	5.43	4.96	4.56

For recommended centre distances see page 437

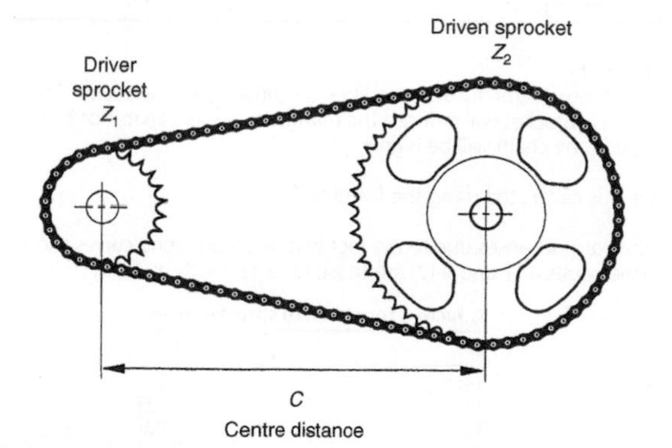

Establish selection factors

The following factors will be used later on to determine the selection power.

Application factor f_1

Factor f_1 takes account of any dynamic overloads depending on the chain operating conditions. The value of factor f_1 can be chosen directly or by analogy using chart 2.

Driven machine characteristics		Characteristics of driver		
		Smooth running Electric motors, steam and gas turbines, internal combustion engines with hydraulic coupling	**Slight shocks Internal combustion engines with six cycles or more with mechanical coupling, electric motors with frequent starts**	**Moderate shocks Internal combustion engines with less than six cycles with mechanical coupling**
Smooth running	Centrifugal pumps and compressors, printing machines, paper calanders, uniformly loaded conveyors, escalators, liquid agitators and mixers, rotary driers, fans	**1**	**1.1**	**1.3**
Moderate shocks	Pumps and compressors (3+ cycles), concrete mixing machines, non uniformly loaded conveyors, solid agitators and mixers	**1.4**	**1.5**	**1.7**
Heavy shocks	Planers, excavators, roll and ball mills, rubber processing machines, presses and shears one and two cycle pumps and compressors, oil drilling rigs	**1.8**	**1.9**	**2.1**

Tooth factor f_2

The use of a tooth factor further modifies the final power selection. The choice of a smaller diameter sprocket will reduced the maximum power capable of being transmitted since the load in the chain will be higher.

Tooth factor f_2 is calculated using the formula $f_2 = \frac{19}{Z_1}$

Note that this formula arises due to the fact that selection rating curves shown in the rating charts (see pages 441 and 442) are those for a 19 tooth sprocket.

f_2 factors for standard sprocket sizes

Z_1	f_2
15	1.27
17	1.12
19	1.00
21	0.91
23	0.83
25	0.76

Calculate the selection power

Multiply the power to be transmitted by the factors obtained from step two.

Selection power = Power to be transmitted $\times f_1 \times f_2$ (kW).

This selection power can now be used with the appropriate rating chart, see pages 441 and 442.

Select chain drive

From the rating chart, select the smallest pitch of simplex chain to transmit the selection power at the speed of the driving sprocket Z_1.

This normally results in the most economical drive selection. If the selection power is now greater than that shown for the simplex chain, then consider a multiplex chain of the same pitch size as detailed in the ratings chart.

Calculate chain length

To find the chain length in pitches (L) for any contemplated centre distance of a two point drive, use the formula below:

$$\text{Length } (L) = \frac{Z_1 + Z_2}{2} + \frac{2C}{P} + \frac{\left(\frac{Z_2 - Z_1}{2p}\right)^2 \times P}{C}$$

The calculated number of pitches should be rounded up to a whole number of even pitches. Odd numbers of pitches should be avoided because this would involve the use of a cranked link which is not recommended. If a jockey sprocket is used for adjustment purposes, two pitches should be added to the chain length (L).

C is the contemplated centre distance in mm and should generally be between 30 and 50 pitches.

For example, for $1\frac{1}{2}$ in. pitch chain $C = 1.5 \times 25.4 \times 40 =$ **1524 mm**.

Calculate exact centre distance

The actual centre distance for the chain length (L) calculated by the method above, will in general be greater than that originally contemplated. The revised centre distance can be calculated from the formula below.

$$C = \frac{P}{8}\left[2L - Z_2 - Z_1 + \sqrt{(2L - Z_2 - Z_1)^2 - \left(\frac{\pi}{3.88}(Z_2 - Z_1)^2\right)}\right]$$

where:
P = chain pitch (mm)
L = chain length (pitches)
Z_1 = number of teeth in driver sprocket
Z_2 = number of teeth in driven sprocket

Drive with multiple sprockets

When designing a drive with multiple sprockets, the chain length calculation becomes more complicated. Most CAD systems, however, can be used to calculate chain length by wrapping a polyline around the pitch circle diameter (PCD's) of each sprocket. A scale manual drawing could also give a fairly accurate result as follows:

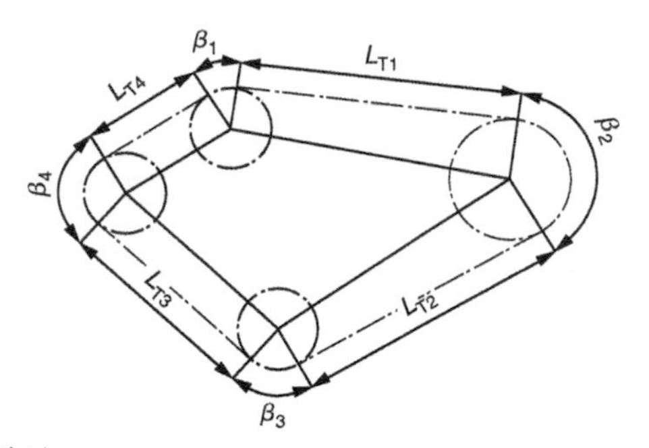

Measure lengths L_{Ti}
Measure angles β_i

The theoretical length in pitches can now be calculated by the addition of all L_T and β values using the following formula.

where:
P = the chain pitch
Z_i = the number of teeth

$$\text{Number of pitches} = \frac{1}{P}\sum_{i=1}^{i=n} L_{Ti} + \sum_{i=1}^{i=n} \frac{\beta_i Z_i}{360°}$$

This calculation method can also be applied on drives where the chain is driven on guide rails or around jockey sprockets. These should be considered as ordinary sprockets.

Sprockets for transmission chain

Renold manufacture a comprehensive range of stock sprockets for European standard chains up to 2 in. pitch.

Other sizes of sprocket, including those to American standard dimensions, are available on request.

Special sprockets are also manufactured on request, in special materials or formats, normally to suit a specific application in harsh or difficult drive situations, examples being:

- Sprockets incorporating shafts.
- Welded or detachable hubs.
- Shear pin devices fitted.
- Necklace sprockets made up of chain plates and individual tooth sections for turning large drums or tables.
- Combination sprockets (two or more sprockets combined having different pitch sizes and numbers of teeth).
- Sprockets in two or more sections, that is, split sprockets or segmental sprockets.

Examples of two typical special sprockets.

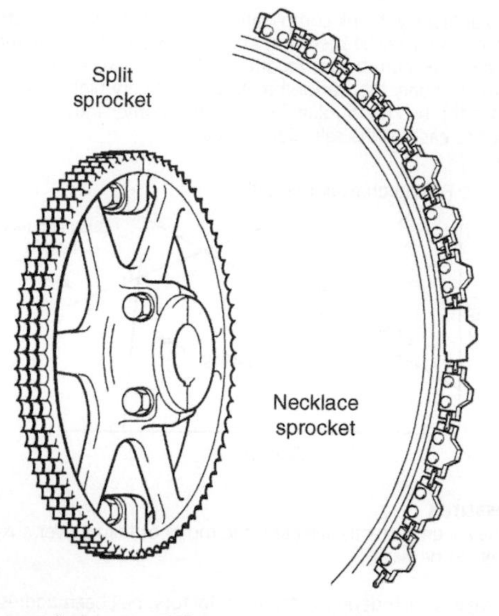

Selection of sprocket materials

Choice of material and heat treatment will depend upon shape, diameter and mass of the sprocket. The table below can be used as a simple guide on the correct selection of sprocket material.

Sprocket	Smooth running	Moderate shocks	Heavy shocks
Up to 29T	EN8 or EN9	EN8 or EN9 hardened and tempered or case hardened mild steel	EN8 or EN9 hardened and tempered or case hardened mild steel
30T and over	Cast iron	Mild steel or meehanite	EN8 or EN9 hardened and tempered or case hardened mild steel

Kilowatt ratings, for European and ANSI chains, shown in the ratings charts on pages 441 and 442 are based on the following conditions:

(a) Service factor of 1.
(b) Wheel centre distance of 30–50 times the chain pitch.
(c) Speed of driver sprocket (Z_1) whether on the driving or driven shaft.
(d) Two sprocket drive arrangement.
(e) Adjustment by centre distance or jockey on unloaded strand.
(f) Riveted endless chain (press fit connector).
(g) Correct lubrication.
(h) Accurate shaft/sprocket alignment.

Under these conditions a service life of approximately 15 000 h can ordinarily be expected when the chain operates under full rating. The kilowatt ratings for multiple strand European chains up to triplex are given respectively in columns 2 and 3, for ANSI chains up to quadruplex in columns 2, 3 and 4.

5.3.11 Rating chart construction

The rating charts at first sight look complicated, however, they are constructed from three simple lines. From this it may be seen that at lower speeds the failure mode is likely to be plate fatigue if the maximum power recommendation is exceeded. However, pin galling will occur due to boundary lubrication break down at very high speeds. At the intersection of these lines the bush and roller fatigue curve comes into a play and accounts for the rounded tops to each of the selection curves.

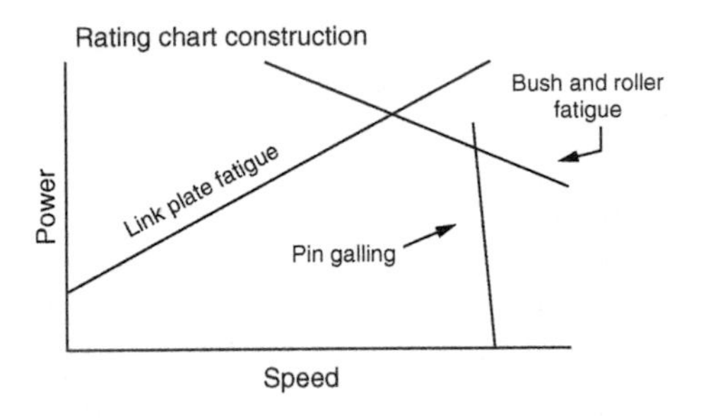

Bearing pressures

When a chain has been correctly selected, the mode of failure over a very long period of time is most likely to be wear.

The subject of wear, which depends on many factors, has been addressed earlier in this guide, however, a very useful indicator of the likely wear performance is the magnitude of pressure between the key mating surfaces, that is, pin and bush.

This pressure is known as the bearing pressure, and is obtained by dividing the working load by the bearing area. Bearing areas for standard chains are quoted in the designer data at the end of this guide.

The following table gives an indication of the implications of various bearing pressures but should not be used without reference to the other chain selection methods given in this guide.

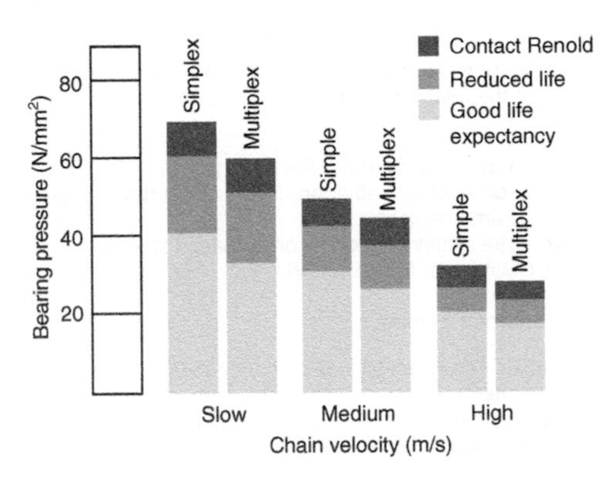

Slow velocity up to 60% of maximum allowable speed.
Medium velocity 60–80% of maximum allowable speed.
High velocity over 80% of maximum allowable speed.

Note: There is some variation between chains, and the above figures should be used as a guide only.

5.3.12 European chain rating chart

European Standard chain drives
Rating chart using 19T driver sprocket

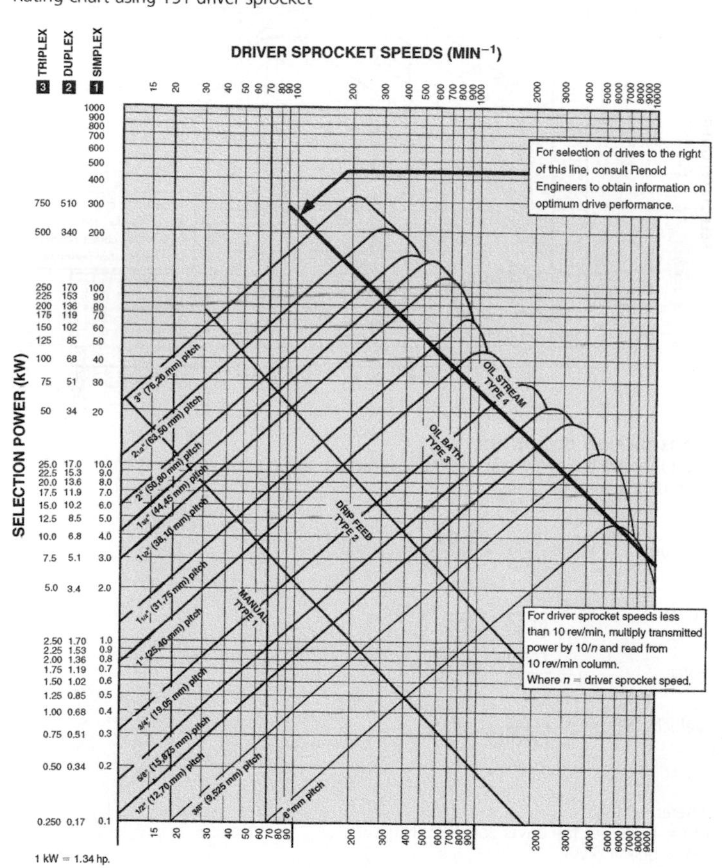

5.3.13 ANSI rating chart

American Standard chain drives
Rating chart using 19T driver sprocket

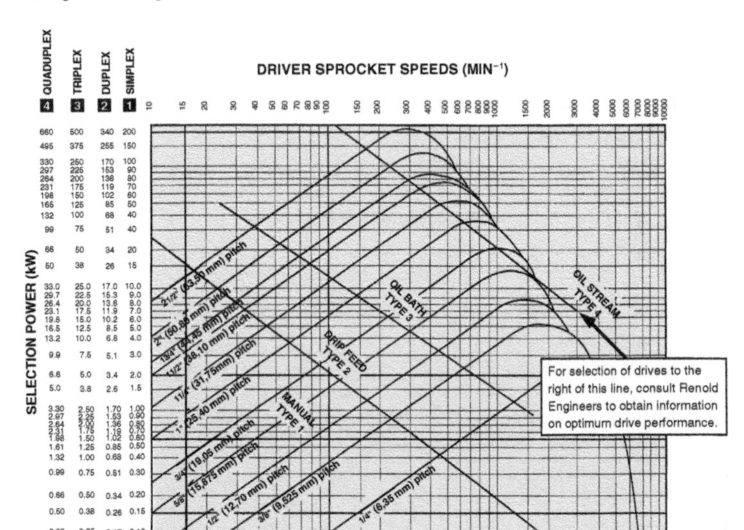

Transmission equations

The following equations give the relationships between power, torque and velocity for various drive arrangements.

Torque $$M = \frac{F_1\ 1}{2000}\ \frac{9550.Pr}{1}$$

Power $$Pr = \frac{M_1}{9550}\ \frac{1}{1000}\ (kW)$$

Force $$_1 = \frac{1000\ Pr}{v}\ \frac{2000}{1}\ (N)$$

Velocity $$v = \frac{1\ 1}{60\ 000}\ (m/s)$$

where:
Md = torque of the driver sprocket (Nm)
Pr = power (kW)
d_1 = pitch circle diameter of the driver sprocket (mm)
n_1 = driver sprocket speed (rev/min)
Z_1 = number of teeth in the driver sprocket
Z_2 = number of teeth in the driven sprocket

v = linear speed of the chain (m/s)
F_1 = chain pull (N)
P = pitch of the chain (mm)

Centripetal acceleration

Centripetal acceleration affecting parts of the chain engaged on the sprockets is determined by:

$$F_2 = q \cdot v^2 \text{ (N)}$$

where:
F_2 = force (N)
q = mass of the chain (kg/m)

From this formula we can see that at high speed, this force is not negligible and is the main reason for speed limitation.

5.3.14 Chain suspension force

The force acting between one link and the next due to the mass of the chain is small and is internally balanced within the chain. This will do no more than cause the chain to adopt a sagging catenery shape between the sprockets.

Allowance will need to be made in the installation for the slightly different postures adopted by the chain between zero and maximum load.

5.3.15 Lubrication

Chain drives should be protected against dirt and moisture and be lubricated with good quality non-detergent mineral based oil. A periodic change of oil is desirable. Heavy oils and greases are generally too stiff to enter the chain working surfaces and should not be used.

Care must be taken to ensure that the lubricant reaches the bearing areas of the chain. This can be done by directing the oil into the clearances between the inner and outer link plates, preferably at the point where the chain enters the sprocket on the bottom strand.

The table below indicates the correct lubricant viscosity for various ambient temperatures.

Ambient temperature	Lubricant rating	
°C	SAE	BS 4231
−5 to +5	20	46 to 68
5 to 40	30	100
40 to 50	40	150 to 220
50 to 60	50	320

For the majority of applications in the above temperature range, a multigrade SAE 20/50 oil would be suitable.

Use of grease

As mentioned above, the use of grease is not recommended. However, if grease lubrication is essential, the following points should be noted:

- Limit chain speed to 4 m/s.
- Applying normal greases to the outside surfaces of a chain only seals the bearing surfaces and will not work into them. This causes premature failure. Grease has to be heated until fluid and the chain is immersed and allowed to soak until all air bubbles cease to rise. If this system is used, the chains need regular cleaning and regreasing at intervals depending on the drives' power and speed. It should also be noted that temperatures above 80°C will cause damage to many greases and reduce their effectiveness.

Abnormal ambient temperatures

For elevated temperatures up to 250°C, dry lubricants such as colloidal graphite or M_OS_2 in white spirit or poly-alkaline glycol carriers are most suitable.

Conversely, at low temperatures between −5°C and −40°C, special low temperature initial greases and subsequent oil lubricants are necessary. Lubricant suppliers will give recommendations.

5.3.16 Lubricating methods

There are four basic methods for lubricating chain drives. The recommended lubrication method is based on the chain speed and power transmitted, and can be found in the rating charts (see pages 441 and 442).

Type 1, Manual operation

Oil is applied periodically with a brush or oil can, preferably once every 8 h of operation. Volume and frequency should be sufficient to just keep the chain wet with oil and allow penetration of clean lubricant into the chain joints.

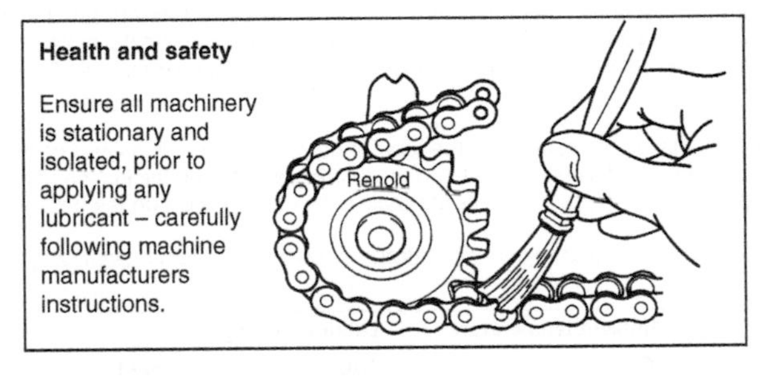

Applying lubricant by aerosol is also a satisfactory method, but it is important that the aerosol lubricant is of an approved type for the application, such as that supplied by Renold. This type of lubricant 'winds' in to the pin/bush/roller clearances, resisting both the tendency to drip or drain when the chain is stationary and centrifugal 'flinging' when the chain is moving.

Type 2, Drip lubrication

Oil drips are directed between the link plate edges from a drip lubricator. Volume and frequency should be sufficient to allow penetration of lubricant into the chain joints.

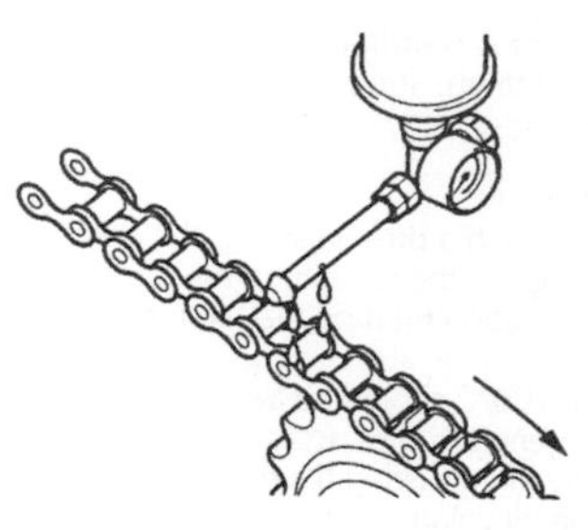

Type 3, Bath or disc lubrication

With oil bath lubrication the lower strand of chain runs through a sump of oil in the drive housing. The oil level should cover the chain at its lowest point whilst operating.

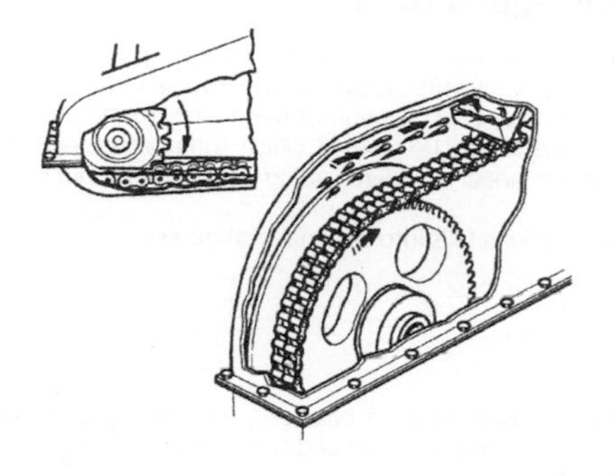

With slinger disc lubrication an oil bath is used, but the chain operates above the oil level. A disc picks up oil from the sump and deposits it on the chain by means of deflection plates. When such discs are employed they should be designed to have peripheral speeds between 180 and 2440 m/min.

Type 4, Stream lubrication

A continuous supply of oil from a circulating pump or central lubricating system is directed onto the chain. It is important to ensure that the spray holes from which the oil emerges are in line with the chain edges. The spray pipe should be positioned so that the oil is delivered onto the chain just before it engages with the driver sprocket.

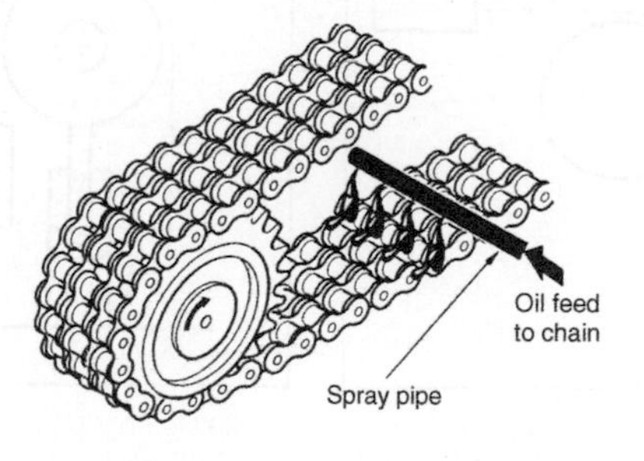

This ensures that the lubricant is centrifuged through the chain and assists in cushioning roller impact on the sprocket teeth. Stream lubrication also provides effective cooling and impact damping at high speeds.

Effect of temperature

An important factor to control in a drive system is the chain and chaincase temperatures during operation. Depending on the severity of the drive service, continuity of use, etc., special attention to the lubrication method may be required.

Chain temperatures above 100°C should be avoided if possible due to lubrication limitations, although chain can generally give acceptable performance up to around 250°C in some circumstances. A way of improving the effectiveness of the lubrication and its cooling effect is to increase the oil volume (up to 4.5 l/min per chain strand) and incorporate a method of external cooling for the oil.

5.3.17 Lifting applications

This section covers applications, such as lifting and moving, where the loads involved are generally static. Obviously, dynamic loads are also involved in most applications and the designer needs to take due consideration of these. The machinery designer should also refer to DTI Publication INDY J1898 40 M which summarises legislation in place from 1st January 1993 to 1st January 1995 regarding machinery product standards.

Chain for lifting applications falls into two main categories:

- Leaf chains.
- Bush and roller chains.

Leaf chain

Leaf chain is generally used for load balancing type lifting applications as illustrated below. They must be anchored at either end since there is no means of geared engagement in the chain itself.

Safety factors

A safety factor of 7:1 is normal for steady duty reciprocating motion, for example, fork lift trucks. For medium shock loads, 9:1 and for heavy shock loads, 11:1.

Operating speed

Applications should not exceed a maximum chain speed of 30 m/min.

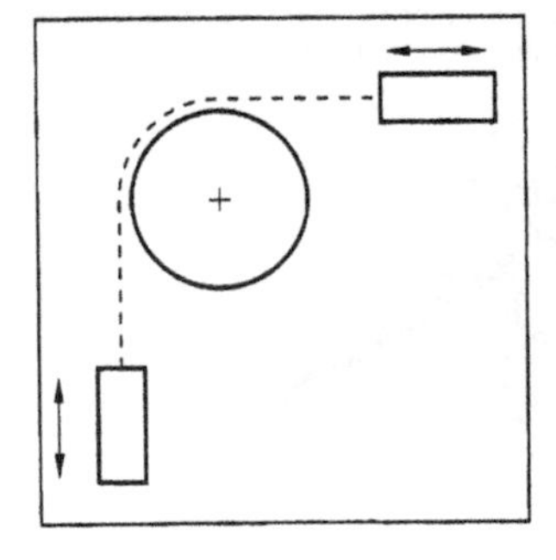

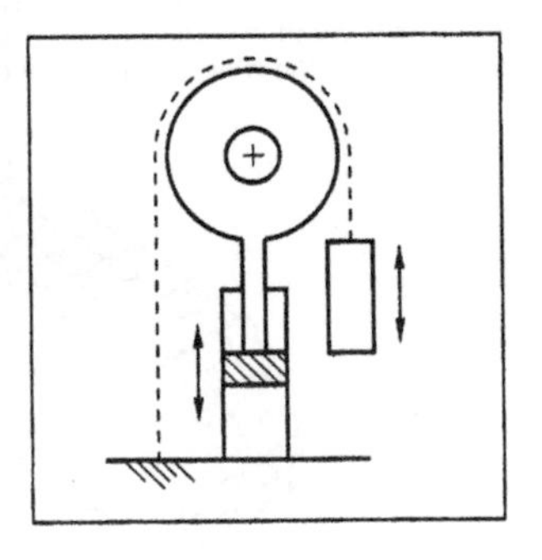

Applications

1. Machine tools – planers, drills, milling heads, machine centres.
2. Fork lift trucks, lifts, hoists.
3. Counterweight balances – jacks, doors, gates, etc.

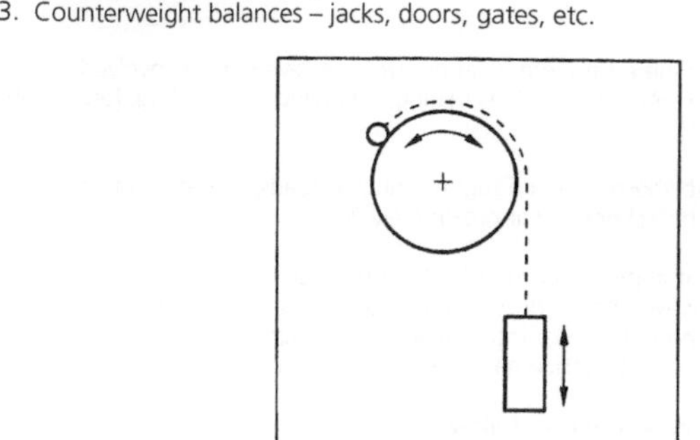

Bush and Roller chains

Bush and roller chains can be used for lifting and moving purposes, and have the advantage over leaf chain in that they may be geared into a suitable driving sprocket. Roller chain has a better wear resistance than leaf chain and may be used at higher speeds.

Safety factors

Applications vary widely in the nature of loads applied and it is therefore recommended that factors of safety are applied which allow for some degree of abuse:

- A factor of safety of 8:1 in non-passenger applications.
- A factor of safety of 10:1 in passenger applications.

Lower factors of safety than these may be used (except for passenger applications), where careful consideration of the maximum loads and health and safety implications have been made. For comments on this see the section 'Influences on chain life'.

Operating speeds

Applications should not normally exceed a maximum chain speed of 45 m/min. For speeds higher than this, consider selection as if the chain were in a power transmission application converting the chain load to power using the following formula:

$$\text{Power} = FV\ (\text{kW})$$

where:
F = Load (kN)
V = Velocity of chain (m/s)

Then apply selection power factors as shown in step two of 'Drive Selection'.

Calculate equivalent rev/min by using the smallest sprocket in the system where

$$\text{speed} = \frac{60\,000\ V}{PZ}$$

where:
P = Chain pitch (mm)
Z = No of teeth in sprocket

Select lubrication methods also from the selection chart.

5.3.18 ANSI Xtra range

Transmission chain is also available in heavy duty versions of the ANSI standard range of chain.

These chains are suitable where frequent or impulsive load reversals are involved. Typical applications are in primary industries, such as mining, quarrying, rock drilling, forestry and construction machinery.

In order to accommodate these higher fatigue inducing loads, material for inner and outer plates is increased in thickness by approximately 20%.

This modification does not improve the tensile strength since the pin then becomes the weakest component. However, heavy duty chains with higher tensile strength are available. This is achieved by through hardening instead of case hardening the pin, but unfortunately this action reduces wear performance due to the lower pin hardness.

Renold ANSI Xtra chains are available as follows:

- Xtra H range Thicker plates
- Xtra V range Through hardened pins
- Xtra HV range Thicker plates and through hardened pins.

The H and HV chains are not suitable or appropriate for high speed transmission applications.

The following points should also be noted:

- The V range of chains are totally interchangeable with standard ANSI chain.
- Simple chains of standard, H or HV designs all have identical gearing dimensions and therefore can operate on the same sprockets as for standard chains. The thicker plates will require a larger chain track and it may be desirable to use sprockets with heat treated teeth. Multiplex chain requires an increased transverse pitch of the teeth but other gearing dimensions are the same.
- The only reason to use H or HV chains is where fatigue life is a problem. We do not make any cranked (offset) links or slip-fit connecting links for this range, since these have a lower fatigue resistance.
- Detachable (cottered) versions can be produced if required as could triplex or wider chains.

5.3.19 Influences on chain life

Factors of safety

All Renold chain is specified by its minimum tensile strength. To obtain a design working load it is necessary to apply a 'factor of safety' to the breaking load. However, before considering this, the following points should be noted:

- Most chain side plates are manufactured from low to medium carbon steel and are sized to ensure they have adequate strength and also ductility to resist shock loading.
- These steels have yield strengths around 65% of their ultimate tensile strength. What this means is that if chains are subjected to loads of greater than this, depending on the material used in the side plates, then permanent pitch extension will occur.
- Most applications are subjected to transient dynamic loads well in excess of the maximum static load and usually greater than the designer's estimate.

- Motors, for example, are capable of up to 200% full load torque output for a short period.

The consequences of these points are that chain confidently selected with a factor of safety of 8:1 on breaking load is, in effect, operating with a factor of safety of around 5:1 on yield and much less than this when the instantaneous overload on the drive is considered.

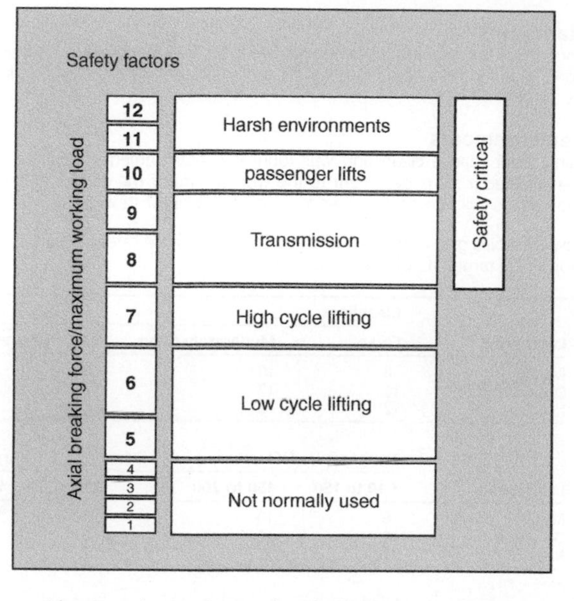

A further consideration when applying a factor of safety to a chain application is the required chain life.

In a properly maintained application a life of 8 000 000 cycles or 15 000 h, whichever comes first, is normal. Wear will be the usual mode of failure.

In applications where low factors of safety are required, the life will reduce accordingly.

The maximum working load is obtained by dividing the chain minimum tensile strength by the factor of safety.

The table below gives a rough indication of life for various factors of safety.

Factor		Cycles maximum	Type of application
Simple	Multiplex		
5.0	6.0	1 000 000	Dynamic load does not exceed working load
6.0	7.2	2 000 000	
8.0	8.0	8 000 000	Dynamic loads can occasionally exceed working load by 20%
10.0	10.0	8 000 000	All passenger lifts

It should be noted that at factors below 8:1, bearing pressures increase above the maximum recommended, with the result that increased wear will arise unless special attention is taken with lubrication, for example:

- More frequent lubrication.
- Higher performance lubricants.
- Better methods of applying lubrication.

Important note

For factors of 5:1 the resulting bearing pressure is 50% higher than recommended and chain working under these conditions will wear prematurely, whatever type of lubrication regime is used.

Harsh environments

In anything other than a clean and well lubricated environment, the factor of safety should be adjusted if some detriment to the working life of the chain is to be avoided. Low temperatures will also decrease working life, especially if shock loads are involved.

The following tables give a general guide to the appropriate safety factors for different applications for a target life of 8 000 000 cycles.

	Cleanliness		
Lubrication	**Clean**	**Moderately clean**	**Dirty/abrasive**
Regular	8	10	12
Occasional	10	12	14
None	12	12	14

	Temperature (°C)		
Lubrication	**+10 to 150**	**150 to 200**	**200 to 300**
Regular	8	10	12
Occasional	10	12	14
None	12	12	14

	Load regime		
Temperature (°C)	**Smooth**	**Moderate shocks**	**Heavy shocks**
+10 to +150	8	11	15
0 to +10	10	15	19
−20 to 0	12	20	25
−40 to −20	15	25	33

5.3.20 Chain extension

When designing lifting applications it can be useful to know how much a chain will extend under a given load.

The approximate elongation of a chain under a given load can be measured by using the following formulae:

- Simplex Chain $$\Delta L = \frac{(14.51) \cdot 10^{-5} \cdot L}{P^2} F_1$$
- Duplex Chain $$\Delta L = \frac{(9.72) \cdot 10^{-5} \cdot L}{P^2} F_1$$

- Triplex Chain $$\Delta L = \frac{(7.26) \cdot 10^{-5} \cdot L}{P^2} F_1$$

where:
ΔL = change in chain length (mm)
L = original length of the chain (mm)
P = pitch of the chain (mm)
F_1 = average load in the chain.

5.3.21 Matching of chain

Any application in which two or more strands of transmission chain are required to operate side by side in a common drive, or conveying arrangement, may involve the need for either pairing or matching. Such applications generally fall into one of the following categories:

Length matching for conveying and similar applications

Wherever length matching of transmission chain is necessary it is dealt with as follows:

- The chains are accurately measured in handling lengths between 3 m and 8 m as appropriate and then selected to provide a two (or more) strand drive having overall length uniformity within close limits. However, such length uniformity will not necessarily apply to any intermediate sections along the chains, but the actual length of all intermediate sections, both along and across the drive, will not vary more than our normal manufacturing limits. However, adapted transmission chains are usually manufactured to specific orders which are generally completed in one production run so that it is reasonable to assume that length differences of intermediate sections will be small.
- Chains are supplied in sets which are uniform in overall length within reasonably fine limits and will be within our normal manufacturing limits. It should be noted that chain sets supplied against different orders at different times may not have exactly the same lengths to those supplied originally, but will vary by no more than our normal tolerance of 0.0%, +0.15%.

Pitch matching transmission drive chains

Pitch matched chains are built up from shorter subsections (usually 300 to 600 mm lengths) which are first measured and then graded for length. All subsections in each grade are of closely similar length and those forming any one group across the set of chains are selected from the same length grade.

The requisite number of groups are then connected to form a pitch matched set of chains, or alternatively, if this is too long for convenient handling, a set of handling sections for customer to assemble as a final set of pitch matched chain. Suitable tags are fixed to the chains to ensure they are connected together in the correct sequence.

Identification of handling lengths

	Handling lengths		
	1	**2**	**3**
A strand	A–A1	A1–A2	A2–A3
B strand	B–B1	B1–B2	B2–B3
C strand	C–C1	C1–C2	C2–C3

Long chains are made up in sections, each section being numbered on end links. Sections should be so joined up that end links with similar numbers are connected. Where chains are to run in sets of two or more strands, each strand is stamped on end links of each section with a letter, in addition to being numbered. Correct consecutive sections for each strand must be identified from the end links and joined up as indicated.

By these means, the actual length of any intermediate portion of one strand (as measured from any one pitch point to any other) will correspond closely with that of the transversely equivalent portion on the other strands, generally within 0.05 mm, depending on the chain pitch size.

Pitch matching adapted transmission chains (when attachments are fitted to chains)

With the sole exception of extended bearing pins, it is not possible to match the pitch of holes in attachments themselves to within very fine limits, due to the additional tolerances to be contended with (bending, holing, etc.).

Colour coding

For customers who wish to match their chains, perhaps in order to fit special attachments *in situ*, Renold colour code short lengths of chain within specified tolerance bands. These will normally be red, yellow or green paint marks to indicate lower, mid and upper thirds of the tolerance band. For even finer tolerance bands additional colours can be used, but normally a maximum of five colours will be more than adequate.

Colour	Red	0.05%
	Yellow	0.10%
	Green	0.15%
	Blue	For finer
	White	Tolerances

5.3.22 To measure chain wear

A direct measure of chain wear is the extension in excess of the nominal length of the chain. the chain wear can therefore be ascertained by length measurement in line with the instructions given below:

- Lay the chain, which should terminate at both ends with an inner link (part No 4), on a flat surface, and, after anchoring it at one end, attach to the other end a turnbuckle and a spring balance suitably anchored.
- Apply a tension load by means of the turnbuckle amounting to:

Simplex chain	$P^2 \times 0.77$ (N)
Duplex chain	$P^2 \times 1.56$ (N)
Triplex chain	$P^2 \times 2.33$ (N)

where P is the pitch in mm.

In the case of double pitch chains (e.g. chains having the same breaking load and twice the pitch) apply measuring loads as for the equivalent short pitch chains.

As an alternative, the chain may be hung vertically and the equivalent weight attached to the lower end.

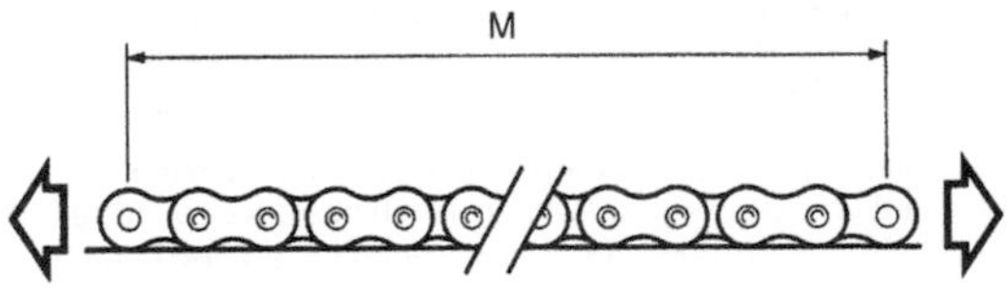

M

- Measure length 'M' (see diagram above) in millimetres from which the percentage extension can be obtained from the following formula:

$$\text{Percentage extension} = M - \frac{(N \cdot P)}{N \cdot P} \times 100$$

where:
N = Number of pitches measured
P = Pitch

- As a general rule, the useful life of the chain is terminated and the chain should be replaced when extension reaches 2% (1% in the case of double pitch chains). For drives with no provision for adjustment, the rejection limit is lower, dependent on the speed and layout. A usual figure is between 0.7% and 1.0% extension.

Renold chain wear guide
A simple-to-use chain wear guide is available from Renold chain for most popular sizes of chain pitch. Please contact your Sales Office for details.

5.3.23 Repair and replacement

Sprockets
Examination of both flanks will give an indication of the amount of wear which has occurred. Under normal circumstances this will be evident as a polished worn strip about the pitch circle diameter of the sprocket tooth.

If the depth of this wear 'X' has reached an amount equal to 10% of the 'Y' dimension, then steps should be taken to replace the sprocket. Running new chain on sprockets having this amount of tooth wear will cause rapid chain wear.

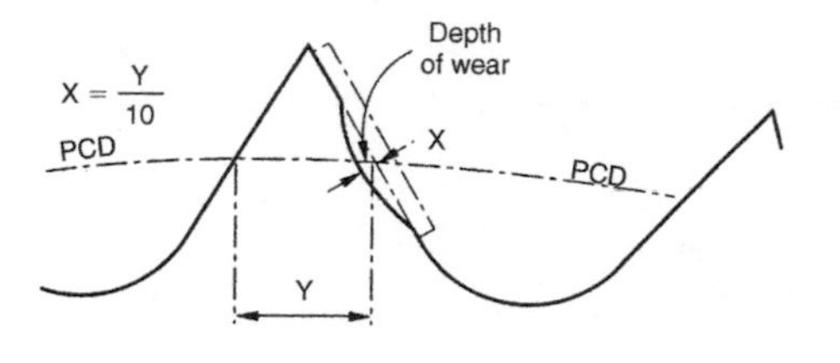

It should be noted that in normal operating conditions, with correct lubrication the amount of wear 'X' will not occur until several chains have been used.

Chain
Chain repair should not as a rule be necessary. A correctly selected and maintained chain should gradually wear out over a period of time (approximately 15 000 h), but it should not fail. Please refer to the Installation and Maintenance section, which gives an indication of the service life remaining.

If a transmission chain sustains damage due to an overload, jam-up, or by riding over the sprocket teeth, it should be carefully removed from the drive and given a thorough visual examination. Remove the lubricating grease and oil to make the job easier.

Depending on the damage, it may be practicable to effect temporary repairs using replacement links. However, it is not a guarantee that the chain has not been over stressed and so made vulnerable to a future failure. The best policy, therefore, is to remove the source of trouble and fit a new chain. This should be done for the following reasons:

1. The cost of down time to the system or machine can often outweigh the cost of replacing the chain.
2. A new or even used portion of chain or joints assembled into the failed chain will cause whipping and load pulsation. This can, and probably will, produce rapid failure of the chain and will accelerate wear in both the chain and its sprockets.

If a chain has failed two or more times, it is certain the chain will fail again in time. If no replacement is immediately available, repair the chain, but replace it at the earliest opportunity.

5.3.24 Chain adjustment

To obtain full chain life, some form of chain adjustment must be provided, preferably by moving one of the shafts. If shaft movement is not possible, an adjustable jockey sprocket engaging with the unloaded strand of the chain is recommended. Generally the jockey should have the same number of teeth as the driver sprocket and care should be taken to ensure the speed does not exceed the maximum shown in the rating charts (see pages 437 and 438).

The chain should be adjusted regularly so that, with one strand tight, the slack strand can be moved a distance 'A' at the mid point (see diagram below). To cater for any eccentricities of mounting, the adjustment of the chain should be tried through a complete revolution of the large sprocket.

A = total movement

C = horizontal centre distance

$$\text{Total movement } 'A' \text{ (mm)} = \frac{C \text{ (mm)}}{K}$$

where:
K = 25 for smooth drives
50 for shock drives

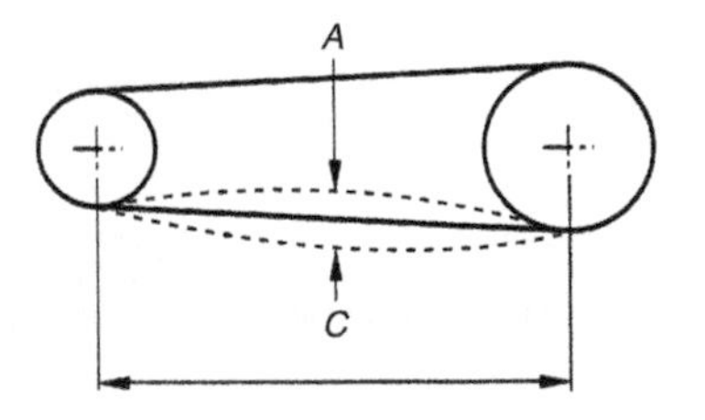

For vertical drives please consult the installation and maintenance section, which gives more details on chain adjustment.

5.3.25 Design ideas

A variety of applications.

Conveying, indexing, lifting and pulling, power transmission and timing.

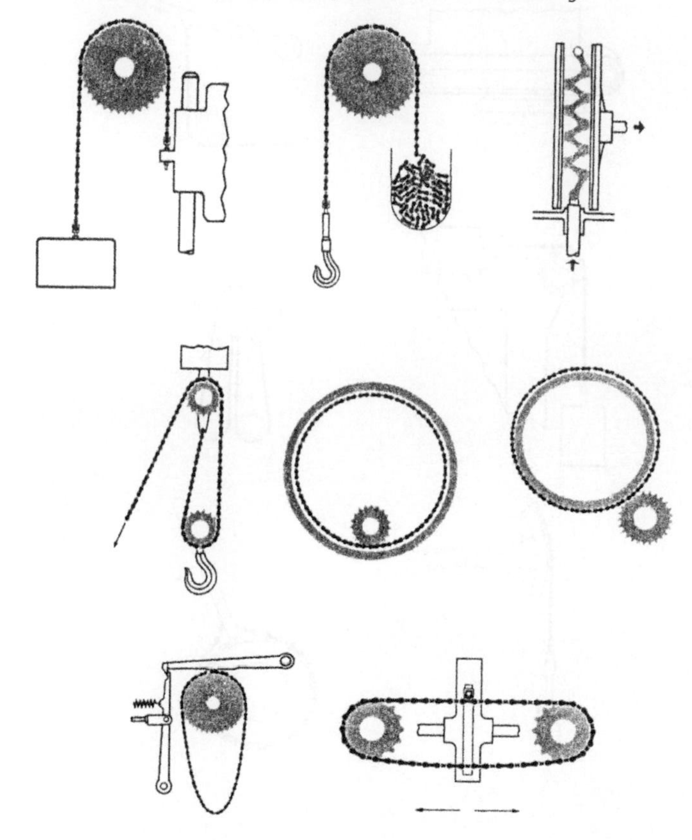

A variety of industries.

Aircraft, automotive, marine, mechanical handling, motorcycle, nuclear oilfield.

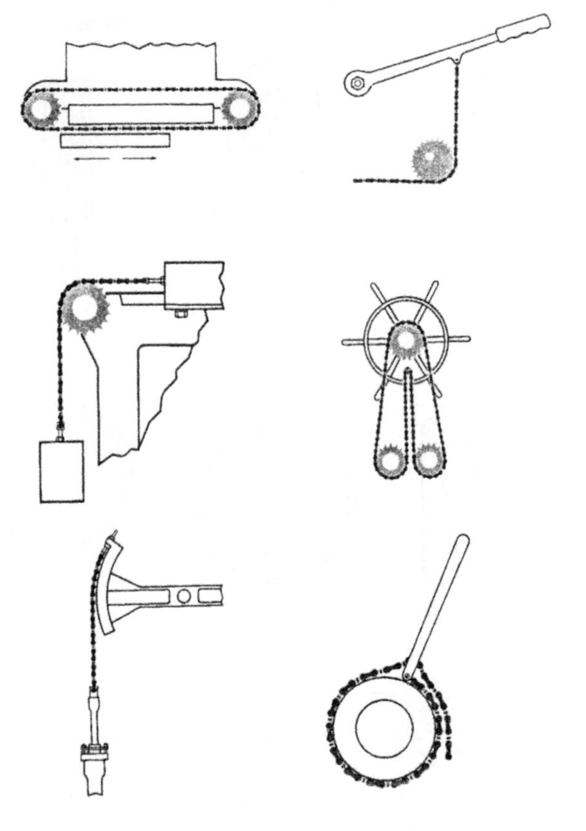

5.3.26 Table of PCD factors

To obtain PCD of any sprocket with 9 to 150 teeth, multiply chain pitch by appropriate factor.

For example, the PCD of a 38T sprocket of 3/4 in. (19.05 mm) pitch = 19.05 × 12.110 = **230.70 mm**.

No. of teeth	PCD factor	No. of teeth	PCD factor	No. of teeth	PCD factor
9	2.924	57	18.153	105	33.428
10	3.236	58	18.471	106	33.746
11	3.549	59	18.789	107	34.064
12	3.864	60	19.107	108	34.382
13	4.179	61	19.426	109	34.701
14	4.494	62	19.744	110	35.019
15	4.810	63	20.062	111	35.337
16	5.126	64	20.380	112	35.655
17	5.442	65	20.698	113	35.974
18	5.759	66	21.016	114	36.292
19	6.076	67	21.335	115	36.610
20	6.392	68	21.653	116	36.928
21	6.709	69	21.971	117	37.247
22	7.027	70	22.289	118	37.565
23	7.344	71	22.607	119	37.883
24	7.661	72	22.926	120	38.202
25	7.979	73	23.244	121	38.520
26	8.296	74	23.562	122	38.838
27	8.614	75	23.880	123	39.156
28	8.931	76	24.198	124	39.475
29	9.249	77	24.517	125	39.793
30	9.567	78	24.835	126	40.111
31	9.885	79	25.153	127	40.429
32	10.202	80	25.471	128	40.748
33	10.520	81	25.790	129	41.066
34	10.838	82	26.108	130	41.384
35	11.156	83	26.426	131	41.703
36	11.474	84	26.744	132	42.021
37	11.792	85	27.063	133	42.339
38	12.110	86	27.381	134	42.657
39	12.428	87	27.699	135	42.976
40	12.746	88	28.017	136	43.294
41	13.063	89	28.335	137	43.612
42	13.382	90	28.654	138	43.931
43	13.700	91	28.972	139	44.249
44	14.018	92	29.290	140	44.567
45	14.336	93	29.608	141	44.885
46	14.654	94	29.927	142	45.204
47	14.972	95	30.245	143	45.522
48	15.290	96	30.563	144	45.840
49	15.608	97	30.881	145	46.159
50	15.926	98	31.200	146	46.477
51	16.244	99	31.518	147	46.795
52	16.562	100	31.836	148	47.113
53	16.880	101	32.154	149	47.432
54	17.198	102	32.473	150	47.750
55	17.517	103	32.791		
56	17.835	104	33.109		

5.3.27 Simple point to point drives: Example one

The following worked examples give simple step-by-step guidance on selecting various types of chain drive systems. Renold technical staff are available to advise on any chain selection problems.

For details of transmission equations see page 442.

Example one: Rotary pump drive

Given:

- Pump speed — 360 rev/min
- Power absorbed — 7.5 kW
- Driver — Electric motor at 1440 rev/min
- Constraints — Centre distance approximately 458 mm; Adjustment by shaft movement

Selection parameters

- Use $Z_1 = 19T$
- No polygonal effect
- Satisfactory for smooth drives

Calculate the drive ratio as follows:

$$\textit{Drive ratio} = i = \frac{Z_2}{Z_1} = \frac{N_2}{N_1} = \frac{1440}{360} = 4$$

Therefore the driven number of teeth

$$Z_2 = 4 \times Z_1 = 4 \times 19 = \mathbf{76T}$$

Selection factors

Application factor $f_1 = 1$ (driver and driven sprockets smooth running):

$$\text{Tooth factor } f_2 = \frac{19}{Z_1} = \frac{19}{19} = 1$$

Selection power = 7.5 × 1 × 1 = **7.5 kW**

Select chain

The chain can now be selected using charts 3 and 4 and cross referencing power to speed, giving the following possibilities:

0.5″ BS Simplex	(approximately 81% of rated capacity)
0.375″ BS Duplex	(approximately 98% of rated capacity)
0.5″ ANSI Simplex	(approximately 83% of rated capacity)
0.375″ ANSI Duplex	(approximately 84% of rated capacity)

0.375″ ANSI Duplex chain is unsuitable as it is a bush chain.

Note: The approximation percentage of rated capacity is calculated by dividing the selection power at 1440 rev/min by the chains maximum capacity at 1440 rev/min.

For this example we will choose 0.5″ European Simplex.

Installation parameters

Lubrication European chain rating chart (see page 441) clearly indicates the chain needs oilbath lubrication. The chain will need to be enclosed and run in a sump of oil.

We now calculate the chain length:

$$L = \frac{Z_1 + Z_2}{2} + \frac{2C}{P} + \frac{P \cdot \left(\frac{Z_2 - Z_1}{2\pi}\right)^2}{C}$$

$$L = \frac{19 + 76}{2} + \frac{2 \times 458}{12.7} + \frac{12.7 \cdot \left(\frac{76 - 19}{2\pi}\right)^2}{458} = \mathbf{121.9}$$

Round up to the nearest number of even pitches, that is, **122**.

Centre distance calculation

The centre distance of the drive can now be calculated using the formula shown below:

$$= \frac{P}{8}\left[2L - Z_2 - Z_1 + \sqrt{(2L - Z_2 - Z_1)^2 - \frac{\pi}{3.88}(Z_2 - Z_1)^2}\right]$$

$$= \frac{12.7}{8}\left[((2 \times 122) - 76 - 19) + \sqrt{((2 \times 122) - 76 - 19)^2 - \frac{\pi}{3.88} \times (76 - 19)^2}\right]$$

$$= \mathbf{458.6\,mm}$$

Adjustment

Provide for chain wear of 2% or two pitches, whichever is smaller, in this case, $(122 \times 1.02) - 122 =$ **2.44 pitches**.

Therefore use two pitches and recalculate using:

$L = 124$ in the above equation. This gives
$C = 471.7\,\text{mm}$

that is, total adjustment of **13.1 mm**.

Note that in practice, some negative adjustment will facilitate assembly and will be essential if it is intended to assemble chain which is pre-joined into an endless loop.

Other data

$$\text{Chain velocity } \frac{N \times P \times Z_1}{60\,000} = \frac{1440 \times 12.7 \times 19}{60\,000} = \mathbf{5.79\,m/s}$$

Load in chain due to power transmitted $= \dfrac{Q \times 1000}{v}$
(where Q = Selection power (kW))

$$= \frac{7.5 \times 1000}{5.79} = \mathbf{1295\,N}$$

Load in chain due to centripetal acceleration = Chain mass/metre × velocity^2
$= 0.68 \times 5.79^2$
$= 23\,N$

Total chain working load $= \mathbf{1318\,N}$

Note the load in the chain due to centripetal acceleration becomes much more significant at higher speeds since the square of the chain velocity is in the equation:

Chain axial breaking force = **19 000 N**

Chain safety factor $= \dfrac{1900}{1318} = \mathbf{14.4}$

Chain bearing area = **50 mm²**

Bearing pressure $= \dfrac{\text{Working load}}{\text{Bearing area}} = \dfrac{1318}{50} = \mathbf{23.36\,N/mm^2}$

5.3.28 Simple point to point drives: Example two

The following worked examples give simple step-by-step guidance on selecting various types of chain drive systems. Renold technical staff are available to advise on any chain selection problems.

For details of transmission equations see page 442.

Example two: 4-cylinder compressor

Given:

- Pump speed 250 rpm
- Power absorbed 250 kW
- Driver Electric motor at 960 rpm
- Constraints Centre distance approx. 1500 mm

Selection parameters

Use a 25T sprocket for an impulsive drive (see page 434 selection of drive ratio and sprockets:

$$\textit{Drive ratio} = \frac{Z_2}{Z_1} = \frac{N_2}{N_1} = \frac{960}{250} = \mathbf{3.84}$$

Number of teeth $Z_2 = 3.84 \times Z_1 = 3.84 \times 25 = \mathbf{95T}$

Selection factors

Application factor $f_1 = 1.5$ (driver and driven sprocket medium impulsive):

Tooth factor $f_2 = \frac{19}{Z_1} = \frac{19}{25} = \mathbf{0.76}$

Selection power = Transmitted power $\times f_1 \times f_2$ (kW)
Selection power = $250 \times 1.5 \times 0.76 = \mathbf{285\,kW}$

Select chain

The chain can now be selected using European Chain Rating Chart (see page 441) by cross referencing the power (285 kW on the vertical axis) and speed (960 rev/min on the horizontal axis).

Two matched strands of 1.25" pitch European triplex chains could be used with a heat treated 25 tooth steel driver and a 95 tooth driven sprocket to give a drive ratio of 3.8:1.

Installation parameters

Lubrication European chain rating chart (see page 441) clearly shows that an oilstream system is required on this drive. The chain should run in an enclosure with a pump and sump arrangement.

We will now calculate the chain length:

$$L = \frac{25+95}{2} + \frac{2 \times 1500}{31.75} + \frac{31.75\left(\frac{95-25}{2\pi}\right)^2}{1500} = \mathbf{157.12}$$

Round up to the nearest number of even pitches that is, **158**.

Centre distance calculation

The centre distance of the drive can now be calculated using the standard formula below:
= 1514.44 mm

$$C = \frac{31.75}{8}\left[\left((2 \times 158) - 95 - 25\right) + \sqrt{\left((2 \times 158) - 95 - 25\right)^2 - \frac{\pi}{3.88}(95-25)^2}\right]$$

Adjustment

Chain velocity $\frac{960 \times 31.75 \times 25}{60\,000} = \mathbf{12.7\,m/s}$

Load in the chain $\frac{285 \times 1000}{12.7} = \mathbf{22\,440\,N}$

Load in the chain due to centripetal acceleration = $11.65 \times 2 \times 12.7 \times 12.7 = \mathbf{3758\,N}$

Total chain working load = **26027 N**

$$\text{Bearing pressure} = \frac{\text{Working load}}{\text{Bearing area}} = \frac{23\,272}{885 \times 2} = \mathbf{14.7\,N/mm^2}$$

$$\text{Chain safety factor} = \frac{\text{Breaking load}}{\text{Working load}} = \frac{294\,200}{26\,027} \times 2 = \mathbf{22.6}$$

Multi-shaft drives

Shafts in series

This arrangement shows the driving of live roller conveyors.

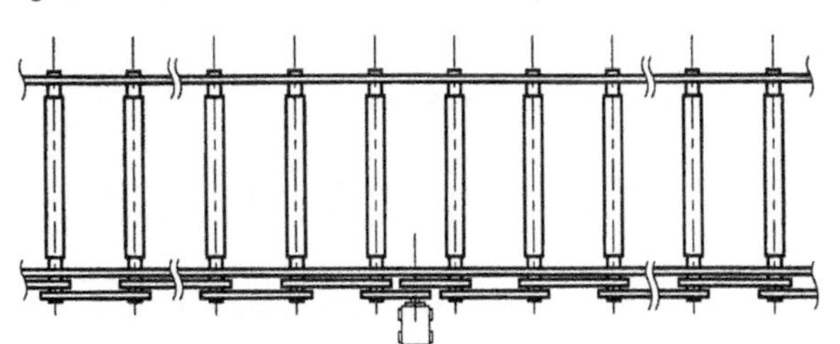

The choice of the chain is based on the slipping torque between the rollers and the material to be transported. The safety factor to be applied for this type of drive is typically:

Safety factor = 5 for one direction drives

Safety factor = 8 for reversible drives

Every roller except the last comprises two simple sprockets, or one special sprocket to be used with two simple chains. At low speeds or in reversible drives, sprockets with hardened teeth should be used.

Roller conveyors with less than 10 rollers can be driven from one of the ends of the track. When the number of rollers is higher, it is recommended that the driving arrangement is in the middle of the conveyor in order to have a better distribution of the power and the highest overall efficiency.

If we assume that a drive operating under ideal conditions such as a clean environment and correct lubrication achieves an efficiency of $R\%$, then the overall efficiency of a roller conveyor with X rollers will be:

$$100\left(\frac{R}{100}\right)^X = \frac{R^X}{100^{(X-1)}}$$

If the individual drive efficiency R is equal to 98%, then the drive of a roller conveyor with 30 rollers will therefore only have an overall efficiency of 55%.

Consequently, it is recommended that no more than 30 rollers per drive are used. For roller conveyors with more than 30 rollers, use multiple drives.

The drive should be able to develop a torque corresponding to the slipping torque of the loaded rollers.

5.3.29 Simple point to point drives: Example three

The following worked examples give simple step-by-step guidance on selecting various types of chain drive systems. Renold technical staff are available to advise on any chain selection problems.

For details of transmission equations see page 438.

Example three

Given:

- Moving a stack of steel plates.
- 20 rollers with a diameter of 150 mm.
- Shafts with a diameter of 60 mm on ball bearings.
- Weight of one roller 1900 N.
- There are two stacks on the conveyor at any one time.
- One stack weighs 17 500 N with a length of 1500 mm.
- Total nett load: 35 000 N (two stacks).
- Centre distance of the rollers: 300 mm.
- Linear speed: 15 m/min.
- PCD of the sprockets: 140 mm.
- Impulsive load: 30 starts per hour, in one direction.

Assumptions

- A drive is placed in the middle with 10 rollers on each side.
- The rolling resistance of the rollers is 0.05.
- The friction resistance between the rollers and the load is 0.25.
- The efficiency per drive is 98%.

Selection calculations

Every stack of steel is 17 500 N and is conveyed by:

$$\frac{\text{Stack length}}{\text{Centre distance of rollers}} = \frac{1500}{300} = \mathbf{5\ rollers}$$

or 10 rollers for the total nett load.

If a nett load of 35 000 N is added to the total weight of 10 bearing rollers (19 000 N), then this gives a gross load of 54 000 N. The tangential force for 10 rollers is: 54 000 × 0.05 = 2700 N and the corresponding torque is:

$$F \times d\,(\text{force} \times \text{distance}) = 2700 \times \frac{0.06}{2} = \mathbf{81\ Nm}$$

Note: where d = shaft diameter.

For each group of 10 rollers the efficiency will be:

$$\frac{98^{10}}{100^{9}} = \mathbf{81.7\%}$$

The effective torque then becomes:

$$\frac{\text{Actual torque}}{\text{Efficiency}} = \frac{81}{0.817} = \mathbf{99\ Nm}$$

For sprockets with a pitch circle diameter of 140 mm, the pull in the chain will be:

$$\frac{2000 \times \text{Md}}{d_1} = \frac{2000 \times 99}{140} = \mathbf{1414\ N}$$

The friction force for a friction coefficient of 0.25 is 35 000 × 0.25 = **8750 N**.

The corresponding torque is equal to:

$$F \times d(\text{force} \times \text{distance}) = 8750 \times \frac{0.15}{2} = \mathbf{656\ Nm}$$

Note: Where d = radius of shaft.

The total drive torque is 656 + 81 = **737 Nm**

The effective torque is therefore:

$$\frac{737}{0.817} = \mathbf{902\ Nm}$$

The pull in the chain then becomes:

$$\frac{2000 \times Md}{d_1} = \frac{2000 \times 902}{140} = \mathbf{12\,886\ N}$$

Per drive we can now evaluate chain ISO 16B-1 or Renold chain 1 10 088 running with two sprockets with 17 teeth and a pitch circle diameter of 138 mm.

In normal use:

$$\text{The safety factor } \frac{\text{Axial breaking force}}{\text{Working load}} = \frac{67\,000}{1414} = \mathbf{47.4}$$

$$\text{Bearing pressure } \frac{\text{Working load}}{\text{Bearing area}} = \frac{1414}{207} = \mathbf{6.83\ N/mm^2}$$

When slipping they are:

$$\text{The safety factor } \frac{\text{Axial breaking force}}{\text{Pull in chain}} = \frac{67\,000}{12\,886} = \mathbf{5.2}$$

$$\text{Bearing pressure } \frac{\text{Chain pull}}{\text{Bearing area}} = \frac{12\,886}{207} = \mathbf{62.26\ N/mm^2}$$

The linear speed of the chain is:

$$\frac{\text{stack speed} \times d_1}{\text{roller dia}}\left(\frac{1}{60}\right) = \frac{15 \times 0.138}{0.15 \times 60} = \mathbf{0.23\ m/sec}$$

Note: Where d_1 = PCD of sprocket in metres.

For each group of 10 rollers the power is: $\frac{F_1 \cdot V}{1000}$

Under normal working conditions

$$\frac{\text{Working load} \times \text{linear speed}}{1000} = \frac{1414 \times 0.23}{1000} = \mathbf{0.33\,kW}$$

$$\frac{\text{Chain pull} \times \text{linear speed}}{1000} = \frac{12\,886 \times 0.23}{1000} = \mathbf{2.96\,kW}$$

when the rollers are slipping.

- Taking the efficiency of the gear until into account and adding a factor of 25% to this total power, 3.7 kW will be necessary.

Note: At higher linear speeds, we should also take into account other additional factors such as the moment of inertia of the rollers and the power needed to accelerate the various components of the system.

Shafts in parallel

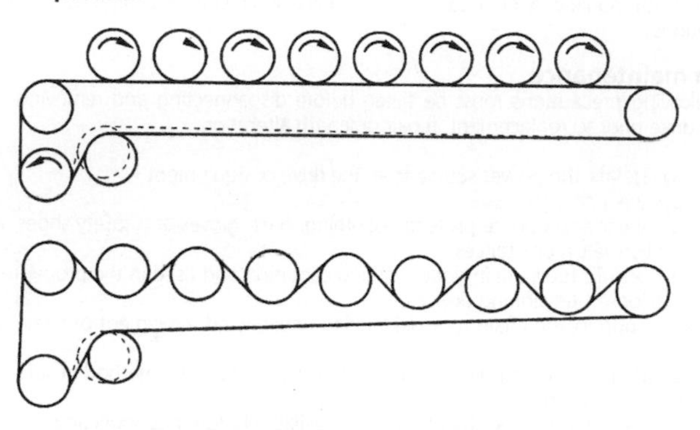

Drives of this type will only be used when:

- There is a steady load, preferably divided evenly over the sprocket system.
- At linear speeds not higher than 1.5 m/sec.
- It is driven in one direction only.

The efficiency of this driving method is higher than for the series drive because there is reduced tooth contact.

Every drive needs special attention with regard to the positioning of the driver sprocket, the jockey and the reversing pinions.

The layout of the sprockets, the support and the guidance of the chain determine to a large extent, the service life of the chain.

The chain in most cases is quite long and a good grip on the driver sprocket is only possible when a degree of pre-tensioning is applied. This should never exceed half the normal pulling load of the application.

The method of selection is the same as for that detailed under **Shafts in Series**.

Drives mounted as in bottom figure have an efficiency under normal conditions of:

- 94% with 5 rollers
- 89% with 10 rollers
- 84% with 15 rollers
- 79% with 20 rollers
- 75% with 25 rollers.

5.3.30 Safety warnings

Connecting links

No. 11 or 26 joints (slip fit) should not be used where high speed or arduous conditions are encountered. In these or equivalent circumstances where safety is essential, a riveting link No. 107 (interference fit) must be used.

Wherever possible, driver should have sufficient overall adjustment to ensure the use of an even number of pitches throughout the useful life of the chain. A cranked link joint (No. 12 or 30) should only be used as a last resource and restricted to light duty, non-critical applications.

Chain maintenance

The following precautions must be taken before disconnecting and removing a chain from a drive prior to replacement, repair or length alteration.

1. Always isolate the power source from the drive or equipment.
2. Always wear safety glasses.
3. Always wear appropriate protective clothing, hats, gloves and safety shoes, as warranted by the circumstances.
4. Always ensure tools are in good working condition and used in the proper manner.
5. Always loosen tensioning devices.
6. Always support the chain to avoid sudden unexpected movement of chain or components.
7. Never attempt to disconnect or reconnect a chain unless the method of safe working is fully understood.
8. Make sure correct replacement parts are available before disconnecting the chain.
9. Always ensure that directions for current use of any tools is followed.
10. Never re-use individual components.
11. Never re-use a damaged chain or chain part.
12. On light duty drives where a spring clip (No. 26) is used, always ensure that the clip is fitted correctly in relation to direction of travel.

For further information on:

- BS and ANSI products and dimensions,
- chain installation and maintenance,
- designer guide,
- industry applications.

Consult the web site of Renold Power Transmission Ltd. www.renold.com

See also their general catalogue obtainable from the address in Appendix 3.

5.4 Power transmission: shafts

5.4.1 Square and rectangular parallel keys, metric series

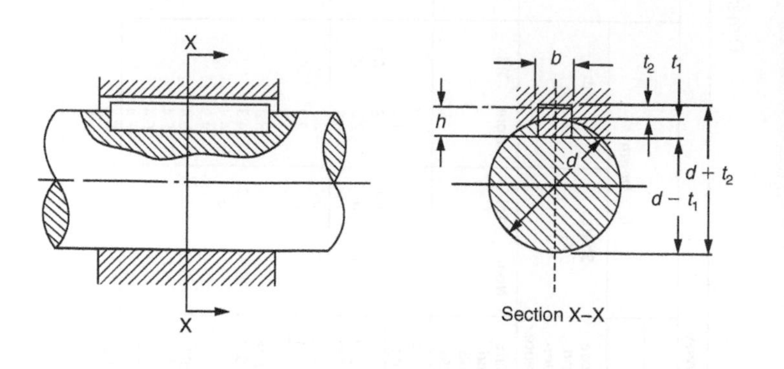

Section X–X

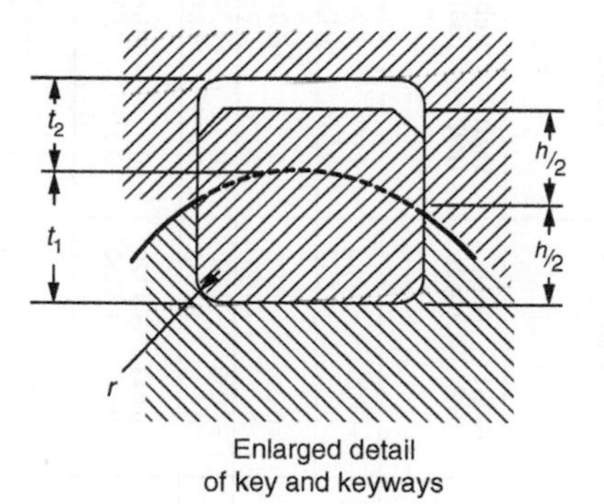

Enlarged detail of key and keyways

continued

Section 5.4.1 (*continued*)

Dimensions in mm

Shaft		Key	Keyway											
Nominal diameter (d)		(see section 5.4.2) section ($b \times h$) width × thickness	Width (b)						Depth				Radius r	
				Tolerance for class of fit					Shaft t_1		Hub t_2			
				Free		Normal		Close and interference						
Over	Incl.		Nom.	Shaft (H9)	Hub (D10)	Shaft (N9)	Hub (Js9)	Shaft and hub (P9)	Nom.	Tol.	Nom.	Tol.	Max.	Min.
6	8	2 × 2	2	+0.025	+0.060	−0.004	+0.012	−0.006	1.2		1.0		0.16	0.08
8	10	3 × 3	3	0	+0.020	−0.029	−0.012	−0.031	1.8		1.4		0.16	0.08
10	12	4 × 4	4						2.5	+0.1	1.8	+0.1	0.16	0.08
12	17	5 × 5	5	+0.030	+0.078	0	+0.015	−0.012	3.0	0	2.3	0	0.25	0.16
17	22	6 × 6	6	0	+0.080	−0.030	−0.015	−0.042	3.5		2.8		0.25	0.16
22	30	8 × 7	8	+0.036	+0.095	0	+0.018	−0.015	4.0	+0.2	3.3	+0.2	0.25	0.16
30	38	10 × 8	10	0	+0.040	−0.036	−0.018	−0.051	5.0	0	3.3	0	0.40	0.25
38	44	12 × 8	12						5.0		3.3		0.40	0.25
44	50	14 × 9	14	+0.043	+0.120	0	+0.021	−0.018	5.5		3.8		0.40	0.25
50	58	16 × 10	16	0	+0.050	−0.043	−0.021	−0.061	6.0	+0.2	4.3	+0.2	0.40	0.25
58	65	18 × 11	18						7.0	0	4.4	0	0.40	0.25
65	75	20 × 12	20						7.5		4.9		0.60	0.40
75	85	22 × 14	22	+0.052	+0.149	0	+0.026	−0.022	9.0		5.4		0.60	0.40
85	95	25 × 14	25	0	+0.065	−0.052	−0.026	−0.074	9.0		5.4		0.60	0.40
95	110	28 × 16	28						10.0		6.4		0.60	0.40
110	130	32 × 18	32						11.0		7.4		0.60	0.40
130	150	36 × 20	36	+0.062	+0.180	0	+0.031	−0.025	12.0		8.4		1.00	0.70
150	170	40 × 22	40	0	+0.080	−0.062	−0.031	−0.088	13.0	+0.3	9.4	+0.3	1.00	0.70
170	200	45 × 25	45						13.0	0	10.4	0	1.00	0.70

For full range and further information see BS 4235: Part 1: 1972.

5.4.2 Dimensions and tolerances for square and rectangular parallel keys

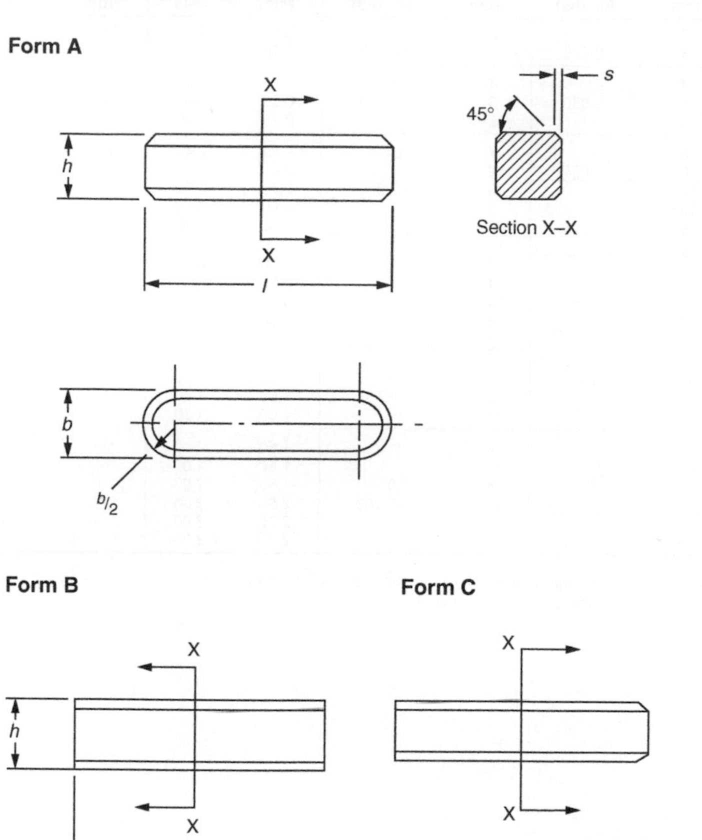

Width (*b*)		Thickness (*h*)		Chamfer (s)		Range of lengths (*l**)	
Nom.	Tol. (h9)	Nom.	Tol. (h9)	Min.	Max.	From	Incl.
2	0	2	0	0.16	0.25	6	20
3	−0.025	3	−0.025	0.16	0.25	6	36
4	0	4	0	0.16	0.25	8	45
5	−0.030	5	−0.030	0.25	0.40	10	56
6		6		0.25	0.40	14	70
8	0	7	Tol. (h11)	0.25	0.40	18	90
10	−0.036	8	0	0.40	0.60	22	110
12		8	−0.090	0.40	0.60	28	140
14	0	9		0.40	0.60	36	160
16	−0.043	10		0.40	0.60	45	180
18		11		0.40	0.60	50	200
20		12		0.60	0.80	56	220
22	0	14	0	0.60	0.80	63	250
25	−0.052	14	−0.110	0.60	0.80	70	280
28		16		0.60	0.80	80	320
32		18		0.60	0.80	90	360
36	0	20		1.00	1.20	100	400
40	−0.062	22	0	1.00	1.20	–	–
45		25	−0.130	1.00	1.20	–	–
50		28		1.00	1.20	–	–
56		32		1.60	2.00	–	–
63	0	32		1.60	2.00	–	–
70	−0.074	36	0	1.60	2.00	–	–
80		40	−0.160	2.50	3.00	–	–
90	0	45		2.50	3.00	–	–
100	−0.087	50		2.50	3.00	–	–

For full range and further information see BS 4235: Part 1: 1972.
*For preferred sizes see BS 4235: Table 9.

5.4.3 Square and rectangular taper keys, metric series

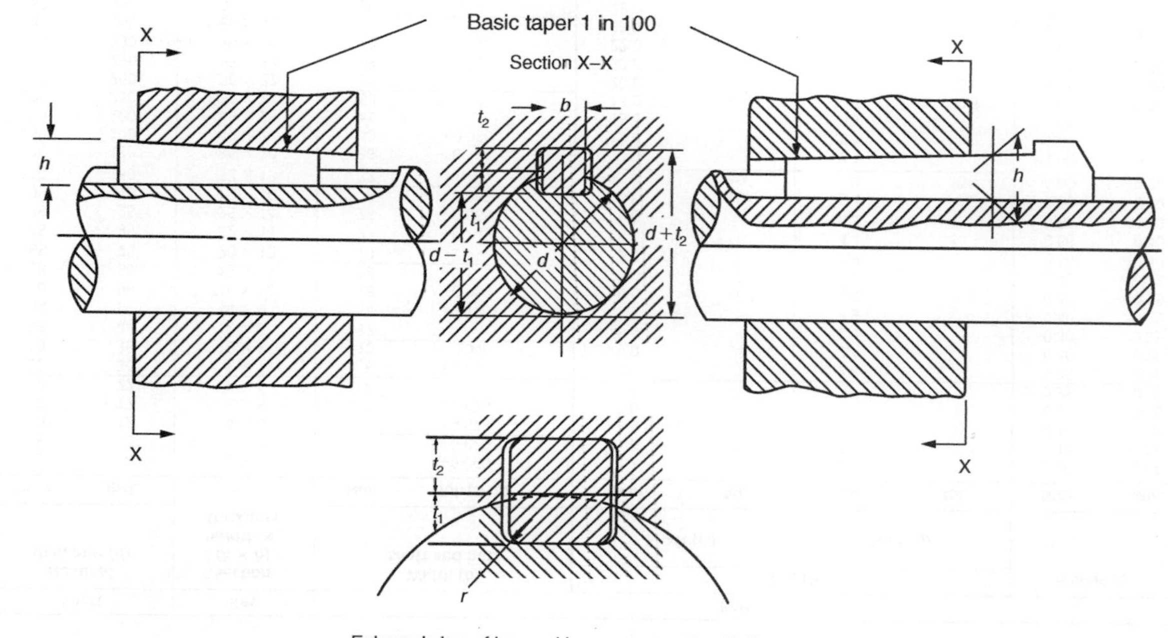

Enlarged view of key and keyway on section X–X

Dimensions in mm

Shaft		Key	Keyway							
Nominal diameter *(d)*		Section *(b × h)* width × thickness	Width *(b)*, shaft and hub		Depth				Radius *(r)*	
					Shaft (t_1)		Hub (t_2)			
Over	Incl.		Nom.	Tol. (D10)	Nom.	Tol.	Nom.	Tol.	Max.	Min.
6	8	2 × 2	2	+0.060	1.2		0.5		0.16	0.08
8	10	3 × 3	3	+0.020	1.8	+0.1	0.9	+0.1	0.16	0.08
10	12	4 × 4	4	+0.078	2.5	0	1.2	0	0.16	0.08
12	17	5 × 5	5	+0.030	3.0		1.7		0.25	0.16
17	22	6 × 6	6		3.5		2.2		0.25	0.16
22	30	8 × 7	8	+0.098	4.0		2.4		0.25	0.16
30	38	10 × 8	10	+0.040	5.0		2.4		0.40	0.25
38	44	12 × 8	12		5.0		2.4		0.40	0.25
44	50	14 × 9	14	+0.120	5.5		2.9		0.40	0.25
50	58	16 × 10	16	+0.050	6.0	+0.2	3.4	+0.2	0.40	0.25
58	65	18 × 11	18		7.0	0	3.4	0	0.40	0.25
65	75	20 × 12	20		7.5		3.9		0.60	0.40
75	85	22 × 14	22	+0.149	9.0		4.4		0.60	0.40
85	95	25 × 14	25	+0.065	9.0		4.4		0.60	0.40
95	110	28 × 16	28	+0.065 +0.149	10.0	+0.2	5.4	+0.2	0.60	0.40
110	130	32 × 18	32		11.0	0	6.4	0	0.60	0.70
130	150	36 × 20	36	+0.180	12.0		7.1		1.00	0.70
150	170	40 × 22	40	+0.080	13.0		8.1		1.00	0.70
170	200	45 × 25	45		15.0		9.1		1.00	0.70
200	230	50 × 28	50		17.0	+0.3	10.1	+0.3	1.00	0.70
230	260	56 × 32	56		20.0	0	11.1	0	1.60	1.20
260	290	63 × 32	63	+0.220	20.0		11.1		1.60	1.20
290	330	70 × 36	70	+0.120	22.0		13.1		1.60	1.20
330	380	80 × 40	80		25.0		14.1		2.50	2.00
380	440	90 × 45	90	+0.260	28.0		16.1		2.50	2.00
440	500	100 × 50	100	+0.120	31.0		18.1		2.50	2.00

For full range and further information see BS 4235: Part 1: 1972.

5.4.4 Dimensions and tolerances for square and rectangular taper keys

Plain key | **Gib-head key**

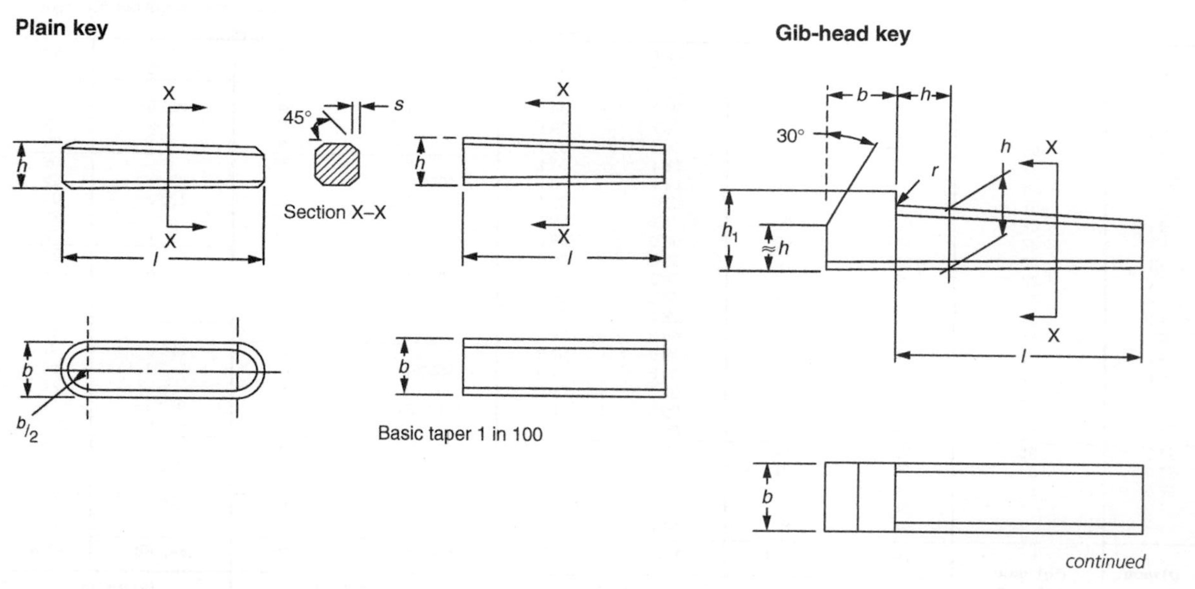

continued

Section 5.4.4 (*continued*)

Dimensions in mm

Width (*b*)		Thickness (*h*)		Chamfer (*s*)		Length (*l**)		Gib-head nom′ (h_1)	Radius nom (*r*)
Nom.	Tol. (h9)	Nom.	Tol. (h9)	Min.	Max.	From	Incl.		
2	0	2	0	0.16	0.25	6	20	–	–
3	−0.025	3	−0.025	0.16	0.25	6	36	–	–
4	0	4	0	0.16	0.25	8	45	7	0.25
5	−0.030	5	−0.030	0.25	0.40	10	56	8	0.25
6		6		0.25	0.40	14	70	10	0.25
8	0	7		0.25	0.40	18	90	11	1.5
10	−0.036	8	Tol. (h11)	0.40	0.60	22	110	12	1.5
12		8	0	0.40	0.60	28	140	12	1.5
14	0	9	−0.090	0.40	0.60	36	160	14	1.5
16	−0.043	10		0.40	0.60	45	180	16	1.5
18		11		0.40	0.60	50	200	18	1.5
20		12		0.60	0.80	56	220	20	1.5
22	0	14	0	0.60	0.80	63	250	22	1.5
25	−0.052	14	−0.110	0.60	0.80	70	280	22	1.5
28		16		0.60	0.80	80	320	25	1.5
32		18		0.60	0.80	90	360	28	1.5
36	0	20		1.00	1.20	100	400	32	1.5
40	−0.062	22	0	1.00	1.20	–	–	36	1.5
45		25	−0.130	1.00	1.20	–	–	40	1.5
50		28		1.00	1.20	–	–	45	1.5
56		32		1.60	2.00	–	–	50	1.5
63	0	32		1.60	2.00	–	–	50	1.5
70	−0.074	36	0	1.60	2.00	–	–	56	1.5
80		40	−0.160	2.50	3.00	–	–	63	1.5
90	0	45		2.50	3.00	–	–	70	1.5
100	−0.087	50		2.50	3.00	–	–	80	1.5

For full range and further information see BS 4235: Part 1: 1972.
* For preferred lengths see BS 4235: Part 1: Table 9.

5.4.5 Woodruff keys and keyways, metric series

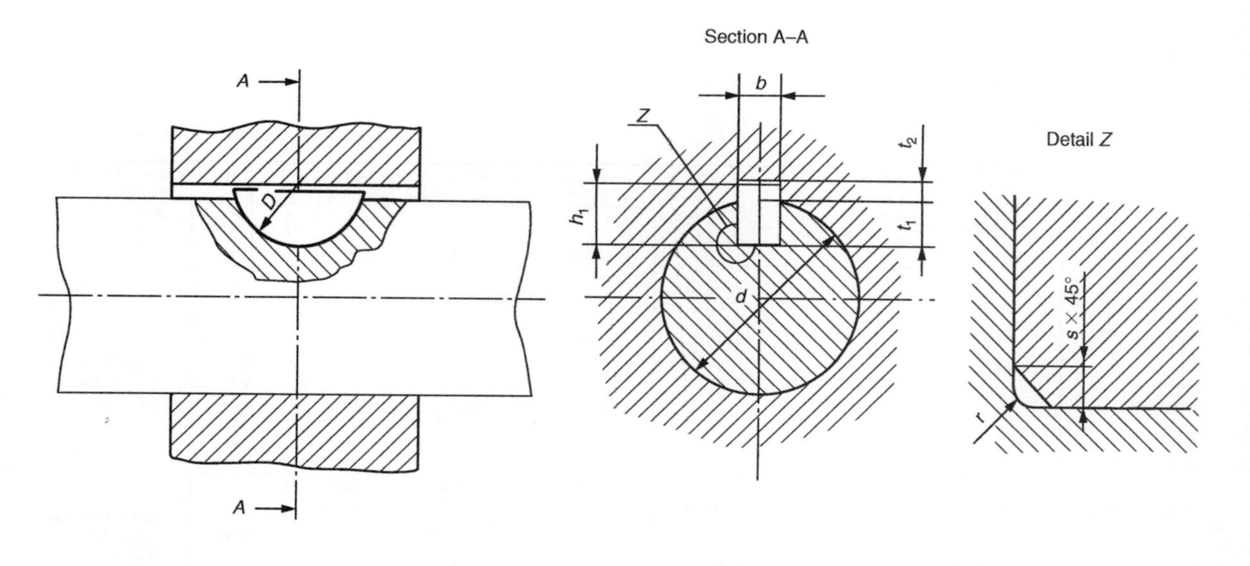

Dimensions in mm

Key size of normal for $b \times h_1 \times D$ or equivalent Whitney form	Width (b)				Depth				Radius (r)	
	Nom.	Tolerance			Shaft (t_1)		Hub (t_2)			
		Normal fit		Close fit						
		Shaft (N9)	Hub (Js9)	Shaft and hub (P9)	Nom.	Tol.	Nom.	Tol.	Max.	Min.
1.0 × 1.4 × 4	1.0	−0.004 −0.029	+0.012 −0.012	−0.006 −0.031	1.0	+0.1 0	0.6	+0.1 0	0.16	0.08
1.5 × 2.6 × 7	1.5				2.0		0.8		0.16	0.08
2.0 × 2.6 × 7	2.0				1.8		1.0		0.16	0.08
2.0 × 3.7 × 10	2.0				2.9		1.0		0.16	0.08
2.5 × 3.7 × 10	2.5				2.7		1.2		0.16	0.08
3.0 × 5.0 × 13	3.0				3.8	+0.2 0	1.4		0.16	0.08
3.0 × 6.5 × 16	3.0				5.3		1.4		0.16	0.08
4.0 × 6.5 × 16	4.0	0 −0.030	+0.015 −0.015	−0.012 −0.042	5.0		1.8		0.25	0.16
4.0 × 7.5 × 19	4.0				6.0		1.8		0.25	0.16
5.0 × 6.5 × 16	5.0				4.5		2.3		0.25	0.16
5.0 × 7.5 × 19	5.0				5.5		2.3		0.25	0.16
5.0 × 9.0 × 22	5.0				7.0	+0.3 0	2.3		0.25	0.16
6.0 × 9.0 × 22	6.0				6.5		2.8		0.25	0.16
6.0 × 11.0 × 28	6.0				7.5		2.8	+0.2 0	0.25	0.16
8.0 × 11.0 × 28	8.0	0 −0.036	+0.018 −0.018	−0.015 −0.051	8.0		3.3		0.40	0.25
10.0 × 13.0 × 32	10.0				10.0		3.3		0.40	0.25

For further information see BS 4235: Part 2: 1977.

5.4.6 Dimensions and tolerances for Woodruff Keys

Normal form

X = Sharp edges removed

Whitney form

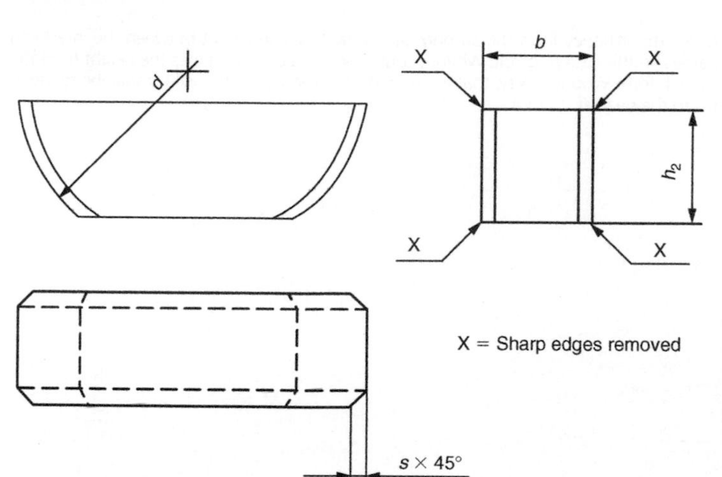

Dimensions in mm

Width b Nom.	Tol. (h9)*	Height (h_1)		Diameter (d)		Chamfer (s)	
		Nom.	Tol. (h11)	Nom.	Tol. (h12)	Min.	Max.
1.0	0 / −0.025	1.4	0 / −0.060	4	0 / −0.120	0.16	0.25
1.5		2.6		7	0 / −0.150	0.16	0.25
2.0		2.6		7		0.16	0.25
2.0		3.7	0 / −0.075	10		0.16	0.25
2.5		3.7		10		0.16	0.25
3.0		5.0		13	0 / −0.180	0.16	0.25
3.0		6.5	0 / −0.090	16		0.16	0.25
4.0	0 / −0.030	6.5		16		0.25	0.40
4.0		7.5		19	0 / −0.210	0.25	0.40
5.0		6.5		16	0 / −0.180	0.25	0.40
5.0		7.5		19	0 / −0.210	0.25	0.40
5.0		9.0		22		0.25	0.40
6.0		9.0		22		0.25	0.40
6.0		10.0		25		0.25	0.40
8.0	0 / −0.036	11.0	0 / −0.110	28		0.40	0.60
10.0		13.0		32	0 / −0.250	0.40	0.60

For further information see BS 4235: Part 2: 1977.
*A tolerance closer than h9 may be adopted subject to agreement between interested parties.

Note: The Whitney form should only be adopted by agreement between the interested parties. In this case h_2 of the Whitney form key shall equal 0.8 times the height h_1 of the normal form Woodruff key, that is, $h_2 = 0.8h_1$. The calculated values shall be rounded off to the nearest 0.1 mm.

5.4.7 Shaft ends types: general relationships

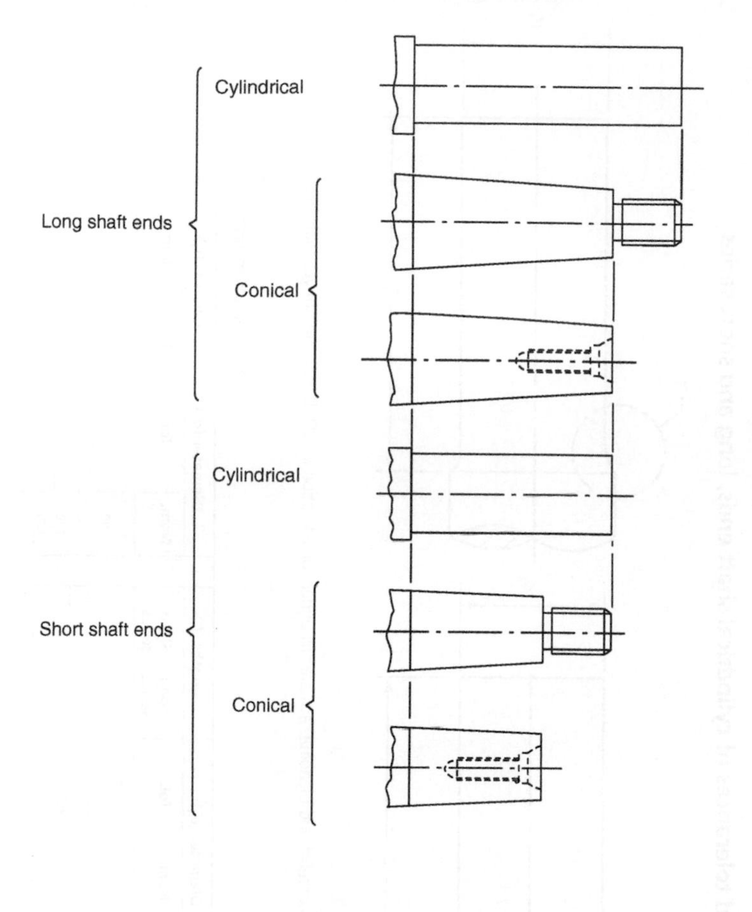

5.4.8 Dimensions and tolerances of cylindrical shaft ends, long and short series

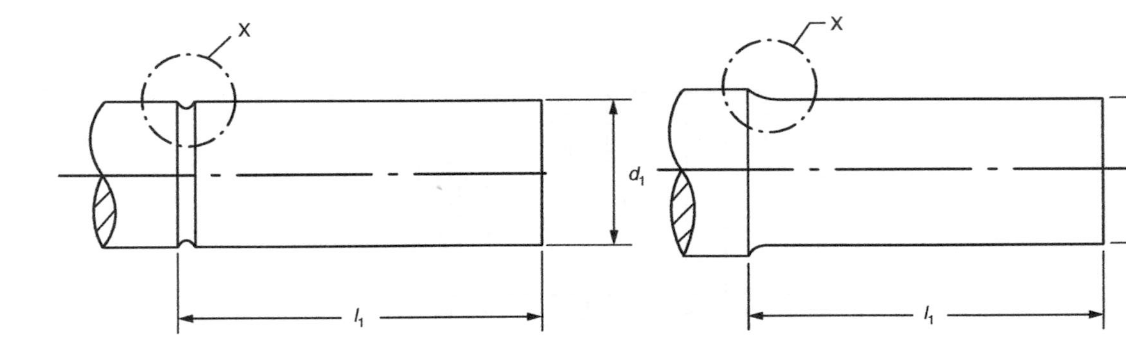

Detail X: undercut grooves for tool relief and blendina radii are still under consideration in ISO. Details are not yet available.

Dimensions in mm

Diameter (d_1)		Lengths (l_1)		Diameter (d_1)		Lengths (l_1)	
Nom.	Tol.	Long series	Short series	Nom.	Tol.	Long series	Short series
6 7	$j6^{+0.006}_{-0.002}$	16	–	100 110 120 125	$m6^{+0.035}_{+0.013}$	210	165
8 9	$j6^{+0.007}_{-0.002}$	20	–				

10		23	20*
11	$j6\,^{+0.008}_{-0.003}$		
12		30	25*
14			
16		40	28
18			
19	$j6\,^{+0.009}_{-0.004}$		
20		50	36
22			
24			
25		60	42
28			
30		80	58
32	$k6\,^{+0.018}_{+0.002}$		
35			
38			
40		110	82
42			
45			
48			
50			
55	$m6\,^{+0.030}_{+0.011}$		
56			
60		140	105
63			
65			
70			
71			
75			
80		170	130
85	$m6\,^{+0.035}_{+0.013}$		
90			
95			

130	$m6\,^{+0.040}_{+0.015}$	250	200
140			
150			
160			
170			
180			
190	$m6\,^{+0.046}_{+0.017}$	300	240
200			
220			
240		350	280
250			
260	$m6\,^{+0.052}_{+0.020}$		
280		410	330
300			
320	$m6\,^{+0.057}_{+0.021}$		
340		550	450
360			
380			
400		650	540
420	$m6\,^{+0.063}_{+0.023}$		
440			
450			
460			
480			
500			
530	$m6\,^{+0.070}_{+0.026}$	800	680
560			
600			
630			

* The dimensions thus indicated are not in agreement with the related dimensions of long series conical shaft ends. See BS4506 Table 2 and Clause 2.

5.4.9 Dimensions of conical shaft ends with parallel keys, long series

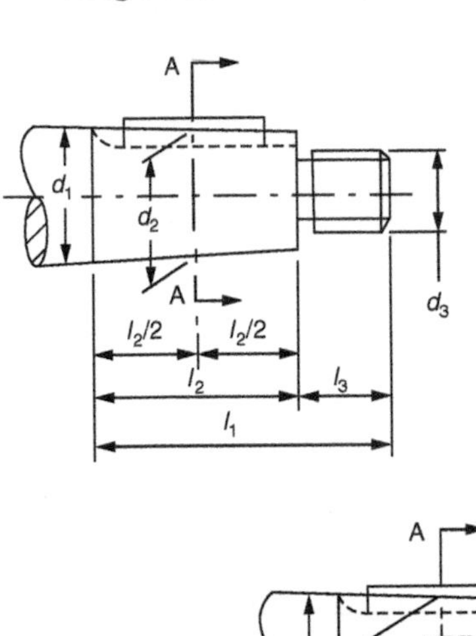

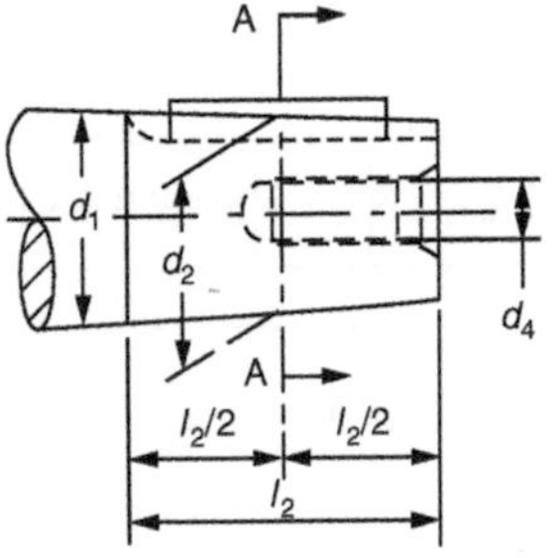

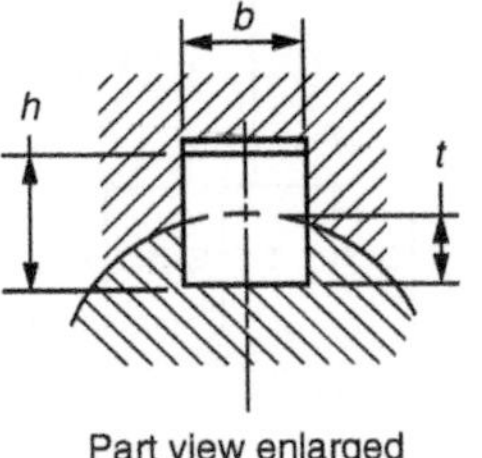

Part view enlarged at sections A–A

Keyway may have forms other than shown.
Conicity of 1:10 corresponds to $(d_1 - d_2)/(l_2/2) = 1/10$.

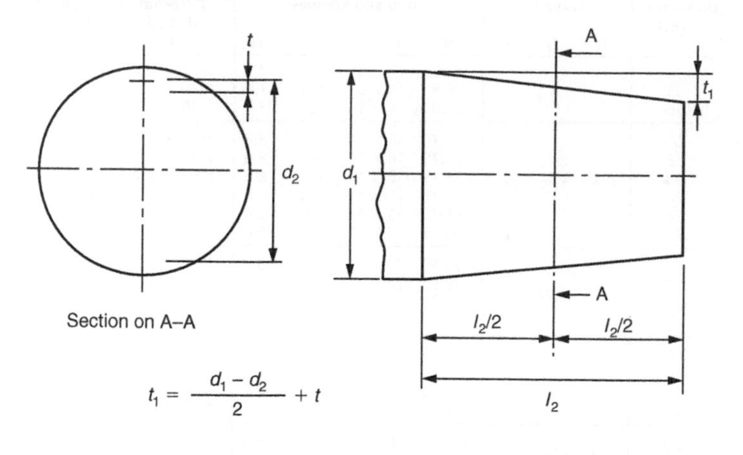

$$t_1 = \frac{d_1 - d_2}{2} + t$$

Dimensions in mm

Diameter (d_1)	Length			Key and keyway				External thread (d_3)	Internal thread (d_4)
	l_1	l_2	l_3	d_2	$b \times h$	t	t_1		
6	16	10	6	5.5	–	–	–	M4 × 0.7	–
7				6.5	–	–	–	M4 × 0.7	–
8	20	12	8	7.4	–	–	–	M6 × 1	–
9				8.4	–	–	–	M6 × 1	–
10	23	15*	8	9.25	–	–	–	M6 × 1	–
11				10.25	2 × 2	1.2	1.6	M6 × 1	–
12	30	18*	12	11.1	2 × 2	1.2	1.7	M8 × 1	M4 × 0.7
14				13.1	3 × 3	1.8	2.3	M8 × 1	M4 × 0.7
16	40	28	12	14.6	3 × 3	1.8	2.5	M10 × 1.25	M4 × 0.7
18				16.6	4 × 4	2.5	3.2	M10 × 1.25	M5 × 0.8
19				17.6	4 × 4	2.5	3.2	M10 × 1.25	M5 × 0.8
20	50	36	14	18.2	4 × 4	2.5	3.4	M12 × 1.25	M6 × 1
22				20.2	4 × 4	2.5	3.4	M12 × 1.25	M6 × 1
24				22.2	5 × 5	3.0	3.9	M12 × 1.25	M6 × 1
25	60	42	18	22.9	5 × 5	3.0	4.1	M16 × 1.5	M8 × 1.25
28				25.9	5 × 5	3.0	4.1	M16 × 1.5	M8 × 1.25
30	80	58	22	27.1	5 × 5	3.0	4.5	M20 × 1.5	M10 × 1.5
32				29.1	6 × 6	3.5	5.0	M20 × 1.5	M10 × 1.5
35				32.1	6 × 6	3.5	5.0	M20 × 1.5	M10 × 1.5
38				35.1	6 × 6	3.5	5.0	M24 × 2.0	M12 × 1.75
40	110	82	28	35.9	10 × 8	5.0	7.1	M24 × 2.0	M12 × 1.75
42				37.9	10 × 8	5.0	7.1	M24 × 2.0	M12 × 1.75
45				40.9	12 × 8	5.0	7.1	M30 × 2.0	M16 × 2.0
48				43.9	12 × 8	5.0	7.1	M30 × 2.5	M16 × 2.0

continued

Section 5.4.9 (*continued*)

Diameter (d_1)	Length l_1	Length l_2	Length l_3	Key and keyway d_2	Key and keyway $b \times h$	Key and keyway t	Key and keyway t_1	External thread (d_3)	Internal thread (d_4)
50				45.9	12 × 8	5.0	7.1	M36 × 3.0	M16 × 2.0
55				50.9	14 × 9	5.5	7.6	M36 × 3.0	M20 × 2.5
56				51.9	14 × 9	5.5	7.6	M36 × 3.0	M20 × 2.5
60	140	105	35	54.75	16 × 10	6.0	8.6	M42 × 3.0	M20 × 2.5
63				57.75	16 × 10	6.0	8.6	M42 × 3.0	M20 × 2.5
65				59.75	16 × 10	6.0	8.6	M42 × 3.0	M20 × 2.5
70				64.75	18 × 11	7.0	9.6	M48 × 3.0	M24 × 3.0
71				65.75	18 × 11	7.0	9.6	M48 × 3.0	M24 × 3.0
75				69.75	18 × 11	7.0	9.6	M48 × 3.0	M24 × 3.0
80	170	130	40	73.50	20 × 12	7.5	10.8	M56 × 4.0	M30 × 3.5
85				78.50	20 × 12	7.5	10.8	M56 × 4.0	M30 × 3.5
90				83.50	22 × 14	9.0	12.3	M64 × 4.0	M30 × 3.5
95				88.50	22 × 14	9.0	12.3	M64 × 4.0	M36 × 4.0
100	210	165	45	91.75	25 × 14	9.0	13.1	M72 × 4.0	M36 × 4.0
110				101.75	25 × 14	9.0	13.1	M80 × 4.0	M42 × 4.5
120				111.75	28 × 16	10.0	14.1	M90 × 4.0	M42 × 4.5
125				116.75	28 × 16	10.0	14.1	M90 × 4.0	M48 × 5.0
130	250	200	50	120.0	28 × 16	10.0	15.0	M100 × 4.0	–
140				130.0	32 × 18	11.0	16.0	M100 × 4.0	–
150				140.0	32 × 18	11.0	16.0	M110 × 4.0	–
160	300	240	60	148.0	35 × 20	12.0	18.0	M125 × 4.0	–
170				158.0	36 × 20	12.0	18.0	M125 × 4.0	–
180				168.0	40 × 22	13.0	19.0	M140 × 6.0	–
190	350	280	70	176.0	40 × 22	13.0	20.0	M140 × 6.0	–
200				186.0	40 × 22	13.0	20.0	M160 × 6.0	–
220				206.0	45 × 25	15.0	22.0	M160 × 6.0	–

For further information see BS 4506.
* The dimensions thus indicated are not in agreement with the related dimensions for long series conical shaft ends. See BS 4506: Table 2 and Clause 2.

5.4.10 Dimensions of conical shaft ends with diameters above 220 mm with the keyway parallel to the shaft surface, long series

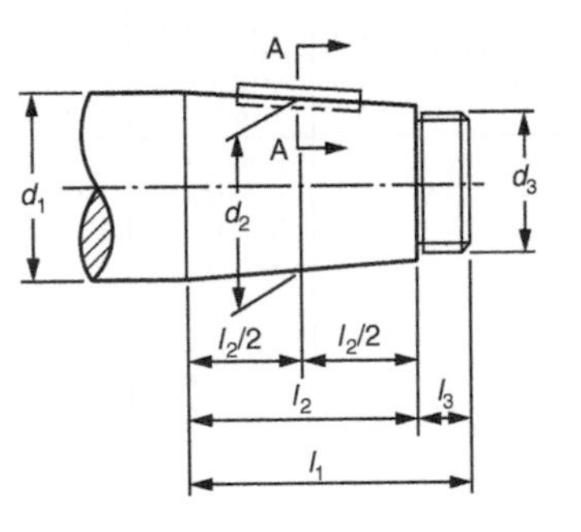

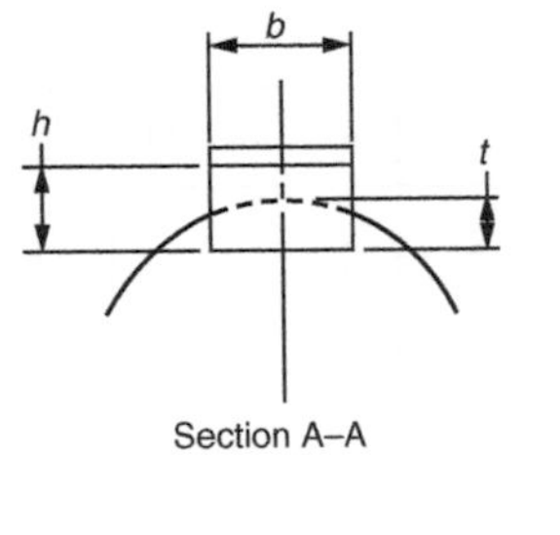

Conicity of 1:10 corresponds to $(d_1 - d_2)/(l_2/2) = 1/10$.

Dimensions in mm

Diameter (d_1)	Length (l_1)	(l_2)	(l_3)	Key and keyway (d_2)	$b \times h$	t	External thread (d_3)
240				223.5	50 × 28	17	M180 × 6
250	410	330	80	233.5	50 × 28	17	M180 × 6
260				243.5	50 × 28	17	M200 × 6
280				261.0	56 × 32	20	M220 × 6
300	470	380	90	281.0	63 × 32	20	M220 × 6
320				301.0	63 × 32	20	M250 × 6
340				317.5	70 × 36	22	M280 × 6
360	550	450	100	337.5	70 × 36	22	M280 × 6
380				357.5	70 × 36	22	M300 × 6
400				373.0	80 × 40	25	M320 × 6
420				393.0	80 × 40	25	M320 × 6
440				413.0	80 × 40	25	M350 × 6
450	650	540	110	423.0	90 × 45	28	M350 × 6
460				433.0	90 × 45	28	M380 × 6
480				453.0	90 × 45	28	M380 × 6
500				473.0	90 × 45	28	M420 × 6
530				496.0	100 × 50	31	M420 × 6
560	800	680	120	526.0	100 × 50	31	M450 × 6
600				566.0	100 × 50	31	M500 × 6
630				596.0	100 × 50	31	M550 × 6

For further information see BS 4506.

5.4.11 Dimensions of conical shaft ends with parallel keys, short series

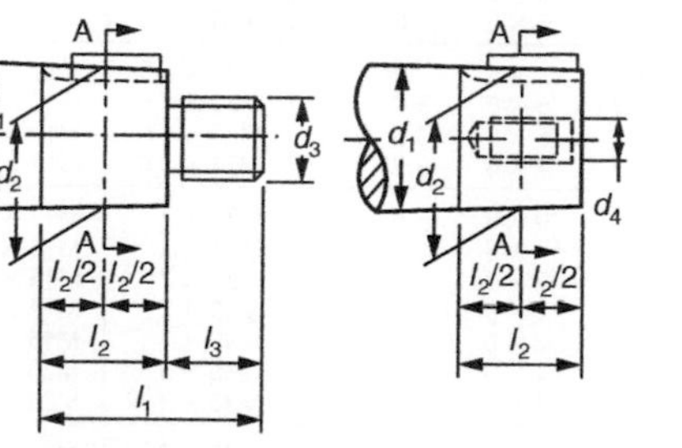

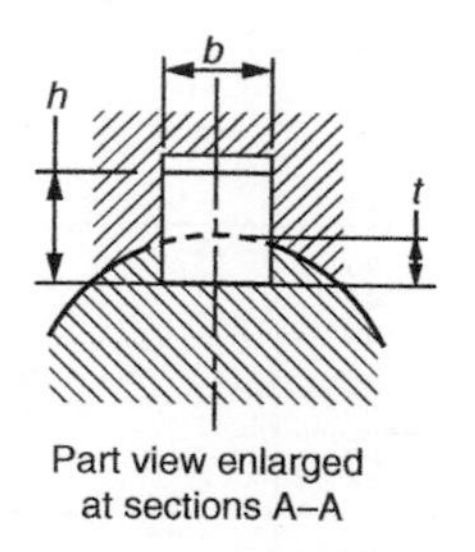

Part view enlarged at sections A–A

Keyway may have forms other than shown.
Conicity of 1:10 corresponds to $(d_1 - d_2)/(l_2/2) = 1/10$.

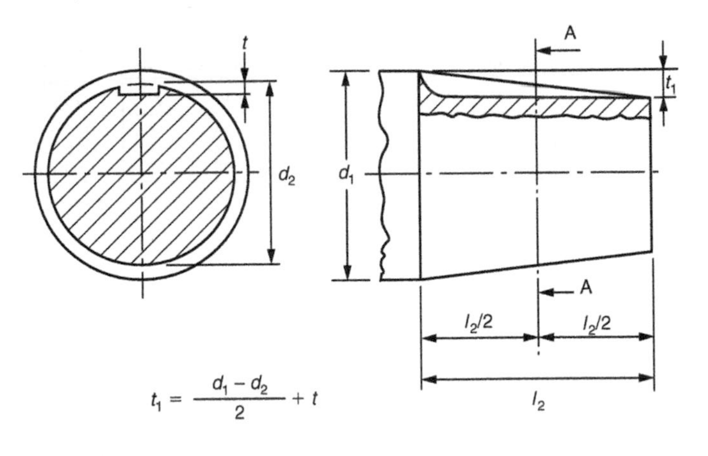

$$t_1 = \frac{d_1 - d_2}{2} + t$$

Dimensions in mm

Diameter (d_1)	Length			Key and key way				External thread (d_3)	Internal thread (d_4)
	(l_1)	(l_2)	(l_3)	(l_2)	$b \times h$	t	t_1		
16	28	16	12	15.2	3 × 3	1.8		M10 × 1.25	M4 × 0.7
18				17.2	4 × 4	2.5		M10 × 1.25	M5 × 0.8
19				18.2	4 × 4	2.5		M10 × 1.25	M5 × 0.8
20	36	22	14	18.9	4 × 4	2.5		M12 × 1.25	M6 × 1.0
22				20.9	4 × 4	2.5		M12 × 1.25	M6 × 1.0
24				22.9	5 × 5	3.0		M12 × 1.25	M6 × 1.0
25	42	24	18	23.8	5 × 5	3.0		M16 × 1.5	M8 × 1.25
28				26.8	5 × 5	3.0		M16 × 1.5	M8 × 1.25
30	58	36	22	28.2	5 × 5	3.0		M20 × 1.5	M10 × 1.5
32				30.2	6 × 6	3.5		M20 × 1.5	M10 × 1.5
35				33.2	6 × 6	3.5		M20 × 1.5	M10 × 1.5
38				36.2	6 × 6	3.5		M24 × 2.0	M12 × 1.75
40	82	54	28	37.3	10 × 8	5.0		M24 × 2	M12 × 1.75
42				39.3	10 × 8	5.0		M24 × 2	M12 × 1.75
45				42.3	12 × 8	5.0		M30 × 2	M16 × 2.0
48				45.3	12 × 8	5.0		M30 × 2	M16 × 2.0
50				47.3	12 × 8	5.0		M36 × 3	M16 × 2.0
55				52.3	14 × 9	5.5		M36 × 3	M20 × 2.5
56				53.3	14 × 9	5.5		M36 × 3	M20 × 2.5
60	105	70	35	56.5	16 × 10	6.0		M42 × 3	M20 × 2.5
63				59.5	16 × 10	6.0		M42 × 3	M20 × 2.5
65				61.5	16 × 10	6.0		M42 × 3	M20 × 2.5
70				66.5	18 × 11	7.0		M48 × 3	M24 × 3.0
71				67.5	18 × 11	7.0		M48 × 3	M24 × 3.0
75				71.5	18 × 11	7.0		M48 × 3	M24 × 3.0

continued

Dimensions in mm

Diameter (d_1)	Length (l_1)	(l_2)	(l_3)	Key and key way (l_2)	$b \times h$	t	t_1	External thread (d_3)	Internal thread (d_4)
80	130	90	40	75.5	20 × 12	7.5		M56 × 4	M30 × 3.5
85				80.5	20 × 12	7.5		M56 × 4	M30 × 3.5
90				85.5	22 × 14	9.0		M64 × 4	M30 × 3.5
95				90.5	22 × 14	9.0		M64 × 4	M36 × 4.0
100	165	120	45	94.0	25 × 14	9.0		M72 × 4	M36 × 4.0
110				104.0	25 × 14	9.0		M80 × 4	M42 × 4.5
120				114.0	28 × 16	10.0		M90 × 4	M42 × 4.5
125				119.0	28 × 16	10.0		M90 × 4	M48 × 5.0
130	200	150	50	122.5	28 × 16	10.0		M100 × 4	–
140				132.5	32 × 18	11.0		M100 × 4	–
150				142.5	32 × 18	11.0		M110 × 4	–
160	240	180	60	151.0	36 × 20	12.0		M125 × 4	–
170				161.0	36 × 20	12.0		M125 × 4	–
180				171.0	40 × 22	13.0		M140 × 6	–
190	280	210	70	179.5	40 × 22	13.0		M140 × 6	–
200				189.5	40 × 22	13.0		M160 × 6	–
220				209.5	45 × 25	15.0		M160 × 6	–

For further information see BS 4506.

5.4.12 Transmissible torque values

Shaft end diameter d_1 (mm)	Transmissible torque T (Nm) (a)	(b)	(c)
6		0.307	0.145
7		0.53	0.25
8		0.85	0.4
9		1.25	0.6
10		1.85	0.875
11		2.58	1.22
12		3.55	1.65
14		6.00	2.8
16		9.75	4.5
18		14.5	6.7
19		17.5	8.25
20		21.2	9.75
22		29.0	13.6
24		40.0	18.5
25		46.2	21.2
28		69.0	31.5
30	206	87.5	40.0
32	250	109.0	50.0
35	325	150.0	69.0
38	425	200.0	92.5
40	487	236.0	112

Shaft end diameter d_1 (mm)	Transmissible torque T (Nm) (a)	(b)	(c)
42	560	280	132
45	710	355	170
48	850	450	212
50	950	515	243
55	1280	730	345
56	1360	775	355
60	1650	975	462
63	1900	1150	545
65	2120	1280	600
70	2650	1700	800
71	2780	1800	825
75	3250	2120	1000
80	3870	2650	1250
85	4750	3350	1550
90	5600	4120	1900
95	6500	4870	2300
100	7750	5800	2720
110	10 300	8250	3870
120	13 700	11 200	5150
125	15 000	12 800	6000
130	17 000	14 500	

continued

Section 5.4.12 (*continued*)

Shaft end diameter d_1 (mm)	Transmissible torque T (Nm)		
	(a)	(b)	(c)
140	21 200	19 000	
150	25 800	24 300	
160	31 500	30 700	
170	37 500	37 500	
180	45 000		
190	53 000		
200	61 500		
220	82 500		
240	1 06 000		
250	1 18 000		
260	1 36 000		
280	1 70 000		
300	2 06 000		
320	2 50 000		

Shaft end diameter d_1 (mm)	Transmissible torque T (Nm)		
	(a)	(b)	(c)
340	3 00 000		
360	3 55 000		
380	4 25 000		
400	4 87 000		
420	5 60 000		
440	6 50 000		
450	6 90 000		
460	7 50 000		
480	8 50 000		
500	9 50 000		
530	11 50 000		
560	13 60 000		
600	16 50 000		
630	19 00 000		

The values of transmissible torque have been calculated from the following formulae and rounded off to the values of the R80 (preferred numbers) series:

(a) Transmission of pure torque: $T = 2.45166\pi \times 10^{-3} \times d_1^3$ Nm.
(b) Transmission of a known torque associated with a bending moment of a known magnitude: $T = 58.8399 \times 10^{-5} \times d_1^{3.5}$ Nm.
(c) Transmission of a known torque associated with an undetermined bending moment: $T = 27.45862 \times 10^{-5} \times d_1^{3.5}$ Nm.

These three formulae assume use of a steel with a tensile strength of 490 to 590 N/mm^2. These values are intended to provide a rapid comparison between shafts of different sizes and *not* as fundamental design criteria. Steady torque conditions are assumed.

5.4.13 Straight-sided splines for cylindrical shafts, metric

Designation: nominal dimensions

The profile of a splined shaft or hub is designated by stating, in the following order:

The number of splines N
The minor diameter d

The outside diameter D.
For example, shaft (or hub) $6 \times 23 \times 26$.

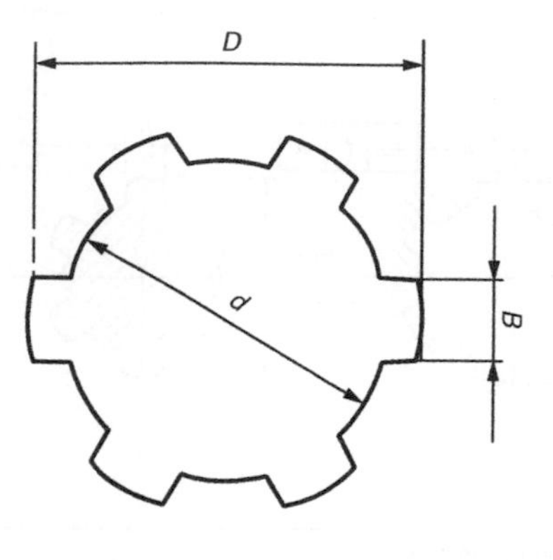

Section 5.4.13 (*continued*)

d (mm)	Light series Designation	N	D (mm)	B (mm)	Medium series Designation	N	D (mm)	B (mm)
11					6 × 11 × 14	6	14	3
13					6 × 13 × 16	6	16	3.5
16					6 × 16 × 20	6	20	4
18					6 × 18 × 22	6	22	5
21					6 × 21 × 25	6	25	5
23	6 × 23 × 26	6	26	6	6 × 23 × 28	6	28	6
26	6 × 26 × 30	6	30	6	6 × 26 × 32	6	32	6
28	6 × 28 × 32	6	32	7	6 × 28 × 34	6	34	7
32	8 × 32 × 36	8	36	6	8 × 32 × 38	8	38	6
36	8 × 36 × 40	8	40	7	8 × 36 × 42	8	42	7
42	8 × 42 × 46	8	46	8	8 × 42 × 48	8	48	8
46	8 × 46 × 50	8	50	9	8 × 46 × 54	8	54	9
52	8 × 52 × 58	8	58	10	8 × 52 × 60	8	60	10
56	8 × 56 × 62	8	62	10	8 × 56 × 65	8	65	10
62	8 × 62 × 68	8	68	12	8 × 62 × 72	8	72	12
72	10 × 72 × 78	10	78	12	10 × 72 × 82	10	82	12
82	10 × 82 × 88	10	88	12	10 × 82 × 92	10	92	12
92	10 × 92 × 98	10	98	14	10 × 92 × 102	10	102	14
102	10 × 102 × 108	10	108	16	10 × 102 × 112	10	112	16
112	10 × 112 × 120	10	120	18	10 × 112 × 125	10	125	18

Tolerances on holes and shafts

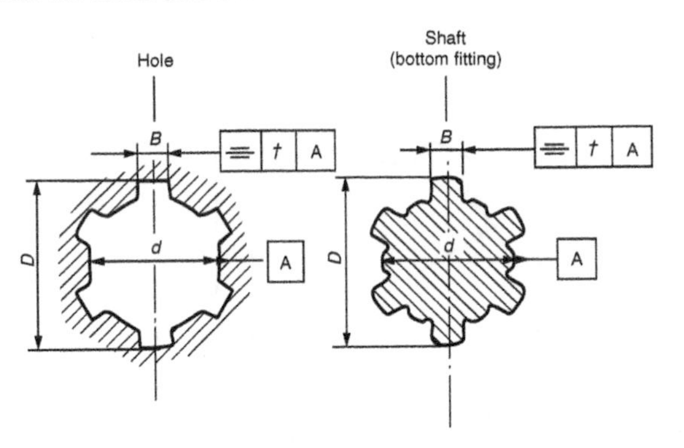

Tolerances on hole						Tolerances on shaft			Mounting type
Not treated after broaching			Treated after broaching						
B	D	d	B	D	d	B	D	d	
H9	H10	H7	H11	H10	H7	d10	a11	f7	Sliding
						f9	a11	g7	Close sliding
						h10	a11	h7	Fixed

Tolerances on symmetry

Dimensions in mm

Spline width	B	3	3.5 4 5 6	7 8 9 10	12 14 16 18
Tolerance of symmetry	t	0.010 (IT7)	0.012 (IT7)	0.015 (IT7)	0.018 (IT7)

The tolerance specified on *B* includes the index variation (and the symmetry variation).

Notes:
(a) With certain milling cutters, it is possible for special applications to produce splines without bottom tool clearance with a very reduced fillet radius between the spline side and the minor diameter d (e.g. milling cutters with fixed working positions).
(b) The dimensional tolerances on holes and shafts relate to entirely finished workpieces (shafts and hubs). Tooling should therefore be different for untreated workpieces, or workpieces treated before machining and for workpieces treated after machining.
(c) For further information on straight sided splines and gauges for checking such splines. See BS 5686: 1986.

Involute splines

These have a similar profile to spur gear teeth. They have a much greater root strength than straight sided splines and can be produced on gear cutting machines with standard gear tooth cutters. However, their geometry and checking is much more complex than for straight sided splines and is beyond the scope of this book. See BS 3550 (inch units) and BS 6186 (metric units).

5.5 Tapers

5.5.1 Self-holding Morse and metric 5% tapers

Dimensions

Tolerances of symmetry are given in millimetres

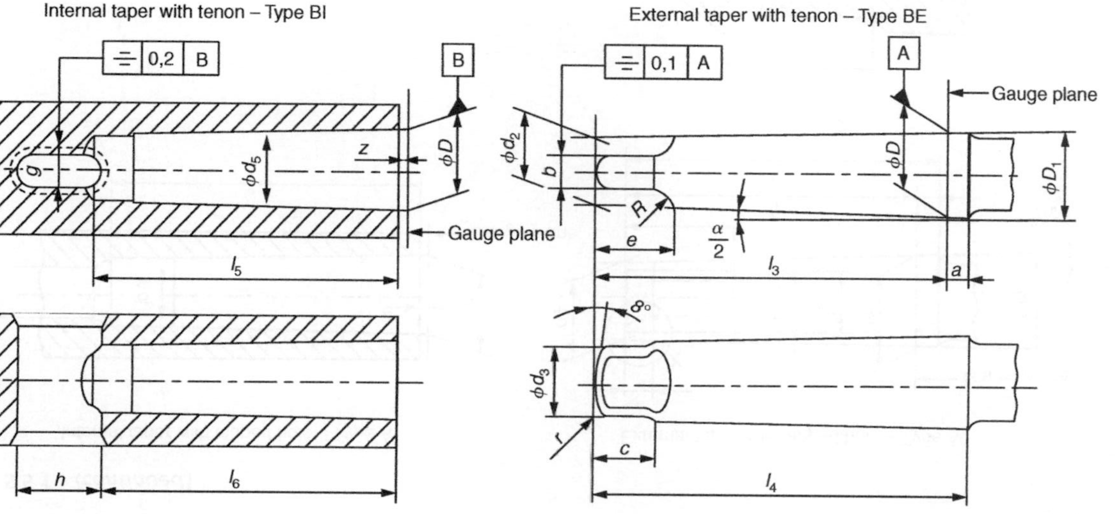

continued

Section 5.5.1 (*continued*)

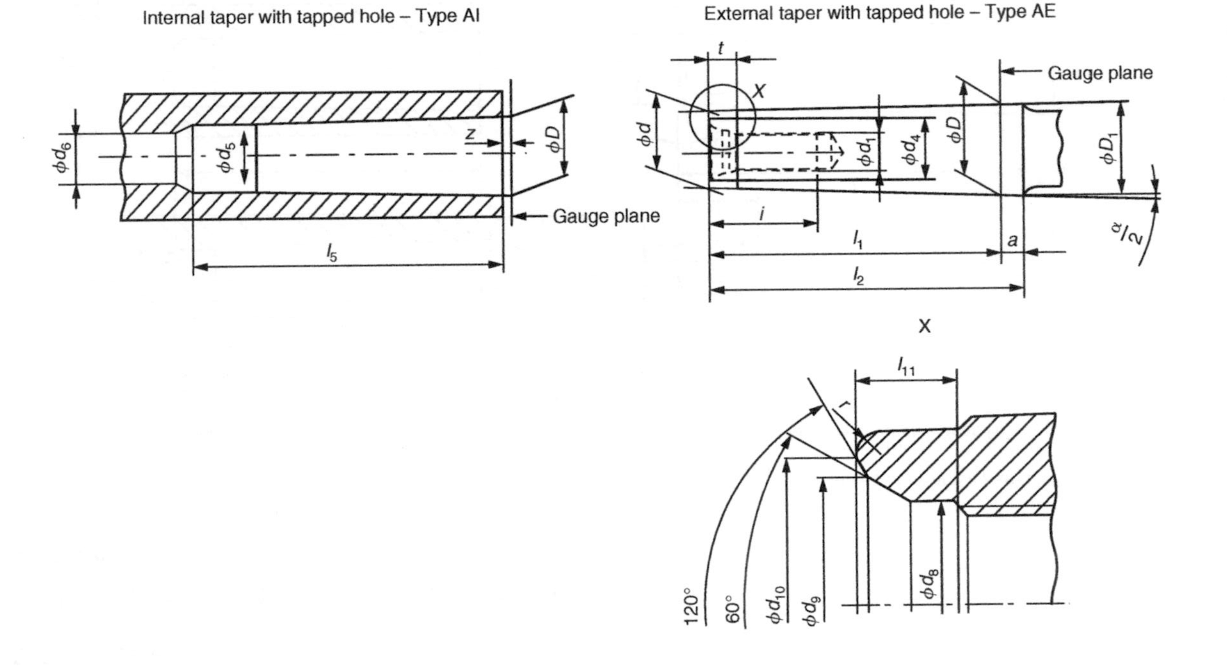

Tolerances of symmetry in millimetres

Internal taper with tenon and coolant supply – Type BIK

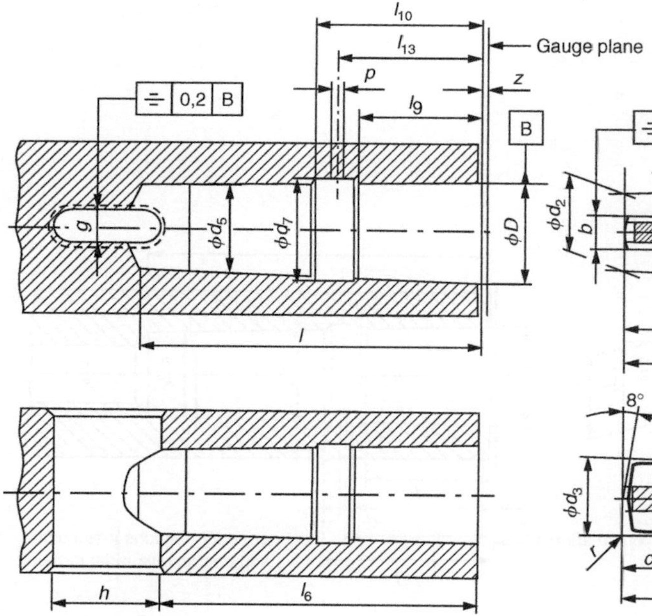

External taper with tenon and coolant supply – Type BEK

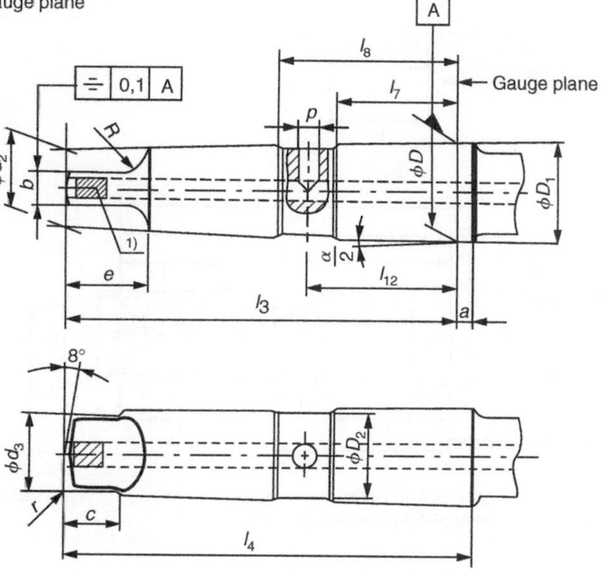

1) Optional gap

continued

Section 5.5.1 (*continued*)

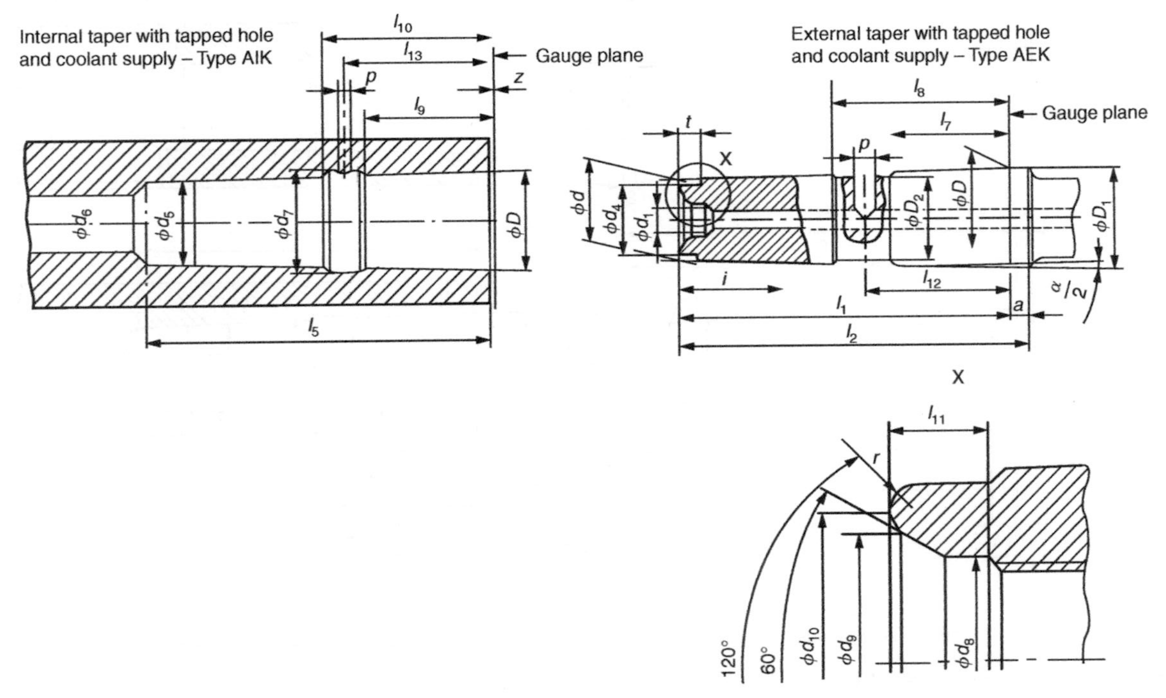

Numbers 0–6 Morse tapers and 5% metric tapers

Dimensions in mm

Designation		Metric tapers		Morse tapers							Metric tapers				
		4	6	0	1	2	3	4	5	6	80	100	120	160	200
Taper ratio		**1:20 = 0,05**		**0.624 6:12 = 1:19.212 = 0.052 05**	**0.598 58:12 = 1:20.047 = 0.049 88**	**0.599 41:12 = 1:20.02 = 0.049 95**	**0.602 35:12 = 1:19.922 = 0.050 2**	**0.623 26:12 = 1:19.254 = 0.051 94**	**0.631 51:12 = 1:19.002 = 0.052 63**	**0,62565:12 = 1:19.18 = 0.052 14**	**1:20 = 0,05**				
External taper	D	4	6	9.045	12.065	17.78	23.825	31.267	44.399	63.348	80	100	120	160	200
	a	2	3	3	3.5	5	5	6.5	6.5	8	8	10	12	16	20
	D_1 (1)	4.1	6.2	9.2	12.2	18	24.1	31.6	44.7	63.8	80.4	100.5	120.6	160.8	201
	D_2	–	–	–	–	15	21	26	40	56	–	–	–	–	–
	d (1)	2.9	4.4	6.4	9.4	14.6	19.8	25.9	37.6	53.9	70.2	88.4	106.6	143	179.4
	d_1 (2)	–	–	–	**M6**	**M10**	**M12**	**M16**	**M20**	**M24**	**M30**	**M36**	**M36**	**M48**	**M48**
	d_2 (1)	–	–	6.1	9	14	19.1	25.2	36.5	52.4	69	87	105	141	177
	d_3 max.	–	–	6	8.7	13.5	18.5	24.5	35.7	51	67	85	102	138	174
	d_4 max.	2.5	4	6	9	14	19	25	35.7	51	67	85	102	138	174
	d_8	–	–	–	6.4	10.5	13	17	21	26	–	–	–	–	–
	d_9	–	–	–	8	12.5	15	20	26	31	–	–	–	–	–
	d_{10} max.	–	–	–	8.5	13.2	17	22	30	11.5	–	–	–	–	–
	l_1 max.	23	32	50	53.5	64	81	102.5	129.5	182	196	232	268	340	412
	l_2 max.	25	35	53	57	69	86	109	136	190	204	242	280	356	432
	l_3 $_{-0.1}^{0}$	–	–	56.5	62	75	94	117.5	149.5	210	220	260	300	380	460
	l_4 max.	–	–	59.5	65.5	80	99	124	156	218	228	270	312	396	480
	l_7 $_{-0.1}^{0}$	–	–	–	–	20	29	39	51	81	–	–	–	–	–
	l_8 $_{0}^{0.1}$	–	–	–	–	34	43	55	69	99	–	–	–	–	–
	l_{11}	–	–	–	4	5	5.5	8.2	10	11.5	–	–	–	–	–
	l_{12}	–	–	–	–	27	36	47	60	90	–	–	–	–	–
	p	–	–	–	3.3	4.2	5	6.8	8.5	10.2	–	–	–	–	–
	b h13	–	–	3.9	5.2	6.3	7.9	11.9	15.9	19	26	32	38	50	62
	c (3)	–	–	6.5	8.5	10	13	16	19	27	24	28	32	40	48
	e max.	–	–	10.5	13.5	16	20	24	29	40	48	58	68	88	108

continued

Dimensions in mm

Designation			Metric tapers		Morse tapers							Metric tapers				
			4	6	0	1	2	3	4	5	6	80	100	120	160	200
Taper ratio			1:20 = 0.05		0.624 6:12 = 1:19,212 = 0,052 05	0.598 58:12 = 1:20.047 = 0.049 88	0.599 41:12 = 1:20,02 = 0.049 95	0.602 35:12 = 1:19.922 = 0.050 2	0.623 26:12 = 1:19.254 = 0.051 94	0.631 51:12 = 1:19.002 = 0.052 63	0.62565:12 = 1:19.18 = 0.052 14	1:20 = 0,05				
Internal taper	i	min.	–	–	–	16	24	24	32	40	47	59	70	70	92	92
	R	max.	–	–	4	5	6	7	8	12	18	24	30	36	48	60
	r		–	–	1	1.2	1.6	2	2.5	3	4	5	5	6	8	10
	t	max.	2	3	4	5	5	7	9	10	16	24	30	36	48	60
	d_5	H11	3	4.6	6.7	9.7	14.9	20.2	26.5	38.2	54.8	71.5	90	108.5	145.5	182.5
	d_6	min.	–	–	–	7	11.5	14	18	23	27	33	39	39	52	52
	d_7		–	–	–	–	19.5	24.5	32	44	63	–	–	–	–	–
	l_5	min.	25	34	52	56	67	84	107	135	188	202	240	276	350	424
	l_6		21	29	49	52	62	78	98	125	177	186	220	254	321	388
	l_9		–	–	–	–	22	31	41	53	83	–	–	–	–	–
	l_{10}		–	–	–	–	32	41	53	67	97	–	–	–	–	–
	l_{13}		–	–	–	–	27	36	47	60	90	–	–	–	–	–
	g	A13	2.2	3.2	3.9	5.2	6.3	7.9	11.9	15.9	19	26	32	38	50	62
	h		8	12	15	19	22	27	32	38	47	52	60	70	90	110
	p		–	–	–	–	4.2	5	6.8	8.5	10.2	–	–	–	–	–
	z	(4)	0.5	0.5	1	1	1	1	1	1	1	1.5	1.5	1.5	2	2

(1) For D_1 and d or d_2, approximate values are given for guidance.
(The actual values result from the actual values of a and l_1 or l_3 respectively, taking into account the taper ratio and the basic size D.)

(2) d_1 is the nominal thread diameter, either a metric thread M with standard pitch or, if expressly stated, a UNC thread (see table 'Numbers 1–6 Morse tapers and Numbers 1–3 (Brown & Sharpe tapers' for inch sizes). In every case, the appropriate symbol M or UNC shall be marked on the component.

(3) It is permissible to increase the length c over which the tenon is turned to diameter d_3, but without exceeding e.

(4) z is the maximum permissible deviation, outwards only, of the position of the gauge plane related to the basic size D from the nominal position of coincidence with the leading face.

Numbers 1–6 Morse tapers and Nos 1–3 Brown & Sharpe tapers

Dimensions in mm

Designation			Brown & Sharpe tapers			Morse tapers					
			1	2	3	1	2	3	4	5	6
Taper ratio			**0.502:12 = 1:23,904 = 0.041 83**	**0.502:12 = 1:23,904 = 0.041 83**	**0.502:12 = 1:23,904 = 0.041 83**	**0.598 58:12 = 1:20,047 = 0.049 88**	**0.599 41:12 = 1:20,02 = 0.049 95**	**0.602 35:12 = 1:19,922 = 0.050 2**	**0.623 26:12 = 1:19,254 = 0.051 94**	**0.631 51:12 = 1:19,002 = 0.052 63**	**0.625 65:12 = 1:19,18 = 0.052 14**
External taper	D		0.239 22	0.299 68	0.375 25	0.475	0.7	0.938	1.231	1.748	2.494
	a		3/32	3/32	3/32	1/8	3/16	3/16	1/4	1/4	1/16
	D_1	(1)	0.243 14	0.303 6	0.379 17	0.481 2	0.709 4	0.947 4	1.244	1.761 2	2.510 3
	D_2		–	–	–	0.393 7	0.590 6	0.826 8	1.102 4	1.574 8	2.204 7
	d	(1)	0.2	0.25	0.312 5	0.369	0.572	0.778	1.02	1.475	2.116
	d_1	(2)	–	–	–	UNC 1/4	UNC 3/8	UNC 1/2	UNC 5/8	UNC 5/8	UNC 1
	d_2	(1)	0.189 54	0.236 93	0.296 81	0.353 4	0.553 3	0.752 9	0.990 8	1.438 8	2.063 9
	d_3	max.	11/64	7/32	9/32	11/32	17/32	23/32	31/32	1 13/32	2
	d_4	max.	11/64	7/32	9/32	11/32	17/32	23/32	31/32	1 13/32	2
	d_8		–	–	–	0.251 97	0.413 38	0.511 81	0.669 29	0.826 77	1.023 62
	d_9		–	–	–	0.314 96	0.492 12	0.590 55	0.787 4	1.023 62	1.220 47
	d_{10}	max.	–	–	–	0.334 64	0.519 68	0.689 29	0.866 14	1.181 1	1.417 32
	l_1	max.	15/16	1 3/16	1 1/2	2 1/8	2 9/16	3 3/16	4 1/16	5 3/16	7 1/4
	l_2	max.	1 1/32	1 9/32	1 19/32	2 1/4	2 3/4	3 3/8	4 5/16	5 7/16	7 9/16
	l_3	0 −0.004	1 3/16	1 1/2	1 7/8	2 7/16	2 15/16	3 11/16	4 5/8	5 7/8	8 1/4
	l_4	max.	1 9/32	1 19/32	1 31/32	2 9/16	3 1/8	3 7/8	4 7/8	6 1/8	8 9/16
	l_7	0 −0.004	–	–	–	19/32	25/32	1 9/64	1 17/32	2	3 3/16
	l_8	0 −0.004	–	–	–	1 3/16	1 11/32	1 19/64	2 3/16	2 23/32	3 29/32
	l_{11}		–	–	–	0.157 48	0.196 85	0.216 53	0.322 83	0.393 7	0.452 75
	l_{12}		–	–	–	–	1.062 99	1.417 32	1.850 39	2.362 2	3.543 3
	p		–	–	–	1/8	11/64	13/64	9/32	21/64	13/32
	b	h12	0.125	0.156 2	0.187 5	0.203 1	0.25	0.312 5	0.468 7	0.625	0.75

continued

Section 5.5.1 (*continued*)

Dimensions in mm

Designation			Brown & Sharpe tapers 1	Brown & Sharpe tapers 2	Brown & Sharpe tapers 3	Morse tapers 1	Morse tapers 2	Morse tapers 3	Morse tapers 4	Morse tapers 5	Morse tapers 6
Taper ratio			**0.502:12 = 1:23,904 = 0.041 83**	**0.502:12 = 1:23,904 = 0.041 83**	**0.502:12 = 1:23,904 = 0.041 83**	**0.598 58:12 = 1:20,047 = 0.049 88**	**0.599 41:12 = 1:20,02 = 0.049 95**	**0.602 35:12 = 1:19,922 = 0.050 2**	**0.623 26:12 = 1:19,254 = 0,051 94**	**0.631 51:12 = 1:19,002 = 0.052 63**	**0.625 65:12 = 1:19,18 = 0.052 14**
Internal taper	c	(3)	1/4	5/16	3/8	11/32	13/32	17/32	5/8	3/4	1 1/16
	e	max.	0.381	0.455	0.532	0.52	0.66	0.83	0.96	1.15	1.58
	i	min.	–	–	–	1/2	3/4	0.944 88	1 1/4	1 1/4	1.850 4
	R	max.	3/16	3/16	3/16	3/16	1/4	9/32	5/16	0.472 44	0.708 66
	r		1/32	1/32	3/64	3/64	1/16	5/64	3/32	1/8	5/32
	t	max.	1/8	1/8	1/8	3/16	3/16	1/4	1/4	5/16	3/8
	d_5	H11	0.203	0.255	0.319	0.378	0.588	0.797	1.044	1.502	2.157 48
	d_6	min.	–	–	–	9/32	7/16	9/16	11/16	11/16	1 1/8
	d_7		–	–	–	17/32	49/64	31/32	1 17/64	1 47/64	2 31/64
	l_5	min.	1	1 1/4	1 9/16	2 3/16	2 21/32	3 9/32	4 5/32	5 5/16	7 3/8
	l_6		29/32	1 1/8	1 13/32	2 1/16	2 1/2	3 1/16	3 7/8	4 15/16	7
	l_9		–	–	–	43/64	7/8	17/32	1 39/64	2 3/32	3 17/64
	l_{10}		–	–	–	1 1/16	1 17/64	1 39/64	2 3/32	2 41/64	3 13/16
	l_{13}		–	–	–	–	1.062 99	1.417 32	1.850 39	2.362 2	3.543 3
	g	H12	0.141	0.172	0.203	0.223	0.27	0.333	0.493	0.65	0.78
	h		13/32	9/16	23/32	3/4	7/8	1 1/8	1 1/4	1 1/2	1 7/8
	p		–	–	–	1/8	11/64	13/64	9/32	21/64	13/32
	z	(4)	0.04	0.04	0.04	0.039 3	0.039 3	0.039 3	0.039 3	0.039 3	0.039 3

(1) For D_1 and d or d_2, approximate values are given for guidance.
(The actual values result from the actual values of a and l_1 or l_3 respectively, taking into account the taper ratio and the basic size D.)

(2) d_1 is the nominal thread diameter: either a UNC thread or, if expressly stated, a metric thread M with standard pitch (see table 'Numbers 1–6 Morse tapers and 5% metric tapers' for metric sizes). In every case, the appropriate symbol UNC or M shall be marked on the component.

(3) It is permissible to increase the length c over which the tenon is turned to diameter d_3, but without exceeding e.

(4) z is the maximum permissible deviation, outwards only, of the position of the gauge plane related to the basic size D from the nominal position of coincidence with the leading face.

5.5.2 Tapers for spindle noses

All dimensions are in mm.

Tapers Nos 30–60

Note:– For the spindle nose No. 60, the tenons can be fixed by two screws, as for the spindle noses Nos 65–80.

continued

Section 5.5.2 (*continued*)

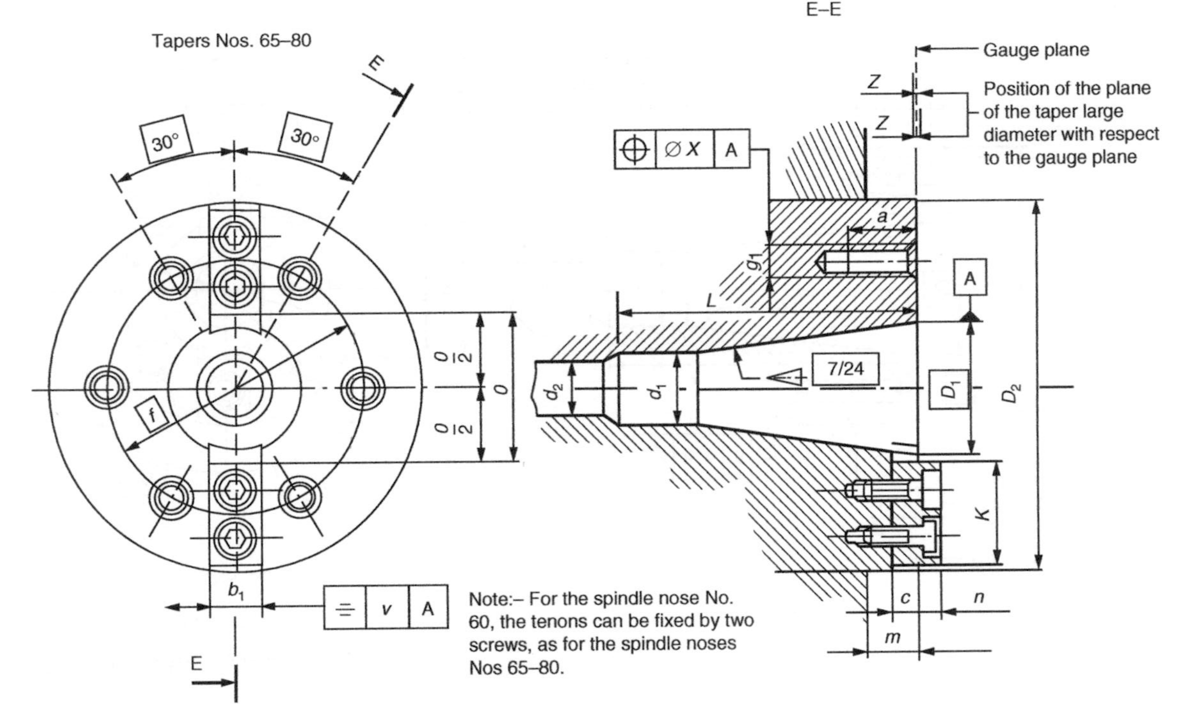

Note:– For the spindle nose No. 60, the tenons can be fixed by two screws, as for the spindle noses Nos 65–80.

Designation and dimensions

Designation No.	Taper		Recess		(2)	Tenon						External centring					
	D_1 (1)	z	d_1 H12	L min.	d_2 min.	b_1 (3)	v	c min.	n max.	O/2 min.	K max.	D_2 h5	m min.	f	g_1 (4)	a min.	x
30	31.75	0.4	17.4	73	17	15.9	0.06	8	8	16.5	16.5	69.832	12.5	54	M10	16	0.15
40	44.45	0.4	25.3	100	17	15.9	0.06	8	8	23	19.5	88.882	16	66.7	M12	20	0.15
45	57.15	0.4	32.4	120	21	19	0.06	9.5	9.5	30	19.5	101.6	18	80	M12	20	0.15
50	69.85	0.4	39.6	140	27	25.4	0.08	12.5	12.5	36	26.5	128.57	19	101.6	M16	25	0.2
55	88.9	0.4	50.4	178	27	25.4	0.08	12.5	12.5	48	26.5	152.4	25	120.6	M20	30	0.2
60	107.95	0.4	60.2	220	35	25.4	0.08	12.5	12.5	61	45.5	221.44	38	177.8	M20	30	0.2
65	133.35	0.4	75	265	42	32	0.1	16	16	75	58	280	38	220	M24	36	0.25
70	165.1	0.4	92	315	42	32	0.1	20	20	90	68	335	50	265	M24	45	0.25
75	203.2	0.4	114	400	56	40	0.1	25	25	108	86	400	50	315	M30	56	0.32
80	254	0.4	140	500	56	40	0.1	31.5	31.5	136	106	500	50	400	M30	63	0.32

(1) D_1: Basic diameter defining the gauge plane.
(2) Opening for traction bar.
(3) Assembly of the tenon in the slot: M6-h5 fit.
(4) Thread diameter g_1: this is either a metric thread M with coarse pitch or, if expressly stated, a UN thread according to table 'on thread specification'. In every case, the appropriate symbol M or UN shall be marked on the component.

Thread specification

Designation No.	30	40	45	50	55	60	65	70	75	80
g_1	UN 0,375-16	UN 0,500-13	UN 0,500-13	UN 0,625-11	UN 0,750-10	UN 0,750-10	UN 1,000-8	UN 1,000-8	UN 1,250-7	UN 1,250-7

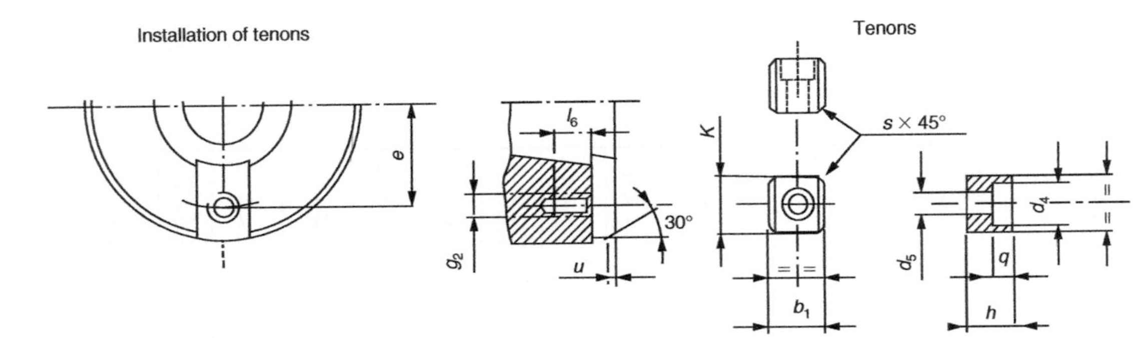

Note:– For spindle nose No. 60, the tenons can be fixed by two screws, as for the spindle noses Nos 65–80.

Noses Nos 65–80

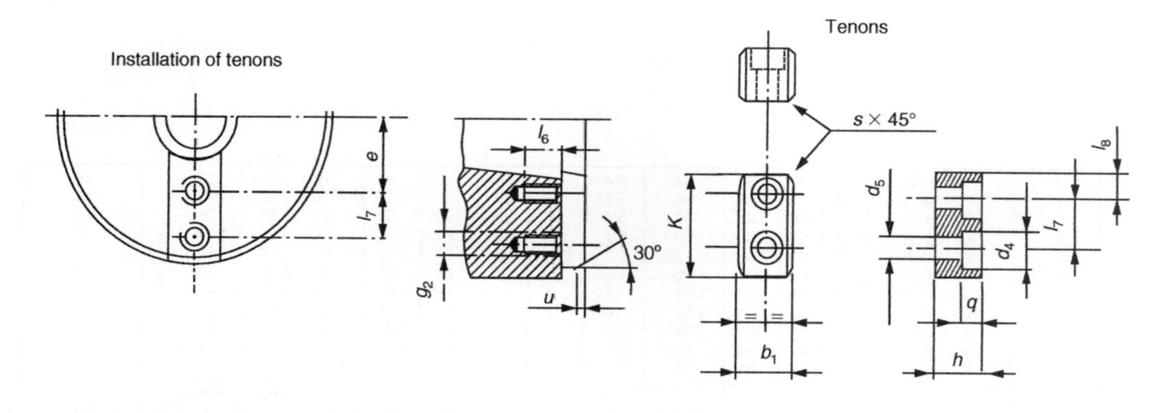

Complementary dimensions

Designation No.	Tenon									Slot				Screws ISO 4762	Chamfer
	b_1	h max.	k max.	d_5	d_4	q	l_7	l_8	s max.	$e \pm 0.2$	g_2	l_6	l_7		u
30	See Table 1(a)	16	16.5	6.4	10.4	7	–	–	1.6	25	M6	9	–	M6 × 16	2
40		16	19.5	6.4	10.4	7	–	–	1.6	33	M6	9	–	M6 × 16	2
45		19	19.5	8.4	13.4	9	–	–	1.6	40	M8	12	–	M8 × 20	2
50		25	26.5	13	19	13	–	–	2	49.5	M12	18	–	M12 × 25	3
55		25	26.5	13	19	13	–	–	2	61.5	M12	18	–	M12 × 25	3
60		25	45.5	13	19	13	–	–	2	84	M12	18	–	M12 × 25	3
		25	45.5	13	19	13	22	11.7	2	73	M12	18	22	M12 × 25	3
65		32	58	17	25	17	28	15	2.5	90	M16	25	28	M16 × 35	4
70		40	68	17	25	17	36	16	2.5	106	M16	25	36	M16 × 45	4
75		50	86	21	31	21	42	22	2.5	130	M20	30	42	M20 × 55	4
80		63	106	21	31	21	58	24	2.5	160	M20	30	58	M20 × 65	4

5.5.3 Tapers for tool shanks

All dimensions are in mm.

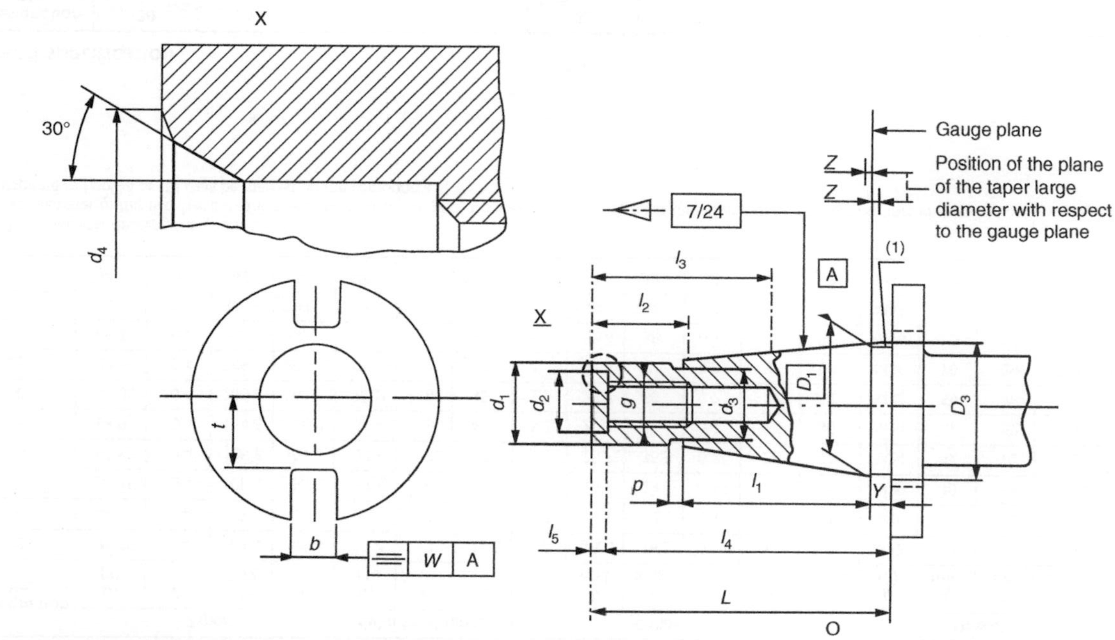

1) Optional groove. Without groove, cylindrical joining surface with diameter $D_3 = D_1 - 0.5$.

Designation No.	Taper				Cylindrical tenon			Collar					Thread					
	D_1 (1)	Z	L h12	l_1	d_1 a10	p	d_3	y	b H12	t max.	w	d_2	d_4 max.	g (2)	l_2 min.	l_3 min.	l_4 0–0.5	l_5
30	31.75	0.4	68.4	48.4	17.4	3	16.5	1.6	16.1	16.2	0.12	13	16	M12	24	34	62.9	5.5
40	44.45	0.4	93.4	65.4	25.3	5	24	1.6	16.1	22.5	0.12	17	21.5	M16	32	43	85.2	8.2
45	57.15	0.4	106.8	82.8	32.4	6	30	3.2	19.3	29	0.12	21	26	M20	40	53	96.8	10
50	69.85	0.4	126.8	101.8	39.6	8	38	3.2	25.7	35.3	0.2	26	32	M24	47	62	115.3	11.5
55	88.9	0.4	164.8	126.8	50.4	9	48	3.2	25.7	45	0.2	26	36	M24	47	62	153.3	11.5
60	107.95	0.4	206.8	161.8	60.2	10	58	3.2	25.7	60	0.2	32	44	M30	59	76	192.8	14
65	133.35	0.4	246	202	75	12	72	4	32.4	72	0.3	38	52	M36	70	89	230	16
70	165.1	0.4	296	252	92	14	90	4	32.4	86	0.3	38	52	M36	70	89	280	16
75	203.2	0.4	370	307	114	16	110	5	40.5	104	0.3	50	68	M48	92	115	350	20
80	254	0.4	469	394	140	18	136	6	40.5	132	0.3	50	68	M48	92	115	449	20

(1) D_1: Basic diameter defining the gauge plane.
(2) Thread diameter g: this is either a metric thread M with coarse pitch or, if expressly stated, a UN thread according to table on 'thread specification'. In every case, the appropriate symbol M or UN shall be marked on the component.

Thread specification

Designation No.	30	40	45	50	55	60	65	70	75	80
g	UN 0,500-13	UN 0,625-11	UN 0,75-10	UN 1,000-8	UN 1,000-8	UN 1,25-7	UN 1,375-6	UN 1,375-6	UN 1,750-5	UN 1,750-5

5.5.4 Tool shank collars

All dimensions are in mm.

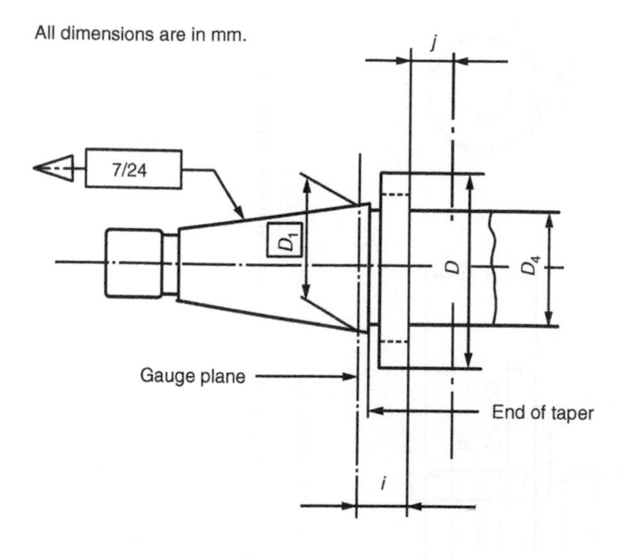

Designation and dimensions

Designation No.	D_1	i[a,b] ± 0,1	D	D_4[b] max.	j[b,c] min.
30	31.75	9.6	50	36	9
40	44.45	11.6	63	50	11
45	57.15	15.2	80	68	13
50	69.85		97.5	78	16
55	88.9	17.2	130	110	
60	107.95	19.2	156	136	
65	133.35	22	195	By agreement between customer and supplier	
70	165.1	24	230		
75	203.2	27	280		
80	254	34	350		

[a] The distance between the front face of the collar and the gauge plane having the basic diameter D_1 (and not the great base plane of the taper).
[b] These values are only prescribed for those tools that are intended for attachment on the collar front face.
[c] Tool fixing area.

5.5.5 Bridgeport R8 taper

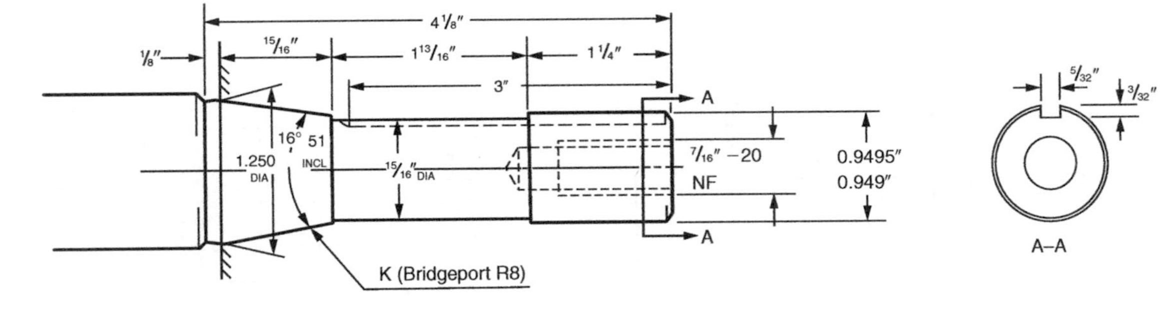

The R8 taper was originally introduced by the Bridgeport Machine Tool Co., for their vertical spindle turret mills.

This taper is now widely adopted for the spindles of similar machines by other manufacturers and it is also used for the spindles of small, low-cost milling machines imported from the Far East.

5.6 Fluid power transmission systems

Mechanical power transmission systems depend on such devices as shafts, universal joints, gears, belts, chains, etc., to transmit energy and motion from one part of a system to another. As such they tend to be relatively inflexible. Fluid power systems, although less efficient in the use of energy, are extremely flexible and controllable. The following table compares electrical, hydraulic and pneumatic systems.

Comparisons of electrical, hydraulic and pneumatic systems

	Electrical	**Hydraulic**	**Pneumatic**
Energy source	Usually from outside supplier	Electric motor or diesel driven	Electric motor or diesel driven
Energy storage	Limited (batteries)	Limited (accumulator)	Good (reservoir)
Distribution system	Excellent, with minimal loss	Limited basically a local facility	Good Can be treated as a plant wide service
Energy cost	Lowest	Medium	Highest
Rotary actuators	AC and DC motors Good control on DC motors AC motors cheap	Low speed Good control Can be stalled	Wide speed range Accurate speed control difficult
Linear actuator	Short motion via solenoid Otherwise via mechanical conversion	Cylinders Very high force	Cylinders Medium force
Controllable force	Possible with solenoid and DC motors Complicated by need for cooling	Controllable high force	Controllable medium force
Points to note	Danger from electric shock	Leakage dangerous and unsightly Fire hazard	Noise

Source: Table 1.1 in *Hydraulics & Pneumatics*, 2nd edn. Andrew Parr: B/Heinemann.

This section is concerned only with fluid power transmission, that is pneumatic (gases: usually air) and hydraulic (liquids: mainly oil or water). The main advantages and disadvantages of such systems arise out of the different characteristics of low density compressible gases and (relatively) high density incompressible liquids. A pneumatic system, for example, tends to have a 'softer' action than a hydraulic system which is more positive. A pneumatic system also exhausts to the atmosphere and this simplifies the pipework since no return circuit is required. A liquid-based hydraulic system can operate at much higher pressures and can provide much higher forces. Hydraulic systems employ water as the operating fluid where large volumes are required as in operating the raising and lowering mechanism of Tower Bridge in London or in the lift bridges employed mainly on the canal systems and waterways of continental Europe. However for most industrial purposes the hydraulic fluid is oil which is self-lubricating and does not cause corrosion.

5.6.1 A typical pneumatic system

The following figure is a schematic diagram of a pneumatic system. Air is drawn from the atmosphere via an *air filter* and raised to the required pressure by an *air compressor* which is usually driven by an electric motor (portable compressors as used by the construction industry have the air compressor driven by a diesel engine). The compression process raises the temperature of the air. Air also contains a significant amount of water vapour. Before passing to the compressed air to the *receiver* (storage reservoir) the air must be cooled and this results in any water vapour present condensing out. This condensate should be removed before the air reaches the receiver. A pressure regulator switch turns the motor on when the pressure in the receiver falls and turns the motor off when the air reaches a pre-determined pressure. The receiver is also fitted with a *safety valve* in case the pressure regulator switch fails. This safety valve must be capable of passing the full output of the compressor. The receiver is usually followed by a *lubricator* which allows an oil mist to enter the air stream and lubricates the control valve and actuator. In the simplest system the air passes to a control valve followed by the *actuator*: generally a piston and cylinder to provide linear motion.

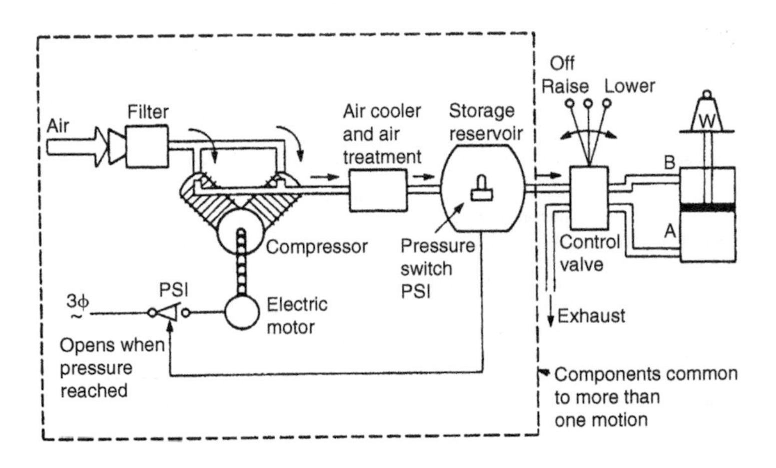

Pneumatic solution.
Source: Figure 1.3 in *Hydraulics & Pneumatics*, 2nd edn. Andrew Parr: B/Heinemann.

Operating pressures in a pneumatic system are much lower than those used in hydraulic systems. A typical pneumatic system pressure being about 10 bar which will provide a force of 98.1 kN cm^{2}cm of piston area. Therefore actuators in pneumatic systems need to be much larger than those used in hydraulic systems to move the same loads. The compressibility of air makes it necessary to store a large volume of air in a receiver to be drawn upon by the actuator as and when required. Without this reservoir of air there would be a slow and pulsating exponential rise in pressure with a corresponding slow and pulsating piston movement when the regulating valve is first opened. Most industrial installations require the compressed and conditioned air to be piped round the factory from the receiver to the various points where it will be required. Plug-in sockets are provided at each outlet point to receive the hose attached to the appliance (drill, rivet gun, nibbler, etc.). The socket is self sealing, so that there is no loss of air and pressure, when the hose is disconnected.

5.6.2 A typical hydraulic system

The following figure shows a typical hydraulic installation. This generally uses oil (water is only used for very large scale installations) as the activating fluid. Unlike the pneumatic system, discussed in Section 5.6.1, a hydraulic system must form a closed loop so that after passing through the actuator it returns to the oil storage tank for re-use. Further, hydraulic installations are usually single purpose, integrated installations that operate a single device, for example, the table traverse of a surface grinding machine or for powering a mobile crane. Hydraulic fluid is never piped around a whole factory to provide a 'supply on demand system' like compressed air. Hydraulic systems work at much higher pressures, typically 150 bar compared to 10 bar for a pneumatic system, therefore the actuator (cylinder and piston or hydraulic motor) can be considerably smaller that its pneumatic counterpart for a given application. Further, since oil or water is virtually incompressible hydraulic systems are much more positive than pneumatic systems.

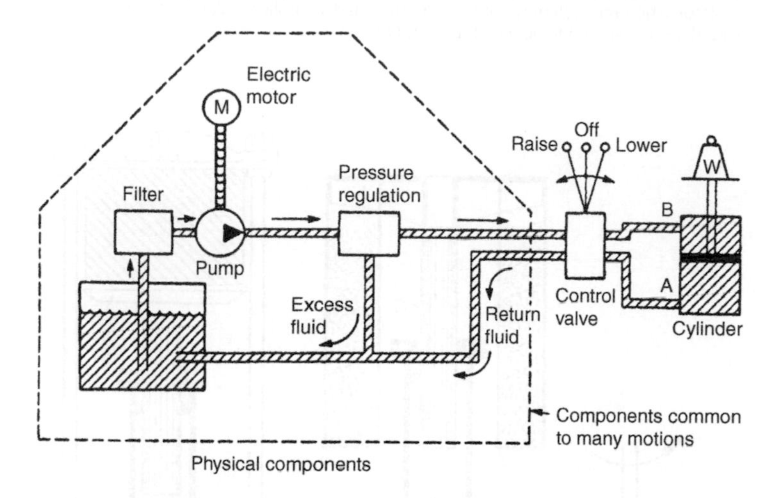

Hydraulic solution.
Source: Figure 1.2 in *Hydraulics & Pneumatics*, 2nd edn. Andrew Parr: B/Heinemann.

Fluid (oil or water) is drawn from the *storage tank* via a *filter* to the motor driven pump. On a machine tool this is driven by an electric motor, whilst on a crane or other mobile device it is driven by a diesel engine. Unlike a compressor the pump runs all the time that the machine is in use. If the oil pressure builds up beyond a safe predetermined value a *pressure regulating valve* returns the surplus oil to the storage tank. If an *oil cooler* is not fitted in the system then the capacity of the storage tank must be such that the oil has time to cool down an does not drop in viscosity. The piston movement is controlled by a *three position changeover valve* as shown. Oil is admitted at a point A to raise the piston (and the load W). Any oil already in the cylinder can escape via B and return to the storage tank via the valve. Oil is admitted to point B to lower the load and any oil already in the cylinder is allowed to escape via point A and return to the cylinder. The speed of movement can be controlled by the volume flow rate; that is the amount the valve ports are opened by the operator. This ability to provide precise control at low speeds is one of the main advantages of hydraulic systems. In the centre position, the valve locks the fluid into the cylinder on each side of the piston so that it cannot move unless leakage occurs.

The regulator bypasses the excess oil still being pumped back to the storage tank. When oil is used as the actuating fluid the system is self-lubricating.

For large scale systems, particularly when using water to reduce costs, very large volumes of water are required at very high pressures for example when raising and lowering Tower Bridge in London. Since this operation is only required at relatively infrequent intervals it would be uneconomical to employ a pump big enough to drive the operating rams directly. The solution is to install a *hydraulic accumulator* as shown in following figure between the pump and the operating valve and rams. The hydraulic accumulator consists of a large cylinder and ram carrying a heavy mass to provide the required pressure. The *size* of the cylinder and ram depends on the *volume* of fluid required to operate the actuating rams of (in this example) the bridge. The *pressure* is controlled by the *mass* bearing down on the accumulator. A relatively small and economic pump providing the hydraulic fluid at the required pressure but at a low delivery volume rate is used to raise the accumulator when the system is not being used. The accumulator, alone, provides the pressure fluid on demand to actuate the operating rams.

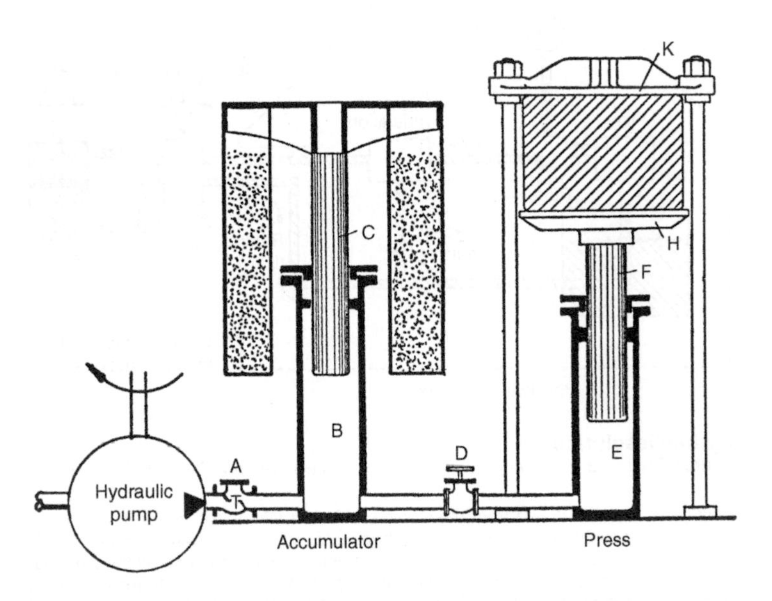

Water is forced by a pump past the valve A into the cylinder B. This raises the heavily loaded ram C. Meantime valve D is closed. The water cannot escape from B by the valve A, for the pressure upon the top closes it. So water under pressure is stored in the accumulator. When it is desired to work the press the valve D is opened and the water is forced from cylinder B into E. This raises the ram F, and compresses the bale of cotton, or whatever it may be, between the faces H and K.

Hydraulic press and accumulator.

5.6.3 Air compressor types

Piston compressors

These can deliver compressed air to suit the requirements of any system. They can also deliver air at higher pressures than vane and screw compressors. On the downside, they are noisier than vane and screw compressors and their output pulsates and requires to be fed into a receiver to smooth out the flow. Single cylinder compressors are the worst in this respect. Multistage piston compressors are more efficient due mainly to *intercooling* between the stages. They are capable of higher maximum pressures.

Diaphragm compressors

These are small scale inexpensive reciprocating type compressors of limited output capable of delivering oil-free air for very small systems. Since the air is free from oil they are widely used for small scale spray painting and air-brush applications. They are usually used for portable, stand-alone compressors, by contractors on site for powering such devices as staple guns.

Rotary vane compressors

These are quiet, inexpensive and deliver a steady stream of compressed air without the pulsations associated with the piston and diaphragm types. Lubrication of the vane edges is required by a recirculating lubrication system, with the oil being purged from the air before delivery. Multistage vane compressors are available with intercooling between the stages where higher pressures and larger volumes of air are required. The principle of operation is shown in following figure.

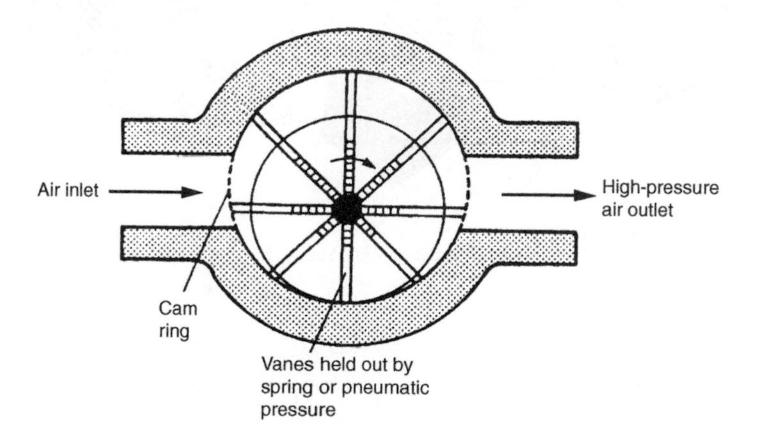

The vane compressor.
Source: Figure 5.27 (adapted) *Hydraulics & Pneumatics*, Andrew Parr: B/Heinemann.

Screw compressors

These are quiet, efficient but expensive and are usually chosen to supply large scale factory installations The principle of operation is shown in figure 'the screw compressor'. It can be seen that a rotor with male lobes meshes with a smaller diameter rotor with female lobes. Oil not only lubricates the bearings but also seals the clearances between the rotors and cools the air before purged from the air at the point of discharge. The figure 'compressor types: typical scope' shows the scope of the main types of compressor described above.

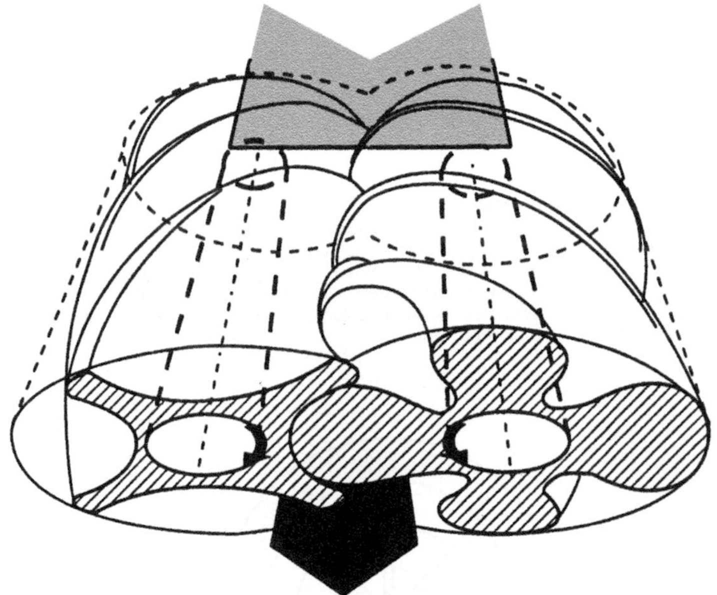

The screw compressor.
Source: Figure 4.6 in *Practical Pneumatics*, Chris Stacey: Newnes.

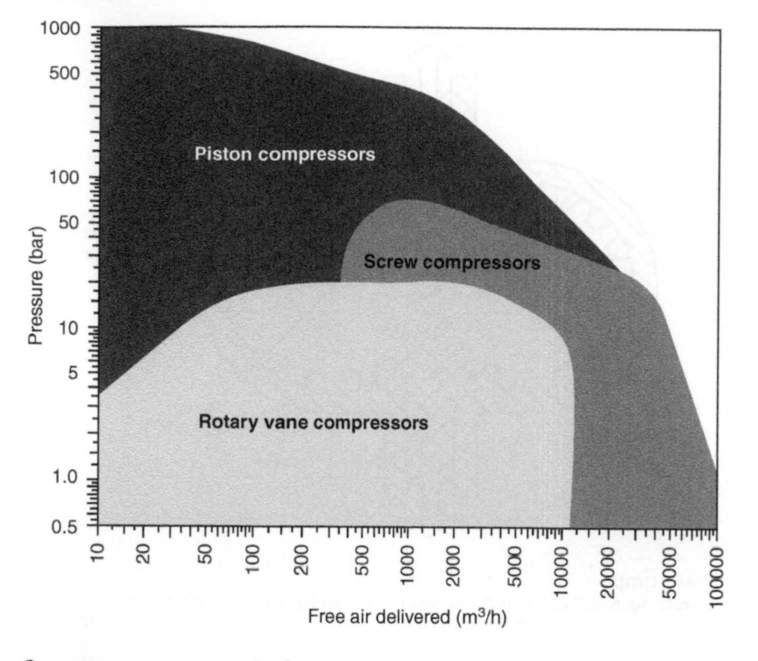

Compressor types – typical scope.
Source: Figure 4.2 in *Practical Pneumatics*, Chris Stacey: Newnes.

5.6.4 Hydraulic pumps

Gear pumps

These are the simplest and most robust positive displacement pump, having just two moving parts as shown in figure 'gear pump'. Only one gear needs to be driven by the power source. The direction of rotation of the gears should be noted since they form a partial vacuum as they come out of mesh, drawing fluid into the inlet chamber. The oil is carried round between the gear teeth and the outer casing of the pump resulting in a continuous supply of oil into the delivery chamber. The pump displacement, the volume of fluid delivered, is determined by the volume of fluid between each pair of teeth and the speed of rotation. The pump delivers a fixed volume of fluid for each rotation and the outlet port pressure is determined by the 'back-pressure' opposing the flow in the rest of the system. This type of pump is used up to pressures of about 150 bar and about 6750 l/min. At 90% the volumetric efficiency is the lowest of the pump types to be described in the section. There are a number of variations on the gear pump available including the *internal gear pump* and the *lobe pump*. A lobe pump is shown in figure 'the lobe pump'.

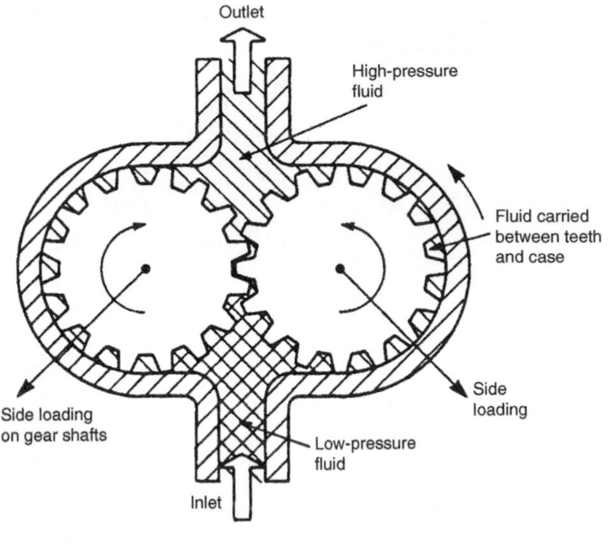

Gear pump.
Source: Figure 2.7 in *Hydraulics & Pneumatics*, Andrew Parr: B/Heinemann.

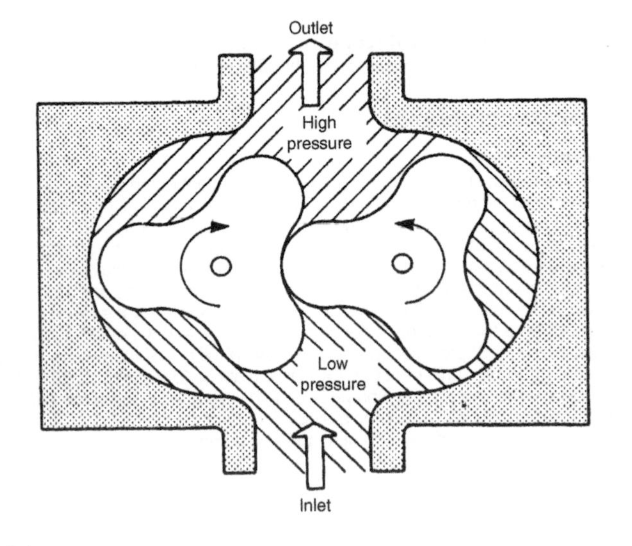

The lobe pump.
Source: Figure 2.8 in *Hydraulics & Pneumatics*, Andrew Parr: B/Heinemann.

Vane pumps

The relatively low volumetric efficiency of the gear type pump stems from clearances between the teeth, and between the gears and the outer casing. These sources of leakage are largely over come in vane type pumps by the use of spring loaded vanes as shown in figure 'unbalanced vane pump'. Vane type hydraulic pumps are similar in principle to the vane type compressor described in Section 5.6.3 except that when delivering oil they are self-lubricating. An alternative design is shown in figure 'balanced vane pump'.

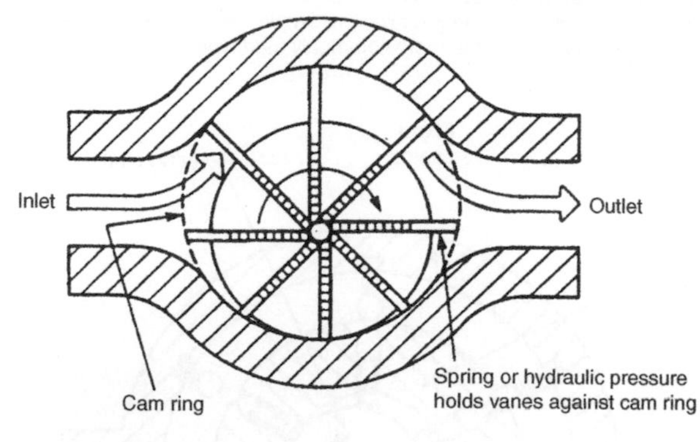

(a) Unbalanced vane pump

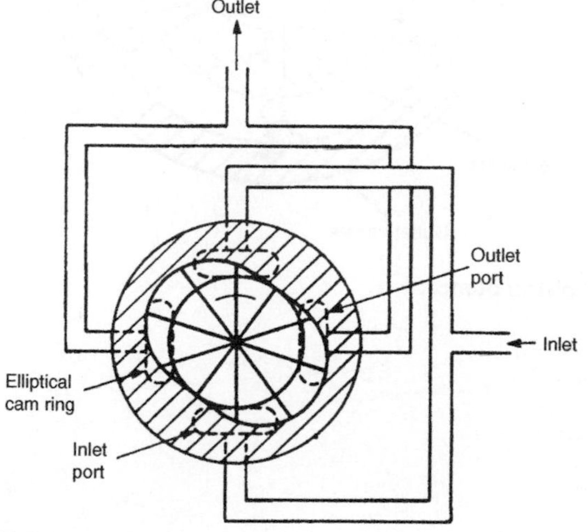

(b) Balanced vane pump

Vane pumps.
Source: Figure 2.10 in *Hydraulics & Pneumatics*, Andrew Parr: B/Heinemann.

Piston pumps

Reciprocating force pumps are only used on very large scale applications in conjunction with an accumulator to smooth out the pulsations. Nowadays, for most applications requiring multiple piston pumps, the pistons and cylinders are arranged radially as shown in figure 'radial piston pump'. The pump consists of several hollow pistons inside a stationary cylinder block. Each piston has spring loaded inlet and outlet valves. As the cam rotates, fluid is transferred relatively smoothly from the inlet port to the outlet port. The pump shown in figure 'piston pump with stationary can and rotating block' is similar in principle but uses a stationary cam and a rotating cylinder block. This arrangement removes the need for multiple inlet and outlet valves and is consequently more simple, reliable and cheaper to manufacture and maintain, hence this is the more commonly used type.

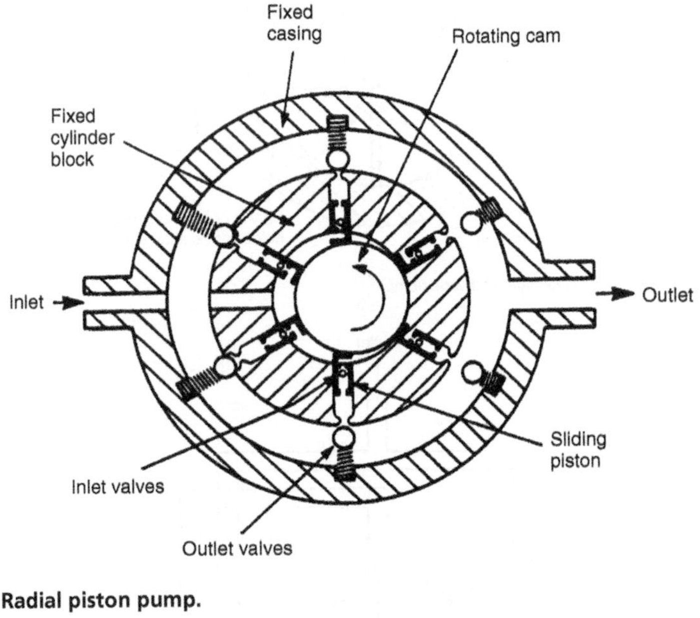

Radial piston pump.

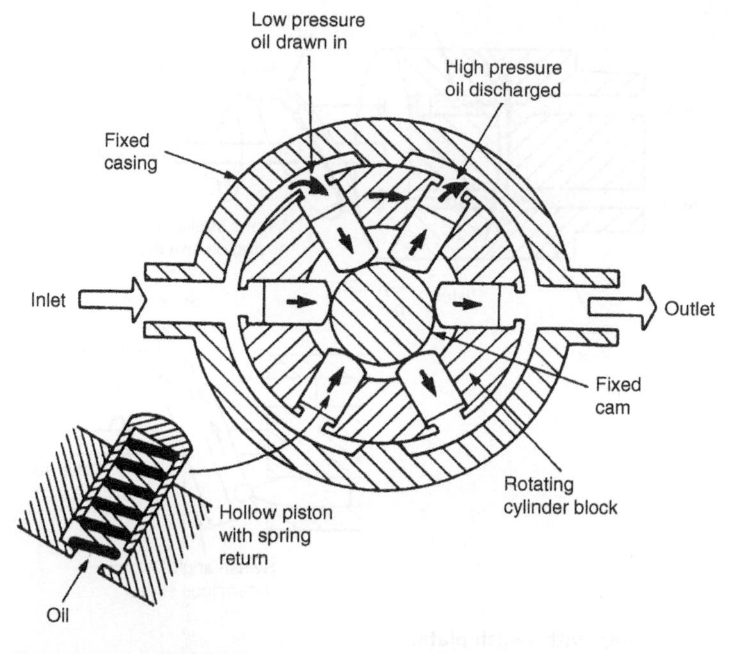

Piston pump with stationary cam and rotating block.
Source: Figures 2.12 and 2.13 in *Hydraulics & Pneumatics*, Andrew Parr: B/Heinemann.

Axial (swash plate) pumps

This is also a multiple piston pump. The pistons are arranged in a rotating cylinder block so that they are parallel to the axis of the drive shaft as shown in following figure. The piston stroke is controlled by an angled swash plate as shown. Each piston is kept in contact with the swash plate by spring pressure or by a rotating shoe plate linked to the swash plate. Pump capacity is controlled by altering the angle of the swash plate. The larger the angle the greater the rate of flow for a given speed of rotation. The maximum swash plate angle is limited by the maximum designed piston stroke length. Zero flow rate is achieved when the swash plate is perpendicular to the drive shaft. Reversing the swash plate angle reverses the direction of fluid flow through the pump.

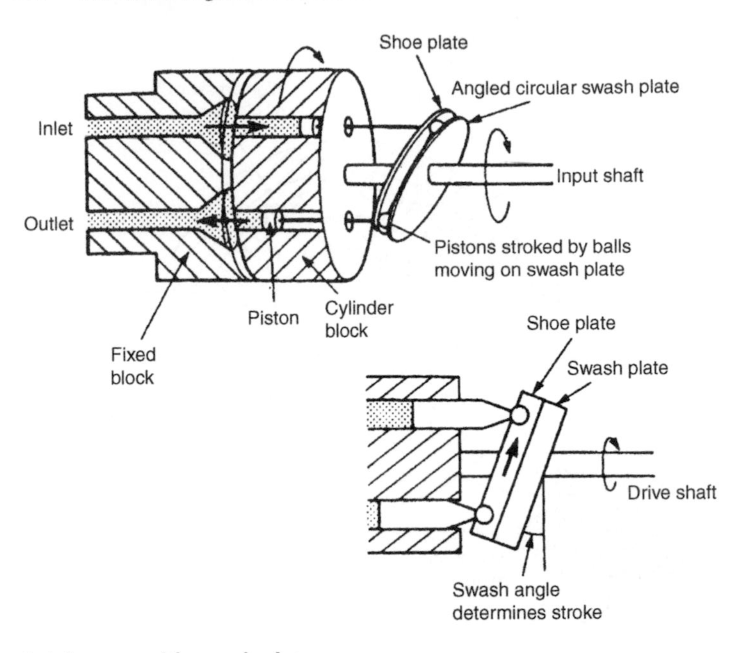

Axial pump with swash plate.
Source: Figure 2.14 in *Hydraulics & Pneumatics*, Andrew Parr: B/Heinemann.

All types of piston pumps have very high volumetric efficiency and can be used at the highest hydraulic pressures. They are more complex than vane and gear pumps and are, therefore more expensive in first cost and to maintain. The following table compares the advantages and limitations of various types of hydraulic pumps.

Comparison of hydraulic pump types

Type	Maximum pressure (bar)	Maximum flow (l/min)	Variable displacement	Positive displacement
Centrifugal	20	3000	No	No
Gear	175	300	No	Yes
Vane	175	500	Yes	Yes
Axial piston (port-plate)	300	500	Yes	Yes
Axial piston (valved)	700	650	Yes	Yes
In-line piston	1000	100	Yes	Yes

Specialist pumps are available for pressures up to about 7000 bar at low flows. The delivery from centrifugal and gear pumps can be made variable by changing the speed of the pump motor with a variable frequency (VF) drive.
Source: Table 2.1 in *Hydraulics & Pneumatics*, Andrew Parr: B/Heinemann.

5.6.5 Actuators (linear)

Both pneumatic and hydraulic systems use linear actuators where motion is required in a straight line as in clamping and operating machine tool work tables. Such devices have a piston and cylinder where short strokes are required. They are essentially the same design for both pneumatic and hydraulic applications except that hydraulic actuators are more heavily constructed to withstand the higher pressures and forces associated with hydraulic systems. They may be single acting as shown in figure 'single-acting cylinder', double acting as shown in figure 'double-acting cylinders' or with adjustable cushioning as shown in figure 'adjustable cushioning'. Hydraulic devices often require long-stroke actuators. These are called *rams* and a typical configuration is shown in figure 'long-stroke hydraulic ram and cylinder'. It is single acting and relies on the back-force of the load being moved for the return stroke. The range of such devices is too great to list in this book and the reader is referred to Appendix 3 for the details of suppliers who can provide comprehensive catalogues. Wherever possible pneumatic and hydraulic clamping devises should be designed so that the clamping force is not reduced or removed (unlocked) in the event of a system failure.

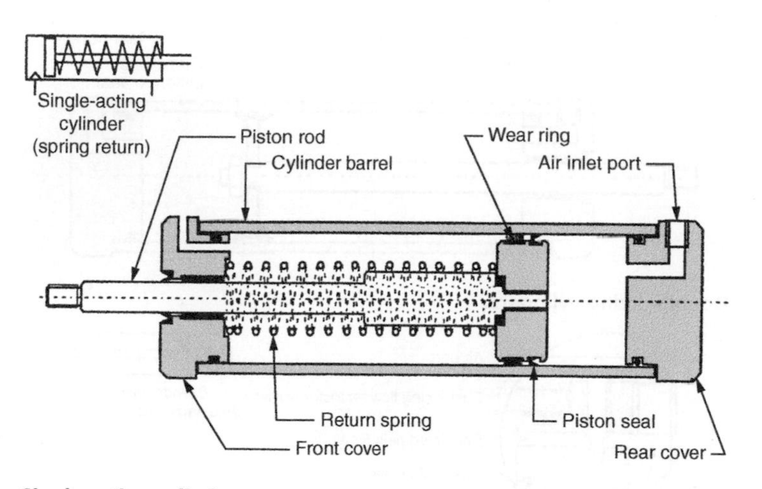

Single-acting cylinder.

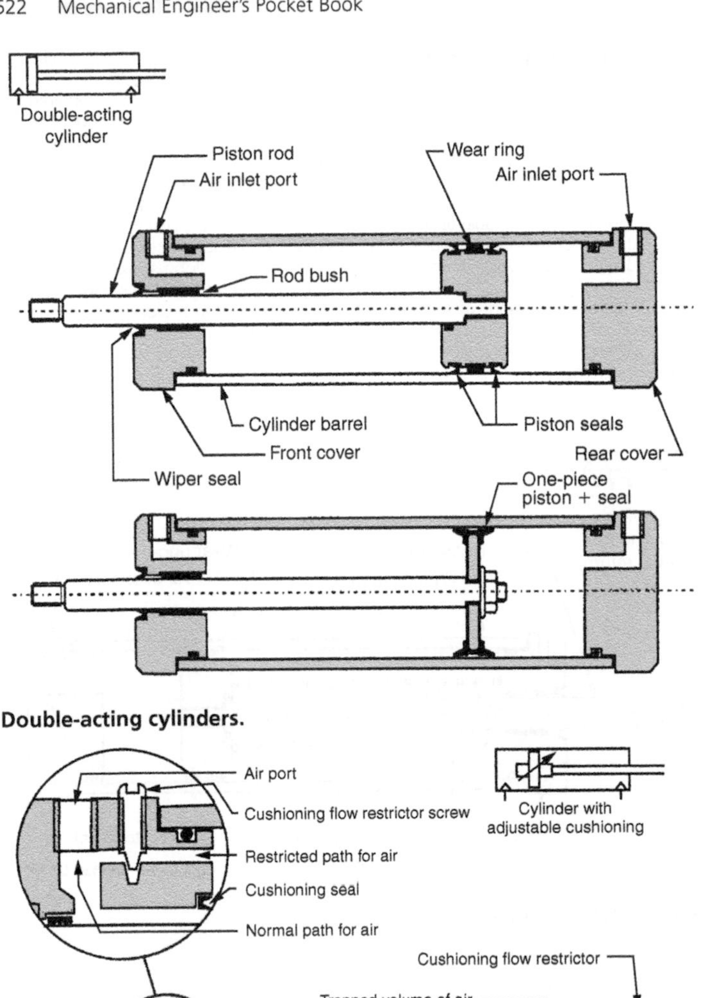

Double-acting cylinders.

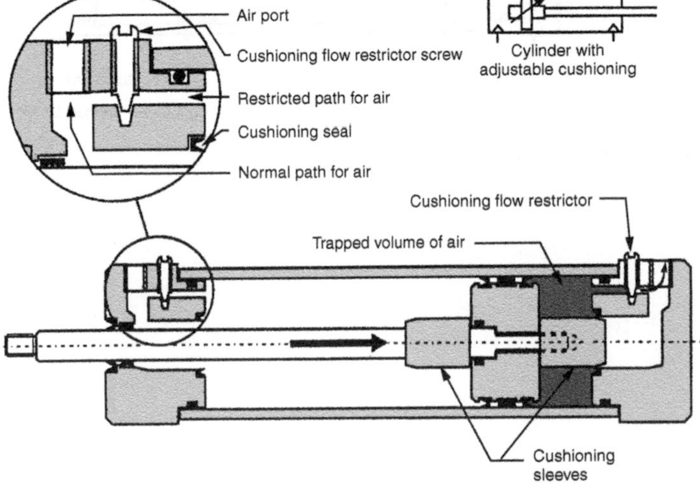

Adjustable cushioning.

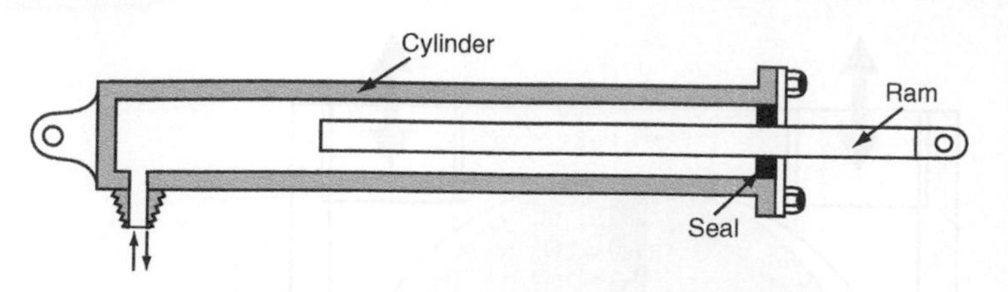

Long-stroke hydraulic ram & cylinder.
Pneumatic & Hydraulic Linear actuators.

5.6.6 Actuators (rotary)

These are similar in construction to the compressors and pumps described earlier. The direction of rotation depends on the direction of the fluid flow.

Pneumatic

For general applications these can be the radial piston and cylinder type as shown in figure 'the piston motor' where high power at moderate speeds is required. They are bulky, costly and noisy and require a silencer fitted to the exhaust system. A cheaper and lighter solution where lower power and higher speeds are required is the vane motor as shown

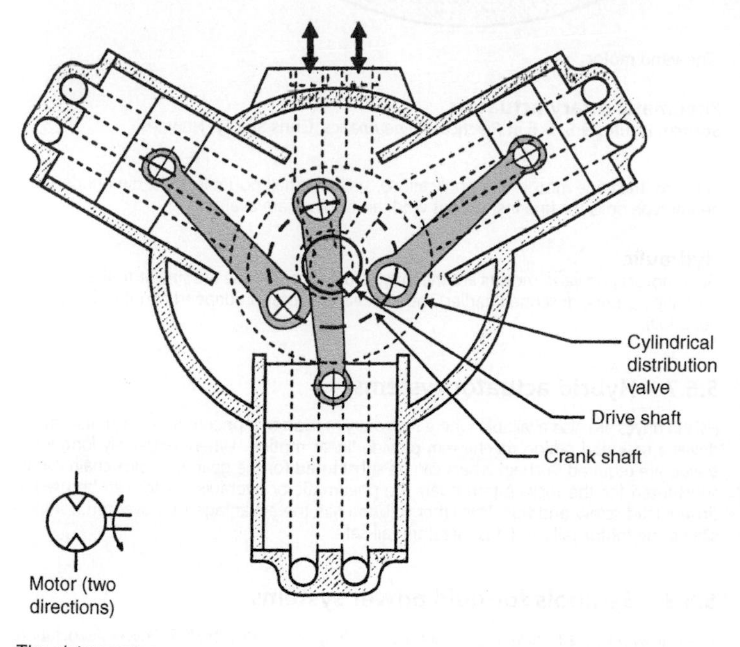

The piston motor

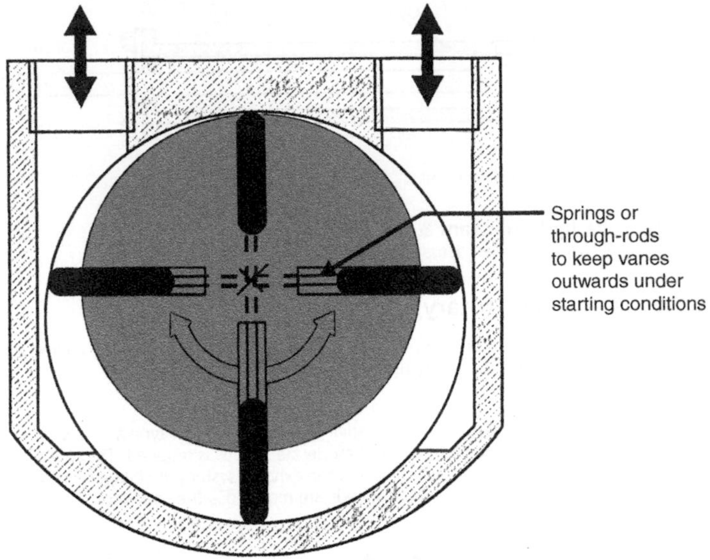

The vane motor

Pneumatic rotary actuators.
Source: Figures 9.5, 9.6 in *Practical & Pneumatics*, Chris Stacey: Newnes.

in figure 'the vane motor'. For applications, such as small portable drills, die-grinders and dental type drills for fine instrument work air turbines are available.

Hydraulic

Gear motors and vane motors are the most widely used. These are similar to the gear and vane type pumps described earlier. The direction of rotation depends on the direction of fluid flow.

5.6.7 Hybrid actuator systems

Hybrid drives are also available where the rotary motion of a pneumatic or hydraulic motor drives a rack and pinion mechanism provide linear motion. Where extremely long linear travels are required, a chain wheel can be substituted for the gear and roller chain can be substituted for the rack. Alternatively the pneumatic or hydraulic motor can be used to drive a lead screw and nut. This latter solution has the advantage that over-run is impossible if the motor fails and it is therefore fail-safe.

5.6.8 Symbols for fluid power systems

The following design data is provided by courtesy of the British Fluid Power Association. Some of the more widely used symbols for fluid power systems are shown in the following

Hydraulic and pneumatic symbols.

Symbol	Description	Symbol	Description
	Source of energy – Hydraulic		Hydraulic cylinder – Double-acting
	Source of energy – Pneumatic		Pneumatic cylinder – Double-acting
	Hydraulic pump Fixed displacement One flow direction		Cylinder – Double-acting – Double-ended Piston rod Hydraulic
	Hydraulic pump Variable displacement Two flow directions		Pneumatic cylinder – Single-acting – Spring return
	Hydraulic motor Fixed displacement One flow direction		Cylinder – Double-acting – Adjustable cushions both ends Pneumatic
	Hydraulic motor Variable displacement Two flow directions of rotation		Pressure intensifier – Single fluid – Hydraulic
	Air compressor		Semi-rotary actuator – Double-acting – Hydraulic
	Pneumatic motor Fixed displacement One flow direction		Semi-rotary actuator – Single-acting – Spring return – Pneumatic
	Accumulator – Gas loaded		Telescopic cylinder – Double-acting – Hydraulic
	Air receiver	M	Electric motor (from IEC 617)

Symbol	Description	Symbol	Description
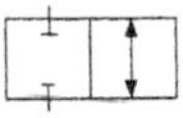	Directional control 2/2 valve	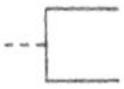	Valve control mechanism – By pressure
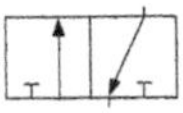	Directional control 3/2 valve	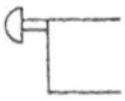	Valve control mechanism – By push button
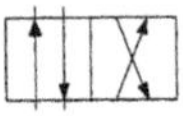	Directional control 4/2 valve	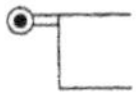	– By roller
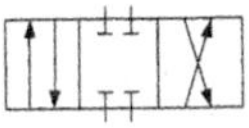	Directional control 4/3 valve – Closed centre	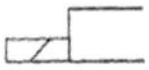	– By solenoid – Direct
	Directional control 5/2 valve	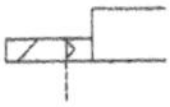	– By solenoid – With pressure pilot – Pneumatic
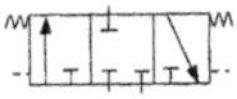	Directional control 3/3 valve – Closed centre – Spring-centred, pilot operated	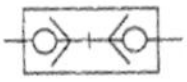	Quick-release coupling – With non-return valves – Connected
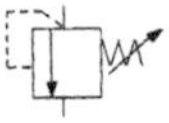	Pressure relief valve – Single stage – Adjustable pressure		Flexible line – Hose
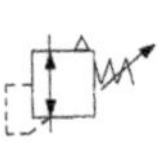	Pressure reducing valve – With relief – Pneumatic	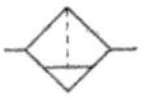	Filter
	Non-return valve	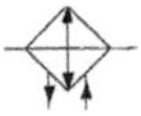	Cooler – With coolant flow line indication
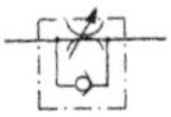	One-way restrictor or flow control valve	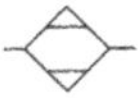	Air dryer

figure. For the full range of graphic symbols the reader is referred to ISO 1219-1 as amended in 1991. For the rules relating to circuit layout see ISO 1219-2. For port identification and operator marking see ISO 9461 (hydraulic) or CETOP RP68P (pneumatic) or ISO 5599 (pneumatic).

5.6.9 Fluid power transmission design data (general formulae)

Hydraulic

(a) Pumps and motors

Flow rate (l/min) $Q = \dfrac{D \cdot n}{1000}$

Shaft torque (Nm) $T = \dfrac{D \cdot p}{20\pi}$

Shaft power (kW) $P = \dfrac{T \cdot n}{9554}$

Hydraulic power (kW) $P = \dfrac{Q \cdot p}{600}$

For a quick calculation $\dfrac{\text{Power (kW) P} = \text{tonnes} \times \text{mm/sec}}{100}$

(b) Cylinders

Pressure (N/m^2) $p = \dfrac{F}{A}$

Flow rate (l/min) $Q = 60.A.v.10^3$

F = force (N)
A = area (m^2)
v = velocity (m/s)
p = pressure (bar)
D = displacement (cm^3/rev)
n = rev/min

(c) Flow

Flow (l/min) $Q \propto \sqrt{\Delta p}$

Δp = pressure change (bar)

That is, if you double the flow you get 4 times the pressure change

Pressure loss in pipes

Flow in l/min	Tube bore size (mm)								
	5	**7**	**10**	**13**	**16**	**21**	**25**	**30**	**36**
1	0.69	0.22							
2	1.38	0.44							
3	2.07	0.66	0.17						
5	4.14	1.24	0.24						
7.5	6.55	1.72	0.31						
10		3.10	0.38	0.14					
15		5.38	0.69	0.21	0.08				
20			1.10	0.30	0.14				
30			2.21	0.69	0.25	0.04			
40				1.17	0.45	0.08	0.04		
50					0.59	0.12	0.07	0.03	
75					1.31	0.23	0.14	0.06	0.02
100						0.41	0.22	0.13	0.03
150							0.45	0.23	0.06
200								0.41	0.10
250									0.16

This chart gives the approximate pressure drop in smooth bore straight pipes, in bar per 3 m length. Bends and fittings will increase the above pressure losses and manufacturers should be consulted for more accurate figures.

Pneumatic

(a) Flow through pipes

$$\Delta p = \frac{1.6 \times 10^8 \times (Q \times 10^{-3})^{1.85} \times L \times 10^{-3}}{d^5 \times p}$$

where:

Δp = pressure drop (bar)
Q = free air flow (m^3/s) = l/s $\times 10^{-3}$
L = pipe length (m)
d = internal diameter of pipe

(b) Velocity through pipes

$$v = \frac{1273Q}{(p+1)d^2}$$

where:
v = flow velocity (m/s)
p = initial pressure (bar)
d = inside pipe diameter (mm)

If the free air flow is known, the minimum inside diameter to keep velocity below 6 m/s, can be found from:

$$d\ (\text{mm}) = \sqrt{\frac{212 \times Q}{(p+1)}}$$

For normal installations, where the pressure is about seven bar gauge, this can be simplified to:

d (mm) should be greater than $5 \times \sqrt{Q}$

Source: Pages 14 &15 BFPA Data books.

5.6.10 Fluid power transmission design data (hydraulic cylinders)

Output force and maximum rod lengths

Example: Knowing the output force required (200 kN) and the pressure of the system (160 bar), connect Output force through pressure to cut cylinder diameter.

Answer: 125 mm.

To find the maximum length of a piston rod. Connect output force required (200 kN) through rod diameter (70 mm) to cut the maximum rod length scale; this gives you the (Lm) dimensions.

Answer: 2800 mm.

To find the actual length stroke (LA) for a specific mounting use formulae below.

Maximum stroke lengths for specific mounting cases

Foot mounted, eye rod end .... LA = Lm × 0.8
Foot mounted, rigidly supported rod .... LA = Lm
Front flange, eye rod end .... LA = Lm × 0.8
Front flange, rigidly supported rod .... LA = Lm
Rear flange, eye rod end .... LA = Lm × 0.4
Real flange, rigidly supported rod .... LA = Lm × 0.8
Rear eye, eye rod end .... LA = Lm × 0.3
Trunnion head end, eye rod end .... LA = Lm × 0.3
Trunnion gland end, eye rod end .... LA = Lm × 0.6
Trunnion gland end, rigidly supported end ..... LA = Lm × 0.8

For intermediate trunnion positions scaled multiplier factors must be taken. Clevis and spherical eye mountings have the same factor as eye mountings.

Example: Having found Lm (2800 mm) for rear flange mount with eye rod end LA = Lm × 0.4 = 2800 × 0.4 = **1120 mm**.

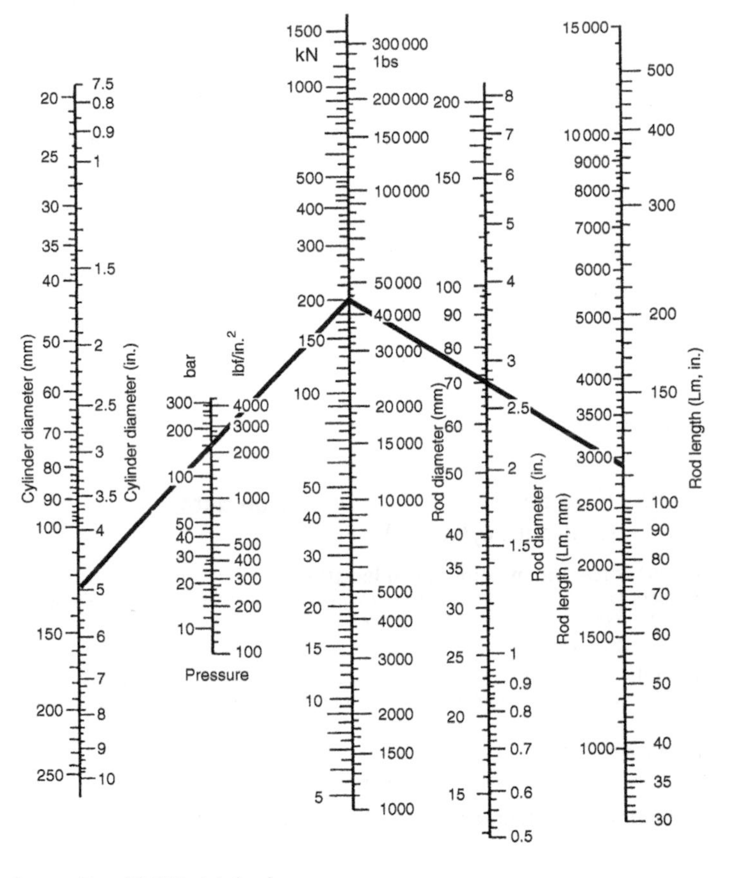

Source: Page 16 BFPA data book.

5.6.11 Fluid power transmission design data (hydraulic pipes and hoses)

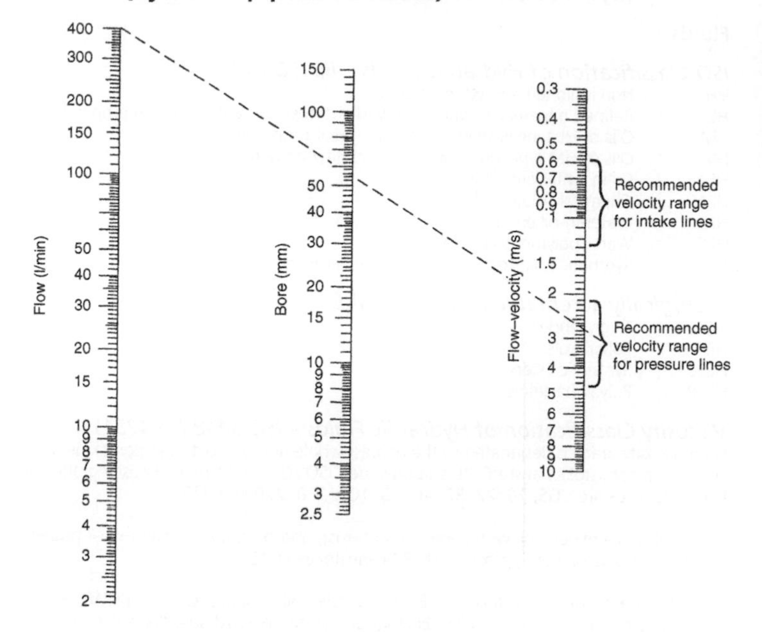

Nomogram for determining pipe sizes in relation to flow rates and recommended velocity ranges.

Based on the formula:

$$\text{Velocity of fluid in pipe (m/s)} = \frac{\text{Flow rate (l/min)} \times 21.22}{d^2}$$

where d = bore of pipe (mm)

Recommended velocity ranges based on oils having a maximum viscosity grade of 70cSt at 40°C and operating between 18°C and 70°C.

Note: For pipe runs greater than 10 m pipe size should be increased correspondingly. Intake line should never exceed 1 m long.

For further information, see:

BFPA/P7 Guidelines to the selection and application of tube couplings for use in fluid power systems.

BFPA/P47 Guidelines for the use of fluid power hose assemblies.

Source: Page 17 BFPA design data book.

5.6.12 Fluid power transmission design data (hydraulic fluids, seals and contamination control)

Fluids

ISO Classification of Hydraulic Fluids – BS ISO 6743-4

HH	Non inhibited refined mineral oils
HL	Refined mineral oils with improved anti-rust and anti-oxidation properties
HM	Oils of HL type with improved anti-wear properties
HV	Oils of HM type with improved viscosity/temperature properties
HFAE	Oil in water emulsions
HFAS	Chemical solutions in water
HFB	Water-in-oil emulsions
HFC	Water polymer solutions
HFDR	Synthetic fluids of the phosphate ester type

Ecologically acceptable hydraulic fluids

HETG	Tryglycerides
HEPG	Polyglycols
HEES	Synthetic Esters
HEPR	Polyalphaclefins

Viscosity Classification of Hydraulic Fluids – ISO 3448 (BS 4231)

Each viscosity grade is designated by the nearest whole number to its mid-point kinematic viscosity in centistokes at 40°C. It is abbreviated ISOVG ... Common viscosity grades of hydraulic fluids are VG5, 10, 22, 32, 46, 68, 100, 150, 220 and 320.

Thus HM32 is a mineral oil with improved anti-rust, anti-oxidation and anti-wear properties having a viscosity of approximately 32 centistokes at 40°C.

For further details of specific fluids see BFPA/P12, Mineral oil data sheets and BFPA/P13, Fire resistant fluids data sheets, BFPA/P67 Ecologically acceptable hydraulic fluids data sheets.

Seals

Seal material		Recommended for
Acrylonitrile butadiene	**(NBR)**	air, oil, water, water/glycol
Polyacrylate rubber	**(ACM)**	air, oil
Polyurethane	**(AU, EU)**	air, oil
Fluorocarbon rubber	**(FKM)**	air, oil water, water/glycol, chlorinated hydrocarbons, triaryl phosphates
Silicone	**(VOM)**	air, oil, chlorinated hydrocarbons
Styrene Butadiene	**(SBR)**	air, water, water/glycol
Ethylene propylene diene	**(EPDM)**	air, water, water/glycol, phosphate ester
Polytetrafluorethylene	**(PTFE)**	air, oil, water, water/glycol, phosphate ester

For full details of seal compatibilities, see ISO 6072: Hydraulic fluid power, compatibility between elastomeric materials and fluids or BS 7714: Guide for care and handling of seals for fluid power applications. For recommendation of O-ring seal standards, see BFPA/P22 'Industrial O-ring Standards – Metric vs Inch.'

Source: Pages 18–20 inclusive, BFPA data book.

Fluid cleanliness

The presence of particulate contamination ('dirt') is the single most important factor governing the life and reliability of fluid power systems. Operating with clean fluids is essential. Advice on contamination control can be gained from guide line document BFPA/P5.

Contamination control

Specification of Degree of Filtration – ISO 4572 (BS 6275/1)

The multipass test, ISO 4572 (BS 6275/1), was introduced to overcome the difficulties in comparing the performance of filters. The element is subjected to a constant circulation of oil during which time fresh contaminant (ISO Test Dust) is injected into the test rig. The contaminant that is not removed by the element under test is recirculated thereby simulating service conditions.

The filtration ratio β of the filter is obtained by the analysis of fluid samples extracted from upstream and downstream of the test filter, thus

$$\beta x = \frac{\text{number of particles larger than } x \text{ upstream of the filter}}{\text{number of particles larger than } x \text{ downstream of the filter}}$$

The rating of a filter element is stated as the micrometer size where βx is a high value (e.g. 100 or 200)

Fluid cleanliness standards

The preferred method of quoting the number of solid contaminant particles in a sample is the use of BS ISO 4406.

The code is constructed from the combination of two range numbers selected from the following table. The first range number represents the number of particles in a millilitre sample of the fluid that are larger than 5 μm, and the second number represents the number of particles that are larger than 15 μm.

Number of particles per millilitre		Scale number
More than	up to and including	
10 000	20 000	21
5 000	10 000	20
2 500	5000	19
1 300	2500	18
640	1300	17
320	640	16
160	320	15
80	160	14
40	80	13
20	40	12
10	20	11
5	10	10
2.5	5	9
1.3	2.5	8
0.64	1.3	7

For example code 18/13 indicates that there are between 1300 and 2500 particles larger than 5 μm and between 40 and 80 particles larger than 15 μm.

For further details and comparisons of ISO 4406 with other cleanliness classes, see BFPA/P5 – Guidelines to contamination control in fluid power systems.

Flushing

Formula for flow required to adequately flush an hydraulic system:

$Q > 0.189\ vd$ (l/min)

where:

Q = flow (l/min)
v = kinematic viscosity (cSt)
d = pipe bore (mm)

For further information on flushing see BFPA/P9 – Guidelines to the flushing of hydraulic systems.

Cleanliness of components

Three methods exist for measuring the cleanliness of components: test rig, flush test, strip and wash. The level of cleanliness required must be agreed between the supplier and customer but the methods are fully described in BFPA/P48 – Guidelines to the cleanliness of hydraulic fluid power components.

5.6.13 Fluid power transmission design data (hydraulic accumulators)

Storage applications

Storage applications	Formula to estimate accumulator volume for storage applications
Slow charge Slow charge	$V_1 = \dfrac{V_a \times \dfrac{p_2}{p_1}}{1 - \dfrac{p_2}{p_3}}$
Fast charge Fast discharge	$V_1 = \dfrac{V_a \times \left(\dfrac{p_2}{p_1}\right)^{\frac{1}{1.4}}}{1 - \left(\dfrac{p_2}{p_3}\right)^{\frac{1}{1.4}}}$
Slow charge Fast discharge	$V_1 = \dfrac{V_a \times \dfrac{p_3}{p_1}}{\left(\dfrac{p_3}{p_2}\right)^{\frac{1}{1.4}} - 1}$

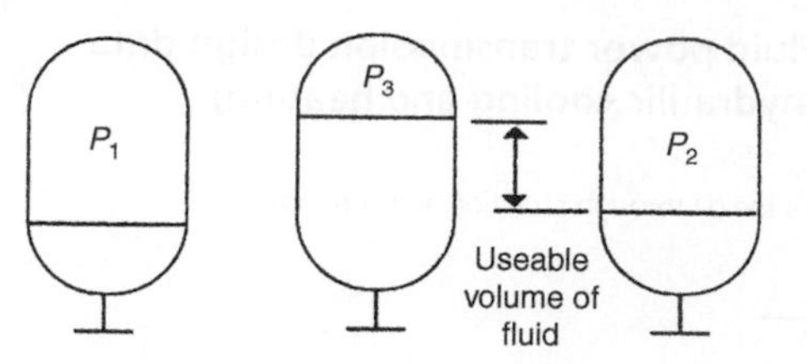

The precharge pressure is chosen to 90% of the min. working pressure. n varies between 1 and 1.4 depending on whether the charge is slow (isothermal) or fast (adiabatic).

Pump pulsation

Pump pulsation and Formula to size accumulator to reduce pump pulsations.

(a) Minimum effective volume (l) $V_1 = \frac{k \cdot Q}{n}$

Note: It is good engineering practice to select an accumulator with port connection equal to the pump port connection.

(b) To check the level of pulsation obtained.
Volume of fluid entering accumulator $= D \cdot C$

For pulsation damping precharge pressure $P_1 = 0.7 \cdot P_2$
and assuming change from P_1 to P_2 is isothermal, then $V_2 = 0.7 \cdot V_1$

$$V_3 = V_2 - (D \cdot C)$$

$$P_3 = P_3 \left(\frac{V_2}{V_3} \right)^{1.4}$$

Hence: Percentage pulsation above and below mean is $\left(\frac{P_2 - P_3}{P_2} \right) 100$

V_1 = effective gas volume
V_2 = min. gas volume
V_3 = max. gas volume
p_1 = pre charge pressure
p_2 = min. working pressure
p_3 = max. working pressure

$V_a = V_3 - V_2$ = working volume (fluid)
k = a constant*
Q = pump flow (l/min)
n = pump speed (rpm) if $n > 100$ use 100
D = pump displacement (l/rev)
C = a constant*

* Dependent on no. of pistons. For multi-piston pumps >3 pistons. $k = 0.45$ and $C = 0.013$.

Source: Page 21 BFPA data book.

5.6.14 Fluid power transmission design data (hydraulic cooling and heating)

Cooling

The tank cools the oil through radiation and convection.

$$P = \frac{\Delta T_1 \cdot A \cdot k}{1000}$$

where:

$k = 12$	at normally ventilated space
$24\,\mathrm{W/m^2\,°C}$	at forced ventilation
6	at poor air circulation

Required volume of water flow through the cooler:

$$Q = 860 \times \frac{\text{power loss}}{\Delta T \text{ water}} \qquad 1/\mathrm{h}$$

Heating

Heating is most necessary if the environmental temperature is essentially below 0°C.

Requisite heating effect:

$$P = \frac{V \cdot \Delta T_2}{35 \cdot \Delta t} \qquad \text{in kW}$$

Energy

$$J = M \cdot C \cdot \Delta T$$

where:
M = Mass (kg)
C = Specific heat capacity J/kg°C
ΔT_1 = temperature difference (°C) fluid/air
ΔT_2 = increase in fluid temperature (°C)
Δt = time (min)
Note: 1 MJ = 0.2777 kW/h.

Heat equivalent of hydraulic power

$$\mathrm{kJ/s} = \frac{\text{flow (l/min)} \times \text{pressure (bar)}}{600} = \mathrm{kW}$$

Change of volume at variation of temperature

Change or volume $\Delta V = 6.3 \times 10^{-4} \cdot V \cdot \Delta T$

Change of pressure at variation of temperature

Note: With an infinite stiff cylinder.
Change of pressure $\Delta p = 11.8 \cdot \Delta T$ (in general, affected by many variables)

Example: The temperature variation of the cylinder oil from nighttime (10°C) to daytime/solar radiation (50°C) gives:

$\Delta P = 11.8 \times 40 =$ **472 bar**

Key

ΔT = temperature change (°C)
P = power (kW)
k = heating coefficient (W/m^2 °C)
A = area of tank excluding base (m^2)
Δt = time change (min)
Δp = change in pressure (bar)
C = specific heat capacity (J/kg°C)
V = volume (l)

Source: Page 22 BFPA data book.

5.6.15 Fluid power transmission design data (pneumatic valve flow)

Valve flow performance is usually indicated by a flow factor of some kind, such as '*C*', '*b*', '*Cv*', '*Kv*', and others.

The most accurate way of determining the performance of a pneumatic valve is through its values of '*C*' (conductance) and '*b*' (critical pressure ratio).

These figures are determined by testing the valve to the CETOP RP50P recommendations.

The tests will result typically in a set of curves as shown below.

From these the critical pressure ratio '*b*' can be found. '*b*' represents the ratio of P_2 to P_1 at which the flow velocity goes sonic (the limiting speed of air). Also the conductance '*C*' which represents the flow 'dm^3/s/bar absolute' at this point.

If a set of curves are not available the value of flow for other pressure drops can be calculated from:

$$Q = CP_1\sqrt{1-\left[\frac{\frac{P_2}{P_1}-b}{1-b}\right]^2}$$

where:
P_1 = upstream pressure bar
C = conductance dm^3/s/bar a
Q = flow dm^3/s
P_2 = downstream pressure bar
b = critical pressure ratio

ISO 6538/CETOP RP50P method

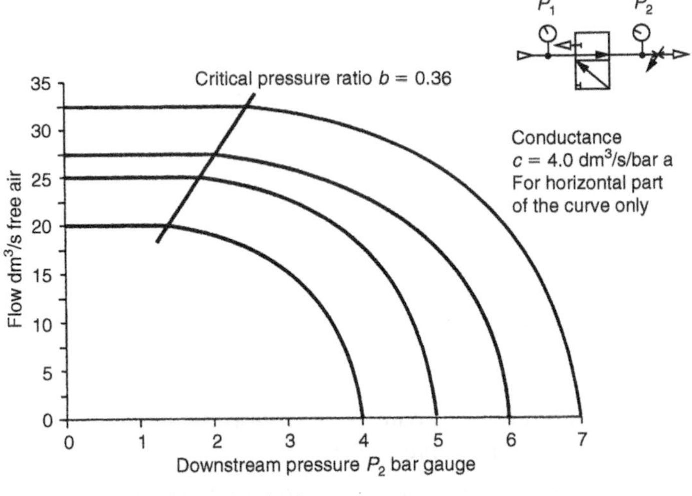

Source: Page 23 (part) BFPA data book.

5.6.16 Fluid power transmission design data (pneumatic cylinders)

Some factors to consider when selecting and using pneumatic linear actuators, air cylinders

Mode of action

Three basic types

- Single-acting, spring return: Movement and force by air pressure in one direction, return movement by internal spring force, usually sprung to outstroke position but occasionally the reverse is available.
- Double-acting: Air pressure required to produce force and movement in both directions of travel.
- Double-acting, through rod: Double-ended piston rod which acts as a normal double-acting cylinder, but mechanical connections can be made to both ends of through rod.

Source: Page 23 (part) BFPA data book.

Rodless cylinders

Where space is at a premium and there are potential loading and alignment problems, a variety of rodless cylinder designs are available. The range of available bore sizes is limited (e.g. 16–100 mm).

Quality classes

Basically three qualities of unit available

- Light duty and compact cylinders: Limited range of bore sizes, up to 100 mm. Not cushioned at stroke ends, or cushion pads only. Check manufacturers data sheets for serviceability and susceptibility to corrosion.
- Medium duty/standard: For normal factory environments. Some degree of corrosion resistance. Serviceable. Cushioned at both ends. Usually double-acting. Bore size range, 32 mm and greater.
- Heavy duty: Rugged construction. Serviceable. Non-corrodible materials. superior cushioning, thicker piston rods and heavy duty mountings. Bore size range 32 mm and greater.

For interchangeability and standard mounting dimensions, see ISO 6432, 8–25 mm bore, ISO 6431, 32–320 bore: standard duty, double-acting, metric dimensions.

Standard bore sizes

- Double-acting: 8, 10, 12, 16, 20, 25, 32, 40 50, 63, 80, 100, 125, 160, 200, 250 and 320 mm.

Standard, stocked strokes

- Double-acting: 25, 50, 80, 100, 125, 160, 200, 250 and 320 mm.

Cylinder thrust

To calculate the theoretical thrust of a double-acting cylinder, use the formula:

- Thrust = $\left(\dfrac{\pi D^2 \times P}{4}\right)$ Newton

  where D = diameter of piston (mm), P = Gauge pressure (bar).

  Pull will be less, due to area of piston rod.

- Pull = $\left[\dfrac{\pi(D^2 - d^2) \times P}{4}\right]$ Newton

  where d = diameter of piston rod (mm).

For static loading, that is, where full thrust is only required when the cylinder comes to rest (e.g. clamping), use the above calculation.

For dynamic loading, that is, where thrust is required throughout the piston travel, allowance has to be made for the exhaust back-pressure, friction, changes in driving pressure, etc., add 30% to the thrust figure required, for normal speeds. For higher speeds add as much as 100%.

Cylinder speeds

With normal loading, valving and pressure: 5–7 bar, the important factor is the relationship between the bore area of the cylinder and the actual bore area of the cylinder inlet ports. Conventionally this is in the order of 100:1 and would result in speeds of 0.3–0.5 m/s. For normal speeds, use a directional control valve and piping of the same size as the cylinder ports. For higher speeds use a cylinder of larger bore size than necessary plus larger valve and pipework but be careful of cushioning problems.

Stroke lengths

For static loading use any convenient standard, stocked stroke length as cushioning is not important. For dynamic loading, order the correct required stroke length as the use of external stops affects cushioning potential.

With long stroke lengths, that is, more than 15 × bore diameter, care must be taken to avoid side-loading on bearing, etc., use pivot type mountings. Check diameter of piston rod to avoid buckling under load. If necessary use a larger bore size cylinder than normal as this will probably have a longer bearing and a thicker piston rod.

5.6.17 Fluid power transmission design data (seals, filtration and lubrication)

Seals

Some miniature pneumatic components and heavy duty valves employ metal to metal sealing. Most equipment uses flexible seals manufactured from synthetic rubber materials. These are suitable for ambient temperatures up to 80°C. Fluorocarbon rubber os silicon rubber (Viton) seals are used for temperatures up to 150°C.

Synthetic seals are resistant to mineral-based hydraulic oils but specific types of oil must be checked with the equipment manufacturer to avoid problems arising from additives (see Section 5.6.12).

Good wear resistance ensures a reasonable performance, even with a relatively wet and dirty air supply. However, to ensure safe operation, with a satisfactory service life, system filtration and some form of lubrication is necessary.

Filtration

Good filtration starts at the compressor with correct siting of the air intake and an intake filter. Error sat this stage cause problems throughout the subsequent installation. After coolers and dryers ensure that the supply enters the ring main in good condition, but condensation and dirt can be picked up on the way to the point of use. Individual filters are necessary at each major application point.

For general industrial purposes, filters with a 40 μm element are satisfactory. For instrument pneumatics, air gauging, spraying, etc., a filter of 5 μm or better is required. High quality filters are often called coalescers. Filters are available with manual, automatic or semi-automatic drain assemblies.

To alleviate the problem of dirt entering open exhaust ports use an exhaust port silencer which also avoids noise problems. Simple exhaust port filters are also available, which offer a reasonable level of silencing, with little flow resistance.

Lubrication/modern equipment is often designed to run without lubrication

However, most industrial air supplies contain a little moisture. All pneumatic components are greased on assembly, unless specifically requested otherwise. This provides significant lubrication and ensures that the equipment performs satisfactorily for several million cycles, particularly if used frequently.

An airline lubricator is included in many industrial applications.

Conditioning units

It is estimated that 90% of failures in pneumatic systems are due to poor quality of the air supply. Contamination is drawn in at the compressor from the atmosphere. The level of contamination is effectively concentrated by the compression. Any contamination will attack the system components.

Modern pneumatic systems will include some form of air conditioning unit comprising a dryer to remove moisture, a filter to remove contamination and perhaps a form of lubrication.

Every application requires careful consideration on the type of conditioning to specify to meet the required operating condition.

Source: Page 26 BFPA data book.

5.6.18 Fluid power transmission design data (air compressors)

As most industrial factory and machine-shop type pneumatic equipment operates at about 6 bar, it is usual to select a compressor installation delivering 7 bar in to the mains, to allow for pipe losses.

Types of compressor

- Displacement compressors: Air is compressed by contracting the space containing air taken in at atmospheric pressure (e.g. reciprocating compressors – piston or diaphragm type; rotary compressors – sliding vane, gear, screw). Roots blower.
- Dynamic compressors: Compression is achieved by converting the air inlet rate into a pressure (e.g. centrifugal compressors – radial impeller, blade type, axial compressors).

Overlap occurs between the various types in terms of capacity and pressure range but some generalisation can be made. Use reciprocating compressors if very high pressures, up to 1000 bar are required. Rotary vane types are used for medium pressures, up to 7 bar and low capacity. Blowers are used for large volumes of low pressure air, up to 1 bar.

Sizes

For industrial applications compressors can be classified:

Small – up to 40 l/s
Medium – 40 to 300 l/s
Large – above 300 l/s

Installation

Three types of installation dependant on application:

- Portable: Paint spraying, tyre inflation, etc.
- Mobile: Road/rock drills, rammers, emergency stand-by sets, etc.
- Fixed: Machines, factory, workshop, etc.

Prime movers

Selection of correct drive unit is essential to obtain efficient and economical supply. Three basic types – **Electric motors** are used for compactness and ease of control; **IC engine** (diesel, petrol, gas) for mobile units, emergency stand-by sets or where an electrical supply is not available; **Turbine** (gas, steam) can be incorporated into the total energy system of a plant using existing steam or gas supplies.

Selection factors

- Delivery pressure: Must be high enough for all existing and potential future requirements. If there is a special requirement for a large volume of either high or low air pressure, it may be better to install a separate unit for that purpose.
- Capacity: Calculate not only the average air consumption but also maximum instantaneous demands, for example, large bore cylinders and air motors, operating at high speeds. Determine use factors. Frequently users add to existing airlines indiscriminately and run out of air.
- Intake siting: Intake air should be as clean and as cold as possible for maximum efficiency.

- Intake filter: High capacity to remove abrasive materials which could lead to rapid wear.
- Air quality: Study air quality requirements throughout the system or plant. The correct combinations of separators, aftercoolers, outlet filters and dryers should be determined. The problem of water removal should not be left to the airline filters associated with individual plant and systems.
- Stand-by capacity: What would happen in an emergency or when an individual compressor requires servicing.
- Air receiver: The system must have adequate storage requirements, not only to meet demand, but also to ensure efficient running of the prime mover.
- Air main capacity: A large bore ring main acts as a useful receiver, reduces pressure drops and operating costs. The cost of larger size of pipework is only a small proportion of the installation costs.

Source: Page 26 BFPA data book.

5.6.19 Fluid power transmission design data (tables and conversion factors in pneumatics)

The tables on this and the subsequent pages can be used to answer frequently asked questions in pneumatics concerning air quality, cylinder forces loading and bending, air consumption plus valve flow and lubrication.

Air quality

ISO 8573-1 specifies quality classes for compressed air. A class number is made up from the individual maximum allowable contents of solid particles, water and oil in air, and can be used to specify air quality for use with valves and other pneumatic applications.

Class	Solids		Water	Oil
	Particle size max. (μm)	Concentration max. (mg/m³)	Max. pressure dew point (°C)	Concentration (mg/m³)
1	0.1	0.1	−70	0.01
2	1	1	−40	0.1
3	5	5	−20	1
4	15	8	+3	5
5	40	10	+7	25
6	–	–	+10	–

For general applications where ambient temperature is between +5°C and +35°C, air quality to ISO8573-1 class 5.6.4 is normally sufficient. This is 40 μm filtration, +10°C maximum pressure dew point and 5 mg/m³ maximum oil content. Pressure dew point is the temperature to which compressed air must be cooled before water vapour in the air starts to condense into water particles.

Air consumption

For cylinders with the bore and rod sizes shown, the values of consumption are for an inlet pressure of 6 bar and only 1 mm of stroke. To find the consumption for a single stroke or one complete cycle take the value from the appropriate column and multiply by the cylinder stroke length in mm. To adjust the value for a different inlet pressure divide by 7 and multiply by the required (absolute) pressure (e.g. gauge pressure in bar +1).

Cylinder forces

The theoretical thrust and pull is related to the effective piston area and the pressure. The tables show the theoretical forces in Newton for single and double-acting cylinders at

Table of air consumption.

Bore (mm)	Rod (mm)	Push stroke consumption (dm^3/mm of stroke at 6 bar)	Pull stroke consumption (dm^3/mm of stroke at 6 bar)	Combined consumption (dm^3/mm of stroke/cycle)
10	4	0.00054	0.00046	0.00100
12	6	0.00079	0.00065	0.00144
16	6	0.00141	0.00121	0.00262
20	8	0.00220	0.00185	0.00405
25	10	0.00344	0.00289	0.00633
32	12	0.00563	0.00484	0.01047
40	16	0.00880	0.00739	0.01619
50	20	0.01374	0.01155	0.02529
63	20	0.02182	0.01962	0.04144
80	25	0.03519	0.03175	0.06694
100	25	0.05498	0.05154	0.10652
125	32	0.08590	0.08027	0.16617
160	40	0.14074	0.13195	0.27269
200	40	0.21991	0.21112	0.43103
250	50	0.34361	0.32987	0.67348

Table of thrust and pulls, single-acting cylinders.

Cylinder bore (mm)	Thrust (N at 6 bar)	Min pull of spring (N)
10	37	3
12	59	4
16	105	7
20	165	14
25	258	23
32	438	27
40	699	39
50	1102	48
63	1760	67
80	2892	86
100	4583	99

Table of thrust and pulls, double-acting cylinders.

Cylinder bore mm (in.)	Piston rod diameter mm (in.)	Thrust (N at 6 bar)	Pull (N at 6 bar)
8	3	30	25
10	4	47	39
12	6	67	50
16	6	120	103
20	8	188	158
25	10	294	246
32	12	482	414
40	16	753	633
44.45 (1.75)	16	931	810
50	20	1178	989
63	20	1870	1681
76.2 (3)	25	2736	2441
80	25	3015	2721
100	25	4712	4418
125	32	7363	6881
152.4 (6)	(1½)	10 944	10 260
160	40	12 063	11 309
200	40	18 849	18 095
250	50	29 452	28 274
304.8 (12)	(2¼)	43 779	42 240
320	63	48 254	46 384
355.6 (14)	(2¼)	59 588	58 049

6 bar inlet to the cylinder. For forces at other pressures divide the figures by 6 and multiply by the required pressure in bar gauge.

Load and buckling

For applications with high side loading, use pneumatic slide actuators or standard cylinders fitted with guide units. Alternatively external guide bearings should be installed.

When a long stroke length is specified, care must be taken to ensure the rod length is within the limits for prevention of buckling. The table shows the maximum stroke length for a variety of installation arrangements.

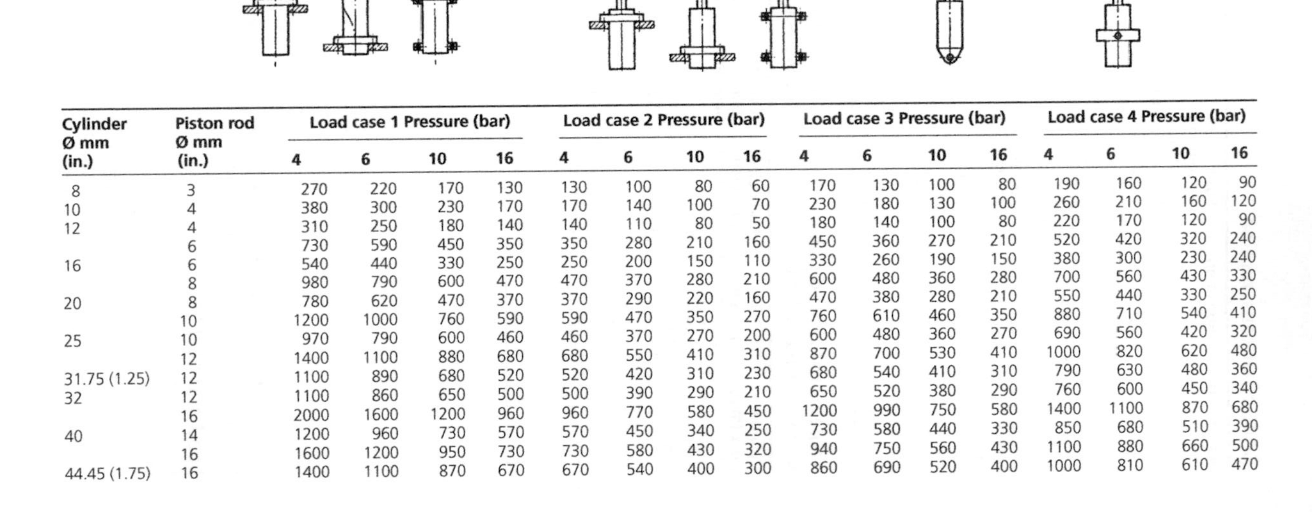

Cylinder Ø mm (in.)	Piston rod Ø mm (in.)	Load case 1 Pressure (bar)				Load case 2 Pressure (bar)				Load case 3 Pressure (bar)				Load case 4 Pressure (bar)			
		4	**6**	**10**	**16**	**4**	**6**	**10**	**16**	**4**	**6**	**10**	**16**	**4**	**6**	**10**	**16**
8	3	270	220	170	130	130	100	80	60	170	130	100	80	190	160	120	90
10	4	380	300	230	170	170	140	100	70	230	180	130	100	260	210	160	120
12	4	310	250	180	140	140	110	80	50	180	140	100	80	220	170	120	90
	6	730	590	450	350	350	280	210	160	450	360	270	210	520	420	320	240
16	6	540	440	330	250	250	200	150	110	330	260	190	150	380	300	230	240
	8	980	790	600	470	470	370	280	210	600	480	360	280	700	560	430	330
20	8	780	620	470	370	370	290	220	160	470	380	280	210	550	440	330	250
	10	1200	1000	760	590	590	470	350	270	760	610	460	350	880	710	540	410
25	10	970	790	600	460	460	370	270	200	600	480	360	270	690	560	420	320
	12	1400	1100	880	680	680	550	410	310	870	700	530	410	1000	820	620	480
31.75 (1.25)	12	1100	890	680	520	520	420	310	230	680	540	410	310	790	630	480	360
32	12	1100	860	650	500	500	390	290	210	650	520	380	290	760	600	450	340
	16	2000	1600	1200	960	960	770	580	450	1200	990	750	580	1400	1100	870	680
40	14	1200	960	730	570	570	450	340	250	730	580	440	330	850	680	510	390
	16	1600	1200	950	730	730	580	430	320	940	750	560	430	1100	880	660	500
44.45 (1.75)	16	1400	1100	870	670	670	540	400	300	860	690	520	400	1000	810	610	470

50	20	2000	1600	1200	930	930	740	550	420	1200	960	720	550	1400	1100	840	640
50.8(2)	20	1900	1600	1200	930	930	740	550	420	1200	960	720	550	1400	1100	840	640
63	20	1500	1200	930	720	720	570	420	310	930	740	550	420	1100	860	650	490
63.5 (2.5)	25	2400	2000	1500	1200	1200	930	700	530	1500	1200	900	690	1700	1400	1100	810
76.2 (3)	25	2000	1600	1200	950	950	760	560	420	1200	980	740	560	1400	1100	860	660
80	25	1900	1500	1100	880	880	700	510	380	1100	910	680	510	1300	1100	800	600
100	25	1500	1200	880	670	670	520	380	270	880	690	510	370	1000	820	600	450
101.6 (4)	32	2400	2000	1500	1100	1100	910	670	500	1500	1200	890	670	1700	1400	1000	790
125	32	2000	1600	1200	910	910	710	520	380	1200	940	690	520	1400	1100	820	620
127 (5)	38.1 (1.5)	2800	2200	1700	1300	1300	1000	760	570	1700	1300	1000	760	2000	1600	1200	900
152.4 (6)	38.1 (1.5)	2300	1800	1400	1100	1100	830	610	440	1400	1100	810	600	1600	1300	950	720
160	40	2400	1900	1500	1100	1100	880	640	480	1400	1200	860	640	1700	1400	1000	760
200	40	1900	1500	1100	860	860	670	480	350	1100	890	650	480	1300	1000	770	580
203.2 (8)	44.45 (1.75)	2300	1900	1400	1100	1100	840	610	440	1400	1100	810	600	1600	1300	960	720
250	50	2400	1900	1400	1100	1100	850	620	440	1400	1100	830	610	1700	1300	980	730
254 (10)	57.15 (2.25)	3100	2500	1900	1400	1400	1100	840	620	1900	1500	1100	830	2200	1700	1300	990
304.8 (12)	57.15 (2.25)	2500	2000	1500	1200	1200	920	660	480	1500	1200	890	660	1800	1400	1100	790
320	63	3000	2400	1800	1400	1400	1100	780	570	1800	1400	1000	780	2100	1700	1200	930
355.6 (14)	57.15 (2.25)	2100	1700	1300	970	970	760	540	380	1300	1000	730	540	1500	1200	870	650

Valve flow

There are a variety of standards and methods for the measurement and display of valve flow performance. These can give rise to confusion and difficulty when comparing the published performance of different valves. The table below provides conversion factors as a guide to expressing valve performance in different units.

Flow factor conversion table

	Factors			Flow*		Orifice Size	
	Cv	Kv	C	m^3/h	l/min	A	S
†Cv	1	0.869	4.08	59.1	985	16.3	21.5
Kv	1.15	1	4.69	67.9	1132	18.7	24.7
C	0.245	0.213	1	14.5	241	4.11	5.27
M^3/h	0.017	0.015	0.069	1	16.67	0.276	0.364
l/min	0.001	0.0088	0.0041	0.06	1	0.016	0.022
A	0.061	0.053	0.243	3.62	60.4	1	1.31
S	0.046	0.040	0.189	2.75	45.8	0.761	1

* Flow parameters are 6 bar inlet and 5 bar outlet at 20°C, 1013 mbar and 65% humidity.
† 'Cv' is specified by ANSI/NFPA.
Source: IMI Norgen Ltd.

How to use

Select the unit of measurement that is known in the left hand column and multiply by the factor given in the column of the required unit of measurement:

'Cv' is specified by ANSI/NFPA.
'Kv' used in Germany and based on water flow.
'C' sonic conductance in $dm^3/s/bar$ specified by ISO 6358.
'A' effective area in mm^2 specified by ISO 6358.
'S' effective area in mm^2 according to the Japanese standard JIS B 8375.

A further measurement is the NW value. This gives the equivalent diameter in mm^2 of the smallest path through a valve. This is non-comparable and not in the table.

Lubricants

When to lubricate, via an oil-fog or micro-fog lubricator, is generally explained in this catalogue. However, the oil recommended is very much dependant on the local conditions and not least availability of various brands and labels.

In each country Norgren can recommend equivalent products, based on the information from the suppliers.

5.6.20 Guideline documents

BFPA/P3	1995	Guidelines for the Safe Application of Hydraulic and Pneumatic Fluid Power Equipment	BFPA/P7	2004	Guidelines to the Design, Installation and Commissioning of Piped Systems Part 1 – Hydraulics
BFPA/P4	1986	Guidelines for the Design of Quieter Hydraulic Fluid Power Systems (Third Edition)	BFPA/P9	1992	Guidelines for the Flushing of Hydraulic Systems
			BFPA/P12	1995	Hydraulic Fluids Mineral Oil Data Sheets
BFPA/P5	1999	Guidelines to Contamination Control in Hydraulic Fluid Power Systems	BFPA/P13	1996	Fire-Resistant Hydraulic Fluids Data Sheets

continued

Section 5.6.20 (*continued*)

Ref.	Year	Title
BFPA/P22	2003	Guidelines on Selection of Industrial O-Rings (Metric & Inch)
BFPA/P27	1993	Guidelines on Understanding the Electrical Characteristics of Solenoids for Fluid Power Control Valves & their Application in Potentially Explosive Atmospheres
BFP/P28	1994	Guidelines for Errors and Accuracy of Measurements in the Testing of Hydraulic & Pneumatic Fluid Power Components
BFPA/P29	1987	General conditions for the Preparation of Terms and Conditions of Sale of UK Fluid Power Equipment Manufacturers and Suppliers
BFPA/P41	1995	Guidelines to Hydraulic Fluid Power Control Components
BFPA/P44	1995	Index of BS/ISO Standards Relating to Fluid Power
BFPA/P47	2004	Guidelines to the Use of Hydraulic Fluid Power Hose and Hose Assemblies
BFPA/P48	1998	Guidelines to the Cleanliness of Hydraulic Fluid Power Components
BFPA/P49	1995	Guidelines to Electro-hydraulic Control Systems
BFPA/P52	1997	Guidelines to the Plugging of Hydraulic Manifolds and Components
BFPA/P53	2002	Fluid Power at the Forefront
BFPA/P54	2003	Guidelines to the Pressure System Safety Regulations 2000 and their Application to Gas-loaded Accumulators
BFPA/P55	1993	Guidelines for the Comparison of Particle Counters and Counting Systems for the Assessment of Solid Particles in Liquid
BFPA/P56	2004	BFPA Fluid Power, Engineer's Data Booklet
BFPA/P57	1993	Guidelines to the Use of Ecologically Acceptable Hydraulic Fluids in Hydraulic Fluid Power Systems
BFPA/P58	2003	The Making of Fluid Power Standards
BFPA/P59	1993	Proceedings of the 1993 BFPA Leak-Free Hydraulics Seminar
BFPA/P60	1994	Leak-Free for Hydraulic Connections
BFPA/P61	1998	A Guide to the Use of CE Mark
BFPA/P65	1995	VDMA 24 568 & 24 569 Rapidly Biologically Degradable Hydraulic Fluids Minimum Technical Requirements & Conversion from Fluids based on Mineral Oils
BFPA/P66	1995	BFPA Survey on Ecologically Acceptable Hydraulic Fluids
BFPA/P67	1996	Ecologically Acceptable Hydraulic Fluids Data Sheets
BFPA/P68	1995	Machinery Directive Manufacturers Declarations
BFPA/P83	2003	The World of Fluid Power 2003 CD Rom (Edition 3)
BFPA/P95	2003	Principles of Hydraulic System Design
BFPA/P100	2003	Guidelines for the Proof & Burst Pressure Testing of Fluid Power Components
BFPDA/D2	1994	Technical Guidelines for Distributors of Hydraulic Fluid Power Equipment

Available from British fluid power association
Source: Page 27 BFPA hand book.

For further information relating to fluid power transmission equipment the reader is referred to the companies and associations listed in Appendix 3.

6 Engineering Materials

6.1 Mechanical properties

6.1.1 Tensile strength

This is the ability of a material to withstand tensile (stretching) loads without rupture occurring. The material is in tension.

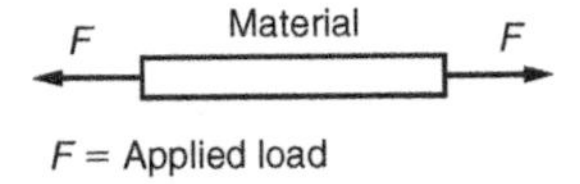

F = Applied load

6.1.2 Compressive strength

This is the ability of a material to withstand compressive (squeezing) loads without being crushed or broken. The material is in compression.

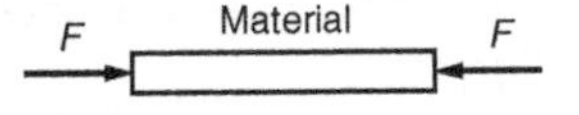

6.1.3 Shear strength

This is the ability of a material to withstand offset or transverse loads without rupture occurring. The rivet connecting the two bars shown is in *shear* whilst the bars themselves are in *tension.* Note that the rivet would still be in *shear* if the bars were in *compression*.

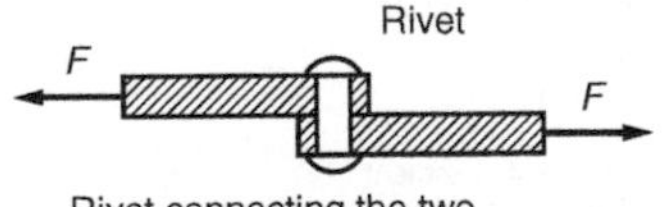

Rivet connecting the two bars is in resisting shear

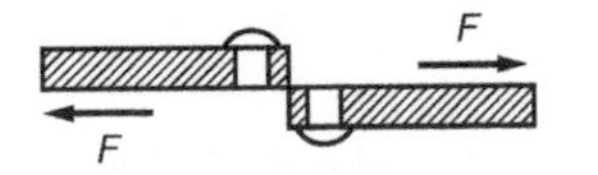

Rivet connecting the two bars has failed in shear

6.1.4 Toughness: impact resistance

This is the ability of a material to resist shatter. If a material shatters it is brittle (e.g. glass). If it fails to shatter when subjected to an impact load it is tough (e.g. rubber). Toughness should not be confused with strength. Any material in which the spread of surface cracks does not occur or only occurs to a limited extent is said to be tough.

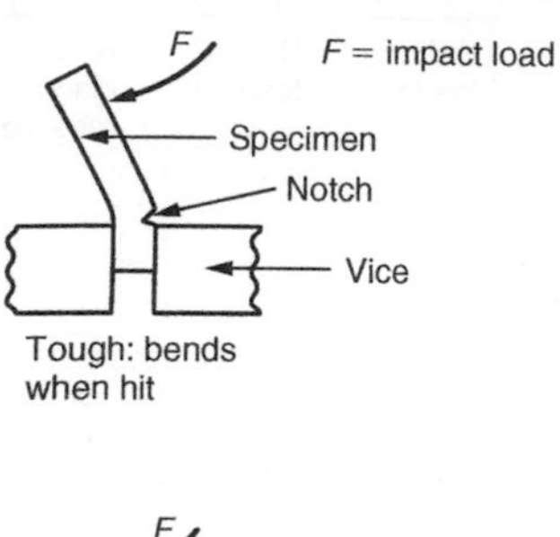

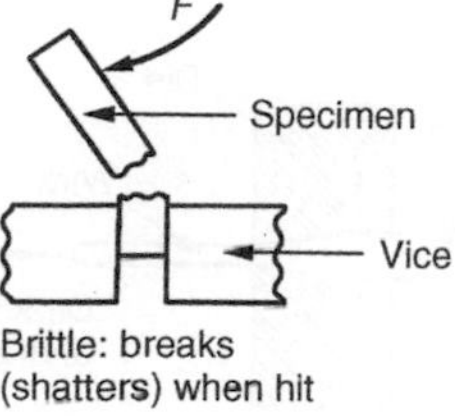

6.1.5 Elasticity

This is the ability of a material to deform under load and return to its original size and shape when the load is removed. Such a material would be required to make the spring as shown.

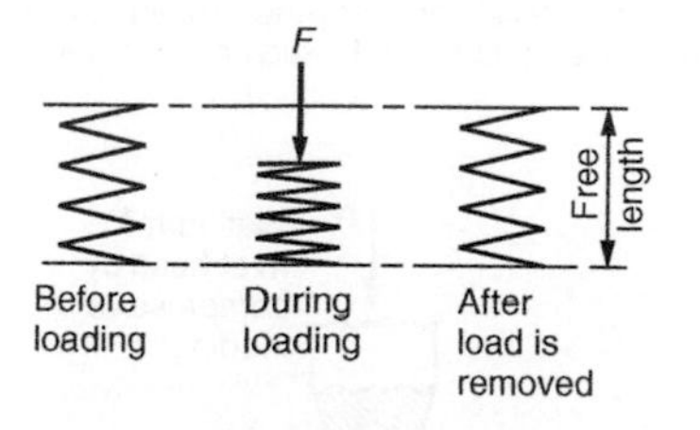

6.1.6 Plasticity

This property is the exact opposite of elasticity. It is the state of a material which has been loaded beyond its elastic state. Under a load beyond that required to cause elastic deformation (the elastic limit) a material possessing the property of plasticity deforms permanently. It takes a *permanent set* and will not recover when the load is removed.

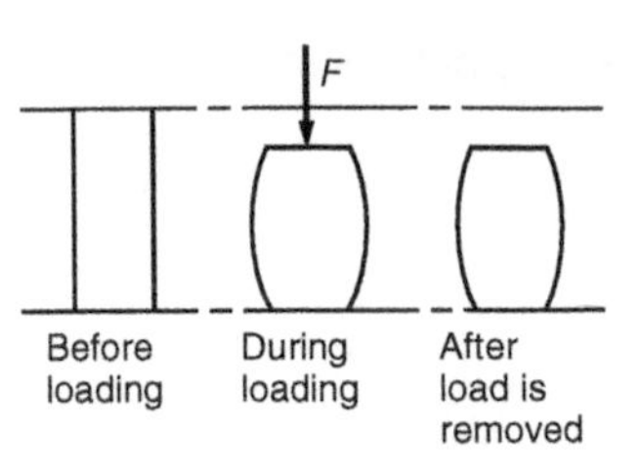

6.1.7 Ductility

This is the term used when plastic deformation occurs as the result of applying a *tensile load*. A *ductile* material combines the properties of plasticity and tenacity (tensile strength) so that it can be stretched or drawn to shape and will retain that shape when the deforming force is removed. For example, in wire drawing the wire is reduced in diameter by drawing it through a die.

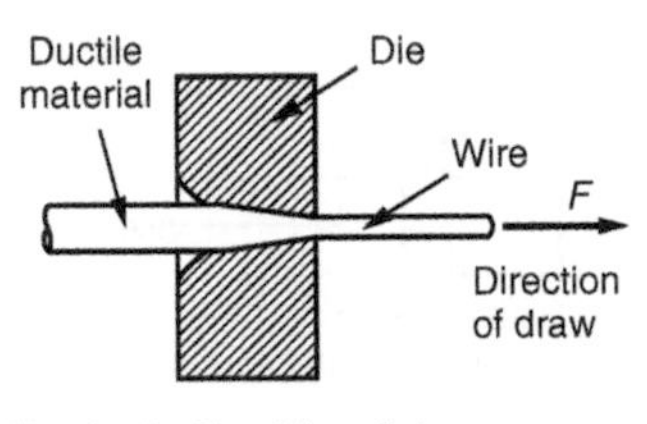

F = Applied load (tensile)

6.1.8 Malleability

This is the term used when plastic deformation occurs as the result of applying a *compressive load*. A *malleable* material combines the properties of plasticity and compressibility, so that it can be squeezed to shape by such processes as forging, rolling and rivet heading.

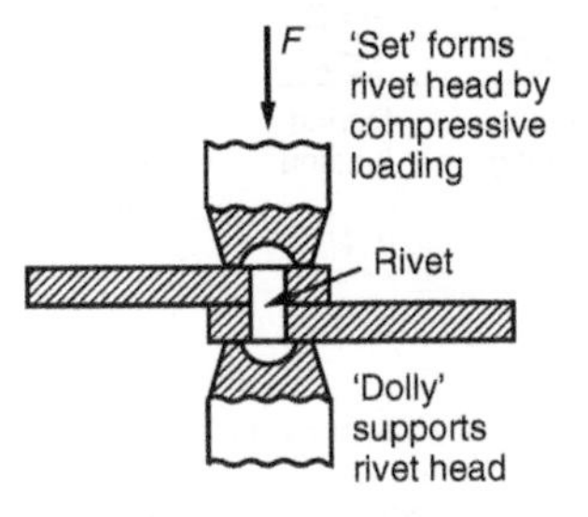

F = Applied load (compressive)

6.1.9 Hardness

This is the ability of a material to withstand scratching (abrasion) or indentation by another hard body. It is an indication of the wear resistance of a material.

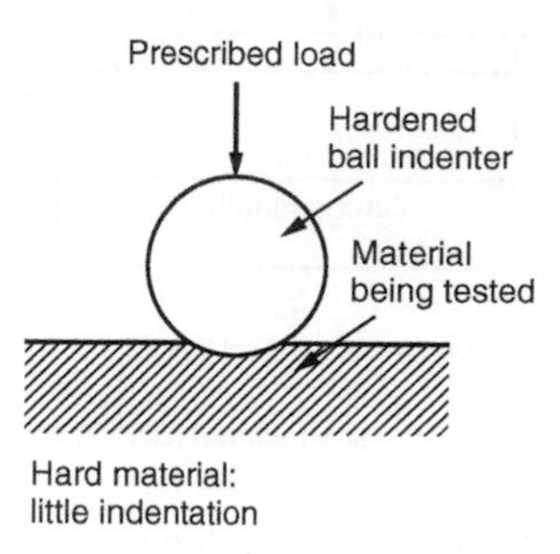

Processes which increase the hardness of materials also increase their tensile strength. At the same time the toughness of the material is reduced as it becomes more brittle.

Hardenability must not be confused with hardness. Hardenability is the ability of a metal to respond to the heat treatment process of quench hardening. To harden it, the hot metal must be chilled at a rate in excess of its *critical cooling rate*. Since any material cools more quickly at the surface than at the centre there is a limit to the size of bar which can cool quickly enough at its centre to achieve uniform hardness throughout. This is the *ruling section* for the material. The greater its hardenability the greater will be its ruling section.

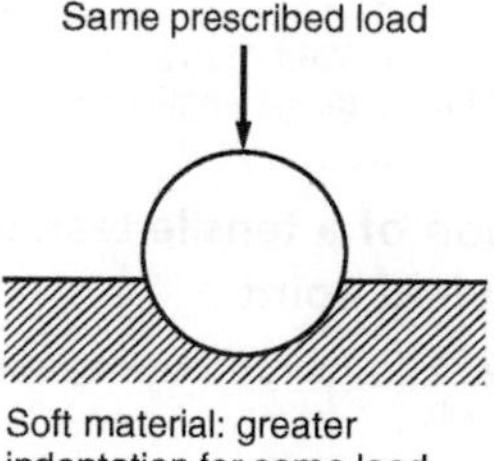

6.1.10 Tensile test

The tensile test is widely used for determining the strength and ductility of a material. The test involves loading a standard specimen axially as shown. The load is increased at a constant rate mechanically or hydraulically. The specimen increases in length until it finally fractures. During the test the specimen is gripped at each end to ensure simple uniaxial loading and freedom of bending. The extension is measured from the *gauge length*. The mid-portion of the specimen is reduced in diameter as shown to ensure fracture occurs within the gauge length.

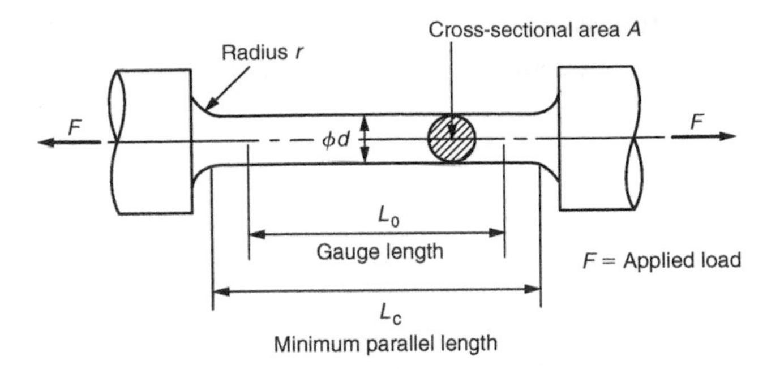

Typical cylindrical tensile test specimen (BS EN 10002-1)

The results of the test are plotted as shown in Section 6.1.11; it is usual to plot the applied load vertically and the resulting extension horizontally. Alternatively, stress and strain may be plotted with the stress vertical and the resulting strain horizontal. For a given specimen similar curves would be produced. The stress and strain relations are:

$$\text{Stress} = \frac{\text{load}}{\text{original cross-sectional area}}$$

$$\text{Strain} = \frac{\text{increase in length under load}}{\text{original length}}$$

Proportional specimens (BS EN 10002-1) are given by the relationship $L_0 = 5.65\sqrt{A}$. Since $A = \pi d^2/4$, then $\sqrt{A} = d\sqrt{(\pi)}/2 = 0.886d$. Thus $L_0 = 5.65 \times 0.886 \approx 5d$. Hence a specimen of 10-mm diameter will have a gauge length of 50 mm.

6.1.11 Interpretation of a tensile test: material showing a yield point

The curve shown is typical for a ductile material with a pronounced yield point (e.g. annealed low carbon steel).

The initial part of the plot from O to A is linear since the material is behaving in an elastic manner (Hooke's law). If the load is released at any point between O and A the specimen will return to its original length. The steeper the slope of OA the more rigid (stiffer) will be the material. The point A is called the *limit of proportionality.*

At the point A some materials may undergo a sudden extension without a corresponding increase in load. This is called the *yield point*, and the yield stress at this point is calculated by dividing the load at yield by the original cross-sectional area.

Beyond the yield point A the plot ceases to be linear since the material is now behaving in a plastic manner. If the load is released at any point in the plastic range, the elastic strain is recovered but the plastic element of the deformation is maintained and the material will have undergone permanent extension. That is, it has taken a *permanent set.*

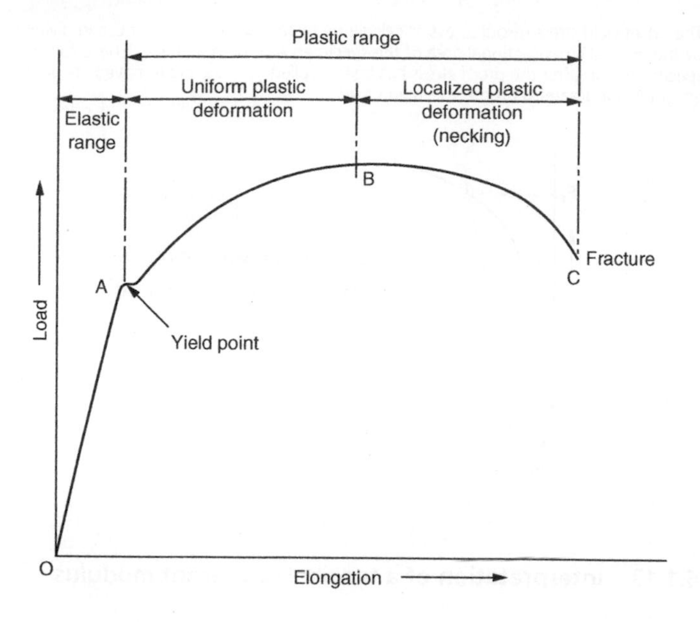

Beyond the point B the material extends with a reducing load. However, since there is a local reduction in cross-sectional area (necking) the stress (load/area) is actually increasing up to the breaking point. The stress calculated at the point B is called the *maximum tensile stress* (or just *tensile strength*) of the material.

The *ductility* of the material is calculated by reassembling the broken specimen and measuring the increase in gauge length.

Then:

$$\text{Elongation (\%)} = \frac{\text{increase in length}}{\text{original length}} \times 100$$

Under service conditions the material is loaded to a value within the OA zone. Usually not more than 50% of the value at A to allow a *factor of safety.*

6.1.12 Interpretation of a tensile test: proof stress

Many materials do not show a marked yield point, and an offset yield stress or *proof stress* is specified instead. This is the stress required to produce a specified amount of plastic deformation.

A line BC is drawn parallel to the elastic part of the plot OA so as to cut the load/elongation curve at C. The offset is specified (usually 0.1% or 0.2% of the gauge length).

The offset yield stress (proof stress) is calculated by taking the load F_1 at C and dividing it by the original cross-sectional area of the specimen A_0. The material will have fulfilled its specification if, after the proof stress has been applied for 15 s and removed, its permanent set is not greater than the specified offset.

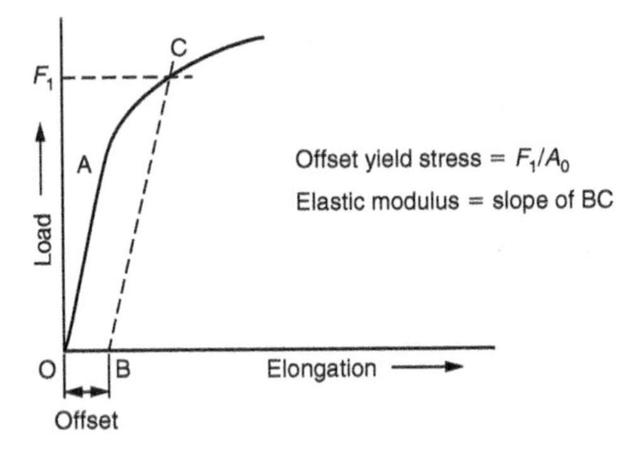

6.1.13 Interpretation of a tensile test: secant modulus

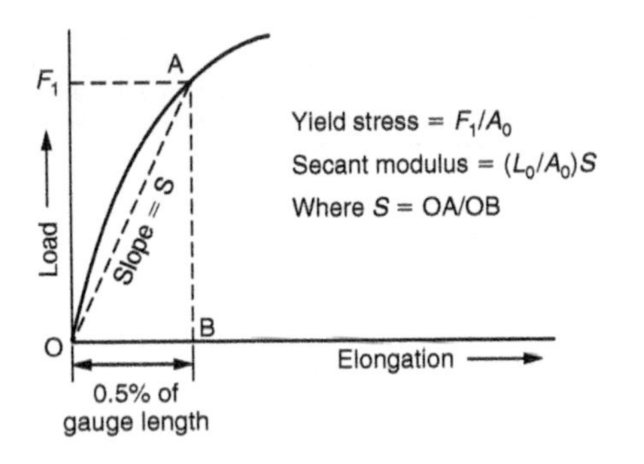

Some highly ductile metals such as annealed copper and certain polymers do not show a linear region on the tensile test plot, and therefore the offset yield stress (proof stress) cannot be determined.

In these cases an appropriate extension is specified (typically 0.5% of gauge length) and the yield stress is specified as the load to produce a total extension of 0.5% of gauge length divided by the original cross-section area A_0.

In place of an elastic modulus the *secant modulus* is used to determine the elasticity of the material.

6.1.14 Impact testing for toughness: Izod test

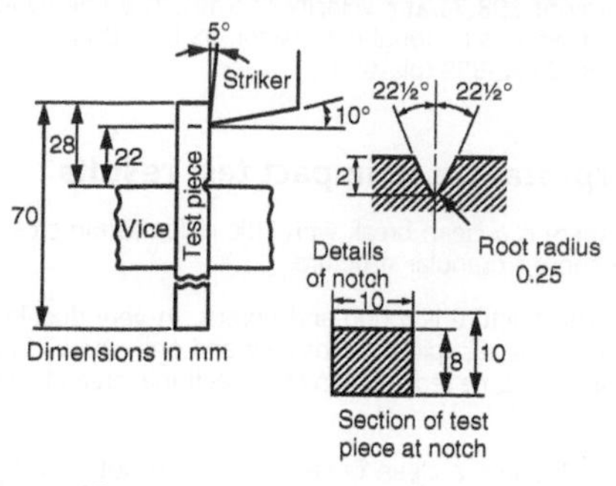

In the Izod impact test a 10-mm square notched test piece is used. It is supported as a *cantilever* in the vice of the testing machine and struck with a kinetic energy of 162.72 J at a velocity of 3.8 m/s. The energy absorbed in deforming or breaking the test piece is its toughness factor. BS 131 (metals), BS 2782-350 (plastics).

6.1.15 Impact testing for toughness: Charpy test

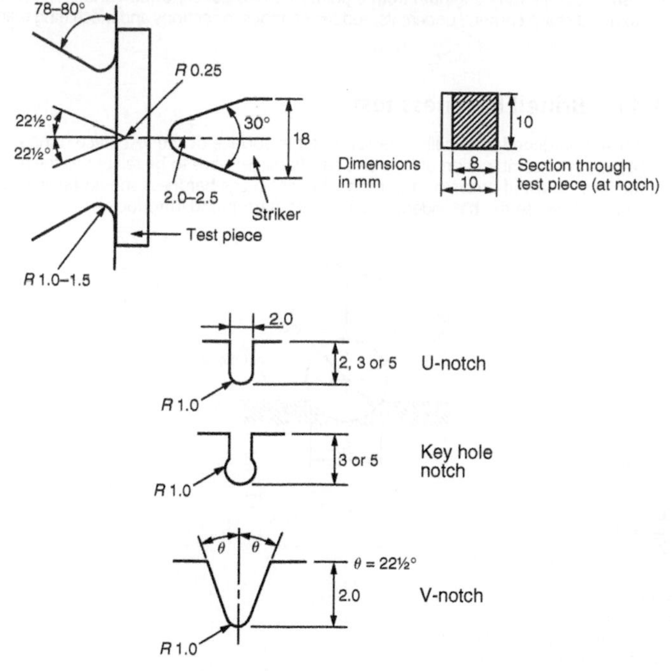

Standard Charpy notches

In the Charpy impact test a 10-mm square specimen is supported as a *beam* and struck with a kinetic energy of 298.3 J at a velocity of 5 m/s. The energy absorbed in bending or breaking the test piece is its toughness factor. BS EN 10045 (metals), BS 131-6 and BS 131-7 (metals), BS 2782-359 (plastics).

6.1.16 Interpretation of impact test results

Brittle metals There is a clean break with little reduction in cross-sectional area. The fractured surfaces show a granular structure.

Ductile metals The fracture is rough and fibrous. In very ductile metals the fracture may not be complete: the test piece bends over and only shows slight tearing from the notch. There will also be some reduction in cross-sectional area at the point of fracture or bending.

Brittle polymers There is a clean break showing smooth, glassy fractured surfaces with some splintering.

Ductile polymers There is no distinctive appearance to the fracture if one occurs at all. There will be a considerable reduction in cross-sectional area and some tearing at the notch.

Crack spread Since the Izod and Charpy tests both use notched test pieces, useful information can be obtained regarding the resistance of a material to the spread of a crack. Such a crack may originate from a point of stress concentration and indicates the need to avoid sharp corners, undercuts, sudden changes in sections and machining marks.

6.1.17 Brinell hardness test

In this test a hardened steel ball is pressed into the surface of the test piece using a prescribed load. The combination of load and ball diameter have to be related to the material under test to avoid errors of distortion. The test piece hardness should be limited to $H_B = 500$, otherwise the ball indenter will tend to flatten and introduce errors (BS 240).

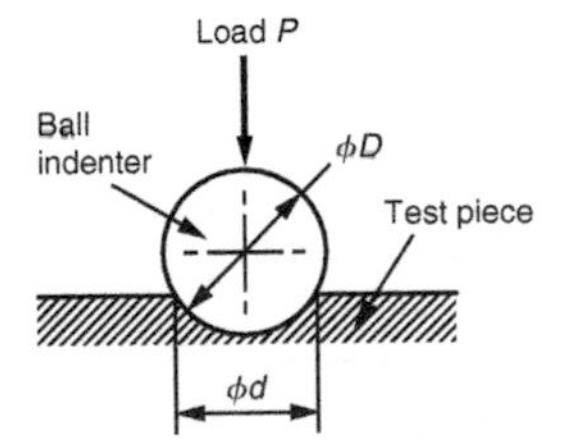

$$\frac{P}{D^2} = k$$

where:
P = load (kg)
D = diameter of indenter (mm)
k = 30 for ferrous metals
= 10 for copper and copper alloys

= 5 for aluminium and aluminium alloys
= 1 for lead, tin, white-bearing metals

$$\text{Hardness number } (H_B) = \frac{\text{load}}{\text{spherical area of indentation}}$$

$$= \frac{P}{\pi(D/2)[D - \sqrt{(D^2 - d^2)}]}$$

6.1.18 Vickers hardness test

In this test a diamond indenter is used in the form of a square-based pyramid with an angle of 136° between facets. Since diamond has a hardness of 6000 H_B, this test can be used for testing very hard materials. Only one size of indenter is used and the load is varied. Standard loads are 5, 10, 20, 30, 50 and 100 kg. It is necessary to state the test load when specifying a Vickers hardness number (e.g. $H_D(50) = 200$).

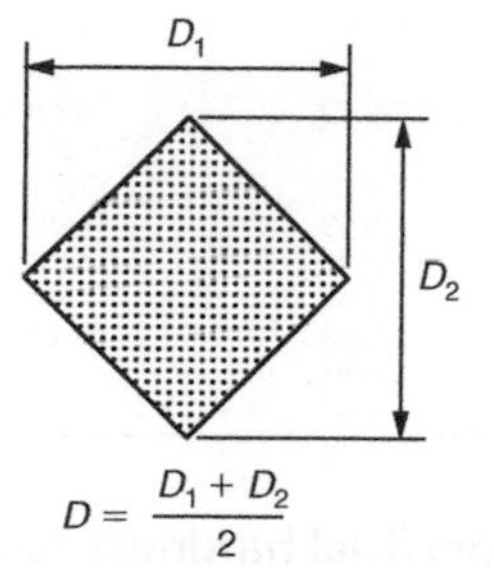

$$D = \frac{D_1 + D_2}{2}$$

Appearance of indentation

$$\text{Hardness value } H_D = 1.844 \frac{P}{D^2}$$

where:
P = load (kg)
D = diagonal length of indentation (average)

BS 427, BS EN 23878 (hard metals), BS 411-6 (metallic coatings) BS 6479 (case-hardened steels), BS 6481 (flame and induction hardened steels), DD ENV 843-4 (monolithic ceramics).

6.1.19 Rockwell hardness test

This test uses either a hard steel ball for softer materials or a 120° diamond cone indenter for harder materials. The test compares the differences in depth of penetration between a minor initial load (98 N) and a major additional load. The result of the test is read directly from the machine scale.

$H_R = E - e$

where:

E = a constant dependent on indenter used

e = the permanent *increase* in penetration due to the application of an additional major load

Scale	Indenter	Additional force (kN)	Applications
A	120° diamond cone	0.59	Sheet steel; shallow case-hardened components
B	Ball, ϕ 1.588 mm	0.98	Copper alloys; aluminium alloys; annealed low-carbon steels
C	128° diamond cone	1.47	Most widely used range; hardened steels; cast irons; deep case-hardened components
D	120° diamond cone	0.98	Thin but hard steel; medium depth case-hardened components
E	Ball, ϕ 3.175 mm	0.98	Cast iron; aluminium alloys, magnesium alloys; bearing metals
F	Ball, ϕ 1.588 mm	0.59	Annealed copper alloys; thin soft sheet metals
G	Ball, ϕ 1.588 mm	1.47	Malleable cast irons; phosphor bronze; gunmetals; cupro-nickel alloys, etc.
H	Ball, ϕ 3.175 mm	0.59	Soft materials; high ferritic alloys; aluminium; lead; zinc
K	Ball, ϕ 3.175 mm	1.47	Aluminium and magnesium alloys
L	Ball, ϕ 6.350 mm	0.59	Thermoplastics
M	Ball, ϕ 6.350 mm	0.98	Thermoplastics
P	Ball, ϕ 6.350 mm	1.47	Thermosetting plastics
R	Ball, ϕ 12.70 mm	0.59	Very soft plastics and rubbers
S	Ball, ϕ 12.70 mm	0.98	–
V	Ball, ϕ 12.70 mm	1.47	–

6.1.20 Rockwell superficial hardness test

In this test the initial force is reduced from 98 to 29.4 N and the additional (major) force is also reduced. This range of tests is used when measuring the hardness of thin sheets and foils.

Scale	Indenter	Additional force (kN)
15-N	120° diamond cone	0.14
30-N	120° diamond cone	0.29
45-N	120° diamond cone	0.44
15-T	Ball, ϕ 1.588 mm	0.14
30-T	Ball, ϕ 1.588 mm	0.29
45-T	Ball, ϕ 1.588 mm	0.44

BS 5600-4-5 (hard metals)
BS 891, 4175 (hard metals)
DD ENV 843-4 (monolithic ceramics)
BS 2782-365C (plastics).

6.1.21 Comparative hardness scales

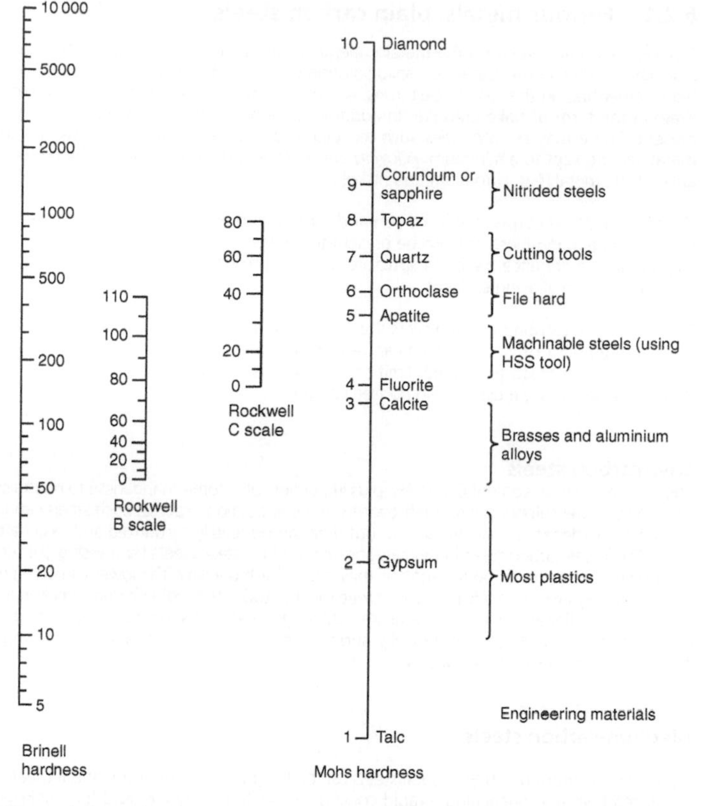

Tables and charts showing comparative hardness figures for various methods of testing should be treated with caution since the tests are carried out under different conditions. For example:

(a) The relatively large diameter ball indenter of the Brinell test and some Rockwell tests displaces the metal of the test piece by plastic flow.
(b) The sharp edged and sharply pointed diamond pyramid of the Vickers test tends to cut its way into the test piece by shear.
(c) The Rockwell test uses yet another form of indenter, namely a 120° diamond cone. This test also compares the increase in depth of penetration when the load is increased, whereas in (a) and (b) the area of indentation is measured for a single stated load.
(d) The Scleroscope is a dynamic test, measuring hardness as a function of resilience.

6.2 Ferrous metals and alloys

6.2.1 Ferrous metals: plain carbon steels

Ferrous metals are based on the metallic element *iron* (Latin *ferrum* = iron). The iron is associated with carbon, either as a solid solution or as the chemical compound iron carbide (cementite). In the case of cast irons, some amount of carbon may be uncombined (free) in the form of flake graphite. In addition to carbon, other elements may also be present. These may be *impurities* such as sulphur and phosphorus which weaken the metal and are kept to a minimum. *Alloying elements* are added to enhance the performance of the metal (e.g. chromium and nickel).

Plain carbon steels consist mainly of iron and carbon, and are the simplest of the ferrous metals. Some manganese will also be present to neutralize the deleterious effects of the sulphur and to enhance the grain structure. It is not present in sufficient quantity to be considered as an alloying element.

The amount of carbon present affects the properties of the steel as shown in 6.2.3. The maximum amount of carbon which can remain combined with the iron at all temperatures is 1.7%. In practice an upper limit of 1.2–1.4% is set to ensure a margin of safety. A steel, by definition, must contain *no* free carbon.

Low-carbon steels

These have a carbon content 0.1–0.3% plus impurities, plus some manganese to neutralize the effect of any sulphur content left over from the extraction process. Such steels cannot be directly hardened by heat treatment, but they can be readily carburized and case hardened. The lower-carbon steels in this category are used for steel sheets for pressing out such components as motorcar body panels as they have a high ductility. The lower-carbon steels in this category are also made into drawn wire rod and tube. The higher-carbon steels in this category are stiffer and less ductile, and are used for general workshop bars, plates and girders. Low-carbon steels are substantially stronger than wrought iron which is no longer considered to be a structural material.

Medium-carbon steels

(a) Carbon content 0.3–0.5%: Such steels can be toughened by heat treatment (heating to red heat and quenching – rapid cooling – in water). They are used for crankshaft and axle forgings where cost is important and the service requirements do not warrant stronger but more expensive alloy steels.
(b) Carbon content 0.5–0.8%: These are used for vehicle leaf springs and garden tools. Such steels can be quench hardened by heat treatment as above.

High-carbon steels

All high carbon steels can be hardened to a high degree of hardness by heating to a dull red heat and quenching. The hardness and application depend on the carbon content, the rate of cooling from the hardening temperature and the degree of tempering after hardening:

(a) Carbon content 0.8–1.0%; used for coil springs and wood chisels.
(b) Carbon content 1.0–1.2%; used for files, drills, taps and dies.
(c) Carbon content 1.2–1.4%; used for fine-edge tools (knives, etc.).

6.2.2 Effect of carbon content on the composition, properties and uses of plain carbon steels

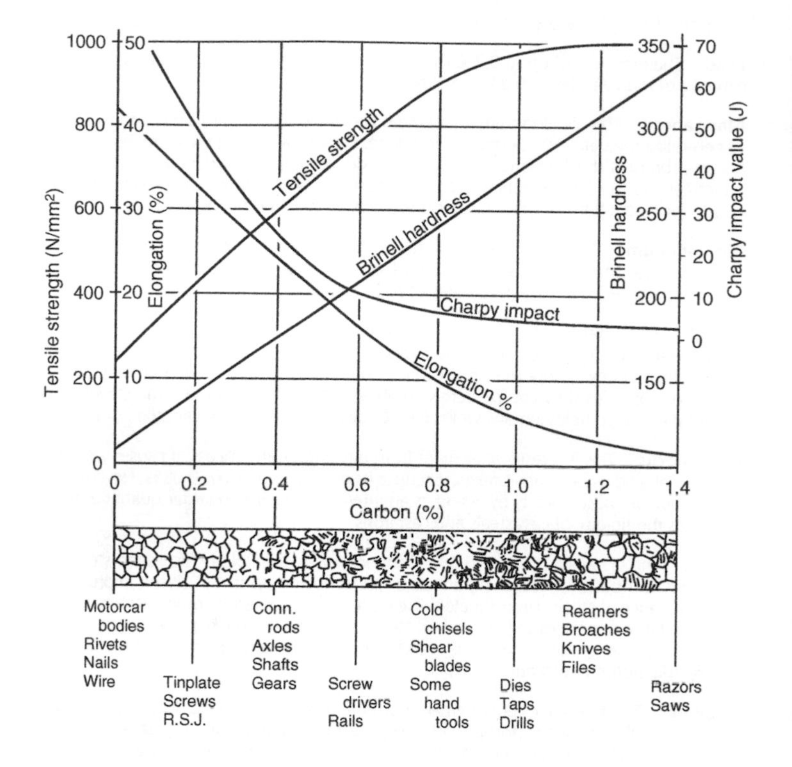

6.2.3 Ferrous metals: alloying elements

Alloy steels are carbon steels containing less than 1% carbon but to which other metals have been added in sufficient quantities to alter the properties of the steel significantly. The more important alloying elements are as follows:

Aluminium Up to 1% aluminium in alloy steels enables them to be given a hard, wear-resistant skin by the process of *nitriding*.

Chromium The presence of small amounts of chromium stabilizes the formation of hard carbides. This improves the response of the steel to heat treatment. The presence of large amounts of chromium improves the corrosion resistance and heat resistance of the steel (e.g. stainless steel). Unfortunately, the presence of chromium in a steel leads to grain growth (see *nickel*).

Cobalt Cobalt induces sluggishness into the response of a steel to heat treatment. In tool steels it allows them to operate at high-level temperatures without softening. It is an important alloying element in some high-speed steels.

Copper Up to 0.5% copper improves the corrosion resistance of alloy steels.

Lead The presence of up to 0.2% lead improves the machinability of steels, but at the expense of reduced strength and ductility.

Manganese This alloying element is always present in steels up to a maximum of 1.5% to neutralize the deleterious effects of impurities carried over from the extraction process. It also promotes the formation of stable carbides in quench-hardened steels. In larger quantities (up to 12.5%) manganese improves the wear resistance of steels by spontaneously forming a hard skin when subject to abrasion (self-hardening steels).

Molybdenum This alloying element raises the high-temperature creep resistance of steels; stabilizes their carbides; improves the high-temperature performance of cutting tool materials; and reduces the susceptibility of nickel–chrome steels to 'temper brittleness'.

Nickel The presence of nickel in alloy steels results in increased strength and grain refinement. It also improves the corrosion resistance of the steel. Unfortunately it tends to soften the steel by graphitizing any carbides present. Since nickel and chromium have opposite properties they are frequently combined together (nickel–chrome steels). Their advantages are complementary, whilst their undesirable effects are cancelled out.

Phosphorus This is a residual element from the extraction process. It causes weakness in the steel, and usually care is taken to reduce its presence to below 0.05%. Nevertheless, it can improve machinability by acting as an internal lubricant. In larger quantities it also improves the fluidity of cast steels and cast irons.

Silicon The presence of up to 0.3% silicon improves the fluidity of casting steels and cast irons without the weakening effects of phosphorus. Up to 1% silicon improves the heat resistance of steels. Unfortunately, like nickel, it is a powerful graphitizer and is never added in large quantities to high-carbon steels. It is used to enhance the magnetic properties of 'soft' magnetic materials as used for transformer laminations and the stampings for electric motor stators and rotors.

Sulphur This is also a residual element from the extraction process. Its presence greatly weakens steel, and every effort is made to refine it out; in addition, manganese is always present in steels to nullify the effects of any residual sulphur. Nevertheless, sulphur is sometimes deliberately added to low-carbon steels to Improve their machinability where a reduction in component strength can be tolerated (sulphurized free-cutting steels).

Tungsten The presence of tungsten in alloy steels promotes the formation of very hard carbides and, like cobalt, induces sluggishness into the response of the steel to heat treatment. This enables tungsten steels (high-speed steels) to retain their hardness at high temperatures. Tungsten alloys form the bases of high-duty tool and die steels.

Vanadium This element enhances the effects of the other alloying elements present and has many and varied effects on alloy steels:

(a) Its presence promotes the formation of hard carbides.
(b) It stabilizes the *martensite* in quench-hardened steels and thus improves hardenability and increases the limiting ruling section of the steel.
(c) It reduces grain growths during heat treatment and hot-working processes.
(d) It enhances the 'hot hardness' of tool steels and die steels.
(e) It improves the fatigue resistance of steels.

6.2.4 British standards relating to ferrous metals

Although BS 970 is now obsolescent, *it is still widely used* but should not be quoted when specifying wrought steels for new designs and products. It has now been superceded by the BS EN 10 000 series of specifications which are English language versions and compatible with the Continental ISO specifications.

One advantage of BS 970: 1991 is that it changed the random numbering system of the earlier version and designates the steels by an alphanumeric code that indicates the actual composition of the material in a logical manner. The designation code of each steel is built up as follows:

(a) The first three symbols are a number code indicating the type of steel:
  000 to 199 are carbon and manganese steels. The numbers represent the manganese content × 100.
  200 to 240 are free-cutting steels. The second and third numbers indicate the sulphur content × 100.
  250 are silicon manganese valve steels.
  300 to 499 are stainless and heat-resisting steels.
  500 to 999 are alloy steels.
(b) The fourth symbol is a letter code:
  A The steel is supplied to a chemical composition determined by chemical analysis of a batch sample.
  H The steel is supplied to a hardenability specification.
  M The steel is supplied to a mechanical property specification.
  S The material is a stainless steel.
(c) The fifth and sixth symbol comprise a number code indicating the actual carbon content × 100

The following are examples of the BS 970: 1991 six-figure code:

070M20 A plain carbon steel with a composition of 0.7% manganese and 0.2% carbon. The M indicates that the steel has a prescribed mechanical property specification.

230M07 A low-carbon, free-cutting steel with a sulphur content of 0.3% and a carbon content of 0.7%.

080A15 A plain carbon steel with a manganese content of 0.8% and a carbon content of 0.15%. The A indicates that the steel must meet a prescribed chemical composition specification.

Unfortunately for alloy steels coded between 500 and 999 the logicality of the first three digits breaks down. In addition to the six-figure grading code, a *condition code* is applied. The code letter indicates the tensile strength range for a given steel after heat treatment. The final factor to be considered in the coding of wrought steels is the *limiting ruling section*. As explained in Section 6.1.9 this is the maximum diameter of bar of a given composition which, after heat treatment, will attain the specified mechanical properties.

For example, a plain carbon steel of composition 070M55(R) containing 0.7% manganese and 0.55% carbon can attain condition (R) after heat treatment providing it is not greater than 100-mm diameter, however if it is to attain condition (S) then its maximum diameter must not exceed 63 mm. In the first example the limiting ruling section is 100 mm and in the second example the limiting ruling section is 63 mm in diameter. This letter/strength relationship code is listed in Table 6.1. The condition code is applied in brackets after the six-symbol code as shown above.

In addition to the standard chemical symbols used to denote the composition of a metal (e.g. C = carbon, Mn = manganese, P = phosphorus, S = sulphur, etc.), the symbols

Table 6.1 Code letter/strength relationship

Condition code letter	Tensile strength (MPa)		Condition code letter	Tensile strength (MPa)	
	min.	max.		min.	max.
P	550	700	V	1000	1150
Q	620	770	W	1080	1240
R	700	850	X	1150	1300
S	770	930	Y	1240	1400
T	850	1000	Z	1540	–
U	930	1080			

used in tables of mechanical properties are as follows:

Tensile properties

R_m Tensile strength
R_e Yield strength
A Percentage elongation after fracture
S_o Original cross-sectional area of the gauge length
R_{p02} 0.2% proof stress (non-proportional elongation)
$R_{p1.0}$ 1.0% proof stress (non-proportional elongation)
$R_{t0.5}$ 0.5% proof stress (total elongation)
$R_{t1.0}$ 1.0% proof stress (total elongation)
R_{eH} Upper yield stress

Impact properties

KCV Charpy V-notch impact value

Hardness

H_B Brinell hardness
H_V Vickers hardness
HRC Rockwell hardness (C scale)

Other

LRS the limiting ruling section.

Note that the specifications for the chemical composition and mechanical properties for plain carbon and alloy steels are now being superceded by the BS EN 10000 series of standards. For example:

- BS EN 10277-2 1999 deals with the composition and mechanical properties of plain carbon bright steel products.
- BS EN 10277-3 deals with the composition and mechanical properties of free-cutting steels.
- BS EN 10277-4 deals with the chemical composition and mechanical properties of case-hardening steels.
- BS EN 10277-5 deals with non-alloy and alloy steels suitable for quench hardening and tempering.

The steels are now designated by a steel name and a steel number. For example a plain carbon steel containing 0.13–0.18% carbon and 0.7–0.9% manganese would be designated 080A15 in BS 970 it is now designated: **steel name C15** and **steel number 1.0401** in BS EN 10277-2.

BS EN 10277 and related standards for steels are highly detailed and comprehensive, and it is not physically possible to reproduce them in this Pocket Book. These standards should be consulted together with BS Guide Lines PD 970:2005 when selecting a suitable steel for a new product design and manufacture. Some examples of typical steels and their applications are listed in Section 6.2.5 and some examples of typical tool and die steels are listed in Section 6.2.6.

6.2.5 Some typical steels and their applications

Plain carbon steels

Type of steel	Typical com-positions (%)	Heat treatment	Typical mechanical properties					Uses
			Yield point (N/mm^2)	Tensile strength (N/mm^2)	Elong-ation (%)	Impact J	Hardness (Brinell)	
Low-carbon steel	0.10 C 0.40 Mn	No heat treatment (except) process annealing to remove the effects of coldwork.	–	2300	28	–	–	Lightly stressed parts produced by cold-forming processes (e.g. deep drawing and pressing).
Structural steels	0.20 C	No heat treatment.	240	450	25	–	–	General structural steel.
	0.20 C 1.50 Mn	No heat treatment.	350	325	20	–	–	High-tensile structural steel for bridges and general building construction – fusion welding quality.
Casting steel	0.30 C	No heat treatment other than 'annealing' to refine grain.	265	500	18	20	150	Castings for a wide range of engineering purposes where medium strength and good machinability are required.
Constructional steels (medium carbon)	0.40 C 0.80 Mn	Harden by quenching from 830°C to 860°C. Temper at a suitable temperature between 550°C and 660°C.	500	700	20	55	200	Axles crankshafts, spindles, etc. under medium stress.
	0.55 C 0.70 Mn	Harden by quenching from 810°C to 840°C. Temper at a suitable temperature between 550°C and 666°C.	550	750	14	–	250	Gears, cylinders and machine tool parts requiring resistance to wear.

continued

Section 6.2.5 *(continued)*

Type of steel	Typical com-positions (%)	Heat treatment	Typical mechanical properties					Uses
			Yield point (N/mm^2)	Tensile strength (N/mm^2)	Elong-ation (%)	Impact J	Hardness (Brinell)	
Tool steels (High carbon)	0.70 C 0.35 Mn	Heat slowly to 790–810°C. and quench in water or brine. Temper at 150–300°C.					780	Hand chisels, cold sets, mason's tools, smith's tools, screwdriver blades, stamping dies, keys, cropping blades, miner's drills, paper knives.
	0.90 C 0.35 Mn	Heat slowly to 760–780°C and quench in water or brine. Temper at 150–350°C.					800	Press tools; punches; dies; cold heading, minting and embossing dies; shear blades; woodworking tools; lathe centres; draw plates.
	1.00 C 0.35 Mn	Heat slowly to 770–790°C and quench in water or brine. Temper at 150–350°C.					800	Taps; screwing dies; twist drills; reamers; counter sinks; blanking tools; embossing, engraving, minting, drawing, needle and paper dies; shear blades, knives; press tools; centre punches; woodworking cutters; straight edges; gouges; pneumatic chisels; wedges.
	1.20 C 0.35 Mn	Heat slowly to 760–780°C and quench in water or brine. Temper at 180–350°C.					800	Engraving tools; files; surgical instruments; taps; screwing tools.

Low-alloy constructional steels

Type of steel	Com-position (%)	Condition	Mechanical properties					Uses
			Yield stress (N/mm^2)	Tensile stress (N/mm^2)	Elong-ation (%)	Izod J	Heat treatment	
Low manganese	0.28 C 1.50 Mn	Normalized.	355	587	20	–	Oil-quench from 860°C (water-quench for sections over 38-mm diameter). Temper as required.	Automobile axles, crankshafts, connecting rods, etc. where a relatively cheap steel is required.
Manganese–chrome	0.40 C 0.90 Mn 1.00 Cr	Quenched and tempered at 600°C.	494	695	25	91	Oil-quench from 850°C; temper between 550°C and 660°C and cool in oil or air.	Crankshafts, axles, connecting rods; other parts in the automobile industry and in general engineering.
Manganese–molybdenum	0.38 C 1.50 Mn 0.50 Mo	28.5-mm bar, o.q. and tempered at 600°C.	1000	1130	19	70	Oil-quench from 830°C to 850°C; temper between 550°C and 650°C and cool in oil or air.	A substitute for the more highly alloyed nickel–chrome–molybdenum steels.
Nickel–chromium	0.31 C 0.60 Mn 3.00 Ni 1.00 Cr	28.5 mm bar, o.q. and tempered at 600°C.	819	927	23	104	Oil-quench from 820°C to 840°C; temper between 550° and 650°C. Cool in oil to avoid 'temper brittleness'.	Highly stressed parts in automobile and general engineering (e.g. differential shafts, stub axles, connecting rods, high-tensile studs, pinion shafts).

continued

Low-alloy constructional steels *(continued)*

Type of steel	Com-position (%)	Condition	Mechanical properties					Uses
			Yield stress (N/mm^2)	Tensile stress (N/mm^2)	Elong-ation (%)	Izod J	Heat treatment	
Nickel–chromium–molybdenum	0.40 C 0.55 Mn 1.50 Ni 1.20 Cr 0.30 Mo	o.q. and tempered at 200°C o.q. and tempered at 600°C.	– 988	2010 1080	14 22	27 69	Oil-quench from 830°C to 850°C; 'light temper' 180–200°C; 'full temper' 550–650°C; cool in oil or air.	Differential shafts, crankshafts and other highly stressed parts where fatigue and shock resistance are important; in the 'light tempered' condition it is suitable for automobile gears; can be surface hardened by nitriding.
	0.30 C 0.55 Mn 4.25 Ni 1.25 Cr 0.30 Mo	Air-hardened and tempered at 200°C.	1470	1700	14	35	Air-harden from 820°C to 840°C; temper at 150–200°C and cool in air.	An air-hardening steel for aero-engine connecting rods, valve mechanisms, gears, differential shafts and other highly stressed parts; suitable for surface hardening by cyanide or carburizing.
Manganese–nickel–chromium–molybdenum	0.38 C 1.40 Mn 0.75 Ni 0.50 Cr 0.20 Mo	28.5 mm bar, o.q. from 850°C and tempered at 600°C.	958	1040	21	85	Oil-quench from 830°C to 850°C; temper at 550–660°C, and cool in air.	Automobile and general engineering components requiring a tensile strength of 700–1000 N/mm^2.

Alloy tool and die steels (see also Section 6.2.6)

Type of steel	Composition (%)	Heat treatment	Uses
'60' carbon-chromium	0.60 C 0.65 Mn 0.65 Cr	Oil-quench from 800°C to 850°C. Temper: (a) For cold-working tools at 200–300°C. (b) For hot-working tools at 400–600°C.	Blacksmith's and boilermaker's chisels and other tools; mason's and miner's tools; vice jaws; hot stamping and forging dies.
1% carbon-chromium	1.00 C 0.45 Mn 1.40 Cr	Oil-quench from 810°C; temperature at 150°C.	Ball and roller bearings; instrument pivots; cams; small rolls.
High carbon, high chromium (HCCR)	2.10 C 0.30 Mn 12.50 Cr	Heat slowly to 750–800°C and then raise to 960–990°C. Oil-quench (small sections can be air cooled). Temper at 150–400°C for 30–60 min.	Blanking punches, dies and shear blades for hard, thin materials; dies for moulding abrasive powders, (e.g. ceramics; master gauges; thread rolling dies).
$\frac{1}{4}$% vanadium	1.00 C 0.25 Mn 0.20 V	Water-quench from 850°C; temper as required.	Cold-drawing dies, etc.
4% vanadium	1.40 C 0.40 Mn 0.40 Cr 0.40 Mo 3.60 V	Water-quench from 770°C; temper at 150–300°C.	Cold-heading dies, etc.
Hot-working die steel	0.35 C 1.00 Si 5.00 Cr 1.50 Mo 0.40 V 1.35 W	Pre-heat to 800°C, soak and then heat quickly to 1020°C and air cool. Temper at 540–620°C for 1½ h.	Extrusion dies, mandrels and noses for aluminium and copper alloys; hot forming, piercing, gripping and heading tools; brass forging and hot pressing dies.

continued

Alloy tool and die steels (see also Section 6.2.6) (*continued*)

Type of steel	Composition (%)	Heat treatment	Uses
High-speed steels 18% tungsten	0.75 C 4.25 Cr 18.00 W 1.20 V	Quench in oil or air blast from 1290°C to 1310°C. Double temper at 565°C for 1 h.	Lathe, planer and shaping tools; millers and gear cutters; reamers; broaches; taps; dies; drills; hacksaw blade; bandsaws; roller bearings at high temperatures (gas turbines).
12% cobalt	0.80 C 4.75 Cr 22.0 W 1.50 V 0.50 Mo 12.0 Co	Quench in oil or air blast from 1300°C to 1320°C. Double temper at 565°C for 1 h.	Lathe, planing and shaping tools, milling cutters, twist drills etc. for exceptionally hard materials; has maximum red hardness and toughness; suitable for severest machining duties (e.g. manganese steels and high tensile steels, close-grained cast irons).
Molybdenum '562'	0.83 C 4.25 Cr 6.50 W 1.90 V 5.00 Mo	Quench in oil or air blast from 1250°C. Double temper at 565°C for 1 h.	Roughly equivalent to the standard 18-4-1 tungsten high speed steel but tougher; drills, reamers, taps, milling cutters, punches, threading dies, cold forging dies.
9% molybdenum 8% cobalt	1.00 C 3.75 Cr 1.65 W 1.10 V 9.50 Mo 8.25 Co	Quench in oil or air blast from 1180°C to 1210°C. Triple temper at 530°C for 1 h.	Similar uses to the 12% Co–22% W high speed steel.

Stainless and heat-resisting steels

Type of steel	Com-position (%)	Condition	Typical mechanical properties				Heat treatment	Uses
			Yield stress (N/mm^2)	Tensile strength (N/mm^2)	Elon-gation (%)	Hardness (Brinell)		
Stainless iron (ferritic)	0.04 C 0.45 Mn 14.00 Cr	Soft	340	510	31	–	Non-hardenable except by cold work.	Wide range of domestic articles, forks, spoons; can be spun, drawn and pressed.
Cutlery steel (martensitic)	0.30 C 0.50 Mn 13.00 Cr	Cutlery temper Spring temper	– –	1670 1470	– –	534 450	Water- or oil-quench (or air cool) from 950 to 1000°C. Temper: for cutlery, at 150–180°C; for springs, at 400–450°C.	Cutlery and sharpedged tools requiring corrosion resistance; circlips, etc; approximately pearlitic in structure when normalized.
18/8 stainless (austenitic)	0.05 C 0.80 Mn 8.50 Ni 18.00 Cr	Softened Cold rolled	278 803	618 896	50 30	170 –	Non-hardening except by cold-work (cool quickly from 1050°C to retain carbon in solid solution).	A highly ductile and easily worked steel that has high corrosion resistance. Widely used for kitchenware, sinks, food processing, brewing and catering equipment. Also used for architectural purposes.
18/8 stainless (weld decay proofed)	0.05 C 0.80 Mn 8.50 Ni 18.00 Cr 1.60 Ti	Softened Cold rolled	278 402	649 803	45 30	180 225	Non-hardening except by cold-work (cool quickly from 1050°C to retain carbon in solid solution).	A weld decay proofed steel (fabrication by welding can be safely employed); used extensively in nitric acid plant and similar chemical processes.

Section 6.2.6 Some typical tool steels

Composition / Designation	C		Si		Mn		P		S		Cr		Mo		Ni		Co		Cu		Sn		V		W		Annealed hardness (max.)*	Hardness after heat treatment (min.)
	min.	max.	min.	max.	min.	max.	min.	max.	min.	max.	min.	max.	min.	max.	min.	max.	min.	max.	min.	max.	min.	max.	min.	max.	min.	max.		
High-speed tool steels	%	%	%	%	%	%	%	%	%	%	%	%	%	%	%	%	%	%	%	%	%	%	%	%	%	%	HB	HV
BM1	0.75	0.85	–	0.40	–	0.40	–	0.035	–	0.035	3.75	4.5	8.0	9.0	–	0.40	–	$1.0^{\dagger}$	–	0.20	–	0.05	1.0	1.25	1.0	2.0	241	823
BM2	0.82	0.92	–	0.40	–	0.40	–	0.035	–	0.035	3.75	4.5	4.75	5.5	–	0.40	–	$1.0^{\dagger}$	–	0.20	–	0.05	1.75	2.05	6.0	6.75	248	836
BM4	1.25	1.40	–	0.40	–	0.40	–	0.035	–	0.035	3.75	4.5	4.25	5.0	–	0.40	–	$1.0^{\dagger}$	–	0.20	–	0.05	3.75	4.25	5.75	6.5	255	849
BM15	1.45	1.60	–	0.40	–	0.40	–	0.035	–	0.035	4.5	5.0	2.75	3.25	–	0.40	4.5	5.5	–	0.20	–	0.05	4.75	5.25	6.25	7.0	277	869
BM35	0.85	0.95	–	0.40	–	0.40	–	0.035	–	0.035	3.75	4.5	4.75	5.25	–	0.40	4.60	5.20	–	0.20	–	0.05	1.75	2.15	6.0	6.75	269	869
BM42	1.0	1.10	–	0.40	–	0.40	–	0.035	–	0.035	3.5	4.25	9.0	10.0	–	0.40	7.5	8.5	–	0.20	–	0.05	1.0	1.3	1.0	2.0	269	897
BT1	0.70	0.80	–	0.40	–	0.40	–	0.035	–	0.035	3.75	4.5	–	0.70	–	0.40	–	$1.0^{\dagger}$	–	0.20	–	0.05	1.0	1.25	17.5	18.5	255	823
BT4	0.70	0.80	–	0.40	–	0.40	–	0.035	–	0.035	3.75	4.5	–	1.0	–	0.40	4.5	5.5	–	0.20	–	0.05	1.0	1.25	17.5	18.5	277	849
BT5	0.75	0.85	–	0.40	–	0.40	–	0.035	–	0.035	3.75	4.5	–	1.0	–	0.40	9.0	$10.0^{\dagger}$	–	0.20	–	0.05	1.75	2.05	18.5	19.5	290	869
BT6	0.75	0.85	–	0.40	–	0.40	–	0.035	–	0.035	3.75	4.5	–	1.0	–	0.40	11.25	12.25	–	0.20	–	0.05	1.25	1.75	20.0	21.0	302	869
BT15	1.40	1.60	–	0.40	–	0.40	–	0.035	–	0.035	4.25	5.0	–	1.0	–	0.40	4.5	5.5	–	0.20	–	0.05	4.75	5.25	12.0	13.0	290	890
BT21	0.60	0.70	–	0.40	–	0.40	–	0.035	–	0.035	3.5	4.25	–	0.7	–	0.40	–	$1.0^{\dagger}$	–	0.20	–	0.05	0.40	0.60	13.5	14.5	255	798
BT42	1.25	1.40	–	0.40	–	0.40	–	0.035	–	0.035	3.75	4.5	2.75	3.50	–	0.40	9.0	10.0	–	0.20	–	0.05	2.75	3.25	8.5	9.5	277	912
Hot work tool steels																												
BH10	0.30	0.40	0.75	1.10	–	0.40	–	0.035	–	0.035	2.8	3.2	2.65	2.95	–	0.40	–	–	–	0.20	–	0.05	0.30	0.50	–	–	229	–
BH10A	0.30	0.40	0.75	1.10	–	0.40	–	0.035	–	0.035	2.8	3.2	2.65	2.95	–	0.40	2.8	3.2	–	0.20	–	0.05	0.30	1.10	–	–	241	–
BH11	0.32	0.42	0.85	1.15	–	0.40	–	0.035	–	0.035	4.75	5.25	1.25	1.75	–	0.40	–	–	–	0.20	–	0.05	0.30	0.50	–	–	229	–
BH12	0.30	0.40	0.85	1.15	–	0.40	–	0.035	–	0.035	4.75	5.25	1.25	1.75	–	0.40	–	–	–	0.20	–	0.05	–	0.50	1.25	1.75	229	–
BH13	0.32	0.42	0.85	1.15	–	0.40	–	0.035	–	0.035	4.75	5.25	1.25	1.75	–	0.40	–	–	–	0.20	–	0.05	0.90	1.10	–	–	229	–
BH19	0.35	0.45	–	0.40	–	0.40	–	0.035	–	0.035	4.0	4.5	–	0.45	–	0.40	4.0	4.5	–	0.20	–	0.05	2.0	2.4	4.0	4.5	248	–
BH21	0.25	0.35	–	0.40	–	0.40	–	0.035	–	0.035	2.25	3.25	–	0.60	–	0.40	–	–	–	0.20	–	0.05	–	0.40	8.5	10.0	235	–
BH21A	0.20	0.30	–	0.40	–	0.40	–	0.035	–	0.035	2.25	3.25	–	0.60	2.0	2.5	–	–	–	0.20	–	0.05	–	0.50	8.5	10.0	255	–
Hammer die steel BH 224/5	0.49	0.57	–	0.35	0.70	1.00	–	0.030	–	0.025	0.70	1.10	0.25	0.40	1.25	1.80	–	–	–	0.20	–	0.05	–	–	–	–	A*444/447 A 401/429 B 363/388 C 331/352 D 302/321 E 269/293	–

Cold-work tool steels																												
BD2	1.40	1.60	–	0.60	–	0.60	–	0.035	–	0.035	11.5	12.5	0.70	1.20	–	0.40	–	–	–	0.20	–	0.05	0.25	1.00	–	–	255	735
BD2A	1.60	1.90	–	0.60	–	0.60	–	0.035	–	0.035	12.0	13.0	0.70	0.90	–	0.40	–	–	–	0.20	–	0.05	0.25	1.00	–	–	255	763
BD3	1.90	2.30	–	0.60	–	0.60	–	0.035	–	0.035	12.0	13.0	–	–	–	0.40	–	–	–	0.20	–	0.05	–	0.50	–	–	255	763
BA2	0.95	1.05	–	0.40	0.30	0.70	–	0.035	–	0.035	4.75	5.25	0.90	1.1	–	0.40	–	–	–	0.20	–	0.05	0.15	0.40	–	–	241	735
BA6	0.65	0.75	–	0.40	1.8	2.1	–	0.035	–	0.035	0.85	1.15	1.2	1.6	–	0.40	–	–	–	0.20	–	0.05	–	–	–	–	241	735
BO1	0.85	1.0	–	0.40	1.1	1.35	–	0.035	–	0.035	0.40	0.60	–	–	–	0.40	–	–	–	0.20	–	0.05	–	0.25	0.40	0.60	229	735
BO2	0.85	0.95	–	0.40	1.5	1.8	–	0.035	–	0.035	–	–	–	–	–	0.40	–	–	–	0.20	–	0.05	–	0.25	–	–	229	735
BS1	0.48	0.55	0.70	1.0	0.30	0.70	–	0.035	–	0.035	1.2	1.7	–	–	–	0.40	–	–	–	0.20	–	0.05	0.10	0.30	2.0	2.5	229	600
BL1	0.95	1.10	–	0.40	0.40	0.70	–	0.035	–	0.035	1.20	1.60	–	–	–	0.40	–	–	–	0.20	–	0.05	–	–	–	–	229	735
BW2	0.95	1.1	–	0.30	–	0.35	–	0.035	–	0.035	–	0.15	–	0.10	–	0.20	–	–	–	0.20	–	0.05	0.15	0.35	–	–	207	790
Plastics moulding steel																												
BP20	0.28	0.40	0.40	0.60	0.65	0.95	–	0.025	–	0.025	1.50	1.80	0.35	0.55	–	0.40	–	–	–	0.25	–	0.05					‡	‡
BP30	0.26	0.34	–	0.40	0.45	0.70	–	0.025	–	0.025	1.10	1.40	0.20	0.35	3.9	4.3	–	–	–	0.20	–	0.05	–	–	–	–	‡	‡

* The hardness values for BH224/5 are for hardened and tempered steel.
† The maximum cobalt levels for BM1, BM2, BM4, BT1 and BT21 have been increased because of the level of cobalt in the scrap used for manufacture is increased.
‡ BP20 and BP30 are supplied to a wide range of hardness.

6.2.7 Flake (grey), cast irons

Cast iron is the name given to those ferrous metals containing more than 1.7% carbon. Since the maximum amount of carbon which can be held in solid solution as austenite (γ-phase) is 1.7%, there will be excess carbon in all cast irons. This can be either taken up by the iron as cementite (combined carbon) or precipitated out as free carbon in the form of graphite flakes (uncombined carbon).

Slow cooling results in coarse gains of ferrite and large flakes of graphite.

More rapid cooling results in both ferrite and pearlite being present together with finer and more uniformly dispersed flakes of graphite. This results in a stronger, tougher and harder cast iron.

Rapid cooling results in very fine flake graphite dispersed throughout a matrix of pearlite. This results in a further increase in strength and hardness.

It is the grey appearance of the freshly fractured surface of cast iron, resulting from the flake graphite, that gives ferritic and pearlitic cast irons the name *grey cast irons.*

Very rapid cooling and a reduction in silicon content results in all the carbon remaining combined as pearlite and cementite. Since no grey carbon is visible in the fractured surface, such cast iron is referred to as *white cast iron.* It is too hard and brittle to be of immediate use, but white iron castings are used as a basis for the malleable cast irons (see 6.2.11).

As well as iron and carbon the following elements are also present in cast irons:

Silicon This softens the cast iron by promoting the formation of uncombined carbon (graphite) at the expense of combined carbon (cementite). The silicon content is increased in small castings, which tend to cool rapidly, to promote the formation of ferrite and pearlite, and prevent the formation of excess cementite.

Phosphorus This is a residual impurity from the extraction process. Its presence causes embrittlement and hardness. However, its presence is desirable in complex, decorative castings, where strength and shock resistance is relatively unimportant, as it increases the fluidity of molten iron.

Sulphur This is also a residual impurity. It stabilizes the cementite and prevents the formation of flake graphite, thus hardening the iron. The presence of iron sulphide (FeS) causes embrittlement.

Manganese This is added in small quantities to neutralize the effects of the sulphur. It also refines the grain of the cast iron and so increases its strength. Since excess manganese stabilizes the cementite and promotes hardness, the manganese content must be balanced with the silicon content.

A typical composition for a grey cast iron could be as follows (see also Section 6.2.8):

Carbon	3.3%
Silicon	1.5%
Manganese	0.75%
Sulphur	0.05%
Phosphorus	0.5%
Iron	remainder

6.2.8 BS EN 1561: 1997 Grey cast irons

BS EN 1561 specifies the requirements of seven grades of grey cast iron. Unlike earlier standards it does not specify the composition or its processing in the foundry. BS EN 1561 specifies the properties, test conditions and quality control of the castings. How these are attained are left to the foundry in consultation with the customer. In addition the customer may specify or require:

(a) a mutually agreed chemical composition;
(b) casting tolerances, machining locations;
(c) test bars and/or test certificates;
(d) whether testing and inspection is to be carried out in the presence of the customer's representative;
(e) any other requirement such as hardness tests and their locations, non-destructive tests, and quality assurance.

The main properties of grey cast irons as specified in BS EN 1561 are given in the table on page 576. Note that for grey cast iron, hardness is not related to tensile strength but varies with casting section thickness and materials composition.

Tensile strength of grey cast irons

Material designation		Relevant wall thickness[a] (mm)		Tensile strength R_m[b] mandatory values (N/mm^2)		Tensile strength R_m[d] anticipated values in casting[e] (N/mm^2)
Symbol	Number	Over	Up to and including	In separately cast sample	In cast-on sample (min.)	(min.)
EN-GJL-**100**	EN-JL1010	5[f]	40	**100** to 200[g]	–	–
EN-GJL-**150**	EN-JL1020	2.5[f]	5	**150** to 250[g]	–	180
		5	10		–	155
		10	20		–	130
		20	40		120	110
		40	80		110	95
		80	150		100	80
		150	300		90[e]	–
EN-GJL-**200**	EN-JL1030	2.5[f]	5	**200** to 300[g]	–	230
		5	10		–	205
		10	20		–	180
		20	40		170	155
		40	80		150	130
		80	150		140	115
		150	300		130[e]	–
EN-GJL-**250**	EN-JL1040	5[f]	10	**250** to 350[g]	–	250
		10	20		–	225
		20	40		210	195
		40	80		190	170
		80	150		170	155
		150	300		160[e]	–
EN-GJL-**300**	EN-JL1050	10[f]	20	**300** to 400[g]	–	270
		20	40		250	240
		40	80		220	210
		80	150		210	195
		150	300		190[e]	–

EN-GJL-**350**	EN-JL1060	10[f]	20	**350** to 450[g]	–	315
		20	40		290	280
		40	80		260	250
		80	150		230	225
		150	300		210[e]	–

[a] If a cast-on sample is to be used the relevant wall thickness of the casting shall be agreed upon by the time of acceptance of the order.
[b] If by the time of acceptance of the order proving of the tensile strength has been agreed, the type of the sample is also to be stated on the order. If there is lack of agreement the type of sample is left to the discretion of the manufacturer.
[c] For the purpose of acceptance the tensile strength of a given grade shall be between its nominal value n (position 5 of the material symbol) and $(n + 100)$ N/mm^2.
[d] This column gives guidance to the likely variation in tensile strength for different casting wall thicknesses when a casting of simple shape and uniform wall thickness is cast in a given grey cast iron material. For castings of non-uniform wall thickness or castings containing cored holes, the table values are only an approximate guide to the likely tensile strength in different sections, and casting design should be based on the measured tensile strength in critical parts of the casting.
[e] These values are guideline values. They are not mandatory.
[f] This value is included as the lower limit of the relevant wall thickness range.
[g] The values relate to samples with an as-cast casting diameter of 30 mm, this corresponds to a relevant wall thickness of 15 mm.

Note:
1. 1 N/mm^2 is equivalent to 1 MPa.
2. For high damping capacity and thermal conductivity, EN-GJL-100 (EN-JL1010) is the most suitable material.
3. The material designation is in accordance with EN 1560.
4. The figures given in bold indicate the minimum tensile strength to which the symbol of the grade is related.

Brinell hardness of castings of grey cast iron, mandatory and anticipated values at the agreed test position

Material designation		Relevant wall thickness (mm)		Brinell hardness[a,b] (H_B 30)	
Symbol	Number	Over including	Up to and	min.	max.
EN-GJL-**HB155**	EN-JL2010	**40**[c]	**80**	–	**155**
		20	40	–	160
		10	20	–	170
		5	10	–	185
		2.5	5	–	210
EN-GJL-**HB175**	EN-JL2020	**40**[c]	**80**	**100**	**175**
		20	40	110	185
		10	20	125	205
		5	10	140	225
		2.5	5	170	260
EN-GJL-**HB195**	EN-JL2030	**40**[c]	**80**	**120**	**195**
		20	40	135	210
		10	20	150	230
		5	10	170	260
		4	5	190	275
EN-GJL-**HB215**	EN-JL2040	**40**[c]	**80**	**145**	**215**
		20	40	160	235
		10	20	180	255
		5	10	200	275
EN-GJL-**HB235**	EN-JL2050	**40**[c]	**80**	**165**	**235**
		20	40	180	255
		10	20	200	275
EN-GJL-**HB255**	EN-JL2060	**40**[c]	**80**	**185**	**255**
		20	40	200	275

[a] For each grade, Brinell hardness decreases with increasing wall thickness.
[b] By agreement between the manufacturer and the purchaser a narrower hardness range may be adopted at the agreed position on the casting, provided that this is not less than 40 Brinell hardness units. An example of such a circumstance could be castings for long series production.
[c] Reference relevant wall thickness for the grade.

Note:
1. Information on the relationship between Brinell hardness and tensile strength is indicated in Figure B.1 and the relationship between Brinell hardness and relevant wall thickness in Figure C.2 of Annexes B and C, respectively (see: BS EN 1561).
2. The material designation is in accordance with EN 1560.
3. The figures given in bold indicate the minimum and maximum Brinell hardness, to which the symbol of the grade is related and the corresponding reference relevant wall thickness range limits.

6.2.9 Malleable cast irons

BS EN 1562: 1997

BS EN 1562 specifies the requirements of malleable cast irons. The type of cast iron is indicated by the symbol thus:

EN-GJMW whiteheart malleable cast iron
EN-GJMB blackheart malleable cast iron

This initial letter is followed by a two-figure code designating the minimum tensile strength in MPa of a 12 mm diameter test piece. The test result is divided by ten to give the code. Finally, there are two figures representing the minimum elongation percentage on the specified gauge length.

Thus a complete designation of a malleable cast iron could be EN-GJMW-350-4 this is a whiteheart malleable cast iron with a minimum tensile strength of 350 N/mm^2 on a 12 mm diameter test piece, and a minimum elongation of 4%.

As for grey irons, the specification is not concerned with the composition of the iron. The composition and manufacturing processes are left to the discretion of the foundry in consultation with the customer.

The melt and the castings made from it will have satisfied the requirements of BS 6681 providing the test results and general quality of the castings meet the specifications laid down therein.

Whiteheart process

White iron castings are heated in contact with an oxidizing medium at about 1000°C for 70–100 h, depending upon the mass and the thickness of the castings. The carbon is drawn out of the castings and oxidized, leaving the castings with a ferritic structure at the surface and a pearlitic structure near the centre of the casting. There will be some residual rosettes of free graphite. Whiteheart castings behave more like steel castings but have the advantage of a much lower melting point and greater fluidity at the time of casting.

Applications: Wheel hubs, bicycle and motor cycle frame fittings; gas, water, and steam pipe fittings.

Blackheart process

White iron castings are heated out of contact with air at 850–950°C for 50–170 h, depending upon the mass and the thickness of the castings. Cementite breaks down into small rosettes of free graphite dispersed throughout a matrix of ferrite. This results in an increase in malleability, ductility, tensile strength and toughness.

Applications: Wheel hubs, brake drums, conduit fittings, control levers and pedals.

Pearlitic process

This is similar to the blackheart process but is accompanied by rapid cooling. This prevents the formation of ferrite and flake graphite and instead, results in some rosettes of graphite dispersed throughout a matrix of pearlite. This results in castings which are harder, tougher and with a higher tensile strength, but with reduced malleability and ductility.

Applications: Gears, couplings, camshafts, axle housings, differential housings and components.

Mechanical properties of whiteheart malleable cast irons

Material designation		Nominal diameter of test piece d (mm)	Tensile strength R_m (N/mm^2) min.	Elongation $A_{3.4}$ (%) min.	0.2% proof stress $R_{p0.2}$ (N/mm^2) min.	Brinell hardness (for information only) H_B max.
Symbol	Number					
EN-GJMW-**350-4**	EN-JM1010	6	270	10	–	230
		9	310	5	–	
		12	**350**	**4**	–	
		15	360	3	–	
EN-GJMW-**360-12**[b]	EN-JM1020[b]	6	280	16	–[a]	200
		9	320	15	170	
		12	**360**	**12**	**190**	
		15	370	7	200	
EN-GJMW-**400-5**	EN-JM1030	6	300	12	–[a]	220
		9	360	8	200	
		12	**400**	**5**	**220**	
		15	420	4	230	
EN-GJMW-**450-7**	EN-JM1040	6	330	12	–[a]	220
		9	400	10	230	
		12	**450**	**7**	**260**	
		15	480	4	280	
EN-GJMW-**550-4**	EN-JM1050	6	–	–	–[a]	250
		9	490	5	310	
		12	**550**	**4**	**340**	
		15	570	3	350	

[a] Due to the difficulty in determining the proof stress of small test pieces, the values and the method of measurement shall be agreed between the manufacturer and the purchaser at the time of acceptance of the order.
[b] Material most suitable for welding.

Note:
1. 1 N/mm^2 is equivalent to 1 MPa.
2. The material designation is in accordance with EN 1560.
3. The figures given in bold indicate the minimum tensile strength and minimum elongation $A_{3.4}$ to which the symbol of the grade is related, and the preferred nominal diameter of the test piece and the corresponding minimum 0.2% proof stress.

Mechanical properties of blackheart malleable cast irons

Material designation		Nominal diameter of test piece[a] d (mm)	Tensile strength R_m (N/mm^2) min.	Elongation $A_{3.4}$ (%) min.	0.2% proof stress $R_{p0.2}$ (N/mm^2) min.	Brinell hardness (for information only) H_B
Symbol	Number					
EN-GJMB-**300-6**[b]	EN-JM1110[b]	12 or 15	300	6	–	150 maximum
EN-GJMB-**350-10**	EN-JM1130	12 or 15	350	10	200	150 maximum
EN-GJMB-**450-6**	EN-JM1140	12 or 15	450	6	270	150 to 200
EN-GJMB-**500-5**	EN-JM1150	12 or 15	500	5	300	165 to 215
EN-GJMB-**550-4**	EN-JM1160	12 or 15	550	4	340	180 to 230
EN-GJMB-**600-3**	EN-JM1170	12 or 15	600	3	390	195 to 245
EN-GJMB-**650-2**	EN-JM1180	12 or 15	650	2	430	210 to 260
EN-GJMB-**700-2**	EN-JM1190	12 or 15	700	2	530	240 to 290
EN-GJMB-**800-1**	EN-JM1200	12 or 15	800	1	600	270 to 320

[a] Where a 6 mm diameter test piece is representative of the relevant wall thickness of a casting, this size of the test piece may be used by agreement between the manufacturer and the purchaser at the time of acceptance of the order. The minimum properties given in this table shall apply.
[b] Material intended particularly for applications in which pressure tightness is more important than a high degree of strength and ductility.

Note:
1. 1 N/mm^2 is equivalent to 1 MPa.
2. The material designation is in accordance with EN 1560.
3. The figures given in bold are related to the minimum tensile strength and minimum elongation $A_{3.4}$ of the grade.

6.2.10 Spheroidal graphite cast irons

Spheroidal graphite cast iron is also known as nodular cast iron, ductile cast iron, high-duty cast iron, etc.

The addition of magnesium or cerium to molten grey cast iron prevents the formation of flake graphite upon cooling and solidification. Instead, the uncombined carbon is distributed as fine spheroids throughout the mass of the casting. This results in a more homogeneous structure having greater strength and ductility and less susceptibility to fatigue failure.

BS EN 1563: 1997: Spheroidal graphite cast irons

BS EN 1563 specifies the requirements for spheroidal or nodular graphite cast irons. Again, the standard does not specify the chemical composition of the iron, its method of manufacture or any subsequent heat treatment. The standard is solely concerned with the properties, testing and quality control of the finished castings. How this is attained is left to the discretion of the foundry in consultation with the customer. It is a very comprehensive standard and it is only possible to review briefly some of its more important points within the scope of this chapter. The standard itself should be consulted for more detailed study.

Mechanical properties measured on test pieces machined from separately cast samples

Material designation		Tensile strength R_m (N/mm^2) min.	0.2% s proof stress $R_{p0.2}$ (N/mm^2) min.	Elongation A (%) min.
Symbol	Number			
EN-GJS-350-22-LT[a]	EN-JS1015	350	220	22
EN-GJS-350-22-RT[b]	EN-JS1014	350	220	22
EN-GJS-350-22	EN-JS1010	350	220	22
EN-GJS-400-18-LT[a]	EN-JS1025	400	240	18
EN-GJS-400-18-RT[b]	EN-JS1024	400	250	18
EN-GJS-400-18	EN-JS1020	400	250	18
EN-GJS-400-15	EN-JS1030	400	250	15
EN-GJS-450-10	EN-JS1040	450	310	10
EN-GJS-500-7	EN-JS1050	500	320	7
EN-GJS-600-3	EN-JS1060	600	370	3
EN-GJS-700-2	EN-JS1070	700	420	2
EN-GJS-800-2	EN-JS1080	800	480	2
EN-GJS-900-2	EN-JS1090	900	600	2

[a] LT for low temperature.
[b] RT for room temperature.

Note:
1. The values for these materials apply to castings cast in sand moulds of comparable thermal diffusivity. Subject to amendments to be agreed upon in the order, they can apply to castings obtained by alternative methods.
2. Whatever the method used for obtaining the castings, the grades are based on the mechanical properties measured on test pieces taken from samples separately cast in a sand mould or a mould of comparable thermal diffusivity.
3. 1 N/mm^2 is equivalent to 1 MPa.
4. The material designation is in accordance with EN 1560.

Minimum impact resistance values measured on V-notched test pieces machined from separately cast samples

Material designation		Minimum impact resistance values (in J)					
		At room temperature (23 ± 5) °C		At (−20 ± 2) °C		At (−40 ± 2) °C	
Symbol	Number	Mean value from 3 tests	Individual value	Mean value from 3 tests	Individual value	Mean value from 3 tests	Individual value
EN-GJS-350-22-LT[a]	EN-JS1015	–	–	–	–	12	9
EN-GJS-350-22-RT[b]	EN-JS1014	17	14	–	–	–	–
EN-GJS-400-18-LT[a]	EN-JS1025	–	–	12	9	–	–
EN-GJS-400-18-RT[b]	EN-JS1024	14	11	–	–	–	–

[a] LT for low temperature.
[b] RT for room temperature.

Note:

1. The values for these materials apply to castings cast in sand moulds of comparable thermal diffusivity. Subject to amendments to be agreed upon in the order, they can apply to castings obtained by alternative methods.
2. Whatever the method used for obtaining the castings, the grades are based on the mechanical properties on test pieces taken from samples separately cast in a sand mould or a mould of comparable thermal diffusivity.
3. The material designation is in accordance with EN 1560.

6.2.11 Alloy cast irons

The alloying elements in cast irons are similar to those in alloy steels:

Nickel is used for grain refinement, to add strength and to promote the formation of free graphite. Thus it toughens the casting.

Chromium stabilizes the combined carbon (cementite) present and thus increases the hardness and wear resistance of the casting. It also improves the corrosion resistance of the casting, particularly at elevated temperatures. As in alloy steels, nickel and chromium tend to be used together. This is because they have certain disadvantages when used separately which tend to offset their advantages. However, when used together the disadvantages are overcome whilst the advantages are retained.

Copper is used very sparingly as it is only slightly soluble in iron. However, it is useful in reducing the effects of atmospheric corrosion.

Vanadium is used in heat-resisting castings as it stabilizes the carbides and reduces their tendency to decompose at high temperatures.

Molybdenum dissolves in the ferrite and, when used in small amounts (0.5%), it improves the impact strength of the casting. It also prevents 'decay' at high temperatures in castings containing nickel and chromium. When molybdenum is added in larger amounts it forms double carbides, increases the hardness of castings with thick sections, and also promotes uniformity of the microstructure.

Martensitic cast irons contain between 4% and 6% nickel and approximately 1% chromium (e.g. Ni-hard cast iron). This is naturally martensitic in the cast state but, unlike alloys with rather less nickel and chromium, it does not need to be quench hardened, thus reducing the possibility of cracking and distortion. It is used for components which need to resist abrasion. It can only be machined by grinding.

Austenitic cast irons contain between 11% and 20% nickel and up to 5% chromium. These alloys are corrosion resistant, heat resistant, tough, and non-magnetic.

Since the melting temperatures of alloy cast irons can be substantially higher than those for common grey cast irons, care must be taken in the selection of moulding sands and the preparation of the surfaces of the moulds. Increased venting of the moulds is also required as the higher temperatures cause more rapid generation of steam and gases. The furnace and crucible linings must also be suitable for the higher temperatures and the inevitable increase in maintenance costs is also a significant factor when working with high alloy cast irons.

The *growth* of cast irons is caused by the breakdown of pearlitic cementite into ferrite and graphite at approximately 700°C. This causes an increase in volume. This increase in volume is further aggravated by hot gases penetrating the graphite cavities and oxidizing the ferrite grains. This volumetric growth causes warping and the setting up of internal stresses leading to cracking, particularly at the surface. Therefore, where castings are called upon to operate at elevated temperatures, alloy cast irons should be used. A low cost alloy is Silal which contains 5% silicon and a relatively low carbon content. The low carbon content results in a structure which is composed entirely of ferrite and graphite with no cementite present. Unfortunately Silal is rather brittle because of the high silicon content. A more expensive alloy is Nicrosilal. This is an austenitic nickel-chromium alloy which is much superior in all respects for use at elevated temperatures.

Three typical alloy cast irons are listed in Section 6.2.12 together with their properties, composition and some uses.

6.2.12 Composition, properties and uses of some typical cast irons

Type of iron	Composition (%)	Representative mechanical properties		
		Tensile strength (N/mm^2)	Hardness (Brinell)	Uses
Grey iron	3.30 C 1.90 Si 0.65 Mn 0.10 S 0.15 P	Strengths vary with sectional thickness but are generally in the range 150–350 N/mm^2.		Motor vehicle brake drums.
Grey iron	3.25 C 2.25 Si 0.65 Mn 0.10 S 0.15 P	Strengths vary with sectional thickness but are generally in the range 1 50–350 N/mm^2.	–	Motor vehicle cylinders and pistons.
Grey iron	3.25 C 1.25 Si 0.50 Mn 0.10 S 0.35 P		–	Heavy machine castings.
Phosphoric grey iron	3.60 C 1.75 Si 0.50 Mn 0.10 S 0.80 P		–	Light and medium water pipes.
Chromidium	3.20 C 2.10 Si 0.80 Mn 0.05 S 0.17 P 0.32 Cr	275	230	Cylinder blocks, brake drums, clutch casings, etc.
Wear and shock resistant	2.90 C 2.10 Si 0.70 Mn 0.05 S 0.10 P 1.75 Ni 0.10 Cr 0.80 Mo 0.15 Cu	450	300	Crankshafts for diesel and petrol engines (good strength, shock resistance and vibration damping capacity).
Ni-resist	2.90 C 2.10 Si 1.00 Mn 0.05 S 0.10 P 15.00 Ni 2.00 Cr 6.00 Cu	215	130	Pump castings handling concentrated chloride solutions: an austenitic corrosion resistant alloy.

6.3 Non-ferrous metals and alloys

6.3.1 Non-ferrous metals and alloys – introduction

Non-ferrous metals are all the known metals other than iron. Few of these metals are used in the pure state by engineers because of their relatively low strengths; two notable exceptions are copper and aluminium. Mostly they are used as the bases and alloying elements in both ferrous and non-ferrous alloys. Some non-ferrous metals are used for corrosion resistant coatings: for example, galvanized iron (zinc coated, low-carbon steel) and tinplate (tin coated, low-carbon steel).

It is not possible within the scope of this book to consider the composition and properties of the very large range of non-ferrous materials available. The following sections are, therefore, only an introduction to the composition and properties of some of the more widely used non-ferrous metals and alloys. For further information the wide range of British Standards relating to non-ferrous metals and alloys should be consulted, as should be the comprehensive manuals published by the metal manufacturers and their trade associations (e.g. Copper Development Association).

Only a limited number of non-ferrous alloys can be hardened by heat treatment. The majority can only be *work-hardened* by processing (e.g. cold rolling). Thus the *condition* of the metal, as the result of processing, has an important effect upon its properties, as will be shown in the following sections.

Other notable non-ferrous alloys, which are not included in this section, but which should be considered are:

Magnesium alloys (Elektron): used for ultra-lightweight castings.
Nickel alloys (Nimonic): high-temperature-resistant alloys, used in jet engines and gas turbines.
Zinc-based alloys (Mazak): used for pressure die-casting alloys.

6.3.2 High copper content alloys

Silver copper
The addition of 0.1% silver to high-conductivity copper raises the annealing temperature by 150°C with minimal loss of conductivity. This avoids hard drawn copper components softening when conductors are being soldered to them.

Cadmium copper
Like silver, cadmium has little effect upon the conductivity of the copper. Cadmium strengthens, toughens, and raises the tempering temperature of copper. Cadmium copper is widely used for medium- and low-voltage overhead conductors, overhead telephone and telegraph wires, and the overhead conductors for electrified railways. In the annealed condition it has high flexibility and is used for aircraft wiring where its ability to withstand vibration without failing in fatigue is an important safety factor.

Chromium copper
A typical alloy contains 0.5% chromium. It is one of the few non-ferrous alloys which can be heat treated. Thus it can be manipulated and machined in the ductile condition and subsequently hardened and strengthened by heating to 500°C for approximately 2 h. It has a relatively low conductivity compared with silver copper and cadmium copper.

Tellurium copper
The addition of 0.5% tellurium makes the copper as machineable as free-cutting brass whilst retaining its high conductivity. It also improves the very high corrosion resistance of copper. Tellurium copper is widely used in electrical machines and switchgear in hostile environments such as mines, quarries and chemical plants. The addition of traces nickel and silicon makes tellurium copper heat treatable, but with some loss of conductivity.

Beryllium copper

This is used where mechanical rather than electrical properties are required. Beryllium copper is softened by heating it to 800°C and quenching it in water. In this condition it is soft and ductile and capable of being extensively cold worked. It can be hardened by reheating to 300–320°C for upwards of 2 h. The resulting mechanical properties will depend on the extent of the processing (cold working) received prior to reheating. Beryllium copper is widely used for instrument springs, flexible metal bellows and corrugated diaphragms for aneroid barometers and altimeters, and for the Bourdon tubes in pressure gauges. Hand tools made from beryllium copper are almost as strong as those made from steel, but will not strike sparks from other metals or from flint stones. Thus tools made from beryllium copper alloy are used where there is a high risk of explosion (e.g. mines, oil refineries, oil rigs and chemical plants).

6.3.3 Wrought copper and copper alloys: condition code

O	Material in the annealed condition (soft)
M	Material in the 'as-manufactured' condition
¼H	Material with quarter-hard temper (due to cold working)
½H	Material with half-hard temper
H	Material with fully hard temper
EH	Material with extra-hard temper
SH	Material with spring-hard temper
ESH	Material with extra-spring-hard temper.

The above are listed in order of ascending hardness.

Materials which have acquired a hard temper due to cold working can have their hardness reduced and their ductility increased by heat treatment (e.g. annealing or solution treatment).

Chemical symbols used in the following tables:

Al	aluminium	Co	cobalt	Ni	nickel	Si	silicon
Ar	arsenic	Cr	chromium	P	phosphorus	Sn	tin
Be	beryllium	Cu	copper	Pb	lead	Te	tellerium
Bi	bismuth	Fe	iron	S	sulphur	Ti	titanium
C	calcium	Mg	magnesium	Sb	antimony	Zn	zinc
Cd	cadmium	Mn	manganese	Se	selenium	Zr	zirconium

6.3.4 British Standards relating copper and copper alloys

British Standard BS 2874 Copper and copper alloy rods and sections (chemical composition and mechanical properties) relating to copper and copper alloys is now obsolescent and should not be specified for new designs although the materials listed within this Standard will continue to be available and be used for some time to come.

The replacement standards are as follows. They are very detailed and comprehensive and it is not physically possible to reproduce them within the confines of this Pocket Book.

- BS EN 12163: 1998 – Copper and copper alloys (rod for general purposes).
- BS EN 12167: 1998 – Copper and copper alloys for profiles and rectangular bar for general purposes.
- BS EN 1652: 1997 – Copper and copper alloys for plate, sheet and circles for general purposes.
- BS EN 1653: 1998 – Copper and copper alloys for plate sheet and circles for boilers, pressure vessels and hot water storage units.
- BS EN 1982: 1998 – Copper and copper alloys for ingots and casting.

The following copper and copper alloy materials listed in Sections 6.3.5–6.3.26 inclusive are for guidance only.

6.3.5 Copper and copper alloy rods and sections

Chemical composition, tolerance group and mechanical properties of coppers

BS designation	Material	Material group for tolerances	Chemical composition (%)													Mechanical properties												Nearest ISO designation
			Cu	Sn	Pb	Fe	Ni	As	Sb	S	P	Se	Te	Bi	Total impurities	Condition	Size		Tensile strength (N/mm^2)						Elongation on $5.65\sqrt{S_o}$			
																	Over (mm)	Up to and Including (mm)	Round		Square and hexagonal		Rectangular		Round (%)	Square and hexagonal (%)	Rectangular (%)	
																			min.	max.	min.	max.	min.	max.	min.	min.	min.	
C 101	Electrolytic tough pitch high-conductivity copper	I	**99.90** min. (including Ag)	–	0.005	–	–	–	–	–	–	–	–	0.0010	0.03 (excluding O and Ag)	O	4	6.3	–	260	–	–	–	–	32	–	–	Cu-ETP
																	6.3	10	–	250	–	–	–	–	32	–	–	
																	10	12	–	240	–	240	–	240	40	40	40	
																	12	50	–	230	–	230	–	230	45	45	45	
																	50	80	–	230	–	230	–	–	45	45	–	
C 102	Fire refined high-conductivity copper	I	**99.90** min. (including Ag)	–	0.005	–	–	–	–	–	–	–	–	0.0025	0.04 (excluding O and Ag)	½H	4	6.3	290	–	–	–	–	–	4	–	–	Cu-FRHC
																	6.3	10	280	–	–	–	–	–	8	–	–	
																	10	12	260	–	260	–	250	–	12	12	12	
																	12	25	250	–	250	–	230	–	18	18	18	
																	25	50	230	–	230	–	230	–	22	22	18	
																	50	80	230	–	230	–	–	–	22	22	–	
C 103	Oxygen-free high-conductivity copper	I	**99.95** min. (including Ag)	–	0.005	–	–	–	–	–	–	–	–	0.0010	0.03 (excluding O and Ag)	H	4	6.3	350	–	–	–	–	–	–	–	–	Cu-OF
																	6.3	10	350	–	–	–	–	–	–	–	–	
																	10	12	320	–	310	–	270	–	6	6	6	
																	12	25	290	–	280	–	260	–	8	8	8	
																	25	50	260	–	250	–	250	–	12	12	10	
C 106	Phosphorus deoxidized non-arsenical copper	I	**99.85** min. (including Ag)	0.01	0.010	0.030	0.10	0.05	0.01	–	**0.013–0.050**	Se + Te 0.020–0.010		0.0030	0.06 (excluding Ag. As Ni and P)	O	6	–	210 min.						33			Cu-DHP
																M	6	–	230 min.						13			

Note: For essential alloying elements, limits are in bold type. Unless otherwise stated, figures in the total impurities column include those that are not in bold type. Unless otherwise indicated, all single limits are maxima.

Chemical composition, tolerance group and mechanical properties of alloyed coppers *(continued)*

BS designation	Material	Material group of tolerances	Chemical composition (%)															Mechanical properties						Nearest ISO designation
			Cu	Ni	P	Te	Cr	Co	Be	Zr	Bi	Fe	Sb	Si	S	Sn	Total impurities	Condition	Size		Tensile strength (N/mm^2)	0.2% proof stress (N/mm^2)	Elongation on $5.65\sqrt{S_o}$	
																			Over (mm)	Up to and including (mm)	min.	min.	min.	
C 109	Copper tellurium	I	**Rem.**	–	–	**0.30–0.70**	–	–	–	–	–	–	–	–	–	–	0.2	O	6	–	210	–	28	CuTe
																		M	6 50	50 –	260 240	– –	8 8	
C 111	Copper–sulphur	I	**Rem.**	–	–	–	–	–	–	–	–	–	–	–	**0.3–0.6**	–	0.2	O	6	–	210	–	28	CuS
																		M	6 50	50 –	260 240	– –	8 8	
C 112	Copper–cobalt–beryllium	II	**Rem.**	Ni + Fe 0.5	–	–	–	**2.0–2.8**	**0.4–0.7**	–	–	0.10	–	0.2	–	–	0.05 (excluding Fe, Ni and Si)	TH	–	–	690	–	9	CuCo2Be
C 113	Copper–nickel–phosphorus	II	**Rem.**	**0.8–1.2**	**0.16–0.25**	–	–	–	–	–	–	–	–	–	0.2	–	0.03 (excluding S)	TH	– 25	25 –	410 390	– –	18 20	
CC 101	Copper–chromium	II	**Rem.**	0.02	0.01	–	**0.3–1.4**	–	–	–	0.001	0.08	0.002	0.2	0.08	0.008	0.05 (excluding Fe, Si and S)	TH	– 25	25 –	410 370	– –	15 15	CuCr1
CC 102	Copper–chromium–zirconium	II	**Rem.**	0.02	0.01	–	**0.5–1.4**	–	–	**0.02–0.2**	0.001	0.08	0.002	0.2	–	0.008	0.05 (excluding Fe and Si)	TH	– 25	25 –	410 370	– –	15 15	CuCr1Zr

Note: For essential alloying elements, limits are in bold type. Unless otherwise stated, figures in the total impurities column include those that are not in bold type. Unless otherwise indicated, all single limits are maxima.

6.3.6 Wrought copper and copper alloys

Chemical composition and mechanical properties of brasses

BS designation	Material	Chemical composition (%)											Mechanical properties						Nearest ISO designation
		Cu	Sn	Pb	Fe	Al	Mn	As	Ni	Si	Zn	Total impurities	Condition	Size		Tensile strength (N/mm^2)	0.2% proof stress (N/mm^2)	Elongation on 5.65 $\sqrt{S_o}$ (%)	
														Over (mm)	Up to and including (mm)	min.	min.	min.	
CZ104	Leaded 80/20 brass	**79.0–81.0**	–	0.1–1.0	–	–	–	–	–	–	**Rem.**	0.6	M	6	40	310	–	22	
CZ109	Lead-free 60/40 brass	**59.0–62.0**	–	0.1	–	–	–	–	–	–	**Rem.**	0.3 (excluding Pb)	M	6	40	340	–	25	CuZn40
CZ112	Naval brass	**61.0–63.5**	1.0–1.4	–	–	–	–	–	–		**Rem.**	0.7	M	6	18	400	–	15	CuZn38Sn1
														18	40	350	–	20	
CZ114	High-tensile brass	**56.5–58.5**	0.2–0.8	0.5–1.5	0.3–1.0	1.5	0.6–2.0	–	–	–	**Rem.**	0.5 (excluding Al)	M	6	18	460	270	12	CuZn39 AlFeMn
														18	40	440	250	15	
														40	80	440	210	18	
													H	6	40	520	290	12	
CZ115	High-tensile brass (restricted aluminium)	**56.5–58.5**	0.2–0.8	0.5–1.5	0.3–1.0	0.1	0.5–2.0	–	–	–	**Rem.**	0.5	M	6	18	460	250	12	CuZn39 AlFeMn
														18	80	440	210	15	
													HS	6	40	520	290	12	
														40	60	500	240	14	
														60	80	450	210	18	
CZ116	High-tensile brass	**64.0–68.0**	–	–	0.25–1.2	4.0–5.0	0.3–2.0	–	–	–	**Rem.**	0.5	M	6	18	650	370	10	
														18	40	620	340	12	
														40	–	580	300	15	

CZ121 Pb3	Leaded brass 58% Cu 3% Pb	**56.5–** **58.5**	–	2.5 – 3.5	0.3	–	–	–	–	–	**Rem.**	0.7	M	6 18 40 80	18 40 80 –	425 400 380 350	– – – –	15 20 20 25	CuZn39 Pb3
CZ121 Pb4	Leaded brass 58% Cu 4% Pb	**56.5–** **58.5**	–	3.5 – 4.5	0.3	–	–	–	–	–	**Rem.**	0.7	M	6 18 40 80	18 40 80 –	425 400 380 350	– – – –	15 18 20 25	CuZn38 Pb4
CZ122	Leaded brass 58% Cu 2% Pb	**56.5–** **58.5**	–	1.5 – 2.5	0.3	–	–	–	–	–	**Rem.**	0.7	M	6 18 40 80	18 40 80 –	425 400 380 350	– – – –	18 22 25 25	CuZn40 Pb2

Note: For essential alloying elements, limits are in bold type. Unless otherwise stated, figures in the total impurities column include those that are not in bold type. Unless otherwise indicated, all angle limits are maxima.

continued

Chemical composition and mechanical properties of brasses (*continued*)

BS designation	Material	Chemical composition (%)					Mechanical properties									Nearest ISO designation
		Cu	Pb	Fe	Zn	Total impurities	Condition	Form	Size				Tensile strength (N/mm²)	0.2% proof stress (N/mm²)	Elongation on $5.65\sqrt{S_o}$ (%)	
									Thickness (mm)		Width or diameter or width A/F					
									Over	Up to and including	Over	Up to and including	min.	min.	min.	
CZ124	Leaded brass 62% Cu 3% Pb	**60.0–63.0**	2.5–3.7	0.3	**Rem.**	0.5 (excluding Fe)	M	Round and hexagonal rod	–	–	6.0	25	330	130	12	CuZn36 Pb3
							M				25	50	300	115	18	
							M				50	–	280	95	22	
									–	–						
									–	–						
							½H		–	–	6.0	12	400	160	6	
							½H		–	–	12	25	380	160	9	
							¼H		–	–	25	50	340	130	12	
							½H		–	–	50	–	310	95	18	
							H		–	–	3.0	5.0	550	290	–	
							H		–	–	5.0	8.0	480	220	3	
							M	Rectangles and squares	6.0	25	–	150	300	115	18	
							M		25	–	–	150	280	95	22	
							½H		6.0	12	–	25	340	160	9	
							½H		6.0	12	25	150	310	105	12	
							½H		12	50	–	50	310	105	18	
							½H		12	50	50	150	280	95	18	
							½H		50	–	50	100	280	95	18	

Note: For essential alloying elements, limits are in bold type. Unless otherwise stated, figures in the total impurities column include those that are not in bold type. Unless otherwise indicated, all single limits are maxima.

Section 6.3.6 (*continued*)

BS designation	Material	Chemical composition (%)											Mechanical properties						Nearest ISO designation
		Cu	Sn	Pb	Fe	Al	Mn	As	Ni	Si	Zn	Total impurities	Condition	Size (mm)		Tensile strength (N/mm^2)	0.2% proof stress (N/mm^2)	Elongation on $5.65\sqrt{S_o}$ (%)	
														Over	Up to and including	min.	min.	min.	
CZ128	Leaded brass 60% Cu 2% Pb	58.5–61.0	–	1.5–2.5	0.2	–	–	–	–	–	Rem.	0.5	M	6	18	380	–	22	CuZn38 Pb2
														18	40	380	–	22	
														40	80	350	–	25	
														80	–	350	–	25	
CZ129	Leaded brass 60% Cu 1% Pb	58.5–61.0	–	0.8–1.5	0.2	–	–	–	–	–	Rem.	0.5	M	6	18	380	–	25	CuZn39 Pb1
														18	40	380	–	25	
														40	80	350	–	28	
														80	–	350	–	28	
CZ130	Leaded brass for sections	55.5–57.5	–	2.5–3.5	–	0.5	–	–	–		Rem.	0.7 (excluding Al)	M	6	–	350	–	20	CuZn43 Pb2
CZ131	Leaded brass 62% Cu 2% Pb	61.0–63.0	–	1.5–2.5	0.2	–	–	–	–	–	Rem.	0.5	M	6	18	350	–	22	CuZn37 Pb2
														18	40	350	–	25	
														40	80	350	–	28	
														80	–	330	–	28	
CZ132	Dezincification resistant brass	Rem.	0.2	1.7–2.8	0.2	–	–	0.08–0.15	–	–	35.0–37.0	0.5	O	6	–	280	–	30	
													M	6	18	380	–	20	
														18	40	350	–	22	
														40	80	350	–	25	

continued

Section 6.3.6 *(continued)*

BS designation	Material	Chemical composition (%)											Mechanical properties						Nearest ISO designation
		Cu	Sn	Pb	Fe	Al	Mn	As	Ni	SI	Zn	Total impurities	Condition	Size (mm)		Tensile strength (N/mm^2)	0.2% proof stress (N/mm^2)	Elongation on $5.65\sqrt{S_o}$ (%)	
														Over	Up to and including	min.	min.	min.	
CZ133	Naval brass (uninhibited)	59.0–62.0	0.50–1.0	0.20	0.10	–	–	–	–	–	Rem.	0.4	M	6 18	18 40	400 350	170 150	20 25	
CZ134	Naval brass (high leaded)	59.0–62.0	0.50–1.0	1.3–2.2	0.10	–	–	–	–	–	Rem.	0.2	M	6 18	18 40	400 350	170 150	15 20	
CZ135	High-tensile brass with silicon)	57.0–60.0	0.3	0.8	0.5	1.0–2.0	1.5–3.5	–	0.2	0.3–1.3	Rem.	0.5 (excluding Sn, Pb, Fe and Ni)	M	6	40	550	270	12	CuZn37 Mn3 Al2Si
CZ136	Manganese brass	56.0–59.0	–	3.0	–	–	0.5–1.5	–	–	–	Rem.	0.7 (excluding Pb)	M	6 18	18 40	380 350	– –	20 25	
CZ137	Leaded brass 60% Cu 0.5% Pb	58.5–61.0	–	0.3–0.8	0.2	–	–	–	–	–	Rem.	0.5	M	6 40	40 –	380 350	– –	25 28	CuZn40Pb
NS101	Leaded 10% nickel brass	44.0–47.0	–	1.0–2.5	0.4	–	0.2–0.5	–	9.0–11.0	–	Rem.	0.3 (excluding Fe)	M	6	80	460	–	8	CuNi10 Zn42Pb2

Note: For essential alloying elements, limits are in bold type. Unless otherwise stated, figures in the total impurities column include those that are not in bold type. Unless otherwise indicated, all single limits are maxima.

6.3.7 Wrought copper and copper alloys

Chemical composition and mechanical properties of bronze alloys

BS designation	Material	Chemical composition (%)													Mechanical properties						Nearest ISO designation
		Cu	Sn	Pb	Fe	Al	Mn	P	Ni	Si	Zn	S	C	Total impurities	Condition	Size (mm)		Tensile strength (N/mm²)	0.2% proof stress (N/mm²)	Elongation on $5.65\sqrt{S_o}$ (%)	
																Over	Up to and including	min.	min.	min.	
CA104	10% aluminium bronze (copper–aluminium–iron-nickel)	Rem.	0.10	0.05	4.0–5.5	8.5–11.0	0.50	–	4.0–5.5	0.2	0.40	–	–	0.5 (excluding Mn)	M	6 18 80	18 80 –	700 700 650	400 370 320	10 12 12	CuAl10Ni5Fe4
CA107	Copper–aluminium–silicon	Rem.	0.10	0.05*	0.5–0.7	6.0–6.4	0.10	–	0.10	2.0–2.4	0.40	–	–	0.5	M	6 40	40 –	520 520	270 230	20 20	CuAl7Si2
															½H	6 18 40	18 40 –	630 600 550	350 310 250	10 12 15	
CN102	90/10 copper–nickel iron	Rem.	–	0.01	1.00–2.00	–	0.50–1.00	–	10.0–11.0	–	–	0.05	0.05	0.30	M	6	–	280	–	27	CuNi10Fe1Mn
CN107	70/30 copper–nickel iron	Rem.	–	0.01	0.40–1.00	–	0.50–1.50	–	30.0–32.0	–	–	0.08	0.06	0.30	M	6	–	310	–	27	CuNi30Mn1Fe

continued

Chemical composition and mechanical properties of bronze alloys *(continued)*

BS design-ation	Material	Chemical composition (%)													Mechanical properties						Nearest ISO designstion
		Cu	Sn	Pb	Fe	Al	Mn	P	Ni	SI	Zn	S	C	Total impurities	Con-dition	Size (mm)		Tensile strength (N/mm^2)	0.2% proof stress (N/mm^2)	Elongation on $5.65\sqrt{S_o}$ (%)	
																Over	Up to and including	min.	min.	min.	
CS101	Copper–silicon	Rem.	–	–	0.25	–	0.75–1.25	–	–	2.75–3.25	–	–	–	0.5 (excluding Fe)	M	6 18 40	18 40 80	470 400 380	– – –	15 20 25	
PB102	5% phosphor bronze (copper–tin–phosphorus)	Rem.	4.0–5.5	0.02	0.1	–	–	0.02–0.40	0.3	–	0.30	–	–	0.5	M	6 18 40 60 80 100 120	18 40 60 80 100 120 –	500 460 380 350 320 280 260	410 380 320 250 200 120 80	12 12 16 18 20 22 25	CuSn5
PB104	8% phosphor bronze (copper–tin–phosphorus)	Rem.	7.5–9.0	0.05	0.1	–	–	0.02–0.40	0.3	–	0.30	–	–	0.3	M	6 18 40	18 40 80	550 500 450	400 360 300	15 18 20	CuSn8

Note: For essential alloying elements, limits are in bold type. Unless otherwise stated, figures in the total impurities column include those that are not in bold type. Unless otherwise indicated, all single limits are maxima.

* For welding, lead to be 0.01 max.

6.3.8 Copper sheet, strip and foil

Designation	Material	Copper including silver (%)	Tin (%)	Lead (%)	Iron (%)	Nickel (%)	Arsenic (%)	Antimony (%)	Bismuth (%)	Phosphorus (%)	Oxygen (%)	Selenium (%)	Tellurium (%)	Total impurities (%)	Complies with or falls within ISO
C101	Electrolytic tough pitch high-conductivity copper	**99.90** (min.)	–	0.005	–	–	–	–	0.001	–	–	–	–	0.03 (excluding oxygen and silver)	Cu-ETP (ISO 1337)
C102	Fine-refined tough pitch high-conductivity copper	**99.90** (min.)	–	0.005	–	–	–	–	0.0025	–	–	–	–	0.04 (excluding oxygen and silver)	Cu-FRHC (ISO 1337)
C103	Oxygen-free high-conductivity copper	**99.95**	–	0.005	–	–	–	–	0.001	–	–	–	–	(excluding oxygen and silver)	Cu-OF (ISO 1337)
C104	Tough pitch non-arsenical copper	**99.85**	0.01	0.01	0.01	0.05	0.02	0.005	0.003	–	0.10	0.02	0.01	0.05 (excluding nickel, oxygen and silver)	Cu-FRTP (ISO 1337)
C106	Phosphorus deoxidized non-arsenical copper	**99.85**	0.01	0.01	0.03	0.10	0.05	0.01	0.003	**0.013/ 0.050**	–	Se + Te 0.20	0.01	0.06 (excluding silver, arsenic, nickel and phosphorus)	Cu-DHP (ISO 1337)

These properties are common to C101, C102, C103, C104 and C106 as listed above

Condition	Thickness		Tensile strength		Elongation on 50 mm (min.) (%)	Hardness HV	Bend test			
							Transverse bend		Longitudinal bend	
	Over (mm)	Up to and including (mm)	Up to and including 450 mm wide (min.) (N/mm²)	Over 450 mm wide (min.) (N/mm²)			Angle degrees	Radius	Angle degrees	Radius
O	0.5	10.0	210	210	35	55 (max.)	180	Close	180	Close
M	3.0	10.0	210	210	35	65 (max.)	180	Close	180	Close
½H	0.5	2.0	240	240	10	70/95	180	*t*	180	*t*
	2.0	10.0	240	240	15	70/95	180	*t*	180	*t*
	0.5	2.0	310	280	–	90 min.	90	*t*	90	*t*
H	2.0	10.0	290	280	–	–	–	–	–	–

Based on BS 2870:1980, which should be consulted for full information.
For essential alloying elements, limits are in **bold type**. Unless otherwise stated, figures in total impurities column include those in lighter type. Unless otherwise indicated, all limits are maxima.
Note: N/mm^2 = MPa.

6.3.9 Brass sheet, strip and foil: binary alloys of copper and zinc

For designations CZ125, CZ101, CZ102, CZ103.

Designation	Material	Copper (%)	Lead (%)	Iron (%)	Zinc (%)	Total impurities (%)	Condition	Thickness		Tensile strength	
								Over (mm)	Up to and including (mm)	Up to and including 450 mm wide (min.) (N/mm²)	Over 450 mm wide (min.) (N/mm²)
CZ125	Cap copper	**95.0/98.0**	0.02	0.05	**Rem.**	0.025	O	–	10	–	–
C2101	90/10	**89.0/91.0**	0.05	0.10	**Rem.**	0.40	O	–	10.0	245	245
							½H	–	3.5	310	380
	Brass						½H	3.5	10.0		
							H	–	10.0	350	325
CZ102	85/15 Brass	**84.0/86.0**	0.05	0.10	**Rem.**	0.40	O	–	10.0	245	245
							½H	–	3.5	325	295
							½H	3.5	10.0		
							H	–	10.0	370	340
CZ103	80/20 brass	**79.0/81.0**	0.05	0.10	**Rem.**	0.40	O	–	10.0	265	265
							½H	–	3.5	340	310
							½H	3.5	10.0		
							H	–	10.0	400	370

Further properties for the materials listed above

Designation	Elongation on 50 mm min. (%)	Vickers hardness (H_V) Up to and including 450 mm wide		Over 450 mm wide		Bend test: Transverse bend		Longitudinal bend		Complies with or falls within ISO
		min.	max.	min.	max.	Angle (degree)	Radius	Angle (degree)	Radius	
CZ125	–	–	75	–	75	180	Close	180	Close	–
CZ101	35	–	75	–	75	180	Close	180	Close	ISO 426/1 Cu Zn 10
	7	95	–	85	–	180	Close	180	Close	
						180	t	180	t	
	3	110	–	100	–	90	$2t$	90	t	
CZ102	35	–	75	–	75	180	Close	180	Close	ISO 426/1 Cu Zn 15
	7	95	–	85	–	180	Close	180	Close	
						180	t	180	t	
	3	110	–	100	–	90	$2t$	90	t	
CZ103	40	–	80	–	80	180	Close	180	Close	ISO 426/1 Cu Zn 20
	10	95	–	85	–	180	Close	180	Close	
						180	t	180	t	
	5	120	–	110	–	90	$2t$	90	t	

For designations CZ106, CZ107, CZ108.

Designation	Material	Copper (%)	Lead (%)	Iron (%)	Zinc (%)	Total impurities (%)	Condition	Thickness		Tensile strength	
								Over (mm)	Up to and including (mm)	Up to and including 450 mm wide (min.) (N/mm²)	Over 450 mm wide (min.) (N/mm²)
CZ106	70/30 Cartridge brass	**68.5/71.5**	0.05	0.05	**Rem.**	0.30	O	–	10.0	280	280
							¼H	–	10.0	325	325
							½H	–	3.5		
							½H	3.5	10.0	350	340
							H	–	10.0	415	385
CZ107	2/1 brass	**64.0/67.0**	0.10	0.10	**Rem.**	0.40	O	–	10.0	280	280
							¼H	–	10.0	340	325
							½H	–	3.5		
							½H	3.5	10.0	385	350
							H	–	10.0	460	415
							EH	–	10.0	525	–
CZ108	Common brass	**62.0/65.0**	0.30	0.20	**Rem.**	0.50 (excluding lead)	O	–	10.0	280	280
							¼H	–	10.0	340	325
							½H	–	3.5		
							½H	3.5	10.0	358	350
							H	–	10.0	460	415
							EH	–	10.0	525	–

Further properties for the materials listed above

Designation	Elongation on 50 mm min. (%)	Vickers hardness (H_V) Up to and including 450 mm wide		Over 450 mm wide		Bend test Transverse bend		Longitudinal bend		Complies with or falls within ISO
		min.	max.	min.	max.	Angle (degrees)	Radius	Angle (degrees)	Radius	
CZ106	50	–	80	–	80	180	Close	180	Close	ISO 426/1
						180	Close	180	Close	Cu Zn 30
	35	75	–	75	–	180	Close	180	Close	
	20	100	–	95	–	180	*t*	180	*t*	
	5	125	–	120	–	90	2*t*	90	*t*	
CZ107	40	–	80	–	80	180	Close	180	Close	ISO 426/1
	30	75	–	75	–	180	Close	180	Close	Cu Zn 33
						180	Close	180	Close	
	15	110	–	100	–	180	*t*	180	*t*	
	5	135	–	125	–	90	2*t*	90	*t*	
	–	165	–	–	–	–		90	2*t*	
CZ108		1	2	3	4	5	6	7	8	ISO 426/1 Cu Zn 37

Based on BS 2870: 1980, which should be consulted for full information.
For essential alloying elements, limits are in **bold type**. Unless otherwise stated, figures in total impurities column include those in lighter type. Unless otherwise indicated, all limits are maxima.
Note: N/mm^2 = MPa.

6.3.10 Brass sheet, strip and foil: special alloys and leaded brasses

Designation	Material	Copper (%)	Tin (%)	Lead (%)	Iron (%)	Zinc (%)	Aluminium (%)	Arsenic (%)	Total impurities (%)	Condition	Thickness up to and including (mm)
CZ110	Aluminium brass	**76.0/78.0**	–	**0.04**	**0.06**	**Rem.**	**1.80/2.30**	**0.02/0.05**	0.30	M O	10.0 10.0
CZ112	Naval brass	**61.0/63.5**	**1.0/1.4**	–	–	**Rem.**	–	–	0.75	M or O H	10.0 10.0
CZ118	Leaded brass 64% Cu 1% Pb	**63.0/66.0**	–	**0.75/1.5**	–	**Rem.**	–	–	0.30	½H H EH	6.0 6.0 6.0
CZ119	Leaded brass 62% Cu 2% Pb	**61.0/64.0**	–	**1.0/2.5**	–	**Rem.**	–	–	0.30	½H H EH	6.0 6.0 6.0
CZ120	Leaded brass 59% Cu 2% Pb	**58.0/60.0**	–	**1.5/2.5**	–	**Rem.**	–	–	0.30	½H H EH	6.0 6.0 6.0
CZ123	60/40 low lead brass: stamping brass	**59.0/62.0**	–	**0.3/0.8**	–	**Rem.**	–	–	0.30	M	10.0

Properties for the materials listed above

Designation	Tensile strength (min.) (N/mm^2)	Elongation on 50 mm (min.) (%)	Vickers hardness (H_V)		Bend test				Complies with or falls within ISO
					Transverse bend		Longitudinal bend		
			min.	max.	Angle (degrees)	Radius	Angle (degrees)	Radius	
CZ110	340 310	45 50	– –	– 80	– 180	– Close	– 180	– Close	ISO 462/2 Cu Zn 20/A12
CZ112	340 400	25 20	– –	– –	180 –	*t* –	180 90	*t* *t*	–
CZ118	370 430 510	10 5 3	110 140 165	140 165 190	– – –	– – –	– – –	– – –	–
CZ119	370 430 510	10 5 3	110 140 165	140 165 190	– – –	– – –	– – –	– – –	–
CZ120	– 510 575	10 5 3	110 140 165	140 165 190	– – –	– – –	– – –	– – –	ISO 426/2 Cu Zn 39/Pb 2
CZ123	370	–	20	–	–	–	–	–	ISO 426/2 Cu Zn 40/Pb

Based on BS 2870: 1980, which should be consulted for full information.
For essential alloying elements, limits are in **bold type**. Unless otherwise stated, figures in total impurities column include those in lighter type. Unless otherwise indicated, all limits are maxima.
Note: N/mm^2 = MPa.

6.3.11 Phosphor bronze sheet, strip and foil

Designation	Material	Copper (%)	Tin (%)	Lead (%)	Phosphorus (%)	Zinc (%)	Total impurities (%)	Condition	Thickness up to and including (mm)	Tensile strength	
										Up to and including 450 mm wide min. (N/mm²)	Over 450 mm wide min. (N/mm²)
PB101	4% phosphor bronze (copper–tin–phosphorus)	**Rem.**	**3.5/4.5**	0.2	**0.02/0.40**	0.30	0.50	O	10.0	295	259
								¼H	10.0	340	340
								½H	10.0	430	400
								H	6.0	510	495
								EH	6.0	620	–
PB102	5% phosphor bronze (copper–tin–phosphorus)	**Rem.**	**4.5/5.5**	0.02	**0.02/0.40**	0.30	0.50	O	10.0	310	310
								¼H	10.0	350	350
								½H	10.0	495	460
								H	6.0	570	525
								EH	6.0	645	–
								SH	0.9	–	–
PB103	7% phosphor bronze (copper–tin–phosphorus)	**Rem.**	**5.5/7.5**	0.02	**0.02/0.40**	0.30	0.50	O	10.0	340	340
								¼H	10.0	385	385
								½H	10.0	525	460
								H	6.0	620	540
								EH	6.0	695	–
								SH	0.9	–	–
								ESH	0.6	–	–

Further properties for the materials listed above

Designation	0.2% proof stress		Elongation on 50 mm (%)	Vickers hardness (H_V)				Bend test				Complies with or falls within ISO
	Up to and including 450 mm wide min. (N/mm²)	Over 450 mm wide min. (N/mm²)		Up to and including 450 mm wide		Over 450 mm wide		Transverse bend		Longitudinal bend		
				min.	max.	min.	max.	Angle (degrees)	Radius	Angle (degrees)	Radius	
PB101	–	–	40	–	80	–	80	180	Close	180	Close	ISO 427 Cu Sn 4
	125	125	30	100	–	100	–	180	Close	180	Close	
	390	340	8	150	–	130	–	90	*t*	180	*t*	
	480	435	4	180	–	150	–	–	–	90	*t*	
	580	–	–	180	–	–	–	–	–	90	*t*	
PB102	–	–	45	–	85	–	85	180	Close	180	Close	
	140	140	35	110	–	110	–	180	Close	180	Close	
	420	385	10	160	–	140	–	90	*t*	180	*t*	ISO 427
	520	480	4	180	–	160	–	–	–	90	*t*	Cu Sn 4
	615	–	–	200	–	–	–	–	–	90	*t*	
		–	–	215	200	–	–	–	–	90	*t*	
PB103	–	–	50	–	90	–	90	180	Close	180	Close	
	200	200	40	115	–	113	–	180	Close	180	Close	
	440	380	12	170	–	150	–	90	*t*	180	*t*	ISO 427
	550	480	6	200	–	165	–	–	–	90	*t*	Cu Sn 6
	650	–	–	215	–	–	–	–	–	90	*t*	
		–	–	220*	240*	–	–	–	–	90	*t*	
		–	–	220*	–	–	–	–	–	–	–	

Based on BS 2870: 1980, which should be consulted for full information.
For essential alloying elements, limits are in **bold type**. Unless otherwise stated, figures in total impurities column include those in lighter type. Unless otherwise indicated all limits are maxima.
Note: N/mm² = MPa.
* Up to 150 mm wide only.

6.3.12 Aluminium bronze alloys – introduction

Despite their name, these are *copper*-based alloys containing up to 10% (nominal) aluminium. They combine relatively high strength with excellent corrosion resistance at high temperatures. These alloys can be grouped into two categories.

Single-phase alloys

The single phase, or α-alloys, contain up to 5% aluminium. They are highly ductile and corrosion resistant. As the colour of the α-alloys resembles 18 carat gold, this alloy is widely used in the manufacture of costume jewellery. This range of alloys are also widely used in engineering, particularly for pipework where corrosion resistance at high temperatures is required.

Duplex-phase alloys

These alloys contain approximately 10% aluminium and can be heat treated in a similar manner to plain carbon steels. Heating followed by slow cooling anneals this alloy giving a primary α-phase (analogous to ferrite) in a eutectoid matrix of $\alpha + \gamma_2$ (analogous to pearlite). This results in the alloy becoming relatively soft and ductile. Heating followed by rapid cooling (quenching) produces a hard β' structure (analogous to martensite). An example of the composition and properties of a typical duplex-phase alloy is given in the following table.

The duplex-phase alloys are also used for both sand- and die-casting where high strength combined with corrosion resistance and pressure tightness is required. However, these alloys are not easy to cast as the aluminium content tends to oxidize at their relatively high melting temperatures. Special precautions have to be taken to overcome this difficulty and these increase the cost of the process.

6.3.13 Aluminium bronze sheet, strip and foil

Designation	Material	Copper (%)	Tin (%)	Lead (%)	Iron (%)	Nickel (%)	Zinc (%)	Aluminium (%)	Silicon (%)
CA104	10% aluminium bronze (copper–aluminium–nickel–iron)	**Rem.**	0.10	0.05	**4.0/6.0**	**4.0/6.0**	0.40	**8.5/11.0**	0.10

Table continued from above

Manganese (%)	Magnesium (%)	Total impurities (%)	Condition	Size up to and including (mm)	Tensile strength (min.) (N/mm^2)	0.2% proof stress (min.) (N/mm^2)	Elongation on 50 mm (min.) (%)	Complies with or falls within ISO
0.50	0.05	0.05 (excluding Mn)	M	10	700	380	10	ISO 428 Cu Al 10 Fe 5 Ni 5

Based on BS 2870: 1980, which should be consulted for full information.
For essential alloying elements, limits are in **bold type**. Unless otherwise stated, figures in total impurities column include those in lighter type. Unless otherwise indicated, all limits are maxima.
Note: N/mm^2 = MPa.

6.3.14 Copper–nickel (cupro–nickel) sheet, strip and foil

Designation	Material	Copper (%)	Zinc (%)	Lead (%)	Iron (%)	Nickel (%)	Manganese (%)	Sulphur (%)	Carbon (%)	Total impurities (%)	Condition
CN102	90/10 copper–nickel–iron	**Rem.**	–	0.01	**1.00/2.00**	**10.0/11.0**	**0.50/1.00**	0.05	0.05	0.30	M O
CN104	80/20 copper–nickel	**79.0/81.0**	–	0.01	0.30	**19.0/21.0**	**0.05/0.50**	0.02	0.05	0.10	O O
CN105	75/25 copper–nickel	**Rem.**	0.20	–	0.30	**24.0/26.0**	**0.05/0.40**	0.02	0.05	0.35	O O H
CN107	70/30 copper–nickel–iron	**Rem.**	–	0.01	**0.40/1.00**	**30.0/32.0**	**0.05/1.50**	0.08	0.06	0.30	O O

Table continued from above

Designation	Thickness		Tensile strength min. (N/mm²)	Elongation on 50 mm (%)	Bend test: longitudinal and transverse bend		Vickers hardness (H_V)	Complies with or falls within ISO
	Over (mm)	Up to and including min. (mm)			Angle (degrees)	Radius		
CN102	–	10.0	310	30	–	–	90 (max.)	ISO 429 Cu Ni 10 Fe 1 Mn
	–	10.0	280	40	–	–		
CN104	0.6	2.0	310	35	180	Close	–	ISO 429 Cu Ni 20
	2.0	10.0	310	38	180	Close		
CN105	0.6	2.0	340	30	180	Close	–	ISO 429 Cu Ni 25
	2.0	10.0	340	35	180	Close	–	
	0.6	10.0	–	–	–	–	155 (min.)	
CN107	0.6	2.0	370	30	180	Close	–	ISO 429 Cu Ni 30 Mn 1 Fe
	2.0	10.0	370	35	180	Close		

Based on BS 2870: 1980, which should be consulted for full information.
For essential alloying elements, limits are in **bold type**. Unless otherwise stated, figures in total impurities column include those in lighter type. Unless otherwise indicated, all limits are maxima.
Note: N/mm^2 = MPa.

6.3.15 Nickel–silver sheet, strip and foil

Designation	Material	Copper (%)	Lead (%)	Iron (%)	Nickel (%)	Zinc (%)	Manganese (%)	Total impurities (%)	Condition
NS103	10% nickel–silver (copper–nickel–zinc)	**60.0/65.0**	0.04	0.25	**9.0/11.0**	**Rem.**	**0.05/0.30**	0.50	O ½H H EH
NS104*	12% nickel–silver (copper–nickel–zinc)	**60.0/65.0**	0.04	0.25	**11.0/13.0**	**Rem.**	**0.05/0.30**	0.50	O ½H H EH
NS105	15% nickel–silver (copper–nickel–zinc)	**60.0/65.0**	0.04	0.30	**14.0/16.0**	**Rem.**	**0.05/0.50**	0.50	O ½H H EH
NS106	18% nickel–silver (copper–nickel–zinc)	**60.0/65.0**	0.03	0.30	**17.0/19.0**	**Rem.**	**0.05/0.50**	0.50	O ½H H EH
NS107*	18% nickel–silver (copper–nickel–zinc)	**54.0/56.0**	0.03	0.30	**17.0/19.0**	**Rem.**	**0.05/0.35**	0.50	–
NS111	Lead 10% nickel–silver (copper–nickel–zinc–lead)	**58.0/63.0**	**1.0/2.0**	–	**9.0/11.0**	**Rem.**	**0.10/0.50**	0.50	O ½H H

Table continued from above

Designation	Bend test				Vickers hardness (H_V)		Thickness up to and including (mm)	Complies with or falls within ISO
	Transverse bend		Longitudinal bend					
	Angle degrees	Radius	Angle degrees	Radius	min.	max.		
NS103	180	t	180	t	–	100	10.0	–
	180	t	180	t	125	–	10.0	
	90	t	90	t	160	–	10.0	
	–	–	90	t	185	–	10.0	
NS104*	180	t	180	t	–	100	10.0	–
	180	t	180	t	130	–	10.0	
	90	t	90	t	160	–	10.0	
	–	–	90	t	190	–	10.0	
NS105	180	t	180	t	–	105	10.0	–
	180	t	180	t	135	–	10.0	
	90	t	90	t	165	–	10.0	
	–	–	90	t	195	–	10.0	
NS106	180	t	180	t	–	110	10.0	ISO 430 Cu Ni 18 Zn 20
	180	t	180	t	135	–	10.0	
	90	t	90	t	170	–	10.0	
	–	–	90	t	200	–	10.0	
NS107*	–	–	–	–	–	–	–	ISO 430 Cu Ni 18 Zn 27
NS111	–	–	–	–	–	100	–	–
	–	–	–	–	150	180	–	
	–	–	–	–	160	–	–	

Based on BS 2870: 1980, which should be consulted for full information.
For essential alloying elements, limits are in **bold type**. Unless otherwise stated, figures in total impurities column include those in lighter type. Unless otherwise indicated, all limits are maxima.
* For special requirements relevant to particular applications, see BS 2870: 1980, Section 4.

6.3.16 (a) Miscellaneous wrought copper alloys

Designation[a,c]	C105	C107	C108	C109	CZ109	CZ111	CZ114	CZ115	CZ122	CZ127	CZ132
Material[b]	**Tough pitch arsenical copper**	**Phosphorus deoxidized arsenical copper**	**Cadmium copper**	**Tellurium copper**	**Lead-free 60/40 brass**	**Admiralty brass**	**High-tensile brass**	**High-tensile brass (soldering quality)**	**Leaded brass 58% Cu 2% Pb**	**Aluminium–nickel–silicon brass**	**Dezincification-resistant brass**
Availability[c]	**P**	**P, T**	**W, P**	**RS**	**FS, RS**	**T**	**FS, RS**	**FS, RS**	**FS, RS**	**T**	**FS, RS**
Copper (%)	**99.20 (min)**	**99.20 (min)**	**Rem.**	**Rem.**	**59.0/62.0**	**70.0/73.0**	**56.0/60.0**	**56.0/59.0**	**56.5/60.0**	**81.0/86.0**	**Rem.**
Tin (%)	0.03	0.01	–	–	–	**1.0/1.5**	**0.2/1.0**	**0.6/1.1**	–	0.10	0.20
Lead (%)	0.02	0.01	–	–	0.10	0.7	**0.5/1.5**	**0.5/1.5**	**1.0/2.5**	0.05	**1.7/2.8**
Iron (%)	0.02	0.03	–	–	–	0.6	**0.5/1.2**	**0.5/1.2**	0.30	0.25	0.30
Nickel (%)	0.15	0.15	–	–	–	–	–	–	–	**0.80/1.40**	**35.0/37.0**
Zinc (%)	–	–	–	–	**Rem.**	**Rem.**	**Rem.**	**Rem.**	**Rem.**	**Rem.**	**0.08/0.15**
Arsenic (%)	**0.3/0.5**	**0.3/0.5**	–	–	–	**0.02/0.06**	–	–	–	–	–
Antimony (%)	0.01	0.01	–	–	0.021	–	0.020	–	0.020	–	–
Aluminium (%)	–	–	–	–	–	–	1.5	0.2	–	**0.70/1.20**	–
Bismuth (%)	0.005	0.003	–	–	–	–	–	–	–	–	–
Cadmium (%)	–	–	**0.5/1.2**	–	–	–	–	–	–	–	–
Manganese (%)	–	–	–	–	–	–	**0.3/2.0**	**0.3/2.0**	–	0.10	–
Sulphur (%)	–	–	–	–	–	–	–	–	–	–	–

continued

Section 6.3.16(a) *(continued)*

Designation[a,c]	**C105**	**C107**	**C108**	**C109**	**CZ109**	**CZ111**	**CZ114**	**CZ115**	**CZ122**	**CZ127**	**CZ132**
Material[b]	**Tough pitch arsenical copper**	**Phosphorus deoxidized arsenical copper**	**Cadmium copper**	**Tellurium copper**	**Lead-free 60/40 brass**	**Admiralty brass**	**High-tensile brass**	**High-tensile brass (soldering quality)**	**Leaded brass 58% Cu 2% Pb**	**Aluminium–nickel–silicon brass**	**Dezincification-resistant brass**
Availability[c]	**P**	**P, T**	**W, P**	**RS**	**FS, RS**	**T**	**FS, RS**	**FS, RS**	**FS, RS**	**T**	**FS, RS**
Silicon (%)	–	–	–	–	–	–	–	–	–	**0.80/1.30**	–
Oxygen (%)	**0.70**	–	–	–	–	–	–	–	–	–	–
Phosphorus (%)	–	**0.013/0.050**	–	–	–	–	–	–	–	–	–
Tellurium (%)	Se + Te 0.03	0.010 Se + Te 0.20	–	**0.3/0.7**	–	–	–	–	–	–	–
Total impurities (%)	–	0.07 (excluding silver, arsenic, nickel, phosphorus)	0.05	–	0.30 (excluding lead)	0.30	0.50	0.50	0.75	0.50 (excluding tin, lead, iron, manganese)	0.60

For essential alloying elements, limits are in **bold type**. Unless otherwise stated, figures in total impurities column include those in lighter type. Unless otherwise indicated, all limits are maxima.

[a] These alloys are not included in BS 2870, except in Appendix A of that standard, but are available in other product forms and other BS specifications.

[b] Oxygen-free high-conductivity copper for special applications (C110) is found in BS 1433, BS 1977, BS 3839 and BS 4608. This material is used for conductors in electrical and electronic applications.

[c] Availability: T = tube, FS = forging stock, W = wire, RS = rods, sections, P = plate.

6.3.16(b) Miscellaneous wrought copper alloys

Designation[a]	**CA102**	**C105**	**CN101**	**CN108**	**NS101**	**NS102**	**NS109**	**NS112**	**CS101**
Material	**7% Aluminium–bronze**	**10% Aluminium–bronze**	**95/5 Copper–nickel–iron**	**66/30/2/2 Copper–nickel–iron–manganese**	**Leaded 10% nickel–brass**	**Leaded 14% nickel–brass**	**25% nickel–silver**	**15% leaded nickel–silver (Cu–Ni–Zn–Pb)**	**Silicon–bronze (Cu–Si)**
Availability[b]	**T, P**	**P**	**P**	**T**	**FS, RS**	**RS**	**W**	**RS**	**FS, W, RS, P**
Copper (%)	**Rem.**	**78.0/85.0**	**Rem.**	**Rem.**	**44.0/47.0**	**34.0/42.0**	**55.0/60.0**	**60.0/63.0**	**Rem.**
Tin (%)	–	0.10	0.01	–	–	–	–	–	–
Lead (%)	–	0.05	0.01	–	**1.0/2.5**	**1.0/2.25**	0.025	0.5/1.0	–
Iron (%)	Ni + Fe + Mn **1.0/2.5** (optional) but between these limits	**1.5/3.5**	**1.05/1.35**	**1.7/2.3**	0.4	0.3	0.3	–	0.25
Nickel (%)	–	**4.0/7.0**	**5.0/6.0**	**29.0/32.0**	**9.0/11.0**	**13.0/15.0**	**24.0/26.0**	**14.0/16.0**	–
Zinc (%)	–	0.40	–	–	**Rem.**	**Rem.**	**Rem.**	**Rem.**	–
Arsenic (%)	–	–	0.05	–	–	–	–	–	–
Antimony (%)	–	–	0.01	–	–	–	–	–	–
Aluminium (%)	**6.0/7.5**	**8.5/10.5**	–	–	–	–	–	–	–
Silicon (%)	–	0.15	0.05	–	–	–	–	–	**3.15/3.25**

continued

Section 6.3.16 (*continued*)

Magnesium (%)	–	0.05	–	–	–	–	–	–	–
Manganese (%)	See iron and nickel	**0.5/2.0**	**0.3/0.8**	**1.5/2.5**	**0.2/0.5**	**1.5/3.0**	**0.05/0.75**	**0.1/0.5**	**0.75/1.25**
Sulphur (%)	–	–	0.05	–	–	–	–	–	–
Carbon (%)	–	–	0.05	–	–	–	–	–	–
Phosphorus (%)	–	–	0.03	–	–	–	–	–	–
Total impurities (%)	0.50	0.50	0.30	0.30	0.30 (excluding iron)	0.30 (excluding iron)	0.50	0.50	0.50 (excluding iron)

For essential alloying elements, limits are in **bold type**. Unless otherwise stated, figures in total impurities column include those in lighter type. Unless otherwise indicated, all limits are maxima.

[a] These alloys are not included in BS 2870, except in Appendix A, but are available in other product forms and other BS specifications.

[b] Availability: T = tube, FS = forging stock, W = wire, RS = rods, sections, P = plate.

6.3.17(a) Copper alloys for casting: group A

Designation	PB4		LPB1		LB2		LB4		LG1		LG2	
Material	Phosphor bronze (copper–tin–phosphorus)		Leaded phosphor bronze		Leaded bronze		Leaded bronze		Leaded gunmetal		Leaded gunmetal	
Nominal composition	Cu Sn 10 Pb P		Cu Sn 7 Pb P		Cu Sn 10 Pb 10		Cu Sn 5 Pb 9		Cu Sn 3 Pb 5 Zn 8		Cu Sn 5 Pb 5 Zn 5	
Elements	min. (%)	max. (%)	min. (%)	max. (%)	min. (%)	max. (%)	min. (%)	max. (%)	min. (%)	max. (%)	min. (%)	max. (%)
Copper	Remainder		Remainder		Remainder		Remainder		Remainder		Remainder	
Tin	9.5	11.0	6.5	8.5	9.0	11.0	4.0	6.0	2.0	3.5	4.0	6.0
Zinc	–	0.5	–	2.0	–	1.0	–	2.0	7.0	9.5	4.0	6.0
Lead	–	0.75	2.0	5.0	8.5	11.0	8.0	10.0	4.0	6.0	4.0	6.0
Phosphorus	0.4	1.0	0.33	–	–	0.1	–	0.1[a]	–	–	–	–
Nickel	–	0.5	–	1.0	–	2.0	–	2.0	–	2.0	–	2.0
Iron	–	–	–	–	–	0.15	–	0.25	–	–	–	–
Aluminium	–	–	–	–	–	0.01	–	0.01	–	0.01	–	0.01
Manganese	–	–	–	–	–	0.20	–	0.20	–	–	–	–
Antimony	–	–	–	0.25	–	0.5	–	0.5	–	–	–	–
Arsenic	–	–	–	–	–	–	–	–	–	–	–	–
Iron + arsenic + antimony	–	–	–	–	–	–	–	–	–	0.75	–	0.5
Silicon	–	–	–	–	–	0.02	–	0.02	–	0.02	–	0.02
Bismuth	–	–	–	–	–	–	–	–	–	0.10	–	0.05
Sulphur	–	0.1	–	0.1	–	0.1	–	0.1	–	0.10	–	–
Total impurities	–	0.5	–	0.5	–	0.5	–	0.5	–	1.0	–	0.80

6.3.17(b) Further copper alloys for casting: group A

Designation	SCB1		SCB3		SCB6		DCB1		DCB3		PCP1	
Material	Brass for sand casting		Brass for sand casting		Brass for brazeable casting		Brass for die casting		Brass for die casting		Brass for pressure die casting	
Nominal composition	Cu Zn 25 Pb 3 Sn 2		Cu Zn 33 Pb 2		Cu 15 As		Cu Zn 40		Cu Zn 40 Pb		Cu Zn 40 Pb	
Elements	min. (%)	max. (%)	min. (%)	max. (%)	min. (%)	max. (%)	min. (%)	max. (%)	min. (%)	max. (%)	min. (%)	max. (%)
Copper	70.0	80.0	63.0	67.0	83.0	88.0	59.0	63.0	58.0	62.0	57.0	60.0
Tin	1.0	3.0	–	1.5	–	–	–	–	–	1.0	–	0.5
Zinc	Remainder		Remainder		Remainder		Remainder		Remainder		Remainder	
Lead	2.0	5.0	1.0	3.0	–	0.5	–	0.25[c]	0.5	2.5	0.5	2.5
Phosphorus	–	–	–	0.05	–	–	–	–	–	–	–	–
Nickel	–	1.0	–	1.0	–	–	–	–	–	1.0[d]	–	–
Iron	–	0.75	–	0.75	–	–	–	0.5	–	0.8	–	0.2
Aluminium	–	0.01	–	0.1[b]	–	–	–	–	–	0.8	–	0.5
Manganese	–	–	–	0.2	–	–	–	–	–	0.5	–	–
Antimony	–	–	–	–	–	–	–	–	–	–	–	–
Arsenic	–	–	–	–	–	–	–	–	–	–	–	–
Iron + arsenic + antimony	–	–	–	–	0.05	0.20	–	–	–	–	–	–

Silicon	–	–	–	0.05	–	–	–	–	–	0.05	–	–
Bismuth	–	–	–	–	–	–	–	–	–	–	–	–
Sulphur	–	–	–	–	–	–	–	–	–	–	–	–
Total impurities	–	1.0	–	1.0	– (including lead)	1.0	–	0.75	– (excluding Ni + Pb + Al)	2.0	–	0.5

For full range of alloys and further information, see BS 1400.
For essential alloying elements, limits are in **bold type**. Unless otherwise stated, figures in total impurities column include those in lighter type. Unless otherwise indicated, all limits are maxima.
[a] For continuous casting, phosphorus content may be increased to a maximum of 1.5% and alloy coded with suffix /L.
[b] For pressure tight castings in SCB3 the aluminium content should not be greater than 0.02.
[c] DCB1: 0.1% lead if required.
[d] DCB3: nickel to be counted as copper.

6.3.18(a) Copper alloys for casting: group B

Designation	HCC1		CC1-TF		PB1		PB2		CT1		LG4	
Material	High-conductivity copper		Copper–chromium		Phosphor bronze (copper + tin + phosphorus)		Phosphor bronze (copper + tin + phosphorus)		Copper–tin		87/7/3/3 Leaded gunmetal	
Nominal composition	–		Cu Cr 1		Cu Sn 10 P		Cu Sn 11 P		Cu Sn 10		Cu Sn 7 PB 3 Zn 3	
Elements	min. (%)	max. (%)	min. (%)	max. (%)	min. (%)	max. (%)	min. (%)	max. (%)	min. (%)	max. (%)	min. (%)	max. (%)
Copper	See note[a]		Remainder		Remainder		Remainder		Remainder		Remainder	
Tin			–	–	10.0	11.5	11.0	13.0	9.0	11.0	6.0	8.0[c]
Zinc			–	–	–	0.05	–	0.30	–	0.03	1.5	3.0
Lead			–	–	–	0.25	–	0.50	–	0.25	2.5	3.5
Phosphorus			–	–	0.5	1.0	0.15	0.60	–	0.15[b]	–	–
Nickel			–	–	–	0.1	–	0.50	–	0.25	–	2.0[c]

Iron			–	–	–	0.1	–	0.1	–	0.20	–	0.20
Aluminium			–	–	–	0.1	–	0.01	–	0.01	–	0.01
Manganese			–	–	–	0.05	–	–	–	0.20	–	–
Antimony			–	–	–	0.05	–	–	–	0.20	–	0.25
Arsenic			–	–	–	–	–	–	–	–	–	0.15
Iron + arsenic + antimony			–	–	–	–	–	–	–	–	–	0.40
Silicon			–	–	–	0.02	–	0.02	–	0.01	–	0.01
Bismuth			–	–	–	–	–	–	–	–	–	0.05
Magnesium			–	–	–	–	–	–	–	–	–	–
Sulphur			–	–	–	0.05	–	0.1	0	0.05	–	–
Chromium			**0.50**	**1.25**	–	–	–	–	–	–	–	–
Total impurities	–	–	–	–	–	0.60	–	0.20	–	0.80	–	0.70

6.3.18(b) Further copper alloys for casting: group B

Designation	AB1		AB2		CMA1		HTB1		HTB3	
Material	Aluminium bronze (copper–aluminium)		Aluminium bronze (copper–aluminium)		Copper–manganese–aluminium		High-tensile brass[d]		High-tensile brass	
Nominal composition	Cu Al 10 Fe 3		Cu Al 10 Fe 5 Ni 5		Cu Mn 13 Al 8 Fe 3 Ni 3		Cu Zn 31 Al Fe Mn		Cu Zn 28 Al 5 Fe Mn	
Elements	min. (%)	max. (%)	min. (%)	max. (%)	min. (%)	max. (%)	min. (%)	max. (%)	min. (%)	max. (%)
Copper	**Remainder**		**Remainder**		**Remainder**		**57.0**	–	**55.0**	–
Tin	–	0.1	–	0.1	–	**0.50**	–	**1.0**	–	**0.20**
Zinc	–	0.5	–	0.5	–	**1.00**	**Remainder**		**Remainder**	
Lead	–	0.03	–	0.3	–	0.05	–	**0.50**	–	**0.20**
Phosphorus	–	–	–	–	–	0.05	–	–	–	–
Nickel	–	**1.0**	**4.0**	**5.5**	**1.5**	**4.5**	–	**1.0**	–	**1.0**
Iron	**1.5**	**3.5**	**4.0**	**5.5**	**2.0**	**4.0**	**0.7**	**2.0**	**1.5**	**3.25**
Aluminium	**8.5**	**10.5**	**8.8**	**10.0**	**7.0**	**8.5**	**0.5**	**2.5**	**3.0**	**6.0**
Manganese	–	**1.0**	–	**3.0**	**11.0**	**15.0**	**0.1**	**3.0**	–	**4.0**
Antimony	–	–	–	–	–	–	–	–	–	
Arsenic	–	–	–	–	–	–	–	–	–	
Iron + arsenic + antimony	–	–	–	–	–	–	–	–	–	

Silicon	–	0.2	–	0.7	–	0.15	–	0.10	–	**0.10**
Bismuth	–	–	–	–	–	–	–	–	–	–
Magnesium	–	0.05	–	0.05	–	–	–	–	–	–
Sulphur	–	–	–	–	–	–	–	–	–	–
Chromium	–	–	–	–	–	–	–	–	–	–
Total impurities	–	0.30	–	0.30	–	0.30	–	0.20	–	0.20

For full range of alloys and further information, see BS 1400.
For essential alloying elements, limits are in **bold type**. Unless otherwise stated, figures in total impurities column include those in lighter type. Unless otherwise indicated, all limits are maxima.
[a]HCC1 castings shall be made from the copper grades Cu-CATH-2, Cu-EPT-2 or Cu-FRHC, as specified in BS 6017.
[b]For continuous casting, phosphorus content may be increased to a maximum of 1.5% and the alloy coded with the suffix /L.
[c]Tin plus half-nickel content shall be within the range 7.0–8.0%.
[d]HTB1 subject also to microstructure requirements, see BS 1400, Clause 6.3.

6.3.19(a) Copper alloys for casting: group C

Designation	LB1		LB5		G1		G3		G3-TF	
Material	Leaded bronze		Leaded bronze		Gunmetal		Nickel gunmetal		Nickel gunmetal fully heat treated	
Nominal composition	Cu Pb 15 Sn 9		Cu Pb 20 Sn 5		Cu Sn 10 Zn 2		Cu Sn 7 Ni 5 Zn 3		Cu Sn 7 Ni 5 Zn 3	
Elements	min. (%)	max. (%)	min. (%)	max. (%)	min. (%)	max. (%)	min. (%)	max. (%)	min. (%)	max. (%)
Copper	Remainder		Remainder		Remainder		Remainder		Remainder	
Tin	8.0	10.0	4.0	6.0	9.5	10.5	6.5	7.5	6.5	7.5
Zinc	–	1.0	–	1.0	1.75	2.75	1.5	3.0	1.5	3.0
Lead	13.0	17.0	18.0	23.0	–	1.5	0.10	0.50	0.10	0.50
Phosphorus	–	0.1	–	0.1	–	–	–	0.02	–	0.02
Nickel	–	2.0	–	2.0	–	1.0	5.25	5.75	5.25	5.75
Iron	–	–	–	–	–	0.15[b]	–	–[b]	–	–[b]
Aluminium	–	–	–	–	–	0.01	–	0.01	–	0.01
Manganese	–	–	–	–	–	–	–	0.20	–	0.20

Antimony	–	**0.5**	–	**0.5**	–	–[b]	–	–[b]	–	–[b]
Arsenic	–	–	–	–	–	–[b]	–	–[b]	–	–[b]
Silicon	–	0.02	–	0.01	–	0.02	–	0.02	–	0.02
Bismuth	–	–	–	–	–	0.03	–	0.02	–	0.02
Sulphur	–	0.1	–	–	–	0.10	–	0.10	–	0.10
Magnesium	–	–	–	–	–	–	–	–	–	–
Niobium + tantalum	–	–	–	–	–	–	–	–	–	–
Carbon	–	–	–	–	–	–	–	–	–	–
Chromium	–	–	–	–	–	–	–	–	–	–
Zirconium	–	–	–	–	–	–	–	–	–	–
Cobalt	–	–	–	–	–	–	–	–	–	–
Total impurities	–	0.30	–	0.30	–	0.50	–	0.50	–	0.50

6.3.19(b) Further copper alloys for casting: group C

Designation	**SCB4**		**CT2**		**AB3**		**CN1**		**CN2**	
Material	**Naval brass for sand casting**		**Copper–tin**		**Aluminium–silicon bronze**		**Copper–nickel–chromium**		**Copper–nickel–niobium**	
Nominal composition	**Cu Zn 36 Sn**		**Cu Sn 12 Ni**		**Cu Al 6 Si 2 Fe**		**Cu Ni 30 Cr**		**Cu Ni 30 NB**	
Elements	**min. (%)**	**max. (%)**	**min. (%)**	**max. (%)**	**min. (%)**	**max. (%)**	**min. (%)**	**max. (%)**	**min. (%)**	**max. (%)**
Copper	**60.0**	**63.0**	**85.0**	**87.0**	**Remainder**		**Remainder**		**Remainder**	
Tin	**1.0**	**1.5**	**11.0**	**13.0**	–	0.10	–	–	–	–
Zinc	**Remainder**	–	0.4	–	0.40	–	–	–	–	–
Lead	–	**0.50**	–	0.3	–	0.03	–	0.005	–	0.005
Phosphorus	–	–	**0.05**	**0.40**[a]	–	–	–	0.005	–	0.005
Nickel	–	–	**1.5**	2.5	–	0.10	**29.0**	**33.0**	**28.0**	**32.0**
Iron	–	–	–	0.20	**0.5**	**0.7**	**0.4**	**1.0**	**1.0**	**1.4**
Aluminium	–	0.01	–	0.01	**6.0**	**6.4**	–	–	–	–
Manganese	–	–	–	0.20	–	0.50	**0.4**	**1.0**	**1.0**	**1.4**
Antimony	–	–	–	–	–	–	–	–	–	–
Arsenic	–	–	–	–	–	–	–	–	–	–
Silicon	–	–	–	0.01	**2.0**	**2.4**	**0.20**	**0.40**	**0.20**	**0.40**
Bismuth	–	–	–	–	–	–	–	0.002	–	0.002
Sulphur	–	–	–	0.05	–	–	–	0.01	–	0.01
Magnesium	–	–	–	–	–	0.05	–	–	–	–

Niobium + tantalum	–	–	–	–	–	–	–	–	**1.20**	**1.40**
Carbon	–	–	–	–	–	–	–	0.02	–	0.02
Chromium	–	–	–	–	–	–	**1.5**	**2.0**	–	–
Zirconium	–	–	–	–	–	–	**0.05**	**0.15**	–	–
Cobalt	–	–	–	–	–	–	–	**0.05**	–	0.05
Total impurities	–	0.75	–	0.80	–	0.80	–	0.20	–	0.30

For full range of alloys and further information, see BS 1400.
For essential alloying elements, limits are in **bold type**. Unless otherwise stated, figures in total impurities column include those in lighter type. Unless otherwise indicated, all limits are maxima.
[a] For continuous casting, phosphorus content may be increased to a maximum of 1.5% and the alloy coded with suffix /L.
[b] Iron + antimony + arsenic 0.20% maximum.

6.3.20 Copper alloys for casting: typical properties and hardness values

Designation	Freezing range category	Tensile strength (N/mm^2)				0.2% proof stress (N/mm^2)			
		Sand[a]	Chill	Continuous[b]	Centrifugal	Sand[a]	Chill	Continuous	Centrifugal
Group A									
PB4	L	190–270	270–370	330–450	280–400	100–160	140–230	160–270	140–230
LPB1	L	190–250	220–270	270–360	230–310	80–130	130–160	130–200	130–160
LB2	L	190–270	270–280	280–390	230–310	80–130	140–200	160–220	140–190
LB4	L	160–190	200–270	230–310	220–300	60–100	80–110	130–170	80–110
LG1	L	180–220	180–270	–	–	80–130	80–130	–	–
LG2	L	200–270	200–280	270–340	220–310	100–130	110–140	100–140	110–140
SCB1	S	170–200	–	–	–	80–110	–	–	–
SCB3	S	190–220	–	–	–	70–110	–	–	–
SCB6	S	170–190	–	–	–	80–110	–	–	–
DCB1	S	–	280–370	–	–	–	90–120	–	–
DCB3	S	–	300–340	–	–	–	90–120	–	–
PCB1	S	–	280–370	–	–	–	90–120	–	–
Group B									
HCC1	S	160–190	–	–	–	–	–	–	–
CC1-TF	S	270–340	–	–	–	170–250	–	–	–
PB1	L	220–280	310–390	360–500	330–420	130–160	170–230	170–280	170–230
PB2	L	220–310	270–340	310–430	280–370	130–170	170–200	170–250	170–200
CT1	L	230–310	270–340	310–390	280–370	130–160	140–190	160–220	180–190
LG4	L	250–320	250–340	300–370	230–370	130–140	130–160	130–160	130–160
AB1	S	500–590	540–620	–	560–650	170–200	200–270	–	200–270
AB2	S	640–700	650–740	–	670–730	250–300	250–310	–	250–310
CMA1	S	650–730	670–740	–	–	280–340	310–370	–	–
HTB1	S	470–570	500–570	–	500–600	170–280	210–280	–	210–280
HTB3	S	740–810	–	–	740–930	400–470	–	–	400–500

Group C									
LB1	L	170–230	200–270	230–310	220–300	80–110	130–160	130–190	130–160
LB5	L	160–190	170–230	190–270	190–270	60–100	80–110	100–160	80–110
G1	L	270–340	230–310	300–370	250–340	130–160	130–170	140–190	130–170
G3	L	280–340	–	340–370	–	140–160	–	170–190	–
G3-TF	L	430–480	–	430–500	–	280–310	–	280–310	–
SCB4	S	250–310	–	–	–	70–110	–	–	–
CT2	L	280–330	–	300–350	300–350	160–180	–	180–210	180–210
AB3	S	460–500	–	–	–	180–190	–	–	–
CN1	S	480–540	–	–	–	300–320	–	–	–
CN2	S	480–540	–	–	–	300–320	–	–	–

continued

Section 6.3.20 (*continued*)

Designation	Freezing range category	Elongation on 5.65 $\sqrt{S_o}$ (%)				Brinell hardness (H_B)			
		Sand[a]	Chill	Continuous	Centrifugal[b]	Sand	Chill	Continuous	Centrifugal
Group A									
PB4	L	3–12	2–10	7–30	4–20	70–95	95–140	95–140	95–140
LPB1	L	3–12	2–12	5–18	4–22	60–90	85–110	85–110	85–110
LB2	L	5–15	3–7	6–15	5–10	65–85	80–90	80–90	80–90
LB4	L	7–12	5–10	9–20	6–13	55–75	60–80	60–80	60–80
LG1	L	11–15	2–8	–	–	55–65	65–80	–	–
LG2	L	13–25	6–15	13–35	8–30	65–75	80–95	75–90	80–95
SCB1	S	18–40	–	–	–	45–60	–	–	–
SCB3	S	11–30	–	–	–	45–65	–	–	–
SCB6	S	18–40	–	–	–	45–60	–	–	–
DCB1	S	–	23–50	–	–	–	60–70	–	–
DCB3	S	–	13–40	–	–	–	60–70	–	–
PCB1	S	–	25–40	–	–	–	60–70	–	–
Group B									
HCC1	S	23–40	–	–	–	40–45	–	–	–
CC1-TF	S	18–30	–	–	–	100–120	–	–	–
PB1	L	3–8	2–8	6–25	4–22	70–100	95–150	100–150	25–150
PB2	L	5–15	3–7	5–15	3–14	75–110	100–150	100–150	100–150
CT1	L	6–20	5–15	9–25	6–25	70–90	90–130	90–130	90–130
LG4	L	16–25	5–15	13–30	6–30	70–85	80–95	80–95	80–95
AB1	S	18–40	18–40	–	20–30	90–140	130–160	–	120–160
AB2	S	13–20	13–20	–	13–20	140–180	160–190	–	140–180
CMA1	S	18–35	27–40	–	–	160–210	–	–	–
HTB1	S	18–35	18–35	–	20–38	100–150	–	–	100–150
HTB3	S	11–18	–	–	13–21	150–230	–	–	150–230

Group C									
LB1	L	4–10	3–7	9–10	4–10	50–70	70–90	70–90	70–90
LB5	L	5–10	5–12	8–16	7–15	45–65	50–70	50–70	50–70
G1	L	13–25	3–8	9–25	5–16	70–95	85–130	90–130	70–95
G3	L	16–25	–	18–25	–	70–95	–	90–130	–
G3-TF	L	3–5	–	3–7	–	160–180	–	160–180	–
SCB4	S	18–40	–	–	–	50–75	–	–	–
CT2	L	12–20	–	8–15	10–15	75–110	–	100–150	100–150
AB3	S	20–30	–	–	–	–	–	–	–
CN1	S	18–25	–	–	–	170–200	–	–	–
CN2	S	18–25	–	–	–	170–200	–	–	–

[a] On separately cast test bars.
[b] Values apply to samples cut from centrifugal castings made in metallic moulds. Minimum properties of centrifugal castings made in sand moulds are the same as for sand castings.

6.3.21 Aluminium and aluminium alloys

Pure aluminium has a low mechanical strength and is usually alloyed with other metals to produce a wide range of useful, lightweight materials. The aluminium alloys can be divided up into four categories:

- Wrought alloys that can be heat treated.
- Wrought alloys that cannot be heat treated.
- Cast alloys that can be heat treated.
- Cast alloys that cannot be heat treated.

Like most non-ferrous metals and their alloys, aluminium and aluminium alloys depend on cold-working processes such as rolling and drawing to enhance their properties. Only a relatively few aluminium alloys containing copper and other alloying elements respond to heat treatment. An example of a general-purpose light-alloy originally developed for highly stressed aircraft components is *duralumin*. This alloy has the following composition:

Copper (Cu)	4.0%
Silicon (Si)	0.2%
Manganese (Mn)	0.7%
Magnesium (Mg)	0.8%
Aluminium (Al)	Remainder

It can be softened by heating the alloy to 480°C and quenching this is known as *solution treatment* since the alloying elements form a solid solution with the aluminium rendering the metal ductile and suitable for cold working without cracking. However, it starts to re-harden immediately by a natural process called *age hardening* and becomes fully hardened only after 4 days. During age hardening the unstable supersaturated solution of the alloying elements breaks down with the precipitation of the intermetallic compound $CuAl_2$. The precipitate will be in the form of fine particles throughout the mass of the metal to give greater strength and hardness.

Another alloy that is even stronger has the following composition:

Copper (Cu	1.6%
Titanium (Ti)	0.3%
Magnesium (Mg)	2.5%
Zinc (Zn)	0.2%
Aluminium (Al)	Remainder

This alloy can also be softened by solution treatment by heating it to 465°C and quenching. Again hardening occurs by natural ageing. If the natural ageing process is speeded up by reheating the alloy to 165°C for about 10 h, the process is called *precipitation age hardening* or *artificial ageing*. The age-hardening process can be retarded after solution treatment by refrigeration.

6.3.22 British Standards

Some useful British Standards relating to wrought and cast aluminium and aluminium alloys – both heat-treatable and non-heat-treatable are as follows:

- BS EN 485-2: 2004 Aluminium and Aluminium alloys – sheet, strip and plate – chemical composition.
- BS EN 485-2: 2004 Aluminium and aluminium alloys – sheet, strip and plate – mechanical properties.

- BS EN 573: 1995 (four parts) – wrought products:
  (a) Part 1: Numerical designation system.
  (b) Part 2: Chemical symbol-based designation system.
  (c) Part 3: Chemical composition.
  (d) Part 4: Forms of products.
- BS EN 1559-4: 1999 Founding – additional requirements for aluminium and aluminium alloy castings.
- BS 1706: 1998 Aluminium and aluminium alloy castings – chemical composition and mechanical properties.

These are highly detailed and comprehensive standards, and it is not physically possible to reproduce them within the confines of this Pocket Book. However, they should be consulted when selecting materials for a new product design and manufacture.

Some typical examples of wrought and cast aluminium and aluminium alloy products are listed in tables in the following sections. They are for guidance only.

6.3.23 Unalloyed aluminium plate, sheet and strip

Composition

Materials designation[a]	Silicon (%)	Iron (%)	Copper (%)	Manganese (%)	Magnesium (%)	Aluminium[c] (%)	Zinc (%)	Gallium (%)	Titanium (%)	Others[b]	
										Each (%)	Total (%)
1080(A)	0.15	0.15	0.03	0.02	0.02	99.80	0.06	0.03	0.02	0.02	–
1050(A)	0.40	0.05	0.05	0.05	0.05	99.50	0.07	–	0.05	0.03	–
1200	1.0 Si + Fe		0.05	0.05	–	99.00	0.10	–	0.05	0.05	0.15

Properties

Materials designation (%)	Temper[d]	Thickness > (mm)	Thickness ⩽ (mm)	Tensile strength min. (N/mm^2)	Tensile strength max. (N/mm^2)	Elongation on 50 mm: materials thicker than 0.5 mm (%)	0.8 mm (%)	1.3 mm (%)	2.6 mm (%)	3.0 mm (%)	Elongation on $5.65\sqrt{S_o}$ over 12.5 mm thick min (%)
1080(A)	F	3.0	25.0	–	–	–	–	–	–	–	–
	O	0.2	6.0	–	90	29	29	29	35	35	–
	H14	0.2	12.5	90	125	5	6	7	8	8	–
	H18	0.2	3.0	125	–	3	4	4	5	–	–
1050(A)	F	3.0	25.0	–	–	–	–	–	–	–	–
	O	0.2	6.0	55	95	22	25	30	32	32	–
	H12	0.2	6.0	80	115	4	6	8	9	9	–
	H14	0.2	12.5	100	135	4	5	6	6	8	–
	H18	0.2	3.0	135	–	3	3	4	4	–	–
1200	F	3.0	25.0	–	–	–	–	–	–	–	–
	O	0.2	6.0	70	105	20	25	30	30	30	–
	H12	0.2	6.0	90	125	4	6	8	9	9	–
	H14	0.2	12.5	105	140	3	4	5	5	6	–
	H16	0.2	6.0	125	160	2	3	4	4	4	–
	H18	0.2	3.0	140	–	2	3	4	4	–	–

[a] Composition in per cent (m/m) maximum unless shown as a range or a minimum.
[b] Analysis is regularly made only for the elements for which specific limits are shown. If, however, the presence of other elements is suspected to be, or in the case of routine analysis is indicated to be, in excess of the specified limits, further analysis should be made to determine that these other elements are not in excess of the amount specified.
[c] The aluminium content for unalloyed aluminium not made by a refining process is the difference between 100.00% and the sum of all other metallic elements in amounts of 0.010% or more each, expressed to the second decimal before determining the sum.
[d] An alternative method of production, designated H2, may be used instead of the H1 routes, subject to agreement between supplier and purchaser, and provided that the same specified properties are achieved.

6.3.24 Aluminium alloy plate, sheet and strip: non-heat-treatable

Composition

Material designation[a]	Silicon (%)	Iron (%)	Copper (%)	Manganese (%)	Magnesium (%)	Chromium (%)	Aluminium (%)	Zinc (%)	Other restrictions (%)	Titanium (%)	Others[b]	
											Each (%)	Total (%)
3103	0.50	0.7	0.10	0.9/1.5	0.30	0.10	Rem.	0.20	0.10 Zr + Ti	–	0.05	0.15
3105	0.6	0.7	0.3	0.3/0.8	0.2/0.8	0.20	Rem.	0.40	–	0.01	0.05	0.15
5005	0.3	0.7	0.2	0.2	0.5/1.1	0.10	Rem.	0.25	–	–	0.05	0.15
5083	0.40	0.40	0.10	0.4/1.0	4.0/4.9	0.05/0.25	Rem.	0.25	–	0.15	0.05	0.15
5154	0.50	0.50	0.10	0.50	3.1/3.9	0.25	Rem.	0.20	0.10/0.50 Mn + Cr	0.20	0.05	0.15
5251	0.40	0.50	0.15	0.1/0.5	1.7/2.4	0.15	Rem.	0.15	–	0.15	0.05	0.15
5454	0.25	0.40	0.10	0.5/1.0	2.4/3.0	0.5/2.0	Rem.	0.25	–	0.20	0.05	0.15

Section 6.3.24 (*continued*)

Material designation[a]	Temper[c] (%)	Thickness		0.2% proof stress (N/mm^2)	Tensile strength		Elongation on 50 mm: materials thicker than					Elongation on $5.65\sqrt{S_o}$ over 12.5 mm thick (min.) (%)
		> (mm)	< (mm)		min. (N/mm^2)	max. (N/mm^2)	0.5 mm (%)	0.8 mm (%)	1.3 mm (%)	2.6 mm (%)	3.0 mm (%)	
3103	F	0.2	25.0	–	–	–	–	–	–	–	–	–
	O	0.2	6.0	–	90	130	20	23	24	24	25	–
	H12	0.2	6.0	–	120	155	5	6	7	9	9	–
	H14	0.2	12.5	–	140	175	3	4	5	6	7	–
	H16	0.2	6.0	–	160	195	2	3	4	4	4	–
	H18	0.2	3.0	–	175	–	2	3	4	4	–	–
3105	O	0.2	3.0	–	110	155	16	18	20	20	–	–
	H12	0.2	3.0	115	130	175	2	3	4	5	–	–
	H14	0.2	3.0	145	160	205	2	2	3	4	–	–
	H16	0.2	3.0	170	185	230	1	1	2	3	–	–
	H18	0.2	3.0	190	215	–	1	1	1	2	–	–
5005	O	0.2	3.0	–	95	145	18	20	21	22	–	–
	H12	0.2	3.0	80	125	170	4	5	6	8	–	–
	H14	0.2	3.0	100	145	185	3	3	5	6	–	–
	H18	0.2	3.0	165	185	–	1	2	3	3	–	–
5083	F	3.0	25.0	–	–	–	–	–	–	–	–	–
	O	0.2	80.0	125	275	350	12	14	16	16	16	14
	H22	0.2	6.0	235	310	375	5	6	8	10	8	–
	H24	0.2	6.0	270	345	405	4	5	6	8	6	–

continued

Section 6.3.24 *(continued)*

Material designation[a]	Temper[c] (%)	Thickness		0.2% proof stress (N/mm²)	Tensile strength		Elongation on 50 mm: materials thicker than					Elongation on $5.65\sqrt{S_o}$ over 12.5 mm thick (min.) (%)
		> (mm)	< (mm)		min. (N/mm²)	max. (N/mm²)	0.5 mm (%)	0.8 mm (%)	1.3 mm (%)	2.6 mm (%)	3.0 mm (%)	
5154	O	0.2	6.0	85	215	275	12	14	16	18	18	–
	H22	0.2	6.0	165	245	295	5	6	7	8	8	–
	H24	0.2	6.0	225	275	325	4	4	8	6	5	–
5251	F	3.0	25.0	–	–	–	–	–	–	–	–	–
	O	0.2	6.0	60	160	200	18	18	18	20	20	–
	H22	0.2	6.0	130	200	240	4	5	6	8	8	–
	H24	0.2	6.0	175	275	275	3	4	5	5	5	–
	H28	0.2	3.0	215	255	285	2	3	3	4	4	–
5454	F	3.0	25.0	–	–	–	–	–	–	–	–	–
	O	0.2	6.0	80	215	285	12	14	16	18	18	–
	H22	0.2	3.0	180	250	305	4	5	7	8	–	–
	H24	0.2	3.0	200	270	325	3	4	5	6	–	–

[a] Composition in per cent (m/m) maximum unless shown as a range or a minimum.
[b] Analysis is regularly made only for the elements for which specific limits are shown. If, however, the presence of other elements is suspected to be, or in the case of the routine analysis is indicated to be, in excess of the specified limits, further analysis should be made to determine that these other elements are not in excess of the amount specified.
[c] Either H1 or H2 production routes may be used, subject to agreement between the supplier and purchaser, and provided that the specified properties are achieved.

6.3.25 Aluminium alloy plate, sheet and strip: heat-treatable

Composition

Material designation[a]	Silicon (%)	Iron (%)	Copper (%)	Manganese (%)	Magnesium (%)	Chromium (%)	Nickel (%)	Zinc (%)	Other restrictions (%)	Titanium (%)	Aluminium (%)	Others[b]	
												Each (%)	Total (%)
2014A	0.5/0.9	0.50	3.9/5.0	0.4/1.2	0.2/0.8	0.10	0.10	0.25	0.20 Zr + Ti	0.15	Rem.	0.05	0.15
Clad 2014A[c]	0.5/0.9	0.50	3.0/5.0	0.4/1.2	0.2/0.8	0.10	0.10	0.25	0.20 Zr + Ti	0.15	Rem.	0.05	0.15
2024	0.50	0.50	3.8/4.4	0.3/0.9	1.2/1.8	0.10	–	0.25	–	0.15	Rem.	0.05	0.15
Clad 2024[c]	0.50	0.50	3.8/4.4	0.3/0.9	1.2/1.8	0.10	–	0.25	–	0.15	Rem.	0.05	0.15
6082	0.7/1.3	0.50	0.10	0.4/1.0	0.6/1.2	0.25	–	0.20	–	0.10	Rem.	0.05	0.15

Properties

Material designation[a]	Temper[d]	Thickness		0.2% proof stress[e] (N/mm²)	Tensile strength		Elongation on 50 mm: materials thicker than					Elongation on $5.65\sqrt{S_o}$ over 12.5 mm thick (min.) (%)
		> (mm)	< (mm)		min. (N/mm²)	max. (N/mm²)	0.5 mm (%)	0.8 mm (%)	1.3 mm (%)	2.6 mm (%)	3.0 mm (%)	
2014A	O	0.2	6.0	110	–	235	14	14	16	16	16	
	T4	0.2	6.0	22	400	–	13	14	14	14	14	
	T6	0.2	6.0	380	440	–	6	6	7	7	8	
	T451	6.0	25.0	250	400	–	–	–	–	–	14	12
		25.0	40.0	250	400	–	–	–	–	–	–	10
		40.0	80.0	250	395	–	–	–	–	–	–	7
	T651	6.0	25.0	410	460	–	–	–	–	–	–	6
		25.0	40.0	400	450	–	–	–	–	–	–	5
		40.0	60.0	390	430	–	–	–	–	–	–	5
		60.0	90.0	390	430	–	–	–	–	–	–	4
		90.0	115.0	370	420	–	–	–	–	–	–	4
		115.0	140.0	350	410	–	–	–	–	–	–	4
Clad 2014A[c]T4	O	0.2	6.0	100	–	220	14	14	16	16	16	–
		0.2	1.6	240	385	–	13	14	14	–	–	–
		1.6	6.0	245	395	–	–	–	–	14	14	–
	T6	0.2	1.6	345	420	–	7	7	8	9	–	–
		1.6	6.0	355	420	–	–	–	–	9	9	–
2024	O	0.2	6.0	110	–	235	12	12	14	14	14	–
	T3	0.2	1.6	290	440	–	11	11	11	–	–	–
		1.6	6.0	290	440	–	–	–	–	12	12	–
	T351	6.0	25.0	280	430	–	–	–	–	–	10	10
		25.0	40.0	280	420	–	–	–	–	–	–	9
		40.0	60.0	270	410	–	–	–	–	–	–	9
		90.0	115.0	270	400	–	–	–	–	–	–	8
		115.0	140.0	260	390	–	–	–	–	–	–	7

Clad 2024[b]	O	0.2	6.0	100	–	235	12	12	14	14	14	–
	T3	0.2	1.6	270	405	–	11	11	11	–	–	–
		1.6	6.0	275	425	–	–	–	–	12	12	–
6082	O	0.2	3.0	–	–	155	16	16	16	15	–	–
	T4	0.2	3.0	120	200	–	15	15	15	15	–	–
		3.0	25.0	115	200	–	–	–	–	–	15	12
	T6	0.2	3.0	255	295	–	8	8	8	8	–	–
		3.0	25.0	240	295	–	–	–	–	–	8	–
	T451	6.0	25.0	115	200	–	–	–	–	–	–	15
		25.0	90.0	115	200	–	–	–	–	–	–	15
	T651	6.0	25.0	240	295	–	–	–	–	–	–	8
		25.0	90.0	240	295	–	–	–	–	–	–	8
		90.0	115.0	230	285	–	–	–	–	–	–	7
		115.0	150.0	220	275	–	–	–	–	–	–	6

[a] Composition in per cent (m/m) maximum unless shown as a range or a minimum.
[b] Analysis is regularly made only for the elements for which specific limits are shown. If, however, the presence of other elements is suspected to be, or in the case of routine analysis is indicated to be, in excess of the specified limits, further analysis should be made to determine that these other elements are not in excess of the amount specified.
[c] Unalloyed aluminium grade 1050A is used as cladding material. The cladding thickness is 4% on each side of the material up to and including 1.6 mm thick and 2% on each side for material over 1.6 mm thick.
[d] The tempers Tx51 (stress relieved by stretching) apply to plate and sheet which have been stretched after solution treatment to give a permanent set of approximately 1.5–3.0%.
[e] The proof stress values given are maxima for those alloys in the 'O' temper condition and minima for the remaining alloys.

6.3.26 Aluminium and aluminium alloy bars, extruded tube and sections for general engineering: non-heat-treatable

Composition

Material designation[a]	Silicon (%)	Iron (%)	Copper (%)	Manganese (%)	Magnesium (%)	Chromium (%)	Zinc (%)	Other restrictions (%)	Titanium (%)	Others[b]		Aluminium
										Each (%)	Total (%)	(%)
1050A	0.25	0.40	0.05	0.05	0.05	–	0.07	–	0.05	0.03	–	99.50[d] min.
1200	1.0 Si + Fe		0.05	0.05	–	–	0.10	–	0.05	0.05	0.15	99.00[d] min.
5083	0.4	0.4	0.10	0.40/1.0	4.0/4.9	0.05/0.25	0.25	–	0.15	0.05	0.15	Rem.
5154A	0.50	0.50	0.10	0.50	3.1/3.9	0.25	0.20	0.10–0.50 Mn + Cr	0.20	0.05	0.15	Rem.
5251	0.40	0.50	0.15	0.10/0.5	1.7/2.4	0.15	0.15	–	0.15	0.05	0.15	Rem.

Properties

Material designation	Temper[c]	Diameter (bar) or thickness (tube/sections)		0.2% proof stress	Tensile strength		Elongation on $5.65\sqrt{S_o}$	Elongation on 50 mm
		> (mm)	⩽ (mm)	min. (N/mm^2)	min. (N/mm^2)	max. (N/mm^2)	min. (%)	min. (%)
1050A	F	–	–	–	(60)	–	(25)	(23)
1200	F	–	–	–	(65)	–	(20)	(18)
5083	O	–	150[e]	125	275	–	14	13
	F	–	150	(130)	(280)	–	(12)	(11)
5154A	O	–	150[e]	85	215	275	18	16
	F	–	150	(100)	(215)	–	(16)	(14)
5251	F	–	150	(60)	(170)	–	(16)	(14)

[a] Composition in per cent (m/m) maximum unless shown as a range or a minimum.
[b] Analysis is regularly carried out for the elements for which specific limits are shown. If, however, the presence of other elements is suspected to be, or in the course of routine analysis is indicated to be, in excess of specified limits, further analysis should be made to determine that these other elements are not in excess of the amount specified.
[c] No mechanical properties are specified for materials in the F condition. The bracketed values shown for proof stress, tensile strength and elongation are typical properties, and are given for information only.
[d] The aluminium content for unalloyed aluminium not made by a refining process is the difference between 100% and the sum of all other metallic elements present in amounts of 0.10% or more each, expressed to the second decimal before determining the sum.
[e] No mechanical properties are specified for tube and hollow sections having a wall thickness greater than 75 mm (see BS 1474: 1987, Clause 6).

6.3.27 Aluminium alloy bars, extruded tube and sections for general engineering: heat-treatable

Composition

Material designation[a]	Silicon (%)	Iron (%)	Copper (%)	Manganese (%)	Magnesium (%)	Chromium (%)	Zinc (%)	Other restrictions (%)	Other Titanium (%)	Others[b]		Aluminium (%)
										Each (%)	Total (%)	
2014A	0.5/0.9	0.50	3.9/5.0	0.4/1.2	0.2/0.8	0.10	0.25	0.20 Zr + Ti	0.15	0.05	0.15	Rem.
6060	0.3/0.6	0.1/0.3	0.10	0.10	0.35/0.6	0.05	0.15	–	0.10	0.05	0.15	Rem.
6061	0.4/0.8	0.7	0.15/0.4	0.15	0.8/1.2	0.04/0.35	0.25	–	0.15	0.05	0.15	Rem.
6063	0.2/0.6	0.35	0.10	0.10	0.45/0.9	0.10	0.10	–	0.10	0.05	0.15	Rem.
6063A	0.3/0.6	0.15/0.35	0.10	0.15	0.6/0.9	0.05	0.15	–	0.10	0.05	0.15	Rem.
6082	0.7/1.3	0.50	0.10	0.40/1.0	0.6/1.2	0.25	0.20	–	0.10	0.05	0.15	Rem.
6463	0.2/0.6	0.15	0.20	0.05	0.45/0.9	–	0.05	–	–	0.05	0.16	Rem.

Properties

Material designation[a]	Temper[c]	Diameter (bar) or thickness (tube/sections)[d]		0.2% proof stress	Tensile strength		Elongation	
		> (mm)	⩽ (mm)	min. (N/mm²)	min. (N/mm²)	max. (N/mm²)	on $5.65\sqrt{S_o}$ min. (%)	on 50 mm min. (%)
2014A	T4	–	20	230	370	–	11	10
		20	75	250	390	–	11	–
		75	150	250	390	–	8	–
	T6	150	200	230	370	–	8	–
		–	20	370	435	–	7	6
		20	75	435	480	–	7	–
	T6510	75	150	420	465	–	7	–
		150	200	390	435	–	7	–
6060	T4	–	150	60	120	–	15	–
	T5	–	150	100	145	–	8	–
	T6	–	150	150	190	–	8	–
6061	T4	–	150	115	190	–	16	14
	T6 T6510	–	150	240	280	–	8	7
6063	O	–	200	–	–	140	15	13
	F	–	200	–	(100)	–	(13)	(12)
	T4	–	150	70	130	–	16	14
		150	200	70	120	–	13	–
	T5	–	25	110	150	–	8	7
	T6	–	150	160	195	–	8	7
		50	200	130	150	–	6	–

Section 6.3.27 (*continued*)

Material designation[a]	Temper[c]	Diameter (bar) or thickness (tube/sections)[d]		0.2% proof stress	Tensile strength		Elongation	
		> (mm)	⩽ (mm)	min. (N/mm^2)	min. (N/mm^2)	max. (N/mm^2)	on $5.65\sqrt{S_o}$ min. (%)	on 50 mm min. (%)
6063A	T4	–	25	90	150	–	14	12
	T5	–	25	160	200	–	8	7
	T6	–	25	190	230	–	8	7
6082	O	–	200	–	–	170	16	14
	F	–	200	–	(110)	–	(13)	(12)
	T4	–	150	120	190	–	16	14
		150	200	100	170	–	13	–
	T5	–	6	230	270	–	–	8
		–	20	255	295	–	8	7
	T6	20	150	270	310	–	8	–
	T6510	150	200	240	280	–	5	–
6463	T4	–	50	75	125	–	16	14
	T6	–	50	160	185	–	10	9

[a] Composition in per cent (m/m) maximum unless shown as a range or a minimum.
[b] Analysis is regularly carried out for the elements for which specific limits are shown. If, however, the presence of other elements is suspected to be, or in the course of routine analysis is indicated to be, in excess of specified limits, further analysis should be made to determine that these elements are not in excess of the amount specified.
[c] No mechanical properties are specified for materials in the F condition. The bracketed values shown for proof stress, tensile strength and elongation are typical properties and are given for information only. The temper T6510 is applicable only to bars (see Section 6.3.20).
[d] No mechanical properties are specified for tube and hollow sections having a wall thickness greater than 75 mm (see BS 1474:1987, Clause 6).

6.3.28 Aluminium alloy castings, group A: general purpose

Designation	LM2		LM4		LM6		LM20		LM24		LM25		LM27	
Nominal composition	Al–Si 10 Cu 2		Al–Si 5 Cu 3 Mn 0.5		Al–Si 12		Al–Si 12		Al–Si 8 Cu 3.5		Al–Si 7 Mg 0.5		Al–Si 7 Cu 2 Mn 0.5	
Nearest alloy(s) in ISO 3522	–		Al–Si 5 Cu 3		Al–Si 12 Al–Si 12 Fe		Al–Si 12 Cu Al–Si 12 Cu Fe		Al–Si 8 Cu 3 Fe		Al–Si 7 Mg		Al–Si 5 Cu 3	
Elements	**min. (%)**	**max. (%)**	**min. (%)**	**max. (%)**	**min. (%)**	**max. (%)**	**min. (%)**	**max. (%)**	**min. (%)**	**max. (%)**	**min. (%)**	**max. (%)**	**min. (%)**	**max. (%)**
Aluminium	**Remainder**		**Remainder**		**Remainder**		**Remainder**		**Remainder**		**Remainder**		**Remainder**	
Copper	**0.7**	**2.5**	**2.0**	**4.0**	–	0.1	–	0.4	**3.0**	**4.0**	–	0.20	**1.5**	**2.5**
Magnesium	–	0.3	–	0.2	–	0.1	–	0.2	–	0.30	**0.2**	**0.6**	–	**0.35**
Silicon	**9.0**	**11.5**	**4.0**	**6.0**	**10.0**	**13.0**	**10.0**	**13.0**	**7.5**	**9.5**	**6.5**	**7.5**	**6.0**	**8.0**
Iron	–	1.0	–	0.8	–	0.6	–	1.0	–	1.3	–	0.5	–	0.8
Manganese	–	0.5	**0.2**	**0.6**	–	0.5	–	0.5	–	0.5	–	0.3	**0.2**	**0.6**
Nickel	–	0.5	–	0.3	–	0.1	–	0.1	–	0.5	–	0.1	–	0.3
Zinc	–	2.0	–	0.5	–	0.1	–	0.2	–	3.0	–	0.1	–	1.0
Lead	–	0.3	–	0.1	–	0.1	–	0.1	–	0.3	–	0.1	–	0.2
Tin	–	0.2	–	0.1	–	0.05	–	0.1	–	0.2	–	0.05	–	0.1
Titanium	–	0.2	–	0.2	–	0.2	–	0.2	–	0.2	–	0.2[a]	–	0.2
Each other element[b]	–	–	–	0.05	–	0.05	–	0.05	–	–	–	0.05	–	0.05
Total other elements	–	0.5	–	0.15	–	0.15	–	0.20	–	0.5	–	0.15	–	0.15

Specified impurities are in light type. Analysis is required to verify that the contents of the specified impurities are less than the limits given in the table. Analysis for other elements is made when their presence is suspected to be in excess of the 'each other element' limit.

[a] If titanium alone is used for grain refining, the amount present shall not be less than 0.05%.

[b] In cases where alloys are required in the modified condition, the level of modifying element(s) present is not limited by the specified maximum value for other elements.

6.3.29 Aluminium alloy castings, group B: special purpose

Designation	LM0		LM5		LM9		LM13		LM16		LM21		LM22	
Nominal composition	99.50 + %Al		Al–Mg 5 Mn 0.5		Al–Si 12 Mg 0.5 Mn 0.5		Al–Si 12 Cu 1 Mg 2		Al–Si 5 Cu 1 Mg 0.5		Al–Si 6 Cu 4 Mn 0.4 Mg 0.2		Al–Si 5 Cu 3 Mn 0.4	
Nearest alloy(s) in ISO 3522	–		Al–Mg 5 Si 1 Al–Mg 6		Al–Si 10 Mg		Al–Si 12 Cu Al–Si 12 Cu Fe		Al–Si 5 Cu 1 Mg		Al–Si 6 Cu 4		Al–Si 5 Cu 3	
Elements	**min. (%)**	**max. (%)**	**min. (%)**	**max. (%)**	**min. (%)**	**max. (%)**	**min. (%)**	**max. (%)**	**min. (%)**	**max. (%)**	**min. (%)**	**max. (%)**	**min. (%)**	**max. (%)**
Aluminium	**99.50**[a]	–	**Remainder**		**Remainder**		**Remainder**		**Remainder**		**Remainder**		**Remainder**	
Copper	–	0.03	–	0.10	–	0.20	**0.7**	**1.5**	**1.0**	**1.5**	**3.0**	**5.0**	**2.8**	**3.8**
Magnesium	–	0.03	**3.0**	**6.0**	**0.2**	**0.6**	**0.8**	**1.5**	**0.4**	**0.6**	**0.1**	**0.3**	–	0.05
Silicon	–	0.03	–	0.30	**10.0**	**13.0**	**10.0**	**13.0**	**4.5**	**5.5**	**5.0**	**7.0**	**4.0**	**6.0**
Iron	–	0.40	–	0.60	–	0.60	–	1.0	–	0.6	–	1.0	–	0.6
Manganese	–	0.03	**0.3**	**0.7**	**0.3**	**0.7**	–	0.5	–	0.5	**0.2**	**0.6**	**0.2**	**0.6**
Nickel	–	0.03	–	0.10	–	0.10	–	1.5	–	0.25	–	0.3	–	0.15
Zinc	–	0.07	–	0.10	–	0.10	–	0.5	–	0.1	–	2.0	–	0.15
Lead	–	0.03	–	0.05	–	0.10	–	0.1	–	0.1	–	0.2	–	0.10
Tin	–	0.03	–	0.05	–	0.05	–	0.1	–	0.05	–	0.1	–	0.05
Titanium	–	–	–	0.20	–	0.02	–	0.2	–	0.2[b]	–	0.2	–	0.20
Each other element[c]	–	–	–	0.05	–	0.05	–	0.05	–	0.05	–	0.05	–	0.05
Total other elements[c]	–	–	–	0.15	–	0.15	–	0.15	–	0.15	–	0.15	–	0.15

Specified impurities are shown in light type. Analysis is required to verify that the contents of the specified impurities are less than the limits given in the table. Analysis for other elements is made when their presence is suspected to be in excess of the 'each other element' limit.

[a] The aluminium content shall be determined by difference; that is, by subtracting the total of all other elements listed.

[b] If titanium alone is used for grain refining, the amount present shall be not less than 0.05%.

[c] In cases when alloys are required in the modified condition, the level of any modifying element present is not limited by the specified maximum value for other elements.

6.3.30 Aluminium alloy castings, group C: special purpose and of limited application

Designation	LM12		LM26		LM28[a]		LM29[a]		LM30		LM31[b]	
Nominal composition	**Al–Cu 10 Mg 0.3**		**Al–Si 10 Cu 3 Mg 1**		**Al–Si 18 Cu 1.5 Mg 1 Ni 1**		**Al–Si 23 Cu 1 Mg 1 Ni 1**		**Al–Si 17 Cu 4.5 Mg 0.5**		**Al–Zn 5 Mg 0.7 Cr 0.5 Ti**	
Nearest alloy(s) in ISO 3522	–		–		–		–		–		**Al–Zn 5 Mg**	
Elements	**min. (%)**	**max. (%)**	**min. (%)**	**max. (%)**	**min. (%)**	**max. (%)**	**min. (%)**	**max. (%)**	**min. (%)**	**max. (%)**	**min. (%)**	**max. (%)**
Aluminium	**Remainder**		**Remainder**		**Remainder**		**Remainder**		**Remainder**		**Remainder**	
Copper	**9.0**	**11.0**	**2.0**	**4.0**	**1.3**	**1.8**	**0.8**	**1.3**	**4.0**	**5.0**	–	0.10
Magnesium	**0.2**	**0.4**	**0.5**	**1.5**	**0.8**	**1.5**	**0.8**	**1.3**	**0.4**	**0.7**	**0.5**	**0.75**
Silicon	–	2.5	**8.5**	**10.5**	**17.0**	**20.0**	**22.0**	**25.0**	**16.0**	**18.0**	–	0.25
Iron	–	1.0	–	1.2	–	0.70	–	0.70	–	1.10	–	0.50
Manganese	–	0.6	–	0.5	–	0.60	–	0.60	–	0.30	–	0.10
Nickel	–	0.5	–	1.0	**0.8**	**1.5**	**0.8**	**1.3**	–	0.10	–	0.10
Zinc	–	0.8	–	1.0	–	0.20	–	0.20	–	0.20	**4.8**	**5.7**
Lead	–	0.1	–	0.2	–	0.10	–	0.10	–	0.10	–	0.05
Tin	–	0.1	–	0.1	–	0.10	–	0.10	–	0.10	–	0.05
Titanium	–	0.2	–	0.2	–	0.20	–	0.20	–	0.20	–	0.25[c]
Chromium	–	–	–	–	–	0.60	–	0.60	–	–	**0.4**	**0.6**
Each other element[e]	–	0.05	–	0.05	–	0.10[d]	–	0.10[d]	–	0.10	–	0.05
Total other elements[e]	–	0.15	–	0.15	–	0.30	–	0.30	–	0.30	–	0.15

Specified impurities are shown in light type. Analysis is required to verify that the contents of the specified impurities are less than the limits given in the table. Analysis for other elements is made when their presence is suspected to be in excess of the 'each other element' limit.

[a] LM28 and LM29 are also subject to metallographical structure requirements (BS 1490, Note 5.4).

[b] LM31 castings in M condition have to be naturally aged for 3 weeks before use, or before determination of mechanical properties.

[c] If titanium alone is used for grain refining, the amount present shall be not less than 0.05%.

[d] Maximum cobalt content.

[e] In cases when alloys are required in the modified condition, the level of any modifying element present is not limited by the specified maximum value for other elements.

6.3.31 Aluminium alloy castings: mechanical properties

Designation[a]	Condition	Tensile strength		Elongation on $5.65\sqrt{S_o}$	
		Sand or investment casting min. (N/mm²)	Chill cast min. (N/mm²)	Sand or investment casting min. (%)	Chill cast min. (%)
Group A					
LM2	M	–	150	–	–
LM4	M	140	160	2	2
	TF	230	280	–	–
LM6	M	160	190	5	7
LM20	M	–	190	–	5
LM24	M	–	180	–	1.5
LM25	M	130	160	2	3
	TE	150	190	1	2
	TB7[b]	160	230	2.5	5
	TF	230	280	–	2
LM27	M	140	160	1	2
Group B					
LMO	M	–	–	–	–
LM5	M	140	170	3	5
LM9	M	–	190	–	3
	TE	170	230	1.5	2
	TF	240	295	–	–
LM13[c]	TE	–	210	–	–
	TF	170	280	–	–
	TF7	140	200	–	–
LM16	TB	170	230	2	3
	TF	230	280	–	–
LM21	M	150	170	1	1
LM22	TB	–	245	–	8
Group C					
LM12	M	–	170	–	–
LM26[c]	TE	–	210	–	–
L28[d]	TE	–	170	–	–
	TF	120	190	–	–
LM29[d]	TE	120	190	–	–
	TF	120	190	–	–
LM30	M	–	150	–	–
	TS	–	160	–	–
LM31[e]	M	215	–	4	–
	TE	215	–	4	–

[a] Properties are obtained on separately cast test samples (see BS 1490: 1988, Note 5.3.1).
[b] After solution treatment, castings have to be heated at a temperature and for a time that will ensure reasonable stability of mechanical properties.
[c] LM13, LM26, LM28 and LM29 are subject to hardness requirements (see BS 1490, Note 5.3.2 and Table 5).
[d] LM28 and LM29 are subject to microstructure requirements (see BS 1490, Note 5.4).
[e] LM31 castings in the M condition have to be naturally aged for 3 weeks before use or before the determination of mechanical properties.

6.3.32 BS EN 29453: 1993

Chemical compositions of tin–lead and tin–lead–antimony solder alloys

Group	Alloy No.	Alloy designation	Melting or solidus/ liquidus temperature (°C)	Chemical composition % (m/m)										
				Sn	Pb	Sb	Cd	Zn	Al	Bi	As	Fe	Cu	Sum of all impurities (except Sb, Bi and Cu)
Tin–lead alloys	1	S-Sn63Pb37	183	62.5–63.5	Rem.	0.12	0.002	0.001	0.001	0.10	0.03	0.02	0.05	0.08
	1a	S-Sn63Pb37E	183	62.5–63.5	Rem.	0.05	0.002	0.001	0.001	0.05	0.03	0.02	0.05	0.08
	2	S-Sn60Pb40	183–190	59.5–60.5	Rem.	0.12	0.002	0.001	0.001	0.10	0.03	0.02	0.05	0.08
	2a	S-Sn60Pb40E	183–190	59.5–60.5	Rem.	0.05	0.002	0.001	0.001	0.05	0.03	0.02	0.05	0.08
	3	S-Pb50Sn50	183–215	49.5–50.5	Rem.	0.12	0.002	0.001	0.001	0.10	0.03	0.02	0.05	0.08
	3a	S-Pb50Sn50E	183–215	49.5–50.5	Rem.	0.05	0.002	0.001	0.001	0.05	0.03	0.02	0.05	0.08
	4	S-Pb55Sn45	183–226	44.5–45.5	Rem.	0.50	0.005	0.001	0.001	0.25	0.03	0.02	0.08	0.08
	5	S-Pb60Sn40	183–235	39.5–40.5	Rem.	0.50	0.005	0.001	0.001	0.25	0.03	0.02	0.08	0.08
	6	S-Pb65Sn35	183–245	34.5–35.5	Rem.	0.50	0.005	0.001	0.001	0.25	0.03	0.02	0.08	0.08
	7	S-Pb70Sn30	183–255	29.5–30.5	Rem.	0.50	0.005	0.001	0.001	0.25	0.03	0.02	0.08	0.08
	8	S-Pb90Sn10	268–302	9.5–10.5	Rem.	0.50	0.005	0.001	0.001	0.25	0.03	0.02	0.08	0.08
	9	S-Pb92Sn8	280–305	7.5–8.5	Rem.	0.50	0.005	0.001	0.001	0.25	0.03	0.02	0.08	0.08
	10	S-Pb98Sn2	320–325	1.5–2.5	Rem.	0.12	0.002	0.001	0.001	0.10	0.03	0.02	0.05	0.08
Tin–lead alloys with antimony	11	S-Sn63Pb37Sb	183	62.5–63.5	Rem.	0.12–0.50	0.002	0.001	0.001	0.10	0.03	0.02	0.05	0.08
	12	S-Sn60Pb40Sb	183–190	59.5–60.5	Rem.	0.12–0.50	0.002	0.001	0.001	0.10	0.03	0.02	0.05	0.08
	13	S-Pb50Sn50Sb	183–216	49.5–50.5	Rem.	0.12–0.50	0.002	0.001	0.001	0.10	0.03	0.02	0.05	0.08
	14	S-Pb58Sn40Sb2	185–231	39.5–40.5	Rem.	2.0–2.4	0.005	0.001	0.001	0.25	0.03	0.02	0.08	0.08
	15	S-Pb69Sn30Sb1	185–250	29.5–30.5	Rem.	0.5–1.8	0.005	0.001	0.001	0.25	0.03	0.02	0.08	0.08
	16	S-Pb74Sn25Sb1	185–263	24.5–25.5	Rem.	0.5–2.0	0.005	0.001	0.001	0.25	0.03	0.02	0.08	0.08
	17	S-Pb78Sn20Sb2	185–270	19.5–20.5	Rem.	0.5–3.0	0.005	0.001	0.001	0.25	0.03	0.02	0.08	0.08

Note:
1. All single figure limits are maxima.
2. Elements shown as 'Rem.' (i.e. Remainder) are calculated as differences from 100%.
3. The temperatures given under the heading 'Melting or solidus/liquidus temperature' are for information purposes and are not specified requirements for the alloys.

Chemical compositions of soft solder alloys other than tin–lead and tin–lead–antimony alloys

Group	Alloy No.	Alloy designation	Melting or solidus/ liquidus temperature (°C)	Chemical composition % (m/m)												
				Sn	Pb	Sb	Bi	Cd	Cu	In	Ag	Al	As	Fe	Zn	Sum of all impurities
Tin–antimony	18	S-Sn95Sb5	230–240	Rem.	0.10	4.5–5.5	0.10	0.002	0.10	0.05	0.05	0.001	0.03	0.02	0.001	0.2
Tin–lead–bismuth and bismuth–tin alloys	19	S-Sn60Pb38Bi2	180–185	59.5–60.5	Rem.	0.10	2.0–3.0	0.002	0.10	0.05	0.05	0.001	0.03	0.02	0.001	0.2
	20	S-Pb49Sn48Bi3	178–205	47.5–48.5	Rem.	0.10	2.5–3.5	0.002	0.10	0.05	0.05	0.001	0.03	0.02	0.001	0.2
	21	S-Bi57Sn43	138	42.5–43.5	0.05	0.10	Rem.	0.002	0.10	0.05	0.05	0.001	0.03	0.02	0.001	0.2
Tin–lead–cadmium	22	S-Sn50Pb32Cd18	145	49.5–50.5	Rem.	0.10	0.10	17.5–18.5	0.10	0.05	0.05	0.001	0.03	0.02	0.001	0.2
Tin–copper and tin–lead–copper alloys	23	S-Sn99Cu1	230–240	Rem.	0.10	0.05	0.10	0.002	0.45–0.90	0.05	0.05	0.001	0.03	0.02	0.001	0.2
	24	S-Sn97Cu3	230–250	Rem.	0.10	0.05	0.10	0.002	2.5–3.5	0.05	0.05	0.001	0.03	0.02	0.001	0.2
	25	S-Sn60Pb38Cu2	183–190	59.5–60.5	Rem.	0.10	0.10	0.002	1.5–2.0	0.05	0.05	0.001	0.03	0.02	0.001	0.2
	26	S-Sn50Pb49Cu1	183–215	49.5–50.5	Rem.	0.10	0.10	0.002	1.2–1.6	0.05	0.05	0.001	0.03	0.02	0.001	0.2
Tin–indium	27	S-Sn50In50	117–125	49.5–50.5	0.05	0.05	0.10	0.002	0.05	Rem.	0.01	0.001	0.03	0.02	0.001	0.2
Tin–silver and tin–lead–silver alloys	28	S-Sn96Ag4	221	Rem.	0.10	0.10	0.10	0.002	0.05	0.05	3.5–4.0	0.001	0.03	0.02	0.001	0.2
	29	S-Sn97Ag3	221–230	Rem.	0.10	0.10	0.10	0.002	0.10	0.05	3.0–3.5	0.001	0.03	0.02	0.001	0.2
	30	S-Sn62Pb36Ag2	178–190	61.5–62.5	Rem.	0.05	0.10	0.002	0.05	0.05	1.8–2.2	0.001	0.03	0.02	0.001	0.2
	31	S-Sn60Pb36Ag4	178–180	59.5–60.5	Rem.	0.05	0.10	0.002	0.05	0.05	3.0–4.0	0.001	0.03	0.02	0.001	0.2
Lead–silver and lead–tin–silver alloys	32	S-Pb98Ag2	304–305	0.25	Rem.	0.10	0.10	0.002	0.05	0.05	2.0–3.0	0.001	0.03	0.02	0.001	0.2
	33	S-Pb95Ag5	304–365	0.25	Rem.	0.10	0.10	0.002	0.05	0.05	4.5–6.0	0.001	0.03	0.02	0.001	0.2
	34	S-Pb93Sn5Ag2	298–301	4.8–5.2	Rem.	0.10	0.10	0.002	0.05	0.05	1.2–1.8	0.001	0.03	0.02	0.001	0.2

Notes:

1. All single figure limits are maxima.
2. Elements shown as 'Rem.' (i.e. Remainder) are calculated as differences from 100%.
3. The temperatures given under the heading 'Melting or solidus/liquidus temperature' are for information purposes and are not specified requirements for the alloys.

6.3.33 Typical uses of soft solders

Group	Alloy no.	Typical uses
Tin–lead alloys	1 1a 2	Soldering of electrical connections to copper; soldering brass to zinc; hand soldering of electronic assemblies; hot dip coating of ferrous and non-ferrous metals; high-quality sheet metal work; capillary joints including light gauge tubes in copper and stainless steel; manufacture of electronic components; machine soldering of printed circuits.
	2a	Hand and machine soldering of electronic components; can soldering.
	3 3a 4 5	General engineering work on copper, brass and zinc; can soldering.
	6 7	Jointing of electric cable sheaths; 'wiped' joints.
	8 9	Lamp solder; dipping solder; for service at very low temperatures (e.g. less than −60°C).
Tin–lead alloys with antimony	11 12	Hot dip coating and soldering of ferrous metals; high-quality engineering; capillary joints of ferrous metals; jointing of copper conductors.
	13 14	General engineering; heat exchangers; general dip soldering; jointing copper conductors.
	15 16 17	Plumbing, wiping of lead and lead alloy cable sheathing; dip soldering.
Tin–antimony alloys	18	High service temperature (e.g. greater than 100°C) and refrigeration equipment; step soldering.
Tin–lead bismuth Bismuth–tin alloys	19 20 21	Low melting point solder for assemblies that could be damaged by normal soldering temperatures; step soldering for thermal cutouts.
Tin–lead cadmium alloys	22	As above, but more ductile.
Tin–copper Tin–lead–copper alloys	23 24 25 26	For capillary joints in all copper plumbing installations and particularly in those installations where the lead content of the solder is restricted.
Tin–indium	27	Very low melting range alloy.
Tin–silver	28 29	High service temperature (e.g. greater than 100°C).
Tin–lead silver alloys	30 31	For capillary joints in all copper plumbing installations and particularly in those installations where the lead content of the solder is restricted.
Lead–silver Lead–tin–silver alloys	32 33 34	For service both at high (e.g. greater than 100°C) and very low (e.g. less than −60°C) temperatures; soldering of silver-coated substrates.

6.3.34 Silver soldering (hard soldering)

Silver solders have a higher melting range, greater ductility and very much greater strength than soft solders. Special fluxes based on boron are required and, because of the high melting temperatures involved, the joint has to be heated with a brazing torch (blowpipe).

6.3.35 Group AG: silver brazing filler metals

Type	Composition (% by mass)																Melting range (approximately)	
	Silver		Copper		Zinc		Cadmium[a]		Tin		Manganese		Nickel		Impurity limits applicable to all types		Solidus	Liquidus
	min.	max.	min.	max.	min.	max.	min.	max.	min.	max.	min.	max.	min.	max.		max.	(°C)	(°C)
AG1	49.0	51.0	14.0	16.0	14.0	18.0	18.0	20.0	–	–	–	–	–	–	Aluminium	0.0010	620	640
AG2	41.0	43.0	16.0	18.0	14.0	18.0	24.0	26.0	–	–	–	–	–	–	Beryllium	0.0005	610	620
AG3	37.0	39.0	19.0	21.0	20.0	24.0	19.0	21.0	–	–	–	–	–	–	Bismuth	0.005	605	650
AG11	33.0	35.0	24.0	26.0	18.0	22.0	20.0	22.0	–	–	–	–	–	–	Lead	0.025	610	670
AG12	29.0	31.0	27.0	29.0	19.0	23.0	20.0	22.0	–	–	–	–	–	–	Phosphorus	0.008	600	690
															Silicon	0.05		
AG14	54.0	56.0	20.0	22.0	21.0	23.0	–	0.025	1.7	2.3	–	–	–	–	Titanium	0.002	630	660
AG20[b]	39.0	41.0	29.0	31.0	27.0	29.0	–	0.025	1.7	2.3	–	–	–	–	Zirconium	0.002	650	710
AG21[b]	29.0	31.0	35.0	37.0	31.0	33.0	–	0.025	1.7	2.3	–	–	–	–			665	755
															Total of all impurities	0.25		
AG5	42.0	44.0	36.0	38.0	18.0	22.0	–	0.025	–	–	–	–	–	–			690	770
AG7[c]	71.0	73.0	27.0	29.0	–	–	–	0.025	–	–	–	–	–	–			780	780
AG9	49.0	51.0	14.5	16.5	13.5	17.5	15.0	17.0	–	–	–	–	2.5	3.5			635	655
AG13	59.0	61.0	25.0	27.0	12.0	16.0	–	0.025	–	–	–	–	–	–			695	730
AG18	48.0	50.0	15.0	17.0	21.0	25.0	–	0.025	–	–	6.5	8.5	4.0	5.0			680	705
AG19	84.0	86.0	–	–	–	–	–	0.025	–	–	14.0	16.0	–	–			960	970

[a]Filler metals containing cadmium produce fumes during brazing that, if inhaled, can be dangerous to health. In view of the low-threshold limit value (TLV) for cadmium oxide, 0.05 mg/m^2, it is recommended that a local extraction system is used during brazing.

[b]Attention is drawn to the fact that it is claimed that AG20 and AG21 are the subject of British patent No. 1436943 and No. 1532879, respectively, copies of which can be obtained from the Patent Office, 25 Southampton Buildings, London WC2A 1AY. BSI takes no position as to the validity of the patents or whether they are still in force. The patents are irrevocably endorsed 'licences of right' under Section 46 of the Patents Act 1977 (or Section 35(2)(a) of the Patents Act 1949 if applicable), which states:

'(3) Where such an entry is made in respect of a patent:

(a) any person shall, at any time after the entry is made, be entitled as of right to a licence under the patent on such terms as may be settled by agreement or, in default of agreement, by the Comptroller on the application of the proprietor of the patent or the person requiring the licence.'

Licence details may be obtained from the registered proprietors of the patents.

[c]Suitable for vacuum applications. For the full range of filler metals (spelters) for brazing, see BS 1845.

6.4 Metallic material sizes

6.4.1 Metallic material sizes: introduction to BS 6722: 1986

British Standard BS 6722 supersedes BS 4229 Parts 1 and 2, BS 4391 and DD5 which are now withdrawn.

In the case of wire, the dimensions recommended are those previously given in BS 4391, since they are firmly and logically established in the wire and associated industries.

For other products where a clear pattern of sizes does not exist, and where availability and demand vary from one sector of industry to another, it was considered that a list of recommended dimensions based on the rounded R20 series (BS 2045) should be established without any attempt to distinguish between product types or between ferrous and non-ferrous materials. Suitable dimensions may then be drawn from this list to establish a range of sizes to meet best the pattern of demand.

The detailed recommendations for bar shapes previously given in BS 4229 have not therefore been included. In a similar way, the recommendations previously given in DD5 for plate and sheet have been rationalized and the recommendations for thickness of sheet and plate are now referenced as for bar dimensions. Details of preferred sizes for hexagon bars are given in BS 3692.

It is recommended that applicable British Standards for metallic materials should include combinations of dimensions recommended in BS 6722: 1986. BS 6722 gives recommended dimensions for use as a basis for establishing the sizes of metallic materials in the form of wires, bars (excluding hexagon bars) and flat products including sheet and plate:

Wires The diameters of wires should be selected from the recommended dimensions given in Section 6.4.2.

Bars The diameters of round bars and the thickness and widths of bars of rectangular or square cross section should be selected from the recommended dimensions given in Section 6.4.3.

Flat products The widths and/or lengths for flat products (sheet, strip and plate) should be selected from the recommended dimensions given in Section 6.4.4. The thicknesses for flat products should be selected from the recommended dimensions given in Section 6.4.3.

6.4.2 Recommended diameters of wires, metric

Choice (values in mm)									
First	Second	Third	First	Second	Third	First	Second	First	Second
0.010			0.100			1.0		10.0	
	0.011			0.112			1.12		11.2
0.012			0.125			1.25		12.5	
		0.013			0.132				
	0.014			0.140			1.40		14.0
		0.015			0.150				
0.016			0.160			1.60		16.0	
		0.017			0.170				
	0.018			0.180			1.80		18.0
		0.019			0.190				
0.020			0.200			2.00		20.0	
		0.021			0.212				
	0.022			0.224			2.25		
		0.024			0.236				
0.025			0.250			2.50			
		0.026			0.265				
	0.028			0.280			2.80		
		0.030			0.300				
0.032			0.315			3.15			
		0.034			0.335				
	0.036			0.355			3.55		
		0.038			0.375				
0.040			0.400			4.00			
		0.042			0.425				
	0.045			0.450			4.50		
		0.048			0.480				
0.050			0.500			5.00			
		0.053			0.530				
	0.056			0.560			5.60		
		0.060			0.600				
0.063			0.630			6.30			
		0.067			0.670				
	0.071			0.710			7.10		
		0.075			0.750				
0.080			0.800			8.00			
		0.085			0.850				
	0.090			0.900			9.00		
		0.095			0.950				

For metric inch wire-gauge equivalent, see Section 6.4.7.

6.4.3 Recommended dimensions for bar and flat products

Choice (values in mm)							
First	Second	First	Second	First	Second	First	Second
0.10		1.0		10.0		100	
	0.11		1.1		11.0		110
0.12		1.2		12.0		120	
	0.14		1.4		14.0		140
0.16		1.6		16.0		160	
	0.18		1.8		18.0		180
0.20		2.0		20.0		200	
	0.22		2.2		22.0		220
0.25		2.5		25.0		250	
	0.28		2.8		28.0		280
0.30		3.0		30.0		300	
	0.35		3.5		35.0		
0.40		4.0		40.0			
	0.45		4.5		45.0		
0.50		5.0		50.0			
	0.55		5.5		55.0		
0.60		6.0		60.0			
	0.70		7.0		70.0		
0.80		8.0		80.0			
	0.90		9.0		90.0		

6.4.4 Recommended widths and lengths of flat products

Width (mm)	Length (mm)
400	2500
500	3000
600	4000
800	5000
1000	6000
1200	8000
1250	10000
1500	
2000	

6.4.5 Mass of metric round and square bars

	Steel		Copper		Brass		Tin bronze		Duralumin	
Size (mm)	Round	Square	Round	Square	Round	Square	Round	Square	Round	Square
1.0	0.0062	0.0079	0.0071	0.0090	0.0065	0.0085	0.0069	0.0088	0.0023	0.0027
1.1	0.0075	0.0095	0.0085	0.0108	0.0081	0.0103	0.0084	0.0106	0.0029	0.0035
1.2	0.0089	0.0113	0.0101	0.0129	0.0096	0.0122	0.0100	0.0127	0.0032	0.0041
1.4	0.0120	0.0154	0.0137	0.0175	0.0130	0.0166	0.0134	0.0173	0.0044	0.0056
1.6	0.0160	0.0201	0.0181	0.024	0.0173	0.0217	0.0179	0.0225	0.0058	0.0073
1.8	0.020	0.0254	0.023	0.029	0.0216	0.0275	0.0224	0.0285	0.0073	0.0093
2.0	0.025	0.0314	0.028	0.036	0.027	0.034	0.028	0.035	0.0091	0.0114
2.5	0.039	0.0491	0.044	0.056	0.042	0.053	0.043	0.055	0.0142	0.0179
3.0	0.055	0.0707	0.063	0.080	0.060	0.076	0.062	0.079	0.0200	0.0257
3.5	0.076	0.0962	0.087	0.109	0.082	0.104	0.085	0.108	0.0277	0.0350
4.0	0.099	0.126	0.113	0.143	0.107	0.136	0.111	0.141	0.0360	0.0459
4.5	0.125	0.159	0.142	0.181	0.135	0.172	0.140	0.178	0.0455	0.0579
5.0	0.154	0.196	0.175	0.223	0.166	0.212	0.173	0.220	0.0560	0.0713
5.5	0.187	0.237	0.213	0.270	0.202	0.256	0.209	0.265	0.0680	0.0863
6.0	0.222	0.283	0.253	0.322	0.240	0.306	0.249	0.317	0.0808	0.1030
7.0	0.302	0.385	0.344	0.438	0.326	0.416	0.338	0.431	0.110	0.140
8.0	0.395	0.502	0.450	0.571	0.427	0.543	0.442	0.562	0.144	0.183
9.0	0.499	0.636	0.568	0.724	0.539	0.688	0.559	0.712	0.182	0.232
10.0	0.617	0.785	0.702	0.893	0.667	0.849	0.691	0.879	0.225	0.286
11.0	0.746	0.950	0.849	1.081	0.806	1.027	0.836	1.064	0.272	0.346
12.0	0.888	1.130	1.010	1.286	0.960	1.222	0.995	1.266	0.323	0.411
14.0	1.208	1.539	1.375	1.751	1.306	1.664	1.353	1.724	0.440	0.560
16.0	1.578	2.010	1.796	2.287	1.706	2.173	1.767	2.251	0.574	0.732

18.0	1.998	2.543	2.274	2.894	2.160	2.749	2.238	2.848	0.727	0.926
20.0	2.466	3.140	2.806	3.573	2.666	3.394	2.762	3.517	0.898	1.143
25.0	3.853	4.906	4.385	5.583	4.165	5.303	4.315	5.495	1.402	1.786
30.0	5.158	7.065	5.870	8.040	5.876	7.637	5.777	7.913	1.878	2.572
35.0	7.553	9.616	8.595	10.943	8.165	10.395	8.459	10.770	2.749	3.500
40.0	9.865	12.56	11.23	14.29	10.66	13.58	11.05	14.07	3.590	4.572
45.0	12.48	15.90	14.20	18.09	13.49	17.19	13.98	17.81	4.542	5.788
50.0	15.41	19.63	17.54	22.34	16.66	21.22	17.26	21.99	5.609	7.145
55.0	18.65	23.75	21.22	27.03	20.16	25.67	20.89	26.60	6.789	8.645
60.0	22.20	28.26	25.26	32.16	24.00	30.55	24.86	31.65	8.081	10.29
70.0	30.21	38.47	34.38	43.78	32.66	41.59	33.84	43.09	11.00	14.00
80.0	39.46	50.24	44.91	57.17	42.66	54.31	44.20	56.27	14.36	18.29
90.0	49.94	63.59	56.83	72.37	53.99	68.74	55.93	71.22	18.18	23.15
100.0	61.65	78.50	70.16	89.33	66.64	84.86	69.05	87.92	22.44	28.57
120.0	88.78	113.04	101.03	128.64	95.97	122.20	99.43	126.60	32.32	41.15
160.0	167.84	200.96	191.00	228.69	181.44	217.24	187.98	225.08	61.09	73.15
200.0	246.60	314.00	280.63	319.36	266.58	339.43	276.19	351.68	89.76	114.30
250.0	385.31	490.63	438.48	558.34	416.52	530.37	431.55	549.51	140.25	178.59
300.0	554.80	706.50	631.36	804.00	599.74	763.73	621.38	791.28	201.95	257.17

Values in kg/m.

6.4.6 Hexagon bar sizes for screwed fasteners, metric

Choice	Nominal size of thread	Hexagon bar sizes (mm)				Mass (kg/m) (max. bar size: steel)
		A/F		A/C		
		max.	min.	max.	min.	
First (preferred)	M1.6	3.2	3.08	3.7	3.48	0.07
	M2	4.0	3.88	4.6	4.38	0.11
	M2.5	5.0	4.88	5.8	5.51	0.17
	M3	5.5	5.38	6.4	6.08	0.21
	M4	7.0	6.85	8.1	7.74	0.33
	M5	8.0	7.85	9.2	8.87	0.44
	M6	10.0	9.78	11.5	11.05	0.68
	M8	13.0	12.73	15.0	14.38	1.15
	M10	17.0	16.73	19.6	18.90	1.97
	M12	19.0	18.67	21.9	21.10	2.45
	M16	24.0	23.67	27.7	26.75	3.62
	M20	30.0	29.67	34.6	33.53	6.12
	M24	36.0	35.38	41.6	39.98	8.81
	M30	46.0	45.38	53.1	51.28	14.40
	M36	55.0	54.26	63.5	61.31	20.52
	M42	65.0	64.26	75.1	72.61	28.67
	M48	75.0	74.26	86.6	83.91	35.15
	M56	85.0	84.13	98.1	95.07	49.00
	M64	95.0	94.13	109.7	106.37	61.22
Second (non-preferred)	M14	22.0	21.67	25.4	24.49	3.29
	M18	27.0	26.67	31.2	30.14	4.95
	M22	32.0	31.61	36.9	35.74	6.96
	M27	41.0	40.38	47.3	45.63	11.43
	M33	50.0	49.38	57.7	55.80	17.00
	M39	60.0	59.26	69.3	66.96	24.42
	M45	70.0	69.26	80.8	78.26	32.23
	M52	80.0	79.26	92.4	89.56	43.41
	M60	90.0	89.13	103.9	100.72	55.00
	M68	100.0	99.13	115.5	112.02	67.84

A/F: across flats; A/C: across corners.

The mass of common non-ferrous hexagon bars can be determined by multiplying the mass of a corresponding steel bar by one of the following conversion factors:

Brass × 1.081
Bronze × 1.120
Duralumin × 0.364.

For example, the mass of a hexagonal steel bar 30 mm across flats (A/F) is 6.12 kg/m. Thus the mass of a hexagonal brass bar 30 mm A/F is 6.12 × 1.081 = 6.616 kg/m.

6.4.7 Gauge sizes and equivalents

ISWG	British (in.)	Metric (mm)	Mass (kg/m) (steel)	ISWG	British (in.)	Metric (mm)	Mass (kg/m) (steel)
50	0.0010	0.025	0.0000044	21	0.032	0.8	0.0041
49	0.0012	0.030	0.0000057	20	0.036	0.9	0.0052
48	0.0016	0.040	0.0000102	19	0.040	1.0	0.0064
47	0.0020	0.050	0.000016	18	0.048	1.2	0.0092
46	0.0024	0.061	0.000023	17	0.056	1.4	0.0125
45	0.0028	0.071	0.000032	16	0.064	1.6	0.0163
44	0.0032	0.081	0.000041	15	0.072	1.8	0.0211
43	0.0036	0.091	0.000052	14	0.080	2.0	0.025
42	0.0040	0.101	0.000064	13	0.092	2.3	0.034
41	0.0044	0.112	0.000077	12	0.104	2.6	0.043
40	0.0048	0.122	0.000092	11	0.116	3.0	0.054
39	0.0052	0.127	0.000108	10	0.128	3.3	0.065
38	0.0060	0.152	0.000144	9	0.144	3.7	0.083
37	0.0068	0.172	0.000185	8	0.160	4.1	0.102
36	0.0076	0.177	0.000231	7	0.176	4.5	0.123
35	0.0084	0.203	0.000256	6	0.192	4.9	0.147
34	0.0092	0.230	0.000338	5	0.212	5.4	0.178
33	0.0100	0.254	0.000397	4	0.232	5.9	0.214
32	0.0108	0.27	0.000463	3	0.252	6.4	0.25
31	0.0116	0.28	0.000527	2	0.276	7.0	0.31
30	0.0124	0.32	0.000613	1	0.300	7.6	0.36
29	0.0136	0.35	0.000735	1/0	0.324	8.2	0.42
28	0.0148	0.37	0.00087	2/0	0.348	8.8	0.48
27	0.0164	0.4	0.00107	3/0	0.372	9.4	0.55
26	0.0180	0.45	0.00129	4/0	0.400	10.2	0.64
25	0.0200	0.5	0.00159	5/0	0.432	11.0	0.74
24	0.022	0.55	0.00192	6/0	0.464	11.8	0.86
23	0.024	0.6	0.0023	7/0	0.500	12.7	1.0
22	0.028	0.7	0.0031				

The mass of common non-ferrous wires can be determined by multiplying the mass of a corresponding steel wire by one of the following conversion factors:

Brass $\times$ 1.081
Bronze $\times$ 1.120
Duralumin $\times$ 0.364.

6.5 Polymeric (plastic) materials

6.5.1 Polymeric (plastic) materials – introduction

Polymeric materials are conventionally referred to as 'plastics'. This is a misnomer since polymeric materials rarely show plastic properties in their finished condition; in fact, many of them are elastic. However, during the moulding process by which they are formed they are reduced to a plastic condition by heating them to a temperature above that of boiling water, and it is from this that they get the generic name of *plastics*.

There are two main groups of polymeric materials:

Thermoplastics These can be softened as often as they are reheated. They are not so rigid as the thermosetting plastic materials but tend to be tougher. Examples of thermoplastic polymeric materials are listed in Section 6.5.3.

Thermosetting plastics (thermosets) These undergo a chemical change during moulding and can never again be softened by reheating. This chemical change, called *curing*, is triggered by the temperature and pressure of the moulding process. These materials are harder and more brittle than the thermoplastic materials (see Section 6.5.2).

Polymers

Polymers are formed by combining together a large number of basic units (monomer molecules) to form long-chain molecules (polymers). These polymer molecules may be one of three types as follows:

Linear polymer chain

Linear polymer chains can move past each other easily, resulting in a non-rigid, flexible, thermoplastic material such as *polyethylene*.

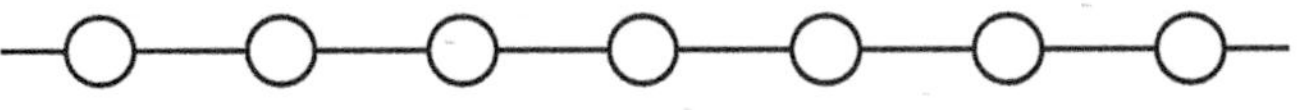

Branched linear polymer chain

It is more difficult for branched linear chains to move past each other, and materials with molecules of this configuration are more rigid, harder and stronger. Such materials also tend to be less dense since the molecule chains cannot pack so closely together. Heat energy is required to break down the side branches so that the chains can flow more easily, and this raises their melting point above that for materials with a simple linear chain.

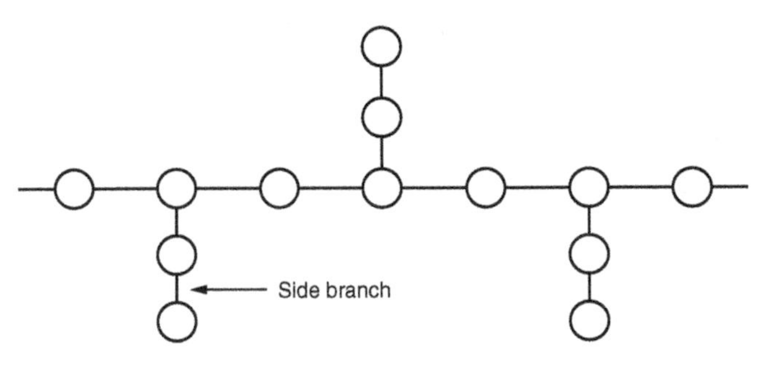

Cross-linked polymer

The cross-linked molecular chain is typical of the thermosetting plastics. Thermosets are rigid and tend to be brittle once the cross-links are formed by 'curing' the material during the moulding process. The *elastomers* are an intermediate group of materials which exhibit the toughness and resilience of rubber. This is achieved by a more limited cross-linking than that of the rigid thermosets.

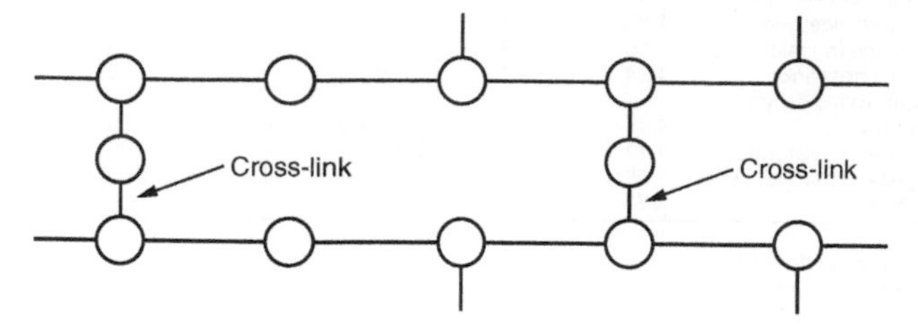

Reinforced polymeric materials

In this group of materials synthetic, polymeric materials are used to bond together strong reinforcing materials to produce high-strength composites.

Glass-reinforced plastics (GRP)

High-strength glass fibres are bonded together using polyester or epoxide resins. The fibres may be in the form of rovings (ropes), woven cloth or chopped strand mat. GRP are used for a wide range of products, including: printed circuit boards for high-quality electronic equipment, safety helmets, and boat hulls and superstructures.

Laminated plastics

Sheets of paper, cotton cloth, woollen cloth or woven glass fibre are impregnated with an appropriate synthetic resin and then stacked between polished metal sheets in a hydraulic press. The combined heat and pressure cause the laminates to bond together and to cure. The moulded sheets, rods, tubes and other sections produced by the process have high strength and can be machined on conventional machined tools into screwed fastenings, bushes, gears, etc. in a manner similar to metals. A typical range of such materials is available under the trade name of *Tufnol*.

Adhesives

For detail, see Section 4.5.

6.5.2 Some important thermosetting polymers

Material	Relative density	Tensile strength (N/mm^2)	Elongation (%)	Impact strength (J)	Maximum service temperature (°C)
Phenol formaldehyde[a]	1.35	35–55	1.0	0.3–1.5	75
Urea formaldehyde[b]	1.50	50–75	1.0	0.3–0.5	75
Melamine formaldehyde[b]	1.50	56–80	0.7	0.2–0.4	100
Casein (cross-linked with formaldehyde)	1.34	55–70	2.5–4.0	1.5–2.0	150
Epoxides[c]	1.15	35–80	5.0–10.0	0.5–1.5	200
Polyesters (unsaturated)	1.12	50–60	2.0	0.7	220
Polyesters (alkyd resins)	2.00	20–30	0.0	0.25	150
Silicones	1.88	35–45	30–40	0.4	450

[a] With wood flour filler.
[b] With cellulose filler.
[c] Rigid, unfilled.

Additives

Plasticizers These reduce the rigidity and brittleness of polymeric materials and improve their flow properties during moulding.

Fillers These are bulking agents which not only reduce the cost of the moulding powder, but have a considerable influence on the properties of a moulding produced from a given polymeric material. Fillers improve the impact strength and reduce shrinkage during moulding. Typical fillers are:

Glass fibre: good electrical properties.
Wood flour, calcium carbonate: low cost, high bulk, low strength.
Aluminium powder: expensive but high strength and wear resistance.
Shredded paper (*cellulose*), *shredded cloth*: good mechanical strength with reasonable electrical insulation properties.
Mica granules: good heat resistance and good electrical insulation properties.

Stabilizers These are added to prevent or reduce degradation, and include antioxidants, antiozonants and ultraviolet ray absorbants.

Colourants These can be subdivided into dyestuffs, organic pigments and inorganic pigments. Dyestuffs are used for transparent and translucent plastics. Pigments have greater opacity, colour stability and heat stability than dyestuffs. They are unsuitable for transparent plastics.

Antistatic agents These provide improved surface conductivity so that static electrical charges can leak away, thus reducing the attraction of dust particles, the risk of electric shock and the risk of explosion in hazardous environments caused by the spark associated with an electrical discharge.

6.5.3 Some important thermoplastic polymers

Material	Crystallinity (%)	Density (kg/m³)	Tensile strength (N/mm²)	Elongation (%)	Impact strength (J)	T_g^a (°C)	T_m^a (°C)	Maximum service temperature (°C)
Polyethylene (polythene)	60[b]	920	11	100–600	No fracture	−120	115	85
	95[c]	950	31	50–800	5–15	−120	138	125
	60	900	30–35	50–600	1–10	−25	176	150
Polypropylene	0	1070	28–53	1–35	0.25–2.5	100	–	65–85
Polystyrene	0	1400[d]	49	10–130	1.5–1.8	87	–	70
Polyvinyl chloride (PVC)	0	1300[e]	7–25	240–380	–	87	–	60–105
Polytetrafluoroethylene (Teflon) (PTFE)	90	2170	17–25	200–600	3–5	−126	327	260
Polymethyl methacrylate (Plexiglass) (Perspex)	0	1180	50–70	3–8	0.5–0.7	0	–	95
Acrylonitrile–butadiene–styrene (ABS) (high-impact polystyrene)	0	1100	30–35	10–140	7–12	−55	–	100
Polyamides (nylon) (properties for nylon 66)	Variable	1140	50–85	60–300	1.5–15.0	50	265	120
Polyethylene teraphthalate (Terylene): one of the polyester group	60	1350	Over 175	60–110	1.0	70	267	69
Polyformaldehyde[f]	70–90	1410	50–70	15–75	0.5–2.0	−73	180	105
Polyoxymethylene[f]	70–90	1410	60–70	15–70	0.5–2.0	−76	180	120
Polycarbonate	0	1200	60–70	60–100	10–20	150	–	130
Cellulose acetate	0	1280	24–65	5–55	0.7–7.0	120	–	70
Cellulose nitrate (highly flammable)	0	1400	35–70	10–40	3–11	53	–	–
Polyvinylidene chloride	60	1680	25–35	Up to 200	0.4–1.3	−17	198	60

[a] T_g = glass transition temperature, T_m = melting temperature.
[b] Low-density polyethylene.
[c] High-density polyethylene.
[d] Unplasticized.
[e] Plasticized.
[f] Members of the polyacetal group of plastics: these are strong and stiff with high creep resistance and resistance to fatigue.

7 Linear and Geometric Tolerancing of Dimensions

7.1 Linear tolerancing

7.1.1 Limits and fits

It is not possible to work to an exact size nor is it possible to measure to an exact size. Therefore dimensions are given limits of size. Providing the dimensions of a part lie within the limits of size set by the designer, then the part will function correctly. Similarly the dimensions of gauges and measuring equipment are given limits of size. As a general rule, the limits of size allocated to gauges and measuring instruments are approximately 10 times more accurate than the dimensions they are intended to check (gauges) or measure (measuring instruments).

The upper and lower sizes of a dimension are called the *limits* and the difference in size between the limits is called the *tolerance*. The terms associated with limits and fits can be summarized as follows:

- *Nominal size*: This is the dimension by which a feature is identified for convenience. For example, a slot whose actual width is 25.15 mm would be known as the 25-mm wide slot.
- *Basic size*: This is the exact functional size from which the limits are derived by application of the necessary allowance and tolerances. The basic size and the nominal size are often the same.
- *Actual size*: The measured size corrected to what it would be at 20°C.
- *Limits*: These are the high and low values of size between which the size of a component feature may lie. For example, if the lower limit of a hole is 25.05 mm and the upper limit of the same hole is 25.15 mm, then a hole which is 25.1 mm diameter is *within limits* and is acceptable. Examples are shown in Fig. 7.1(a).
- *Tolerance*: This is the difference between the limits of size. That is, the upper limit minus the lower limit. Tolerances may be bilateral or unilateral as shown in Fig. 7.1(b) and (c).
- *Deviation*: This is the difference between the basic size and the limits. The deviation may be symmetrical, in which case the limits are equally spaced above and below the basic size (e.g. 50.00 ± 0.15 mm). Alternatively, the deviation may be asymmetrical, in which case the deviation may be greater on one side of the basic size than on the other (e.g. $50.00 + 0.25$ or -0.05).
- *Mean size*: This size lies halfway between the upper and lower limits of size, and must not be confused with either the nominal size nor the basic size. It is only the same as the basic size when the deviation is symmetrical.
- *Minimum clearance (allowance)*: This is the clearance between a shaft and a hole under maximum metal conditions. That is, the largest shaft in the smallest hole that the limits will allow. It is the tightest fit between shaft and hole that will function correctly. With a *clearance fit* the allowance is positive. With an *interference fit* the allowance is negative. These types of fit are discussed in the next section.

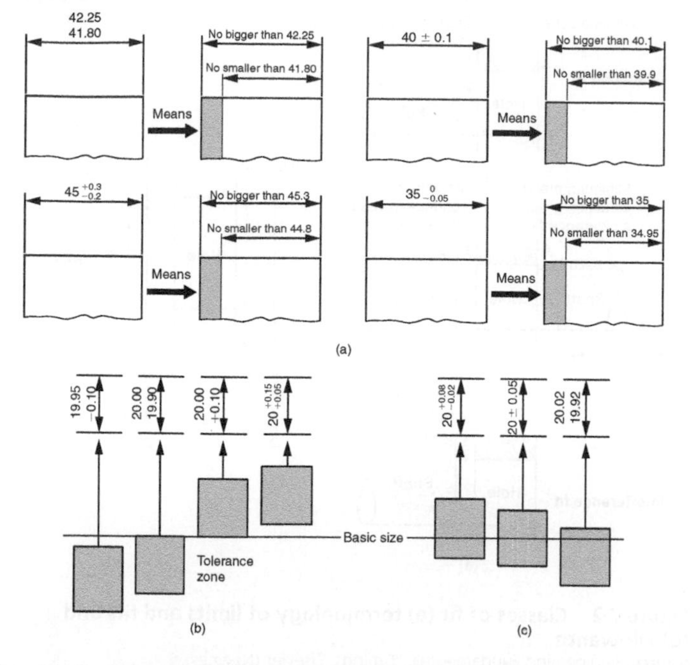

Figure 7.1 Toleranced dimensions (a) methods of tolerancing; (b) unilateral tolerance and (c) bilateral tolerance. Dimensions in mm.

Source: Engineering Fundamentals: Timings: Elsevier (Newnes).

7.1.2 Classes of fit

Figure 7.2(a) shows the classes of fit that may be obtained between mating components. In the *hole basis system* the hole size is kept constant and the shaft size is varied to give the required class of fit. In an *interference fit* the shaft is always slightly larger than the hole. In a *clearance fit* the shaft is always slightly smaller than the hole. A *transition fit* occurs when the tolerances are so arranged that under maximum metal conditions (largest shaft: smallest hole) an interference fit is obtained, and that under minimum metal conditions (largest hole: smallest shaft) a clearance fit is obtained. The hole basis system is the most widely used since most holes are produced by using standard tools such as drills and reamers. It is then easier to vary the size of the shaft by turning or grinding to give the required class of fit.

In a *shaft basis system* the shaft size is kept constant and the hole size is varied to give the required class of fit. Again, the classes of fit are *interference fit transition fit* and *clearance fit*. Figure 7.2(b) shows the terminology relating to limits and fits.

7.1.3 Accuracy

The greater the accuracy demanded by a designer, the narrower will be the tolerance band and the more difficult and costly will it be to manufacture the component within the limits

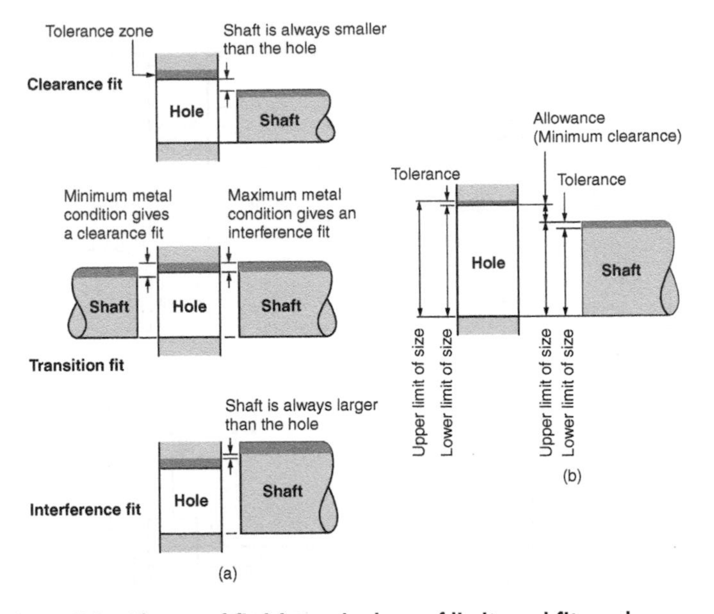

Figure 7.2 Classes of fit (a) terminology of limits and fits and (b) allowance.
Source: Engineering Fundamentals: Timings: Elsevier (Newnes).

specified. Therefore, for ease of manufacture at minimum cost, a designer never specifies an accuracy greater than is necessary to ensure the correct functioning of the component. The more important factors affecting accuracy when measuring components are as follows.

7.2 Standard systems of limits and fits (introduction)

This section is based on BS EN 20286-2: 1993 and is compatible with ISO 286-2: 1988 tables of standard tolerance grades and limit deviations for holes and shafts. These tables in these standards are suitable for all classes of work from the finest instruments to heavy engineering. It allows for the size of the work, the type of work, and provides for both hole basis and shaft basis systems as required.

7.2.1 Application of tables of limits and fits

The tables provide for 28 types of *shaft* designated by lower-case letters, a, b, c, etc., and 28 types of *hole* designated by upper-case (capital) letters A, B, C, etc. To each type of shaft or hole the grade of tolerance is designated by a number 1–18 inclusive. The letter indicates the position of the tolerance relative to the basic size and is called the *fundamental deviation*. The number indicates the magnitude of the tolerance and is called the *fundamental tolerance*. A shaft is completely defined by its basic size, letter and number (e.g. 75 mm h6). Similarly a hole is completely defined by its basic size, letter and number, for example: 75 mm H7. Table 7.1 shows a selection of limits and fits for a wide range of hole and shaft combinations for a variety of applications.

Table 7.1 Primary selection of fits.

Normal sizes		Loose clearance		Average clearance		Close clearance		Precision clearance		Transition		Interference	
Over (mm)	Up to (mm)	H9	e9	H8	f7	H7	g6	H7	h6	H7	k6	H7	p6
–	3	+25	−14	+14	−6	+10	−2	+10	−0	+10	+6	+10	+12
		+0	−39	+0	−16	+0	−8	+0	−6	+0	+0	+0	+6
3	6	+30	−20	+18	−10	+12	−4	+12	−0	+12	+9	+12	+20
		+0	−50	+0	−22	+0	−12	+0	−8	+0	+1	+0	+12
6	10	+36	−25	+22	−13	+15	−5	+15	−0	+15	+10	+15	+24
		+0	−61	+0	−28	+0	−14	+0	−9	+0	+1	+0	+15
10	18	+43	−32	+27	−16	+18	−6	+18	−0	+18	+12	+18	+29
		+0	−75	+0	−34	+0	−17	+0	−11	+0	+1	+0	+18
18	30	+52	−40	+33	−20	+21	−7	+21	−0	+21	+15	+21	+35
		+0	−92	+0	−41	+0	−20	+0	−13	+0	+2	+0	+22
30	50	+62	−50	+39	−25	+25	−9	+25	−0	+25	+18	+25	+42
		+0	−112	+0	−50	+0	−25	+0	−16	+0	+2	+0	+26
50	80	+74	−60	+46	−30	+30	−10	+30	−0	+30	+21	+30	+51
		+0	−134	+0	−60	+0	−29	+0	−19	+0	+2	+0	+32
80	120	+87	−72	+54	−36	+35	−12	+35	−0	+35	+25	+35	+59
		+0	−159	+0	−71	+0	−34	+0	−22	+0	+3	+0	+37
120	180	+100	−85	+63	−43	+40	−14	+40	−0	+40	+28	+40	+68
		+0	−185	+0	−83	+0	−39	+0	−25	+0	+3	+0	+43
180	250	+115	−100	+72	−50	+46	−15	+46	−0	+46	+33	+46	+79
		+0	−215	+0	−96	+0	−44	+0	−29	+0	+4	+0	+50
250	315	+130	−110	+81	−56	+52	−17	+52	−0	+52	+36	+52	+88
		+0	−240	+0	−108	+0	−49	+0	−32	+0	+4	+0	+56
315	400	+140	−125	+89	−62	+57	−18	+57	−0	+57	+40	+57	+98
		+0	−265	+0	−119	+0	−54	+0	−36	+0	+4	+0	+62
400	500	+155	−135	+97	−68	+63	−20	+63	−0	+63	+45	+63	+108
		+0	−290	+0	−131	+0	−60	+0	−40	+0	+5	+0	+68

Source: Abstract from BS 4500 (data compatible with BS EN 20286-2: 1993).

Consider a precision clearance fit as given by a 75 mm H7 hole and a 75 mm h6 shaft as shown in Fig. 7.3(a). This hole and shaft combination is high-lighted in Table 7.1.

Hole

Enter the table along diameter band 50–80 mm, and where this band crosses the column H7 the limits are given as +30 and +0. These dimensions are in units of 0.001 mm [0.001 mm = 1 micrometre (μm)]. Therefore, when applied to a basic size of 75 mm, they give working limits of 75.030 and 75.000 mm as shown in Fig. 7.3(b).

Shaft

Enter the table along diameter band 50–80 mm, and where this band crosses the column h6 the limits are given as −0 and −19. Again these dimensions are in units of 0.001 mm (1 μm). Therefore, when applied to a basic size of 75 mm, they give working limits of 75.000 and 74.981 mm as shown in Figure 7.3(b).

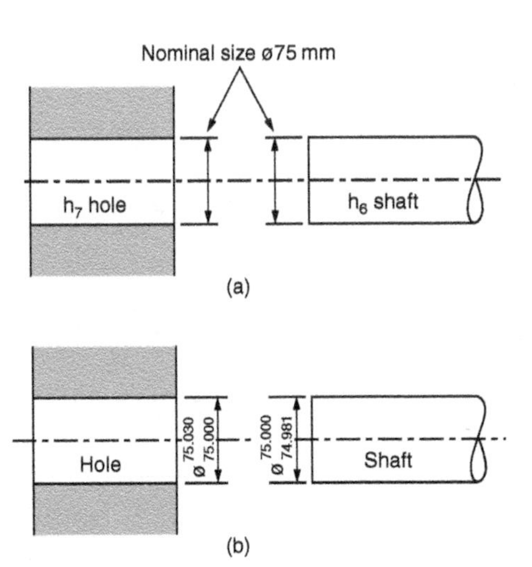

Figure 7.3 Application of limits and fits (a) tolerance specification – precision clearance fit and (b) dimensional limits to give precision of the clearance fit as derived from BS 4500. Dimensions in mm.

7.2.2 Selection of tolerance grades

From what has already been stated, it is obvious that the closer the limits (smaller the tolerance) the more difficult and expensive it is to manufacture a component. It is no use specifying very small (close) tolerances if the manufacturing process specified by the designer cannot achieve such a high degree of precision. Alternatively there is no point in choosing a process on grounds of low cost if it cannot achieve the accuracy necessary for the component to function correctly. Table 7.2 based on BS EN 20286-1: 1993 shows the standard tolerances from which the tables of limits and fits are derived.

Table 7.2 Numerical values of standard tolerance grades IT for basic sizes up to 3 150 mm[1]

Basic size mm		Standard tolerance grades																	
		IT1[2]	IT2[2]	IT3[2]	IT4[2]	IT5[2]	IT6	IT7	IT8	IT9	IT10	IT11	IT12	IT13	IT14[3]	IT15[3]	IT16[3]	IT17[3]	IT18[3]
Above	Up to and including	Tolerances																	
		µm											mm						
–	3[3]	0.8	1.2	2	3	4	6	10	14	25	40	60	0.1	0.14	0.25	0.4	0.6	1	1.4
3	6	1	1.5	2.5	4	5	8	12	18	30	48	75	0.12	0.18	0.3	0.48	0.75	1.2	1.8
6	10	1	1.5	2.5	4	6	9	15	22	36	58	90	0.15	0.22	0.36	0.58	0.9	1.5	2.2
10	18	1.2	2	3	5	8	11	18	27	43	70	110	0.18	0.27	0.43	0.7	1.1	1.8	2.7
18	30	1.5	2.5	4	6	9	13	21	33	52	84	130	0.21	0.33	0.52	0.84	1.3	2.1	3.3
30	50	1.5	2.5	4	7	11	16	25	39	62	100	160	0.25	0.39	0.62	1	1.6	2.5	3.9
50	80	2	3	5	8	13	19	30	46	74	120	190	0.3	0.46	0.74	1.2	1.9	3	4.6
80	120	2.5	4	6	10	15	22	35	54	87	140	220	0.35	0.54	0.87	1.4	2.2	3.5	5.4
120	180	3.5	5	8	12	18	25	40	63	100	160	250	0.4	0.63	1	1.6	2.5	4	6.3
180	250	4.5	7	10	14	20	29	46	72	115	185	290	0.46	0.72	1.15	1.85	2.9	4.6	7.2
250	315	6	8	12	16	23	32	52	81	130	210	320	0.52	0.81	1.3	2.1	3.2	5.2	8.1
315	400	7	9	13	18	25	36	57	89	140	230	360	0.57	0.89	1.4	2.3	3.6	5.7	8.9
400	500	8	10	15	20	27	40	63	97	155	250	400	0.63	0.97	1.55	2.5	4	6.3	9.7
500	630[2]	9	11	16	22	32	44	70	110	175	280	440	0.7	1.1	1.75	2.8	4.4	7	11
630	800[2]	10	13	18	25	36	50	80	125	200	320	500	0.8	1.25	2	3.2	5	8	12.5
800	1000[2]	11	15	21	28	40	56	90	140	230	360	560	0.9	1.4	2.3	3.6	5.6	9	14
1000	1250[2]	13	18	24	33	47	66	105	165	260	420	660	1.05	1.65	2.6	4.2	6.6	10.5	16.5
1250	1600[2]	15	21	29	39	55	78	125	195	310	500	780	1.25	1.95	3.1	5	7.8	12.5	19.5
1600	2000[2]	18	25	35	46	65	92	150	230	370	600	920	1.5	2.3	3.7	6	9.2	15	23
2000	2500[2]	22	30	41	55	78	110	175	280	440	700	1100	1.75	2.8	4.4	7	11	17.5	28
2500	3150[2]	26	36	50	68	96	135	210	330	540	860	1350	2.1	3.3	5.4	8.6	13.5	21	33

Note: This table, taken from ISO 286-1, has been included in this part of ISO 286 to facilitate understanding and use of the system.

[1] Values for standard tolerance grades IT01 and IT0 for basic sizes less than or equal to 500 mm are given in ISO 286-1, Annex A, Table 5.

[2] Values for standard tolerance grades IT1 to IT5 (inclusive) for basic sizes over 500 mm are included for experimental use.

[3] Standard tolerance grades IT14 to IT18 (inclusive) shall not be used for basic sizes less than or equal to 1 mm.

It can be seen that as the *International Tolerance* (IT) number gets larger, so the tolerance increases. The recommended relationship between process and standard tolerance is as follows:

IT16 Sand casting, flame cutting (coarsest tolerance)
IT15 Stamping
IT14 Die-casting, plastic moulding
IT13 Presswork, extrusion
IT12 Light presswork, tube drawing
IT11 Drilling, rough turning, boring
IT10 Milling, slotting, planing, rolling
IT9 Low grade capstan and automatic lathe work
IT8 Centre lathe, capstan and automatic lathe work
IT7 High-quality turning, broaching, honing
IT6 Grinding, fine honing
IT5 Machine lapping, fine grinding
IT4 Gauge making, precision lapping
IT3 High quality gap gauges
IT2 High quality plug gauges
IT1 Slip gauges, reference gauges (finest tolerance).

Example 7.1
Determine a process that is suitable for manufacturing the shaft shown in Fig. 7.4.

From Table 7.2 it can be seen that for a shaft of 40 mm nominal diameter and a tolerance of 0.03 mm (30 μm) the IT number lies between IT7 and IT8. Therefore a good turned finish would be sufficiently accurate. However to avoid wear the shaft would probably receive some form of heat treatment to harden the bearing surfaces, in which case the tolerance could easily be achieved by commercial quality cylindrical grinding.

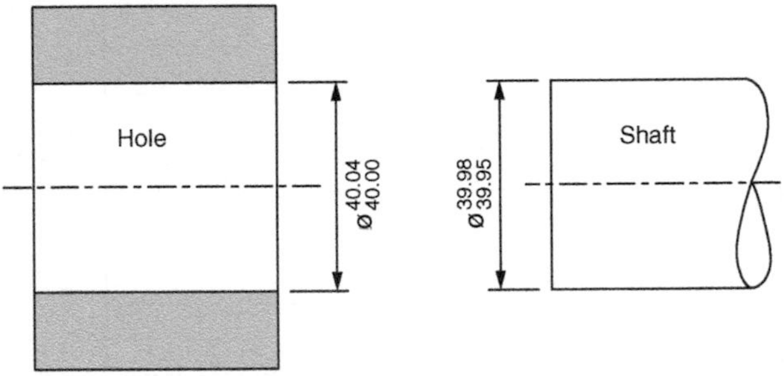

Figure 7.4 Average clearance fit: toleranced dimensions. Dimensions in mm.

7.3 Geometric tolerancing

Geometric tolerancing is a complex subject and there is only room to consider the basic principles in this volume. The reader is referred to BS ISO 1101: 2004 for a detailed explanation and examples of applications. The linear tolerancing of component dimensions is only concerned with the limits of size within which the component must be manufactured. Geometric tolerancing takes the dimensioning of components further and is concerned with the correct geometric shape of the component.

The importance of correct geometric accuracy as well as correct dimensional accuracy is being increasingly recognized.

Figure 7.5(a) shows a shaft whose dimensions have been given *limits of size*. The difference between the given limits is called the *tolerance*. For many applications, if the process used can achieve the dimensional tolerance, it will also produce a geometrically correct form. For example, the shaft shown in Fig. 7.5(a) would most likely be cylindrically ground to achieve the dimensional accuracy specified. Cylindrical grinding between centres should give a satisfactory degree of roundness. However, the shaft could also be mass-produced by centreless grinding and some *out-of-roundness* may occur such as *ovality* and *lobing*. Figure 7.5(b) shows how the shaft can be out-of-round yet still remain within its limits of size.

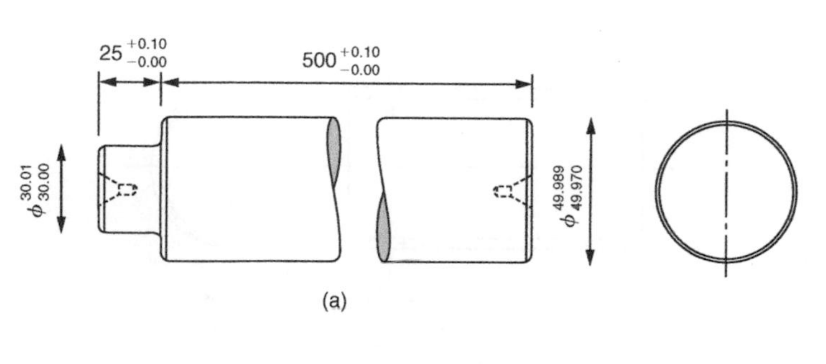

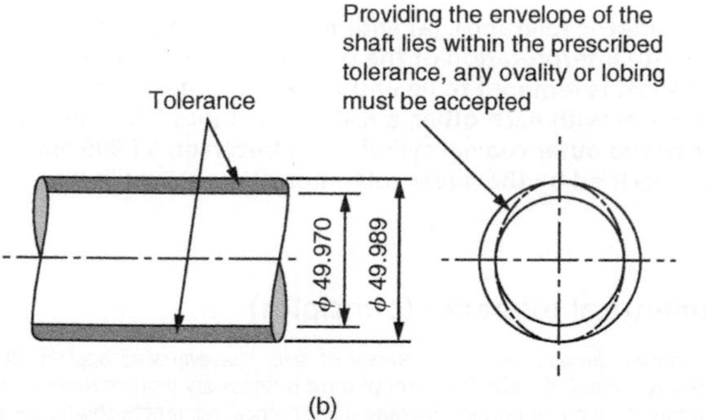

Figure 7.5 Need for geometric tolerance: (a) shaft and (b) geometrical error. Dimensions in mm.

If cylindricity is important for the correct functioning of the component, then additional information must be added to the drawing to emphasize this point. This additional information is a *geometrical tolerance* which, in this case, would be added as shown in Fig. 7.6(a).

The symbol in the box indicates cylindricity, and the number indicates the geometrical tolerance. Figure 7.6(b) shows the interpretation of this tolerance.

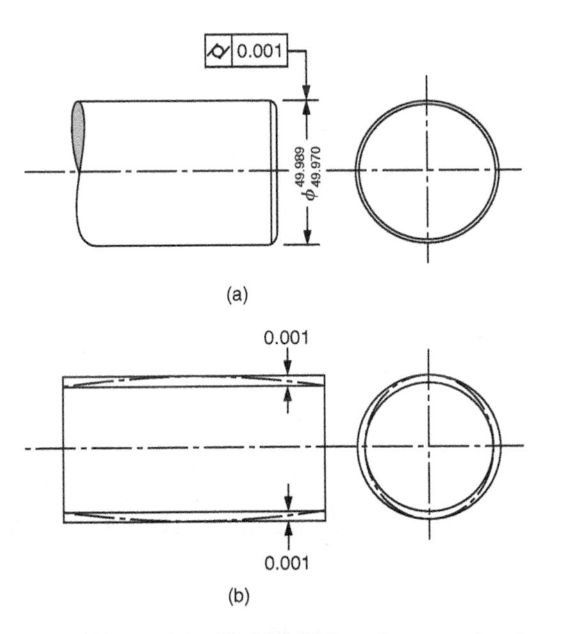

Figure 7.6 Geometric tolerance: (a) addition of geometric tolerance for cylindricity and (b) interpretation of the geometric tolerance. The curved surface of the shaft is required to lie within two cylinders whose surfaces are coaxial with each other, a RADIAL distance 0.001 mm apart. The diameter of the outer coaxial cylinder lies between 49.989 and 49.970 mm as specified by the linear tolerance. Dimensions in mm.

7.3.1 Geometrical tolerance (principles)

In some circumstances, dimensions and tolerance of size, however well applied, do not impose the necessary control of form. If control of form is necessary then geometrical tolerances must be added to the component drawing. Geometrical tolerances should be specified for all requirements critical to the interchangeability and correct functioning of components. An exception can be made when the machinery and techniques used can be relied upon to achieve the required standard of form. Geometrical tolerances may also need to be specified even when no special size tolerance is given. For example, the thickness of a surface plate is of little importance, but its accuracy of flatness is of fundamental importance.

Figure 7.7 shows the standard tolerance symbols as specified in BS ISO 1101: 1983 and it can be seen that they are arranged into groups according to their function. In order to apply geometrical tolerances it is first necessary to consider the following definitions.

Features and tolerances		Toleranced characteristics	Symbols	Subclauses
Single features	Form tolerances	Straightness	⏤	14.1
		Flatness	⏥	14.2
		Circularity	○	14.3
		Cylindricity	⌭	14.4
Single or related features		Profile of any line	⌒	14.5
		Profile of any surface	⌓	14.6
Related features	Orientation tolerances	Parallelism	//	14.7
		Perpendicularity	⊥	14.8
		Angularity	∠	14.9
	Location tolerances	Position	⌖	14.10
		Concentricity and coaxiality	◎	14.11
		Symmetry	⌯	14.12
	Run-out tolerances	Circular run-out	↗	14.13
		Total run-out	⌰	14.14

Figure 7.7 Geometrical tolerancing symbols (a) symbols for toleranced characteristics.

Descriptions		Symbols	Clauses
Toleranced feature indications	Direct	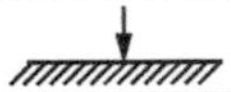	6
	By letter	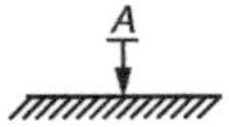 A	7.4
Datum indications	Direct		8
	By letter	 A A	
Datum target		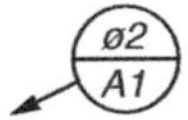 ø2 A1	ISO 5459
Theoretically exact dimension		50	10
Projected tolerance zone		Ⓟ	11
Maximum material condition		Ⓜ	12

Figure 7.7 Geometrical tolerancing symbols (b) additional symbols.

7.3.2 Tolerance frame

The tolerance frame is a rectangular frame (box) that is divided into two or more compartments containing the requisite information. Reading from left to right, this information is as shown in Fig. 7.8(a) and includes:

- The tolerance symbol.
- The tolerance value in the same units as the associated linear dimension.
- If required, a letter or letters identifying the datum feature or features.
- If required, additional remarks may be added as shown in Fig.7.8(b).

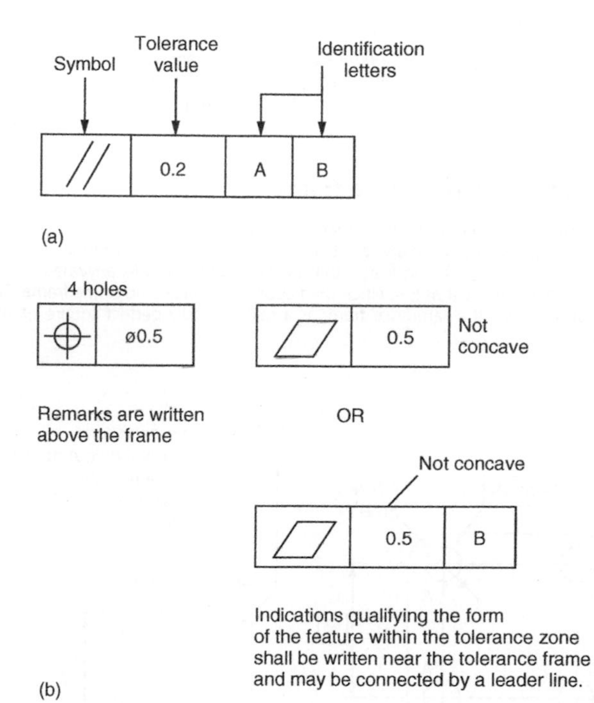

Figure 7.8 The tolerance frame (a) tolerance frame and (b) qualifying remarks.

7.3.3 Geometrical tolerance

This is the maximum permissible overall variation of form or of position of a feature. That is, it defines the size and shape of a tolerance zone within which the surface, median plane or axis of the feature is to lie. It represents the full indicator movement it causes where testing with a dial test indicator (DTI) is applicable, for example, the 'run-out' of a shaft rotated about its own axis.

7.3.4 Tolerance zone

This is the zone within which the feature has to be contained. Thus, according to the characteristic that is to be toleranced and the manner in which it is to be dimensioned, the tolerance zone is one of the following:

- A circle or a cylinder.
- The area between two parallel lines or two parallel straight lines.
- The space between two parallel surfaces or two parallel planes.
- The space within a parallelepiped.

Note
The tolerance zone, once established, permits any feature to vary within that zone. If it is considered necessary to prohibit sudden changes in surface direction, or to control the rate of change of a surface within this zone, this should be additionally specified.

7.3.5 Geometrical reference frame

This is the diagram composed of the constructional dimensions that serve to establish the true geometrical relationships between positional features in any one group as shown in Fig. 7.9. For example, the symbol indicates that the hole centre may lie anywhere within a circle of 0.02 mm diameter that is itself centred upon the intersection of the frame. The dimension indicates that the reference frame is a geometrically perfect square of side length 40 mm.

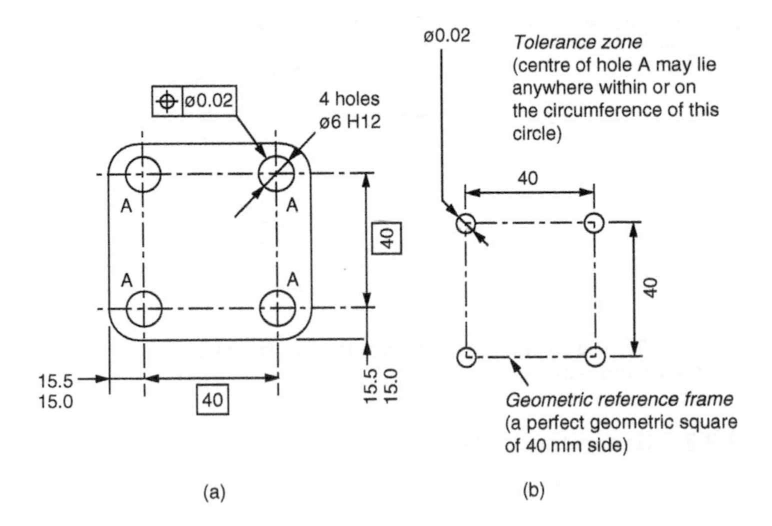

Figure 7.9 The geometric tolerance diagram (a) drawing requirement and (b) reference frame. Dimensions in mm.

7.3.6 Applications of geometrical tolerances

Figure 7.9 shows a plate in which four holes need to be drilled. For ease of assembly, the relationship between the hole-centres is more important than the group position on the plate. Therefore a combination of dimensional and geometrical tolerances has to be used.

Using the dimensional tolerances and the boxed dimension that defines true position, it is possible to determine the pattern locating tolerance zone for each of the holes as shown in Fig. 7.10. That is, the reference frame (which must always by a perfect square of side length 40 mm in this example) may lie in any position providing its corners are within the shaded boxes. The shaded boxes represent the *dimensional tolerance zones*. The hole centres may then (in this example) lie anywhere within the 0.02 mm circles that are themselves centred upon the corners of the reference frame.

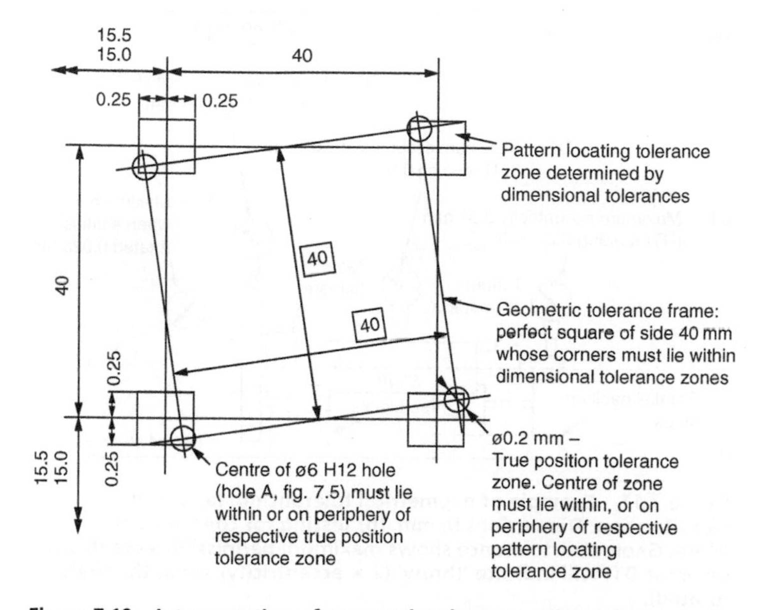

Figure 7.10 Interpretation of geometric tolerances. Dimensions in mm.

Figure 7.11 shows a stepped shaft with four concentric diameters and a flange that must run true with the datum axis. The shaft is to be located within journals at *X* and *Y*, and these are identified as datums. Note the use of leader lines terminating in solid triangles resting upon the features to be used as datums. The common axis of these datum diameters is used as a datum axis for relating the tolerances controlling the concentricity of the remaining diameters and the true running of the flange face. Figure 7.11(a) also shows that the dependent diameters and face carry geometric tolerances as well as dimensional tolerances. The tolerance frames carry the concentricity symbol, the concentricity tolerance and the datum identification letters. These identification letters show that the concentricity of the remaining diameters is relative to the common axis of the datum

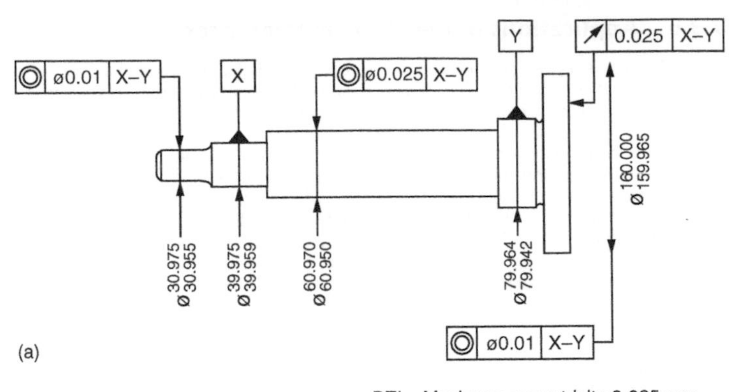

Figure 7.11 Example of geometrical tolerancing. (a) Datum identification. Dimensions in mm. (b) Testing for concentricity. (Note: Geometric tolerance shows maximum permissible eccentricity, whereas DTI will indicate 'throw' (2 × eccentricity) when the shaft is rotated).

diameters X and Y. The tolerance frame whose leader touches the flange face indicates that this face must run true, within the tolerance indicated when the shaft rotates in the vee-blocks supporting it at X and Y, as shown in Fig. 7.11(b).

These two examples are in no way a comprehensive resume of the application of geometric tolerancing but stand as an introduction to the subject and as an introduction to BS ISO 1101:1983. This British Standard details the full range of applications and their interpretation, and includes a number of fully dimensioned examples.

7.4 Virtual size

To understand this section, it is necessary to introduce the terms *maximum metal condition* and *minimum metal condition*.

- The *maximum metal condition* exists for both the hole and the shaft when the least amount of metal has been removed but the component is still within its specified limits of size, that is, the largest shaft and smallest hole that will lie within the specified tolerance.
- The *minimum metal condition* exists when the greatest amount of metal has been removed but the component is still within its limits of size, that is the largest hole and the smallest shaft that will lie within the specified tolerance.

Figure 7.12 shows the effect of a change in geometry on the fit of a pin in a hole. In Figure 7.12(a), an ideal pin is used with a straight axis. It can be seen that, as drawn, there is clearance between the pin and the hole. However, in Fig. 7.12(b) the pin is distorted and now becomes a tight fit in the hole despite the fact that its dimensional tolerance has not changed.

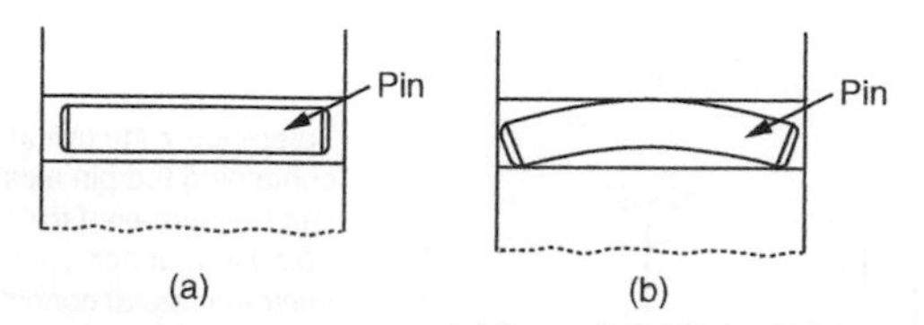

Figure 7.12 Effect of geometry on fit (a) pin with straight axis: clearance fit in hole and (b) pin with bowed axis: tight fit in hole despite no change in size.

Figure 7.13(a) shows the same pin and hole, only this time dimensional and geometrical tolerances have been added to the pin. For simplicity the hole is assumed to be geometrically correct. The tolerance frame shows a straightness tolerance of 0.05 mm. This means that the *axis of the pin* can bow within the confines of an imaginary cylinder 0.05 mm diameter. The dimensional limits indicate a tolerance of 0.08 mm. The worst conditions for assembly (tightest fit) occurs when the pin and the hole are in their respective maximum metal conditions and, in addition, the maximum error permitted by the geometrical tolerancing is also present. Figure 7.13(b) shows the fit between the pin and the hole under these conditions.

Under such conditions the pin will just enter a truly straight and cylindrical hole equal to the pin diameter under maximum metal conditions (25.00 mm) plus the geometrical tolerance of 0.05 mm, that is a hole diameter of 25.05 mm. This theoretical hole diameter of 25.05 mm is referred to as the *virtual size for the pin*.

Under minimum metal conditions, the geometrical tolerance for the pin can be increased without altering the fit of the pin in the hole. As shown in Fig. 7.13(c), the geometric tolerance can now be increased to 0.13 mm without any increase in the virtual size or the change of fit.

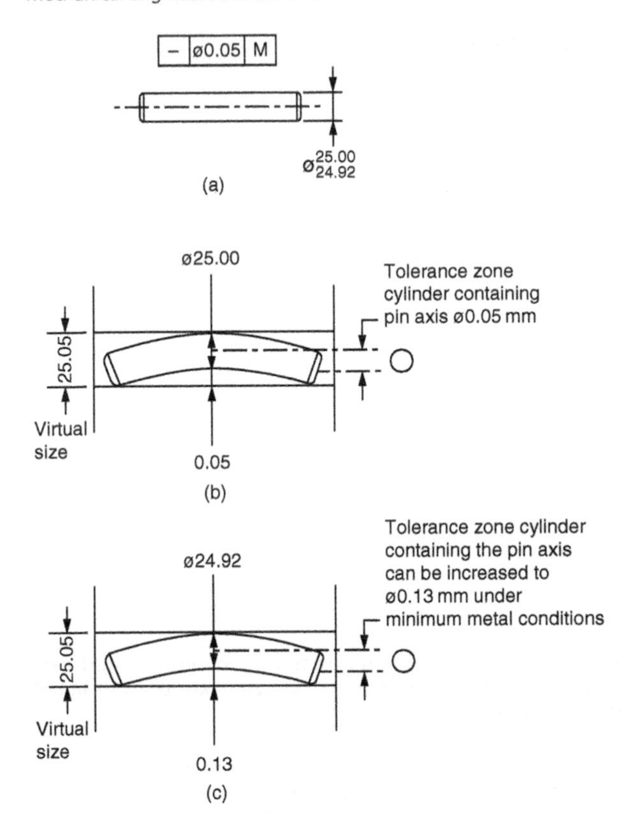

Figure 7.13 Virtual size (a) pin with geometric and dimensional tolerances; (b) virtual size under maximum metal conditions (pin diameter = 25.00mm) and (c) virtual size under minimum metal conditions (pin diameter = 24.92). Dimensions in mm.

7.5 The economics of geometrical tolerancing

Figure 7.14 shows a typical relationship between tolerance and manufacturing costs for any given manufactured component part. As tolerance is reduced the manufacturing costs tend to rise rapidly. The addition of geometrical tolerances on top of the dimensional tolerances aggravates this situation still further. Thus geometric tolerances are only applied if the function and the interchangeability of the component is critical and absolutely dependent upon a tightly controlled geometry. To keep costs down, it can often be assumed that the machine or process will provide adequate geometric control and only dimensional tolerances are added to the component drawing. For example, it can generally be assumed that a component finished on a cylindrical grinding machine

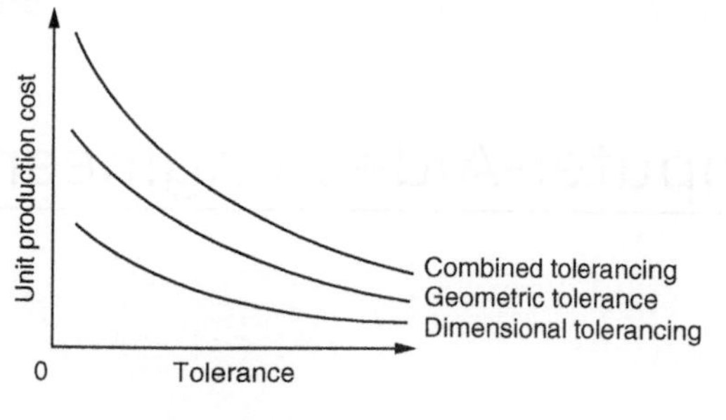

Figure 7.14 Cost of tolerancing.

between centres will be straight and cylindrical to a high degree of geometrical accuracy. However, current trends in quality control and just-in-time (JIT) scheduling tends to increase the demand for closer dimensional and geometrical control to ensure no hold ups occur during assembly and that any quality specifications are met first time without the need for corrective action or rejection by the customer.

8

Computer-Aided Engineering

8.1 Computer numerical control

8.1.1 Typical applications of computer numerical control

Computer numerical control (CNC) is applied to a wide range of production processes in many industries. In the engineering industries it is applied to such processes as:

(a) Machine tools, including
- Milling machines and machining centres
- Centre lathes and turning centres
- Drilling machines
- Precision, grinding machines
- Electro-discharge machining (EDM) (spark erosion) machines
- Die sinking machines

(b) Sheet metal working machines, including
- Turret punching machines
- Riveting machines
- Forming machines

(c) Fabrication equipment, including
- Flame cutting machines
- Welding machines
- Tube bending machines

(d) Inspection machines for checking three-dimensionally (3D) contoured components.

8.1.2 Advantages and limitations of CNC

It is evident from the wide and increasing use of computer numerically controlled machines in manufacturing industry that the advantages substantially outweigh the limitations.

Advantages

High productivity Although the cutting speeds and feeds for CNC machines are the same as for manually operated machines, much time is saved by rapid traversing and positioning between operations. Also a wider range of operations is possible on a CNC machine, avoiding the necessity to pass the work from one machine to another for, say, each of drilling, milling and boring. This reduces the need for expensive jigs and fixtures, and avoids reserves of work in progress between operations. In addition, CNC machines do not become tired and they maintain a constant rate of productivity. If the work is robot fed, they can work 'lights out' through the night.

Design flexibility Complex shapes are easily produced on CNC machines. In addition, contoured solids can be produced on CNC machines which cannot be produced on conventional, manually operated machines.

Management control Production rates, reduced scrap and improved quality comes under management control with CNC machines and is not influenced by operator performance.

Quality CNC machines have a higher accuracy and better repeatability than conventional machines. If the machine is fitted with adaptive control it will even sense tool failure or other variations in performance and either stop the machine or, if fitted with automatic tool changing, select backup tooling from the tool magazine before scrap is produced or the machine damaged.

Reduced lead time The lead time for CNC machines is much less than for other automatic machines. There are no complex and expensive cams and form tools; it is necessary simply to write a part program and load it into the machine memory. Complex profiles are generated using standard tooling.

Limitations

Capital cost The initial cost of CNC machines is substantially higher than for manually operated machines of the same type. However, in recent years, the cost differential has come down somewhat.

Tooling cost To exploit the production potential of CNC machine tools, specialized cutting tools are required. Although the initial cost is high, this largely reflects the cost of the tool shanks and tool holding devices which do not have to be replaced. The cost of replacement tool tips is no higher than for conventional machines.

Maintenance Due to the complexity of CNC machine tools, few small and medium companies will have the expertise to carry out more than very basic maintenance and repairs. Therefore, maintenance contracts are advisable. These are expensive, approximately 10% of the capital cost per annum.

Training Programmer/operator training is required. This is usually provided by the equipment manufacturer but can be time consuming and costly.

Depreciation As with all computer-based devices, CNC controllers rapidly become obsolescent. Therefore, CNC machines should be amortized over a shorter period than is usual with manually operated machines and should be replaced approximately every 5 years.

8.1.3 Axes of control for machine tools

There are a number of axis configurations for CNC machine tools; the most common are used with vertical/horizontal milling machines and lathes as shown. Note that:

(a) The *Z*-axis is always the main spindle axis.
(b) The *X*-axis is always horizontal and perpendicular to the *Z*-axis.
(c) The *Y*-axis is perpendicular to both the *X- and Z-axis.*

Vertical milling machine: axes of control

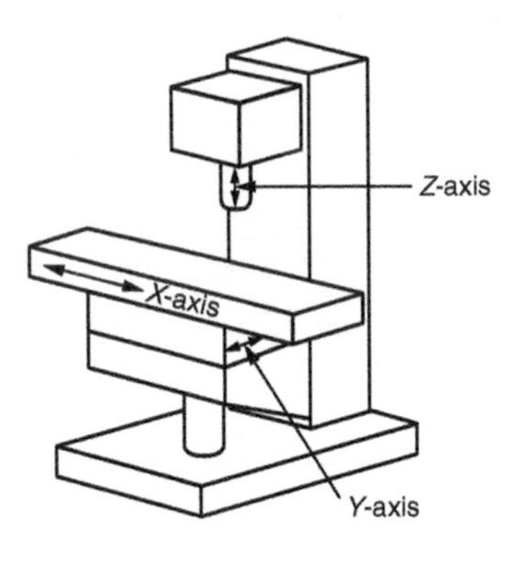

Horizontal milling machine: axes of control

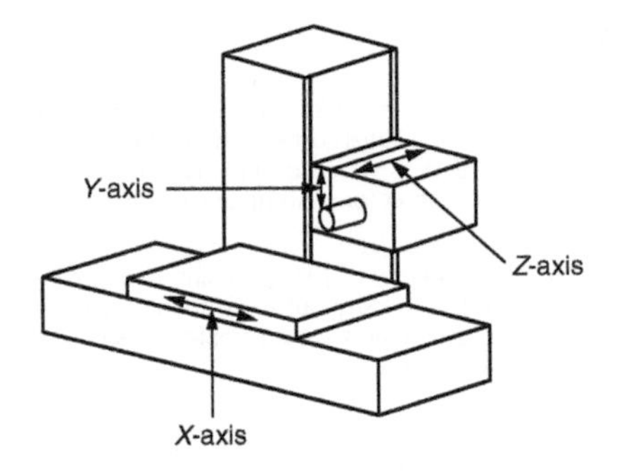

Centre lathe: axes of control

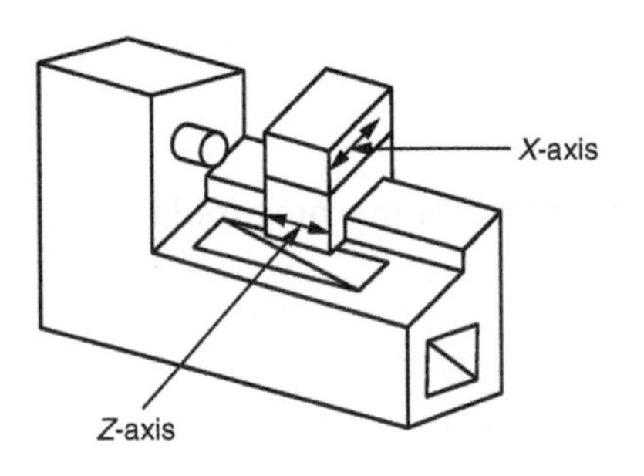

8.1.4 Control systems

Control systems for CNC machine tools broadly fall into two categories as follows.

Open-loop control

This system derives its name from the fact that there is no feedback in the system, and hence no comparison between input and output. The system uses stepper motors to drive the positioning mechanism, and these have only limited torque compared with more conventional servo motors. Further, if the drag of the mechanism causes the motor to stall and miss a step, no corrective action is taken by the system.

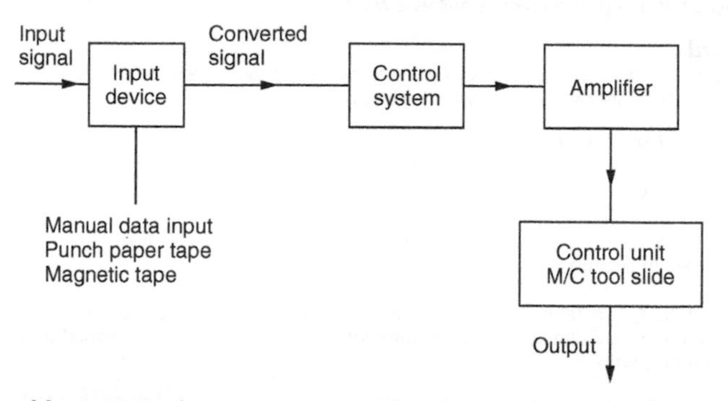

Closed-loop control

The following figure explains closed-loop control system:

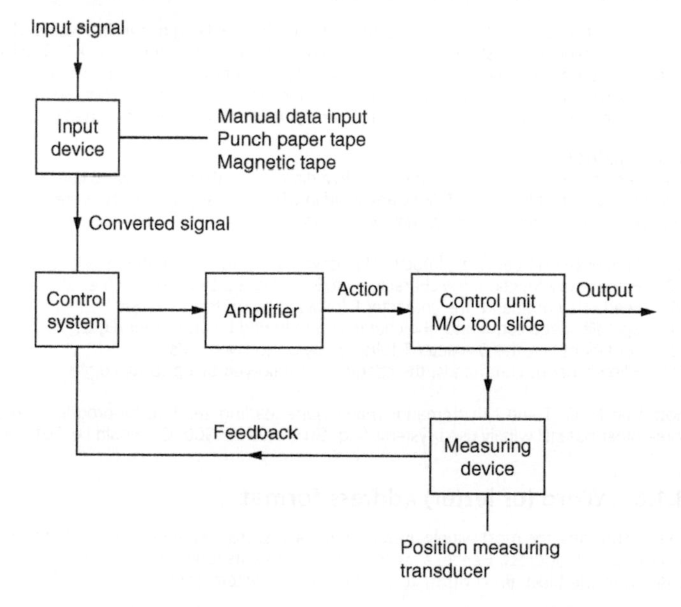

In a closed-loop control system the feedback continuously influences the action of the controller and corrects any positional errors. This system allows the use of servo motors to drive the positioning mechanism, and these have a much higher torque than the stepper motors used with the open-loop control system.

8.1.5 Program terminology and format

Character

A character is a number, letter or other symbol which is recognized by the controller. An associated group of characters makes a word.

Word

A word is a group of characters which defines one complete element of information (e.g. N100). There are two types of word as follows.

Dimensional words

These are any words related to a linear dimension; that is, any word commencing with the characters X, Y, Z, I, J, K or any word in which these characters are inferred.

The letters X, Y, Z refer to dimensions parallel to the corresponding machine axes, as explained in Section 8.1.3.

The letters I, J, K refer to arcs of circles. The start and finish positions of the arcs are defined by X, Y, Z dimensions, whilst the centre of radius of the arc is defined using I, J, K dimensions, with:

I dimensions corresponding to X dimensions
J dimensions corresponding to Y dimensions
K dimensions corresponding to Z dimensions.

Current practice favours the use of the decimal point in specifying dimension words. Thus, a machine manual may stipulate that an X dimension word has the form X4, 3, which means that the X dimension may have up to 4 digits in front of the decimal point and up to 3 digits behind the decimal point. Some older systems, still widely in use, do *not* use the decimal point but use leading and trailing zeros; for example, 25.4 would be written 0025400.

Management words

These are any words which are not related to a dimension; that is, any word commencing with the characters N, G, F, S, T, M or any word in which the above characters are inferred. Examples of management words may be as follows:

N4 Sequence number: the character N followed by up to four digits (i.e. N1 … N9999).
G2 Preparatory function: the character G followed by up to two digits (i.e. G0 … G99).
F4 Feed rate command: the character F followed by up to four digits.
S4 Spindle speed command: the character S followed by up to four digits.
T2 Tool identifier: the character T followed by up to two digits.
M2 Miscellaneous commands: the character M followed by up to two digits.

Note that N, G, T and M commands *may require* leading zeros to be programmed on some older but still widely used systems (e.g. G0 would be G00, G1 would be G01, etc.).

8.1.6 Word (or letter) address format

This is, currently, the most widely used format. Each word commences with a letter character called an *address*. Hence, a word is identified by its letter character and *not* by its position in the block (in contrast to the fixed block system described below). Thus in a word (or letter) address format, instructions which remain unchanged from a previous block may be omitted from subsequent blocks.

A typical letter address format, as given by a maker's handbook, could be:

Metric	N4 G2 X4, 3 Y4, 3 Z4, 3 I4, 3 J4, 3 K4, 3 F3 S4 T2 M2
Inch	N4 G2 X3, 4 Y3, 4 Z3, 4 I3, 4 J3, 4 K3, 4 F3 S4 T2 M2

8.1.7 Coded information

A CNC program contains the information for the manufacture of a component part. The CNC controller regulates the signals and sequence to the various drive units.

Codes

Block numbers	N
Preparatory functions	G
Dimensional data	X, Y, Z, I, J, K
Feed rates	F
Spindle speeds	S
Tool numbers	T
Miscellaneous functions	M

A sample of a CNC program could look like this:

```
N5   G90     G71     G00    X25.0  Y25.0  T01  M06
N10  X100.0  Y100.0  S1250
N15  G01     Z25.0   F125   M03
```

Block numbers

The block number is usually the first word which appears in any block. Blocks are numbered in steps of 5 or 10 so that additional blocks can be easily inserted in the event of an omission.

Preparatory functions (G)

These are used to inform the machine controller of the functions required for the next operation. Standardized preparatory functions are shown in the following table. **In practice, the actual codes used will depend on the control system and the machine type. The codes that one system uses can vary from those of another, so reference to the relevant programming manual is essential.**

Code number	Function[a]	
G00	Rapid positioning, point to point	(M)
G01	Positioning at controlled feed rate, normal dimensions	(M)
G02	Circular interpolation, normal dimensions	(M)
G03	Circular interpolation counter clockwise (CCW), normal dimensions	(M)
G04	Dwell for programmed duration	
G05	Hold: cancelled by operator	
G06	Reserved for future standardization: not	
G07	normally used	
G08	Programmed slide acceleration	
G09	Programmed slide deceleration	
G10	Linear interpolation, long dimensions	(M)
G11	Linear interpolation, short dimensions	(M)
G12	3D interpolation	(M)
G13–G16	Axis selection	(M)
G17	*XY* plane selection	(M)
G18	*ZX* plane selection	(M)
G19	*YZ* plane selection	(M)
G20	Circular interpolation clockwise (CW), long dimensions	(M)
G21	Circular interpolation CW, short dimensions	(M)

continued

Table (*continued*)

Code number	Function[a]	
G22	Coupled motion positive	
G23	Coupled motion negative	
G24	Reserved for future standardization	
G25–G29	Available for individual use	
G30	Circular interpolation CCW, long dimensions	(M)
G31	Circular interpolation CCW, short dimensions	(M)
G32	Reserved for future standardization	
G33	Thread cutting, constant lead	(M)
G34	Thread cutting, increasing lead	(M)
G35	Thread cutting, decreasing lead	(M)
G36–G39	Available for individual use	
G40	Cutter compensation, cancel	(M)
G41	Cutter compensation, left	(M)
G42	Cutter compensation, right	(M)
G43	Cutter compensation, positive	
G44	Cutter compensation, negative	
G45	Cutter compensation +/+	
G46	Cutter compensation +/−	
G47	Cutter compensation −/−	
G48	Cutter compensation −/+	
G49	Cutter compensation 0/+	
G50	Cutter compensation 0/−	
G51	Cutter compensation +/0	
G52	Cutter compensation −/0	
G53	Linear shift cancel	(M)
G54	Linear shift *X*	(M)
G55	Linear shift *Y*	(M)
G56	Linear shift *Z*	(M)
G57	Linear shift *XY*	(M)
G58	Linear shift *XZ*	(M)
G59	Linear shift *YZ*	(M)
G60	Positioning exact 1	(M)
G61	Positioning exact 2	(M)
G62	Positioning fast	(M)
G63	Tapping	
G64	Change of rate	
G65–G79	Reserved for future standardization	
G80	Fixed cycle cancel	(M)
G81–G89	Fixed cycles	(M)
G90–G99[b]	Reserved for future standardization	

(M) indicates modal G commands which remain in force from line to line until cancelled.

[a] Variations on standard codings occur not only between the different makes of controller but between different models by the same maker and even between different types of the same model. Thus again it must be stressed that the programmer/operator *must* work from the manufacturer's manual for a given machine/controller combination.

[b] Most control units use G90 to establish the program in *absolute* dimensional units, and G91 to establish the program in *incremental* dimensional units, however. FANUC control units use G20 in place of G90, and G21 in place of G92.

Dimensional words

A CNC machine tool will have axes which are addressed by letters (*X*, *Y*, *Z*, etc.). The CNC program will instruct the controller to drive the appropriate machine elements to a required position parallel to the relevant axes using dimensional words (X25.0 Y55.0) which consists of the axis letter plus a dimension. These dimensional words are sign sensitive (− or +). If there is no sign, the number is assumed positive by the controller. This adds built-in safety to a CNC program, ensuring that should a CNC programmer fail to input a minus value to a *Z*-axis command the tool should move to 'safe' as shown below. That is, it moves away from the workpiece, as shown in the following figure.

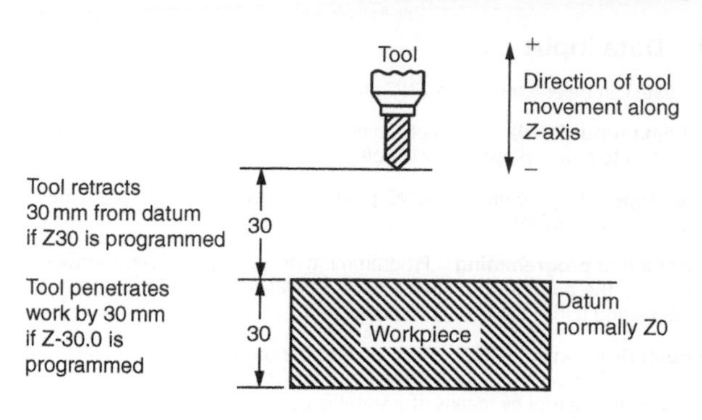

Consider a line of program N10 G01 X25.0 Y25.0 Z30.0. This would result in the tool moving to the coordinates X25, Y25, Z30. However, since the Z word is positive the tool will go up and not down 30 mm from the datum. To drill the component to a depth of 30 mm the line of program should read N10 G01 X25.0 Y25.0 Z-30.0.

Feed rate (F)

Words commencing with F indicate to the controller the desired feed rate for machining. There are a number of ways of defining the feed rate:

(a) Millimetres/minute: F30 = 30 mm/min.
(b) Millimetres/revolution: F0.2 = 0.2 mm/rev.
(c) Feed rate number 1–20 indicates feed rates predetermined by the manufacture in rev/min or mm/min as appropriate. To use this system the feed rate command is prefaced by a G code. Only a few older machines use this system.

Spindle speed (S)

Words commencing with the letter S indicate to the controller the desired spindle speed for machining. Once again there are a number of ways of defining the spindle speed, each selected by use of the appropriate G code:

(a) Spindle speed in revolutions/minute (r.p.m.).
(b) Cutting speed in metres/minute.
(c) Constant cutting speed in metres/minute.

Tool numbers (T)

Each tool used during the machining of a part will have its own number (e.g. T01 for tool 1, T19 for tool 19). Most CNC machine controllers can hold information about 20 tools or more. This tool information will take the form of a tool length offset (TLO) and a tool diameter/radius compensation (see Sections 8.1.9 and 8.1.10). Where the machine is fitted with automatic tool changing it will also identify the position of the tool in the tool magazine.

Miscellaneous functions (M)

Apart from preparatory functions there are a number of other functions that are required throughout the program (e.g. 'switch spindle on', 'switch coolant on', etc). These functions have been standardized and are popularly known as M codes. The address letter M is followed by two digits. The most commonly used M codes are as follows:

M00	Program stop	M04	Spindle on CCW	M08	Flood coolant on
M01	Optional stop	M05	Spindle off	M09	Coolant off
M02	End program	M06	Tool change	M30	End of tape
M03	Spindle on CW	M07	Mist coolant on		

8.1.8 Data input

Program data can be entered into a CNC machine by a number of methods:

Manual data input (MDI) This method requires data input by an operator key boarding in the data to a tape punch (now obsolescent).

Magnetic tape This requires the CNC program to be 'copied' to a magnetic tape or cartridge (now obsolescent).

Computer aided programming Programming on a computer using software that not only generates the program but simulates its operation to test its accuracy and safety before up-loading it into the machine control unit.

Direct numerical control (DNC) This requires the prepared CNC program to be stored in the memory of a remote computer. The CNC data is then transferred directly down line to the CNC machine tool by means of a suitable communications link.

Conversational data input This enables the operator to load the program directly into the machine control unit by responding to 'prompts' & 'menus' appearing on the control unit screen. The controller has a split memory so that a previous program can continue operating the machine whilst a new program is being loaded.

8.1.9 Tool length offsets: milling

Most machining operations require the use of a number of tools, which vary in length and diameter. In order to allow for these variables, TLO and tool diameter compensation facilities can be used.

TLO allow for a number of tools with varying length to be programmed to a common datum or zero, as shown in the figure.

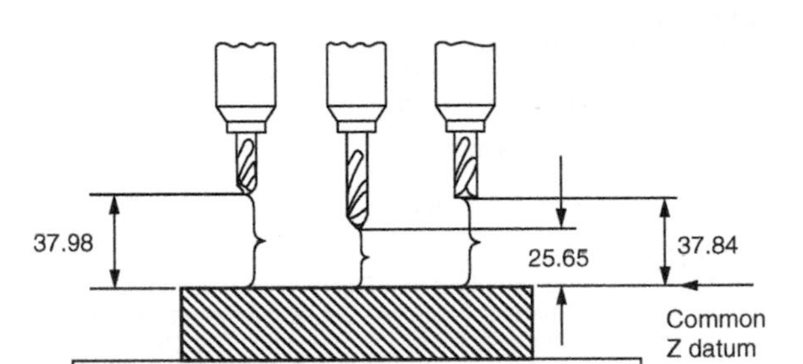

8.1.10 Cutter diameter compensation: milling

When using diameter or radius compensation, normal practice would be to program for the actual size of the component and then use the diameter compensation facility to allow for the radius of the cutter. Another use of this feature allows for the changing of tools of different diameters, without altering the CNC part program. Diameter compensation is activated by the following preparatory codes (see Section 8.1.7):

G41 Compensates, cutter to the left
G42 Compensates, cutter to the right

G41 Compensation to the left

The cutter is always to the *left* of the surface being machined when looking along the path of the cutter, as indicated by the arrows in the figure.

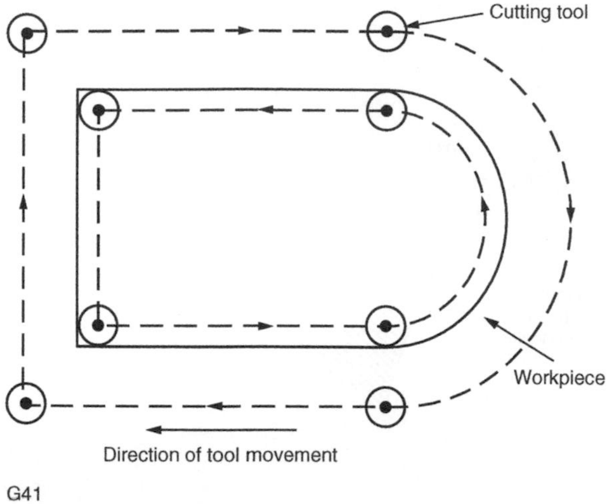

G41

G42 Compensation to the right

The cutter is always to the right of the surface being machined when looking along the path of the cutter, as indicated by the arrows in the figure.

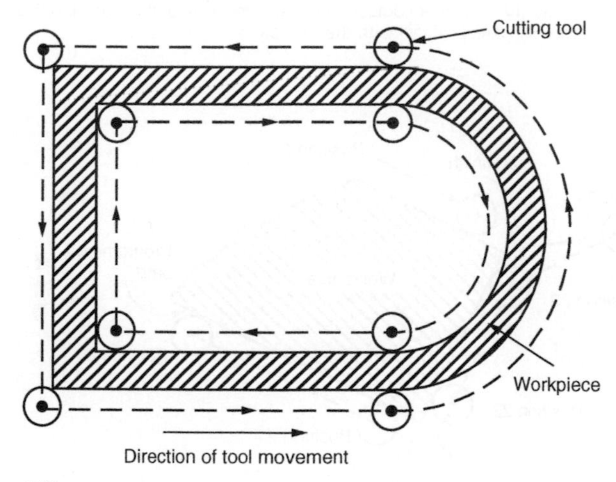

G42

G40 Turning compensation on and off

Whenever G41 or G42 is activated, the cutter diameter is compensated during the next move in the X and Y axes, as shown. The preliminary movement of the cutter before cutting commences is called *ramping on*.

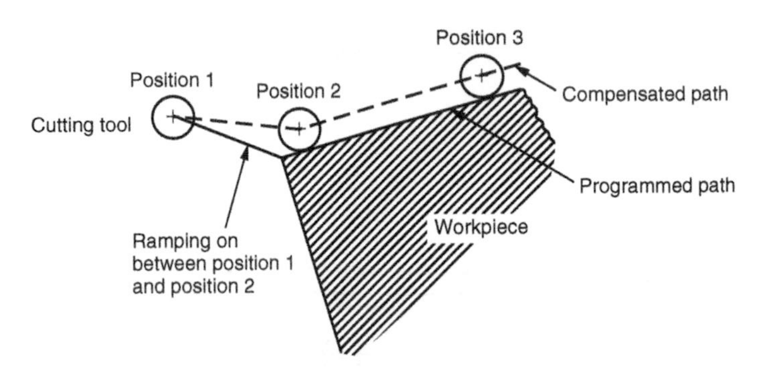

Similarly G40 causes the compensation to be cancelled on the next X and Y movements as shown, and is known as *ramping off*.

Ramping on and ramping off allow the feed servo motor to accelerate to the required feed rate, and to decelerate at the end of the cut whilst the cutter is clear of the work. It therefore follows that diameter compensation can never be applied or cancelled when the cutting tool is in direct contact with the workpiece.

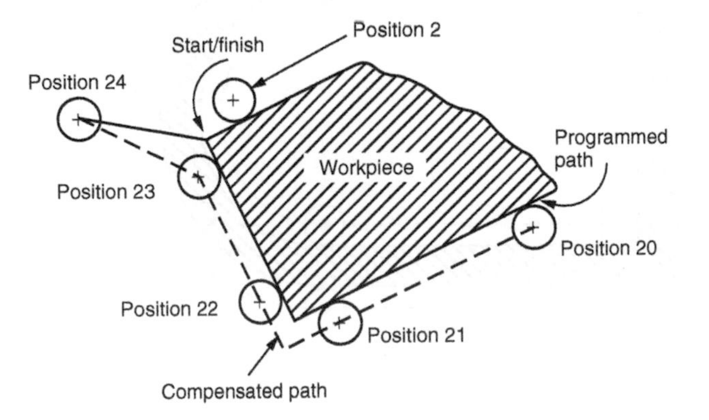

8.1.11 Programming techniques: milling and drilling

Canned cycles

To save repetitive programming on similar operations, CNC controllers offer a range of canned or fixed cycles. It is only necessary to provide dimensional data with the required G code. Some examples of canned cycles for milling are as follows:

G80	Cancels fixed or canned cycles
G81	Drilling cycles
G82	Drilling cycles with dwell
G83	Deep hole drilling cycle
G84	Tapping cycle
G85	Boring cycle
G86	Boring cycle with dwell
G87/88	Deep hole boring cycle

G81 Drilling cycle

One of the most popular canned cycles is the drilling cycle G81. The machine movements involved in drilling a series of holes are shown and are as follows:

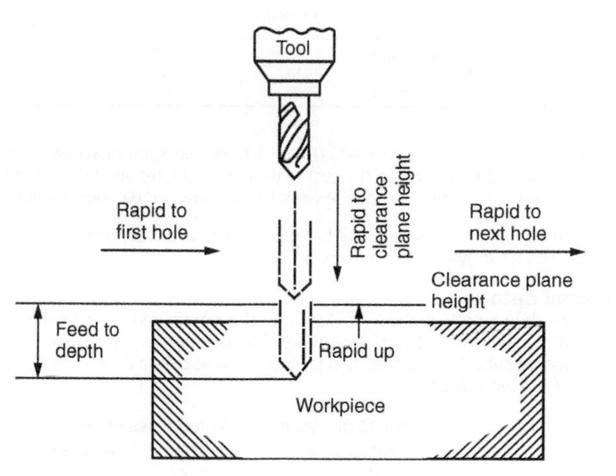

(a) Rapid to centre of first hole.
(b) Rapid to clearance plane height.
(c) Feed to depth of hole.
(d) Rapid up to clearance plane height.
(e) Rapid to centre of next hole.
(f) Repeat for as many holes as required.

The only data which needs to be programmed is:

Tool number
Hole position (X, Y)
Hole depth (Z)
Spindle speed (S)
Feed rate (F).

8.1.12 Programming example: milling

Component

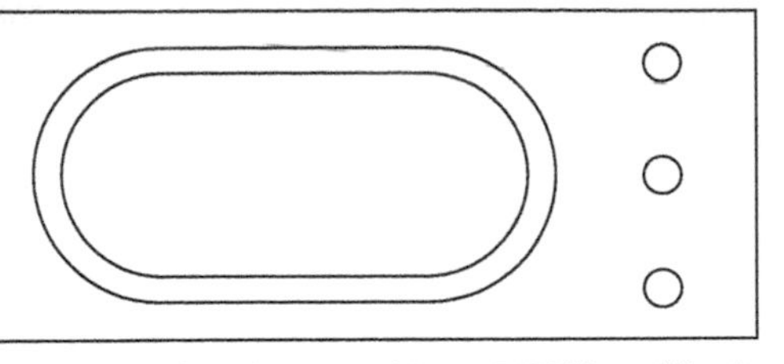

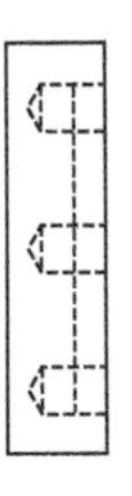

Material: aluminium alloy BS 1474: 1987 type 5154(A) condition 0.

Machining data

Tool no.	Tool description	Spindle speeds (rev/min)	Feed rates (mm/min)	
			Vertical	Horizontal
T01	10 mm diameter slot drill	1500	70	100
T02	No. 2 centre drill	2500	40	–
T03	8 mm diameter twist drill	1760	60	–

The tool change position will be X – 100.0, Y – 100.0. The appropriate M code in the program for tool changing will move the tool to these coordinates and retract the tool up the *Z*-axis to a 'home' position clear of the work to facilitate manual tool change.

Centre line path/point programming will be used, and to avoid complexity no diameter compensation will be shown in this example.

Dimensional data

The dimensions shown provide the *X*, *Y*, *Z* data for the 10 mm wide and 8 mm deep slot, and the 8 mm diameter and 12 mm deep holes. The numbers in the circles refer to the tool positions as described in the specimen program to follow. The *Z* datum (Z0) would be the top face of the workpiece.

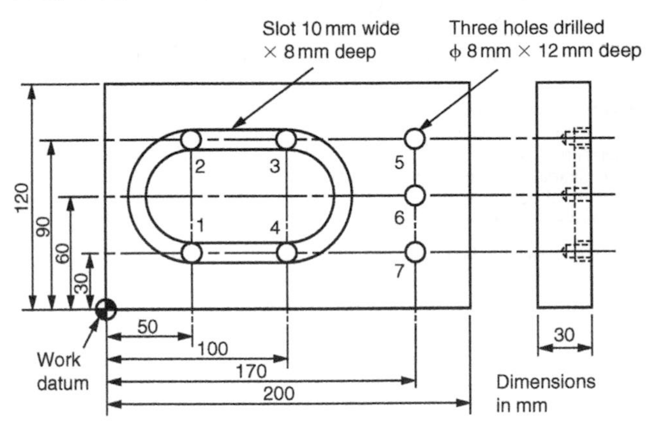

Specimen program

The format of the following program example would be suitable for the Bridgeport BOSS 6 software. When evaluating the program, reference should be made to the coded information given in Section 8.1.7.

CNC listing	Description
%	Per cent symbol = start of program/tape.
N5 G90 G71 G00 G75	Set up default preparatory codes. G75 is special to BOSS 6 software.
N10 X − 100.0 Y − 100.0 S1500 T01 M06	Rapid move to tool change (home) position. Set spindle speed (rev/min). Select tool required and insert first tool (slot drill). Stop control for tool change.
N15 X50.0 Y30.0	Restart control. Rapid move to position 1.
N20 Z1.0 M03	Rapid tool move to 1.0 mm above workpiece. Start up spindle (speed set in block N10).
N25 G01 Z − 8.0 F70 M07	Linear feed rate move to required depth (70 mm/min). Mist coolant on.
N30 G02 X50.0 Y90.0 150.0 J60.0 F100	Circular interpolation to *X* and *Y* coordinates at position 2, using *I* and *J* coordinates for interpolation centers. Increase feed rate (horizontal) to 100 mm/min.
N35 G01 X100.0 Y90.0	Linear feed rate move as set in block N30 to position 3.
N40 G02 X100.0 Y30.0 100.0 J60.0	Circular interpolation to *X* and *Y* coordinates at position 4 at feed rate set in block N30.
N45 G01 X50.0 Y30.0	Linear feed rate move as set in block N30 to position 1.
N50 G00 Z1.0	Rapid move tool clear of workpiece along *Z*-axis.
N55 X − 100.0 Y − 100.0 S2500 T02 M06	Rapid move to tool change (home) position and stop spindle. Remove tool 1 and insert tool 2 (center drill). Set spindle speed to 2500 rev/min ready for drilling.
N60 X170.0 Y90.0	Restart control. Rapid move to position 5.
N65 Z1.0 M03	Rapid move to 1.0 mm above workpiece. Start up spindle (speed set to 2500 rev/min in block N55).
N70 G81 X170.0 Y90.0 Z − 5.0 F40 M07	Start drilling cycle (see Section 8.1.11) and centre drill first hole (position 5) to depth at 40 mm/min feed rate. Mist coolant on.
N75 X170.0 Y60.0	C/drill second hole (position 6).
N80 X170.0 Y30.0	C/drill third hole (position 7).
N85 G80	Cancel drilling cycle.
N90 G00 − X100.0 Y − 100.0 S1760 T03 M06	Rapid move to tool change (home) position and stop spindle. Remove tool 2. Insert tool 3 (twist drill). Set spindle speed to 1760 rev/min ready for drilling.
N95 X170.0 Y90.0	Rapid move to position 5.
N100 Z1.0 M03	Rapid move tool to 1 mm above workpiece. Start spindle.
N105 G81 X170.0 Y90.0 Z − 12.0 F60 M07	Start drilling cycle and drill first hole to depth (12 mm) at position 5 with 60 mm/min feed rate. Mist coolant on.
N110 X170.0 Y60.0	Drill second hole (position 6).
N115 X170.0 Y30.0	Drill third hole (position 7).
N120 G80	Turn spindle off.
N125 G00 X − 100.0 Y − 100.0 M06	Return tool to tool change (home) position and stop spindle. Ready for repeat of program.
N130 M02	End or program.
N135 M30	End of tape.

8.1.13 Tool offsets: lathe

On a CNC lathe, to keep a common datum with all the tools on the turret, TLO are required in both the X- and Z-axis as shown.

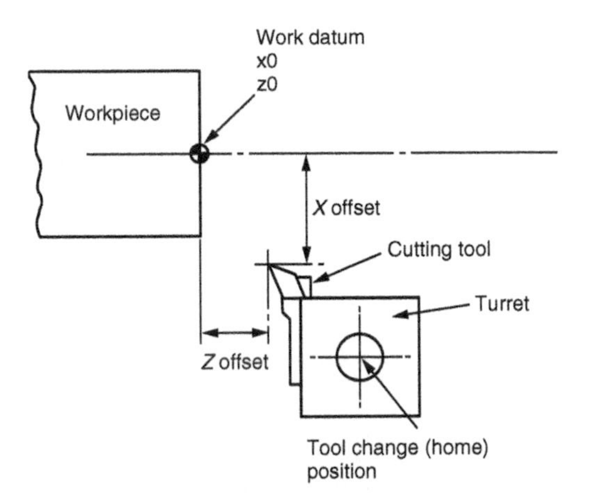

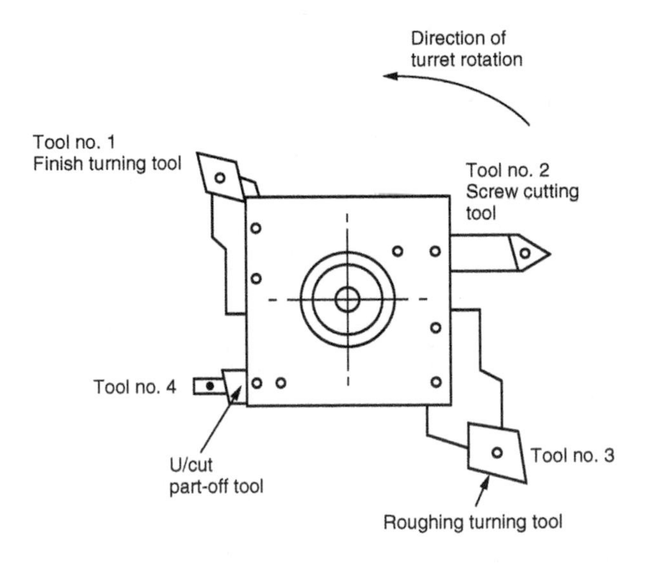

On a CNC centre lathe the turret will hold a number of different tools. Each tool will have its own *X* and *Z* offset, which will become operative when the tool is called into the program. When the turret is indexed all the tools will take up a different position relative to the workpiece. As can be seen in the diagram, since the various tools protrude from the turret by different distances in terms of *X* and *Z*, different TLO will need to be set for each tool.

8.1.14 Tool nose radius compensation: lathe

Tool nose radius compensation (TNRC) is used in a similar manner to diameter compensation when milling. Consequently the same preparatory codes (G codes) are used to activate and deactivate TNRC. TNRC reduces the complex and repetitive calculations required for tool paths. The programmer can program as if sharp pointed tools were being used. The codes are as follows:

G41 Compensates tool nose radius to the left
G42 Compensates tool nose radius to the right
G40 Cancels tool nose radius compensation

G41 Compensation to the left

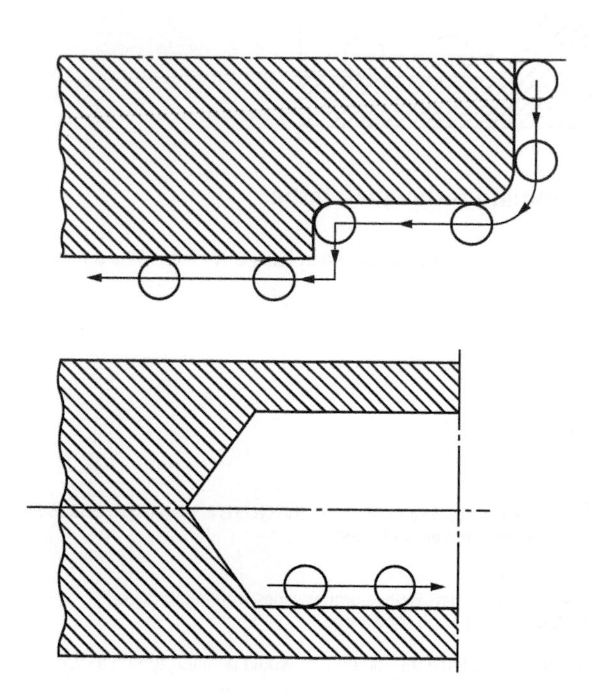

G42 Compensation to the right

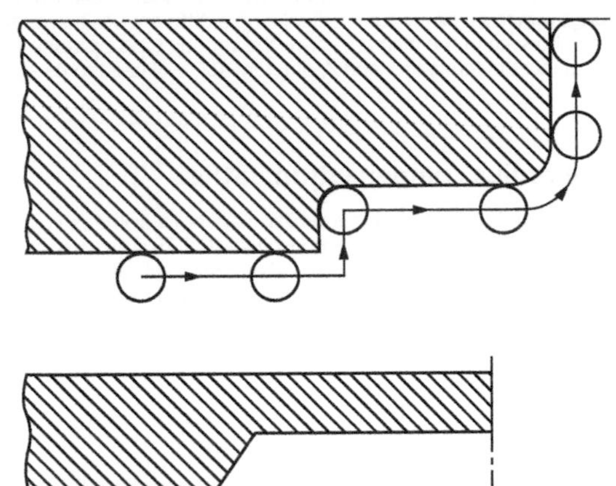

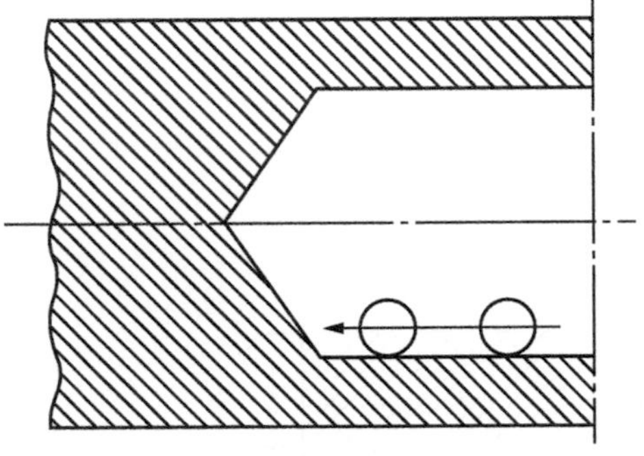

8.1.15 Programming techniques: lathe

Canned cycles

Examples of some of the canned cycles available when using CNC centre lathe controllers are as follows:

G66/67	Contouring cycles
G68/69	Roughing cycle
G81	Turning cycle
G82	Facing cycle
G83	Deep hole drilling cycle
G84/85	Straight threading cycles
G88	Auto grooving cycle

G68 Roughing cycle

This cycle is used for large amounts of stock removal. The cycle is activated by a G68 code. The machine moves are shown and are as follows:

(a) Rapid move to start point.
(b) A number of linear roughing passes (the number varies with depth of cut).
(c) A profiling pass, leaving on a finish allowance.

After the roughing cycle is completed the component would be finish turned to size.

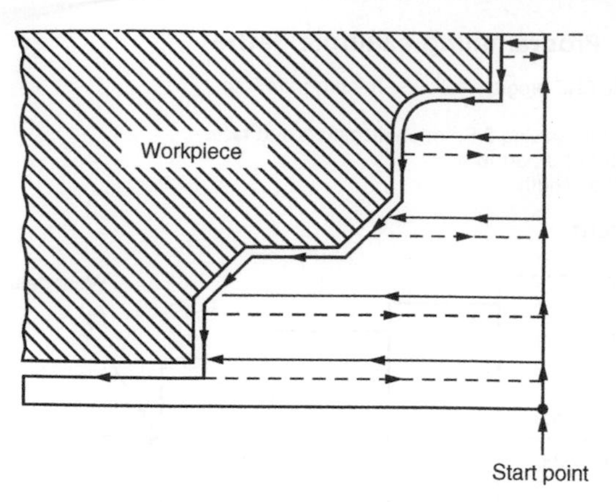

G83 Deep hole drilling cycle

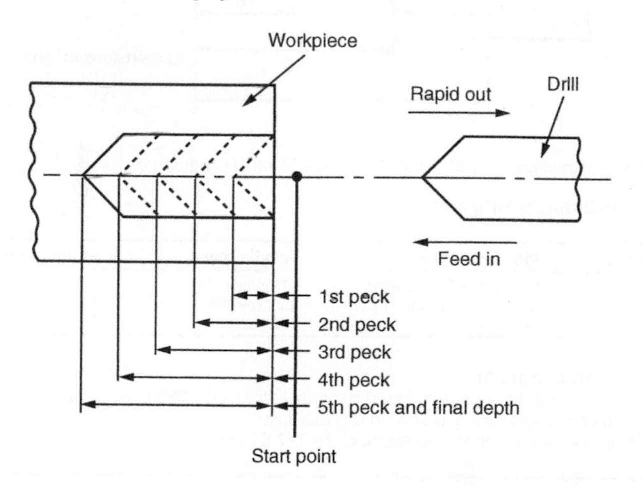

The deep hole drilling cycle is used to drill holes with a high depth/diameter ratio, and provides 'flute clearance' during the cycle. The cycle is activated by a G83 code. The moves involved are shown and are as follows:

(a) Rapid move to start point.
(b) Feed to first peck depth.
(c) Rapid move out to start point.
(d) Rapid in just short of first peck.
(e) Feed to second peck depth.
(f) Repeat movements (c)–(e) for each successive peck depth until required depth is achieved.

8.1.16 Programming example: lathe

An example CNC program follows. The program is to:

(a) Rough turn leaving 0.5 mm on diameters and faces
(b) Finish turn the part to size
(c) Part off to length.

Component

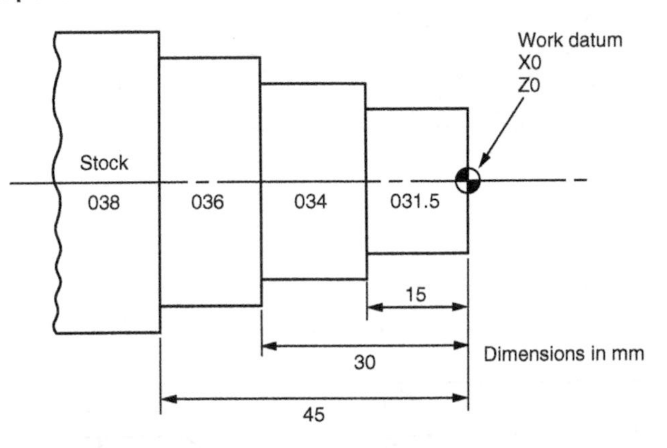

Material: aluminium alloy BS 1474: 1987 type 5154(A) condition 0.

Tool and machining data

Tool no.	Tool description	Spindle speed	Feed rates (mm/rev)
T101	Turning tool for roughing	130 m/min	0.2
T202	Turning tool for finishing	2500 rev/min	0.05
T303	Part off tool	120 m/min	0.05

Specimen program

(a) The format of the program would be suitable for a GE 1050 CNC controller.
(b) Diameter programming is used in this example.
(c) Tool change position (home position) is X177.8 Z254.

CNC program listing	Description
%	Start of program/tape.
N0010 G71	Metric input.
N0020 G95	Feed rate, mm/rev.
N0030 G97 S1000 M03	Direct rev/min spindle speed. 1000 rev/min switch on spindle.
N0040 G00 M08	Rapid move set. Switch on coolant.
N0050 G53 X177.8 Z254.0 T0	Cancels work offset. Moves turret to home position. Cancels tool offsets.
N0060 T100	Rotates turret to position 1.

continued

Table (*continued*)

CNC program listing	Description
N0070 G54 X00 Z2.0 T101	Activates work offset. Rapid move to coordinates. Activates tool offset number 1 with tool number 1.
N0080 G92 R0 S2000	Sets up parameters for constant surface cutting speed (CSS) (maximum 2000 rev/min).
N0090 G96 S130	Activates CSS at 130 m/min.
N0100 G01 X0.0 Z0.5 F0.2	Linear feed rate move at 0.2 mm/rev.
N0110 X36.5	Linear feed rate move at 0.2 mm/rev.
N0120 Z50.0	Turning beyond end of component to reduce diameter for parting off. Linear feed rate moves at 0.2 mm/rev.
N0130 G00 X37.0	Rapid move of tool off workpiece.
N0140 Z0.5	Rapid move of tool off workpiece.
N0150 X34.5	Rapid move of tool off workpiece.
N0160 G01 Z − 29.5	Linear feed rate move.
N170 G00 X35.0	Rapid move.
N0180 Z0.5	Rapid move.
N0190 X32.0	Rapid move.
N0200 G01 Z − 14.5	Linear feed rate move.
N0120 G00	Rapid move.
N0220 G53 X177.8 Z254.0 T0	Cancels work and tool offsets and moves turret to home position.
N0230 T200	Rotates turret to position 2.
N0240 G54 X10.0 Z10.0 T202	Activates work offset. Rapid move to coordinates. Activates tool offset number 2 with tool number 2.
N0250 G41 X0.0 Z0.05	Moves to coordinates and activates TNRC.
N0260 G97 S2500	Direct rev/min, 2500 rev/min.
N0270 G01 Z0.0 F0.05	Linear feed rate move at 0.05 mm/rev.
N0280 X31.5	Linear feed rate moves to profile part to size.
N0290 Z − 15.0	
N0300 X34.0	
N0310 Z − 30.0	
N0320 X36.0	
N0330 Z − 45.0	
N0340 G40	Cancels TNRC.
N0350 G00	Rapid move.
N0360 G53 X177.8 Z254.0 T0	Cancels work offset. Rapid move to home position. Cancels tool offsets.
N0370 T300	Rotates turret to position 3.
N0380 G54 X38.0 Z − 45 T303	Activates work offset. Rapid move to coordinates. Activates tool offset number 3 with tool number 3.
N0390 G92 R38 S1000	Sets up parameters for CSS.
N0400 G96 S120	Activates CSS at 120 m/min.
N0410 G01 X − 2.0 F0.05	Linear feed rate move.
N0420 G00 X38.0 M05	Rapid clear of work. Switch off spindle.
N0430 G53 X177.8 Z254.0 T0	Cancels work and tool offsets. Rapid to home position.
N0440 M02	End of program.

8.1.17 Glossary of terms

Absolute programming A system of programming where all positional dimensions are related to a common datum.

Adaptive control A system of sensors which changes speeds/feeds in response to changes in cutting loads. Also provides for in-process gauging and size correction.

Address In programming, a symbol which indicates the significance of the information immediately following that symbol.

Backlash 'Wasted' movements between interacting mechanical parts due to wear and/or manufacturing tolerance.

Binary coded decimal A method of representing decimal numbers by a series of binary numbers.

Binary number A number system with a base of 2.

Bit An abbreviation of 'binary digit'.

Block A line or group of words which contains all the instructions for one operation.

Byte A group of 8 bits.

CAM Computer-aided manufacture.

Canned cycle An operation which has been preprogrammed and which can be called up by a single instruction.

Circular interpolation A type of contouring control which uses the information contained in a single instruction to produce movement along the arc of a circle.

CLF Cutter location file.

Closed-loop control A system in which a signal depending on the output is fed back to a comparing device so that the output can be compared with the input, and a corrective signal generated if necessary.

Continuous path system A system in which the tool path results from the coordinated simultaneous motion of the work and/or cutter along two or more axes.

Cutter compensation An adjustment which compensates for the difference between the actual and programmed cutter diameters. See also *Tool length offset* (*TLO*).

Datum The reference position from which all absolute coordinates are taken.

DNC Direct numerical control.

DO loop A loop back to a repeated instruction in a program.

Encoder A device which is used to convert the position of a moving device into an electronic signal.

Feed rate The rate at which the cutting tool is advanced into the workpiece.

Incremental system A system by which each positional dimension is measured from the preceding position.

In-process gauging A system of gauging built into the machine which gauges the work whilst it is being manufacture so that any errors due to tool wear can be compensated for automatically. In the event of tool failure the machine will automatically change to back-up tooling. Thus the machine can be used unattended in a 'lightsout' environment. Used in conjunction with adaptive control.

Interpolation The joining up of programmed points to make a smooth curve by computation.

Linear path system A point-to-point system in which the tool or work moves only in straight lines.

Manual data input (MDI) A means of manually entering instructions to a controller, usually via a key panel.

Numerical control Control of machine movements by the direct insertion of numerical data.

Open-loop system A system where there is a single forward path between the input instruction and the output signal with no feedback.

Parallel path system A point-to-point system in which the tool or work moves only in straight lines parallel to any axis.

Parity A checking method to help determine if the tape has been prepared correctly. The standard for even parity is that of the International Standards Organization (ISO), whereas the standard for odd parity is that of the Electronic Industries Association (EIA).

Part program A program in machine control language and format required to accomplish a given task.

Point-to-point system A system in which the controlled motion is required only to reach a given end point, with no path control during the movements from one point to the next.

Post-processor A set of computer instructions which transforms tool centre line data into machine movement commands using the tape code and format required by a specific machine tool control system.

Rapid traverse Movements at a high traverse rate, normally used to save time when positioning the cutter before cutting commences.

Servo system An automatic control system, incorporating power amplification and feedback, designed to make a machine table follow the programmed route.

Spindle speed The rotational speed of the cutting tool or workpiece.

Subroutine A sequence of instructions or statements used to perform an operation which is frequently required. It is prepared by the part programmer and temporarily stored in the controller memory.

Tape punch A device for transferring instructions to a punched tape by perforating it with holes.

Tape reader A device for sensing and transmitting the instructions recorded on a punched or magnetic tape.

Tool length offset (TLO) An instruction which adjusts the position of the spindle along the *Z*-axis of a milling machine to take account of different lengths of cutting tool. On a lathe, offset is parallel to both the *X- and Z-axis*.

Word A set of characters which give a single complete instruction to a machine's control system.

Word address A block of characters preceded by a literal symbol to identify them.

8.2 Computer-aided design

8.2.1 An introduction to computer-aided design

Computer-aided design (CAD) is the application of a computer system which can be used to solve or enhance the solution of design and draughting problems. It not only eases the work of the draughtsperson in preparing drawings in both 2D and 3D, but it can be extended to provide:

- Geometric modelling: wire frame and solid modelling.
- Finite element analysis techniques to solve thermal and mechanical stress problems.
- Testing.
- Documentation.
- CNC programming.

For the architect and civil engineer, the ability to draw not only individual buildings but whole estates and complexes in 3D and then change the viewpoint at the touch of a button is of inestimable value. The use of virtual reality techniques enables the designer to 'walk through' an estate or complex.

By far the most widely used application of CAD in small- and medium-engineering companies is as a draughting aid. The capabilities of a typical draughting system could be:

- The production of 'library symbols' such as BS 308 conventions, pneumatic symbols, hydraulic symbols, electrical symbols and electronic symbols.
- The production of three view orthographical drawings using these symbols, to avoid having to repeatedly draw small details (e.g. screwed fastenings).
- The automatic production of isometric an oblique drawings from orthographical drawings and the ability to change the viewpoint at will.
- Drawing can be speeded up by such commands as: mirror imaging, rotation, copying, move, zoom, scaling, fillets, rectangles and polygons.
- The automatic shading or hatching of section drawings.
- The facility to make major or minor changes to a drawing without redrawing.
- The production of printed circuit board (PCB) layouts.
- The production of plant layouts.
- The production of parts lists, bills of materials, etc., directly from the digitized drawing data.
- Ease of storage on floppy disks.

Some typical examples of drawings are shown below. These were produced on *DesignCAD* and are reproduced by courtesy of *P.M.S (Instruments) Ltd.*

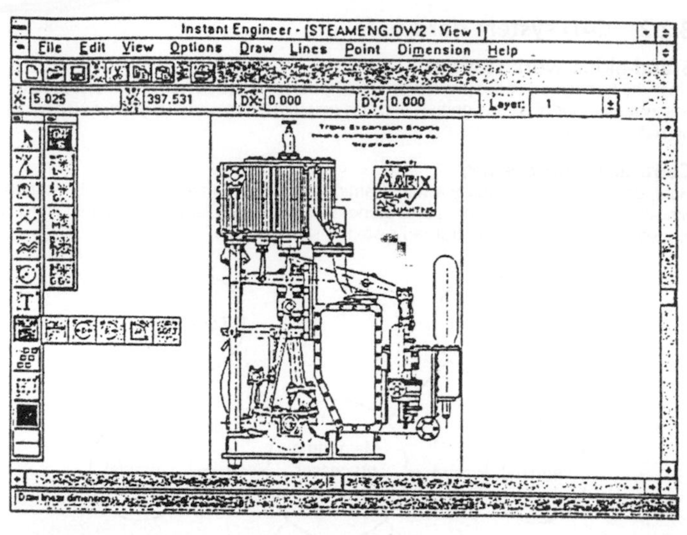
Instant Engineer - [STEAMENG.DW2 - View 1]
File Edit View Options Draw Lines Point Dimension Help
X: 5.025
Y: 397.531
DX: 0.000
DY: 0.000
Layer: 1

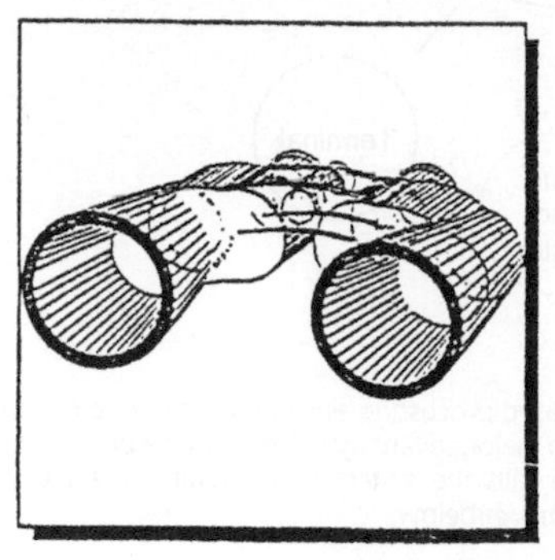

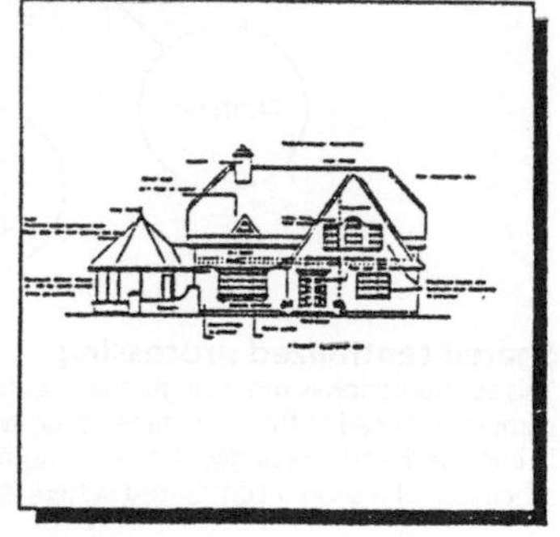

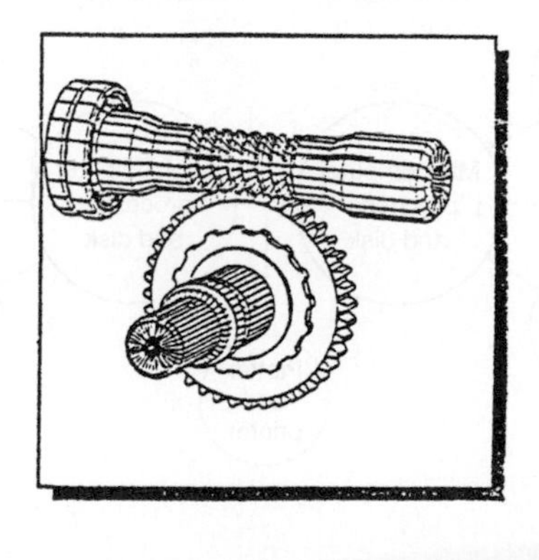

8.2.2 CAD system hardware

Stand-alone work stations

Since the first edition of this pocket book was published, the power of desktop computers has increased out of all recognition and it is now possible to run full-scale CAD packages on such machines.

Centralized processing

Here, one computer (normally a mainframe) supports a number of terminals. Response times can be slower than for 'stand-alone' systems, particularly if the system is heavily loaded. A schematic of a centralized system is shown below.

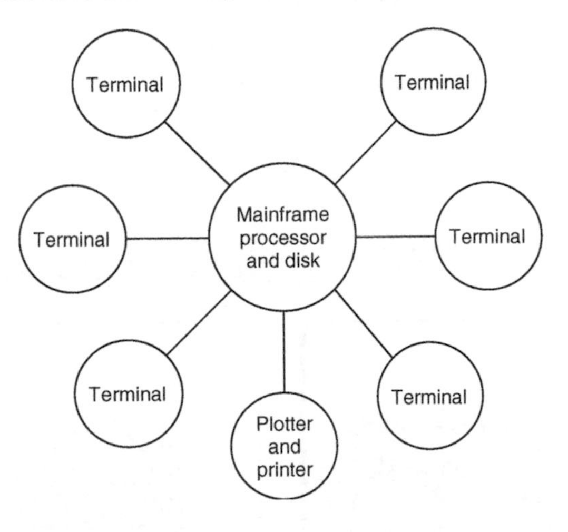

Shared centralized processing

This configuration is more costly than centralized processing since more than one central computer is used at the same time. It has two major advantages. Firstly the response time is improved and, secondly, if one computer fails the system is not totally disabled. A schematic of a shared centralized system is shown below.

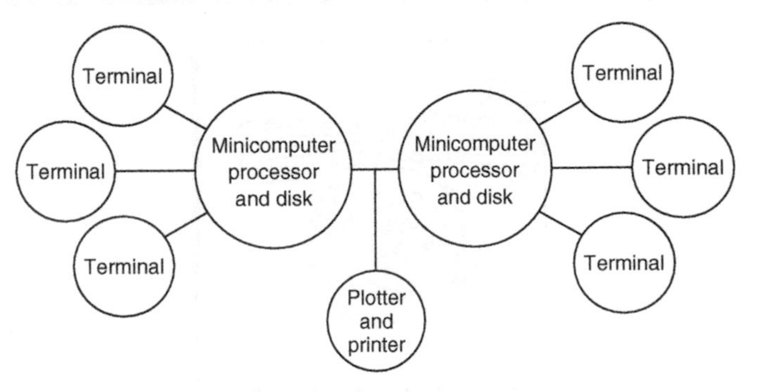

Integrated system

With this configuration, a number of computers are networked together so that the system files can be shared. This arrangement also allows the plotter, printers and disk drives required to become a shared resource, resulting in more economical loading of the system, faster response time and less chance of total disablement. The initial capital outlay is lower, as the system can be started up with only a few workstations. Additional workstations can then be added to the network as and when required. A schematic of an integrated system is shown below.

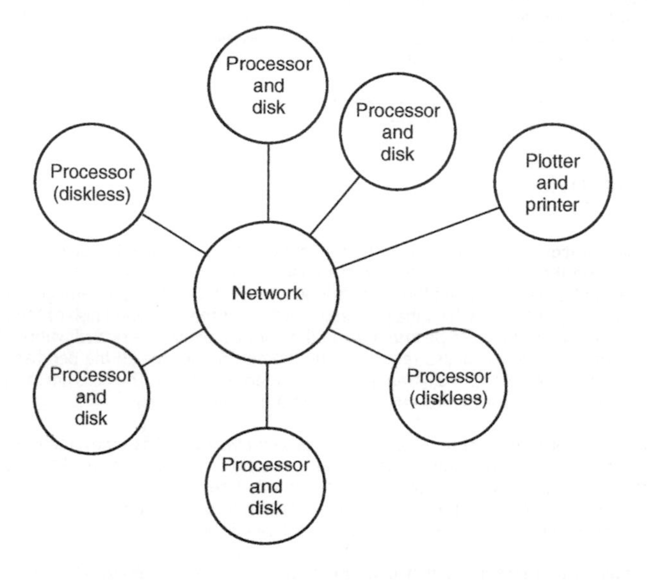

Visual display units

Nowadays these have to be colour monitors since modern CAD packages have 'layering' facilities. For example, an architect can produce the ground floor plan of a building in one colour, and superimpose the first floor plan in another colour to ensure structural integrity. The plumbing and central heating layout can then be superimposed in another colour an the electrical installation in yet another colour to make sure that everything is in the correct place. The layers can then be printed individually or in combination as required.

When producing orthographical engineering drawings it is useful to use one colour layer for the construction lines an another colour layer for the final outline.

When producing PCB layouts, one colour can be used for the tracks on the top of the board and another colour layer for the tracks on the underside of the board.

Colour presentations are also used when producing publicity material, charts and graphs for inclusion in reports, also for overhead projector transparencies.

Input devices

Keyboards These are familiar to all computer operators and are user friendly. These are used to input commands, text and parameter values.

Digitizer tablet The data tablet detects the position of a mechanical scanning or hand-held cursor. The tablet contains an electromagnetic, electrostatic or electroresistive grid which corresponds to the resolution of the visual display unit (VDU). Thus, *X* and *Y* coordinates on the tablet can be mapped directly on to the screen.

Joystick This is a two-axes control for providing a rapid movement of the cursor. It is familiar to players of video games. When used with CAD systems it allows more rapid cursor movement than is possible from the keyboard. However, the positional accuracy is low and is dependent on operator skill.

Mouse The mouse is now widely used in place of the joystick and is essential when using *Windows* versions of CAD software. The 'click' keys on the mouse allow the command to be entered quickly and easily once the cursor has been positioned.

Output devices

Hard copy can be printed from digitized data in a number of ways.

Flat bed plotter The paper is held to the flat bed of the machine by suction. The pen moves along the *X- and Y-axis* to draw the image. The carriage moves along the *X*-axis and the penhead moves along the *Y*-axis bridge which is located on the carriage. There is also a 'home' position to which the penhead is parked at the start and finish of the drawing. This allows the paper to be inserted and the finished drawing removed without fouling the pen. The penhead also returns to the 'home' position when the pen has to be changed to accommodate a different line thickness and/or colour. The pens are located in a magazine and are selected on command from the computer.

Drum or roller bed plotter The bed rotates to provide the *X*-axis movement and the pen carriage moves along the *Y*-axis parallel to the drum axis of rotation. This arrangement allows large drawings to be produced, yet little floor space is required. These plotters are not quite so accurate as flat bed plotters due to paper stretch and slip. They are also more limited in the range of materials that they will accept.

For small-scale drawings, conventional dot matrix, laser jet and electrostatic printers as used for normal text hard copy printing are suitable.

8.2.3 CAD system software

There are now many CAD packages available. These may be in 2D or 3D. Generally, a 2D package will do anything that can be done manually using a drawing board and T-square together with conventional drawing instruments. You can also produce isometric and oblique 'pictorial' representations. To save time it is possible to purchase 'libraries' of standard symbols. Some companies provide library information of their products on disk for their customers to use. For example, manufacturers of bathroom and kitchen fittings provide scale drawings of their product range on disk for architects to import directly into their designs. Some users will build up their own libraries of frequently used components using 'block' commands.

3D software is necessary when you require:

- **Wire frame modelling** This described the corners and edges of a model as shown below. It can lead to a view looking ambiguous.

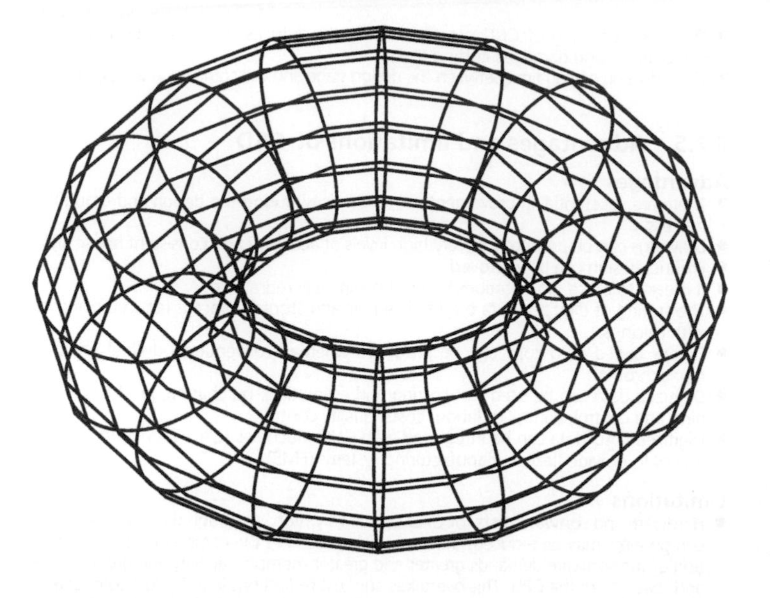

- **Surface models** These are usually used when components have a complex surface, such as a car or an aircraft panel. Information relating to the surface can be extracted from the shaded surface model.
- **Solid models** These provide detailed information concerning areas, volumes and centre of gravity of a component part. Solid models can be built up from standard 3D shapes and modified to get the desired result.

All CAD software in current use is configured to run in a *Windows* environment but some versions are available for use with Mackintosh computers. There is a wide range of software packages available configured for use by engineering designers, architects, interior designers, nautical architects and others. They invariably have provision for customising by the user to suit specific applications and are supported by 'libraries' of data and symbols that can be selected as required. Although requiring massive computer power compared with earlier versions, the latest, modern desk-top and lap-top computers can still support most of the packages in common use.

8.2.4 Computer-aided design and manufacture

The integration of CAD with computer-aided manufacture (CAM) involves the sharing of a database between these distinct elements. The advantages of such a system are:

- There is consistent reliability of information relating to design and manufacture without the possibility of copying errors.
- CNC programmes can be constructed directly from the digitized data of the drawing in much shorter times.
- Complex parts are programmed without the need for lengthy calculations.

- On-screen proofing of CNC program tool paths reduces the possibility of scrapped components and damaged cutting tools.
- The reduction of lead time between the design stage and the completion of manufacture.

8.2.5 Advantages and limitations of CAD

Advantages
- There are substantial improvements in productivity during the design and draughting process.
- Drawings can be produced to very high levels of accuracy and consistent house styles.
- Product design can be improved.
- It is easy to make modifications without completely redrawing.
- The storing of digitized data on disk is simple and storage space is reduced compared with tracings.
- Colour coded hard copy can be produced whereas conventional dyeline prints are monochrome.
- Digitized data can be used for: costing and estimating, purchasing, production planning and control, control of labour costs, quality control.
- Digitized CAD data can be integrated with CNC machines (CAD/CAM) and also with robots to provide flexible manufacturing systems (FMS).

Limitations
- Hardware and software costs for CAD can be very high. Although the cost of individual components may be reduced, the continual increase in the complexity and sophistication of the software demands greater and greater memory capacity and greater power and speed from the CPU. This overtakes any cost reductions in individual components.
- Powerful CAD systems require regular maintenance, although they are becoming very much more reliable and maintenance contracts are available.
- Initial set-up and staff training can be expensive whilst libraries of conventions an symbols are prepared and digitized to suit company products.
- A suitable ergonomic environment is required to avoid operator fatigue and ensure economic use of the system.

8.3 Industrial robots

8.3.1 An introduction to robotics

An industrial robot is a general purpose, programmable machine possessing certain anthropomorphic (human-like) characteristics. The most typical anthropomorphic characteristic of a robot is its arm. This arm, together with the robot's capability to be programmed, makes it ideally suited to a variety of production tasks, including machine loading, spot welding, spray painting and assembly. The robot can be programmed to perform a sequence of mechanical motions, and it can repeat that motion sequence indefinitely until reprogrammed to perform some other job.

Manufacturers use robots mostly to reduce manning levels. Robots, used either with other robots or with other machines, have two major advantages compared with traditional machines. First, they allow almost total automation of production processes, leading to increased rates of production better quality control and an increased response to varying demand. Second, they permit the adaptability, at speed, of the production unit. The production line can be switched rapidly from one product to another similar product, for example from one model of car to another. Again, when a breakdown immobilizes one element in the production unit, that element's function can be replaced quickly.

Adaptable production units are known as FMS. A flexible unit would comprise a small number of robots and computer-controlled machine tools working together to produce or part-produce a particular component or assembly. The logical development of such systems is to integrate a number of these flexible cells into a fully automated workshop or factory.

It must be remembered that the robot, although a highly sophisticated device, cannot by itself solve all the problems that can be solved by the human operator. It must therefore be associated with additional techniques (e.g. CAM).

The term *robotics* has two currently accepted meanings:

(a) In the strictest sense of the word, it implies the further development of automation by improving the robot as we know it at present.
(b) In the broader sense, it involves the development not only of the robot itself, but also processes associated with the robot (e.g. CAD) or the consideration of the robot as a machine with special properties in association with other machines (e.g. flexible manufacturing cells or systems).

8.3.2 Robot control

An industrial robot shares many attributes in common with a CNC machine tool. The type of technology used to operate CNC machine tools is also used to actuate the robot's mechanical arm. However, the uses of the robot are more general, typically involving the handling of work parts. Robots can be programmed from a computer on-line or off-line. Alternatively, the robot can be guided through a sequence of movements from a remote key pad coupled to the robot by a trailing lead. Thirdly, the robot can be guided manually through a sequence of movements as when spray painting, with the human operator holding and directing the robot's spray gun. The instructions are retained in the robot's electronic memory. In spite of these differences, there are definite similarities between robots and CNC machine tools in terms of power drive technologies, feedback systems, the use of computer control and even some of the industrial applications.

8.3.3 Robot arm geometry

A robot must be able to reach workpieces and tools. This requires a combination of an arm and a wrist subassembly, plus a 'hand', usually called the *end effector*. The robot's sphere of influence is based on the volume into which the robot's arm can deliver the wrist subassembly. A variety of geometric configurations have been researched and their relative kinematic capabilities appraised. To date, the robot manufacturers have specialized in the following geometric configurations:

(a) Cartesian coordinates: three linear axes.
(b) Cylindrical coordinates: two linear axes and one rotary axis.
(c) Spherical coordinates: one linear axis and two rotary axes.
(d) Revolute coordinates: three rotary axes.

Cartesian coordinate robots

Cartesian coordinate robots consist of three orthogonal linear sliding axes, the manipulator hardware and the interpolator. The control algorithms are similar to those of CNC machine tools. Therefore, the arm resolution and accuracy will also be of the order of magnitude of machine tool resolution.

An important feature of a cartesian robot is its equal and constant spatial resolution; that is, the resolution is fixed in all axes of motions and throughout the work volume of the robot arm. This is not the case with other coordinate systems.

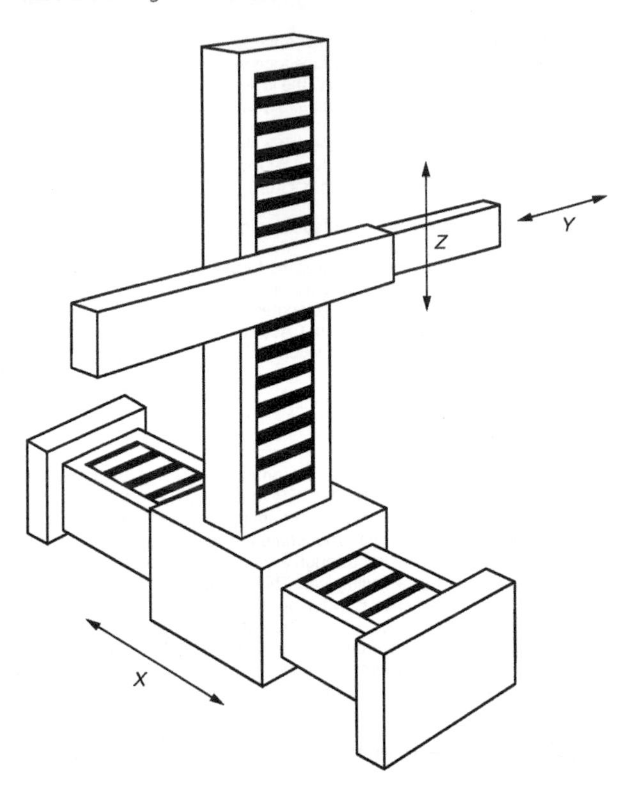

Cylindrical coordinate robots

Cylindrical coordinate robots consist of a horizontal arm mounted on a vertical column, which in turn is mounted on a rotary base. The horizontal arm moves in and out, the carriage moves vertically up and down on the column, and these two units rotate as a single assembly on the base. The working volume is therefore the annular space of a cylinder.

The resolution of the cylindrical robot is not constant and depends on the distance between the column and the wrist along the horizontal arm. Given the standard resolution of an incremental digital encoder on the rotary axis and arm length of only 1 m, then the resolution at the hand at full arm extension will be of the order of 3 mm. This is two orders of magnitude larger than is regarded as the state of the art in machine tools (0.01 mm). This is one of the drawbacks of cylindrical robots as compared to those with cartesian frames. Cylindrical geometry robots do offer the advantage of higher speed at the end of the arm provided by the rotary axis, but this is often limited by the moment of inertia of the robot arm.

In robots which contain a rotary base, good dynamic performance is difficult to achieve. The moment of inertia reflected at the base drive depends not only on the weight of the object being carried but also on the distance between the base shaft and the manipulated object. This is regarded as one of the main drawbacks of robots containing revolute joints.

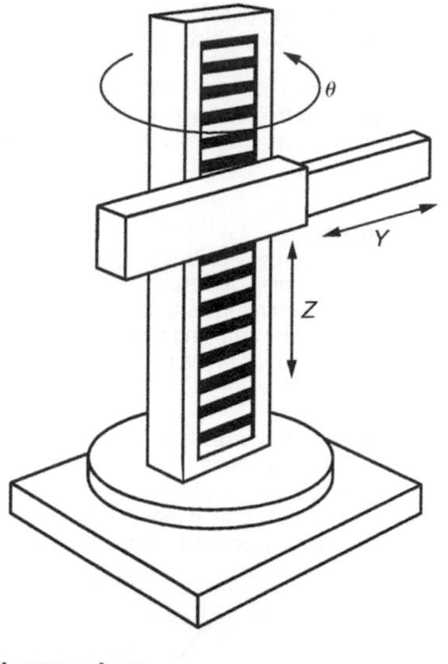

Spherical coordinate robots

The kinematic configuration of spherical coordinate robot arms is similar to the turret of a tank. These arms consist of a rotary base, an elevated pivot and a telescopic arm which moves in and out. The magnitude of rotation is usually measured by incremental encoders mounted on the rotary axes. The working envelope is a thick spherical shell.

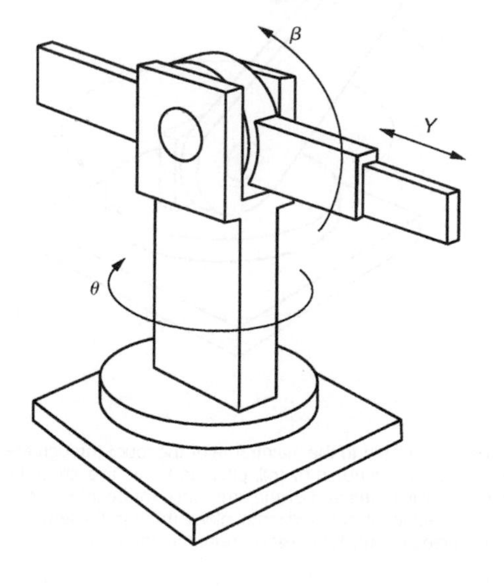

The disadvantage of spherical robots compared with their cartesian counterparts is that there are two axes having a low resolution, which varies with the arm length.

The main advantage of spherical robots over the cartesian and cylindrical ones is a better mechanical flexibility. The pivot axis in the vertical plane permits convenient access to the base or under-the-base level. In addition, motions with rotary axes are much faster than those along linear axes.

Revolute coordinate robots

The revolute, or articulated, robot consists of three rigid members connected by two rotary joints and mounted on a rotary base. It closely resembles the human arm. The end effector is analogous to the hand, which attaches to the forearm via the wrist. The elbow joint connects the forearm and the upper arm, and the shoulder joint connects the upper arm to the base. Sometimes a rotary motion in the horizontal plane is also provided at the shoulder joint.

Since the revolute robot has three rotary axes it has a relatively low resolution, which depends entirely on the arm length. Its accuracy is also the poorest since the articulated structure accumulates the joint errors at the end of the arm. The advantage of such a structure and configuration is that it can move at high speeds and has excellent mechanical flexibility, which has made it the most popular medium-sized robot.

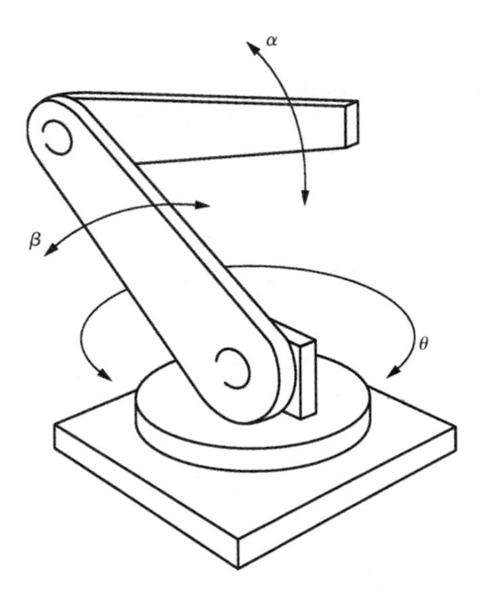

The wrist

The end effector is connected to the mainframe of the robot through the wrist. The wrist includes three rotary axes, denoted by roll, pitch and yaw. The roll (or twist) is a rotation in a plane perpendicular to the end of the arm; pitch (or bend) is a rotation in a vertical plane and yaw is a rotation in the horizontal plane through the arm. However, there are applications which require only two axes of motion in the wrist.

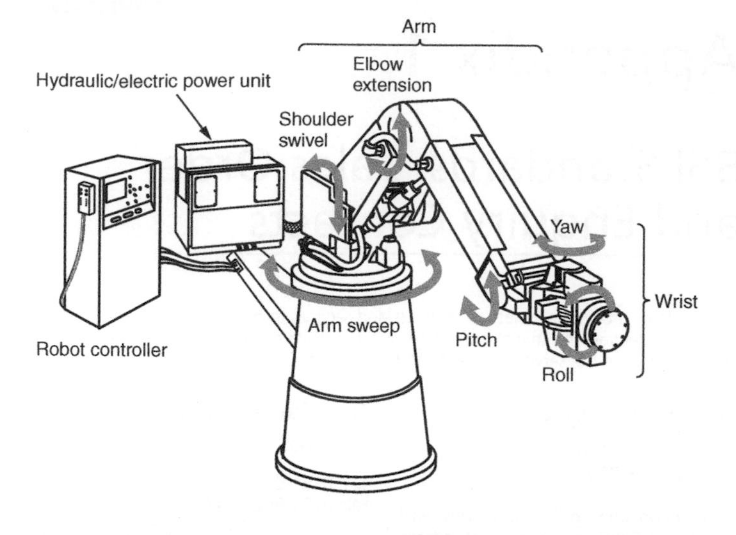

In order to reduce weight at the wrist, the wrist drives are sometimes located at the base of the robot, and the motion is transferred with rigid links or chains. Reduction of weight at the wrist increases the maximum allowable load and reduces the moment of inertia, which in turn improves the dynamic performance of the robot arm. The pitch, roll and yaw movements of the wrist can be seen on the robot shown.

Appendix 1

BSI Standards: Sales Order and Enquiry Contacts

BSI standards provides a variety of products and services to help standards users to manage their collection and to make standards work for their organization. The following is a brief summary of the services available and appropriate contact numbers.

Customer services

Tel.: 0181 996 7000
Fax: 0181 996 7001

- *for identifying, price quotations and ordering British and Foreign standards and other publications*
- *"PLUS" – private list updating service*

Information centre

Tel.: 0181 996 7111
Fax: 0181 996 7048

- *for detailed information and searches on British and overseas standards*
- *technical help to exporters*
- *certification and testing requirements overseas*
- *EC standardization developments*

Membership services

Tel.: 0181 996 7002
Fax: 0181 996 7001

- *members help desk*
- *membership administration*

Translations and language services

Tel.: 0181 996 7222
Fax: 0181 996 7047

- *for technical, standards and commercial translations*

Copyright

Tel.: 0181 996 7070
Fax: 0181 996 7400

- *copyright licences and enquiries*

Library services

Tel.: 0181 996 7004
Fax: 0181 996 7005

- *library services and enquiries*

Electronic products help desk

Tel.: 0181 996 7333
Fax: 0181 996 7047

- *perinorm*
- *electronic product development*

Ordering BSI publications

Orders can be placed by post, phone, fax or telex through BSI customer services.

Post: BSI customer services
BSI Standards
389 Chiswick High Road
London W4 4AL

Tel.: 0181 996 7000
Fax: 0181 996 7001

The hours of opening are 8.30 am to 5.30 pm Mondays to Fridays except public holidays

PLUS
Private list updating service

Contact: Tel.: 0181 996 7398
Fax: 0181 996 7001

Electronic media and databases – Help desk

Contact: Fax: 0181 996 7047

Copyright licences
Contact: Pamela Danvers

Tel.: 0181 996 7070
Fax: 0181 996 7001

Non-members

Non-members should send remittance with order, based on the prices given, or pay directly by credit card.

Members

Members will be invoiced in the usual way and will receive the appropriate discounts.

Prices

The group (Gr) number of each entry, in conjunction with the key below, indicates the UK price of the document. Postage and packing is included

Group no.	Non-members	Members
0	2.60	1.30
1	2.60	1.30
2	6.70	3.35
3	12.00	6.00
4	17.50	8.75
5	26.00	13.00
6	32.60	16.30
7	47.00	23.50
8	61.50	30.75
9	72.50	36.25
10	86.50	43.25
11	92.00	46.00
12	104.00	52.00
13	116.50	58.25
14	127.50	63.75
15	130.00	65.00

The above prices are for guidance and may be subject to variation.

Members' order hotline

Tel.: 0181 996 7003

For members who know exactly what they want to order.

Speeds up the placing of telephone orders.

Priority service

Urgent orders received before 12.00 h, by phone, fax or telex can be sent by priority service and will be despatched on the same day by first class mail (*orders for the priority service should be clearly marked*).

The charge for this service is 10% of invoice value, with a minimum charge of £1.00 and a maximum charge of £50.00.

Sales outlets

See the BSI catalogue for a full list of BSI sales outlets

Information centre

For detailed information and searches on British and overseas standards;

- certification and testing requirements overseas;
- technical help to exporters;
- EC development re. standardization

Tel.: 0181 996 7021
Electrical
Tel.: 0181 996 7022
Consumer products
Tel.: 0181 996 7023
Construction
Tel.: 0181 996 7024
Mechanical
Fax: 0181 996 7048

- Library loans International and foreign standards and related technical documents are available for loan to BSI members. The current price of tokens is £35.00 per book for orders or 10 or more books (please quote token numbers when requesting items for loan). Library contact:

Tel.: 0181 996 7004
Fax: 0181 996 7005

BSi News Update
compiled and edited by
Kay Westlake
Tel.: 0181 996 7060
Fax: 0181 996 7089

Membership services

Tel.: 0181 996 7002
Fax: 0181 996 7001

Translations and language services

Tel: 0181 996 7222
Fax: 0181 996 7047

- for technical, standards and commercial translations

Copyright

Copyright subsists in all BSI publications. **BSI also holds the copyright, in the UK**, of the publications of the international standardization bodies. Except as permitted under the Copyright, Design and Patents Act 1988 no extract may be reproduced, stored in a retrieval system or transmitted in any form or by any means – electronic, photocopying, recording or otherwise – without prior written permission from BSI. If permission is granted, the terms may include royalty payments or a licensing agreement.

Details and advice can be obtained from the:

Copyright Manager,
BSI, 389 Chiswick High Road,
London W4 4AL

Tel.: 0181 996 7070

BSI print on demand policy

- BSI has revised its production processes following intensive investment and process redesign enabling the rapid production of products on a variety of media from paper to electronic books.
- Increasingly, all British, European and International standards you order will be printed on demand from images stored in electronic files.
- By providing standards as looseleaf, hole-punched documents, amendments can be integrated more easily as replacement pages to provide improved, up-to-date and complete working documentation. Eventually the messy and time-consuming cut-and-paste methods will no longer be necessary to update your standards.
- During the transition phase from existing processes the product may arrive in various formats. We apologize for this and also for the quality of original material we have to use until the changeover is complete. We hope to start receiving live files from our international colleagues over the next 12 months which will significantly improve the print quality of BS, ISOs and BS EN ISOs.

- All of the new-format standards are printed on special watermarked paper with the words 'licenced copy' inside the paper, so that you can show that you own the official standard supplied by BSI.

BSI standards

389 Chiswick High Road
London W4 4AL

Tel.: 0181 996 9000
Fax: 0181 996 7400

Customer services

Tel.: 0181 996 7000
Fax: 0181 996 7001

Membership administration

Tel.: 0181 996 7002
Fax: 0181 996 7001

Information centre

Tel.: 0181 996 7111
Fax: 0181 996 7048

BSI quality assurance

Tel.: 01908 220908
Fax: 01908 220671

BSI testing

Tel.: 01442 230442
Fax: 01442 321442

BSI product certification

Tel.: 01908 312636
Fax: 01908 695157

BSI training services

Tel.: 0181 996 7055
Fax: 0181 996 7364

- Materials and chemicals
- Health and environment
- Consumer products
- Engineering
- Electrotechnical
- Management systems
- Information technology
- Building and civil engineering

- British standards
- Corresponding international standards
- European standards
- Handbooks and other publications

BSI global

Web site: www.bsi.global.com
E-mail: info@bsi.global.com

Appendix 2

Library Sets of British Standards in the UK

The following libraries hold full sets of British standards in either paper, CD-ROM or microfiche formats. The names of public libraries are shown in italics. For other libraries, it is advisable to make prior written application in order to ascertain the hours and conditions for access. **These sets are for reference only and attention is drawn to Copyright Law.**

ENGLAND

Avon

Bath — *Central Library*; University of Bath
Bristol — *Commercial Library*

Bedfordshire

Bedford — *Central Library*
Cranfield — Cranfield University
Leighton Buzzard — *Public Library*
Luton — *Central Library*; University of Luton

Berkshire

Reading — *Central Library*; University of Reading; Reading College
Slough — *Central Library*

Buckinghamshire

Aylesbury — *County Hall*
High Wycombe — College of H & E
Milton Keynes — *Central Library*

Cambridgeshire

Cambridge — *Central Library*
Peterborough — *Central Library*

Cheshire

Crewe — *Central Library*
South Wirral — Ellesmere Port Library
Stockport — *Central Library*
Warrington — *Central Library*

Cleveland

Cleveland — *County Library*
Hartlepool — *Reference Library*
Middlesbrough — University of Teesside

Cornwall

Redruth — Cornwall College of Further & Higher Education
Truro — *Reference Library*

Cumbria

Barrow-in-Furness — *County Library*
Carlisle — *County Library*
Kendal — *County Library*

Derbyshire

Chesterfield — *Central Library*
Derby — University of Derby

Devon

Barnstaple — *Central Library*
Exeter — *Central Library*
Plymouth — *Reference Library*

Dorset

Poole — *Reference Library, The Dolphin Centre*

Durham

Darlington — College of Technology
Durham — *County Library*; University Library; New College

Essex

Barking — *Central Library*

Chelmsford	*Central Library*
	Anglia Polytechnic University
Colchester	*Library*
Ilford	*Central Library*
Romford	*Central Library*
Southend-on-Sea	College of Arts & Technology
	Southend Library

Gloucestershire

Cheltenham	*County Library*

Hampshire

Basingstoke	*Public Library*
Farnborough	*Public Library*
Portsmouth	*Central Library*
	Highbury College of Technology
	University Library of Portsmouth
Southampton	*Central Library*
Winchester	*County Library*

Hereford & Worcester

Redditch	*Redditch Library*

Hertfordshire

Hatfield	*Central Library*
	University of Hertfordshire
Stevenage	*Central Library*
Watford	*Central Library*

Humberside

Grimsby	*Central Library*
Hull	*Central Library*
Scunthorpe	*Central Library*

Jersey

St Helier	*The Jersey Library*

Kent

Bexley Heath	*Central Library*
Bromely	*Central Library*
Chatham	*Central Library*
Maidstone	*County Library*
Margate	*Public Library*
Tonbridge	*Central Library*

Lancashire

Blackburn	*Central Library*
Bolton	*Central Library*
	Institute of Higher Education
Preston	*Central Library*
	Lancashire Polytechnic

Leicestershire

Leicester	De Montfort University
	Information Centre
	University of Leicester
Loughborough	University of Technology

Lincolnshire

Lincoln	*Central Library*

Greater London

Battersea	*Reference Library*
Chiswick	**BSI Library**
City University	Reference Library
Gower Street	University College London
Hammersmith	*Central Library*
Haringey	Middx University, Bounds Green
	Central Library
Hendon	*The Burroughs*
Holborn	*The British Library*
Islington	*Central Library*
Kensington	Imperial College of Science
	Educational Libraries
Kingston	Kingston University
Palmers Green	*Reference Library*
Southwark	South Bank University
Stratford	Newham Community College
	Reference Library
Swiss Cottage	*Reference Library*
Waltham Forest	Waltham College
Westminster	University of Westminster
	Westminster Libraries
Woolwich	*Woolwich Central Library*

Greater Manchester

Ashton-under-Lyne	*Public Library*
Manchester	John Rylands University Library
	Metro University Library
	Public Library
	UMIST Library
	University of Manchester
Oldham	*Reference Library*
Salford	College of Technology
	University of Salford

Wigan — Wigan & Leigh College

Merseyside
Birkenhead — *Central Library*
Liverpool — *Central Reference Library*
John Moores University
University, Harold Cohen Library
St Helens — Gamble Institute

Middlesex
Uxbridge — Brunel University

W Midlands
Birmingham — Aston University
Chamberlain Square
University of Birmingham
University of Central England
Coventry — *Central Library*
Coventry University
Lanchester University Library
University of Warwick
Dudley — *Reference Library*
Solihull — *Central Library*
Walsall — *Central Library*
West Bromwich — *Central Library*
Wolverhampton — *Central Library*

Norfolk
Norwich — *County Hall*

Northamptonshire
Northampton — *Central Library*

Northumberland
Ashington — College of Arts & Technology

Nottinghamshire
Nottingham — *County Library*
Trent University Library
University of Nottingham

Oxfordshire
Didcot — Rutherford Appleton Laboratory
Oxford — *Central Library*
Oxford Brookes University

Shropshire
Telford — *St Quentin Gate*

Somerset
Bridgewater — *County Library*

Staffordshire
Hanley — *Library*

Suffolk
Ipswich — *County Library*
Lowestoft — *Central Library*

Surrey
Croydon — *Central Library*
Guildford — College of Technology
Sutton — *Central Library*
Woking — *Public Library*

E Sussex
Brighton — *Reference Library*

W Sussex
Brighton — University of Brighton
Crawley — *Public Library*

Tyne and Wear
Gateshead — *Central Library*
Newcastle — *Central Library*
Polytechnic University of Newcastle
North Shields — *Central Library*
South Shields — *Central Library*
South Tyneside College
Sunderland — *County Library*
University of Sunderland
Washington — *Central Library*

Warwickshire
Rugby — *Central Library*

Wiltshire
Trowbridge — *Public Library*

N Yorkshire
Northallerton — *County Library*
York — Askham Bryan College
Central Library

S Yorkshire
Barnsley — *Central Library*
Doncaster — *Central Library*

Rotherham *Central Library*
Sheffield *Central Library*
City Polytechnic Library
University of Sheffield

W Yorkshire
Bradford *Public Library*
University Library
Huddersfield *Central Library*
University of Huddersfield
Leeds *Central Library*
Metropolitan University
University, Edward Boyle Library
Wakefield *Library Headquarters*

N IRELAND
Antrim
Ballymena *Area Library*
Belfast *Central Library*
College of Technology
Science Library, Queens University
Newtonabbey University of Ulster

Armagh
Armagh *Southern Education & Library Board*

Down
Ballynahinch *South Eastern Education & Library Board*

Tyrone
Omagh *County Library*

SCOTLAND
Grampian
Aberdeen *Central Library*

Lanarkshire
Hamilton Bell College of Technology

Lothian
Edinburgh *Central Library*
Heriot-Watt University
Napier Polytechnic
University, Engineering Library

Strathclyde
East Kilbride *Central Library*
Glasgow Glasgow University Library
The Mitchell Library
Strathclyde University Library

Tayside
Dundee *Central Library*
Institute of Technology
Forfar *Forfar Central Library*

WALES
Clwyd
Mold *County Civic Centre*

Dyfed
Llanelli *Public Library*

S Glamorgan
Cardiff *Central Library*

W Glamorgan
Swansea *Central Library*
University Library

Appendix 3

Contributing Companies

- **British Fluid Power Association**
  Cheriton House
  Cromwell Park
  Chipping Norton
  Oxfordshire OX7 5SR

  Tel.: 01608-647900
  Fax: 01608-647919
  e-mail: enquiries@bfpa.co.uk
  web site: www.bfpa.co.uk

- **ContiTech United Kingdom Ltd**
  Power Transmissions
  PO Box 26,
  Wigan, WN2 4WZ

  Tel.: 01942-52500
  Fax: 01942-524000
  web site: www.contitech.de

- **David Brown Engineering Ltd**
  Park Gear Works
  Lockwood
  Huddersfield HD4 5DD

  Tel.: 01484-465500
  Fax: 01484-465501
  e-mail: ringleby@davidbrown.textron.com
  web site: www.tyextron PT.com

- **Emhart Teknologies, Tucker Fasteners Ltd**
  Walsall Road
  Birmingham B42 1BP

  Tel.: 0121-356-4611
  Fax: 0121-356-1598
  web site: www.emhart.com

- **IMI Norgren Ltd**
  Brookside Business Park
  Green gate
  Middleton
  Manchester M24 1GS

  Tel.: 0161-655-7300
  Fax: 0161-655-770
  web site: www.norgren.com

- **Henkel Loctite Adhesives Ltd**
  Technologies House
  Wood Lane End
  Hemel Hemstead
  Hertfordshire HP2 4RQ

  Tel.: 01442-278000
  Fax: 01442-278071
  web site: www.loctite.com

- **National Broach and Machine Co**
  17500 Twenty three Mile Road
  Mt Clemens
  Michigan 48044
  USA

  Tel.: 313-263-0100
  Fax: 313-263-4571

- **Renold plc**
  Renold House
  Styal Road
  Wythenshawe
  Manchester M22 5WL

  Tel.: 0161-498-4500
  Fax: 0161-437-7782
  e-mail: enquiry@renold.com
  web site: www.renol.com

- **Sandvik Coromant UK**
  Manor Way
  Halesowen
  West Midlands B62 8QZ

  Tel.: 0121-550-4700
  Fax: 0121-550-0977

- **Tucker Fasteners Ltd**
  Emhart Fastening Technologies
  Walsall Road
  Birmingham B42 1BP

  Tel.: 0121-356-4811
  Fax: 0121-356-1598

Appendix 4

Useful References

(a) Carbide tooling: Sandvik Coromant UK, Manor Way Halesowen, West Midlands, B62 8QZ. Tel: 0121-504-5000, Fax: 0121-504-5555. Web site: www.sandvik.com

(b) Highspeed steel tooling: British Standards:
Twist drill sizes – metric: BS 328
Hand reamers – metric: BS 328: Pt 4
Long flute machine reamers: BS 328: Pt 4
Machine chucking reamers with MT shanks: BS 328: Pt 4
Shell reamers with taper bores and arbors for shell reamers: BS 328: Pt 4: Appendix B
Taper pin reamers: BS 328: Pt 4: Appendix B
Counterbores with parallel shanks and integral pilots: BS 328
Counterbores with MT shanks and detachable pilots: BS 328: Pt 5
Countersinks with parallel shanks and MT shanks: BS 328: Pt 5
Single point cutting tools: butt welded HSS: BS 1296: Pts 1 to 4
Milling cutters for horizontal milling machines: BS 4500
T-slot milling cutters with MT shanks: BS 122: Pt 3
Shell-end milling cutters: BS 122: Pt 3
Screwed shank end mills and slot drills: BS 122: Pt 4

(c) Bonded abrasive (grinding) wheels: BS 4481: Pts 1 to 3

INDEX

Printed and bound by CPI Group (UK) Ltd, Croydon, CR0 4YY

03/10/2024

01040432-0002